Advanced Optical
Instruments and Techniques
Volume 2

T0174245

OPTICAL SCIENCE AND ENGINEERING

Founding Editor
Brian J. Thompson
University of Rochester
Rochester, New York

RECENTLY PUBLISHED

Fundamentals and Basic Optical Instruments, *edited by Daniel Malacara Hernández*
Advanced Optical Instruments and Techniques, *edited by Daniel Malacara Hernández*
Laser Beam Shaping Applications, Second Edition, *Fred M. Dickey and Todd E. Lizotte*
Tunable Laser Applications, Third Edition, *edited by F. J. Duarte*
Laser Safety: Tools and Training, Second Edition, *edited by Ken Barat*
Optical Materials and Applications, *edited by Moriaki Wakaki*
Lightwave Engineering, *Yasuo Kokubun*
Handbook of Optical and Laser Scanning, Second Edition, *Gerald F. Marshall*
    and Glenn E. Stutz
Computational Methods for Electromagnetic and Optical Systems, Second Edition,
    John M. Jarem and Partha P. Banerjee
Optical Methods of Measurement: Wholefield Techniques, Second Edition, *Rajpal S. Sirohi*
Optoelectronics: Infrared-Visible-Ultraviolet Devices and Applications, Second Edition,
    edited by Dave Birtalan and William Nunley
Photoacoustic Imaging and Spectroscopy, *edited by Lihong V. Wang*
Polarimetric Radar Imaging: From Basics to Applications, *Jong-Sen Lee and Eric Pottier*
Near-Earth Laser Communications, *edited by Hamid Hemmati*
Slow Light: Science and Applications, *edited by Jacob B. Khurgin and Rodney S. Tucker*
Dynamic Laser Speckle and Applications, *edited by Hector J. Rabal and Roberto A. Braga Jr.*
Biochemical Applications of Nonlinear Optical Spectroscopy, *edited by Vladislav Yakovlev*
Optical and Photonic MEMS Devices: Design, Fabrication and Control, *edited by Ai-Qun Liu*
The Nature of Light: What Is a Photon?, *edited by Chandrasekhar Roychoudhuri,*
    A. F. Kracklauer, and Katherine Creath
Introduction to Nonimaging Optics, *Julio Chaves*
Introduction to Organic Electronic and Optoelectronic Materials and Devices,
    edited by Sam-Shajing Sun and Larry R. Dalton
Fiber Optic Sensors, Second Edition, *edited by Shizhuo Yin, Paul B. Ruffin, and Francis T. S. Yu*
Terahertz Spectroscopy: Principles and Applications, *edited by Susan L. Dexheimer*
Photonic Signal Processing: Techniques and Applications, *Le Nguyen Binh*
Smart CMOS Image Sensors and Applications, *Jun Ohta*
Organic Field-Effect Transistors, *Zhenan Bao and Jason Locklin*

*Please visit our website **www.crcpress.com** for a full list of titles*

Advanced Optical
Instruments and Techniques
Volume 2

Edited by
Daniel Malacara-Hernández
Brian J. Thompson

CRC Press
Taylor & Francis Group
Boca Raton London New York

CRC Press is an imprint of the
Taylor & Francis Group, an **informa** business

CRC Press
Taylor & Francis Group
6000 Broken Sound Parkway NW, Suite 300
Boca Raton, FL 33487-2742

First issued in paperback 2019

ISBN-13: 978-1-4987-2067-0 (hbk)
ISBN-13: 978-0-367-87295-3 (pbk)

Library of Congress Cataloging-in-Publication Data

Names: Malacara, Daniel, 1937- editor. | Thompson, Brian J., editor.
Title: Fundamentals and basic optical Instruments ; Advanced optical instruments and techniques / [edited by] Daniel Malacara Hernandez and Brian J. Thompson.
Other titles: Handbook of optical engineering. | Advanced optical instruments and techniques
Description: Second edition. | Boca Raton : CRC Press, 2017-2018. | Series: Optical science and engineering | "The second edition of the Handbook of Optical Engineering has been enlarged and it is now divided in two volumes. The first volume contains thirteen chapters and the second volume contains twenty one, making a total of 34 chapters"--Volume 1, preface. | Includes bibliographical references and index. Contents: volume 1. Fundamentals and basic optical instruments -- volume 2. Advanced optical instruments and techniques.
Identifiers: LCCN 2017021177| ISBN 9781498720748 (hardback : v. 1) | ISBN 9781315119984 (ebook : v. 1) | ISBN 9781498720670 (hardback : v. 2) | ISBN 9781315119977 (ebook : v. 2)
Subjects: LCSH: Optics--Handbooks, manuals, etc. | Optical instruments--Handbooks, manuals, etc.
Classification: LCC TA1520 .H368 2018 | DDC 621.36--dc23
LC record available at https://lccn.loc.gov/2017021177

Visit the Taylor & Francis Web site at
http://www.taylorandfrancis.com

and the CRC Press Web site at
http://www.crcpress.com

Contents

Preface..vii

Contributors ix

1   Optics of Biomedical Instrumentation...1
     Shaun Pacheco, Zhenyue Chen, and Rongguang Liang

2   Wavefront Slope Measurements in Optical Testing........................ 37
     Alejandro Cornejo-Rodríguez, Alberto Cordero-Davila, and Fermín S. Granados Agustín

3   Basic Interferometers 71
     Daniel Malacara-Hernández

4   Modern Fringe Pattern Analysis in Interferometry....................... 101
     Manuel Servín, Malgorzata Kujawinska, and José Moisés Padilla

5   Optical Methods in Metrology: Point Methods.............................153
     Zacarías Malacara-Hernández and Ramón Rodriguez-Vera

6   Optical Metrology of Diffuse Objects: Full-Field Methods.........213
     Marija Strojnik, Malgorzata Kujawinska, and Daniel Malacara-Hernández

7   Active and Adaptive Optics 245
     Daniel Malacara-Hernández and Pablo Artal

8   Holography.. 259
     Pierre-Alexandre Blanche

9   Fourier Optics and Image Processing 299
     Francis T.S. Yu

10   Light-Sensitive Materials: Silver Halide Emulsions,
     Photoresist, and Photopolymers 353
     Sergio Calixto, Daniel J. Lougnot, and Izabela Naydenova

11   Electro-Optical and Acousto-Optical Devices 409
     Mohammad A. Karim

12   Radiometry.. 459
     Marija Strojnik and Michelle K. Scholl

13 Color and Colorimetry...517
    Daniel Malacara-Doblado

14 The Human Eye and Its Aberrations...543
    Jim Schwiegerling

15 Incoherent Light Sources 561
    Zacarías Malacara-Hernández

16 Lasers...579
    Vicente Aboites and Mario Wilson

17 Spatial and Spectral Filters 603
    Angus Macleod

18 Optical Fibers and Accessories..633
    Andrei N. Starodoumov

19 Isotropic Amorphous Optical Materials 677
    Luis Efrain Regalado and Daniel Malacara-Hernández

20 Anisotropic Materials 695
    Dennis H. Goldstein

21 Optical Fabrication 727
    David Anderson and Jim Burge

Index..765

Preface

The Second Edition of the Handbook of Optical Engineering has been enlarged and is now divided into two volumes. The first volume contains thirteen chapters and the second volume contains twenty-one, making a total of thirty-four chapters. In the first volume, Chapter 4, "Ray Tracing," by Ricardo Florez-Hernández and Armando Gómez-Vieyra; Chapter 5, "Optical Design and Aberrations," by Armando Gómez-Vieyra and Daniel Malacara-Hernández; Chapter 9, "Polarization and Polarizing Optical Devices," by Rafael Espinosa-Luna and Qiwen Zhan; and Chapter 12, "Microscopes," by Daniel Malacara-Doblado and Alejandro Téllez-Quiñones, have been added. In Volume 2, Chapter 1, "Optics of Biomedical Instrumentation," by Shaun Pacheco, Zhenyue Chen, and Rongguang Liang; Chapter 7, "Active and Adaptive Optics," by Daniel Malacara-Hernández and Pablo Artal; Chapter 13, "Color and Colorimetry," by Daniel Malacara-Doblado; and Chapter 14, "The Human Eye and Its Aberrations," by Jim Schwiegerling, are new.

Most of the rest of the chapters were also updated, some with many changes and others with slight improvements. The contributing authors of some chapters have been modified to include a new collaborator.

The general philosophy has been preserved to include many practical concepts and useful data.

This work would have been impossible without the collaboration of all authors of the different chapters and many other persons. The first editor is grateful to his assistant, Marissa Vásquez, whose continuous and great help was of fundamental importance. Also, the first editor is especially grateful to his family, mainly his wife, Isabel, who has always encouraged and helped him in many ways.

Daniel Malacara-Hernández
Brian J. Thompson

Contributors

Vicente Aboites
Centro de Investigaciones en Óptica, AC
León, Gto., México

David Anderson
Rayleigh Optical Corporation
Tucson, Arizona, USA

Pablo Artal
Laboratorio de Óptica, Instituto
    Universitario de
Investigación en Óptica y Nanofísica
Universidad de Murcia
Murcia, Spain

Pierre-Alexandre Blanche
College of Optical Sciences
University of Arizona
Tucson, Arizona, USA

Jim Burge
College of Optical Sciences
University of Arizona
Tucson, Arizona, USA

Sergio Calixto
Centro de Investigaciones en Óptica, AC
León, Gto., México

Zhenyue Chen
College of Optical Sciences
University of Arizona
Tucson, Arizona, USA

Alberto Cordero-Dávila
Facultad de Ciencias Físico-Matemáticas
Benemérita Universidad Autónoma de Puebla
Puebla, Pue., México

Alejandro Cornejo-Rodríguez
Instituto Nacional de Astrofísica, Óptica y
    Electrónica
Puebla, Pue., México

Dennis H. Goldstein
Air Force Research Laboratory
Eglin AFB, Florida, USA

Fermín S. Granados Agustín
Instituto Nacional de Astrofísica, Óptica y
    Electrónica
Puebla, Pue., México

Mohammad A. Karim
City College of the City University of
    New York
New York, NY, USA

Malgorzata Kujawinska
Institute of Precise and Optical Instruments
Technical University
Warsaw, Poland

Rongguang Liang
College of Optical Sciences
University of Arizona
Tucson, Arizona, USA

Daniel J. Lougnotand
UMR CNRS
Mulhouse, France

Angus Macleod
Thin Film Center Inc.
Tucson, Arizona, USA

Daniel Malacara-Doblado
Centro de Investigaciones en Óptica, AC
León, Gto., México

Daniel Malacara-Hernández
Centro de Investigaciones en Óptica, AC
León, Gto., México

Zacarías Malacara-Hernández
Centro de Investigaciones en Óptica, AC
León, Gto., México

Izabela Naydenova
College of Sciences and Health
Dublin Institute of Technology
Kevin St, Dublin

Shaun Pacheco
College of Optical Sciences
University of Arizona
Tucson, Arizona, USA

José Moisés Padilla
Centro de Investigaciones en Óptica, AC
León, Gto., México

Luis Efrain Regalado
Siguiendo al Sol SA de CV
León, Gto., México

Ramón Rodríguez-Vera
Centro de Investigaciones en Óptica, AC
León, Gto., México

Michelle K. Scholl
Alenka Associates
Tempe, Arizona, USA

Jim Schwiegerling
College of Optical Sciences
University of Arizona
Tucson, Arizona, USA

Manuel Servín
Centro de Investigaciones en Óptica, AC
León, Gto., México

Andrei N. Starodoumov
Coherent Inc
Santa Clara, California, USA

Marija Strojnik
Centro de Investigaciones en Óptica, AC
León, Gto., México

Mario Wilson
CICESE
Ensenada, BC, México

Francis T. S. Yu
Pennsylvania State University
University Park, Pennsylvania, USA

1

Optics of Biomedical Instrumentation

1.1 Wide-Field Microscopy.. 1
    Optical Layout • Resolution
1.2 Fluorescence Microscope 5
    Introduction to Fluorescence Process • Fluorescence Imaging Systems
1.3 Confocal Microscopy..9
    Principle • Components • Types of Confocal Microscopes
1.4 Optical Sectioning Structured Illumination Microscopy (OS-SIM) 16
    Principle • Optical Sectioning Strength • Optical Sectioning Algorithm • Problem of Speed and Solution
1.5 Super-Resolution Structured Illumination Microscopy (SR-SIM)..20
    Principle • SR-SIM Instrumentation • Reconstruction Algorithm • Nonlinear SIM • Combining OS-SIM and SR-SIM
1.6 Endoscopy..26
    Introduction • Basic Optics for Endoscopes • Objective Lenses • Relay Lenses
References 33

Shaun Pacheco,
Zhenyue Chen, and
Rongguang Liang

1.1 Wide-Field Microscopy

The goal of a microscope is to produce a magnified image of a microscopic sample without degrading the image quality. Microscopes have become an essential tool for many biomedical applications. They are used in investigating biological processes, diagnosing diseases, and quantitatively measuring biological processes *in vitro* and *in vivo*. This section introduces the key components in an optical microscope and how diffraction-limited resolution is defined.

1.1.1 Optical Layout

A typical optical layout for a microscope with an infinity corrected objective is shown in Figure 1.1. The object plane is at the front focal plane of the objective, and the output for an infinity-corrected objective is a collimated beam for every object point. A tube lens is used to form an intermediate image, which can be directly imaged onto an electronic sensor or observed by the human eye through the eyepiece. The exit pupil of the objective lens is typically set at the rear focal plane to make the objective object space telecentric. In a telecentric system, the chief rays are parallel to the optical axis and the system magnification is constant even if the object is displaced from the focal plane. Microscope objectives are well corrected for aberrations, and thus produce diffraction-limited imaging.

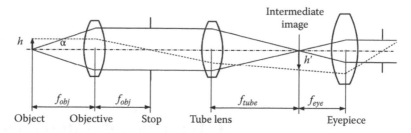

FIGURE 1.1 A typical microscope design for an infinity-corrected objective.

Two important properties of a microscope are the numerical aperture (NA) and magnification. NA of the microscope objective is defined as

$$NA = n \sin \alpha \tag{1.1}$$

where n is the refraction index of the medium between the front lens of the objective and the object, α is half acceptance angle of the objective. The magnification of the objective is defined as

$$M_{obj} = \frac{f_{tube}}{f_{obj}}, \tag{1.2}$$

where f_{tube} is the focal length of the tube lens and f_{obj} is the focal length of the objective. The total magnification of the microscope with an eyepiece is the product of the magnification of the objective and the magnification of the eyepiece

$$M_{microscope} = M_{obj} M_{eye}. \tag{1.3}$$

The magnification of the eyepiece is approximately

$$M_{eye} \approx \frac{250 \text{ mm}}{f_{eye}}, \tag{1.4}$$

where f_{eye} is the focal length of the eyepiece.

1.1.2 Resolution

The complex exit pupil of the objective is defined as

$$P(x, y) = A(x, y) \exp\left(i \tfrac{2\pi}{\lambda} W(x, y)\right), \tag{1.5}$$

where A is the amplitude function and W is the aberration function of the objective at the exit pupil for wavelength λ. Diffraction from the exit pupil to the plane of focus yields the impulse response

$$h(x_i, y_i) = \int\int_{-\infty}^{\infty} P(x, y) \exp\left(-i 2\pi \left(\frac{x}{\lambda d_i} x_i + \frac{y}{\lambda d_i} y_i\right)\right) dx dy, \tag{1.6}$$

where d_i is the distance from the exit pupil to the image plane, and x_i and y_i are the spatial coordinates at the image plane. The impulse response is the Fourier transform of the complex pupil function.

For an aberration free objective with a circular aperture, the pupil function $P(x, y)$ is unity, the point spread function (PSF) is:

$$\left|h(r_i)\right|^2 = \left(\frac{\pi R^2}{\lambda z}\right)^2 \left(2\frac{J_1\left(\frac{2\pi NA r_i}{\lambda}\right)}{\frac{2\pi NA r_i}{\lambda}}\right)^2,$$
(1.7)

where r_i is the radial coordinate in image space, R is the radius of the exit pupil, and J_1 is the Bessel function of the first kind. The normalized irradiance distribution of the diffraction-limited PSF is shown in Figure 1.2. The first zero of the Airy pattern is at a radial distance

$$r_i = \frac{0.61\lambda}{NA}.$$
(1.8)

Rayleigh's criterion is often used as a measure of the resolution in microscopy. Rayleigh's criterion states that two incoherent point sources are barely resolvable when the center of one falls exactly on the null of the second, as shown in Figure 1.3. This corresponds to a dip between the two peaks of approximately 74% of the maximum. All sources closer than Rayleigh's criterion cannot be resolved, and are assumed to come from the same point source.

For incoherent illumination, the intensity of the image is given by

$$I_i(x, y) = I_g(x, y) * \left|h(x, y)\right|^2,$$
(1.9)

where $*$ is the convolution operator, $|h|^2$ is the incoherent PSF of the optical system and I_g is the image as predicted by geometrical optics. The resulting image is a blurring of the geometrical optics image by the point spread function.

Since an optical system can be modeled as a linear shift invariant system, the imaging equation can be written in the frequency domain. The imaging equation in frequency space is

$$G_i(k_x, k_y) = OTF(k_x, k_y)G_g(k_x, k_y),$$
(1.10)

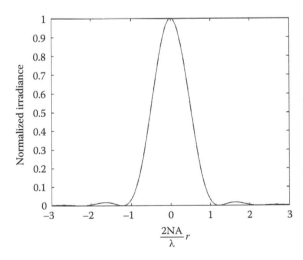
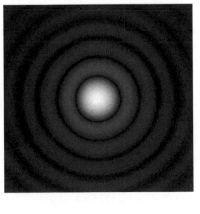

FIGURE 1.2 The lateral and 2D profile of the diffraction-limited point spread function.

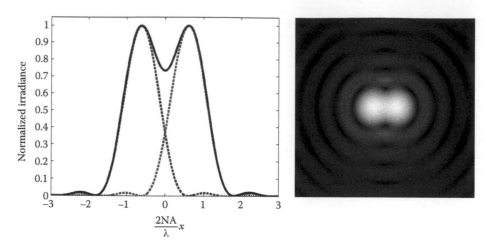

FIGURE 1.3   The lateral and 2D profile of Rayleigh's criterion.

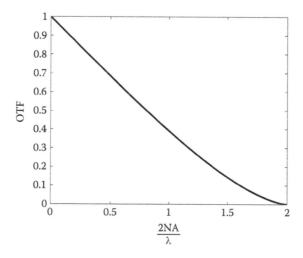

FIGURE 1.4   Diffraction-limited OTF.

where k_x and k_y are the spatial frequency coordinates, G_i is the normalized Fourier transform of the image, G_g is the normalized Fourier transform of the image predicted by geometrical optics, and OTF is the optical transfer function of the imaging system. OTF is the normalized Fourier transform of the PSF:

$$\text{OTF}(k_x, k_y) = \frac{\iint |h(x,y)|^2 \exp\left(-i2\pi(k_x x + k_y y)\right) dxdy}{\iint |h(x,y)|^2 dxdy}.$$

(1.11)

Some important properties of the OTF are

1. $\text{OTF}(0,0) = 1$
2. $\text{OTF}(k_x, k_y) = \text{OTF}^*(-k_x, -k_y)$
3. $|\text{OTF}(k_x, k_y)| \leq |\text{OTF}(0,0)| = 1$

The OTF for a diffraction-limited system is shown in Figure 1.4; it has a cutoff frequency at $2NA/\lambda$. Objects with a higher spatial frequency cannot be resolved, unless super-resolution techniques are used.

1.2 Fluorescence Microscope

Fluorescence microscope plays a major role in the fields of cell and molecular biology, due to its intrinsic selectivity that can provide high contrast between objects of interest and background. This is important in biomedical imaging, since biological structures of interest can be fluorescently labeled to more easily study biological phenomenon. Over the past several decades, different microscope designs have appeared with the aim of increasing image contrast, penetration depth, and spatial resolution. This section discusses the principle of fluorescence microscope and related instrumental techniques.

1.2.1 Introduction to Fluorescence Process

When illuminated with light in a suitable spectrum, some specimen, living or non-living, organic or inorganic, absorb light and then radiate light in a different wavelength from the illumination light. As shown in Figure 1.5a, when molecules absorb light with a suitable wavelength λ_{ex}, electrons may be raised from ground state S_0 to a higher energy and vibrationally excited state S_2. This process may only take 10^{-15} seconds. Within 10^{-14}–10^{-11} seconds, the excited electrons may lose some vibrational energy to the surrounding environment in the form of heat, and they relax to the lowest vibrational energy S_1 level within the electronically excited state from which the fluorescence emission originated. When electrons relax from the excited state to the ground state, light is often emitted at a longer wavelength λ_{em}. This emission process may take 10^{-9} seconds. The wavelength of the emitted light is determined by E_{em}, the energy difference between the energy levels of the two states during the emission of light

$$\lambda_{em} = \frac{hc}{E_{em}}, \tag{1.12}$$

where h is Planck's constant and c is the speed of light.

The entire fluorescence process cycles when the fluorescent specimen is illuminated, unless the fluorophore is irreversibly destroyed in the excited state. If the exciting radiation is stopped, the fluorescence process ends. Therefore, the same fluorophore can be repeatedly excited and detected. This is the fundamental process of the fluorescence detection technique.

Fluorescent molecules, which are capable of emitting fluorescence when illuminated with suitable light, are known as fluorescent probes, fluorochromes, or fluorescent dyes. When the fluorochromes

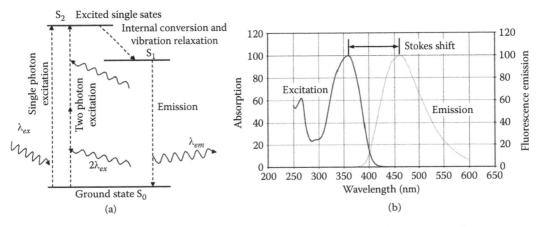

FIGURE 1.5 (a) Jablonski energy diagram illustrating fluorescence processing. (b) Excitation and emission spectra. Curves are normalized to the same peak height.

are conjugated to a larger macromolecule, they are known as fluorophores. There are two types of fluorophores, intrinsic and extrinsic. Intrinsic fluorophores are those that occur naturally, such as aromatic amino acids, nicotinamide adenine dinucleotide (NADH), flavins, porphyrins, and green fluorescent protein. Extrinsic fluorophores are synthetic dyes or modified biochemically.

Each fluorescent molecule has unique and characteristic spectra for excitation and emission. The relative intensities of these spectra are usually plotted side by side in the same graph, as shown in Figure 1.5b. Curves are normalized to the same peak height. An excitation spectrum describes the relative probability that a fluorophore can be excited by a given wavelength of an illuminating light. An emission spectrum is a plot of the relative intensity of emitted light as a function of wavelength. It shows the relative probability of an emitted photon in a particular wavelength. The difference in the wavelength between the emission peak and excitation peak is called Stokes shift. Stokes shift is fundamental to the sensitivity of fluorescence techniques. By completely filtering out the excited light without blocking the emitted fluorescence, it is possible to see only the objects that are fluorescent.

Another commonly used fluorescence imaging technique is multiphoton imaging. As shown in Figure 1.5a, in multiphoton imaging two or three absorbed photons must have a wavelength about twice or triple that required for one-photon excitation. If two or three photons reach the fluorophore at the same time (within an interval of about 10^{-18} seconds), the fluorophore can be excited and emits fluorescence. The probability of the near-simultaneous absorption of two photons is extremely low. Therefore, a high flux of excitation photons is typically required, usually from a femtosecond laser.

Basically, there are two types of fluorescence: intrinsic fluorescence (also called autofluorescence and endogenous fluorescence) and extrinsic fluorescence (also called exogenous fluorescence). Autofluorescence emission, arising from intrinsic fluorophores, is an intrinsic property of cells, while extrinsic fluorescence is obtained by adding exogenous fluorophores, such as FITC, GFP, and PE. The exogenous fluorophore offers an alternative for fluorescence imaging. It is a molecule that is naturally occurring or specially designed and can be used as a probe to label cells or tissues. Probes can be designed to localize the tissues, cells, or proteins within a cell, as well as to respond to a specific stimulus and to monitor the production of a gene product. Fluorophores can be bound to an antibody for delivery to specific targets. Green fluorescence protein (GFP), red fluorescence protein (RFP), and rhodamine are some examples of naturally occurring fluorophore proteins that emit fluorescence light in the green or red wavelengths.

1.2.2 Fluorescence Imaging Systems

Most fluorescence imaging systems require the following key elements: a light source to excite fluorescence, illumination optics, light collection or imaging optics, fluorescence filters to isolate emission photons from excitation photons, and a detector to capture the fluorescence signal. Florescence filters generally include an excitation filter, emission filter, and a dichroic beamsplitter, depending on the application and configuration.

The wavelength of the light source determines which type of fluorophore can be excited and how well that fluorophore can be excited. Besides the light source power, the wavelength is the major factor in determining the probed depth given that the penetration depth of light in tissue strongly depends on the wavelength. The light source power affects the SNR of the fluorescence image. Different applications have very different requirements for light sources, as well as the illumination optics associated with them.

Imaging optics collects and delivers the fluorescence light to the sensor. Because the fluorescence signal is weak, the light collection optics should have a high light collection efficiency. The detector is one of the most critical components in fluorescence imaging because it determines at what level the fluorescence signal can be detected, what relevant structures can be resolved, and/or the dynamics of the process that is visualized and recorded.

1.2.2.1 Fluorescence Filters

Fluorescence filters are required in fluorescence imaging systems to control the spectra of the excitation and emission lights. Without filters, the detector would not be able to distinguish between the desired fluorescence from scattered excitation light and the autofluorescence from the sample, substrate, and other optics in the system. Fluorescence filters are the key components to achieving this goal. The challenge is that the excitation light is usually about 10^5 to 10^6 times brighter than the fluorescence emission.

Generally, there are three filters in a fluorescence imaging system: an excitation filter, a dichroic beamsplitter, and an emission filter, as shown in Figure 1.6. The excitation filter selects a range of wavelengths from a broadband source to excite the sample. The emission filter transmits the emission and rejects the excitation wavelengths. A dichroic beamsplitter serves a dual function, reflecting the excitation wavelengths to the sample through the microscopic objective lens, or other imaging system, and transmitting the emission to the detector or the eyepiece.

It is crucial that optical filters be chosen to give the best performance, both in brightness and contrast, for a given application and fluorophore. The ratio of emitted fluorescence intensity to excitation light intensity in a typical application is between 10^{-4} (for highly fluorescent samples) and 10^{-6}. Therefore, the system must attenuate the excitation light by as much as 10^7 (weak fluorescence) without diminishing the fluorescence signal. A good combination of optical filters can reduce excitation light noise, or stray and scattered light, from the excitation source, as well as instrument autofluorescence outside the emission band.

Both shortpass and bandpass filters can be used in the illumination path to condition the illumination spectrum, but a bandpass filter is usually preferred because the shortpass filter also passes the UV light, which may cause photobleaching and fluorescence noise. Because the observed background autofluorescence roughly increases in proportion to the bandwidth of the excitation spectrum, narrowing the bandwidth of the excitation filter can reduce the background noise and enhance the image contrast. Emission filters can be either longpass or bandpass filters. A longpass filter may be preferred when the application requires a maximum emission signal and when spectral discrimination is not necessary. The longpass filter transmits fluorescence from all fluorophores with an emission spectrum longer than the cut-on wavelength. It is also useful for simultaneous detection of spectrally distinct multiple emissions. The dichroic mirror is mounted at a 45° angle to the optical axis of the microscope objective lens to reflect light in the excitation band and to transmit light in the emission band. The transmittance cutoff of the dichroic mirror lies between the fluorophore's excitation spectrum and its emission spectrum such that the excitation and emission wavelengths are separated effectively. The sharper the slope

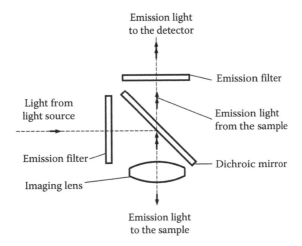

FIGURE 1.6 The basic configuration of fluorescence filters in a fluorescence microscope.

of the transition from transmission to blocking or from blocking to transmission, the higher the light efficiency and image contrast.

Brightness and contrast are the two aspects to evaluating a selected set of fluorescence filters. Ideally, the selection of fluorescence filters should maximize both brightness and contrast. Bandpass filters with a bandwidth of 20–40 nm are optimal for most fluorescence imaging. Filters with a bandwidth greater than 40 nm allow for the collection of light at a wider spectral range and give a higher total signal. Filters with a bandwidth narrower than 20 nm transmit less signal and are the most useful with fluorophores with very narrow emission spectra.

1.2.2.2 System Consideration

The design parameters for a fluorescence imaging system include the power and homogeneity of the excitation light, the FOV, and the NA of the detection system that collects the emission light, the sensitivity and noise characteristics of the detector, and the transmission and blocking capabilities of the fluorescence filters.

Fluorescence signals are usually very weak and require long exposure times due to the low concentration of fluorophores. Therefore, the design of an optical system for fluorescence analysis must consider the entire optical path. Low fluorescence light requires a highly efficient optical system to improve light-capturing abilities, thus increasing sensitivity and throughput, to provide a higher dynamic range to accommodate the vast differences in fluorophore concentrations across a sample array and to reduce crosstalk between sample spots through improved optical resolution.

To optimize the detection of a fluorescence signal, fluorescence filters are selected to maximize blocking in the transmission passband of the emission filter in the illumination path and to maximize blocking in the corresponding transmission passband of the excitation filter in the detection path. In general, it is preferable to block out-of-band light with an excitation filter instead of an emission filter so that the specimen will be exposed to less radiation and fewer components and less-complicated coating in the detection path. A simpler detection path always helps to improve image quality given that the quality requirement of the components is higher in the imaging path.

The optimal position of the excitation filter is where the range of the ray angle is small and away from the light source to reduce the angular effect and autofluorescence in the illumination path. To reduce the autofluorescence from the components in the detection path, the emission filter should be placed in front of other optical components in the detection path to reduce the autofluorescence. However, in many applications, it is not practical to place the emission filter as the first element. For example, in fluorescence microscopy, there is not enough room for the emission filter in front of the objective lens. The next optimal location for the emission filter is where the range of the ray angle is small.

For systems whose excitation and emission paths share common optical elements, such as a microscope's objective lens, and for systems whose excitation and emission paths do not share the same elements, but it is not appropriate to place the emission filter in front of the detection path, special attention should be paid when selecting optical materials in the optical design of fluorescence imaging systems.

As a general guideline, the optical elements in the detection path should be as small as possible to increase light transmission, increase the SNR, and minimize the autofluorescence of the optical components. The same requirement is also desired for the excitation path.

1.2.2.3 Multiphoton Imaging

Although wide field fluorescence microscopy can provide submicron resolution of biochemical events in living systems, it is limited in sensitivity and spatial resolution by background noise caused by fluorescence above and below the focal plane. Confocal fluorescence technique circumvents this problem by rejecting out-of-focus background fluorescence by using pinhole apertures, producing thin unblurred optical sections from deep within thick specimens. Multiphoton fluorescence technique is an alternative to confocal microscopy through selective excitation coupled to a broader range of detection choices, dramatically increasing the efficiency of emitted fluorescence signals.

Excitation in multiphoton microscopy occurs only at the focal point of a diffraction-limited microscope, optically sectioning thick biological specimens to obtain three-dimensional resolution. The advantage of multiphoton imaging is that it can probe selected regions beneath the specimen surface because the position of the focal point can be accurately determined and controlled. The highly localized excitation energy can minimize photobleaching of fluorophores attached to the specimen and can reduce photodamage, increasing cell viability and the subsequent duration of experiments that investigate the properties of living cells. In addition, the application of longer excitation wavelengths permit deeper penetration into biological materials and reduce the high degree of light scattering that is observed at shorter wavelengths. These advantages enable researchers to study thick living tissue samples, such as brain slices and developing embryos.

1.3 Confocal Microscopy

One disadvantage of wide-field microscopy is that the light from the out-of-focus region of the image is overlapped with the focused region at the image plane. This results in poor contrast in the final image. Confocal microscopy was developed by Minksy in 1957 [1]. Confocal microscopes place a pinhole in front of a detector to reject the out-of-focus light, which leads to optical sectioning and improved contrast over wide-field microscopy. Due to the optical sectioning, a confocal microscope is used to image a volume noninvasively. This section discusses the principle of confocal microscopy and its improvement over conventional wide-field microscopes. The components used in a confocal microscope and some modifications to the design are also discussed.

1.3.1 Principle

The principle of a confocal microscope is shown in Figure 1.7. Light travels through an illumination pinhole and is focused onto a sample on the focal plane of the objective. The reflected and/or fluorescent light at the focal plane returns through the objective and is focused at the image plane, where a detection pinhole is placed, as shown in Figure 1.7a. The pinhole has the function of preventing the majority of the out-of-focus light from reaching the detector. The returning light from the focal point passes through the small detection pinhole, where it is finally detected by the detector. However, as shown in

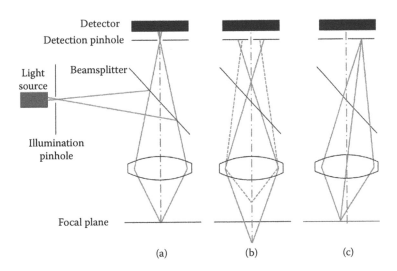

FIGURE 1.7 Concept of confocal microscope. (a) Focus beam is transmitted through the detection pinhole. (b) The majority of the out-of-focus light is blocked by the detection pinhole. (c) Light that is laterally shifted from the focused beam is also blocked by the pinhole.

Figure 1.7b, the light from the axial plane above and below the focal plane focuses either above or below the detection pinhole and the majority of the out-of-focus light is blocked by the pinhole. Additionally, the pinhole blocks the light from points on the focal plane that are laterally shifted from the illumination spot as shown in Figure 1.7c. Since the out-of-focus light is rejected, confocal microscopes have optical sectioning capabilities and have increased contrast over wide-field microscopes. Due to the pinhole, confocal microscopes only image a single point onto a detector at a time. Therefore, a scanning system is required to scan the focused beam across the object of interest.

The PSF behind the detection pinhole for a confocal microscope is the product of the PSFs of the illumination path and the detection path:

$$\text{PSF}_{tot}(r) = \text{PSF}_{ill}(r)\text{PSF}_{det}(r),\tag{1.13}$$

where PSF_{ill} and PSF_{det} are the PSFs for the illumination and detection path, respectively. Consider a diffraction-limited confocal microscope, the total PSF is proportional to

$$\text{PSF}_{tot}(r) \propto \text{somb}^2\!\left(\frac{2\text{NA}_{ill}}{\lambda_{ill}}r\right)\text{somb}^2\!\left(\frac{2\text{NA}_{det}}{\lambda_{det}}r\right),\tag{1.14}$$

where NA_{ill} and λ_{ill} are the NA and wavelength for the illumination path, and NA_{det} and λ_{det} are the NA and wavelength for the detection path. For a reflectance confocal system with epi-illumination, the illumination and detection path utilize the same objective and wavelength, so the PSF of the confocal system is

$$\text{PSF}_{tot}(r) \propto \text{somb}^4\!\left(\frac{2\text{NA}}{\lambda}r\right).\tag{1.15}$$

Therefore, the PSF of a confocal microscope is the square of the Airy disk. A comparison of the lateral profile of the PSF for a wide-field and confocal microscope is shown in Figure 1.8a. While the first zeros of the PSFs are at the same location, Figure 1.8 shows the full width at half maximum (FWHM) for a confocal microscope is smaller than conventional microscopes. The FWHM for a confocal microscope is

$$\text{FWHM} = \frac{0.37\lambda}{\text{NA}}.\tag{1.16}$$

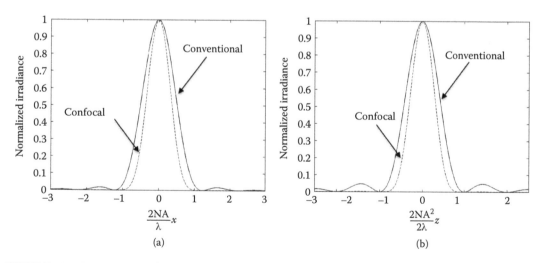

FIGURE 1.8   A comparison of the (a) lateral PSF and (b) axial PSF for conventional microscopy with confocal microscopy.

Note if the detection and illumination wavelength are different, the mean wavelength is used for the calculation of the FWHM. The mean wavelength is calculated by

$$\bar{\lambda} = \sqrt{2}\, \frac{\lambda_{ill}\lambda_{det}}{\sqrt{\lambda_{ill}^2 + \lambda_{det}^2}}. \tag{1.17}$$

The resolution according to Rayleigh's criterion, which is when the dip between the two peaks is approximately 74% of the maximum, is

$$\Delta r_{confocal} = \frac{0.44\lambda}{\lambda}. \tag{1.18}$$

Furthermore, the axial PSF of a confocal microscope is [1]

$$\text{PSF}(z) \propto \text{sinc}^2\left(\frac{NA_{ill}^2}{2\lambda_{ill}}z\right)\text{sinc}^2\left(\frac{NA_{det}^2}{2\lambda_{det}}z\right) \tag{1.19}$$

Figure 1.8b compares the axial PSF of the confocal microscope to conventional microscopes. The FWHM of the axial PSF is

$$\text{FWHM} \cong \frac{1.28\bar{\lambda}}{NA^2} \tag{1.20}$$

and the axial resolution from Rayleigh's criterion is

$$\Delta z_{confocal} = \frac{1.5\lambda}{NA^2}. \tag{1.21}$$

Figure 1.9 shows $x\,z$ slices of the lateral and axial resolution. Note the 3D PSF for the confocal microscope is narrower than the PSF for conventional microscopes in all 3 dimensions, therefore the resolution is improved laterally and axially. Additionally, the pinhole blocks the out-of-focus light, so confocal microscopes have optical sectioning capabilities. The optical sectioning allows the confocal microscope

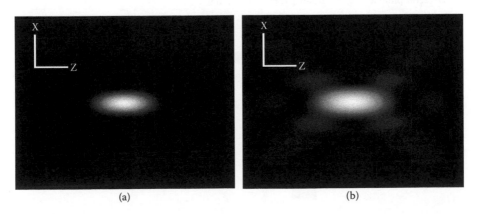

(a)   (b)

FIGURE 1.9 The PSF in the *x-z* plane for (a) confocal microscopy and (b) conventional microscopy.

to capture 3D images of an object of interest. This discussion assumes the pinhole is infinitely small, when the pinhole is increased, the lateral and axial resolution decreases [2].

1.3.2 Components

Although the system design for a confocal microscope varies depending on the specific application, a representative example layout for a confocal microscope is shown in Figure 1.10. Confocal microscopes consist of a light source, an illumination pinhole, illumination optics, a beamsplitter, a scanning mechanism, one or two relay systems, a microscope objective, detection optics, a detection pinhole and a detector.

1.3.2.1 Illumination

The light source and related optics in a confocal microscope are responsible for uniformly illuminating the entrance pupil of the objective. The light source must pass through an illumination pinhole. Additional illumination optics is used to ensure the entrance pupil of the objective is uniformly illuminated. One common configuration is to collimate the illumination beam directly after the illumination pinhole as shown in Figure 1.10.

Lasers are common light sources for confocal microscopes. Lasers are advantageous since they are high power, are stable, and can be focused to a small spot on the sample. If the laser has multiple modes, it has to be spatially filtered. The pinhole used during spatial filtering can simultaneously act as the illumination pinhole. Some incoherent light sources that are used in confocal microscopes are arc lamps and LEDs. The light source should be chosen based on the confocal system design. Some designs are more suited to coherent illumination than incoherent illumination.

1.3.2.2 Beamsplitter

The type of beamsplitter used in a confocal microscope depends on imaging modalities. For reflectance confocal microscopy, a 50/50 beamsplitter may be used. By using this beamsplitter half the illumination light and half the detection light are lost upon reflection. If the reflected signal is low, a beamsplitter with a different transmission/reflection ratio can be used. For instance, a 90/10 beamsplitter, which transmits 90% of the light, is better suited for applications in which the reflected signal is low. Using a high power light source can compensate the loss of 90% of the illumination light upon reflection.

For fluorescence confocal microscopy, a dichroic mirror is used as a beamsplitter. The dichroic mirror typically reflects the excitation light and transmits emission light with high efficiency. In addition to a dichroic mirror, an excitation filter and emission filter are needed to further enhance image contrast.

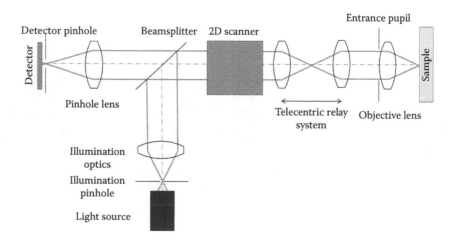

FIGURE 1.10   Example of a typical set-up for a confocal microscope.

The configuration with dichroic mirror, emission filter and excitation filter is especially desirable for fluorescence microscopy, since the fluorescence signal is often very weak.

1.3.2.3 Scanner

The easiest way to scan a confocal microscope is to implement stage scanning. This allows the optical system to remain stationary, so there are no complications added to the system. The major disadvantage of stage scanning is that the scanning speed is too slow for real-time imaging.

There are multiple methods that are used to increase the speed of scanning to video rate. The most common is the use of galvanometer scanners. Galvanometer scanners consist of the galvanometer, a mirror, and a servo driver that controls the system. The galvanometer mirrors can operate either in closed loop or resonant scanning. The resonant scanning mirrors are advantageous since they yield extremely high scanning speeds. However, at the resonant frequency there is no control over the speed or the angular range of the mirror. The scanning path is often sinusoidal for resonant scanning galvanometers. For this reason closed loop scanning mirrors are used when the scanning speed and scanning range needs to be modified regularly. Closed loop scanning mirrors cannot achieve as high speeds as resonant mirrors, but the scanning path and speed are easily controlled. The scanning mirrors should be optically conjugate to the entrance pupil of the objective.

Polygonal scanners utilize a rotating polygonal optical element with three or more reflective facets. By rotating this polygon at high speeds, the output beam is scanned in a sawtooth pattern. If each facet of the polygon is slightly tilted, the beam position fluctuates from facet to facet.

Acousto-optical scanners utilize a surface acoustic grating to deflect a beam. They are capable of extremely high scan rates, up to 20,000 sweeps per second [3]. However, they provide small angular deflection, limited angle resolution, and low light efficiency. Furthermore, the scan angle is wavelength dependent. Due to the wavelength dependence, this is not suitable for fluorescence imaging.

1.3.2.4 Objective Lens

The microscope objective determines the resolution of the confocal microscope. NA of the objective not only determines the resolution of the microscope, but also determines the light collection efficiency and optical sectioning ability. Inherent in microscope objectives is the tradeoff between NA and field of view (FOV). As the NA increases, FOV and the working distance of the microscope shrink.

For good optical performance, the microscope objective should be well-corrected for aberrations. Since confocal microscopes image details inside the tissue, spherical aberration, caused by the tissue thickness, significantly reduces the resolution. To correct for the spherical aberration induced by the tissue, liquid-immersion objective lenses are often used in confocal microscopy. Objective lenses are typically designed to be telecentric for confocal systems to maintain constant magnification as the microscopes image deep into tissue. The telecentricity also ensures the collection angle for every point on the sample is constant.

1.3.2.5 Pinhole

The pinhole in the confocal microscope determines both the optical sectioning capability and the resolution. The derivation of the PSF for a confocal microscope assumes an infinitely small pinhole. There is an inherent tradeoff in the choice of a pinhole. A smaller pinhole yields better resolution and better optical sectioning; however, there is less signal. In light starved samples, a small pinhole may not yield an appropriate signal-to-noise ratio (SNR). If the pinhole is increased, the signal is higher, but more light is detected outside the focal point. If the pinhole is increased too much, the system is no longer confocal. A further discussion of the pinhole size on the resolution is found in Refs. [3–5].

Typically a pinhole size of 1 Airy unit is chosen, since it has been shown optical sectioning does not considerably improve if the pinhole is smaller [3]. An Airy unit is the radial distance to the first zero of the Airy disk. Since the Airy disk size changes with wavelength, the optimal pinhole size for different wavelengths varies. The performance can be increased for a multi-wavelength confocal microscope if each wavelength has its own pinhole aperture and detector.

1.3.2.6 Relay Systems

In addition to the relay system coupling two scan mirrors optically, another relay system is usually needed to image the scan mirror at the entrance pupil of the objective lens. The basic requirement of a relay system in confocal imaging is that the aberrations of the relay system should be well-controlled and the relay lens should be telecentric.

The most straightforward relay system is a 4f system consisting of two identical doublets assembled afocally. The first scan mirror is located at the front focal plane of the first lens, and the second scan mirror is at the rear focal plane of the second lens. The odd aberrations, coma, distortion and lateral chromatic aberration are compensated by the symmetrical configuration. The residual aberrations are spherical aberration, field curvature, astigmatism, and axial chromatic aberration. To address the chromatic aberrations inherent in refractive relay lenses, reflective relay lenses have been developed. Given that the optical properties of a reflective surface only depend on the radius of the mirror, chromatic aberrations disappear.

1.3.2.7 Detector

Common detectors for confocal microscopes are photomultiplier tubes or avalanche photodiodes. Both allow for detection of weak signal with a high SNR. For parallelized confocal systems, the detector may either consist of an array of point detectors (photomultiplier tubes and avalanche photodiodes) or use a high-sensitivity sensor. This high sensitivity sensor can be a cooled charge coupled device (CCD), electron multiplying charge coupled device (EMCCD) or a scientific complementary metal-oxide semiconductor (sCMOS).

1.3.3 Types of Confocal Microscopes

The confocal microscope is basically a point scanning imaging system. The major issue with point scanning systems is the imaging speed. This section will discuss various scanning methods for improved imaging speed.

1.3.3.1 Point Scanning

The most common confocal system is the point scanning confocal microscope. The illumination pinhole and detection pinhole are confocal on the sample, allowing for improved resolution and optical sectioning capabilities. The excitation light has to be scanned in two dimensions over the entire sample; its major disadvantage is the time it requires to scan over the entire 2D FOV of interest. Figure 1.10 is a typical configuration for a point scanning confocal microscope.

1.3.3.2 Line Scanning

A line scanning confocal microscope scans a focused line across the sample, and the detection pinhole is now a detection slit. The line scanning approach can increase the imaging speed since only one dimension on the sample has to be scanned. In fact, the confocal principle using a slit was described by Goldman in 1940 before Minksy developed the point-scanning confocal microscope [6]. The major limitation is that the resolution is only enhanced in the dimension perpendicular to the slit. The resolution parallel to the slit is not enhanced by the slit.

1.3.3.3 Nipkow Disk

Instead of using just a single pinhole for the confocal imaging, a Nipkow disk utilizes a rotating disk with a large number of pinholes. This is essentially a massively parallel confocal microscope. By rotating the disk at high speeds, this allows for extremely fast acquisition of confocal images. Using a Nipkow disk, confocal images can be directly observed with the naked eye.

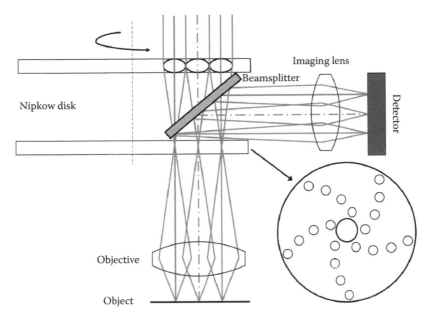

FIGURE 1.11 Set-up for a Nipkow disk confocal microscope that utilizes microlenses to increase illumination efficiency.

The Nipkow disk was invented by Nipkow in 1884 as a way to transfer 2D spatial information into a temporal electrical signal by placing a series of holes on a scanning disk. The Nipkow disk was modified by Petran and Hadravsky in 1968 to demonstrate confocal imaging using a tandem-scanning reflected-light microscope (TSRLM) [7]. A major disadvantage of the Nipkow disk confocal microscope is low illumination efficiency, since only a small percentage of the light passes through the small pinholes. If more pinholes are added to the Nipkow disk, there is a greater chance of crosstalk between each pinhole.

While a few methods exist to improve illumination efficiency, the most commonly used is to place a disk with microlenses above the Nipkow disk. The microlens disk is aligned with the Nipkow disk and they rotate together. Each of the microlenses focuses the light onto the corresponding pinhole, as shown in Figure 1.11 This improves the light efficiency from ~2% to almost 70% [8].

1.3.3.4 Confocal Microscopes Using Spatial Light Modulators

Another method for parallel confocal scanning is to use a spatial light modulator as the scanning mechanism [9]. A digital micro-mirror device (DMD) is one type of spatial light modulator that has been used for confocal imaging. A DMD is a device that consists of hundreds of thousands of micro-mirrors that are independently switched "on" and "off" at extremely high speeds. The "on" state typically rotates the mirror by 12 degrees, whereas the "off" state rotates the mirror by -12 degrees. Each mirror in the DMD can serve as a pinhole for the illumination and detection path. By temporally switching the micro-mirrors "on" and "off," the sample can be scanned in high speed with multiple beams simultaneously. Since multiple beams are typically scanned simultaneously, a CCD or CMOS sensor is typically used to capture the emission signal.

1.3.3.5 Fiber Confocal

Due to the high resolution imaging capability, confocal imaging have many potential clinical applications. However, the size of traditional confocal microscopes makes it problematic to be used for *in vivo* applications. In order to overcome the size constraints, there are multiple designs that create a small portable confocal microscope that can be used for clinical applications. One such design is to use fibers to create a flexible system with a miniature confocal head. For fiber confocal systems, the fiber acts as both the illumination and detection pinholes. When using fibers for confocal systems, special care should be taken to efficiently couple light into the fibers and to remove Fresnel reflections at the fiber ends, since this acts as background noise for the system. Two designs that utilize fibers for a confocal microscope are single-fiber confocal microscopes and fiber bundle confocal microscopes.

For single-fiber confocal microscopes, light is coupled into a single fiber and the coupled light is output to a miniature confocal imaging probe. The probe typically consists of method to scan the fiber and a small objective lens [10–17]. The fiber can be mechanically scanned using piezoelectric actuators, electromagnetic actuators, or electromagnetism [18–20]. Another promising method is to use MEMS mirrors to scan the light output from the fiber. A challenge in using MEMS mirrors is maintaining a small volume for the confocal probe.

Instead of using a single fiber for a small, confocal microscope, a fiber bundle can be used to transfer the light to a small confocal probe [21, 22]. Using a fiber bundle eliminates the need for a scanning mechanism in the confocal probe. Since the beam scanning mechanism can be placed at the proximal end of the fiber, a high speed scanning mechanism can be used to scan the beam into individual fibers at the proximal end of the fiber bundle. Each fiber at the distal end acts as the illumination and detection pinhole for the confocal probe. Due to the spacing between the fibers in the fiber bundle, the acquired image is pixelated, which is one limitation of this method. One variant of this design is to use a DMD to couple light into individual fibers [23]. There are two advantages of using a DMD, the scanning pattern does not have to be a raster pattern and multiple fibers can be illuminated simultaneously in parallel, which increases the acquisition speed. Since the entire DMD is illuminated while only a few pixels in the DMD are "on," this design has low light efficiency. Furthermore, if the DMD is not well aligned with the fiber bundle, some pixels on the DMD will image onto the spaces between the fibers in the fiber bundle.

1.3.3.6 Spectral Confocal Imaging

The number of spectral bands measured in a confocal system depends on the number of spectral filters used in the confocal microscope. To distinguish between the signal from multiple fluorescent dyes, multiple optical filters need to be used before to differentiate the fluorescent dyes. However, the emission spectra from multiple dyes may begin to overlap. The optical filters may be unable to separate the signal from multiple dyes due to this overlap, which leads to spectral crosstalk in the measurements.

To overcome the limitations of using multiple optical filters in a confocal microscope, a hyperspectral confocal microscope records the emission spectrum for every voxel in the imaged volume by placing a spectrometer after the detection pinhole [24]. Hyperspectral imaging is a natural extension of confocal microscopy, since the emission spectrum is detected after passing through a pinhole, which acts as the entrance pinhole for a spectrometer. The measured spectra are used to identify and distinguish multiple fluorescent dyes in the sample [25]. Even if the emission spectra overlap, the fluorescent dyes can be distinguished through analysis. Hyperspectral imaging has numerous medical applications [26].

1.4 Optical Sectioning Structured Illumination Microscopy (OS-SIM)

In order to perform three-dimensional (3D) imaging, optical microscopes need to be capable of optical sectioning, which is the ability to distinguish an in-focus signal from the out-of-focus background. Common techniques for fluorescence optical sectioning are confocal laser scanning microscopy and two-photon microscopy. Optical sectioning SIM is an interesting alternative in optical sectioning techniques, since it has the potential for high imaging speeds, large fields of view or acquisition for long periods of time. OS-SIM illuminates the sample object with sinusoidal patterns, and removes the out-of-focus blur computationally to achieve thin optical sections with a capacity similar to confocal microscopy [17–19]. This section discusses OS-SIM, including its principles, the optical sectioning ability and reconstruction algorithms, as well as its instrumentation problems.

1.4.1 Principle

OS-SIM was developed by Wilson et al. in 1997 [27]. The typical experimental set-up is shown in Figure 1.12. Usually, a diffraction grating with ±1 diffraction orders or a digital micro-mirror device (DMD) is employed to generate the illumination pattern. Using an incoherent light source, the periodic pattern is projected in the focal plane of a conventional microscope and three images are acquired, each at a different phase shift; that is, 0, $2\pi/3$, $4\pi/3$. OS-SIM can be implemented for both fluorescence imaging and white light imaging. For fluorescence imaging, the fluorescent signal is modulated in the direction perpendicular to the long axis of the grid lines of the illumination pattern, reproducing the projected variations in intensity. When defocused the pattern contrast is reduced, resulting in axially displaced features being effectively Kohler illuminated and weaker than those in the focal plane. Consequently, out-of-focus fluorophores do not exhibit modulated fluorescence when comparing the three raw images. The grid pattern mask is removed on a pixel-by-pixel basis using

$$I_{\text{sec}}(x,y)=[(I_1(x,y)-I_2(x,y))^2+(I_1(x,y)-I_3(x,y))^2+(I_2(x,y)-I_3(x,y))^2]^{1/2}, \qquad (1.22)$$

where I_1, I_2, I_3 represent the three phase shifted images, (x, y) represents the pixel coordinate in the images. Figure 1.1 shows the comparison of optical sectioned image and conventional wide-field image of lily pollen grain [27]. Figure 1.13a represents an autofocused image obtained by displaying the maximum image intensity at each pixel throughout a 30-μm axial scan with the 50x, 0.75 NA objective lens. Figure 1.13b, shows a conventional image taken at a mid-plane through the grain. Figure 1.13a shows that the optical sectioned image dramatically removes the out-of-focus light. It should be noted that techniques used in OS-SIM are limited to optical sectioning, and are not able to achieve super-resolution.

1.4.2 Optical Sectioning Strength

Karadaglić and Wilsoned showed that the sectioned image is described by [28]

$$I_{\text{sec}}(x,y)=|S_p(x,y)| \qquad (1.23)$$

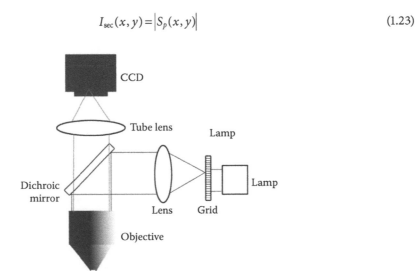

FIGURE 1.12 Typical experimental set-up for OS-SIM. (From Benjamin, T. et al., *J. Opt.*, 15, 94004, 2013.)

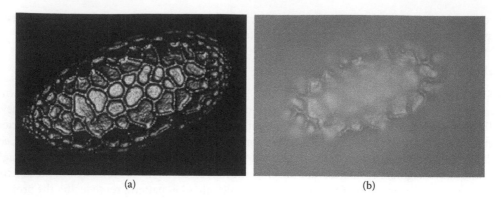

(a)   (b)

FIGURE 1.13   Comparison of optically sectioned image and wide-field image. (a) Optically sectioned image of lily pollen grain. The field size is 100 μm × 70 μm. (b) Conventional image of the lily pollen grain. (From Neil, M. A. A. et al., *Opt. Lett.* 22, 1905–1907, 1997.)

where

$$S_p(x,y) = \exp(-i2\pi px)F^{-1}\left\{\tilde{S}(k_x,k_y)\tilde{H}(k_x+p,k_y)\right\}. \tag{1.24}$$

$\tilde{S}(k_x,k_y)$ refers to the Fourier transform of the sample, $\tilde{H}$ is the optical transfer function of the microscope (OTF), p is the spatial frequency of the illumination pattern and F^{-1} is the inverse Fourier transform operator. Since $\tilde{H}(k_x,k_y;dz)$ drops sharply with defocus dz, this method has optical sectioning ability. Stokseth [29] has developed an analytic approximation for the unaberrated OTF. The full width between zero values of the OTF in the axial direction is described by

$$\Delta = \frac{1.22\lambda}{(2-s)s}, \tag{1.25}$$

where $s = \sqrt{k_x^2 + k_y^2}$ and is measured in units of NA/λ. Therefore, the minimum axial width is at $s = $ NA/λ. Thus, the optimal p for optical sectioning is NA/λ [30].

The optical sectioning strength is represented by the axial full width at half maximum (FWHM) of the grid pattern, which is estimated using [31]

$$\text{FWHM}_z = \frac{(3.63/16\pi)(\lambda \cdot 10^{-3})}{\left[\eta \sin^2(\alpha/2)v_g(1-v_g/2)\right]} \tag{1.26}$$

where λ is the excitation wavelength in nm, η is the refractive index of the sample medium, and α is the objective aperture angle. v_g is effective grid frequency in the CCD sensor plane and is described by

$$v_g = \frac{\beta\lambda v}{\text{NA}} \tag{1.27}$$

where β is the magnification of the illumination pattern between the grid plane and the sample plane in the illumination path, and v is the actual grid frequency in the grid plane. NA is the numerical aperture of the objective. For the system in Figure 1.14, β can be calculated by

$$\beta = f_{projection}/f_{obj} = f_{projection}M/f_{tube} \tag{1.28}$$

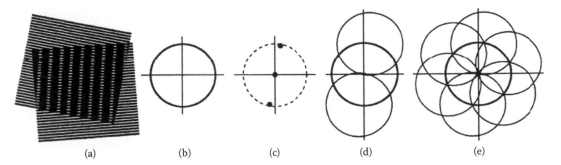

(a)   (b)   (c)   (d)   (e)

FIGURE 1.14 Concept of resolution enhancement by structured illumination. (a) If two line patterns are superposed, their product will contain Moiré fringes (seen here as the apparent vertical stripes in the overlap region). (b) A conventional microscope is limited by diffraction. The set of low-resolution information that it can detect defines a circular "observable region" of Fourier space. (c) A sinusoidal illumination pattern has only three Fourier components. The possible positions of the two-side components are limited by the same circle that defines the observable region (dashed). If the sample is illuminated with such structured light, Moiré fringes will appear, which represent information that has changed position in Fourier space. The amounts of that movement correspond to the three Fourier components of the illumination. The observable region will thus contain, in addition to the normal information, moved information that originates in two offset regions (d). From a sequence of such images with different orientation and phase of the pattern, it is possible to recover information from an area twice the size of the normally observable region, corresponding to twice the normal resolution (e). (From Chang B. J. et al., *Opt. Exp.* 17, 14710–14721, 2009.)

where $f_{projection}$ is the focal length of the projection lens, f_{tube} is the focal length of the tube lens and M is the nominal magnification of the objective. Note that $v_g = 0$ corresponds to conventional wide-field microscopy, while $v_g = 1$ corresponds to the maximum sectioning strength [27].

1.4.3 Optical Sectioning Algorithm

In the original implementation [27], the root-mean-square (RMS) method employed in Equation (1.22) is used to achieve the optical sectioned images. There are two drawbacks associated with this RMS method [32]. The first is that one cannot properly define a point spread function (PSF), meaning that the imaging properties are, theoretically, sample-dependent. However, this sample-dependence varies slightly in practice. A second drawback is that any noise in the images, such as shot noise, when processed by a nonlinear algorithm, leads to a bias in the results. Extra care must be taken to correct for this bias. An alternative processing algorithm [33, 34] circumvents this nonlinearity problem and even leads to the possibility of super-resolution. Benjamin et al. proposed a new algorithm for optical sectioning using 3 raw images as the RMS method to create an optically sectioned image with better resolution, higher contrast, and better image fidelity [35].

1.4.4 Problem of Speed and Solution

In SIM, the optical sectioning strength increases as the spatial frequency of the illuminating fringes increases. Sinusoidal fringe illumination with a spatial frequency near the cut-off frequency of the microscope's OTF is essential to maximize the optical sectioning strength. The phase shifted sinusoidal fringe was originally produced by mechanical motion of a diffraction grating, which had a low speed and low precision of phase shifts [34]. The SIM frame-rate can be increased drastically by using a spatial light modulator (SLM), which has the advantage of generating and controlling the fringe patterns accurately and quickly [36]. State-of-art ferroelectric liquid crystal on silicon spatial light modulators (LCOS-SLM) allows higher reflectivity and frame-rates [37]. SIM using a fringe projection scheme that combines DMD with LED illumination achieved a lateral resolution of 90 nm and an optical sectioning depth of 120 μm [38].

Since several raw images (at least three) are required to calculate the optical sectioning image, this restricts the speed for real time imaging. An alternative structured illumination technique called HiLo microscopy can increase the frame rate since it requires only one structured illumination image and one standard uniform illumination image [39]. The demodulation of contrast is performed spatially with the single structured image rather than temporally with a sequence of structured images. The two images required for HiLo can be acquired sequentially, largely eliminating motion artifacts. Moreover, HiLo microscopy is insensitive to defects or sample-induced distortions in the illumination structure. A wide-field fluorescence microscope set-up, which combines HiLo microscopy technique with the use of a two-color fluorescent probe, was proposed by Muro et al. [40]. It allows one-shot fluorescence optical sectioning of thick biological moving samples, which is illuminated simultaneously with a flat and a structured pattern at two different wavelengths. It can achieve a frame rate of 25 images per second.

1.5 Super-Resolution Structured Illumination Microscopy (SR-SIM)

Confocal microscopy, multiphoton fluorescence microscopy, and OS-SIM allow the observation of live sample structures in real time with improved optical sectioning and contrast. However, all these approaches are diffraction limited, which means they cannot achieve a resolution better than that defined by the diffraction limit. This is not enough for observing cellular and subcellular level phenomena.

The ability to surpass the diffraction limit has been an ongoing goal in microscopy. Ambrose introduced the concept of total internal reflection fluorescence microscopy in 1956, which is one of the first instances of super-resolution fluorescence microscopy. Over the years, more super-resolution techniques have been developed. In order to collect information from a sample in three dimensions, a far-field fluorescence super-resolution imaging technique (STED) was proposed in the 1990s [41]. In mid-2000, other techniques, such as single-molecule localization based super-resolution microscopy (STORM and PALM), were developed [42, 43]. In this section, we will focus on super-resolution structured illumination microscopy (SR-SIM) [34]. SIM commonly uses the interference between two beams to generate a series of sinusoidal illumination patterns. In 2005, Gustafsson demonstrated saturated structured illumination microscopy (SSIM) [44]. In 2008, he described how SIM can be applied in three dimensions to double the axial as well as the lateral resolution with true optical sectioning [33]. In 2009, he and his colleagues demonstrated a high-speed SIM that is capable of 100 nm resolution at frame rates up to 11 Hz for several hundred time points. By using a ferroelectric liquid crystal on silicon spatial light modulator (SLM) with 1,024 × 768 pixels to produce the patterns, the pattern-switching time is decreased by three orders of magnitude [37]. In 2011, they used a system similar to that of the live TIRF-SIM for live 3D SIM [45]. In the same year, they demonstrated whole-cell super-resolution imaging by nonlinear SIM [46]. With reversible photoswitch of a fluorescent protein, it requires nonlinearity at light intensities six orders of magnitude lower than those needed for saturation. They experimentally demonstrated approximately 40 nm resolution on purified microtubules labeled with the fluorescent photoswitchable protein Dronpa.

SR-SIM uses the projection of a high-frequency sinusoidal pattern with the excitation light. When such a sinusoidal wave interferes with the sample, Moiré patterns arise and allow spatial details below the diffraction limit to become visible. A maximum resolution enhancement of two can be achieved when the emitted fluorescence intensity is linear to that of the exciting light. By introducing non-linear effects into the illumination pattern, non-linear SIM is capable of greater resolution improvement than linear SIM [47]. Recent developments of SIM allow fast, multicolor, and three-dimensional high-resolution live-cell imaging [48–50]. To meet the needs of biological research, SIM aims at faster speed, higher resolution, and more robust algorithms.

1.5.1 Principle

As shown in Figure 1.14a, if two high frequency patterns are superposed, Moiré fringes appear and high frequency information becomes apparent due to Moiré effects. Similarly, when the unresolved, unknown high frequency sample structure is superposed with the structured illumination, the Moiré pattern arises in which spatial details below the diffraction limit become visible. A diffraction limited optical system can only detect information below the highest spatial frequencies that the objective can transmit ($2NA/\lambda$), as shown in Figure 1.14b. With structured illumination consisting of a sinusoidal stripe pattern, its Fourier transform has three non-zero components as shown in Figure 1.14c. The origin point corresponds to the zeroth order component and the other two correspond to the first order component, which are offset from the origin in a direction orthogonal to the stripe direction and by a distance proportional to the inverse line spacing of the pattern. When the sample is illuminated with the structured light, the captured image contains Moiré fringes corresponding to information in which the position in Fourier space has been offset. As shown in Figure 1.14d, due to frequency mixing, the observable regions also contain, in addition to the normal image of spatial frequencies (center circle), two new offset frequency images that are centered on the edge of the original field. These offset images contain higher spatial frequencies that are not observed using traditional microscope optical systems. Figure 1.15 illustrates how Figure 1.14d works. In Figure 1.15, the black curve represents the Fourier transformation of the sample [51]. Since the multiplication of the illumination pattern and the sample information in real space corresponds to a convolution in Fourier space, the same sample information will be "attached" to each of these three peaks introduced by the sinusoidal illumination pattern. Consequently, previously unresolvable sample information outside the detection OTF is thereby shifted into the observable region, making it now resolvable to optical imaging. If the illumination is achieved through the objective, the maximally possible illumination frequency will be close to the edge of the observable region of the detection OTF as shown in Figure 1.14c. However, it is impossible to separate these three parts as shown in Figure 1.14d using a single image. The coefficients by which different Fourier transformed components are added together depend on the phase provided by the structured illumination pattern. Since the illumination pattern is known and controllable, they can be separated through simple arithmetic by recording three or more phase shifted images. These Fourier transformation components can be restored to its proper position. The highest spatial frequency possible is the cutoff frequency of the microscope, which is also the largest offset possible. With this offset, it is possible to access Fourier transformation coefficients out to double the previous observable region in the phase shifted direction. By repeating this with the patterns oriented in different directions, one can gather essentially all the Fourier transformation coefficients with a circle twice as large as the physically observable region, as shown in Figure 1.14e.

Note that if data is acquired with at least N different phases, the number of equations is at least equal to the number of unknowns. In this way, it allows the N information components to be separated by applying a simple $N \times N$ matrix to solve it. If phases are badly chosen, it may lead to an ill-conditioned or even singular matrix. The natural choice of equally spaced phases on the $0-2\pi$ interval produces a well-conditioned separation matrix, which in fact performs a discrete Fourier transform of the data with respect to the phase-shift variable [33]. Usually, three phase shifting patterns with phase 0, $2\pi/3$, $4\pi/3$ in one orientation and three orientations 0°, 60°, 120° are chosen in the image acquisition.

1.5.2 SR-SIM Instrumentation

The most important component in a SR-SIM instrument is the generation and modulation of the structured illumination pattern. The original SR-SIM experimental set-up by Gustafsson [33, 34] employs a phase grating to generate two beams. The ±1 orders are selected to generate the spatial varying sinusoidal patterns. The microscope objective lens projects a demagnified image of this grating onto the sample surface. The fringe spatial frequency is close to the cut-off frequency of the objective lens. The orientation and phase of the stripe illumination pattern are controlled through rotation and lateral translation of the grating.

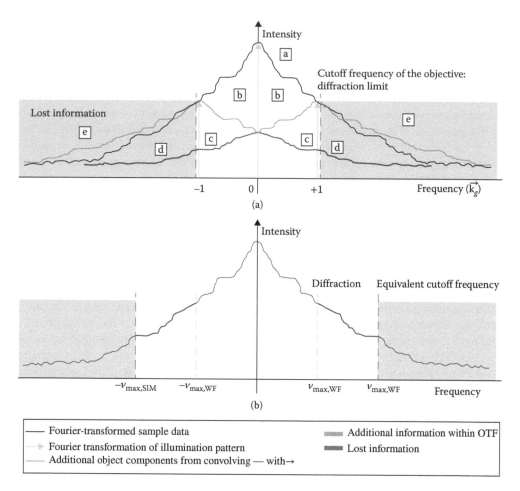

FIGURE 1.15 Lateral resolution enhancement. (a) Imaging model: The Fourier-transformed sample informa-tion (a) is convolved with the Fourier transformation of the illumination pattern (e). This results in additional object components (b) at the position of the ±first orders. Additional information lies within the support region of the detection optical transfer function (OTF) (c). Some information is still lost (d). (b) Image reconstruction: The equivalent cutoff frequency is twice as big as that in the wide-field (WF) case. This figure does not show the decay of the OTF with frequency and shows only the cutoff outside its support region. (From Jost, A. and Heintzmann, R., *Ann. Rev. Mater. Res.* 43, 261–282, 2013.)

The ability to use this for real-time imaging is limited, since traditional structured illumination pattern generation and modulation takes on the order of seconds. Recently, alternative pattern pro-jection systems have overcome the problem and become popular. Liquid-crystal spatial light modula-tors (SLMs) [52, 53] and digital mirror devices (DMDs) [38, 54] have been demonstrated as powerful methods for creating structured illumination. These devices provide great flexibility in terms of pattern generation, as they can be used either as a patterned mask for optical sectioning SIM applications or as a grating to generate interference illumination patterns for SR-SIM.

1.5.3 Reconstruction Algorithm

Since each acquired image contains information in several orders, the standard reconstruction algo-rithm for SIM and non-linear SIM separates the orders from the raw data and combines them in one final super-resolution image [33, 55–58].

The observed data $D(\vec{r})$ is a convolution of the object emission distribution $E(\vec{r})$ with the PSF of the imaging system

$$D(\vec{r}) = E(\vec{r}) * |h(\vec{r})|^2, \tag{1.29}$$

where $\vec{r}$ represents the spatial coordinate of the object. If the object S is illuminated with an intensity distribution $I(\vec{r})$, the resulting emission distribution is

$$E(\vec{r}) = S(\vec{r})I(\vec{r}). \tag{1.30}$$

For structured illumination, the pattern is a sum of a finite number of components and is described by

$$I(\vec{r}_{xy}, z) = \sum_m I_m(z) J_m(\vec{r}_{xy}), \tag{1.31}$$

where $\vec{r}_{xy}$ denotes the lateral coordinates (x, y) and z denotes the axial coordinate. J_m is the lateral function with a single spatial frequency $\vec{p}_m$ and it is described by

$$J_m(\vec{r}_{xy}) = \exp\left(i\left(2\pi\vec{p}_m \cdot \vec{r}_{xy} + \varphi_m\right)\right), \tag{1.32}$$

where φ_m is the phase shift of the pattern on the mth order. Therefore, the acquired data $D(\vec{r})$ is written as

$$D(\vec{r}) = \sum_m \left[\left(|h|^2 I_m\right) * (SJ_m) \right](\vec{r}). \tag{1.33}$$

Let D_m denote the mth term of the above sum, its Fourier transform will take the form

$$D_m(k) = O_m(k)[\tilde{S}(k) * \tilde{J}_m(k_{xy})], \tag{1.34}$$

where O_m is the Fourier transform of $(|h|^2 I_m)$; $\tilde{S}(k)$ indicates the Fourier transform of the object; $\tilde{J}_m$ is the Fourier transform of J_m. Since J_m is a harmonic wave with only one spatial frequency $\vec{p}_m$, its Fourier transform is described by

$$\tilde{J}_m(k_{xy}) = \delta(k_{xy} - \vec{p}_m)e^{i\varphi_m}. \tag{1.35}$$

The observed data is re-written as

$$\tilde{D}(k) = \sum_m \tilde{D}_m(k) = \sum_m O_m(k)e^{i\varphi_m}\tilde{S}(k - p_m). \tag{1.36}$$

The natural choice of equally spaced phases on the 0-2π interval produces a well-conditioned separation matrix. The total phase shift φ_m is calculated as $\varphi_m = m\varphi$ and $\vec{p}_m = m\vec{p}$ for harmonic patterns. So observed data is then revised as

$$\tilde{D}(k) = \sum_m O_m(k)e^{im\varphi}\tilde{S}(k - m\vec{p}) \tag{1.37}$$

Because O_m and $\vec{p}_m$ are known, the separated information components are computationally moved back (by a distance $\vec{p}_m$) to their true positions in Fourier space, recombined into a single extended-resolution data set, and finally retransformed to real space to acquire the super-resolution image. The final reconstructed super-resolution image in frequency domain is described by

$$\hat{\tilde{S}}(k) = \frac{\sum_{d,m} O_m^*(k+m\vec{p}_d)\tilde{D}_{d,m}(k+m\vec{p}_d)}{\sum_{d',m'} \left|O_{m'}(k+m'\vec{p}_{d'})\right|^2 + w^2} A(k),\tag{1.38}$$

where d represents the orientation, primed parameters refer to the specimen reference frame, the corresponding unprimed parameters refer to the data set reference frame, w^2 is the Wiener parameter, which is taken to be a constant and adjusted empirically, and $A(k)$ is an apodization function.

1.5.4 Nonlinear SIM

Although its resolution is limited, linear SIM is still the commonly used high-resolution microscopy method for live cell imaging due to its speed [15]. Nonlinear SIM offers an attractive way to improve resolution without modifying the SIM set-up. High-frequency harmonics are introduced into the effective pattern to increase its resolution enhancement power. Utilization of non-linear effects to increase the resolution in SR-SIM was theorized in 2002 by Heintzmann with the concept of saturated pattern excitation microscopy (SPEM) [59]. In 2005, Gustafsson proposed the first practical implementation of non-linear SIM (NL-SIM), where 2D lateral resolution smaller than 50 nm using fluorescent beads is achieved [60].

The resolution extension limit can be exceeded if the emission rate can be made to depend nonlinearly on the illumination intensity. If such a non-linearity can be described by a sinusoid raised to the nth power, it will give rise to $n-1$ new harmonics. If the nonlinearity has an infinite number of harmonics, it can give rise to theoretically unlimited resolution. Ultimately, the number of accessible harmonics N is limited by more practical considerations, such as photobleaching and signal-to-noise ratio. The contribution of each harmonic is determined in the same manner as SR-SIM, requiring one to image N different phases of the pattern at each orientation. One of the fundamental difficulties with saturated structured illumination microscopy (SSIM) is the required illumination intensities for saturating fluorophores in the excited state. Unfortunately, because saturation requires extremely high light intensities that are likely to accelerate photobleaching and damage even fixed tissue, this implementation is of limited use for studying biological samples. SSIM also requires a much larger set of raw wide-field images to create a reconstruction. Using nine total harmonics (three additional orders created by non-linearities) to extend resolution requires one to image nine different phases of the pattern. In addition to photobleaching, sample drift can become problematic due to the extended imaging time and sensitivity of the technique.

Reversible photoswitching of a fluorescent protein provides the required nonlinearity at light intensities six orders of magnitude lower than those needed for saturation. In general, photoswitchable fluorescent molecules can be reversibly switched between two spectrally distinct states using light; saturating either of these population states results in a nonlinear relationship between the fluorescence emission and the illumination intensity. This method has been implemented on purified protein absorbed to non-biological structures and protein-filled bacteria in one dimension with a wide-field configuration [61, 62], or in two dimensions with a point-scanning, donut-mode configuration [63]. Rego et al. experimentally demonstrate approximately 40 nm resolution on purified microtubules labeled with the fluorescent photoswitchable protein Dronpa [63]. Figure 1.16 shows part of their experimental results. They imaged the microtubules with the nonlinear structured-illumination microscope in two dimensions with a 10 saturated level. Figure 1.16a shows the comparison among conventional total internal reflection fluorescence

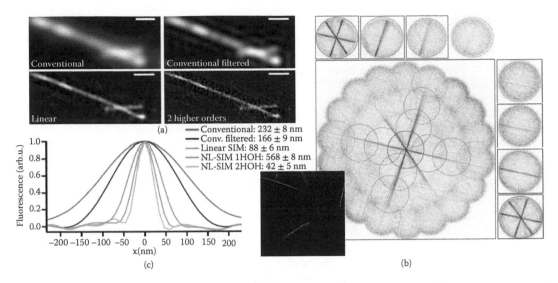

FIGURE 1.16 Comparison of conventional TIRF, linear SIM-TIRF, and NL-SIM-TIRF. (a) Two Dronpa-coated microtubules imaged by conventional TIRF, conventional filtered TIRF, linear SIM-TIRF, and NL-SIM-TIRF with two higher-order harmonics. Scale bar is 500 nm. (b) Fourier transforming a different subset of the data (Inset) reveals strong features in frequency space. Two microtubules display features that are visible in the second higher-order information component, which is most obvious when the orders are separated above and to the right. (c) Line profiles were taken through 10 microtubules and subsequently aligned and averaged. The filtered conventional image and associated profile were generated by processing the data and then discarding all orders except for the conventional component and as such represents the conventional Wiener filtered data. As expected, the final resolution of the NL-SIM data with two additional higher-order harmonics (2HOH) is 4 times that of conventional microscopy, or 42 nm. (From Rego, E. H. et al., *Proc. Natl. Acad. Sci. USA*, 109, E135, 2011.)

(TIRF) image and linear SIM-TIRF, filtered TIRF and NL-SIM-TIRF images. Figure 1.16b shows the Fourier transform of a different subset of the data (Inset) in frequency space. Two microtubules display features that are visible in the second higher-order information component, which is most obvious when the orders are separated above and to the right. Figure 1.16c shows the corresponding inverted spatial super-resolution image of Figure 1.16b. Averaging line profiles of 10 microtubules, a resolution of 42 nm, or four times the resolution of a conventional microscope is acquired.

1.5.5 Combining OS-SIM and SR-SIM

OS-SIM and SR-SIM have many similarities. They have similar experimental set-up, both the optical sectioning ability and super-resolution ability depend on the spatial frequency of the projected fringes. Three-dimensional SIM is a wide-field technique that can achieve lateral and axial resolution of 100 nm and 300 nm using spatially structured illumination light [33]. The lateral resolution, axial resolution, and true optical sectioning are achieved by moving high-resolution sample information into the normal pass-band of the microscope through frequency mixing. Combining OS-SIM and SR-SIM to achieve 3D resolution enhancement involves a modification to two-dimensional SR-SIM. Instead of maximizing the frequency of the pattern to equal $2NA/\lambda$, it is set to NA/λ. The result is a greater overlap of OTF support regions, filling in the missing cone and allowing for optical sectioning with modest gains in lateral resolution [64].

Holleran et al. developed different approaches for reconstructing 2D structured illumination images in order to combine SR-SIM and OS-SIM to allow fast, optically sectioned, super-resolution imaging [58]. Linear reconstruction method is employed to maximize the axial frequency extent of the combined 2D structured illumination passband along with an empirically optimized approximation to this scheme.

For sinusoidal excitation at half the incoherent cutoff frequency, it was found that removing the zero order passband except for a small region close to the excitation frequency enables optimal reconstruction of optically sectioned images with enhanced spatial resolution. The zero order passband, except for a small region close to the excitation frequency, is replaced by the complementary information from the displaced first order passband.

1.6 Endoscopy

1.6.1 Introduction

The earliest endoscope using lenses was invented by Maximilian Nitze in 1877 and was used to examine the interior of the urinary bladder through the urethra [65]. In 1959, Harold H. Hopkins invented rod lenses for image transmission; and in 1963, Karl Storz combined rod lenses for image transmission with fiber bundles for illumination [66]. A video endoscope with a camera on the eyepiece was first introduced in 1987, and a video endoscope with a camera at the distal end of the endoscope was demonstrated in 1992 [65]. In the late 1990s, with the advances in light sources, sensors, and electronics, a wireless endoscope was invented for the investigation of the small bowel [67].

Generally, endoscopes comprise of an airtight and waterproof elongated tube having a distal end with an objective lens for imaging and a proximal end with an eyepiece for viewing. The elongated tube includes a relay lens system, or a fiber bundle, to transmit the image formed by the objective lens to the proximal end of the tube. The function of the eyepiece is to magnify the image at the proximal end for the observer. For endoscopes used in clinical applications, they additionally consist of a work channel and an irrigation channel.

1.6.2 Basic Optics for Endoscopes

The basic elements of an endoscope include an illumination system, imaging system, image transmission system (relay system), and a viewing system (eyepiece or electronic sensor). Figure 1.17 is a diagram of a conventional optical system within a rigid endoscope with only one relay stage for simplicity. It consists of three basic and separate optical components in sequence of the direction of the traveling light: an objective lens forms the first inverted intermediate or final image of the object, a relay lens system re-images the intermediate image to the final image, and an eyepiece or camera lens.

According to the definition of optical transmission systems, endoscopes are classified into three groups: rigid, fiber optic, and video. In recent years, new types of endoscopes, such as wireless, scanning, and stereo, have been developed.

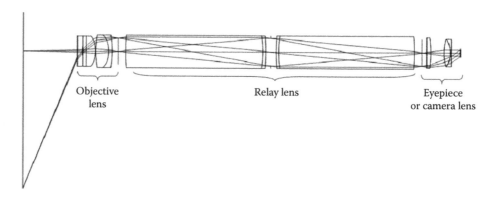

Objective lens   Relay lens   Eyepiece or camera lens

FIGURE 1.17 Typical optical layout of a rigid endoscope with a one-stage relay lens. It consists of three lenses: the objective lens, relay lens, and eyepiece or camera lens.

Rigid endoscopes have a rigid tube to house the refractive relay lenses as shown in Figure 1.17. The relay lenses transfer the image at the distal end to the proximal end of the tube so that the image can be directly viewed through an eyepiece or a digital camera. The depth of field of the rigid endoscope is determined by the viewing configuration. When an eyepiece is mounted at the proximal end of the endoscope, the accommodation of the human eye helps to extend the depth of field. When a CCD or CMOS sensor is used at the proximal end of an endoscope, the depth of field is smaller than that of an endoscope using an eyepiece. The depth of focus can be calculated from the requirement of the resolution and the size of the pixel in the CCD or CMOS. One requirement for objective lenses used in rigid endoscopes is the telecentricity that prevents light loss during image transmission from the distal end to the proximal end.

Flexible endoscopes use imaging fiber bundles to transmit the image from the distal end of the endoscope to the imaging detector or eyepiece, as shown in Figure 1.18. The advantages of fiber-optic endoscopes include: (1) transmitting the image over long distances, which is not achievable with a lens relay, and (2) observing around corners, which is achieved through excellent flexibility. An imaging fiber bundle typically contains about 3000 or more optical fibers packed into a hexagonal array. The spatial arrangements of the fibers at both the proximal and distal ends are identical, resulting in a spatially coherent transmission of the image. The optical fibers are multimode, the core diameter is typically 3 μm, and the overall diameter of the imaging bundle may vary between 0.2 and 3 mm. The resolution of flexible endoscopes is limited by the core diameter of the fiber bundle. Fiber bundles have a smaller depth of field than in conventional endoscopes with refractive lens relays, which is due to the pixilated nature of the fiber bundle. Unlike endoscopes with lens relays where the eyepiece can be designed to compensate for the residual aberrations accumulated from the objective and relay lenses, the eyepiece in fiber endoscopes can only give a sharp image of what appears at the proximal end of the fiber bundle, and it cannot correct the blur of the image formed on the distal end by the objective lens and transmitted by the fiber bundle.

As shown in Figure 1.19, the video endoscope has an image sensor in the distal end, therefore the image relay is not necessary. The imaging head of a video endoscope generally consists of an electronic sensor, optics, illumination, and electrical wiring. Video endoscope can be more flexible than the

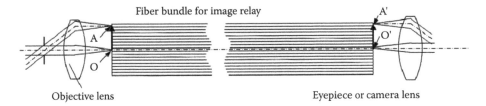

Fiber bundle for image relay

Objective lens

Eyepiece or camera lens

FIGURE 1.18 Optical layout of a typical fiber-optic endoscope.

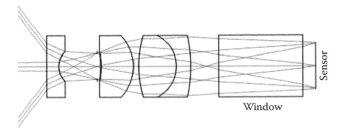

Window

Sensor

FIGURE 1.19 Optical layout of a typical video endoscope.

flexible endoscope with fiber bundle and can provide a sharper image because the number of the pixels in a sensor is greater than the number of fibers in a fiber bundle. The objective lens for a video endoscope is required to have a short structural length, a small diameter, and a wide field angle. Compared to the objective lenses used in rigid and flexible endoscopes employing fiber bundles, telecentricity is not a requirement. When the size of a sensor does not allow the objective lens to be placed perpendicular to the optical axis, a folding mirror or prism is required so that the sensor can be placed parallel to the optical axis of the objective lens. The resolution and sensitivity of image sensors are constantly improving; smaller sensors with higher resolutions will provide for the opportunity to design smaller endoscopic probe heads and to place two or more sensors into probe heads for stereoscopic viewing of an object.

Wireless endoscopy, also referred to as capsule endoscopy, is a significant technical breakthrough in endoscopy. Figure 1.20 is a basic configuration of a capsule endoscope. Inside the cylindrical tube and dome, there is an objective lens, a light source (and illumination optics if necessary), an image sensor, electronics to operate the endoscope and digitize the image, a battery to supply power to the image sensor and electronics, and an antenna/transmitter. The capsule is typically one inch long with a diameter of a half inch. The viewing window usually has a spherical or nearly spherical shape in order to enable smooth insertion into a body cavity. The imaging system, consisting of an objective and dome, typically has a wide angle FOV, preferably 100° or more and a large depth of field, for example, from 0 to 30 mm. Due to space constraints, both the illumination light source and the imaging system are located within the same optical dome, light reflected by the dome surface may reach the image sensor thereby deteriorating the imaging quality. Therefore, it is necessary to optimize the shape of the viewing dome, the locations of the LEDs and the imaging system to prevent the stray light from reaching the image sensor.

Another key component in an endoscope is illumination optics. It is one of the major obstacles in designing an endoscope having a small diameter. Before the introduction of optical fibers into endoscopy in the early 1960s, the illumination of the endoscope relied on a small tungsten filament lamp installed at the distal end of the endoscope. Fiber bundles are typically used in the modern endoscope to provide sufficient and uniform illumination.

The efficiency of a fiber bundle for illumination depends on the core diameter and the NA of the individual fibers. Usually the fiber with a large NA is used in illumination systems in order to convey more light from the light source down to the distal end of the endoscope. In recent years, LEDs are used in endoscopes. LEDs are ideal light sources for endoscopes because of their compact size and low power consumption. A portable system is possible with LEDs as the illumination light source because they can be operated by a battery. LEDs can be either mounted in the proximal end or mounted directly in the distal end of the endoscope. When LEDs are mounted outside the endoscope, a fiber light guide or a plastic light pipe is required to convey the light to the distal end.

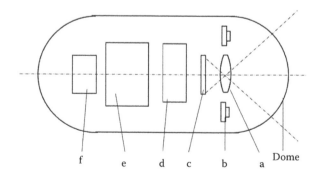

FIGURE 1.20 Basic configuration of a capsule endoscope. (a) Imaging optics, (b) light source, (c) image sensor, (d) battery, (e) electronics, and (f) antenna/transmitter. (From Liang, R., *Optical Design for Biomedical Imaging*, SPIE Press, Bellingham, WA, 2010.)

1.6.3 Objective Lenses

The objective lens is essential for all types of endoscopes. The objective lens typically has a large field of view (FOV), for example, 100°, and a large depth of field so that objects at a distance between 3 mm and 50 mm can be observed without using the focus adjustment.

The diameter of the endoscope objective lens should be small enough so that it can be used through small openings. In addition, a high degree of telecentricity is required for the objective lenses used in rigid endoscopes with relay lens systems and flexible endoscopes with image fiber bundles.

With the above requirements, a retrofocus objective lens is generally used in endoscopic imaging systems. The retrofocus objective lens primarily consists of two lens groups that are separated by an aperture stop, as shown in Figure 1.21a. The back focal length (BFL) is longer than the effective focal length (EFL). The front lens group has a negative refractive power to achieve a large field angle. The rear group usually consists of several lenses and has a positive refractive power as a whole. In a typical objective lens for an endoscope, the power mainly relies upon a front lens group that is located closer to the object to be observed. If a front lens group is eliminated, such as a landscape lens in Figure 1.21b, the overall length of the lens can be reduced. However, this leads to deterioration in optical performance. In order to correct chromatic aberrations, the objective lens commonly uses one or more cemented lenses.

For objective lenses used in rigid endoscopes with relay lenses or flexible endoscopes with fiber bundles, the aperture stop is placed in the front of the focal point of the rear group with positive power so that the chief rays in the image space are parallel to the optical axis. On the other hand, telecentricity is not required for the objective lens used in video or wireless endoscopes.

Due to the configuration with a negative refractive power on the object side of the aperture stop and a positive refractive power on the image side of the aperture stop, the correction for lateral chromatic aberration and distortion is particularly difficult. The correction of distortion in a telecentric objective lens system for endoscopes may influence other off-axis aberrations. For a compact, wide-angle objective lens optical system, it becomes difficult to correct aberrations, especially the off-axis aberrations. Therefore, correction of aberrations other than distortion is the key to designing a wide angle and compact objective lens for endoscopes.

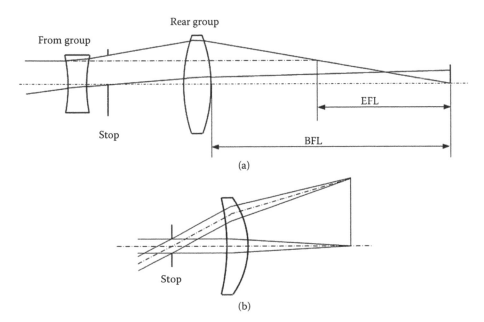

FIGURE 1.21 (a) The retrofocus lens is characterized by a long back focal length, and (b) landscape lens with aperture in front. (From Liang, R., *Optical Design for Biomedical Imaging*, SPIE Press, Bellingham, WA, 2010.)

One example of a simple objective lens is the one shown in Figure 1.22a. It consists of a negative singlet, a positive singlet, and an aperture stop placed between the two lenses. This type of objective lens has a relatively long back working distance, however the NA is limited. Another drawback is that it provides nearly no freedom to correct chromatic aberrations. Figure 1.22b shows the landscape lens with a stop at the front; it has only two elements. It is an all-positive configuration, and therefore, it has an inherent field curvature, which can be reduced by using glass with a very high refractive index. However, glass with a high refractive index usually has low Abbe numbers, resulting in a relatively large lateral chromatic aberration. Similar to the objective lens in Figure 1.22a, the correction of spherical aberration, field curvature, astigmatism, and lateral chromatic aberration is limited as a result of the simple configuration. Since there is no negative lens in front of the aperture stop, field dependent aberrations, such as astigmatism, field curvature, distortion, and later chromatic aberration are relatively larger than the objective lens in Figure 1.22a.

The chief rays of the objective lenses discussed above are not parallel to the optical axis, which means the exit pupil is not at infinity. They are only suitable for video endoscopes with electronic sensors, such as CCDs and CMOSs, at the distal end. Figure 1.22c is a three-element telecentric objective lens. The image surface is at the last surface of the objective lens so that it can be glued or joined to the incident end of the fiber bundle. If this objective lens is used together with a relay lens system, the image plane should be away from the lens surface to avoid a tight requirement on the last surface. Figure 1.22d is an objective lens with a doublet in the rear group instead of a singlet to correct the chromatic aberrations. The cemented doublet is an effective element to correct chromatic aberrations.

Lateral chromatic aberration not only generates color blurring of the peripheral image, but also reduces the resolution of the peripheral region. For the objective lens used with a fiber bundle, the lateral

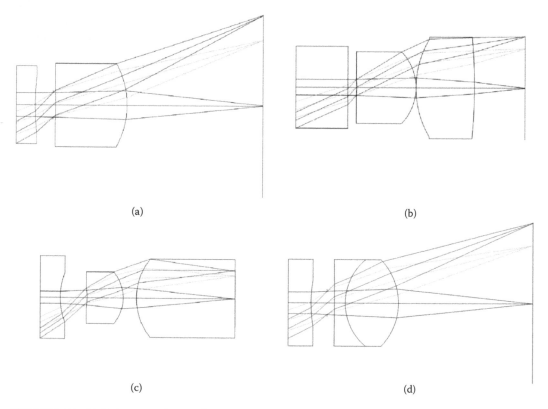

(a)

(b)

(c)

(d)

FIGURE 1.22   (a) Endoscopic objective lens with two singlets, (b) objective lens with a landscape lens, (c) three-element, telecentric objective lenses, and (d) objective lens with a doublet to correct chromatic aberrations.

chromatic aberration should be minimized with reference to the distance between the optical fiber cores. If the lateral chromatic aberration is not corrected sufficiently, the transmitted image will produce noticeable color diffusion. There are two approaches to address lateral chromatic aberration. One is to improve the degree of symmetry by inserting an additional lens on the object side of the aperture stop. Another approach is to use cemented lenses.

The symmetrical configuration is effective in reducing the lateral, but not axial, chromatic aberration. Generally, it is preferable to dispose of optical components for correcting the lateral color at a position that is distant from the aperture stop. Figure 1.23a is one design using this approach. The lateral chromatic aberration can be well balanced; coma and astigmatism are reasonably controlled. A cemented doublet is an effective way to reduce axial and lateral chromatic aberrations. Conventionally, a cemented doublet is placed on the image side of the aperture stop to accomplish achromatism in an endoscopic objective lens. To correct for the lateral color, it is preferable to place the cemented doublet at a position that is a distance from the aperture stop where the ray height is large.

Figure 1.22d is an objective lens with a doublet as the last element close to the image plane. The major aberrations remaining in the objective lens include coma, field curvature and distortion. Typically, the two approaches are used together to design a high performance endoscopic objective with better control on axial and lateral chromatic aberrations, as shown in Figure 1.23b.

The imaging system in the wireless endoscope consists of the dome and objective lens. Due to the fact that the illumination LEDs are in the dome as well, the reflected light from the dome surface may reach the detector and degrade image quality. Therefore the dome and objective lens are always considered together when designing an imaging system.

The viewing dome is usually made of a plastic material; the material should have a low moisture absorption rate, low residual metals, and adequate hardness. In addition, the material for optical domes should be able to withstand the extreme chemical environments within the stomach and intestine, where the pH value ranges from 2.0 to 9.0, respectively. The optical quality and transparency of the viewing dome need to be preserved during the entire endoscopic process.

The optical system shown in Figure 1.24a includes a spherical viewing dome, a singlet aspherical imaging lens, and a solid-state sensor. The aperture stop is placed in front of the imaging lens and is at the center of the viewing dome. Since it is a single element imaging system, there are considerable chromatic aberrations. The other major uncorrected aberrations include distortion and astigmatism. Both aberrations are difficult to correct because the imaging system only consists of a single lens, and the aperture is in front of the lens. In order to improve system performance, another lens can be added in front of the aperture stop. Figure 1.24b is an objective lens consisting of two elements. The odd-ordered aberrations, such as coma, distortion, and lateral chromatic aberration, are reduced because of the symmetric configuration.

While the viewing dome can be optimized to improve the image quality of the objective lens, one additional consideration is the illumination because the imaging and illumination systems are in the

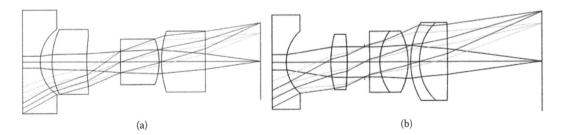

(a)   (b)

FIGURE 1.23   (a) Objective lens with a symmetrical lens configuration to reduce lateral chromatic aberrations and (b) objective lens with a positive lens in the front group and a doublet in the rear group for chromatic aberration correction. (From Liang, R., *Optical Design for Biomedical Imaging*, SPIE Press, Bellingham, WA, 2010.)

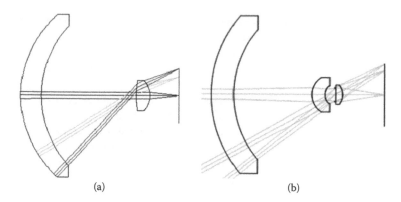

FIGURE 1.24   (a) Objective lens with a single lens and (b) objective lens with two elements arranged symmetrically to the aperture stop.

same compartment. The basic requirement of an illumination system in a wireless endoscope includes a large FOV and good illumination uniformity. Another important consideration is whether the illumination light from the TIR and the Fresnel reflection on the surface of the viewing window will reach the solid-state image sensor.

1.6.4 Relay Lenses

Rigid endoscopes require the use of a relay system to transfer an image through a distance inside a tube of limited diameter to the proximal end of the tube where the eyepiece or imaging lens is located. Depending on the tube length, generally four or more relay stages are necessary to transfer the image from the distal end to the proximal end. The brightness of an image in a rigid endoscope is ultimately determined by the optical invariant of the relay section, which is proportional to the NA and the diameter of the relay lenses.

For endoscopes used in a broad spectrum, such as in white light observation, the relay lens needs to be achromatized. In a conventional relay system, achromatization is best achieved by splitting the relay objective lens into a doublet. A doublet can also reduce the spherical aberration and compensate for the small amount of axial chromatic aberration introduced by the field lenses if they are displaced from the intermediate image planes.

Figure 1.25a is a conventional relay lens design. It consists of lenses of positive power only, generating significant positive field curvature. The coma, distortion, and lateral chromatic aberration are canceled completely by the symmetry of the relay system.

The remaining aberrations are spherical aberration, field curvature, astigmatism, and axial chromatic aberration. Due to the difficulties in loading thin lenses into a long, rigid tube, the lenses are generally mounted in a thin-walled tube that slides into a long, rigid outer tube. Therefore, the clear aperture of the conventional relay lens system is relatively small, resulting in a low light throughput.

In late 1950s and early 1960s, H. H. Hopkins developed a rod lens relay system as shown in Figure 1.25b [66, 68]. In contrast to the conventional relay system, where the medium between the field lens and the relay imaging lens is air, a Hopkins relay system has a glass rod. The roles of the glass and air are interchanged. Compared to the conventional relay, where $n = 1$, the light throughput of the rod lens relay increases by a factor of n^2. Another advantage of the rod lens relay system is that the rod lens permits a greater diameter for a given outer diameter of a rigid tube, increasing the throughput dramatically. To improve the manufacturability, a number of new designs of rod lenses have been in development [69].

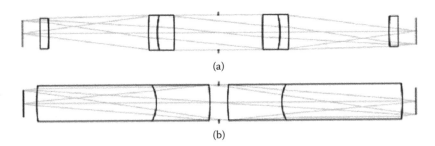

FIGURE 1.25 (a) Conventional relay lens and (b) Hopkins rod lens relay system.

References

1. Minksy M. Microscopy apparatus. *U.S. Patent No. 2,013,467*; 1961.
2. Wilhelm S, Gröbler B, Gulch M, Heinz H. *Confocal Laser Scanning Microscopy: Principles.* Carl Zeiss: Jena, Germany, 2003.
3. Drazic V. Dependence of two- and three-dimensional optical transfer functions on pinhole radius in a coherent confocal microscope. *J. Opt. Soc. Am. A* 9, 725–731; 1992.
4. Wilson T, Carlini AR. Size of detector in confocal imaging system. *Opt. Lett.* 12, 227–229; 1987.
5. Sheppard CJR, Rehman S. Confocal microscopy. In *Biomedical Optical Imaging Technologies Design and Applications*, R. Liang, ed. Springer Berlin Heidelberg, 2013.
6. Goldman H. Spaltlampenphotographie und –photometrie. *Ophthalmologica* 98, 257–270; 1940.
7. Petran M, Hadravsky M, Egger MD, Galambos. Tandem-scanning reflected-light microscope. *J. Opt. Soc. Am.* 58, 661–664; 1968.
8. Tanaami T, Otsuki S, Tomosada N, Kosugi Y, Shimizu M, Ishida H. High-speed 1-frame/ms scanning confocal microscope with a microlens and Nipkow disks. *Appl. Opt.* 41, 4704–4708; 2002.
9. Botvinick EL, Li F, Cha S, Gough DA, Fainman Y, Price JH. In vivo confocal microscopy based on the Texas Instruments digital micromirror device. *Proc. SPIE* 3921, 12–19; 2000.
10. Kanai M. Condensing optical system, confocal optical system, and scanning confocal endoscope. *U.S. Patent Application No. 20050052753*; 2005.
11. Dickensheets DL, Kino GS. Micromachined scanning confocal optical microscope. *Opt. Lett.* 21, 764–766; 1996.
12. Seibel EJ, Smithwick QYL. Unique features of optical scanning single fiber endoscopy. *Lasers Surg. Med.* 30(3), 177–183 (2002).
13. D. L. Dickensheets and G. S. Kino, Silicon-micromachined scanning confocal optical microscope. *J. Microelectromech. Syst.* 7, 38–47; 1998.
14. Ra H, Taguchi Y, Lee D, Piyawattanametha W, Solgaard O. Two-dimensional MEMS scanner for dual-axes confocal in vivo microscopy. *IEEE Int. Conf. Microelectromech. Syst*, 862–865; 2006.
15. Maitland KC, Shin HJ, Ra H, Lee D, Solgaard O, Richards-Kortum R. Single fiber confocal microscope with a two-axis gimbaled MEMS scanner for cellular imaging. *Opt. Exp.* 14, 8604–8612; 2006.
16. Miyajima H, Asaoka N, Isokawa T, Ogata M, Aoki Y, Imai M, Fujimori O, Katashiro M, Matsumoto K. A MEMS electromagnetic optical scanner for a commercial confocal laser scanning microscope. *J. Microelectromech. Syst.* 12, 243–251; 2003.
17. Seibel EJ, Smithwick QYJ, Brown CM, Reinhall PG. Single fiber flexible endoscope: General design for small size, high resolution and wide field of view. *Proc. SPIE* 4158, 29–39; 2001.
18. Rosman GE, Jones BC, Pattie RA, Byrne CG, Optical fiber scanning apparatus. *U.S. Patent 7 920 312*, Apr. 5, 2011.
19. Li Z, Fu L. Note: A resonant fiber-optic piezoelectric scanner achieves a raster pattern by combining two distinct resonances. *Rev. Sci. Instrum.* 83, 086102; 2012.

20. Mansoor H, Zeng H, Tai IT, Zhao J, Chiao M. A handheld electromagnetically actuated fiber optic raster scanner for reflectance confocal imaging of biological tissues. *IEEE Trans. Biomed. Eng.* 60, 1431–1438.

21. Gmitro AJ, Aziz D. Confocal microscopy through a fiber-optic imaging bundle. *Opt. Lett.* 18, 565–567; 1993.

22. Lin CP, Webb RH. Fiber-coupled multiplexed confocal microscope. *Opt. Lett.* 25, 954–956; 2000.

23. Lane PM, Dlugan ALP, Richards-Kortum R, MacAulay CE. Fiber-optic confocal microscopy using a spatial light modulator. *Opt. Lett.* 25, 1780–1782; 2000.

24. Sinclair MB, Haaland DM, Timlin JA, Jones HDT. Hyperspectral confocal microscope. *Appl. Opt.* 45, 6283–6291; 2006.

25. Haaland DM, Jones HDT, Van Benthem MH, et al. Hyperspectal confocal fluorescence imaging: Exploring alternative multivariate curve resolution approaches. *Appl. Spectrosc.* 63(3), 271–279; 2009.

26. Lu G, Fei B. Medical hyperspectral imaging: A review. *J. Biomed. Opt.* 19(1), 010901; 2014.

27. Neil MAA, Juskaitis R, Wilson T. Method of obtaining optical sectioning by using structured light in a conventional microscope. *Opt. Lett.* 22, 1905–1907; 1997.

28. Karadaglić D, Wilson T. Image formation in structured illumination wide-field fluorescence microscopy. *Micron* 39, 808–818; 2008.

29. Stokseth PA. Properties of a defocused optical system. *J. Opt. Soc. Am.* 59, 1314–1321; 1969.

30. Benjamin T, Momany M, Kner P. Optical sectioning structured illumination microscopy with enhanced sensitivity. *J. Opt.* 15, 94004; 2013.

31. Allen JR, Ross ST, Davidson MW. Structured illumination microscopy for superresolution. *Chemphyschem* 15, 566–576; 2014.

32. Mertz J. Optical sectioning microscopy with planar or structured illumination. *Nat. Meth.* 8, 811; 2011.

33. Gustafsson MGL, Shao L, Carlton PM, et al. Three-dimensional resolution doubling in wide-field fluorescence microscopy by structured illumination. *Biophys. J.* 94, 4957–4970; 2008.

34. Gustafsson MGL. Surpassing the lateral resolution limit by a factor of two using structured illumination microscopy. *J. Microsc.* 198, 82–87; 2000.

35. Benjamin T, Momany M, Kner P. Optical sectioning structured illumination microscopy with enhanced sensitivity. *J. Opt.* 15, 94004; 2013.

36. Chang BJ, Chou LJ, Chang YC, Chiang SY. Isotropic image in structured illumination microscopy patterned with a spatial light modulator. *Opt. Exp.* 17, 14710–14721; 2009.

37. Kner P, Chhun BB, Griffis ER, Winoto L, Gustafsson MGL. Super-resolution video microscopy of live cells by structured illumination. *Nat. Meth.* 6, 339–342; 2009.

38. Dan D, Lei M, Yao B, et al. DMD-based LED-illumination super-resolution and optical sectioning microscopy. *Sci. Rep.* 3, 2013.

39. Lim D, Chu KK, Mertz J. Wide-field fluorescence sectioning with hybrid speckle and uniform-illumination microscopy. *Opt. Lett.* 33, 1819–1821; 2008.

40. Muro E, Vermeulen P, Ioannou A, et al. Single-shot optical sectioning using two-color probes in HiLo fluorescence microscopy. *Biophys. J.* 100, 2810–2819 (2011).

41. S. W. Hell and J. Wichmann, Breaking the diffraction resolution limit by stimulated emission: Stimulated-emission-depletion fluorescence microscopy. *Opt. Lett.* 19, 780–782; 1994.

42. Rust MJ, Bates M, Zhuang X. Sub-diffraction-limit imaging by stochastic optical reconstruction microscopy (STORM). *Nat. Meth.* 3(10), 793–796; 2006.

43. Betzig E, Patterson GH, Sougrat R, et al. Imaging intracellular fluorescent proteins at nanometer resolution. *Science* 313(5793), 1642–1645; 2006.

44. Gustafsson MGL. Nonlinear structured-illumination microscopy: Wide-field fluorescence imaging with theoretically unlimited resolution. *Proc. Natl. Acad. Sci. USA* 102, 13081–13086; 2005.

45. Shao L, Kner P, Rego EH, Gustafsson MGL. Super-resolution 3D microscopy of live whole cells using structured illumination. *Nat. Meth.* 8, 1044–1046; 2011.

46. Rego EH, Shao L, Macklin JJ, et al. Nonlinear structured-illumination microscopy with a photo-switchable protein reveals cellular structures at 50-nm resolution. *Proc. Natl. Acad. Sci. USA* 109, E135; 2011.

47. Gustafsson MGL. Nonlinear structured-illumination microscopy: Wide-field fluorescence imaging with theoretically unlimited resolution. *Proc. Natl. Acad. Sci. USA* 102, 13081–13086; 2005.

48. Krzewina LG, Kim MK. Single-exposure optical sectioning by color structured illumination microscopy. *Opt. Lett.* 31, 477–479; 2006.

49. Schermelleh L, Carlton PM, Haase S, et al. Subdiffraction multicolor imaging of the nuclear periphery with 3D structured illumination microscopy. *Science* 320(5881), 1332–1336; 2008.

50. Kner P, Chhun BB, Griffis ER, Winoto L, Gustafsson MGL. Super-resolution video microscopy of live cells by structured illumination. *Nat. Meth.* 6(5), 339–342; 2009.

51. Jost A, Heintzmann R. Superresolution multidimensional imaging with structured illumination microscopy. *Ann. Rev. Mater. Res.* 43, 261–282; 2013.

52. Hirvonen LM, Wicker K, Mandula O, Heintzmann R. Structured illumination microscopy of a living cell. *Eur. Biophys. J.* 38, 807–812; 2009.

53. Shao L, Kner P, Rego EH, Gustafsson MGL. Super-resolution 3D microscopy of live whole cells using structured illumination. *Nat. Meth.* 8, 1044–1046; 2011.

54. Muro E, Vermeulen P, Ioannou A, et al. Single-shot optical sectioning using two-color probes in HiLo fluorescence microscopy. *Biophys. J.* 100, 2810–2819; 2011.

55. Shroff SA, Fienup JR, Williams DR. Lateral superresolution using a posteriori phase shift estimation for a moving object: Experimental results. *J. Opt. Soc. USA* 27, 1770–1782; 2010.

56. Shroff SA, Fienup JR, Williams DR. Phase-shift estimation in sinusoidally illuminated images for lateral superresolution. *J. Opt. Soc. USA* 26, 413–424; 2009.

57. Wicker K, Mandula O, Best G, Fiolka R, Heintzmann R. Phase optimisation for structured illumination microscopy. *Opt. Exp.* 21, 2032–2049; 2013.

58. Holleran KO Shaw M. Optimized approaches for optical sectioning and resolution enhancement in 2D structured illumination microscopy. *Biomed Opt. Exp.* 5, 2580–2590; 2014.

59. Heintzmann R, Jovin TM, Cremer C. Saturated patterned excitation microscopy—A concept for optical resolution improvement. *J. Opt. Soc. Am.* 19, 1599–1609; 2002.

60. Gustafsson MGL. Nonlinear structured-illumination microscopy: Wide-field fluorescence imaging with theoretically unlimited resolution. *Proc. Natl. Acad. Sci. USA* 102, 13081–13086; 2005.

61. Schwentker MA, Bock H, Hofmann M, et al. Wide-field subdiffraction RESOLFT microscopy using fluorescent protein photoswitching. *Microsc. Res. Tech.* 70, 269–280; 2007.

62. Hofmann M, Eggeling C, Jakobs S, and Hell SW. Breaking the diffraction barrier in fluorescence microscopy at low light intensities by using reversibly photoswitchable proteins. *Proc. Natl. Acad. Sci. USA* 102, 17565–17569; 2005.

63. Rego EH, Shao L, Macklin JJ, et al. Nonlinear structured-illumination microscopy with a photo-switchable protein reveals cellular structures at 50-nm resolution. *Proc. Natl. Acad. Sci. USA* 109, E135; 2011.

64. Benjamin T, Momany M, Kner P. Optical sectioning structured illumination microscopy with enhanced sensitivity. *J. Opt.* 15, 94004; 2013.

65. Berci G, Forde KA. History of endoscopy: What lessons have we learned from the past? *Surg. Endosc.* 14, 5–15; 2000.

66. Hopkins HH., Optical principles of the endoscope. In *Endoscopy*. G. Berci, ed., pp. 3–26. New York, NY: Appleton-Century-Crofts, 1976.

67. Iddan G, Meron G, Glukhovsky A, Swain P. Wireless capsule endoscopy. *Nature* 405, 417; 2000.

68. Hopkins HH., Optical system having cylindrical rod-like lenses. *U.S. Patent No. 3,257,902*; 1966.

69. Liang R. *Optical Design for Biomedical Imaging*. Bellingham, WA: SPIE Press 2010.

2

Wavefront Slope Measurements in Optical Testing

2.1 Introduction and Historical Review...37
2.2 Knife Test 38
    Foucaultgram Simulations
2.3 Wire Test 41
2.4 Ronchi Test 42
    Ronchigram Simulations • Substructured Ronchi Gratings
2.5 Hartmann Test..53
    Screen Design and Hartmanngram Evaluation • Shack-Hartmann Test
2.6 Null Tests: Ronchi, Hartmann, and Shack-Hartmann.................56
    Null Shack-Hartmann Test • Null Screens for Convex Surfaces
2.7 Gradient Tests..60
    Platzeck–Gaviola Test • Irradiance Transport Equation • Roddier Method
Acknowledgments...66
References..66

Alejandro Cornejo
Rodríguez, Alberto
Cordero-Davila,
and Fermín S.
Granados Agustín

2.1 Introduction and Historical Review

In this chapter, we analyze a group of methods for testing optical surfaces and systems whose main characteristics are the measurements of the rays' slopes at certain planes. Auxiliary optics are not required, which allows us to have a direct measurement of the wavefront under test that comes from the optical system or surface under test, which is not possible in other measuring methods.

Historically, the first test that can be classified into this group here described is the so-called knife edge or Foucault test [1]. Subsequently, other techniques were described in the first half of the last century, including the wire [2], Ronchi [3–5], Hartmann [6], and Platzeck–Gaviola tests [7]. More recently, one technique derived from the Hartmann method is the Shack–Hartmann [8] test; two more of the tests were developed by Ichikawa et.al. [9] and Roddier et al. [10], both based on the theory of the irradiance transport equation [11]. The Shack–Hartmann and Roddier tests are normally applied to wavefront testing of working astronomical telescopes, and mainly function under an adapative optics technique.

In general, each one of the tests mentioned has its own physical and mathematical description [12]. However, in this chapter, easier explanations of the methods are given, and a common step-by-step description of the older tests is presented. For the case of the Ronchi and Hartmann methods, a unified theory has also been developed [13]. Even though in some cases geometrical and physical optics can be used to describe the tests, as mentioned, in this chapter geometrical optics is mainly used to explain the methods applied to measuring the slope rays of the wavefronts coming from optical surfaces or systems.

The presentation of each one of the tests contains comments on some practical aspects rather than the theoretical foundations of the methods. Of course, theoretical and practical aspects are considered elsewhere [12, 14].

2.2 Knife Test

It is assumed, in the first place, that the optical system under test is illuminated by a point source, or equivalently by a perfect spherical wavefront; hence, if a perfect spherical wavefront is leaving the exit pupil of an optical system, and you place the knife edge into the light beam, see Figure 2.1a, then the border of the shade that you will see in any observation plane corresponds to a straight line, independently of the orientation and position of knife edge, and the wavelength of the light.

Figure 2.1b shows the observed pattern. From the geometrical point of view, the border between the dark and illuminated zones is defined by the plane that passes through the $\overline{BC}$ border of the knife for an image point A′. Despite the diffraction effects, the border will be a plane that, when it is intersected with

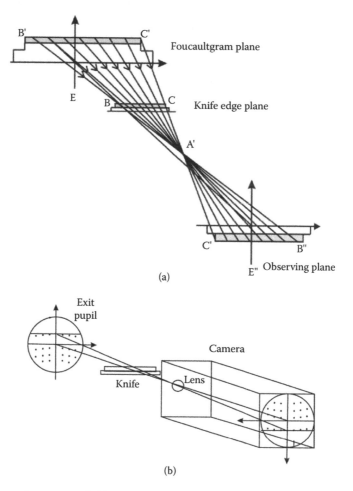

(a)

(b)

FIGURE 2.1 (a) Foucaultgram recorded by means of a camera focused at the exit pupil of the system under test. (b) Foucaultgram of a spherical wavefront.

any other observation plane, E″, a straight line $\overline{B''C'}$ is being defined and therefore, corresponds to the border of the shade that will be observed.

In the usual form of observation, see Figure 2.1b, the camera lens is placed near the image point A′ and you focus the plane of the exit pupil E. In this case, the virtual border (in similarity with the virtual images) is the observed shade and will be a straight line. A real image will be formed, with the aid of a camera lens, on the detector plane (charge-coupled device [CCD], photographic plate), or the retina of the eye in the case of a direct observation. It could also be recorded, in any further plane from the knife edge, without a lens. This last type of observation is very useful when the distance between the camera lens (or the eye) and the observation plane is very short and it is not achieved as an acceptable focus, or when the F# of the system under test is very small. In this last case, the pupil of the eye obstructs the passage of the rays, and only some parts of the pattern are observed with certain circular shape.

When the light rays do not converge toward a unique point, then the projected and observed shadow will not be a straight line; this means that the wavefront is not perfectly spherical. In such case, the border of the shadow is given by the intersection points of the rays that will pass exactly on the border of the knife; that is, the shadow will not follow a straight line.

In this test, the comparison between the experimental and the real patterns are very important; for this reason, in the next section an algorithm will be presented, such that simulated Foucaultgrams can be obtained if skew rays can be traced through the optical system.

2.2.1 Foucaultgram Simulations

In order to simulate the Foucaultgrams, it is assumed that in the optical systems (see Figure 2.2), there are several planes with their respective coordinates as follows: the entrance pupil plane (X, Y), the exit pupil plane (X_o, Y_o), the observation plane (X_{ob}, Y_{ob}), and the knife edge plane (T_X, T_Y).

If it is assumed that the knife edge is parallel to the X-axis, then all the points that belong to the semiplane of the knife edge could be expressed mathematically through the next inequality

$$T_y - T_{yn} \leq 0 \tag{2.1}$$

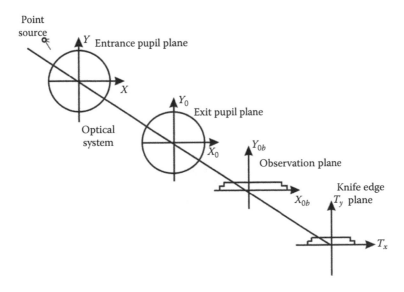

FIGURE 2.2 Planes used in the Foucaultgram simulations.

where T_{yn} is the distance from the knife edge to the origin of the coordinate system. Equation 2.1 is equal to zero for the points that belong to the border of the knife; negative values are obtained for the points over the knife; while for the points that are not covered by the knife edge, values will be greater than zero. This property is important because it serves to distinguish, numerically, between a ray that falls on the knife and another that arrives to the transparent zone.

At once (X_{fi}, Y_{fi}) coordinates have been calculated, see Figure 2.3, where t_k is the shortest distance of the knife edge to the optical axis and θ is the angle of the knife edge with the X_{fi} axis.

The Foucaultgram simulation has its basis in the previous idea and allows us to show that it is not necessary to trace all the rays but only a very limited number. However, several rays are traced from the point source to the knife plane, passing through the entrance pupil, on which the incident rays are located at equally spaced points and in one straight line parallel to the Y-axis, similar to the ray tracing in the optical design.

The important data for each traced ray are the pairs of coordinates (X, Y), (X_{ob}, Y_{ob}), and (T_X, T_Y). If T_y of this last pair satisfies the equality in Equation 2.1, then the corresponding (X_{ob}, Y_{ob}), belongs to the border of the shade. However, when the T_y values are found that do not satisfy Equation 2.1, then it is necessary to identify the successive values for pairs of the coordinates T_y, for which a change of sign according to the inequality (2.1) is obtained. Once a first interval is identified and satisfied, the bisection method is used until the limit of an established required precision. To have as many points as possible for the pattern, the previous procedure is repeated for a family of parallel straight lines within the entrance pupil.

When the edge is rotated an angle q, then it is applied a rotation to the coordinates (TX, TY) for obtaining the new coordinate T_y^*, which must be substituted in Equation 2.1.

Two practical remarks are important. First, if we increase the distance between two consecutive points on the scanned line, then the time for obtaining the pattern diminishes; however, we could not detect some points that would appear when the patterns are closed. Secondly, if you carry out the scan only on parallel straight lines to the axes, then it is difficult to locate some points of the pattern; this occurs mainly for the case when the straight lines of the scan are nearly parallel to the knife edge. To avoid this last problem, it is important to carry out two scans on perpendicular straight lines.

We left out the simulation of Foucaultgrams for systems affected by Seidel aberrations, which can be calculated analytically [13].

In the development of this section, it has been assumed that the illumination of the system under test is with a point source; however, it is possible to substitute it with a linear source, under the basic condition that the linear source is parallel to the border of the knife. In this case, the patterns are almost

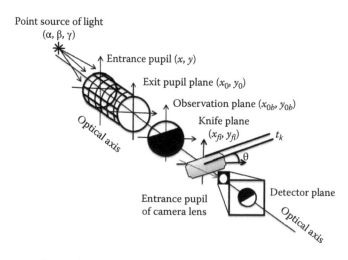

FIGURE 2.3   Experimental setup for the Foucault test.

identical for each point of the linear source. When the knife and the linear source are perpendicular to each other, astigmatic effects will appear that show tilts and displacements of the border of the shades; in this case, the patterns show a slow change of intensities in the border and therefore a poorer definition of the Foucaultgram. It is evident that the use of white light has no effects when the mirrors are tested, but for no corrected refractive systems the chromatic aberration becomes evident and appears as colored borders.

The Foucault test has been analyzed only from a simplified geometrical point of view, but it could also be analyzed using physical optics. For such an analysis, it is demonstrated that if an obstruction in the Fourier plane exists (image plane), then the observed intensity will be uniform. However, when the knife edges are in the Fourier plane, then some changes of intensity are observed, which are proportional to the derivate of the amplitude of the field in the image plane [15]. By using this approach, Landgrave [16] built a phase knife edge over a thin and transparent glass. Rodríguez-Zurita et al. [17] have shown that a system consisting of a polarizer, an Amici prism and an analyzer may act like a phase amplitude spatial filter capable of producing Foucault test-like images.

2.3 Wire Test

The Foucault test is often used to analyze spherical wavefronts, and particularly to test spherical surfaces with the point source placed in the plane of the center of curvature of the surface. However, in testing aspherical mirrors, the Foucault test is not very sensitive, since details are lost. Ritchey [2] developed the wire test, which is more sensitive as a zonal test and is analyzed in this section.

The wire test can be considered as an extension of the Foucault test, since a wire can be assumed to be a double-edged knife. Thus, a wiregram can be obtained from two simulated Foucaultgrams for which the two edge knifes are parallel, and at heights Y_a and Y_b from the optical axis. In the particular case of a perfect spherical wavefront, the zonal pattern will be formed by a straight band.

When the wire is located along the optical axis of an axisymmetrical mirror, at a distance L^* from the vertex of the mirror, then a dark ring can be observed on the mirror, of radius S^*, corresponding to the points from the zone S^* with radius of curvature at the distance L^*. Then, the rays coming from zone S^* cross the wire at position L^*.

With this idea, Ritchey [2] proposed to test parabolic surfaces by comparing the experimental and theoretical longitudinal spherical aberrations. He measured several values of L^* and S^* and compared them using theoretical calculus. The same idea is presented in what follows, applied to any conic mirror, but using the mathematical formulations of Sherwood [18–19] and Malacara [20]. These authors demonstrated independently that, as illustrated in Figure 2.4, the equation of the reflected ray is given by

$$T = \frac{(l+L-2z)\left[1-\left(\dfrac{dz}{ds}\right)^2\right]+2\left(\dfrac{dz}{ds}\right)\left[s-\dfrac{(l-z)(L-z)}{s}\right]}{\dfrac{(l-z)}{s}\left[1-\left(\dfrac{dz}{ds}\right)^2\right]+2\left(\dfrac{dz}{ds}\right)}, \tag{2.2}$$

where T, the transverse aberration, is the height of the reflected ray at the distance L, l is the position of the point source, Figure 2.4, and z is the function that describes the saggita of the mirror, given by

$$z = \frac{cs^2}{1+\sqrt{1-(K+1)c^2s^2}}. \tag{2.3}$$

In this last equation s is the distance to the axis where the incident ray is reflected, c is the paraxial curvature, and K is the conic constant of the mirror under test. From Equation 2.2, it is evident that if

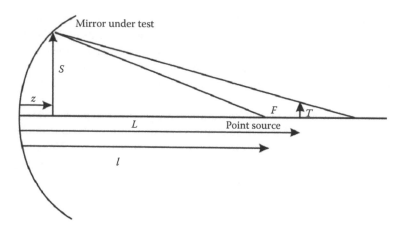

FIGURE 2.4 Geometry for the transverse aberration, *T*, of the reflected ray in the wire test.

we take an incident cone of rays with vertex in the point source and whose base is a circle of radius *s* on the mirror, then the reflected rays will also form a cone whose base is the same zone *s*, and whose vertex will be located in *L**, corresponding to the zero of Equation 2.2 and given by

$$L^* = \frac{(l-2z)\left[1-\left(\dfrac{dz}{ds}\right)^2\right]+2\left(\dfrac{dz}{ds}\right)\left[s-\dfrac{z(l-z)}{s}\right]}{2\left(\dfrac{dz}{ds}\right)\dfrac{(l-z)}{s}\left[1-\left(\dfrac{dz}{ds}\right)^2\right]}. \tag{2.4}$$

From Equation 2.4, it is clear that for a given value of *s* we will have an only value of *L** and since *s* defines a circumference on the mirror, then, if the rays are obstructed precisely in *L**, then at least the rays coming from the circumference of radius *s* will be blocked. Besides, the point of the wire is that it will be observed as a straight line parallel to the wire. Figure 2.5 and Figure 2.6 show how this pattern is formed.

The application of Ritchey's idea begins with the designing of a screen, for testing a parabolic mirror that allows illumination of different zones as the paraxial, intermediate ($0.7071\,S_{max}$) and at the border of the mirror, see Figure 2.7a. The screen is placed in front of the mirror and the wire is displaced along the axis to find the positions corresponding to the different ring zones of the screen on the mirror. In each case they become a series of measurements of *L** and they are compared with the theoretical calculus made with aid of the Equation 2.4.

There are more versatile screens, see Figure 2.7b; however, it is not possible to detect the asymmetrical defects [15]. This type of test is highly recommended for use only in the first phases of fabrication of optical surfaces.

As it was pointed out, a linear source can be used when the wire and the linear source are parallel to each other. In this case, the advantages and disadvantages mentioned in the knife test are retained.

From the physical optics point of view, it could be possible to consider that the wire is a space filter located in the Fourier plane with the advantage, over the knife test, in that here the position and the thickness of the wire could be selected.

2.4 Ronchi Test

As well as the wire test being considered as conceptually, like an extension of the knife test, the Ronchi test [3–5] can be considered, see Figure 2.8a, as an extension of the wire test, if we consider a Ronchi

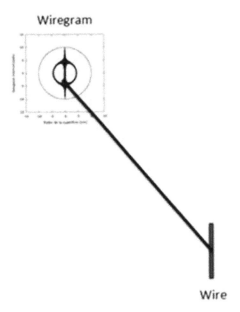

FIGURE 2.5 Pattern and filter for a parabolic mirror (k = –1.0), of 20 cm in diameter, 100.0 cm in curvature radius, and source and filter are located at (0,0,100.0 cm) and 110.2 cm from mirror vertex respectively.

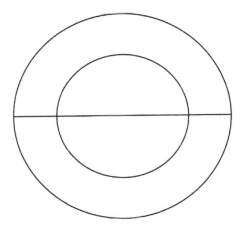

FIGURE 2.6 Simulated Wiregrams.

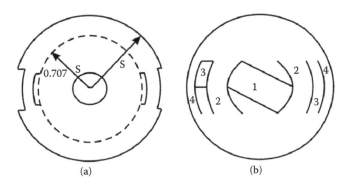

FIGURE 2.7 (a) Zonal screen and (b) Couder screen fov the wire test.

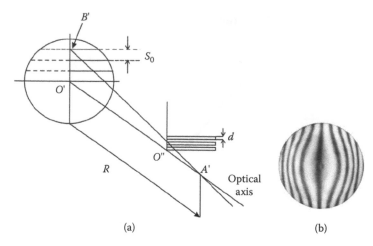

FIGURE 2.8 (a) Ronchigram for a spherical wavefront. A' is the paraxial center of curvature; O" is the ruling plane located along the optical axis; O' is the exit pupil plane or surface under test plane. (b) Experimental Ronchigram for spherical wavefront.

ruling as formed of several wires equidistant and parallel. In the Ronchi ruling, the slits are alternated clear and dark, with the same width, and are assumed to be parallel to the X-axis. The extension for the Ronchi test also gives us global information of the surface; in this test, the information is obtained at the same time from several wires and, therefore, Ronchigram fringes "cover" the exit pupil of the system under test.

From the wire test it is clear that if you are testing a spherical wavefront with a Ronchi ruling, a Ronchigram of parallel and equally spaced fringes will be observed (see Figure 2.8b).

The separation, S_b, between two consecutive fringe borders in the Ronchigrams is equal to

$$S_b = \frac{R}{D_f} d, \qquad (2.5)$$

where R is the radius of curvature of the wavefront, D_f is the separation A'O" between the ideal image point A' and the intersection point O" of the ruling plane with the optical axis, and d is the distance between two consecutive borders in the Ronchi ruling, which is the half period of the ruling.

From Equation 2.5, we obtain a well-known result in the application of the Ronchi test. If D_f is increased, that is, the ruling moves away, before or after the image point A', then the width of the observed fringes diminishes and you will see an increase in the number of fringes in the exit pupil plane. If D_f becomes zero, then S_b becomes infinite and, therefore, a field totally brilliant or dark is observed experimentally, depending upon whether the light arrives to a clear or dark slit on the Ronchi ruling. This result is important if you want to locate the image point of an optical system A'; since in this case, when the ruling is placed at the plane that contains the image point, then the fringes will become, theoretically, infinitely wide. However, in practice, you cannot measure a width greater than the diameter of the exit pupil of the system under test; therefore, you will have an uncertainty in the localization of the image point A'.

For a spherical mirror, another well-known result of this test comes from Equation 2.5, which is the dependence of the width of the fringes S_b with the width d of the ruling slits. In this case, if a ruling of greater frequency is used (a minor d), the frequency of the fringes will also increase; this means that S_b will diminish.

An interesting paper presenting a modification to the Ronchi Test was proposed by Liang and Sasian [21].

2.4.1 Ronchigram Simulations

Up to now we have analyzed the Ronchigrams that could be obtained with spherical surface or wavefronts; in this section, an algorithm will be described to simulate Ronchigrams of optical systems by just doing ray tracing.

The procedure has its theoretical basis in the simulation of several wiregrams. In the Ronchigrams, each wire corresponds to a dark slit of the grating. In practice, the simulation is based in the ray tracing for two neighboring rays (in the entrance pupil), through the optical system and by assuming that these two rays are neighbors also in the ruling plane. The rays at the entrance pupil for the positions (X, Y) and (X', Y') will correspond to the positions (T_X, T_Y) and (T'_X, T'_Y) at the Ronchi ruling (Figure 2.8b). An important case for this pair of rays corresponds to when in the Ronchi ruling one ray falls in a dark zone and the other ray falls in an illuminated one. If the slits of the ruling are assumed parallel to the X-axis, then the borders can be described by the equation

$$T_y = md, \tag{2.6}$$

where $m = 0, \pm1, \pm2, \pm3. \ldots$ Then, in order to identify the existence of one border between two points (T_X, T_Y) and (T'_X, T'_Y) it is possible to calculate

$$M = \mathrm{int}\left(\frac{T_Y}{d}\right) \tag{2.7}$$

and

$$M' = \mathrm{int}\left(\frac{T'_Y}{d}\right), \tag{2.8}$$

where [int] is the computer instruction by means of which the integer part of the quantity that appears between parentheses is calculated.

For two neighboring rays, two possibilities exist: the two integers are equal or different. In the first case, we have not passed over any border and in the second case, it is an indication that we have crossed some border. In this latter case, the separation between the neighboring rays in the entrance pupil can be diminished by using an intermediate point between (X, Y) and (X', Y'); then, the procedure is repeated in order to elect the new subinterval that we refine as in the bisection method. This procedure is followed until the desired limit is reached. As in the case of the Foucaultgrams, two scans are required: one parallel to the X-axis and the other parallel to the Y-axis.

In Figure 2.9, some simulated Ronchigrams are shown.

When the Ronchi ruling is rotated at an angle α, then the coordinates (T_X, T_Y) must be transformed through a rotation with the same angle α, in order to obtain a new coordinate T_y^*, which is substituted in Equations 2.7 and 2.8.

In a paper by Malacara [22] an alternative point of view was developed that can be used exclusively for systems with symmetry of revolution, with respect to the optical axis, and with the point source located on that axis. Under these conditions, the computer programming is easier and elegant, and the procedure is sufficient to simulate the majority of the Ronchigrams needed in an optical shop.

The simplified procedure developed by Sherwood [18–19], Malacara [22] can be interpreted as a ray tracing from a source, on axis, to a point located on the entrance pupil of the system (Figure 2.10), to a certain distance, S, from the optical axis. Such a ray tracing represents all the traced rays over the circle with radius S; that is, they are the traced rays that belong to a cone whose vertex is the point source and whose base is the circle with radius S, on the entrance pupil. Similarly, all the rays that left the system will belong to a

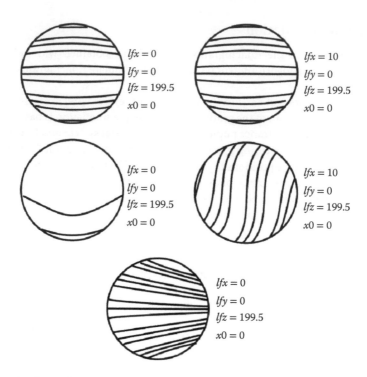

FIGURE 2.9 Simulated Ronchigrams for a parabolic mirror of 15 cm diameter and 200 cm radius of curvature was used. The position of the Ronchi ruling ($LR = lfz$), point source (lfx, lfy, lfz), and center of the conic section ($x0$) are indicated in each case.

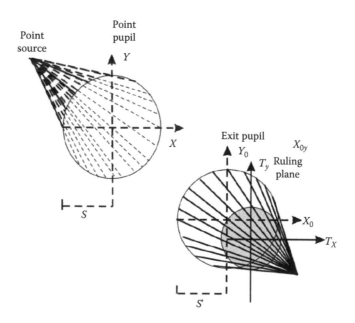

FIGURE 2.10 Geometry used in the simulations of the Ronchigrams for axis symmetrical systems.

cone of rays whose intersection points at the Ronchi ruling plane will define a circle whose points of intersection could be easily calculated. The idea of Malacara [20] has been programmed in order to simulate Ronchigrams for any reflecting conic surface, and a refractive system with centered spherical surfaces.

Other point of view for simulate Ronchigrams was developed by Aguirre-Aguirre et al. [23], the Ronchigrams is obtained by lateral shear interferometry theory, considering fringes with cosine profile increasing, with this procedure, the spatial resolution, Figure 2.11.

To test aspherical surfaces with Ronchi test the simulated Ronchigram are needed. Sherwood [18–19] and Malacara [24] developed an algorithm for axisymetrical systems and with point source on axis. This computer program requires a meridional ray tracing only, then it is easy, elegant and fast. Adding to this, it can be enough to simulate the majority of the Ronchigrams needed in optical shop. This approach has been applied to simulate Ronchigram for any axisymmetric (refractor or reflecting) system, Malacara [20]; another developed algorithm for sections of conic mirrors and source at any position was done by Cordero-Dávila et al. [13]. This one is an iterative method and then the results are good only for borders of the Ronchigram and/or bi-Ronchigram.

As was mentioned previously, the aim of the application of these tests is to know quantitatively the wavefront deformations of any surface or, in general, any optical system. In order to know W (the wavefront deformations) using the Ronchi test, two crossed Ronchigrams are required; since the fringes of a Ronchigram can be interpreted as the level curves of only one of the components of the transverse aberration $\vec{T}$ and for the transverse aberration components (T_X, T_Y), they are related with W by the equation

$$\vec{T} = -\vec{\nabla}W \qquad (2.9)$$

Therefore, it is necessary to carry out an integration to evaluate W from the measurements of the transverse aberrations. Thus, if we have only one Ronchigram, and then only one partial derivative, then we cannot evaluate the line integral; given such situations, we need two crossed Ronchigrams in order to carry out the line integral.

In order to obtain two perpendicular Ronchigrams with only one grating, the first Ronchigram is obtained with the ruling oriented along one direction and, afterwards, the grating is rotated 90 degrees and then the second Ronchigram is recorded [25]. In this case, it is not possible to simultaneously record

Ronchigrams with ray tracing	Simulated Ronchigrams	Experimental Ronchigrams	Correlation coefficient	
			Ray tracing vs. Experimental	Proposed algorithm vs. Experimental
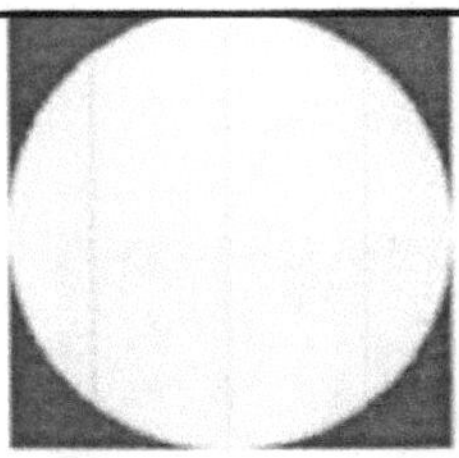	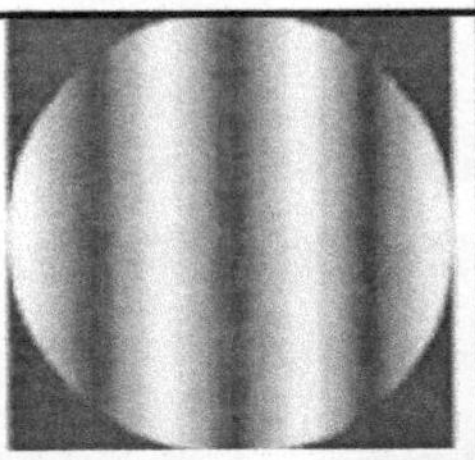	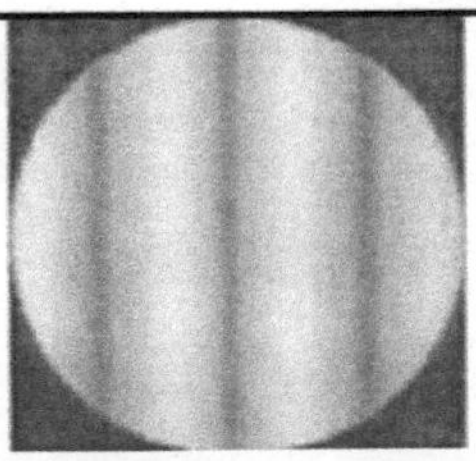	0.1596	0.8126
			0.1886	0.7648

FIGURE 2.11 Ronchigrams is obtained by lateral shear interferometry theory, considering fringes with cosine profile increasing, with this procedure, the spatial resolution.

the two patterns; and then precision is lost if there are temporal variations of W. A second option is achieved by amplitude division of the wavefront (by using a beamsplitter) into two similar channels, each one having a lens, a grating and a detector array. A third option [26,27] is by means of a squared grating, instead of the classical Ronchi ruling. This kind of grating could be considered like the intersection of two crossed Ronchi rulings; and then the obtained pattern could be considered like the intersection points of two crossed Ronchigrams, that in general, they are required in order to calculate W (Figure 2.12).

For the Ronchigram evaluation (Figure 2.13), with the slits of the Ronchi ruling parallel to the X-axis, N vectors (X_i, Y_i, M_{Xi}) are considered, where (X_i, Y_i) are the coordinates of the point associated to the maximums and/or minimums of intensity of the Ronchigram fringes, and M_{Xi} is the interference order associated to the fringe. In the next step, the least-square method is applied to the mentioned vectors in order to estimate the polynomial coefficients of degree $K - 1$, for the bidimensional polynomial transverse aberration component T_X. In a similar way, the polynomial coefficients for T_Y are calculated for the crossed Ronchigram [28]. If it is assumed that W could also be written like another polynomial function but now of degree K, then the integration is carried out using the relationship among the coefficients of the polynomial expressions of W, T_X, and T_Y.

An important aspect to start a pattern evaluation within a unitary circle is the knowledge of the center coordinates and the radius of the exit pupil. Cordero-Dávila et al. [29] have proposed an algorithm that allows us to evaluate both the center coordinates and the radius of a pattern starting with the border points that are supposed to be affected by Gaussian errors.

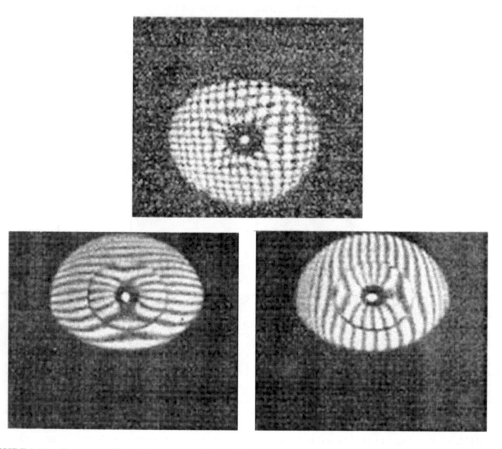

FIGURE 2.12 Two crossed Ronchigrams used for evaluating a mirror.

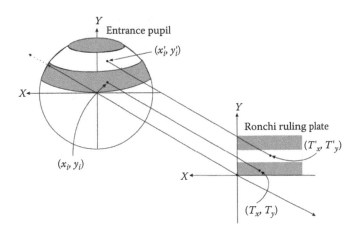

FIGURE 2.13 Ronchigram obtained with the square Ronchi ruling.

a. The Ronchigram evaluations using polynomial fittings have the disadvantage of introducing an overfitting through the election of the polynomial degree. Therefore, Cordero-Dávila et al. [30–31] have proposed a new algorithm by means of which the fitting is carried out assuming that the Gaussian errors are on the fringe coordinates. Another algorithm [27] was developed in order to avoid the overfitting and the incorrect interpretation at the borders. In this case, numerical integration is used, as in the Hartmann test. Unfortunately, this procedure could only be used in the Ronchi test with the squared grating.

b. The Ronchigrams could be analyzed from the point of view of the physical optics predicting the intensities that one should observe [32]. In this last case, the Ronchigrams could also be considered as lateral shearing interferograms. The analysis can be simplified if it is assumed that only two of the diffracted orders interfere, and it is possible to drop the third derivative of W; hence, both interpretations, the geometrical and physical theory, coincide [26].

In the experimental setup of the Ronchi test, it is common to use a point light source or a slit parallel to the ruling lines. Anderson and Porter [33] suggested allowing the grating to extend over the lamp, instead of employing a slit source. In practice, a LED (light emission diode) source for illuminating the grating from behind can be used; this simplifies the experimental array. More recently, Patorsky and Cornejo-Rodríguez [34] found that the setup could be further greatly simplified by illuminating the grating with daylight and setting a strip of aluminum foil just behind this part.

An interesting application of the Ronchi test, as lateral shearing interferometer, was the work by Hegeman [35]. They used the Ronchi test for testing EUV optics for nanolithography.

Another way for evaluating the Ronchigrams was proposed by Aguirre-Aguirre et al. [36]. They present the validation for Ronchigram recovery with the random aberrations coefficients algorithm (ReRRCA), Figure 2.14. This algorithm was proposed to obtain the wavefront aberrations of synthetic Ronchigrams, using only one Ronchigram without the need for polynomial fits or trapezoidal integrations. The validation is performed by simulating different types of Ronchigrams for on-axis and off-axis surfaces. In order to validate the proposed analysis, the polynomial aberration coefficients that were used to generate the simulated Ronchigrams were retrieved. Therefore, it was verified that the coefficients correspond to the retrieved ones by the algorithm with a maximum error of 9%.

To calculate from two crossed Ronchigrams, several methods have been followed:

Polynomial fitting: If a polynomic description (monomials, Zernikes, etc.) of function is assumed, that is, is defined by a coefficients vector, $\vec{B}$. Then each component of the transversal aberration, $[t_x(x,y) \text{ and } t_y(x,y)]$, is described by another polynomial defined respectively by the vectors $\vec{C}$

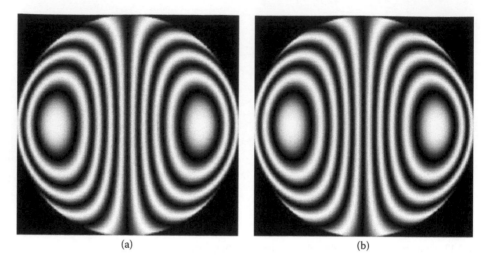

(a)   (b)

FIGURE 2.14   (a) Synthetic Ronchigram of an on-axis hyperbolic surface (outside of focus). (b) Ronchigram recovered by Ronchigram recovery with the random aberrations coefficients (ReRRCA) algorithm[36].

and $\vec{D}$. In this case, two least square fittings on the fringe coordinates (x_i, y_i) and the respective interference orders, M_i, are applied to estimate the two vectors $\vec{C}$ and $\vec{D}$. Cornejo-Rodríguez and Malacara-Hernández [28] used the analytic relations among $\vec{C}$, $\vec{D}$ and $\vec{B}$ to finally calculate this last vector. This method is described, step by step, by Cornejo [37]. It is possible to record only one pattern if a square grid is used. In this case, Cordero-Dávila et al. [27,38] have shown that a numerical or analytical integration can be done.

The Ronchigram evaluations using polynomial fittings have the disadvantage to introduce overfitting through the election of the polynomial degree. Therefore, Cordero-Dávila et al. [30–31] proposed to use a new algorithm by means of which the fitting is carried out by assuming that the Gaussian errors affect the fringe coordinates.

Synchronous phase detection: The synchronous phase detection technique was introduced in traditional Ronchi test by Yatagai [39] in one direction and by Omura and Yatagai [40] in 2-D. The authors moved the Ronchi ruling sideways, and then a periodic phase shift in the Ronchigram was introduced for synchronous phase detection. Following the known methods of phase shifting, they obtained a sectional distribution of the first derivative of the wavefront aberration along the Ronchi ruling displacements (x-axis). By relatively rotating the lens under test, the y-direction derivative of the wavefront is then measured. Next, a wavefront aberration profile for each direction is calculated from the phase distribution corresponding to the first derivative by numerical spatial integration. The final step is a reconstruction of the 2-D distribution of the wavefront aberration under test from the two-direction profiles.

Intensity phase conversion: Der-Shen and Ding-Tin [41] developed a new phase reduction algorithm for the Ronchi test. This method includes finding the fringe centers, normalization, and a linearized intensity phase conversion. These calculi have a systematic error that is eliminated by averaging phases generated by two Ronchigrams shifted by $\pi/2$.

Profilometric measurements: Curvature radiuses of toroidal surfaces were measured by Arasa et al. [42–44] and Royo et al. [45]. Authors recorded several ronchigrams by rotating surface in 30° steps. The Rochigrams are initially smoothed and binarized with a threshold procedure. This leaves a set of wide bright lines.

Ronchi rulings using phase-locked loop: Servin Guirado et al. [46] developed a method for phase detection of modulated Ronchi rulings based on a digital phase-locked loop (PLL). The PLL is used to extract the transverse aberrations from a modulated Ronchi ruling. The use of a PLL as a demodulator is that the detected phase is demodulated continuously (no unwrapping process is

 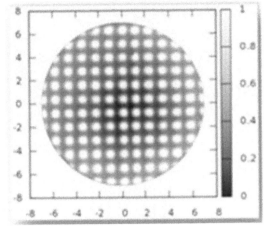

FIGURE 2.15 Example of a bi-Ronchigram.

required). Finally, the wave aberration is obtained by applying the Equation 2.10. Due to the high sensitivity of the PLL, this method allows us to extract information from Ronchigrams formed by almost straight fringes.

$$w_B = w_A - \frac{1}{R} \int\limits_{(x_A,\, y_A)}^{(x_B,\, y_B)} t_x(x,y)dx + t_y(x,y)dy \qquad (2.10)$$

Correlating bi-Ronchigram images: Several efforts were made to evaluate wavefront aberrations by matching Ronchigram images. Hassani and Hooshmand Ziafi [47] matched experimental Ronchigram coming from a lens versus simulated Ronchigrams coming from conic mirrors. Aguirre-Aguirre et al. [20] developed an algorithm to recover the aberration coefficients from a simulated Ronchigram by matching with another simulated Ronchigram.

Recently, Cordero-Davila et al. [38] developed a computer program to simulate Ronchigrams and bi-Ronchigrams for free surfaces described by means of bi-cubic splines. Figure 2.15 shows a bi-ronchigram calculated with the aid of the algorithm described.

2.4.2 Substructured Ronchi Gratings

Murty and Cornejo [48], Cornejo et al. [49], and Robledo Sánchez [50] introduced the idea of substructured Ronchi gratings, with the idea of sharpening and increasing the number of fringes in the Ronchigrams; thereby increasing their spatial resolution and allowing greater accuracy in the evaluation of a surface under test. In a further step, Yaoltzin [51], Cornejo-Rodríguez [52], and Aguirre-Aguirre et al. [53] introduced the implementation of the substructured Ronchi gratings with a LCD; the principal advantage is to have a simple method for generating substructured Ronchi gratings and calculating the intensity pattern produced by this type of grating. An alternative way to produce those substructured Ronchi rulings is by means of the linear combination of classical gratings; the pattern of irradiance produced by these Ronchi gratings will be a linear combination of the intensity patterns produced by each combined classical grating. To verify the results, a comparison between theoretical and experimental Ronchigrams was done for a concave conic mirrors as seen in Figure 2.16.

One advanced application of the Ronchi test is to evaluate optical surfaces with high slopes, using two wavelengths, and the concept of equivalent wavelength, García-Arellano et al. [54]. A spatial modulator

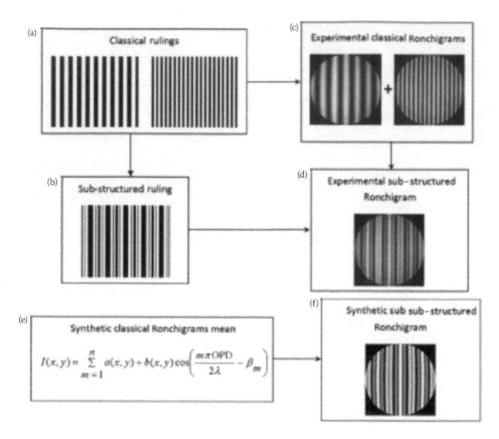

FIGURE 2.16 Flowchart representing the different ways for obtaining a substructured Ronchigram, (a) to (d) refer to the combination of classical gratings; (e) and (f) refer to a synthetic ruling [50].

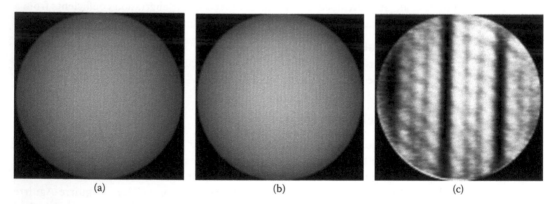

FIGURE 2.17 Generation of a Ronchigram with equivalent wavelength and 7 bit substructured Ronchi rulings. (a) Ronchigram with λ_1; (b) Ronchigram with λ_2; (c) Ronchigram with λ_{eq}.

is used in the implementation of the Ronchi test; in order to generate different wavelengths, a white LED and different color filters were used. Two Ronchigrams with incoherent light, each one for a different color, are registered and computationally processed; thus generating a third one with an equivalent wavelength. The results show that it is possible to generate patterns with traditional and substructured sequences of Katyl rulings, as demonstrated by Figure 2.17. Additionally, they discuss some of the limitations of employing different rulings, and find that appropriate image enhancing algorithms can improve the visibility of the resulting fringes for a better analysis.

2.5 Hartmann Test

The exclusive use of Schlieren's tests in testing optical surfaces could be insufficient, as in the case of the Foucault test in that it is a little sensitive to slow variations of the wavefront [52]. That is not an important problem in surfaces of small dimensions; however, it could be a severe drawback for large surfaces. On the other hand, with the wire test, as has been explained, only radial zones of the surface under test can be carried out.

It is evident that the interferometric tests are much more precise than the geometrical tests. However, interferometric tests could be a very expensive method for testing surfaces and/or optical systems of large aperture. However, the Hartmann test could prove to be an important and economical alternative for testing optics, mainly if large surfaces are under test.

The principal advantage of the Hartmann test is that it does not require the inclusion of wavefront compensators to convert the reflected wavefront into a spherical wavefront, which is required in the use of other methods. It is obvious that correctors can be a source of additional errors.

The basic hypothesis, in the evaluation of the Hartmann test, is that the slopes of the wavefront under test do not change abruptly, but rather in a slow manner; such an assumption is important because the surface or wavefront is sampled in a few zones, where the holes of the Hartmann screen are located, and continuous slow variations are considered.

In this test, the Hartmann screen is a kind of filter, at a certain location plane, and is usually placed at the exit pupil of the system under test. The observation-registering plane is located near the image point of the point source illuminating the screen of the surface. If the coordinates of the centers of each hole in the Hartmann screen are well known and the positions of the centers of the dots in the Hartmanngram are measured, then the director cosines of the rays joining corresponding holes and dots, and then the slopes of the rays of the wavefront can be calculated. Alternatively, two Hartmanngrams recorded with the same Hartmann screen, at two different distances, could be used for evaluation in the Hartmann test, and to avoid the measurement of the distance from the vertex mirror to the observation or Hartmanngram plane; however, you should measure with precision the distance between the observed and registered Hartmanngram planes.

The evaluation of the wavefront is carried out by using Equation 2.9, which gives us the relationship between the transverse aberration and the optical path difference W. In testing a surface, you could suppose the errors, h, are related to W by

$$W = 2h. \tag{2.11}$$

In Equation 2.11, it is assumed normal incidence and, therefore, h is measured along the same direction.

One of the main advantages of the Hartmann method is that the evaluation of W can be done by using polynomial fittings or by means of numerical integration. Theoretically, it is possible to think that it is better to use polynomial fittings, because they avoid the errors due to the numerical integration methods. However, polynomial fittings usually introduce overfitting, which means artificiality in the surface under test or problems of interpretation in the borders.

When surfaces of large dimensions are tested, systematic errors are produced as a result of the laminar turbulence produced by local gradients of temperature, or mirror deformations due to the weight of the mirror, which can be seen as astigmatic aberrations of W.

The use of illuminated screens for the Hartmann test was proposed by Bautista-Elivar et al. [56]. More recently new and powerfull methods using the Hartmann test had been developing by Su et al. [57–59] and Huang et al. [60]. Their methods can be applied either in the laboratory or optical shops (see Figure 2.18).

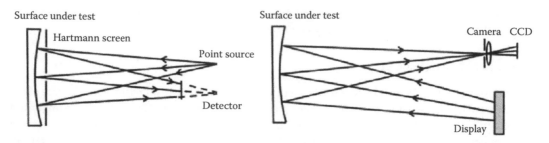

FIGURE 2.18 Schematic setup of (a) a Hartmann test and (b) a software configurable optical test system (SCOTS) test. SCOTS trace rays in reverse.

There is a paper by Malacara-Hernández, and Malacara-Doblado [61], where there is a clear explanation about the Hartmann test and its relation with the Ronchi and Shack-Hartmann methods, with a descriptive connection between the normal and classical arrangements and those call null techniques, either for convex and concave surfaces.

2.5.1 Screen Design and Hartmanngram Evaluation

The holes distribution in the Hartmann screen has changed from the radial distribution used by Hartmann, to the helical one and, finally, to a squared-hole array that has been commonly employed for the testing of large primary mirrors of astronomical telescopes. The main disadvantage of the radial and helical screens is that sampling is not uniform, since the density of holes diminishes radially. Another important disadvantage of these screens is that concentric scratches, or other general defects, produced during the lapping working could not be detected with such a type of screen. Thus, in order to overcome those kinds of problems, a square screen is used, which produces a uniform sampling. Another advantage of this screen is that it could become rigid and then, with this structure, it could achieve high precision in the hole positions. This can be accomplished by making holes of the largest size in the rigid structure and, later on, other holes, with the final size are made on small badges. Finally, the holes in the badges can be aligned and fixed with a higher precision in the rigid structure.

An analysis of the conditions that limit the applicability of the Hartmann test was made by Vitrichenko [62] and Morales and Malacara [63]. Three important factors limiting the use of the Hartmann test are as follows:

- Diffraction effects of the holes of the screen and its mechanical strength, fixed some limit for the diameters of the screen holes and the distance between their centers.
- The total number of holes is limited by the accuracy that must be developed to obtain results of a given reliability.
- The adequacy of the description of the surface of an optical component is limited by the degree of smoothness of the surface.

Some important practical recommendations when the Hartmann test is applied are as follows:

- The screen must be centered accurately [64], since a decentering of the screens leads to an apparent presence of coma.
- The point light source used to illuminate must be centered properly to prevent the introduction of off-axis aberrations.
- The photographic plate or CCD array should be perpendicular to the optical axis.

Once the Hartmanngram has been recorded, the location of the dots on the photographic plate (or CCD) must be measured to a high accuracy. The Hartmanngram coordinates (X_{dei}, Y_{dei}) can be measured by means of a microdensitometer or a measuring microscope having an X–Y traveling stage, or as is usual, if a CCD is used, the locations of the dots can be found after the Hartmanngram has been digitized and stored in a computer. For each dot, the coordinates are measured and a set of numbers (M_{Xi}, M_{Yi}) are assigned. They are related to the center of a hole in the Hartmann screen, with coordinates (X_i, Y_i), by means of the equation

$$(X_i, Y_i) = e(M_{Xi}, M_{Yi}), \tag{2.12}$$

where e is the distance between two nearest hole's centers.

Starting with the N data vectors $(X_{dei}, Y_{dei}, M_{Xi}, M_{Yi})$, it is required to know the coordinates of each one of the dots of the Hartmanngram, with a coordinate system whose origin is at the intersection of the Hartmanngram plane with the optical axis of the system under test. An estimated origin could be found by supposing that the point source is located on the optical axis, and the surface under test is near to an axisymmetrical surface. In this case, the optical axis crosses the Hartmanngram plane at the point (X_{av}, Y_{av}), given by the average of all the coordinates of the dots in the Hartmanngram; and, therefore, the N vectors referred to the optical axis can be calculated from

$$(X_{oai}, Y_{oai}, M_{Xi}, M_{Yi}) = (X_{dei} - X_{av}, Y_{dei}Y_{av}, M_{Xi}, M_{Yi}). \tag{2.13}$$

In the next step, the ideal dot coordinates (X_{eri}, Y_{eri}) are calculated with the aid of an exact ray tracing. From these ideal coordinates, the transverse aberration values can be calculated by means of

$$(X_{abi}, Y_{abi}, M_{Xi}, M_{Yi}) = (X_{oai} X_{eri}, Y_{oai} - Y_{eri}, M_{Xi}, M_{Yi}) \tag{2.14}$$

The new N vectors describe the transverse aberration values of the wavefront that leaves the mirror or system under test. In this data, a defocus error can be present since the hypothetical location of the Hartmanngram plane in the exact ray tracing can be different from the actual Hartmanngram plane. The optimum focus can be found by subtraction of the linear term that can be calculated by means of two least-square fits applied to the data vectors given in Equation 2.14. In this case the focus error data $\left(X_{abi}^*, Y_{abi}^*\right)$, can be calculated as a function of the hole centers with the equation

$$\left(X_{abi}^*, Y_{abi}^*\right) = \left(A + BX_i, A' + B'Y\right), \tag{2.15}$$

where A, B, A', and B' can be evaluated through two least-square independent fittings.

If we apply the two independent fits described above, we can eliminate some astigmatic terms in the linear dependence, which can be interpreted as a defocusing term, and therefore one can have an incorrect interpretation of the form of the surface. In order to avoid the previous problem only one least-squares fitting should be carried out [65].

With the new transverse aberration coordinates $\left(X_{abi} - X_{abi}^*, Y_{abi}^* - Y_{abi}^*\right)$, a numerical or polynomical integration is carried out to get the values of the optical path differences. Once these evaluations are concluded, we can do a new fitting with the terms to fourth degrees but special care must be taken to not subtract them without the quantitative knowledge of the origin of each term; this last factor could be known after evaluating the expected flexions of the mirror and/or the decentering of the source of light.

2.5.2 Shack-Hartmann Test

An important and novelty development of this classical Hartmann test was presented and developed by Platt and Shack [6,66], Shack and Platt [8]; at the present time, the testing method has received the name of Shack-Hartmann Sensor.

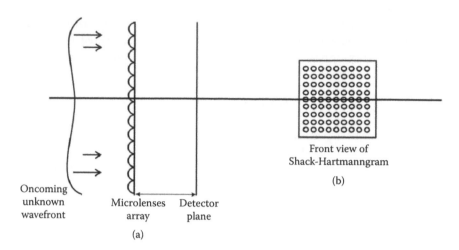

FIGURE 2.19 Schematic diagram for the Shack-Hartmann sensor. (a) Incoming wavefront crossing sensor, producing a series of circular images at the detector. (b) Circular images produced by the sensor on the detector.

Basically, the proposal done by Platt and Shack was locating a small Hartmann screen, close to the center of curvature of a surface under test, with the bright and wonderful idea that in each hole of the screen, small lenses with short focal length were placed, as it is shown in Figure 2.19.

This described Shack-Hartmann sensor or plate, in a short period became one powerful method with applications in optical testing, astronomy, human eye studies, adaptive optics, and other fields where sensing wavefronts are required.

Some examples of the widespread of applications of the Shack-Hartmann method are described in "Robert Shanon and Roland Shack, *Legends in Applied Optics*", J.E. Harvey and R.B. Hooker, Editors. SPIE, Press, Bellingham, Washington, USA, 2005. In particular from this book, there are the papers by Schwiegerling and Neal [67], and Neal [68].

2.6 Null Tests: Ronchi, Hartmann, and Shack-Hartmann

Even with the technological advances in the recording and evaluation of the patterns, important efforts have been done recently in order to design and construct modifications to the traditional tests in such a way that—although still not for spherical wavefronts—we have a direct interpretation as in the case of a pattern of right fringes in the Ronchi test or of a Hartmanngram whose dots are distributed on a squared array.

The previous idea has particular importance in testing aspherical conic sections and/or in the case of production in series since the criterion of acceptance/refusal is very direct. In order to achieve a simplified analysis, an additional optical system could be introduced and, with this, the original wavefront can be modified and converted into a spherical one, as in the case of the optical compensators, whose principal problem is the increment of the number of possible errors, such as defects in their construction and/or assembling of their components. An inexpensive alternative for the Ronchi and Hartmann tests consists in the modification of the Ronchi ruling or the Hartmann screen. In the Ronchi test, this idea was developed qualitatively for conic surfaces by Pastor [69] and with the theory of third order by Popov [70] and Mobsby [71]; and with a more precise solution by Malacara and Cornejo [72]. Finally, the idea was developed for any optical system and using exact ray tracing by Hopkins and Shagan [73].

The null Hartmann test has been studied by Cordero-Dávila et al. [74]. With an analysis based on a common treatment of both the Ronchi and Hartmann tests, Cordero-Dávila et al. [13] have demonstrated that if the centers of the holes in the Hartmann screen are distributed at the intersection points of two crossed Ronchigrams then the dots in the Hartmanngram will be located on the intersection points of the two crossed rulings. And for the Ronchi, if we desire to get a Ronchigram of right- and same-spaced fringes, then the lines of the null Ronchi ruling should contain the Hartmanngram points.

2.6.1 Null Shack-Hartmann Test

In a very recent published paper by Granados-Agustín et al. [75], an analysis of the similitude between the Hartmann, Ronchi, and the novelty Shack-Hartmann test was described by means of Table 2.1. An interesting proposal from such table, in the lowest row, is to have a null Shack-Hartmann method. From a calculated null Ronchi grating, for certain conic surface, along the grating strips, a series of holes with small lenses are located. Therefore, the observed Shack-Hartmann image contains a squared array of image points because the conic surface shape was already compensated with the null Shack-Hartmann plate.

Table 2.1, contains two main columns, one for the kind of filter used (grating or screen), and the other for the observed patterns. Each main column is divided into three sections: (a) the filter column is describing the filters names, a picture of the used filter, and the size and location; (b) the observed patterns column, in a similar way, is describing the name of the pattern, a picture of them, as well as the size and position of the pattern. The letters c.c. and c.s. mean close to the center of curvature and close to the surface, respectively.[75].

TABLE 2.1 Common characteristics of the Hartmann, Ronchi, and Shack-Hartmann tests

Filter			Observed Pattern		
Type		Size position	Type	Tested surface	Size position
Ronchi Ruling		Small c.c.	Ronchigram	 Parabolic	Big c.s.
Hartmann Screen		Big c.s.	Hartmanngram	 Hyperbolic	Small c.c.
Null Ronchi Grating		Small c.c.	Null Rochigram	 Hyperbolic	Big c.s.
Null Hartmann Screen		Big c.s.	Null Hartmanngram	 Parabolic	Small c.c.

(Continued)

TABLE 2.1 (Continued)   Common characteristics of the Hartmann, Ronchi, and Shack-Hartmann tests

Filter			Observed Pattern		
Type		Size position	Type	Tested surface	Size position
Big Hartmann-Ronchi Screen		Big c.s.	Hartmanngram	 Parabolic	Small c.c.
Big Null Hartmann-Ronchi Screen		Big c.s.	Null Hartmanngram	 Parabolic	Small c.c.
Hartmann Screen with Lenses Shack-Hartmann		Small c.c.	Shack Hartmanngram	 Human Eye	Small c.c.
Null Shack-Hartmann Screen		Small c.c.	Null Shack Hartmanngram	 Parabolic	Small c.c.

An alternative method to generating null Ronchi gratings, considering that the period of the Ronchi grating was proposed by Aguirre-Aguirre et al. [76], Figures 2.20 and 2.21, as a function of the x and y positions over the grating, compare to ray tracing or spot diagrams for the null Ronchi gratings calculus. They show results in comparison with the null Ronchi gratings reported in the literature against gratings calculated with their method. They have found that our method can calculate the Ronchi gratings for any conical surface that is satisfactory.

2.6.2 Null Screens for Convex Surfaces

As it is well known, the convex surfaces present some special characteristics to test them.

In order to solve the problem for testing convex surfaces with diameters less than 10 cm, a pioneering paper was published by Díaz-Uribe [77–78], Carmona-Paredes and Díaz-Uribe [79]; screens are located around the surface as it is shown in Figure 2.22. Further proposal has been published since those first papers, with studies about cylindrical screens for concave surfaces, as shown in Figure 2.23 [80–81], or shifting the spots in the screen to produce close images to be analyzed or tilting the null screen [82]. Some of these cylindrical null screens has been applied for the analysis of the human eye by Estrada-Molina et al. [83], and Rodríguez-Rodríguez et al. [84]. These cylindrical screens are also analyzed in the paper by Malacara-Hernández and Malacara-Doblado [85].

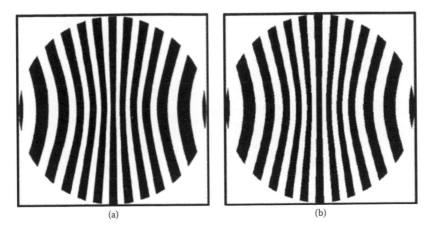

FIGURE 2.20 Grating calculated by (a) D. Malacara and A. Cornejo [67] and (b) our proposal.

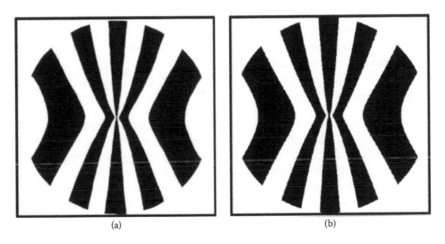

FIGURE 2.21 Grating calculated by (a) G. H. Hopkins and R. H. Shagam [68] and (b) our proposal.

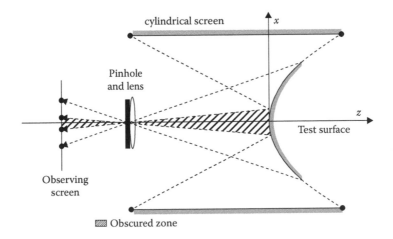

FIGURE 2.22 Layout of the testing configuration for convex surface.

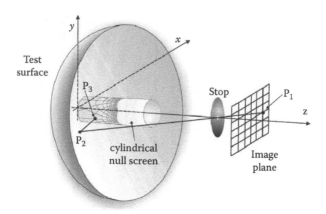

FIGURE 2.23   Layout of the testing configuration for concave surface.

2.7 Gradient Tests

In this section we present two techniques that solve the problem of finding the wavefront coming from a system or surface, mainly by sensing the irradiance for several different planes [86], and at two planes [11], coming from the surface or wavefront under test.

2.7.1 Platzeck–Gaviola Test

A different and interesting approach to the testing of optical surfaces was presented by Platzeck and Gaviola in 1939. In such a technique, the different zones of an aspheric or conic surface are identified by means of the so-called caustic coordinates. These coordinates correspond to the center of curvature of each one of the zones in which the surface or system under test is divided. In Figure 2.24a the meaning of caustic coordinates is illustrated [46]; they physically correspond to the coordinates (x, y) of the par-axial center of curvature, zone a; as well as of the observed $b - b'$, $c - c'$,..., $l - l'$ zones, located at certain distances from the optical axis, and correspond to a certain sagitta value Z_k. For each zone, a spherical region is considered with certain values for the radius of curvature R of the zone, and the corresponding localized center of curvature with coordinates (x, y). The values for the local radius of curvature can be obtained by means of the following equation, derived from calculus,

$$R = \frac{\left[1+\left(\partial z / \partial s\right)^2\right]^{3/2}}{\left(\partial^2 z / \partial^2 s\right)};$$ (2.16)

after obtaining the first and second derivatives of the sagitta z, then

$$R = \frac{1}{c}\left(1 - Kc^2 s^2\right)^{3/2}.$$ (2.17)

For the case of the conic constant $K = 0$, a sphere, the usual relation $R = 1/c$ is obtained. In order to obtain a set of equations for the caustic coordinates x, y, from Figure 2.25, it is possible to write

$$\frac{y}{2(x+Kz)} = \frac{s}{(1/c)-(K+1)^2}$$ (2.18)

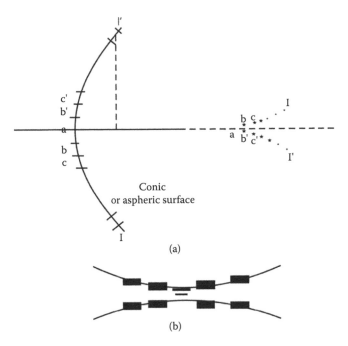

(a)

(b)

FIGURE 2.24 (a) Center of curvature for the different considered spherical zones of a general conic or aspheric surface that corresponds to the paraxial center of curvature, taken as a reference point for the other zones. (b) Best focusing image for certain zones a; b = b', c = c', ..., I = I' of an aspherical or conic surface.

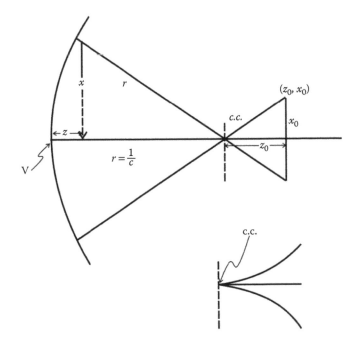

FIGURE 2.25 Caustic coordinates of an aspheric surface.

and

$$R^2 = \left(s + \frac{y}{2}\right)^2 + \left(Hc + h - 2\right)^2.$$

(2.19)

By means of Equations 2.17 through 2.19, equations for x and y can be obtained as

$$x = -Kz\left\{3 + cz(k+1)\left[cz(k+1) - 3\right]\right\}$$

(2.20)

and

$$y = -2scKz\left\{\frac{2 + cz(K+1)\left[cz(K+1) - 3\right]}{1 - cz(K+1)}\right\}.$$

(2.21)

It is important to notice that the caustic coordinates (x, y) from Equations 2.20 and 2.21 are measured taking as a origin the corresponding paraxial center of curvature for the zone a of the surface.

Experimentally what is necessary to do is to register, in a photographic film or with the use of some other modern detector, the position of a wire or a slit that moves along the optical axis producing a sharp focusing image for the different zones of the surface. A detailed procedure for this experiment can be read in Schroader [87] paper, or in the same Platzeck and Gaviola [81] work.

In Figure 2.24b, it can be seen how a complete set of focal points for the different zones of the surface under test are registered in a photographic plate.

2.7.2 Irradiance Transport Equation

The problem with finding the phase of a wavefront from irradiance measurements has been studied intensively by several authors in the last 25 years. Some of them have developed algorithms—for example, Gerchberg and Saxton [88] and Teague [89]—or established more comprehensive theories based in Helmholtz and irradiance transport equations, as those works by Teague [11], Streibl [90], and Ichikawa et al. [9]; interesting review works are those by Fienup [91], Campos-García and Díaz-Uribe [92], Fernández-Guasti et al. [93].

Before establishing a particular application of the solutions of the irradiance transport equations to the field of testing optical surfaces or systems, a general and brief review will be given about the development of obtaining information for the phase wavefront by means of intensity measurements. Finally, the presentation of the work by Teague [11] will be described.

According to a paper by Gerchberg and Saxon, some of the first trials to obtain the wavefront phase from irradiance measurements were done by Schiske [94], Hoppe [95], and Erickson and Klug [96]. In all these works they abandon the Gabor proposals to add a reference wave, in order to find the wavefront phase. In their first paper, Gerchberg and Saxon [97–98], they developed a method for the determination of the phase from intensity recordings in the imaging and diffraction planes. One of the most important characteristics of such a paper was that it was not limited to small phase deviations. In a second paper, the same authors, Gerchberg and Saxon [97–98] improved the computing time, recognizing the wave relation in the imaging and diffraction planes by means of the Fourier transform.

Following the work of many authors to solve the retrieval of phase from irradiance measurements, a crucial problem was always the uniqueness of the solution [(see for example, Gonzalves [99], Devaney and Chidlaw [100], Fienup [91]. A step forward to find the solution was given by Teague [84], where for the first time he established that mathematically it is sufficient to retrieve the phase from irradiance data in two optical planes, and that solution was deterministic. With this last result, the previous problem of the uniqueness of the solution was finally solved. A year later Teague [9] described an alternative method based on Green's functions solution to the propagation equations of phase and irradiance; and Fienup made a comparison among the different algorithms to the retrieval of phase. More recently, Gureyev et al. [101], and Salas [102] found other possible solutions to the radiation transfer equation.

In the next paragraphs, we describe briefly how Teague [11] derived the propagation equation for irradiance.

Starting from the Helmholtz equation

$$\left(\nabla^2 + k^2\right)u\left(x,y,z\right) = 0, \psi, \tag{2.22}$$

where $\nabla^2 = \left(\partial^2/\partial x^2\right)+\left(\partial^2/\partial y^2\right)+\left(\partial^2/\partial z^2\right)$, and $k = 2\pi/\lambda$, it will be assumed that any wave depending only of the position must obey Equation 2.22. For a wave traveling in the z-positive direction and considering only the spatial component, then

$$\psi(x,y,z) = u\left(x,y,z\right)\exp\left(-ikz\right). \tag{2.23}$$

Substituting $\psi(x, y, z)$ of Equation 2.23 into Equation 2.22, we obtain

$$\left[\nabla_T^2 + \frac{\partial}{\partial z^2} - 2ik\left(\frac{\partial}{\partial z}\right)\right]u\left(x,y,z\right) = 0, \tag{2.24}$$

where $\nabla_T^2 = \left(\partial^2/\partial x^2\right)+\left(\partial^2/\partial y^2\right)$. Assuming that the amplitude u varies slowly along the z-direction implies that the term $\partial^2 u/\partial z^2$ can be dropped from Equation 2.24, and the so-called paraxial wave equation is obtained:

$$\left[\nabla_T^2 - 2ik\left(\frac{\partial}{\partial z}\right)\right]u\left(x,y,z\right) - 0. \tag{2.25}$$

If a Fresnel diffraction theory solution is now proposed, without the term $\exp(ikz)$ for Equation 2.25, then the so-called parabolic equation can be derived:

$$\left[\frac{\left(\nabla_T^2\right)}{2k} - k - i\frac{\partial}{\partial z}\right]u_F\left(x,y,z\right) = 0. \tag{2.26}$$

Following the Teague paper [11], where $w_z(x, y, z)$ is normalized such that $|U_F(x, y, z)|^2 = I_F$, with I_F as the irradiance at the point (x, y, z), and writing

$$U_F\left(x,y,z\right) = \left|I_F\right|^{1/2}\exp\left[\ if\left(x,y,z\right)\right], \tag{2.27}$$

and after some algebraic manipulations and with $\phi = (2\pi/\lambda)W\left(x,y,z\right)$, where $W(x, y, z)$ is the wavefront, the irradiance transport equation is obtained as

$$\frac{\partial I}{\partial z} + \nabla_T W \cdot \nabla_T I + I\nabla_T^2 W = 0, \tag{2.28}$$

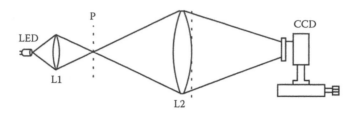

FIGURE 2.26   Experimental setup.

or in a compact form

$$\nabla_{\mathrm{T}} \cdot (I\nabla_{\mathrm{T}} W) = \frac{\partial I}{\partial z}. \tag{2.29}$$

A similar result for the irradiance transport equation was obtained by Streibl, following a different and more simplified approach.

With the irradiance transport equation well established, Teague proved its validity and possible solution by means of Green's functions, carrying out a numerical simulation. Streibl applied it for thin phase structures, obtaining mainly qualitative results. The first quantitative experimented results were obtained by Ichikawa et al. [9]. In order to obtain experimental results, those last authors solved the ITE by the Fourier-transform method, and explained in detail the physical meaning of the different terms of Equation 2.28. In their experiment, Ichikawa et al. used as a phase object a lens and, with the help of a grating and a CCD camera, the irradiance in two planes, separated by 0.7 mm, was registered and the wavefront phase was found with the experimental setup proposed (Figure 2.26). Figure 2.27 shows the results obtained by A.P. Magaña et al. [103], using a Nodal bench, following Ichikawa et al.'s technique.

2.7.3 Roddier Method

For testing astronomical telescopes or for using them with adaptive optics devices, Roddier [104] developed a method to find the phase retrieval from wavefront irradiance measurements. Figure 2.28 shows what Roddier called the curvature sensor, because the aim was to obtain irradiance data at the two off-focus planes, P_1 and P_2, and obtain the phase of the wavefront coming from the optical system.

For his method, Roddier derived the next equation

$$\frac{I_1 - I_2}{I_1 + I_2} = \frac{f(f-1)}{l}\left[\frac{\partial}{\partial n} WS(r-a) - P\left(\frac{f}{l}\bar{r}\right)\nabla_{\mathrm{T}}^2 W\right]. \tag{2.30}$$

On the left side of Equation 2.30 are the irradiance measurements at the two planes P_1 and P_2. On the right side of Equation 2.30, f is the focal length of the system under test; l is the distance where the planes P_1 and P_2 are located symmetrically from the focal point F; and S is the circular Dirac distribution, representing the outward-pointing derivative at the edge of the pupil where it is different to zero.

Equation 2.30 can be derived in a more or less straightforward way from the ITE, Equation 2.28, as has been explained by Roddier et al. [10]. Assuming a plane wavefront at the pupil plane ($Z = 0$), then everywhere except at the pupil border

$$\nabla I\big|_{20} = \begin{cases} 0 \\ -I\big|_{z=0} \end{cases} \hat{n}\partial(r-a). \tag{2.31}$$

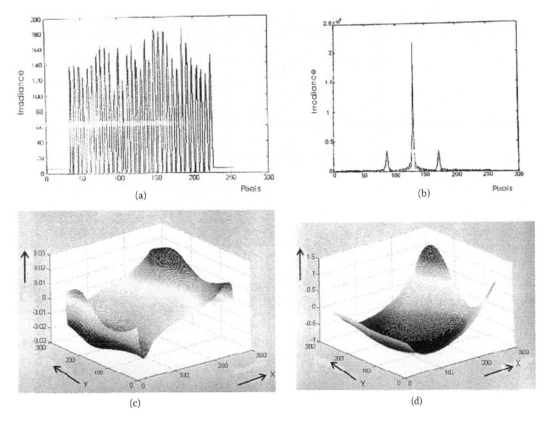

FIGURE 2.27 Wavefront of a lens tested using the method developed by Ichikawa et al. (a) One scan irradiance measurement. (b) Spatial frequency spectra of (a). (c) Derivative of the wavefront. (d) Wavefront shape [98].

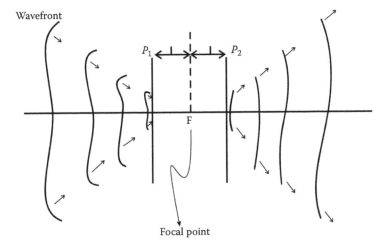

FIGURE 2.28 Roddier's method for measuring the irradiance at the two symmetrical planes, P_1 and P_2, from the focal point F.

Substituting Equation 2.31 into Equation 2.28, we obtain

$$\left.\frac{\partial I}{\partial z}\right|_{z=0} = \left\{ IS(r-a)\frac{\partial}{\partial n}W + P(\bar{r})I\nabla^2 W \right\}\Big|_{z=0}, \tag{2.32}$$

where $P(\bar{r})$ is the pupil transmittance and $\hat{n}\cdot\vec{\nabla}W = (\partial W/\partial n)$.
The irradiance measurements at planes I_1, and I_2 can be written as

$$I_1 = I_{z=0} + \frac{\partial}{\partial z}I\Big|_{z=0}\Delta Z_1$$

and

$$I_2 = I_{z=0} + \frac{\partial}{\partial z}I\Big|_{z=0}\Delta Z_2; \tag{2.33}$$

an important condition for the Roddier method is that $|\Delta Z_1| = |\Delta Z_2| = \Delta Z$. With this last condition, substituting Equation 2.32 into Equation 2.30, the next expression can be derived:

$$\frac{I_1 - I_2}{I_1 + I_2} = \left\{ \partial(r-a)\frac{\partial}{\partial n}W - P(\bar{r})\nabla_T^2 W \right\}\Delta Z. \tag{2.34}$$

Using the thin lens equation, it can be proved that $\Delta Z \cong f(f-l)/l$; with this result, Equation 2.34 can be written as Equation 2.30, which Roddier used for his technique. It is worth mentioning that it is advised that the planes P_1 and P_2 must be outside the caustic in order to avoid mixing the same rays and their irradiance.

Roddier developed an algorithm to find the solution for W of Roddier's equation (Equation 2.30), from the irradiance measurements I_1 and I_2. The Roddier algorithm solves the Poisson equation with Neumann boundary conditions, and the iterative method to find the solution is called overrelaxation. About the problem for the two planes where the irradiance must be measured, Soto et al. [100] analyzed the optimum measurement planes.

Acknowledgments

The authors are grateful to the editorial work by Ana María Zárate Rivera from the Optics Department of INAOE.

References

1. Foucault L.M., Description des Procédés Employés pour Reconnaitre la Configuration des Surfaces Optiques, *C. R. Academ. Sci. Paris*, **47**, 958; (1858); reprinted in *Classiques de la Science*, Vol. **II**, By Armand Colin.
2. Ritchey G.H., On the modern reflecting telescope and the making and testing of optical mirrors, *Smithson. Contrib. Knowl.*, **34**, 3, 1904.
3. Ronchi V., Due Nuovi Metodi per lo Studio delle Superficie e dei Sisemi Occi, *Ann. Soc. Norm. Super. Pisa*, **15**; 1923.
4. Ronchi V., La Frange di Combinazione Nello Studio delle Superficie e dei Sistem Ottics, *Riv. Ottica Mecc. Precis*, **2**, 9, 1923.
5. Ronchi V., Forty years of history of a grating interferometer, *Appl. Opt.* **3**, 4, 1964.

6. Hartmann J., Bemerkungen uber den Bau und die Justirung von Spoktrographen, *Zt., Instrumentenkd* **20**, 47, 1900.

7. Platzeck R, Gaviola E., On the errors of testing a new method for surveying optical surfaces and systems, *J. Opt. Soc. Am.,* **29**, 484, 1939.

8. Shack R.V., Platt B.C., Production and use of a lenticular Hartmann screen, *J. Opt. Soc. Am.,* **61**, (5), 656, 1971.

9. Ichikawa K., Lohmann A., Takeda M., Phase retrieval based on the irradiance transport equation experiments, *Appl. Opt.,* **27**, 3433–3436, 1988.

10. Roddier F., Roddier C., Roddier N., Curvature sensing: A new wavefront sensing method., *Proc. SPIE,* **976**, 203–209, 1988.

11. Teague M.R., Deterministic phase retrieval: A Green's function solution, *J. Opt. Soc. Am.,* **73**, 1434–1441, 1983.

12. Malacara D., *Optical Shop Testing.* Editor, New York, NY: John Wiley & Sons, 1st Ed. 1979; 2nd Ed. 1992.

13. Cordero-Dávila A., Cardona-Núñez O., Cornejo-Rodríguez A., The Ronchi and Hartmann tests with the same mathematical theory, *Appl. Opt.,* **31**, 2370–2376, 1992.

14. De Vany A.S., Some aspects of Interferometric testing and optical figuring. *Appl. Opt.,* **4**, 831–833, 1965.

15. Ojeda-Castañeda J., Foucault, wire, and phase modulation test. In D. Malacara, *Optical Shop Testing,* Editor, New York, NY: John Wiley & Sons, pp. 265–320, 1992.

16. Landgrave J.E.A., Phase Knife Edge Testing, M. Sc. Report Imp. Coll. Sc. And Tech., London, 1974.

17. Rodríguez-Zurita G., Díaz-Uribe R., Fuentes-Madariaga B., Schlieren effect in Amici Prism, *Appl. Opt.,* **29**, 4, 1990.

18. Sherwood A.A., A quantitative analysis of the Ronchi test in term od ray optics, *J. Br. Astronomy Assoc.* **68**, 180, 1958.

19. Sherwood A.A., Ronchi test charts for parabolic mirrors, *J. Proc. R. Soc. New South Wales,* **43**, 19, 1959; reprinted in *Atti. Fond. Giorgio Ronchi Contrib. Ist. Naz. Ottica,* **15**, 340–346, 1960.

20. Malacara D., Ronchi test and transversal spherical aberration, *Bol. Obs. Tonantzintla Tacubaya,* **4**, 73, 1965.

21. Liang C.W., Sasian J., Geometrical optics modeling of the grating-slit test *Opt. Express,* **15**, 1738–1744, 2007.

22. Malacara D., Ronchi test and transversal spherical aberration, *Bol. Obs. Tonantzintla Tacubaya* **4**, 73, 1966.

23. Aguirre-Aguirre D., Izazaga-Perez R., Granados-Agustin F., Percino-Zacarias M., Cornejo-Rodríguez, A., Simulation algorithm for ronchigrams of spherical and aspherical surfaces, with the lateral shear interferometry formalism, *Opt. Rev.,* **20**, 271–276, 2013.

24. Malacara D. Geometrical Ronchi test of aspherical mirrors, *Appl. Opt.,* **4**, 1371–1374, 1965.

25. Cornejo-Rodríguez A., Ronchi test. In D. Malacara, *Optical Shop Testing,* Editor, New York, NY: John Wiley & Sons, pp. 321–3165, 1992.

26. Meyers, W.S, Philip Stahl, H. Contouring of a free oil surface, *Proc. SPIE,* **1755**, 84–94, 1993.

27. Cordero-Dávila A., Luna-Aguilar E., Vaquez Montiel S., Zarate-Vazquez S., Percino Zacarias M.E., Ronchi test with a square grid, *Appl Opt,* **37**, 4, 1998.

28. Cornejo-Rodríguez A., Malacara D., Wavefront determination using Ronchi and Hartmann tests, *Bol. Int. Tonantzintla,* **2**, 127–129, 1976.

29. Cordero-Dávila A., Cardona-Núñez O., Cornejo-Rodríguez A., Least-squares estimators for the center and radius of circular patterns, *Appl. Opt.,* **32**, 5683–5685, 1993.

30. Cordero-Dávila A., Cardona-Núñez O., Cornejo-Rodríguez A., Polynomial fitting of interferograms with Gaussian errors on the fringe coordinates. I: computer simulations, *Appl.Opt.,* **33**, 7339–7342, 1994.

31. Cordero-Dávila A., Cardona-Núñez O., Cornejo-Rodríguez A., Polynomial fitting of interferograms with Gaussian errors on the fringe coordinates. II: Analytical study, *Appl. Opt.,* **33**, 7343–7348, 1994.

32. Malacara D. Analysis of the interferometric Ronchi test, *Appl. Opt.*, **29**, 3633, 1990.
33. Anderson J. A., Porter R.W., Ronchi's method of optical testing, *Astrophys. J.* 70, 175–181, 1929.
34. Patorsky K., Cornejo-Rodríguez A., Ronchi test with daylight illumination, *Appl. Opt.* **25**, 2031–2032, 1986.
35. Hegeman P., Christmann X., Visser M., Braat J. Experimental study of a shearing interferometer concept for at-wavelength characterization of extreme-ultraviolet optics, *Appl. Opt.*, **40**(25), 4526–4533; 2001.
36. Aguirre-Aguirre D., Izazaga-Perez R., Granados-Agustin F., Villalobos-Mendoza B., Percino-Zacarias M., Cornejo-Rodríguez A., Obtaining the wavefront in the Ronchi test using only one Ronchigram with random coefficients of aberration, *Opt. Eng.*, **52**, 053606, 2013.
37. Cornejo A., Ronchi Test. In D. Malacara, *Optical Shop Testing*, Editor, New York, NY: John Wiley & Sons, 3rd edition, 2007.
38. Cordero-Dávila A., Núñez-Alfonso J.M., Luna-Aguilar E., Robledo-Sánchez C.I., Only one fitting for bironchigrams, *Appl. Opt.*, **40**, 31, 5600–560, 2001.
39. Yatagai T., Fringe scanning Ronchi test for aspherical surfaces, *Appl. Opt.*, **23**, 20, 1984.
40. Omura K., Yatagai T., Phase measuring Ronchi test, *Appl. Opt.*, **27**, 3, 1988.
41. Der-Shen W., Ding-Tin L., Ronchi test and a new phase reduction algorithm, *Appl. Opt.*, **29**, 3255–3265, 1990.
42. Arasa J., Royo S., Pizarro C., Toroidal surface profilometries thorugh Ronchi deflectometry: Constancy under rotation of the sample, *Proc. SPIE*, **3491**, 909–915, 1998.
43. Arasa J., Royo S., Tomas N., Simple method for improving the sampling in profile measurements by use of the Ronchi test, *Appl. Opt.*, **39**, 4529–4534, 2000.
44. Arasa J., Royo S., Pizarro C., Profilometry of toroidal surfaces with an improved Ronchi test, *Appl. Opt.*, **39**, 5721–5731, 2000.
45. Royo S., Arasa J., Caum J., Sub-micrometric profilometry of non-rotationally symmetrical surfaces using the Ronchi test, *Proc. SPIE*, **4101**, 153–164, 2000.
46. Servin-Guirado M., Malacara-Hernández D., Cuevas-de-la-Rosa F.J., Direct-phase detection of modulated Ronchi rulings using a phase-locked loop, *Opt. Eng.*, **33**, 1193–1199, 1994.
47. Hassani K.H. Hooshmand Ziafi H., Modified matching Ronchi test to visualize lens aberrations, *Eur. J. Phys.*, **32**, 1385–1390, 2011.
48. Murty MVRK, Cornejo A., Sharpering the fringes in the Ronchi test, *Appl. Opt.*, **12**, 2230, 1973.
49. Cornejo A., Altamirano H., Murty M.V.R.K., Experimental results in the sharpening of the fringes in the Ronchi test, *Boletín del Instituto de Tonantzintla*, **2**, 313, 1978.
50. Robledo Sánchez C., Camacho Basilio G., Jaramillo Núñez A., Cornejo Rodríguez A., Binary grating with variable bar/space ratio following a geometrical progression, *Opt. Commun.*, **119**, 465–470, 1995.
51. Yaoltzin L.Z., Granados-Agustin F., Cornejo-Rodríguez A., Ronchi-test with sub-structured gratings. *20th Congress of International Commission for Optics (ICO). Proc. SPIE*, **6034**, 100–94, 2005.
52. Cornejo-Rodigúez A., Granados-Agustin F., Luna-Zayas Y., The Ronchi test and the use of structured gratings for sharpening the fringes, *Symposium of Optical Fabrication and Testing*, OSA, Rochester, NY, 2006.
53. Aguirre-Aguirre D., Villalobos-Mendoza B., Granados-Agustin F., Izazaga-Pérez R., Campos-García M., Percino-Zacarías M.E., Cornejo-Rodríguez A., Substructured Ronchi gratings from the linear combination of classical gratings, *Opt. Eng.*, **53**, (11), 114111, 2014.
54. García-Arellano A., Granados-Agustín F., Campos-García M., Cornejo-Rodríguez A., Ronchi test with equivalent wavelength, *Appl. Opt.*, **51**, 3071–3080, 2012.
55. Ghozeil I., *Hartmann and other screen tests*. In D. Malacara, *Optical Shop Testing*, Editor, pp. 367–396. New York, NY: John Wiley & Sons, 1992.
56. Bautista-Elivar N., Robledo-Sánchez C.I., Cordero-Dávila A., Cornejo Rodríguez A., Sensing a wave front by use of a diffraction grating, *Appl. Opt.*, **42**, 3737–3741, 2003.

57. Su P., Parks R.E., Wang L., Angel R.P., Burge J.H., Software configurable optical test system: A computerized reverse Hartmann test, *Appl. Opt.,* **49**, 4404–4412, 2010.

58. Su P., Wang Y., Burge J.H., Kaznatcheev K., Idir M., Non-null full field x-ray mirror metrology using SCOTS: A reflection deflectometry approach, *Opt. Express,* **20**, 12393–12407, 2012.

59. Su, T., Wang, S., Parks R.E., Su P., Burge J.H., Measuring rough optical surfaces using scanning long-wave optical test system. 1. Principle and implementation, *Appl. Opt.,* **52**, 7117–7126, 2013.

60. Huang, R., Su P., Horne T., Brusa G., Burge, J. H., Optical metrology of a large deformable aspherical mirror using software configurable optical test system, *Opt. Eng.,* **53**, 2014.

61. Malacara-Hernández D., Malacara-Doblado D., What is a Hartmann test, *Appl. Opt.,* **54** (9), 2296–2300 2015.

62. Vitrichenko E.A. Methods of studying astronomical optics. Limitations of the Hartmann method, *Sov. J. Opt. Technol.,* **20**, 3, 1976.

63. Morales A., Malacara D., Geometrical parameters in the Hartmann test of aspherical mirrors, *Appl. Opt.,* **22**, 3957, 1983.

64. Landgrave J.E.A, Moya J.R., Effect of a small centering error of the Hartmann screen on the computed wavefront aberration, *Appl. Opt.,* **25**, 533, 1986.

65. Zverev V.A., Rodionov S.Λ., Sokolskii M.N., Usoskin V.V., Mathematical principles of Hartmann test of the primary mirror of the large azimuthal telescope, *Sov. J. Opt. Technol.,* **44**, 78, 1977.

66. Platt B.C., Shack R., History and principles of Shack-Hartmann wavefront sensing, *J. Refract. Surg.,* **17**, S573–S577, 2001.

67. Schwiegerling T., Neal D.R. Historical development of the Shack-Hartmann sensor. In J.E. Harvey and R.B. Hooker, Editors, *Robert Shanon and Roland Shack, Legends in Applied Optics,* Bellingham, WA: SPIE Press, p. 132, 2005.

68. Neal D.R., Shack-Hartmann sensor engineered for commercial measurement applications. In Harvey J.E. and Hooker, R.B., Editors, *Robert Shanon and Roland Shack, Legends in Applied Optics,* Bellingham, WA: SPIE Press, p. 140, 2005.

69. Pastor, J., Hologram interferometry and optical technology, *Appl. Opt.,* **8**, 525–531, 1969.

70. Popov G.M., Methods of calculation and testing of Ritchey-Chretien systems, *Izv. Krym. Astrofiz. Obs.,***45**, 188, 1972.

71. Mobsby E., A Ronchi null test for paraboloids, *Sky Telesc.,* **48**, 325–330, 1974.

72. Malacara D., Cornejo A., Null Ronchi test for aspherical surfaces, *Appl. Opt.,* **13**, 1778–1780, 1974.

73. Hopkins G.H., Shagan R.H., Null Ronchi gratings from spot diagram, *Appl. Opt.,* **16**, 2602–2603, 1977.

74. Cordero-Dávila A., Cardona-Núñez O., Cornejo-Rodríguez A., Null Hartmann and Ronchi–Hartmann tests, *Appl. Opt.,* **29**, 4618, 1990.

75. Granados-Agustín F.S., Cardona- Núñez O., Díaz-Uribe R., Percino-Zacarías E., Zarate-Rivera A.M., Cornejo-Rodríguez A., Analysis of the common characteristics of the Hartmann, Ronchi, and Shack–Hartmann tests. *Optik,* **125**, 667–670, 2014.

76. Aguirre-Aguirre D., Granados-Agustin F.S., Izazaga-Perez R., Villalobos-Mendoza B., Cornejo-Rodríguez A., Null Ronchi Gratings as a function of period., *J. Phys.: Conf. Ser.,* **605**, (1), 2015.

77. Díaz-Uribe R., Campos-García M., Null screen testing of fast convex aspheric surfaces, *Appl. Opt.,* **39**, 2670–2677, 2000.

78. Díaz-Uribe R., Medium precision null screen testing of off-axis parabolic mirrors for segmented primary telescope optics: The case of the Large Millimetric Telescope, *Appl. Opt.,* **39**, 2790–2804, 2000.

79. Carmona-Paredes L., Díaz-Uribe R., Geometric analysis of the null screens used for testing convex optical surfaces, *Rev. Mex. Fís.,* **53**, 421–430, 2007.

80. Moreno-Oliva VI, Campos-García M., Bolado-Gómez R., Díaz-Uribe R., Point-shifting in the optical testing of fast aspheric concave surfaces by a cylindrical null-screen, *Appl. Opt.,* **47**, 644–651; 2008.

81. Campos-García M., Bolado-Gómez R., Díaz-Uribe R., Testing fast aspheric concave surfaces with a cylindrical null screen, *Appl. Opt.,* **47**, 849–859, 2008.

82. Avendaño-Alejo M., Moreno-Oliva VI, Campos-García M., Díaz-Uribe R., Quantitative evaluation of an off-axis parabolic mirror by using a tilted null screen, *Appl. Opt.*, **48**, 1008–1015, 2009.

83. Estrada-Molina A., Campos-García M., Díaz-Uribe R., Sagittal and meridional radii of curvature for a surface with symmetry of revolution by using a null-screen testing method, *Appl. Opt.*, **52**, 625–634, 2013.

84. Rodríguez-Rodríguez M.I., Jaramillo-Núñez A., Díaz-Uribe R., Dynamic point shifting with null screens using three LCDs as targets for corneal topography, *Appl. Opt.*, **54**, 6698–6710, 2015.

85. Malacara-Hernández D., Malacara-Doblado D., What is a Hartmann test, *Appl. Opt.*, **54**, 2296–2300, 2015.

86. Platzeck, R., Gaviola E., On the errors of testing a new method for surveying optical surfaces and systems, *J. Opt. Soc. Am.*, **29**, 484–500, 1939.

87. Schroader, I.H., The Caustic Test, *Amateur Making Telescope*, 3; A.G. Ingalls Editor, Scientific American, New York, 429, 1953.

88. Gerchberg R.W., Saxton W.O., A practical algorithm for the determination of phase from image and diffraction plane pictures. *Optik*, **35**, 237–246, 1972.

89. Teague M.R., Irradiance moments: Their propagation and use for unique retrieval of phase, *J. Opt. Soc. Am.*, **72**, 1199–1209, 1982.

90. Streibl N., Phase measuring by the transport equation of intensity, *Opt. Comm.* **49**, 6–10, 1984.

91. Fienup J.R., Phase retrieval algorithms: A comparison. *Appl. Opt.*, **21**, 2758–2769, 1982.

92. Campos-García M., Díaz-Uribe R., Irradiance transport equation from geometrical optics considerations, *Rev. Mex. Fís.*, **52**, 546–549, 2006.

93. Fernández Guasti M., Jiménez J.L., Granados-Agustín F., Cornejo-Rodríguez A., Amplitude and phase Representation of monochromatic fields in physical optics, *J. Opt. Soc. Am.*, **20**, 1629–1634, 2003.

94. Schiske P., Resolution sing Conventional Electron Microscopes for the Measurement of Amplitudes and Phases, *Proceedings 4th European Conference on Electron Microscopy, Rome*, 1968.

95. Hoppe W., Principles of electron structure research at atomic resolution using conventional electron microscopes for the measurement of amplitudes and phases, *Acta Cryst A.*, **26**, 414, 1970.

96. Erickson H, Klug A., The Fourier transforms of an electron monograph: Effects of defocusing and aberrations and implications for the use of underflows contrast enhancement, *Berichte der Bunsen-Gesell Schaft* **74**, 1129; 1970.

97. Gerchberg R.W., Saxton W.O., Phase determination for image and diffraction plane pictures in the electron microscope, *Optik*, **34**, 275–284,1971.

98. Gerchberg R.W., Saxton W.O., A practical algorithm for the determination of phase from image and diffraction plane pictures, *Optik*, **35**, 237–246, 1972.

99. Gonsalves R.A., Phase retrieval from modulus data, *J. Opt. Soc. Am.*, **66**, 961–964, 1976.

100. Devaney A.J., Chidlaw R., On the uniqueness question in the problem of phase retrieval from intensity measurements, *J. Opt. Soc. Am.*, **68**, 1352–1354, 1978.

101. Gureyev T.E., Roberts A., Nugent R.A., Phase retrieval with the transport-of-intensity equation: Matrix solution with use of Zernike polynomials, *J. Opt. Soc. Am.*, A **12**, 1932–1941, 1995.

102. Salas L., Variable separation in curvature sensing: Fast method for solving the irradiance transport equation in the context of optical telescopes, *Appl. Opt.*, **35**, 10, 1593–1596, 1996.

103. Alonso Magaña P., Granados Agustín F., Cornejo Rodríguez A., Medición de la Fase o Frente de Onda con un Banco Nodal, *Rev. Mex. Fís.*, **46** (Suplement No. 2), 54–58, 2000.

104. Roddier F., Curvature sensing: A diffraction theory, *NOAO R&D Notes*, 1982.

105. Soto M., Acosta E., Rios S., Performance analysis of curvature sensors: Optimum positioning of the measurements planes, *Opt. Express*, **11**, 2577–2588, 2003.

3

Basic Interferometers

3.1 Introduction 71
3.2 Coherence of Light Sources for Interferometers 72
3.3 Young's Double Slit 74
    Coherence in Young's Interferometer • Stellar Michelson Interferometer
3.4 Michelson Interferometer 77
3.5 Fizeau Interferometer 80
    Laser Fizeau and Shack Interferometers
3.6 Newton Interferometer 82
3.7 Twyman–Green Interferometer 82
    Laser Twyman–Green Interferometer • Mach–Zehnder Interferometer
3.8 Common Path Interferometers 86
    Burch and Murty Interferometers • Point Diffraction Interferometer
3.9 Wavefronts Producing the Interferograms 87
3.10 Lateral Shearing Interferometers 89
3.11 Talbot Interferometer and Moiré Deflectometry 92
3.12 Foucault Test and Schlieren Techniques 94
3.13 Multiple Reflection Interferometers 96
    Cyclic Multiple Reflection Interferometers
3.14 Cyclic Interferometers 98
3.15 Sagnac Interferometer 99
References 100

Daniel Malacara-
Hernández

3.1 Introduction

Interferometers have been described with detail in many textbooks [1]. They produce the interference of two or more light waves by superimposing them on a screen or the eye. If the relative phase of the light waves is different for different points on the screen, constructive and destructive interference appears at different points, forming interference fringes. Their uses and applications are extremely numerous. In this chapter, only their basic configurations will be described.

To begin, let us consider the interference of two light waves, one having a flat wavefront (constant phase on a plane in space at a given time) and the other a distorted wavefront with deformations $W(x,y)$, as in Figure 3.1. Thus, the amplitude $E_1(x, y)$ in the observing plane is given by the sum of the two waves, with amplitudes $A_1(x, y)$ and $A_2(x, y)$, given by

$$E_1(x,y) = A_1(x,y) \exp\left[ikW(x,y) \right] + A_2(x,y) \exp\left[ikx \sin\theta \right], \tag{3.1}$$

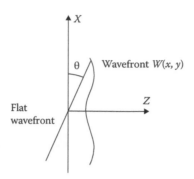

FIGURE 3.1   Two interfering wavefronts.

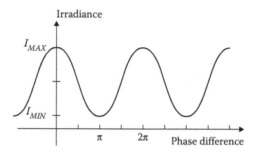

FIGURE 3.2   Irradiance as a function of phase difference for two-wave interference.

where $k = 2\pi/\lambda$ and θ is the angle between the wavefront. The irradiance function $I(x, y)$ may then be written as

$$E_1(x,y) \cdot E_1^*(x,y) = A_1^2(x,y) + A_2^2(x,y)$$
$$+ 2A_1(x,y)A_2(x,y)\cos k\left[x\sin\theta - W(x,y)\right], \tag{3.2}$$

where the symbol * denotes the complex conjugate. This function is plotted in Figure 3.2. We see that the resultant amplitude becomes a maximum when the phase difference is a multiple of the wavelength and a minimum when the phase difference is an odd multiple of half the wavelength. These two conditions are constructive and destructive interference, respectively.

3.2 Coherence of Light Sources for Interferometers

If light source has a single spectral line, we say that it is monochromatic. Then, it is formed by an infinite sinusoidal wavetrain or, equivalently, it has a long coherence length. On the other hand, a light source with several spectral lines or a continuous spectrum is non-monochromatic. Then, its wavetrain or coherence length is short. A light source with a short wavetrain is said to be temporally incoherent and a monochromatic light source is temporally coherent.

The helium–neon laser has a large coherence length and monochromaticity. For this reason, it is the most common light source in interferometry. However, this advantage can sometimes be a problem, because many undesired spurious fringes are formed. Great precautions must be taken to avoid this noise on top of the fringes.

With laser light sources extremely large, optical path differences (OPDs) can be introduced without appreciably losing fringe contrast. Although almost perfectly monochromatic, the light emitted by a gas laser consists of several spectral lines, called longitudinal modes. They are equally spaced but very close together, with a frequency spacing Δv given by

$$\Delta v = \frac{c}{2L},$$ (3.3)

where L is the length of the laser cavity. If this laser cavity length L is modified due to thermal expansion or contraction or to mechanical vibrations, the spectral lines change their frequency, approximately preserving their separation, but with their intensities inside a Gaussian dotted envelope called a power gain curve, as shown in Figure 3.3.

Helium–neon lasers with a single mode or frequency can be constructed, thus producing a perfectly monochromatic wavetrain. However, if special precautions are not taken, because of instabilities in the cavity length, the frequency may be unstable. Lasers with stable frequencies are commercially produced, making possible extremely large OPDs without reducing the fringe contrast. A laser with two longitudinal modes can also be frequency stabilized, if desired, to avoid contrast changes. When only two longitudinal modes are present and they are orthogonally linearly polarized, one of them can be eliminated with a linear polarizer. This procedure greatly increases the temporal coherence of the laser.

With a multimode helium–neon laser the fringe visibility in an interferometer is a function of the OPD, as shown in Figure 3.4. In order to have a good fringe contrast, the OPD has to be an integral multiple of $2L$.

A laser diode can also be used as a light source in interferometers. Creath [2] and Ning et al. [3] have described the coherence characteristics of laser diodes. Their coherence length is of the order of 1 mm, which is a great advantage in many applications, besides the common advantage of their low price and small size.

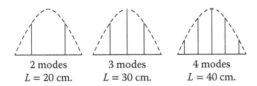

2 modes	3 modes	4 modes
$L = 20$ cm.	$L = 30$ cm.	$L = 40$ cm.

FIGURE 3.3 Longitudinal modes in a He–Ne laser.

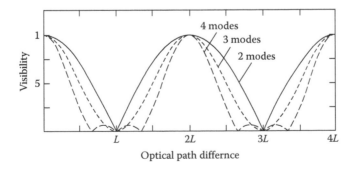

FIGURE 3.4 Contrast as a function of the optical path difference in a two beam interferometer.

3.3 Young's Double Slit

The typical interference experiment is the Young's double slit (Figure 3.5). A line light source emits a cylindrical wavefront that illuminates both slits. The light is diffracted on each slit, producing cylindrical waves diverging from these slits. Any point on the screen is illuminated by the waves emerging from these two slits. Since the total paths from the point source to a point on the observing screen are different, the phases of the two waves are not the same. Then, constructive or destructive interference takes place at different points on the screen, forming interference fringes.

The amplitude E at the observing point D on the screen is

$$E = ae^{i\phi_1} + ae^{i\phi_2}, \tag{3.4}$$

where a is the amplitude at D due to each one of the two slits alone and f_i are the phases whose difference is given by

$$\phi_2 - \phi_1 = kOPD, \tag{3.5}$$

where

$$OPD = (AC + CD) - (AB + BD) \tag{3.6}$$

From Equation 3.2, the irradiance I at the point D is

$$I = EE^* = 2a^2(1 + \cos(kOPD)). \tag{3.7}$$

The minima of the irradiance occurs when

$$OPD = m\lambda, \tag{3.8}$$

where m is an integer; thus

$$(AC + CD) - (AB + BD) = m\lambda, \tag{3.9}$$

which is the expression for a hyperbola. Thus, the bright fringes are located at hyperboloidal surfaces, as in Figure 3.6. If a plane screen is placed a certain distance in front of the two slits, the fringes are straight and parallel with an increasing separation as they separate from the optical axis.

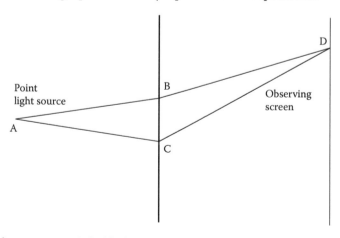

FIGURE 3.5 Interference in Young's double slit.

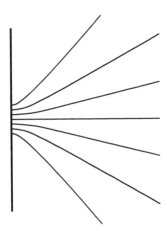

FIGURE 3.6   Locus of fringes in Young's double slit.

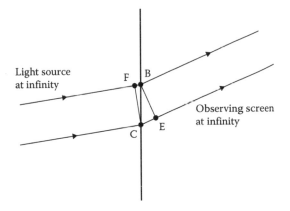

FIGURE 3.7   Young's experiment with light source and observing plane at infinite distances.

For a screen located at infinity, OPD is equal to CE–FB, as in Figure 3.7. Then, Equation 3.4 becomes

$$I = 2a^2\left[1 + \cos\left(kd\left(\sin\theta' - \sin\theta\right)\right)\right],\tag{3.10}$$

where d is the slits' separation and θ is the angle of observation with respect to the optical axis. For small values of θ, these fringes are sinusoidal. The peaks of the irradiance (center of the bright fringes) are given by

$$d\sin\theta = m\lambda,\tag{3.11}$$

so that their separation $\delta\theta$ has to be much larger than the eye resolution, which is about 1 arc minute ($d < 3500\lambda \sim 2$mm).

3.3.1 Coherence in Young's Interferometer

An ideal light source for many optics experiments is a point source with only one pure wavelength (color). However, in practice, most light sources are not a point but have a certain finite size and emit several wavelengths simultaneously. A point source is said to be spatially coherent if, when used to illuminate a system of two slits, interference fringes are produced. If an extended light source is used to illuminate the two slits and no interference fringes are observed, the extended light source is said to be spatially incoherent.

All proceeding theory for the two slits assumes that the light source is a point and also that it is monochromatic. Let us now consider the cases when the light source does not satisfy these conditions. If the light source has two spectral lines with different wavelengths, two different fringe patterns with different fringe separation will be superimposed on the observing screen, as shown in Figure 3.8. The central maxima coincide but they are out of phase for points far from the optical axis. If the light source is white, the fringes will be visible only in the neighborhood of the optical axis.

Let us assume that the two slits are illuminated by two point light sources aligned in a perpendicular direction to the slits. Then, two identical fringe patterns are formed, but one displaced with respect to the other, as in Figure 3.9. If the angular separation between the two light sources, as seen from the slits plane, is equal to half the angular separation between the fringe, the contrast is close to zero. With a single large pinhole the contrast is reduced. More details will be given in the next section, when studying the stellar Michelson interferometer.

3.3.2 Stellar Michelson Interferometer

Let us consider a double-slit interferometer with the light source and the observing screen at infinite distance from the slits' plane. If we have an extended light source, each element with apparent angular dimensions $d\theta_x d\theta_y$ will generate a fringe pattern with irradiance dI, given from Equation 3.10 by

$$dI = \left[1 + \cos\left(kd\left(\sin\theta_y' - \sin\theta_y\right)\right) \right] d\theta_x d\theta_y, \tag{3.12}$$

which is a general expression; if we assume that the angular size of the light source is small and that the fringes are observed in the vicinity of the optical axis, we have

$$dI = \left[1 + \cos\left(kd\left(\theta_y' - \theta_y\right)\right) \right] d\theta_x d\theta_y. \tag{3.13}$$

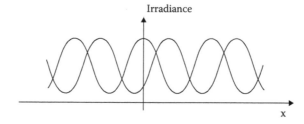

FIGURE 3.8 Superposition of two diffraction patterns with the same frequency.

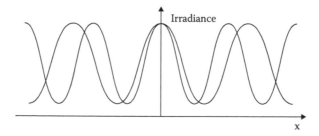

FIGURE 3.9 Superposition of two diffraction patterns with different frequencies.

This is a valid expression for any shape of the light source. If we assume that it is square with angular dimensions equal to $2\alpha \times 2\alpha$, we have

$$I = \alpha^2 + 2\cos\left(kd\alpha'_y\right)\sin\left(kd\theta'_y\right). \tag{3.14}$$

The fringe visibility V or contrast, defined by Michelson, is

$$V = \frac{I_{max} - I_{max}}{I_{max} + I_{min}}. \tag{3.15}$$

Thus, in this case, we have

$$V = \frac{\sin\left(kd\alpha'_y\right)}{\left(kd\alpha'_y\right)} = \sin c\left(kd\alpha'_y\right). \tag{3.16}$$

We can see that the fringe visibility V is a function of the angular size of the light source, as shown in Figure 3.10a, with a maximum when it is a point source ($\alpha = 0$) or the slit separation d is extremely small. The first zero of this visibility occurs when $kd\alpha'_y = \pi$; that is, if $2d\alpha = \lambda$. Using this result, the apparent angular diameter of a square or rectangular light source can be found by forming the fringes with two slits with variable separation. The slits are separated until the fringe visibility becomes zero.

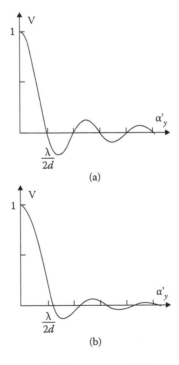

FIGURE 3.10   Contrast variation for (a) two slit light sources and (b) two circular light sources.

If the light source is a circular disk, the visibility can be shown to be an Airy distribution:

$$V = \frac{J_1\left(kd\alpha_y'\right)}{\left(kd\alpha_y'\right)},$$ (3.17)

which is plotted in Figure 3.10b. This interferometer has been used to measure the angular diameter of some stars.

3.4 Michelson Interferometer

A Michelson interferometer (Figure 3.11) is a two-beam interferometer, illuminated with an extended light source. The beam of light from the light source is separated into two beams with smaller amplitudes, at the plane parallel glass plate (beamsplitter). After reflection on two flat mirrors, the beams are reflected back to the beamsplitter, where the two beams are recombined along a common path.

The observing eye (or camera) sees two virtual images of the extended light source, one on top of the other, but separated by a certain distance. The reason is that the two arms of the interferometer may have different lengths. Thus, the optical path difference is given by

$$OPD = 2\left[L_1 - L_2 - nT\right],$$ (3.18)

where T is the effective glass thickness traveled by the light rays on one path through the beamsplitter. On the other hand, from geometrical optics we can see that the virtual images of the extended light source are separated along the optical axis by a distance s, given by

$$s = 2\left(L_1 - L_2 - \frac{T}{n}\right);$$ (3.19)

we can see that these two expressions are different. Either the optical path difference or the two images separation can be made equal to zero.

To observe interference fringes with a non-monochromatic or white light source, the OPD must be zero for all wavelengths. This is possible only if the optical path is the same for the two interfering

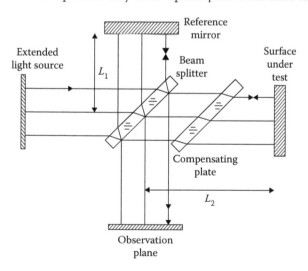

FIGURE 3.11   A Michelson interferometer.

beams at all wavelengths. We can see that the OPD can be made equal to zero by adjusting L_1 and L_2 for any desired wavelength but not for the entire spectrum, unless T is zero or n does not depend on the wavelength. Only if a compensating glass plate is introduced, as in Figure 3.11, can the OPD be made equal to zero for all wavelengths if $L_1 = L_2$. In this manner, white light fringes can be observed. In an uncompensated Michelson interferometer, the optical path difference can also be written as

$$\text{OPD} = s - 2T\left(\frac{n-1}{n}\right), \tag{3.20}$$

but in a compensated interferometer the second term is not present.

If the light source is extended, but perfectly monochromatic, clearly spaced fringes can be observed if the two virtual images of the light sources are nearly at the same plane.

When the two virtual images of the light source are parallel to each other and the observing eye or camera is focused at infinity, circular equal inclination fringes will be observed, as in Figure 3.13a. We see in Figure 3.12a that the OPD in a compensated interferometer is given by

$$\text{OPD} = s \cos \theta; \tag{3.21}$$

thus, the larger the images separation s, the greater the number of circular fringes in the pattern. The diameter of these fringes tends to infinity when the images separation becomes very small.

If there is a small angle between the two light source images, as in Figure 3.12b, the fringes appear curved, as if the center of the fringe had been shifted to one side, as in Figure 3.13b. In this case the fringes are not located at infinity, but close to the light source images. These are called localized fringes.

The fringes may appear to be in front of the two images of the light source, as in Figure 3.12b, between the two images, as in Figure 3.12c, or at the back of the two images, as in Figure 3.12d, depending on the relative position of these two images.

When the angle between the light sources is large, the optical path difference is nearly equal to the local separation between the two images of the light source. The fringes will be almost straight and parallel, as in Figure 3.13c. Their separation decreases when the angle increases. These are called equal thickness fringes.

A final remark about this interferometer, which is valid for all other amplitude division interferometers, is that there are two outputs with complementary interference patterns. In other words, a dark point in one of them corresponds to a bright point on the other and vice versa. The second interferogram is one with the wavefronts going back to the light source. These two interferogram patterns are exactly complementary if there are no energy losses in the system, as can be easily proved with Stokes' relations.

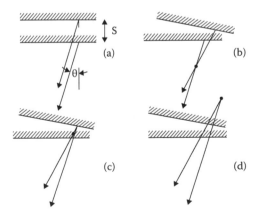

FIGURE 3.12 Images in space of two extended light sources in a Michelson interferometer for four different relative positions.

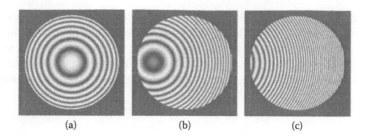

(a)   (b)   (c)

FIGURE 3.13   Fringe patterns in a Michelson interferometer: (a) equal inclination fringes, (b) localized fringes, and (c) equal thickness fringes.

3.5 Fizeau Interferometer

The Fizeau interferometer, illustrated in Figure 3.14, is quite similar to the Michelson interferometer described in the preceding section, producing the two interfering beams by means of an amplitude beamsplitter. Unlike in the Michelson interferometer, the illuminating light source is a monochromatic point, producing a spherical wavefront, which becomes flat after being collimated by a converging lens. This wavefront is reflected back on the partially reflecting front face of the beamsplitter plate. The transmitted beam goes to the optical element to be measured and is then reflected back to the beamsplitter.

The quality of many different optical elements can be evaluated with this interferometer: for example, a glass plate, which also serves here as the reference beamsplitter, as in Figure 3.14. The optical path difference in this interferometer, when testing a single plane parallel plate, is given by

$$\text{OPD} = nt, \tag{3.22}$$

where t is the glass plate thickness. A field without interference fringes is produced when nt is a constant, but n and t cannot be determined, only its product.

In order to test a convex optical surface, the reference surface can be either flat or concave, as in Figure 3.15a and b. The quality of a flat optical surface can be measured with the setup in Figure 3.14. In this case, the OPD is equal to $2d$. If we laterally displace the point light source by a small amount s, the refracted flat wavefront would be tilted at angle θ:

$$\theta = \frac{s}{f}, \tag{3.23}$$

where f is the effective focal length of the collimator. With this tilted flat wavefront, the optical path difference is given by

$$\text{OPD} = 2d \cos \theta. \tag{3.24}$$

The OPD with a small angle θ from the OPD on axis can be approximated by

$$\text{OPD} = 2d\left(1 - \frac{\theta^2}{2}\right) = 2d\left(1 - \frac{s^2}{2f}\right). \tag{3.25}$$

If a small extended light source with semidiameter s is used, the fringes have good contrast, as long as the condition

$$\Delta\text{OPD} = \frac{ds^2}{f} \leq \frac{\lambda}{4} \tag{3.26}$$

is satisfied. The light source can increase its size s only if the air gap d is reduced.

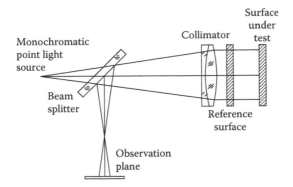

FIGURE 3.14 Fizeau interferometer.

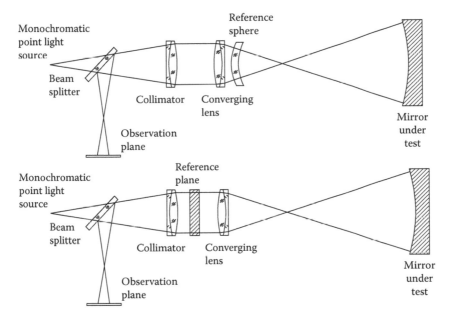

FIGURE 3.15 Fizeau interferometer to test a concave surface with (a) a reference sphere and (b) a reference plane.

If the collimator lens has spherical aberration, the collimated wavefront would not be flat. The maximum transverse aberration (TA) in this lens can be interpreted as the semidiameter s of the light source. Thus, the quality requirements for the collimator lens increase as the OPD is increased. When the OPD is zero, the collimator lens can have any value of spherical aberration without decreasing its precision.

3.5.1 Laser Fizeau and Shack Interferometers

An He–Ne gas laser can be used as a light source for a Fizeau interferometer with the great advantage that a large OPD can be used due to its light temporal coherence. This large OPD is highly necessary when testing concave optical surfaces with a long radius of curvature. However, the high coherence of the laser also brings some problems, such as undesired interference fringes from several optical surfaces in the system. With the introduction of a wedge in the beamsplitter and the use of pinholes acting as spatial filters, undesired reflections can be blocked out. Also, polarizing devices and antireflecting coating can be used for this purpose.

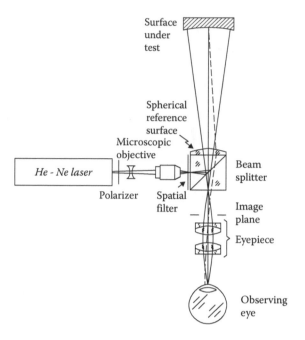

FIGURE 3.16   A Shack–Fizeau interferometer.

A Shack interferometer, as illustrated in Figure 3.16, is an example of a Fizeau interferometer using an He–Ne laser.

The light from an He–Ne laser is focused on a spatial filter in contact with a non-polarizing cube beamsplitter. Since the OPD is large, a high temporal coherence is needed. The gas laser is of such a length and characteristics that it contains two longitudinal modes with linear orthogonal polarizations. One of the two spectral lines is isolated by means of a polarizer.

The reference wavefront is reflected at the spherical convex surface of a plane convex lens cemented to the cube beamsplitter. This cube with the lens can be considered as a thick lens that forms a real image of the surface under test at the image plane; then, this image is visually observed with an eyepiece.

3.6 Newton Interferometer

The Newton interferometer can be considered as a Fizeau interferometer in which the air gap is greatly reduced to less than 1 mm, so that a large extended source can be used. This high tolerance in the magnitude of the angle θ also allows us to eliminate the need for the collimator, if a reasonably large observing distance is desired. Figure 3.17 shows a Newton interferometer, with a collimator, so that the effective observing distance is always infinite. The quality of this collimator does not need to be high. If desired, it can even be taken out, as long as the observing distance is not too short.

A Newton interferometer is frequently used in manufacturing processes, to test planes, concave spherical, or convex spherical optical surfaces by means of measuring test plates with the opposite curvature, placed on top of the surface under test.

3.7 Twyman–Green Interferometer

A Twyman–Green interferometer, designed by Twyman [4] as a modification of the Michelson interferometer is shown in Figure 3.18. The basic modification is to replace the extended light source by a point source and a collimator, as in the Fizeau interferometer. Thus, the wavefront is illuminated with a flat wavefront. Hence, the fringes in a Twyman–Green interferometer are of the equal-thickness type.

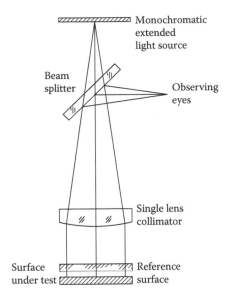

FIGURE 3.17   Newton interferometer.

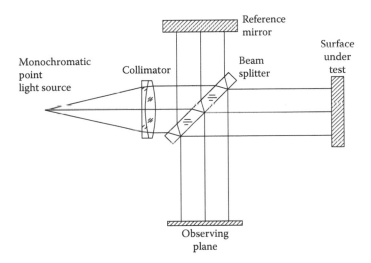

FIGURE 3.18   Twyman–Green interferometer.

As in the Michelson interferometer, white light fringes are observed only if the instrument is compensated with a compensating plate. However, normally, a monochromatic light source is used, eliminating the need for the compensating plate.

The beamsplitter must have extremely flat surfaces and its material must be highly homogeneous. The best surface must be the reflecting one. The nonreflecting surface must not reflect any light, to avoid spurious interference fringes. Thus, the nonreflecting face must be coated with an antireflection multilayer coating. Another possibility is to have an incidence angle on the beamsplitter with a magnitude equal to the Brewster angle and properly polarizing the incident light beam.

The size of the light source can be slightly increased to a small finite size if the optical path difference between the two interferometer arms is small, following the same principles used for the Fizeau interferometer.

A glass plate can be tested as in Figure 3.19a or a convergent lens as in Figure 3.19b. When testing a glass plate, the optical path difference is given by

$$\text{OPD} = (n-1)d. \tag{3.27}$$

When no fringes are present, we can conclude that $(n-1)d$ is a constant, but not independent of n or d. If we compare this expression with the equivalent for the Fizeau interferometer (Equation 3.22), we see that n and t can be measured independently if both Fizeau and Twyman–Green interferometers are used.

A convex spherical mirror with its center of curvature at the focus of the lens is used to test convergent lenses, as in Figure 3.20a, or a concave spherical mirror can be used to test lenses with short focal lengths, as in Figure 3.20b. The small, flat mirror at the focus of the lens can also be employed. The small region being used on the flat mirror is so small that its surface does not need to be very accurate. However, the wavefront is rotated 180°, making the spatial coherence requirements higher and canceling odd aberrations like coma.

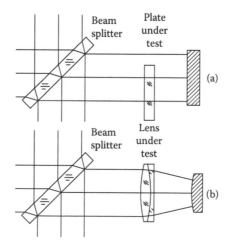

FIGURE 3.19   Testing (a) a glass plate and (b) a lens in a Twyman–Green interferometer.

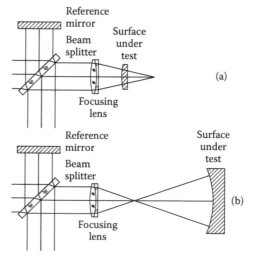

FIGURE 3.20   Testing (a) concave and (b) convex surfaces in a Twyman–Green interferometer.

3.7.1 Laser Twyman–Green Interferometer

Large astronomical mirrors can also be tested with a Twyman–Green unequal-path interferometer, as in Figure 3.21, and described by Houston et al.[5]. However, there are important considerations to take into account because of the large OPD.

1. As in the Fizeau interferometer, when the OPD is large, the collimator as well as the focusing lens must be almost perfect, producing a flat and a spherical wavefront, respectively.
2. The laser must have a large temporal coherence. Ideally, a single longitudinal mode has to be present.
3. The concave mirror under test must be well supported, in a vibration and atmospheric turbulence-free environment.

3.7.2 Mach–Zehnder Interferometer

The Twyman–Green or Michelson configurations are sometimes unfolded to produce the optical arrangement shown in Figure 3.22. An important characteristic is that any sample located in the

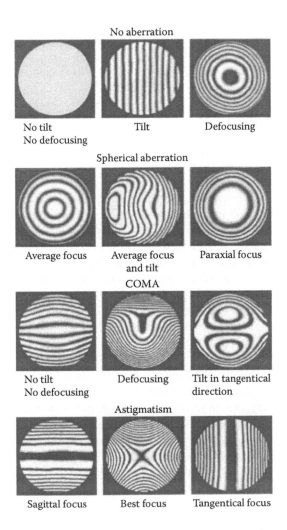

FIGURE 3.21 Interferograms of primary aberrations in a Twyman–Green interferometer.

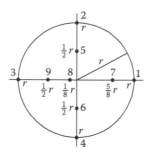

FIGURE 3.22 Mach–Zehnder interferometer.

interferometer is traversed by the light beam only once. Another important feature is that since there are two beamsplitters, the interferometer is compensated if their thicknesses are exactly equal.

3.8 Common Path Interferometers

In common path interferometers, the two interfering wavefronts travel along the same path from the light source to the observing plane. The advantages are that the fringes are quite stable and also that the OPD is nearly zero, thus producing white light fringes.

There are many different types of common path interferometers. Here, a few of the most important will be described.

3.8.1 Burch and Murty Interferometers

The Burch interferometer, also called the scattering interferometer, is illustrated in Figure 3.23a. The real image of a small tungsten lamp is formed at the center of a concave surface under test. This light forming the image passes through a scattering glass plate SP_1 that can be made in several different manners, but the most common is with a half-polished glass surface. The light after the scattering plate can be considered as formed by two beams, one just transmitted undisturbed and another being scattered in a wide range of directions. The direct beam forms the image of the lamp on the central region of the concave surface and the diffracted one illuminates the whole surface of the mirror.

A second identical scattering plate SP_2 is located on the image of the plate SP_1 but rotated 180°. These scattering plates have to be identical point to point. This is the most critical condition, but one possible solution is to make a photographic copy of the first plate SP_1. Both beams passing through the first scattering plate arrive at the second one. Here, the light beam not scattered on the first plate can go through the second plate, again, without being scattered. With this light, the observing eye sees a bright image of the lamp on the surface under test.

This direct beam from the SP_1 beam can also be scattered on the plate SP_2, producing many spherical wavefronts originating at each of the scattering points on the plates SP_2.

The scattered beam from SP_1 can also be considered to be formed by many spherical beams, with center on curvature on each scattering point in SP_1. Each of these spherical wavefronts illuminates the whole concave surface under test. If this surface is spherical, the reflected wavefront is also spherical and convergent to SP_1. However, if the concave surface is not spherical but contains deformations, the convergent wavefronts will also be deformed with twice the value present in the mirror. When these convergent deformed wavefronts pass through the plate SP_2 without being scattered, they interfere with the spherical wavefronts being produced there. The interference pattern is observed, projected over the concave surface.

To avoid the need for two identical scattering plates, a small flat mirror can be placed at the image of the scattering plate SP, as in Figure 3.23b. Then, the light goes back to the scattering plates after being reflected on this mirror and twice on the concave surface.

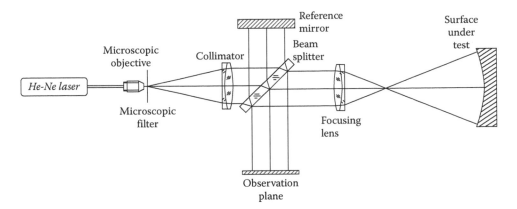

FIGURE 3.23 Scattering interferometer.

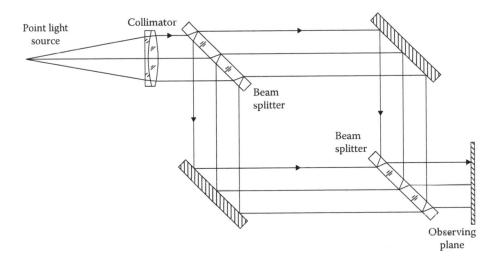

FIGURE 3.24 Point diffraction interferometer.

This interferometer is simpler to construct and is more insensitive to mechanical vibrations of the concave mirror. Another advantage is that the sensitivity to the deformations in the concave mirror is duplicated due to the double reflection here. There are two disadvantages: first, the concave surface has to be coated to increase its reflectivity because the double reflection here reduces the amount of light too much; secondly, the interferometer has no sensitivity to anti-symmetric wavefront deformations. Thus, coma-like aberrations cannot be detected.

3.8.2 Point Diffraction Interferometer

In a point diffraction interferometer, first described by Linnik [6] and later rediscovered independently by Smartt and Strong [7], the aberrated wavefront passes through a specially designed plate, as in Figure 3.24. This plate has a semi-transparent small pinhole with a diameter equal to the Airy disk or smaller to that produced by a perfect spherical surface under test. The aberrated wavefront produces an image much greater than the pinhole size on this plate. The light from this aberrated wavefront goes through the semitransparent plate.

The small pinhole diffracts the light passing through it, producing a spherical wavefront with the center of curvature at this pinhole. After the plate with the pinhole, the two wavefronts, one being aberrated and the second being spherical, produce the interferogram.

3.9 Wavefronts Producing the Interferograms

The Twyman–Green, Fizeau, and common path interferometers, like many other two-wave interferometers, produce the same interferogram if the same aberration is present. The interferograms produced by the Seidel primary aberrations have been described by Kingslake [8], and their associated wavefront deformations can be expressed by

$$W(x, y) = A + Bx + Cy + D(x^2 + y^2)$$
$$+ E(x^2 - y^2) + F(x^2 + y^2)y + G(x^2 + y^2)^2 \tag{3.28}$$

where these aberration coefficients are

> A = Piston or constant term
> B = Tilt about the y axis (image displacement along the x axis)
> C = Tilt about the x axis (image displacement along the y axis)
> D = Defocusing
> E = Astigmatism
> F = Coma
> G = Spherical aberration

This expression applies to axially symmetric optical systems, whose wavefront has symmetry about the z axis. A more general expression is convenient for optical testing, where more complicated wavefront shapes may appear. Also, it is frequently more convenient to write this expression in polar coordinates. The angle θ can be measured from the x or y axis. The common practice in optical design is to measure the angle from the y axis. However, in optical testing this symmetry is not common and therefore it is more natural to measure the angle counter-clockwise from the x axis. Thus, we have

$$x = \rho \cos\theta \quad \text{and} \quad y = \rho \sin\theta \tag{3.29}$$

The terms can be of the form $\rho^n \cos^m \theta$ and $\rho^n \sin^m \theta$ or of the form $\rho^n \cos m\theta$ and $\rho^n \sin m\theta$. The two forms are equivalent. Here, the second form will be used. On the other hand, Hopkins [9] has pointed out that when using polar coordinates, in order to have a single valued function, the following conditions must be satisfied in polar coordinates: (1) the value of m should be smaller than or equal to the value of n, and (2) the sum $n + m$ should be even. In other words, n and m should be both odd or both even. These aberration terms, up to the fourth power are in Table 3.1.

The interferograms produced by these Seidel primary aberrations are illustrated in Figure 3.25. To determine these eight constants from measurements in the interferogram, the eight sampling points shown in Figure 3.26 can be used.

3.10 Lateral Shearing Interferometers

Lateral shear interferometers produce two identical wavefronts, one laterally sheared with respect to the other, as shown in Figure 3.27. The advantage is that a perfect reference wavefront is not needed. The optical path difference in these interferometers can be written as

$$\text{OPD} = W(x, y) - W(x - S, y) + \text{OPD}_0, \tag{3.30}$$

TABLE 3.1 Aberration Terms in $W(\rho, \theta)$. The Angle θ Is Measured Counterclockwise from the x Axis

n	m	r	l	Polar Coordinates	Cartesian Coordinates	Name
0	0	0	0	1	1	Piston
1	0	1	1	$\rho \sin \theta$	y	Tilt about x axis
	1	2	1	$\rho \cos \theta$	x	Tilt about y axis
2	0	3	0	ρ^2	$x^2 + y^2$	Defocusing
	1	4	2	$\rho^2 \sin 2\theta$	$2xy$	Astigmatism, axis at $\pm 45°$
	2	5	2	$\rho^2 \cos 2\theta$	$x^2 - y^2$	Astigmatism, axis at $0°$ or $90°$
3	0	6	1	$\rho^3 \sin \theta$	$(x^2 + y^2)y$	Coma, along y axis
	1	7	1	$\rho^3 \cos \theta$	$(x^2 + y^2)x$	Coma, along x axis
	2	8	3	$\rho^3 \sin 3\theta$	$(3x^2 - y^2)y$	Triangular astigmatism, semi-axes at $30°$, $150°$, $270°$
	3	9	3	$\rho^3 \cos 3\theta$	$(x^2 - 3y^2)x$	Triangular astigmatism, semi-axes at $0°$, $120°$, $240°$
4	0	10	0	ρ^4	$(x^2 + y^2)^2$	Spherical aberration
	1	11	2	$\rho^4 \sin 2\theta$	$2(x^2 + y^2)xy$	Fifth order astigmatism at $\pm 45°$
	2	12	2	$\rho^4 \cos 2\theta$	$x^4 - y^4$	Fifth order astigmatism at $0°$ or $90°$
	3	13	4	$\rho^4 \sin 4\theta$	$4(x^2 - y^2)xy$	Ashtray at $22.5°$ or $67.5°$
	4	14	4	$\rho^4 \cos 4\theta$	$(x^2 - y^2)^2 - 4x^2y^2$	Ashtray at $0°$ or $45°$

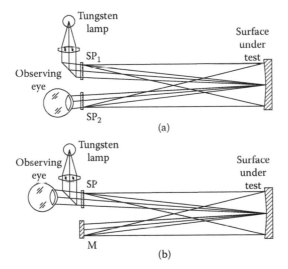

FIGURE 3.25 Unequal path Twyman–Green interferometer.

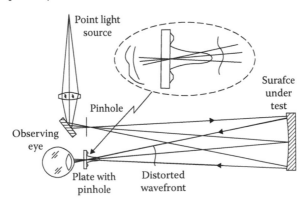

FIGURE 3.26 Points selected to evaluate the primary aberrations in a Twyman–Green interferogram.

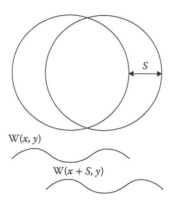

$W(x, y)$

$W(x + S, y)$

FIGURE 3.27 Wavefronts in a lateral shear interferometer.

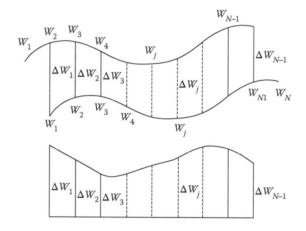

FIGURE 3.28 Saunders' method to find the wavefront in a lateral shear interferometer.

where S is the lateral shear and OPD_0 is the optical path difference with two undistorted wavefronts. If this lateral shear is small compared with the aperture diameter, the smallest spatial wavelength of the Fourier components of the wavefront distortions is much smaller than S. Thus, we may obtain

$$OPD = S \frac{\partial W(x, y)}{\partial x}. \tag{3.31}$$

This interferometer can be quite simple, but a practical problem is that the interferogram represents the wavefront slopes in the shear direction, not the actual wavefront shape. Thus, to obtain the wavefront deformations a numerical integration of the slopes has to be performed; in addition, two laterally sheared interferograms in mutually perpendicular directions are needed.

If the shear is not small enough, the interferogram does not represent the wavefront slope. Then, to obtain the wavefront deformation, a different method has to be used. One of the possible procedures has been proposed by Saunders [10] and is described in Figure 3.28. To begin, let us assume that $W_1 = 0$. Then, we may write

$$
\begin{aligned}
W_1 &= 0 \\
W_2 &= \Delta W_1 + W_1 \\
W_3 &= \Delta W_2 + W_2 \\
W_n &= \Delta W_{n-1} - W_{n-1}
\end{aligned}
\tag{3.32}
$$

A disadvantage of this method is that the wavefront can be evaluated only at points separated by a constant distance S. Intermediate values have to be estimated by interpolation.

An extremely simple lateral shear interferometer was described by Murty [11] and is shown in Figure 3.29. The practical advantages of this instrument are its simplicity, low price, and fringe stability. The only disadvantage is that it is not compensated and, thus, it has to be illuminated by laser light.

The lateral shear interferograms for the Seidel primary aberrations may be obtained as follows. The interferogram for a defocused wavefront is given by

$$2DxS = m\lambda. \tag{3.33}$$

This is a system of straight, parallel, and equidistant fringes. These fringes are perpendicular to the lateral shear direction. When the defocusing is large, the spacing between the fringes is small. When there is no defocus, there are no fringes in the field.

For spherical aberration, the interferogram is given by

$$4A(x^2 + y^2)xS = m\lambda. \tag{3.34}$$

If this aberration is combined with defocus, we have

$$\left[4A(x^2 + y^2)x + 2Dx\right]S = m\lambda. \tag{3.35}$$

The interference fringes are cubic curves. When coma is present, the interferogram is given by

$$2BxyS = m\lambda \tag{3.36}$$

if the lateral shear is S in the *sagittal* direction. When the lateral shear is T in the tangential y-direction, the fringes are described by

$$B(x^2 + 3y^2)T = m\lambda. \tag{3.37}$$

For astigmatism, if the lateral shear is S in the sagittal x-direction, the fringes are described by

$$(2Dx + 2Cx)S = m\lambda \tag{3.38}$$

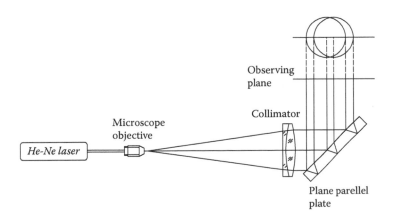

FIGURE 3.29 Murty's lateral shear interferometer.

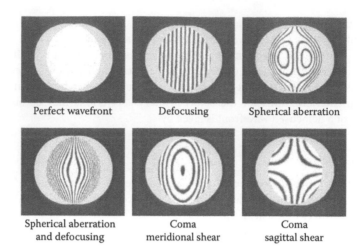

| Perfect wavefront | Defocusing | Spherical aberration |

| Spherical aberration and defocusing | Coma meridional shear | Coma sagittal shear |

FIGURE 3.30   Interferograms in a lateral shear interferometer.

and, for lateral shear T in the tangential y-direction, we have

$$(2Dy - 2Cy)T = m\lambda. \tag{3.39}$$

Then, the fringes are straight and parallel as for defocus, but with a different separation in both interferograms. Typical interferograms for the Seidel primary aberrations are illustrated in Figure 3.30.

The well-known and venerable Ronchi test [17], illustrated in Figure 3.31, can be considered as a geometrical test but also as a lateral shear interferometer. In the geometrical model, the fringes are the projected shadows of the Ronchi ruling dark lines. However, the interferometric model assumes that several laterally sheared wavefronts are produced by diffraction. Thus, the Ronchi test can be considered as a multiple wavefront lateral shear interferometer.

3.11 Talbot Interferometer and Moiré Deflectometry

Projecting the shadow of a Ronchi ruling with a collimated beam of light, as in Figure 3.32, the shadows of the dark and clear lines are not clearly defined due to diffraction. Sharp and well-defined shadows are obtained for extremely short distances from the ruling to the observing screen. When the observing distance is gradually increased, the fringe sharpness decreases, until, at a certain distance, the fringes completely disappear. However, as discovered by Talbot [12], by further increasing the observing distance, the fringes become sharp again and then disappear in a sinusoidal manner. With negative contrast, a clear fringe appears where there should be a dark fringe and vice versa. Talbot was not able to explain this phenomenon but it was later explained by Rayleigh [13]. The period of this contrast variation is called the Rayleigh distance L_r, which can be expressed by

$$L_r = \frac{2d^2}{\lambda}, \tag{3.40}$$

where d is the spatial period (lines separation) of the ruling and λ is the wavelength of the light.

When an aberrated glass plate is placed in the collimated light beam, the observed projected fringes will also be distorted, instead of straight and parallel.

A simple interpretation is analogous to the Ronchi test, with both the geometrical and the interferometric models. The geometrical model interprets the fringe deformation as due to the different local wavefront slopes producing different illumination directions, as in Figure 3.33a; then, the method is

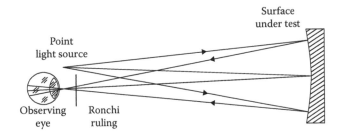

FIGURE 3.31 Ronchi test.

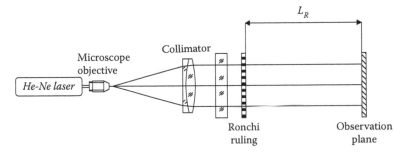

FIGURE 3.32 Observation of the Talbot effect.

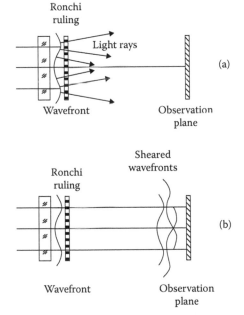

FIGURE 3.33 Interferometric and geometrical interpretations of Talbot interferometry.

frequently known as deflectometry [14]. The interferometric model interprets the fringes as due to the interference between multiple diffracted and laterally sheared wavefronts, as illustrated in Figure 3.33b. Talbot interferometry and their multiple applications have been described by many authors: for example, by Patorski [15] and Takeda and Kobayashi [16].

The fringes being produced have a high spatial frequency and, thus, the linear carrier has to be removed by a Moiré effect with another identical Ronchi ruling at the observation plane.

3.12 Foucault Test and Schlieren Techniques

Leon Foucault [18] proposed an extremely simple method to evaluate the shape of concave optical surfaces. A point light source is located slightly off axis, near the center of curvature. If the optical surface is perfectly spherical, a point image will be formed by the reflected light also near the center of curvature, as shown in Figure 3.34.

A knife edge then cuts the converging reflected beam of light. Let us consider three possible planes for the knife edge:

1. If the knife is inside of focus, the project shadow of the knife will be projected on the optical surface on the same side as the knife.
2. If the knife is outside of focus, the shadow will be on the opposite side to the knife.
3. If the knife is at the image plane, nearly all light is intercepted with even a small movement of the knife.

If the wavefront is not spherical, the shadow of the knife will create a light pattern on the mirror where the darkness or lightness will be directly proportional to the wavefront slope in the direction perpendicular to the knife edge. The intuitive impression is a picture of the wavefront topography. With this test, even small amounts of wavefront deformations of a fraction of the wavefront can be detected.

If a transparent fluid or gas is placed in the light optical path between the lens or mirror producing a spherical wavefront and the knife edge, a good sensitivity to the refractive index gradients in the direction perpendicular to the knife edge is obtained. For example, any air turbulence can thus be detected and measured. This is the working principle of the Schlieren techniques used in atmospheric turbulence studies.

3.13 Multiple Reflection Interferometers

A typical example of a multiple reflection interferometer is the Fabry–Perot [19] interferometer illustrated in Figure 3.35. An extended light source optically placed at infinity by means of a collimator

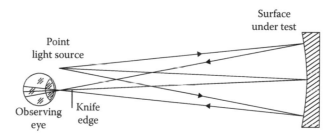

FIGURE 3.34 Foucault test.

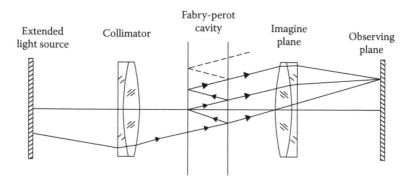

FIGURE 3.35 Fabry–Perot interferometer.

illuminates the interferometer, formed by a pair of plane and parallel interfaces. These two interfaces can be the two highly reflecting (coated) faces of a single plane parallel plate or two reflecting plane faces oriented front-to-front of a pair of glass plates. Then, the observed plane is optically placed at infinity by a focusing lens. Here, circular fringes will be observed.

As shown in Figure 3.36, a ray emitted from the extended light source follows a path with multiple reflections. Then if the amplitude of this ray is a, the resultant transmitted amplitude $E_T(\phi)$ at a point on the observing screen located at an infinite distance is

$$E_T(\phi) = at_1 t_2 + at_1 t_2 r_1 r_2 e^{i\phi} + at_1 t_2 r_1^2 r_2^2 e^{2i\phi} + at_1 t_2 r_1^3 r_2^3 e^{3i\phi} + \cdots \qquad (3.41)$$

thus, obtaining

$$E_T(\phi) = \frac{at_1 t_2}{1 - r_1 r_2 e^{i\phi}}. \qquad (3.42)$$

Assuming now that the two faces are equally reflecting faces and dielectric (non-absorbing), we can consider the Stokes' relations to apply as follows:

$$r^2 + tt' = 1;$$
$$r = -r', \qquad (3.43)$$

where t and r are for a ray traveling towards the interface from vacuum and t' and r' are for a ray traveling to the interface inside the glass. Thus, for this case we can write

$$E_T(\phi) = a\frac{1 - r^2}{1 - r^2 e^{i\phi}}. \qquad (3.44)$$

The irradiance $I_T(\phi)$ of the transmitted interference pattern as a function of the phase difference ϕ between two consecutive rays is then given by the square of the amplitude of this complex amplitude:

$$I_T(\phi) = I_0 \frac{1}{1 + \frac{4r^2}{(1 - r^2)^2} \sin^2\left(\frac{\phi}{2}\right)}, \qquad (3.45)$$

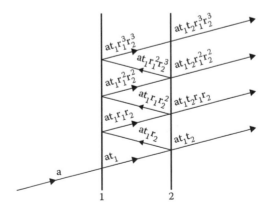

FIGURE 3.36 Multiple reflections in a Fabry–Perot interferometer.

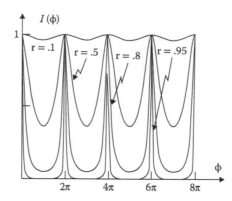

FIGURE 3.37 Irradiance as a function of the phase difference θ in a Fabry–Perot interferometer.

where $I_0 = a^2$ is the irradiance of the incident light beam. This irradiance is plotted in Figure 3.37 for several values of the reflectivity r of the faces. The interesting result is that the fringes become very narrow for high values of this reflectivity; then the position and shape of each fringe can be measured with a high precision.

As in any amplitude-division interferometer without energy losses, there are two complementary interference patterns. The sum of the energy in the reflected and the transmitted patterns must be equal to the incident energy. Thus, we can write

$$I_0 = I_T(\phi) + I_R(\phi), \tag{3.46}$$

where the reflected irradiance is

$$I_R(\phi) = I_0 \frac{\dfrac{4r^2}{\left(1-r^2\right)^2} \sin^2\left(\dfrac{\phi}{2}\right)}{1 + \dfrac{4r^2}{\left(1-r^2\right)^2} \sin^2\left(\dfrac{\phi}{2}\right)}. \tag{3.47}$$

There are a large number of multiple reflection interferometers whose principle is based on this narrowing of the fringes.

3.13.1 Cyclic Multiple Reflection Interferometers

Cyclic multiple reflection interferometers have been described by Garcia-Márquez et al. [20], and are shown in Figure 3.38. The amplitude $E(\phi)$ at the output can be found with a similar method to that used for the Fabry–Perot interferometer, obtaining

$$E(\phi) = a\frac{r + \sigma e^{i\phi}}{1 + \sigma r e^{i\phi}}, \tag{3.48}$$

where σ is the coefficient for the energy loss of the system for one cyclic travel around the system ($\sigma = 1$ if no energy and $\sigma = 0$ if all the energy is lost). This coefficient can be the transmittance or absorbance of

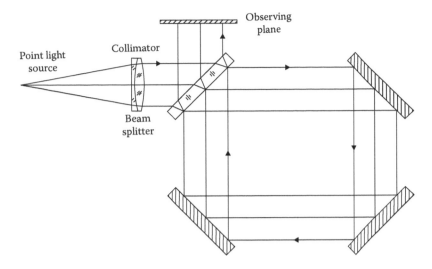

FIGURE 3.38 Cyclic multiple reflections interferometer.

one of the mirrors when it is not 100% reflective. The irradiance $I(\phi)$ as a function of the phase difference ϕ between two consecutive passes through the system is given by

$$I(\phi) = I_0 \frac{(r+\sigma)^2 - 4\sigma r \sin^2\left(\dfrac{\phi}{2}\right)}{(1+\sigma r)^2 - 4\sigma r \sin^2\left(\dfrac{\phi}{2}\right)}, \tag{3.49}$$

where $I_0 = a^2$ is the incident irradiance. It is interesting to consider three particular cases:

1. When the coefficient $\sigma = 1$, i.e., if no energy is lost, then $I(\phi) = a^2$. In other words, all energy arriving to the interferometer is in the output. An interesting consequence is that, then, no fringes can be observed.
2. When the coefficient $\sigma \neq 1$ because the mirror M_2 is a semitransparent mirror with reflectance r, then $\sigma = -r$ and the irradiance $I(\phi)$ becomes equal to the irradiance $I_R(\phi)$ in the Fabry–Perot interferometer. In this case, there is a transmitted interference pattern in the semitransparent mirror, which acts as a second beamsplitter.
3. When the coefficient $\sigma = 1$ due to energy absorption in the mirrors or because an absorbing material is introduced in the interferometer, then there is only one output in the interferometer, but it contains visible fringes. The complementary interference pattern is hidden as absorption.

3.14 Cyclic Interferometers

The basic arrangements for a cyclic interferometer are either square, as in Figure 3.39, or triangular, as in Figure 3.40. The two interfering wavefronts travel in opposite directions around the square or triangular path.

In both the square and the triangular configurations, the two interfering wavefronts keep their relative orientations on the output beam. Also, it can be observed in Figures 3.39 and 3.40 that these interferometers are compensated, so that their optical path difference OPD = 0 for all wavelengths. Thus, white light illumination can be used.

Any transparent object or sample located inside the interferometer will be traversed twice, in opposite directions, by the light beams. In the square configuration these beams pass through the sample with a reversal orientation (not 180° rotation), making it sensitive to anti-symmetric aberrations, such as coma or tilt in one of the mirrors.

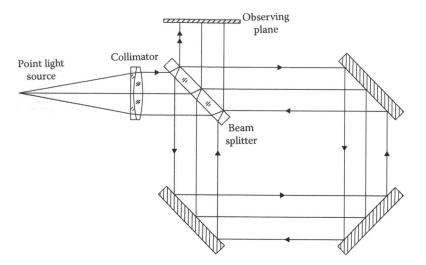

FIGURE 3.39   Square cyclic interferometer.

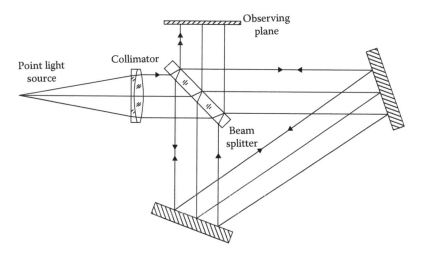

FIGURE 3.40   Triangular cyclic interferometer.

In the triangular configuration the two interfering beams have an even number of reflections (two and four) going from the light source to the observing plane. Therefore, tilt fringes cannot be introduced by tilting any of the two mirrors. However, any tilt or displacement of the mirrors in a perpendicular direction to their surfaces produces a relative lateral shear of the two interfering beams on the observing plane. This property has been used to make lateral shear interferometers.

By introducing a telescopic afocal system in the interferometer, a radial shear interferometer can also be made using this configuration.

3.15 Sagnac Interferometer

The Sagnac interferometer [21] is a cyclic interferometer with a typical square configuration, as in Figure 3.39; it can also be made circular, with a coiled optical fiber, but the working principle is the same. The Sagnac interferometer was used as an optical gyroscope, to sense slow rotations.

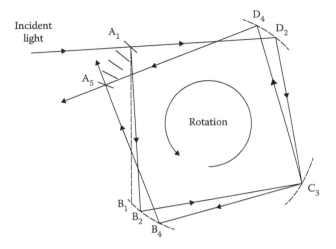

FIGURE 3.41   Sagnac interferometer.

Figure 3.41 shows the working principle. The beamsplitter A and the mirrors B, C, and D form the interferometer. The whole interferometer system rotates, including the light source and observer. Then, for a single travel of the light around the cyclic path in opposite directions, the beamsplitter and the mirrors have consecutive positions, labeled with subscripts 1, 2, 3, 4, and finally 5. When there is no rotation, the path length *s* from one mirror to the next is the same for both beams:

$$s = \sqrt{2}r, \tag{3.50}$$

where *r* is half the diagonal of the square arrangement. However, when the system is rotating, it can be observed in this figure that these paths have different lengths for the two beams, given by

$$s = \sqrt{2}r\left(1 \pm \frac{\theta}{2}\right), \tag{3.51}$$

where θ is the angle rotated between two consecutive positions in Figure 3.41. Thus the OPD at the output for both interfering beams is

$$\text{OPD} = 2\sqrt{2}\, r\theta \tag{3.52}$$

Hence it can be shown that

$$\frac{\text{OPD}}{\lambda} = \frac{4wr^2}{c\lambda} = \frac{4wA}{c\lambda} \tag{3.53}$$

where *A* is the area of the interferometer square.

This small shift in the fringes is constant given a fixed speed of rotation. If the interferometer plane is turned upside down, the fringe shifts in the opposite direction. This optical gyroscope has been used to detect and measure the earth's rotation.

References

1. Malacara D. ed., *Optical Shop Testing*, 3rd edn. New York, NY: John Wiley & Sons, 2007.
2. Creath K. Interferometric Investigation of a Laser Diode. *Appl. Opt.*, **24**, 1291–1293; 1985.
3. Ning Y, Grattan KTV, Meggitt BT, Palmer AW. Characteristics of laser diodes for interferometric use. *Appl. Opt.*, **28**, 3657–3661; 1989.
4. Twyman F. Interferometers for the experimental study of optical systems from the point of view of the wave theory. *Philos. Mag.*, Ser. 6, **35**, 49; 1918.
5. Houston JB Jr, Buccini CJ, O'Neill PK. A laser unequal path interferometer for the optical shop. *Appl. Opt.*, **6**, 1237–1242; 1967.
6. Linnik W. Simple interferometer to test optical systems. *Comptes Rendus del'Académie des Sciences d l'U.R.S.S.*, **1**, 208; 1933.
7. Smartt RN, Strong J. Point diffraction interferometer (abstract only). *J. Opt. Soc. Am.*, **62**, 737; 1972.
8. Kingslake R. The interferometer patterns due to the primary aberrations. *Trans. Opt. Soc.*, **27**, 94; 1925–1926.
9. Hopkins H.H., Wave Theory of Aberrations, Clarendon Press, Oxford, **48**; 1950
10. Saunders JB. Measurement of wavefronts without a reference standard: The wavefront shearing interferometer. *J. Res. Nat. Bur. Stand.*, **65B**, 239; 1961.
11. Murty MVRK. The use of a single plane parallel plate as a lateral shearing interferometer with a visible gas laser source. *Appl. Opt.*, **3**, 331–351; 1964.
12. Talbot WHF. Facts relating to optical science. *Phil. Mag.*, **9**, 401; 1836.
13. Rayleigh L. On Copying Diffraction-gratings, and on some Phenomena contained therewith, *Philos. Mag.* **11**, 196; 1881.
14. Glatt I, Kafri O. Moiré deflectometry—Ray tracing interferometry. *Opt. and Lasers Eng.*, **8**, 277–320; 1988.
15. Patorski K. Moiré methods in interferometry. *Opt. Lasers Eng.*, **8**, 147–170; 1988.
16. Takeda M, Kobayashi S. Lateral aberration measurements with a digital Talbot interferometer. *Appl. Opt.*, **23**, 1760–1764; 1984.
17. Cornejo, A., Ronchi test. In *Optical Shop Testing*, D. Malacara, ed. New York, NY: John Wiley & Sons, 2007.
18. Foucault, L. M., Description des Procédés Employés pou Reconnaitre la Configuration des Surfaces Optiques. *C. R. Acad. Sci. Paris*, **47**, 958, 1858.
19. Fabry, C. and A. Perot, Sur les Franges des Lames Minces Argentées et Leur Application a la Measure de Petites Epaisseurs d'air. *Ann. Chim. Phys.*, **12**, 459, 1897.
20. Garcia-Mairquez, Interferometers Without Observable Fringes, *Opt. Eng.*, **36**, 2863–2867 (1997).
21. Sagnac G. L'ether Lumineux Demontré; por L'Effet due Vent Relatif d'ether Dans un Interferometre en Rotation Uniforme. *Comptes Rendus Academie Science Paris*, **157**, 361–362; 1913.

4

Modern Fringe Pattern Analysis in Interferometry

4.1   Introduction 101
4.2   The Fringe Pattern 102
      Information Content • Fringe Pattern Preprocessing and
      Design • Classification of the Analysis Methods
4.3   Smoothing Techniques 107
      Introduction • Convolution Methods • Regularization
      Methods • Classical Regularization • Frequency Response of
      Regularized Linear Low-Pass Filters
4.4   Temporal Phase-Measuring Methods 111
      Introduction • General Theory of PSI • Analysis of Commonly Used
      Phase-Shifting Algorithms • N-Step Least-Squares Phase-Shifting
      Algorithm • Synthesis of Robust Quadrature Filters for Phase-Shifting
      Interferometry • Design of FTFs by Means of First-Order Building
      Blocks • Six-Step PSA with Robustness against Background Rejection
      and Detuning Error • Seven-Step PSA with a Second-Order Spectral
      Zero at $\omega = -\omega_0$ • Eight-Step Broadband PSA Tuned at $\omega_0 = \pi / 2$
4.5   Spatial Phase-Measuring Methods 125
      The Fourier-Transform Method for Open-Fringes Interferograms •
      Spatial Carrier Phase-Shifting Method • Synchronous Demodulation
      of Open-Fringes Interferograms • Robust Quadrature
      Filters • Regularized Phase-Tracking Technique
4.6   Phase Unwrapping...133
      Unwrapping Using Least-Squares Integration of Gradient
      Phase • Unwrapping Using the Regularized Phase-Tracking
      Technique • Temporal Phase Unwrapping
4.7   Extended Range Fringe Pattern Analysis 137
      Phase Retrieval from Gradient Measurement Using Screen-Testing
      Methods • Wavefront Slope Analysis with Linear Gratings (Ronchi
      Test) • Moiré Deflectometry • Wavefront Analysis with Lateral
      Shearing Interferometry • Wavefront Analysis with Hartmann
      Screens • Wavefront Analysis by Curvature Sensing • Sub-Nyquist
      Analysis
4.8   Applicability of Fringe Analysis Methods....................................147
Acknowledgments..149
References...149

Manuel Servín,
Malgorzata
Kujawinska, and
José Moisés Padilla

4.1 Introduction

Optical methods of testing with the output in the form of an interferogram, or more general fringe pattern, have been used since the early 1800s. However, the routine quantitative interpretation of the information coded into a fringe pattern was not practical in the absence of computers. In the late 1970s the advances in video CCD (charge-coupled device) cameras and image-processing technology coupled with the development of the inexpensive but powerful desktop computer provided the means for the birth and rapid development of automatic fringe pattern analysis. This caused a major resurgence of interest in interferometric metrology and related disciplines and formed an excellent basis for their industrial, medical, civil, and aeronautical engineering applications.

4.2 The Fringe Pattern

4.2.1 Information Content

A fringe pattern can be considered as a sinusoidal signal fluctuation in two-dimensional space (Figure 4.1) given by

$$I(x,y) = a(x,y) + b(x,y)\cos\phi(x,y) + n(x,y), \tag{4.1}$$

or

$$I(x,y) = a(x,y)[1 + V(x,y)\cos\phi(x,y)] + n(x,y), \tag{4.2}$$

where $a(x,y)$, $b(x,y)$, $V(x,y)$ are background, local contrast, and fringe visibility functions, respectively, $n(x,y)$ represents a random noise field, and $\phi(x,y)$ is the phase function obtained when an interferometer, Moiré system, or other device produces a continuous map, which is an analogue of the physical quantity being measured (shape, displacement, deformation, strain, temperature, etc.).

Fringe pattern analysis (fringe analysis for short) refers to full reconversion to the original feature represented by a fringe pattern [1]. In this process, the only measurable quantity is intensity $I(x,y)$. The unknown phase $\phi(x,y)$ should be extracted from Equation 4.1 or 4.2, although it is screened by two other functions $a(x,y)$ and $b(x,y)$. Moreover, $I(x,y)$ depends periodically on the phase, which causes additional problems:

- Due to periodicity, the phase is only determined $\mod 2\pi$ (2π phase ambiguity).
- Due to even character of the cosine function $\cos(\phi) = \cos(-\phi)$, the sign of ϕ cannot be extracted from a single measurement of $I(x,y)$ without a priori knowledge (sign ambiguity).
- In all practical cases some random noise $n(x,y)$ is introduced in an additive and/or multiplicative way.

Additionally, the fringe pattern may suffer from a number of distortions degrading its quality further screening the phase information [2,3].

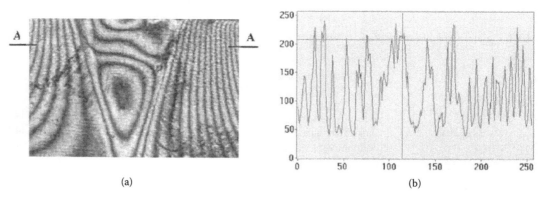

(a)   (b)

FIGURE 4.1 A fringe pattern as a sinusoidal signal fluctuation in space: (a) its cross section A–A and (b) its intensity distribution.

The background and contrast functions contain the intensities of interfering (superposed) fields and the various disturbances. Generally one can say that $a(x, y)$ contains all additive contributions—that is, varying illumination and changing object reflectivity, time-dependent electronic noise due to electronic components of the image-capturing processing, diffraction of dust particles in the optical paths—while $b(x, y)$ comprises all multiplicative influences, including the ratio between the reference and object beams, speckle decorrelation and contrast variations caused by speckles.

For computer-aided quantitative evaluation, the fringe pattern is usually recorded by a CCD camera and stored in the computer memory in a digital format; that is, the recorded intensity is digitized into an array of $M \times N$ image points (pixels) and quantized into G discrete gray values. The numbers M and N set an upper bound to the density of the fringe pattern to be recorded. The sampling theorem demands more than two detection points per fringe; however, due to dealing with finite-sized detector elements, charge leakage to neighboring pixels, and noise in the fringe pattern, one has to supply at least 3–5 pixels per fringe period to yield a reliable phase estimation.

When intensity frames are acquired, the analog video signal is converted to a digital signal of discrete levels. In practice, a quantization into 8 bits (corresponding to 256 gray values) or into 10 bits (giving 1024 values) are the most common. Usually, 8 bits are sufficient for reliable evaluation of fringe patterns. The quantization error is affected by the modulation depth of the signal, as the effective number of quantization levels equals the modulation of the $\times$ signal the number of quantization levels.

4.2.2 Fringe Pattern Preprocessing and Design

A fringe pattern obtained as the output of a measuring system may be modified by the optoelectronic–mechanical hardware (sensors and actuators) and software (virtual sensors and actuators) of the system (Figure 4.2) [4]. These modifications refer to both phase and amplitude (intensity) of the signal produced in space and time, so the most general form of the fringe pattern is given by

$$I(x, y, t) = a_0(x, y) + \sum_{m=1}^{\infty} a_m(x, y) \cos m 2\pi \left[f_{0x} x + f_{0y} y + \upsilon_0 t + \alpha(t) + \phi(x, y) \right] + n(x, y), \qquad (4.3)$$

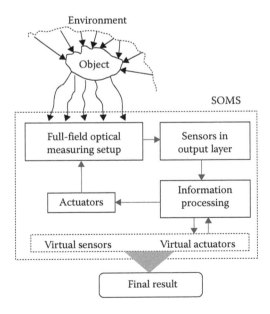

FIGURE 4.2 The scheme of a smart optical measuring system.

where (x, y, t) represent the spatial and temporal coordinates of the signal, respectively, $a_m(x, y)$ is the amplitude of mth harmonic of the signal, f_{0x}, f_{0y} are the fundamental spatial frequencies, v_0 is the fundamental temporal frequency, $\alpha(t)$ is the phase-shift value, the measurand is coded in the phase $\phi(x, y)$, and $n(x, y)$ represents additionally extracted random high-frequency noise.

Assuming a purely sinusoidal signal, Equation 4.3 becomes

$$I(x, y, t) = a(x, y) + b(x, y)\cos 2\pi \left[f_{0x}x + f_{0y}y + v_0t + \alpha(t) + \phi(x, y) \right] + n(x, y), \qquad (4.4)$$

where $a(x, y) = a_0(x, y)$ and $b(x, y) = a_1(x, y)$.

The fringe pattern has to be modified by hardware actuators in the measuring system in order to fulfill the demands of a priori selected analysis method and resistance to environmental conditions. The sensors within the system enable it to determine and control the fringe pattern features ($f_{0x}, f_{0y}, v_0, a(x, y), b(x, y)$) and in this way allow operational parameters of the actuators to be set. Table 4.1 shows the most commonly used actuators in modern research and commercial measuring systems. Special attention is paid to the new possibilities connected with the application of laser diodes [5], fiber optics, spatial light modulators (LCD, DMD) [6], and piezoelectric transducer (PZT) micro-positioning devices. These devices not only allow one to design properly an output fringe pattern, but also help to stabilize fringes in the presence of vibrations [7–9].

However, in a given technical measurement, there is always a certain limit to which the appearance of a fringe pattern can be controlled. Real images are often noisy, low contrast, and with significant variations of background. The fringe pattern analysis method described in the next sections should be designed to handle these problems. However, some general purpose image preprocessing techniques are often used to improve the original data prior to fringe analysis. Two main groups of operations are applied [10,11]:

1. Arithmetic (pixel-to-pixel) operations including normalization, gamma correction, adding/ subtractions, and multiplication/divisions performed directly on the images. These operations lead to production of a fringe pattern, which looks better to human perception and is based on manipulations of the histogram of a digital image.
2. Filtering operations that may be performed alternatively: directly on the image by convolution with a local operator or in Fourier space by multiplying the image spectrum with a filter window. In general, high-pass filtering weakens the influence of non-homogeneous background $a(x, y)$, while a low-pass filter removes high-frequency noise $n(x, y)$.

After correctly performed hardware modifications of the features of fringe patterns and their software preprocessing, the images are ready for further analysis.

4.2.3 Classification of the Analysis Methods

The main task that has to be performed by fringe pattern analysis methods is to compute the phase $\phi(x, y)$ from the measured intensity values (Figure 4.3). This means that an inverse problem has to be solved with all its difficulties:

- The regularization problem (an ill-posed problem due to unknown a, b, ϕ)
- The sign ambiguity problem
- The 2π phase ambiguity problem

The first two difficulties may be overcome by two alternative approaches:

1. Intensity methods in which we work passively on an image intensity distribution captured by a detector. These include fringe extrema localization methods (skeletoning, fringe tracking) and regularization methods.

TABLE 4.1 Actuators in Optical Measuring Systems

Actuator	Signals	Parameters	f_{0x}, f_{0y}	v_0	α	a	b	N_R	m	Reference
Laser diode	Injection current Temperature	Output power Wavelength Degree of coherence	×	×	×	–	×	×	×	Kozlowska and Kujawinska (1997), Takeda and Kitoh (1992)
Non-coherent light source	Current Voltage	Output power	–	–	–	×	×	×	–	Van der Heijden (1994), Joenathan (1992)
Fiber optics with accessories variable couplers	Depending on accessories	Optical path difference	–	–	×	–	×	–	–	Olszak and Patorski (1997), Takeda and Kitoh (1992), Jones (1994)
PZT	Voltage	Displacement Angle	×	–	×	–		–	–	Van der Heijden (1994), Ai (1987), Efron et al. (1989)
SLM (LCD, DMD)	Voltage	Intensity of each pixel or line	×	–	×	–	×	×	×	Efron et al. (1989), Frankowski (1997), Patorski (1993)
Polarizing optics + M[b]	Voltage	Angle Displacement	×	–	×	–	×	–	–	Robinson and Reid (1993), Shagam and Wyant (1978), Asundi and Yung 1991
Gratings +M	Voltage	Displacement Angle	×	–	×	–	×	–	–	Robinson and Reid (1993), Shagam and Wyant (1978), Huntley (1994a)
Optical wedges +M parallel plates	Voltage	Displacement Angle	×	–	×	–		–	–	Robinson and Reid (1993), Schwider (1990)
Acousto-optic devices	Voltage	Wavelength Angle Degree of coherence	–	×	×	–		×	–	Shagam and Wyant (1978), Robinson (1993), Shagam (1978)
Ground glass +M	Voltage	Degree of coherence	–	–	–	–	–	×	–	Robinson and Reid (1993), Schwider (1990), Kujawinska and Kosinski (1997)
CCD cameras Frame-grabber	Voltage Digital signal	Amplification Offset γ correction Exposure time	–	–	–	×	×	×	×	Joenathan and Khorana (1992), Yamaguchi et al. (1996)

[a] FP, fringe pattern.
[b] M, constant or alternate current motor.

2. Phase methods for which we actively modify fringe pattern(s) in order to provide additional information to solve the sign ambiguity problem [12–18]. These include
   - Temporal heterodyning (introducing running fringes) [13]
   - Spatial heterodyning (Fourier-transform method [14], PLL [15], and spatial carrier phase-shifting method [16]
   - Temporal [17] and spatial phase-shifting [18], which are discrete versions of the above methods, where the time or spatially varying interferogram is sampled over a single period

The third difficulty, the 2π phase ambiguity, coming from the sinusoidal nature of the signal is common to fringe pattern analysis methods (Figure 4.3). The only method that measures nearly directly

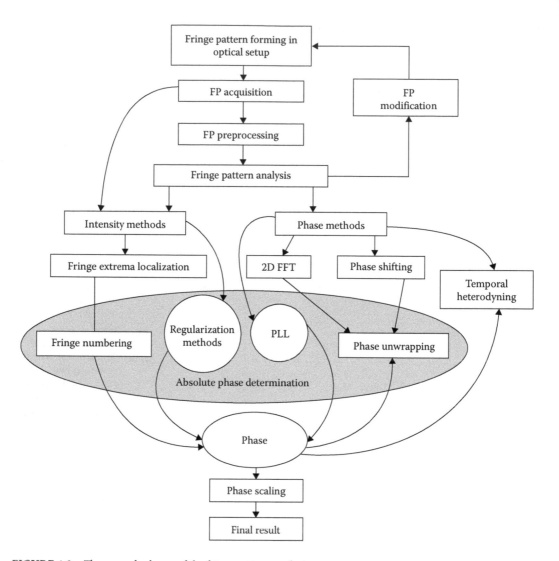

FIGURE 4.3   The general scheme of the fringe pattern analysis process.

absolute phase with no 2π ambiguity is temporal heterodyning [13]. The other fringe pattern analysis methods determine absolute phase $\phi(x, y)$ by

- Fringe numbering and phase extrapolation [1]
- Phase unwrapping [29,20]
- Hierarchical unwrapping [21]
- Regularized phase-tracking techniques [22]

These procedures finalize the fringe measurement stage, which reduces a fringe pattern to a continuous phase map. To solve a particular engineering problem, the stage of phase scaling has to be implemented. It converts the phase map into the physical quantity to be measured in the form which enables further information processing and implementation system, finite element modeling, and machine vision systems [23,11]. This stage is specific application-oriented and is developing rapidly due to the implementation of fringe measurement to a vast range of different types of interferometers, Moiré, and fringe projection systems and due to the increased quality of the phase data obtained.

4.3 Smoothing Techniques

4.3.1 Introduction

It is very common that the fringe pattern, as captured by the video digitizing device, contains excessive noise. Generally speaking, fringe patterns contain mostly a low-frequency signal along with a degrading white noise (multiplicative or additive); therefore, a low-pass filtering (smoothing) of the fringe pattern may remove a substantial amount of this noise, making the demodulation process more reliable. We are going to discuss two basic and commonly used low-pass filters, the averaging convolution window and the regularized low-pass filters.

4.3.2 Convolution Methods

The convolution averaging window is by far the most used low-pass filter in fringe analysis. The discrete impulse response of this filter may be represented by

$$h(x,y) = \frac{1}{9} \begin{pmatrix} 1 & 1 & 1 \\ 1 & 1 & 1 \\ 1 & 1 & 1 \end{pmatrix}. \tag{4.5}$$

The frequency response of this convolution matrix is

$$H(\omega_x, \omega_y) = (1/9)[1 + 2\cos\omega_x + 2\cos\omega_y + 4\cos(\omega_x)\cos(\omega_y)], \tag{4.6}$$

where ω_x and ω_y are the angular frequency in the x- and y-direction, respectively. This convolution filter may be used several times to decrease the band-pass frequency. Using a low-pass convolution filter several times changes the shape of the filter as well as its low-pass frequency. The frequency response of a series of identical low-pass filters will approach a Gaussian-shaped response, as can be seen in Figure 4.4, which shows how rapidly the frequency response's shape of the 3×3 averaging filter changes as it is convolved with itself 1, 2, and 3 times.

4.3.3 Regularization Methods

The main disadvantage of convolution filters as applied to fringe pattern processing is their undesired effect at the edges of the interferogram. This distortion arises because at the boundary of the fringe pattern, convolution filters mix the background illumination with that of the fringe pattern, raising an estimated phase error in that zone. This undesired edge distortion may be so important that some people shrink the interferogram's area to avoid those unreliable pixels near the edge. Also, convolution filters cannot preserve fast gray-level variations while removing substantive amounts of noise. For these reasons, it is more convenient to formulate the smoothing problem in the ways described below.

4.3.4 Classical Regularization

The filtering problem may be stated as follows [24]: find a smooth (or piecewise smooth) $f(.)$ function defined on a two-dimensional field L, given observations $g(.)$ that may be modeled by

$$g(x,y) = f(x,y) + n(x,y), \quad (x,y) \in S, \tag{4.7}$$

where $n(.)$ is a noise field (for example white-Gaussian noise) and S is the region where well-defined fringe data is available. In other words, the filtering problem may be seen as an optimizing problem in

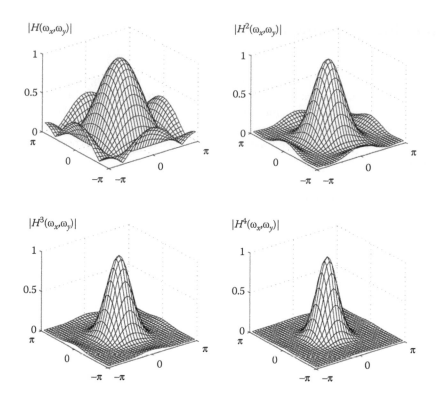

FIGURE 4.4 Frequency response of a 3 × 3 averaging window. As the number of convolution increases, the frequency response tends to a Gaussian shape.

which one has a compromise between obtaining a smooth filtered field $f(x, y)$, while keeping a good fidelity to the observed data $g(x, y)$. In the continuous domain, a common mathematical form for the smoothing problem may be stated as the field $f(x, y)$, which minimizes the following cost or energy functional:

$$U[f(x,y)] = \iint\limits_{(x,y)\in S} \left\{ [f(x,y) - g(x,y)]^2 + \eta \left(\frac{\partial f(x,y)}{\partial x} \right)^2 + \eta \left(\frac{\partial f(x,y)}{\partial y} \right)^2 \right\} \partial y \partial x \qquad (4.8)$$

As the above equation shows, the first term is a measure of the fidelity between the smoothed field $f(x, y)$ and the observed data $g(x, y)$ in a least-squares sense. The second term (the regularizer) penalizes the departure from smoothness of the filtered field $f(x, y)$. The first-order regularizer is also known as a membrane regularizer. That is because the cost functional to be minimized corresponds to the mechanical energy of a two-dimensional membrane $f(x, y)$ attached by linear springs to the observations $g(x, y)$. The parameter η measures the stiffness of the membrane model. A high stiffness value will lead to a smoother filtered field. One may also smooth $f(x, y)$ using higher-order regularizers such as the second-order or thin-plate regularizer. In the continuous domain, the energy functional to be minimized for the filtered field $f(x, y)$ may be stated as

$$U[f(x,y)] = \iint\limits_{(x,y)\in S} \left\{ \frac{1}{\eta}[f(x,y) - g(x,y)]^2 + \left(\frac{\partial^2 f(x,y)}{\partial x^2} \right)^2 + \left(\frac{\partial^2 f(x,y)}{\partial y^2} \right)^2 + \left(\frac{\partial^2 f(x,y)}{\partial y \partial x} \right)^2 \right\} \partial y \partial x \quad (4.9)$$

In this case the smoothed field $f(x, y)$ corresponds to the height of a metallic thin plate attached to the observations $g(x, y)$ by linear springs. Again, the parameter η measures the stiffness of the thin-plate model or conversely (as in the last equation) the looseness of the linear spring connecting the thin plate (filtered field) to the observed data. To optimize these cost functionals using a digital computer, one needs first to discretize the cost functional. Therefore, the functions $f(x, y)$ and $g(x, y)$ are now defined on the nodes of a regular and discrete lattice L, so the integrals become sums over the domain of interest:

$$U[f(x,y)] = \sum_{(x,y) \in S} \sum \left\{ [f(x,y) - g(x,y)]^2 + \eta R_1[f(x,y)] \right\}, \tag{4.10}$$

As mentioned, S is the subset of L where well-defined observations are available. The discrete version of the first-order regularizer $R_1[f(x, y)]$ may be approximated by

$$R_1[f(x,y)] = [f(x,y) - f(x-1,y)]^2 + [f(x,y) - f(x,y-1)]^2 \tag{4.11}$$

and the second-order regularizer $R_2[f(x, y)]$ may be approximated by

$$\begin{aligned} R_2[f(x,y)] = &[f(x+1,y) - 2f(x,y) + f(x-1,y)]^2 \\ &+ [f(x,y+1) - 2f(x,y) + f(x,y-1)]^2 \\ &+ [f(x+1,y+1) - f(x-1,y-1) + f(x-1,y+1) - f(x+1,y-1)]^2. \end{aligned} \tag{4.12}$$

By considering only the first two terms of the second-order regularizer, one may reduce significantly the computational load of the filtering process, while preserving a thin-plate-like behavior.

A simple way to optimize the discrete cost functions stated in this section is by gradient descent, that is,

$$f^0(x,y) = g(x,y),$$
$$f^{k+1}(x,y) = f^k(x,y) - \tau \frac{\partial U[f(x,y)]}{\partial f(x,y)}, \tag{4.13}$$

where k is the iteration number and $\tau \approx 0.1$ is the step size of the gradient search. Although this is a simple optimizing technique, it is a slow procedure especially for high-order regularizers. One may use instead conjugate gradient methods to speed up the optimizing process.

Let us point out a possible practical way of implementing in a digital computer the derivative of the cost function $U[f(x, y)]$ using an irregularly-shaped domain S. Let us define an indicator function $m(x, y)$ in the lattice L having $N \times M$ nodes. The indicator function $m(x, y)$ equals one if the pixel is inside S (valid observations) and zero otherwise. Using this indicator field, the filtering problem with a first-order regularizer may be rewritten as

$$U[f(x,y)] = \sum_{x=0}^{N-1} \sum_{y=0}^{M-1} \left\{ [f(x,y) - g(x,y)]^2 m(x,y) + \eta R_1[f(x,y)] \right\}, \tag{4.14}$$

where

$$\begin{aligned} R_1[f(x,y)] = &[f(x,y) - f(x-1,y)]^2 m(x,y)m(x-1,y) \\ &+ [f(x,y) - f(x,y-1)]^2 m(x,y)m(x,y-1); \end{aligned} \tag{4.15}$$

then, the derivative may be found as

$$\frac{\partial U\left[f(x,y)\right]}{\partial f(x,y)} = \left[f(x,y)-g(x,y)\right]m(x,y)+$$

$$+\eta\left[f(x,y)-f(x-1,y)\right]m(x,y)m(x-1,y)$$

$$+\eta\left[f(x-1,y)-f(x,y)\right]m(x+1,y)m(x,y)$$

$$+\eta\left[f(x,y)-f(x,y-1)\right]m(x,y)m(x,y-1)$$

$$+\eta\left[f(x,y+1)-f(x,y)\right]m(x,y+1)m(x,y). \tag{4.16}$$

As we can see, only the difference terms lying completely within the region of valid fringe data marked by $m(x,y)$ survive. In other words, the indicator field $m(x,y)$ is the function that actually decouples valid fringe data from its surrounding background.

Finally, one may consider linear regularization filters as being more robust than convolution filters in the following sense:

- Unlike convolution filters, the background outside the filtering area S has minimum affect on the filtering process inside S; i.e., minimum cross-talking occurs between the filtered field $f(x,y)$ inside S and its surrounding background in L. In other words, the edge effect of regularized filters at the boundary of S is minimized, in this way the modulating phase near the boundary of S is little affected. This is especially important when dealing with irregular-shaped interferometric regions.
- Regularized linear filters tolerate missing observations due to the capacity of these filters to interpolate over regions of missing data with a well-defined behavior. The interpolating behavior of the filter is given by the mathematical form of the regularization term.
- By modifying the potentials in the cost function, one may obtain many different types of filters, such as quadrature complex-valued quadrature band-pass filters (QFs), which are very important in fringe analysis as phase demodulators.

4.3.5 Frequency Response of Regularized Linear Low-Pass Filters

The filtered field that minimizes the cost functions seen in the previous section smooth out the observation field $g(x,y)$ within the region of well-defined fringe data S. To have a quantitative idea of the amount of smoothing achieved, one may find the frequency response of the regularizer. To find the frequency response of the first-order low-pass filter [25,26], consider an infinite two-dimensional lattice L. Setting the gradient of the cost function to 0 one obtains the following set of linear equations:

$$f(x,y)-g(x,y)+\eta\left[4f(x,y)-f(x-1,y)-f(x+1,y)-f(x,y-1)-f(x,y+1)\right]=0, \tag{4.17}$$

and taking the discrete Fourier transform of the last equation one obtains

$$G(\omega_x,\omega_y)=F(\omega_x,\omega_y)\left[1+2\eta(2-\cos\omega_x-\cos\omega_y)\right]; \tag{4.18}$$

This leads to the following frequency transfer function (FTF),

$$H(\omega_x,\omega_y)=\frac{F(\omega_x\omega_y)}{G(\omega_x,\omega_y)}=\frac{1}{1+2\eta\left[2-\cos(\omega_x)-\cos(\omega_y)\right]}, \tag{4.19}$$

which represents a low-pass filter with a bandwidth controlled by the parameter η.

4.4 Temporal Phase-Measuring Methods

4.4.1 Introduction

Phase-stepping or phase-shifting interferometry (PSI) is the most common technique used to detect the modulating phase of interferograms. Given the high popularity of this method, there has been a lot of research contributing to this area of interferometry [1,3,8,10,27–32]. The PSI method was first introduced by Bruning et al. [17] for testing optical components using a video CCD array to map, over a large number of points, the optical wavefront under analysis. In the PSI technique [33] an interference pattern is phase-stepped under computer control and spatially digitized over several phase steps. The digitized intensity values may then be linearly combined to detect the optical phase at every pixel in the interferogram.

Many algorithms have been proposed to recover the wavefront of interference patterns, for which the emphasis is laid on using a small number of phase-shifted samples. All phase-shifting algorithms (PSAs) proposed work fine whenever the following conditions are met:

- The light-intensity is within the linear range of the CCD.
- The phase-shifted interferograms are taken at exactly the right phase shift.
- The device used to move the reference beam moves linearly in a purely piston fashion.
- All mechanical perturbations (vibrations, air turbulence, and so on) are small during the capture of the digital interferograms.

Sometimes one or several conditions mentioned above are not met, so we need a robust phase-shifting demodulation formula. To evaluate the merits of different PSAs, one needs to analyze their frequency response. Freischlad and Koliopoulos [34] were the first researchers that studied the spectral behavior of commonly used PSAs: they demonstrated that plotting the Fourier transform of two real filters (found respectively on the numerator and denominator of a PSA) without constant or common phase factors one can see whether this algorithm is robust against experimental errors, such as detuning and nonlinear harmonics distortion. This frequency analysis method was later generalized by the works of Surrel [35], and Servin et al.[36,37], which also introduced methods to synthesize robust PSAs. In particular, Surrel [35] pointed out graphically the importance of the complex zeroes and their behavior near them, but lead the behavior outside the zerocs neighbors largely ignored. This was highlighted and remedied by Servin et. al. [37] by using the FTF of the complex quadrature filters used in phase-shifting interferometry.

4.4.2 General Theory of PSI

The phase estimation by phase stepping is achieved by convolving in the time-axis a digital temporal quadrature-filter with the temporal interferograms for each pixel of the interferogram image. Here we will only consider PSAs involving uniformly spaced data with fixed phase-steps values among them (for an analysis of tunable and self-tunable PSAs, see Servin et al., [38]).

Suppose that we have the N phase-stepped interferograms, given by

$$I(x,y,t)=a(x,y)+b(x,y)\cos[\phi(x,y)+\omega_0 t]; \quad t=\{0,1,...,N-1\}. \tag{4.20}$$

As before, $a(x,y)$ represents the background illumination, $b(x,y)$ is the contrast of the interference fringes, and $\phi(x,y)$ is the modulated phase to be determined. The parameter ω_0 is the relative phase-step between successive (temporal) interferograms. Figure 4.5 shows an illustrative example of four phase-shifted fringe patterns.

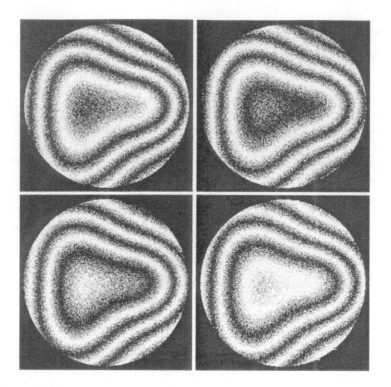

FIGURE 4.5   Four phase-shifted interferograms with a relative phase-step among them given by 2pi/4 radians.

Taking the (temporal) Fourier transform of the space-temporal signal $I(x,y,t)$, in the range $t \in (-\infty, \infty)$ we obtain:

$$I(x,y,\omega) = a(x,y)\delta(\omega) + (1/2)b(x,y)\exp[i\phi(x,y)]\delta(\omega - \omega_0)$$

$$+ (1/2)b(x,y)\exp[-i\phi(x,y)]\delta(\omega + \omega_0), \quad\quad (4.21)$$

where $i = \sqrt{-1}$. In order to isolate one of the spectrally displaced Dirac deltas in the above equation, we need to apply quadrature filter (that is, a one-sided bandpass filter) tuned at the temporal frequency of ω_0 radians/sample. In general, an N-step quadrature linear filter is given by

$$h(t) = \sum_{n=0}^{N-1} c_n \delta(t-n); \quad H(\omega) = F\{h(t)\} = \sum_{n=0}^{N-1} c_n \exp(-i\omega n), \quad\quad (4.22)$$

c_n are complex numbers. These complex coefficients c_n must be chosen in such a way that the FTF, $H(\omega) = F\{h(t)\}$, fulfills at least the following quadrature conditions at ω_0:

$$H(0) = 0, \quad H(-\omega_0) = 0, \quad H(\omega_0) \neq 0. \quad\quad (4.23)$$

Therefore, applying this N-step quadrature linear filter $I(x,y,\omega)$ to results on

$$I(x,y,\omega)H(\omega) = \frac{1}{2}H(\omega_0)b(x,y)\exp[i\phi(x,y)]\delta(\omega - \omega_0). \quad\quad (4.24)$$

In the temporal domain the searched analytic signal is obtained by taking the convolution product $I(x,y,t) * h(t)$. The resulting sequence of $I(x,y,t) * h(t)$ has a temporal support of $2N-1$ samples but we are only interested in the middle term at $t = N-1$, where all available data is overlapped with the filter coefficients. In other words, the most reliable phase estimation is obtained from the analytic signal at $t = N-1$[36], given by

$$A_0(x,y)\exp[i\phi(x,y)] = \sum_{n=0}^{N-1} c_n I(x,y,n), \tag{4.25}$$

where $A_0(x,y) = (1/2)H(w_0)\exp[i(N-1)w_0]b(x,y)$. Finally, the principal value of the estimated phase is computed as the argument of this analytic signal:

$$\phi(x,y) \bmod 2\pi = \arctan\left\{ \frac{\mathrm{Im}[A_0\exp(i\phi)]}{\mathrm{Re}[A_0\exp(i\phi)]} \right\} = \arctan\left\{ \frac{\displaystyle\sum_{n=0}^{N-1} b_n I(x,y,n)}{\displaystyle\sum_{n=0}^{N-1} a_n I(x,y,n)} \right\}, \tag{4.26}$$

where $c_n = a_n + ib_n$. We want to emphasize that these coefficients are complex-valued and we call this the analytic formulation (Equation 4.25) of the PSA. Separating the real and the imaginary parts one obtains the arctangent formulation (Equation 4.26) of the same PSA. These complex numbers c_n fully define the FTF $H(\omega)$. In turn, the FTF allows us to assess the full frequency response of the phase-demodulation formula within the stop-band as well as in the pass-band region of the quadrature filter. From the FTF we can also estimate the PSA's robustness against systematic errors such as detuning, additive random noise degradation, and high-order distorting harmonics, which normally corrupt the interferometric signal.

Detuning is a common systematic error in phase-shifting interferometry. It arises when one applies a PSA tuned at ω_0 but the interferometric data actually has a relative phase-step given by $\omega_0 + \Delta$, where Δ is some unknown (but fixed) value in radians per sample. As demonstrated by Schwider et al. [29], at first-order approximation, this mismatch between the interferometric data and the phase-demodulation formula causes the estimated phase $\varphi(x,y)$ to be distorted by a spurious double-frequency signal into the demodulated phase:

$$\varphi(x,y) = \phi(x,y) + D(\Delta)\sin[2\phi(x,y)], \quad \text{for } |\Delta| \ll 1. \tag{4.27}$$

where $\varphi(x,y)$ is the estimated phase, and $\phi(x,y)$ is the result one would obtain without detuning error. Nowadays we know that the maximum amplitude of the distorting signal, the so-called detuning sensitivity function $D(\Delta)$, depends mainly on the quadrature filter's FTF behavior and the amount of phase-step miscalibration [38]:

$$D(\Delta) = \frac{|H(-\omega_0 - \Delta)|}{|H(\omega_0 + \Delta)|}, \quad \text{for } |\Delta| \ll 1. \tag{4.28}$$

Assuming the phase-shifted interferometric data is corrupted by additive white-Gaussian noise, another important figure of merit that we can evaluate from the FTF (or directly from the PSA's coefficients) is the signal-to-noise (S/N) power ratio gain given by [38]

$$G_{S/N}(\omega_0) = \frac{|H(\omega_0)|^2}{(1/2\pi)\displaystyle\int_{-\pi}^{\pi} H(\omega)H^*(\omega)\,d\omega} = \left|\sum_{n=0}^{N-1} c_n \exp(in\omega_0)\right|^2 \Big/ \sum_{n=0}^{N-1} |c_n|^2. \tag{4.29}$$

For $G_{S/N}(\omega_0) > 1$, the output has a higher S/N ratio than the input data; this is the standard case. If $G_{S/N}(\omega_0) = 1$, the output analytic signal has the same S/N power as the interferogram raw data. Finally, when $G_{S/N}(\omega_0) < 1$, the output has a lower S/N ratio than the input data; this is a highly undesirable situation. For more details, see Servin et al. [37,38]. One may also use the continuous plot of $G_{S/N}(\omega_0)$, $\forall \omega_0 \in (0, \pi)$, to assess the S/N power-ratio for any tuning frequency. This continuous plot $G_{S/N}(\omega_0)$ $\forall \omega_0 \in (0, \pi)$ is called the meta-frequency response of the PSA because it gives us the S/N response of $H(\omega_0)$ for all possible values $\omega_0 \in (0, \pi)$, and it allows us to choose the tuning frequency, which maximizes the S/N power ratio gain.

Whenever the spatio-temporal phase-shifted interferogram is recorded with amplitude distortion (for instance, under gain saturation in the CCD camera), we obtain amplitude distorted interferogram fringes. The non-sinusoidal fringes can be mathematically modeled as

$$I(x,y,t) = a(x,y) + \sum_{n=0}^{\infty} b_n(x,y)\cos\{n[\phi(x,y)+\omega_0 t]\}, \tag{4.30}$$

where $b_n(x,y)$ is the contrast function for the n-th harmonic. Applying a general quadrature filter $h(t)$ to the above equation and taking the (temporal) Fourier transform we have

$$F\{I(x,y,t) * h(t)\} = \sum_{n=-\infty}^{\infty} (b_n/2)\exp(in\phi)H(n\omega_0)\delta(\omega - n\omega_0). \tag{4.31}$$

Since high-quality estimations of the modulating phase require isolating the analytic signal $(b_1/2)H(\omega_0)\exp(i\phi)\delta(\omega - \omega_0)$, the FTF of a robust-to-harmonics quadrature filter must also fulfill

$$H(-n\omega_0) = 0, \quad H(n\omega_0) = 0, \quad \text{for} \quad n = \{2,3,...,k\}, \tag{4.32}$$

assuming that the energy of the k-th harmonic is still significant [38].

To summarize, the most important figures of merit for a linear phase-shifting algorithm can be assessed from its frequency transfer function, which in turn is univocally defined by the complex-valued coefficients $c_n = a_n + ib_n$ of the PSA's analytic formulation.

4.4.3 Analysis of Commonly Used Phase-Shifting Algorithms

In this section we will apply the FTF formalism as assessment tool for some commonly used few-steps PSAs. For the readers' convenience, we will start with the arctangent formulation (most commonly found in the previous literature), from where one can straightforwardly deduce the analytic formulation as well as the spectral response $H(\omega)$. For each PSA to be analyzed, we will plot $|H(\omega)|$ versus ω/ω_0 to assess from a single figure the spectral behavior at the stopbands (all valid PSAs must produce at least first-order spectral zeroes at $\omega = 0$ and at $\omega = -\omega_0$), the amplification value at the passband region (around $\omega = \omega_0$), and the harmonics rejection capabilities (gauged by the spectral zeroes at $\omega/\omega_0 = \pm n$, for $n = \{2,3,...\}$).

The phase-demodulation formulas to be analyzed are the minimum 3-step PSA [33,39], the four-frame PSA [40], and the symmetric five-step PSA with robustness against detuning-error [29,41]. For ease of comparison, we will overlap the continuous plot of $D(\Delta)$ from the Schwider-Hariharan's PSA into the plots corresponding to the three-step and four-step PSAs.

4.4.3.1 Three-Step PSA

Given that in a fringe pattern one normally has three unknowns, three interferograms is the practical minimum to make a PSI measurement:

$$I(x,y,t) = \sum_{n=0}^{2} I(x,y,n\omega_0) = \sum_{n=0}^{2} \left\{ a(x,y) + b(x,y)\cos[\phi(x,y) + \omega_0 t] \right\} \delta(t-n), \tag{4.33}$$

where $\omega_0 \in (0,\pi)$ is the phase-step among the fringe patterns; please note that the endpoints are excluded from the interval. In this equation we have three unknowns: $\{a,b,\phi\}$. Solving algebraically for ϕ one obtains [33,39]:

$$\phi(x,y) \bmod 2\pi = \arctan\left(\frac{1-\cos(\omega_0)}{\sin(\omega_0)} \frac{(I_2 - I_0)}{(I_0 - 2I_1 + I_2)} \right); \tag{4.34}$$

where the spatial dependence was omitted from the irradiances for clarity purposes. By direct comparison with Equation 4.26 we found that $a_n = \{\sin\omega_0, -2\sin\omega_0, \sin\omega_0\}$ and $b_n = \{-1+\cos\omega_0, 0, 1-\cos\omega_0\}$. Taking $c_n = a_n + ib_n$ and substituting in Equation 4.22, after some algebraic manipulation we found the following FTF:

$$H(\omega) = \sum_{n=0}^{2} c_n \exp(-i\omega n) = i[1 - \exp(-i\omega_0)][1 - \exp(-i\omega)]\{1 - \exp[-i(\omega + \omega_0)]\}. \tag{4.35}$$

Clearly, $H(\omega)$ fulfills the fundamental quadrature conditions $H(0) = H(-\omega_0) = 0$, $H(\omega_0) \neq 0$ for all $\omega_0 \in (0,\pi)$. Nevertheless, the maximum S/N power gain is found using the temporal phase-shift of $\omega_0 = 2\pi/3$, for which $G_{S/N}(\omega_0) = 3$ and in turn results in the best harmonics rejection capabilities for a three-step PSA (Servin et al., 2014). Figure 4.6 shows that this three-step PSA fails to reject the $\{...,-5,-2,+4,+7,...\}$ components of the distorting harmonics. Finally, the detuning robustness of this PSA is quite bad due to its first-order spectral zero at $\omega = -\omega_0$. In conclusion, if you are not sure of

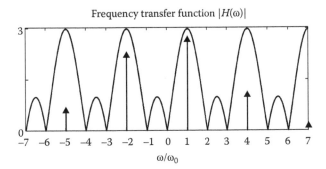
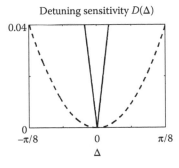

FIGURE 4.6 Frequency response of the minimum three-step PSA for $\omega_0 = 2\pi/3$ radians per sample. The left-side panel shows a normalized-frequency plot of $|H(\omega)|$ where the vertical $\omega/\omega_0 = 1$ arrow at represents the searched analytic signal $A_0 \exp(i\phi)\delta(\omega - \omega_0)$. The continuous line on the right-side panel corresponds to the phase-error due to detuning; the dashed line represents the detuning robustness of the Schwider-Hariharan's PSA included for reference.

knowing the exact phase-step value ω_0 or the linear range of the CCD camera, you must move on to a higher-order and more robust PSA.

4.4.3.2 Four-Step PSA

Another widely used formula for phase stepping (PSA) is the so-called four-frame method (Wyant, 1982). In this case the four recorded interferograms are given by

$$I(x,y,t)=\sum_{n=0}^{3}\left\{a(x,y)+b(x,y)\cos\left[\phi(x,y)+(\pi/2)t\right]\right\}\delta(t-n).\tag{4.36}$$

Solving for the principal value of the modulating phase, we have

$$f(x,y)\bmod 2\pi = \arctan\left(\frac{I_3 - I_1}{I_0 - I_2}\right), \quad (\omega_0 = \pi/2).\tag{4.37}$$

Again, the spatial dependence of the irradiances has been removed for notation clarity. Now, the complex-valued coefficients of the quadrature filter are $c_n = \{1,-i,-1,i\}$. Thus, in analytic form the four-step PSA is given by

$$A_0(x,y)\exp\left[i\phi(x,y)\right]=I_0-iI_1-I_2+iI_3, \quad (\omega_0 = \pi/2).\tag{4.38}$$

To find $H(\omega)$ we need to introduce the numerical values c_n for in Equation 4.22, and after some algebraic manipulation results in

$$H(\omega)=\sum_{n=0}^{3}c_n \exp(-in\omega)=\left[1-\exp(-i\omega)\right]\left\{1-\exp\left[-i(\omega+\pi/2)\right]\right\}\left\{1-\exp\left[-i(\omega+\pi)\right]\right\}.\tag{4.39}$$

This FTF clearly fulfills the fundamental quadrature conditions $H(0)=H(-\omega_0)=0$, $H(\omega_0)\neq 0$ for $\omega_0 = \pi/2$. Moreover, the additional spectral zero at $\omega=\pi$ ensures that both components of the second-order distorting harmonic will be rejected; this is good since in general this distorting harmonic is the one with the highest energy. Figure 4.7 shows that the four-step PSA fails to reject the $\{\ldots,-7,-3,+5,\ldots\}$ components of the distorting harmonics; also the detuning robustness of this PSA is bad due to its first-order spectral zero at $\omega=-\omega_0$. In conclusion, the additional fourth phase-shifted sample improves the harmonics rejection capabilities (respect to the minimum three-step PSA); but the exact phase-step value $\omega_0 = \pi/2$ must be used in the experiment to avoid phase-distortion due to detuning error.

4.4.3.3 Five-Step (4+1) PSA

Another popular formula is the five-step PSA by Schwider et al. [29] and Hariharan et al. [40]. For this demodulation formula, we need five phase-shifted interferograms, given by

$$I(x,y,t)=\sum_{n=0}^{4}\left\{a(x,y)+b(x,y)\cos\left[\phi(x,y)+(\pi/2)t\right]\right\}\delta(t-n).\tag{4.40}$$

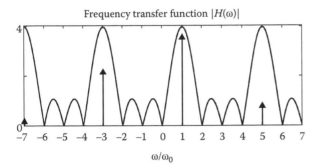
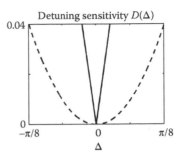

FIGURE 4.7 Frequency response of Wyant's four-step PSA, with $\omega_0 = \pi/2$. The left-side panel shows the normalized-frequency plot of $|H(\omega)|$, where the vertical arrow $\omega/\omega_0 = 1$ at represents the searched analytic signal $A_0 \exp(i\phi)\delta(\omega - \omega_0)$. The continuous line on the right-side panel corresponds to the maximum detuning phase-error; the dashed line represents the detuning robustness of the Schwider-Hariharan's PSA for reference.

The arctangent formula to obtain the modulating phase is

$$\phi(x, y) \bmod 2\pi = \arctan\left(\frac{2I_3 - 2I_1}{I_0 - 2I_2 + I_4} \right), \quad (\omega_0 = \pi/2), \tag{4.41}$$

where the space dependence of the fringe pattern has been omitted for notation simplicity. Proceeding as before we may find its analytic formulation as

$$A_0(x, y)\exp[i\varphi(x, y)] = I_0 - i2I_1 - 2I_2 + i2I_3 + I_4, \quad (\omega_0 = \pi/2), \tag{4.42}$$

and after some algebraic manipulation the FTF is found to be given by

$$H(\omega) = \sum_{n=0}^{4} c_n \exp(-in\omega) = [1 - \exp(-i\omega)]\{1 - \exp[-i(\omega + \pi/2)]\}^2 \{1 - \exp[-i(\omega + \pi)]\}. \tag{4.43}$$

The spectral response of the Schwider-Hariharan's algorithm is shown in Figure 4.8. Its harmonics rejection capability is similar to that of Wyant's four-step PSA: it rejects both components of the second-order distorting harmonic (the one with highest energy) but fails to reject the $\{\ldots, -7, -3, +5, \ldots\}$ components. However, unlike the previously analyzed PSAs, this FTF has a smooth spectral zero at $\omega = -\omega_0$, which provides to this PSA its characteristic robustness against detuning error. In other words, it can tolerate small phase-step miscalibrations (around $\omega_0 = \pi/2$) and still give a reliable phase-estimation.

4.4.4 N-Step Least-Squares Phase-Shifting Algorithm

The N-step least-squares phase-shifting algorithm (LS-PSA) formula was deduced by Bruning et al. [17] following a synchronous demodulation technique for temporal phase-shifted interferometry. Nowadays is arguably the most popular PSA since it has the higher signal-to-noise power ratio gain for any given number of phase-shifted interferograms [28] and it has been demonstrated to provide the best estimation of the searched phase under non-uniform phase-steps [30].

In the arctangent formulation, the N-step LS-PSA is given by

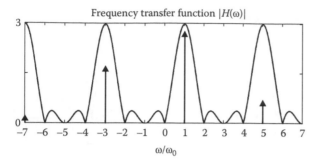

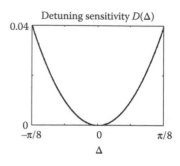

FIGURE 4.8 Frequency response of Schwider-Hariharan's five-step PSA, with $\omega_0 = \pi/2$. The left-side panel shows a normalized-frequency plot of $\left|H(\omega)\right|$, where the vertical arrow at $\omega/\omega_0 = 1$ represents the searched analytic signal $A_0 \exp(i\phi)\delta(\omega - \omega_0)$. The continuous line on the right-side panel corresponds to the maximum phase-error due to detuning.

$$\phi(x,y) \bmod 2\pi = \arctan \frac{\sum_{n=0}^{N-1} \sin(\omega_0 n) I_n(x,y)}{\sum_{n=0}^{N-1} \cos(\omega_0 n) I_n(x,y)}; \quad \omega_0 = \frac{2\pi}{N}\left(\frac{radians}{sample}\right). \tag{4.44}$$

Rewriting the above formula in analytic form, we have

$$A_0(x,y)\exp\left[i\phi(x,y)\right] = \sum_{n=0}^{N-1} \exp(i\omega_0 n) I_n(x,y); \quad \omega_0 = 2\pi/N. \tag{4.45}$$

And according to the FTF formalist, given the complex-valued coefficients $c_n = \exp(i\omega_0 n)$, the FTF for the N-step LS-PSA formula is given by (after some algebraic manipulation)

$$H(\omega) = \sum_{N}^{N-1} \exp\left[-in(\omega - \omega_0)\right] = \prod_{n=0}^{N-2}\left\{1 - \exp\left[-i(\omega + n\omega_0)\right]\right\}; \quad \omega_0 = \frac{2\pi}{N}. \tag{4.46}$$

These $N-1$ spectral zeroes evenly distributed in the FTF (see Figure 4.9) give the N-step LS-PSA formula highest S/N power ratio gain, $G_{S/N}(\omega_0) = N$, and the best harmonics rejection capability for any given number of phase-shifted interferograms [28,30]. On the downside, the N-step LS-PSA formula has no-robustness against detuning error. Also the searched analytic signal $A_0 \exp(i\phi)\delta(\omega - 2\pi/N)$ becomes closer to the baseband as N increases; this may cause an overlap with the background signal at $\omega = 0$ because in real life the bandwidth of their Fourier spectra is not infinitely narrow.

Here ends our review of the general theory for the analysis of PSAs. Next we will discuss the synthesis of new phase-demodulation formulas.

4.4.5 Synthesis of Robust Quadrature Filters for Phase-Shifting Interferometry

Given a number of phase-shifted fringe patterns with a constant phase shift among them, there is an infinite number of quadrature filters that may estimate the desired phase. Many robust PSAs have

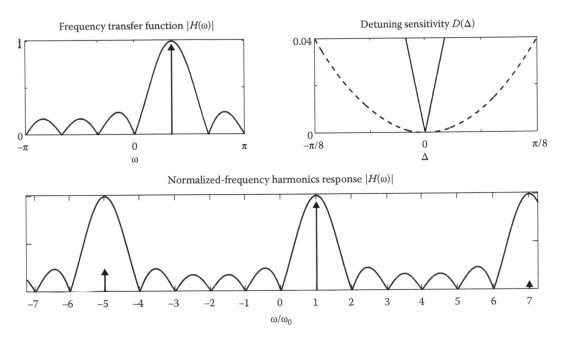

FIGURE 4.9 Frequency response of Bruning's LS-PSA for $N = 6$. The first panel shows five first-order spectral zeroes evenly distributed around $\omega_0 = 2\pi/6$ for $\omega \in [0, \pi)$. The S/N power gain and harmonics rejection capabilities are the best among six-step PSAs: $G_{S/N}(\omega_0) = 6$, and it fails to reject only the $(-5, +7)$ components within the range shown.

been reported over the years, such as the ones made by Morgan [28], Greivenkamp [30], Groot [42], Hibino et al. [43], Schmit and Creath [44,45], and Hibino and Yamaguchi [46], just to mention a few. To find the best PSA for your particular needs (in terms of the number of phase-shifted interferograms, phase-shifting value, detuning robustness, harmonic rejection capability, or noise removal), you may review the specialized literature and test several demodulation formulas until a satisfactory result is found. Or, alternatively, you may design your own PSI quadrature filter.

An efficient design paradigm of PSAs started with the "characteristic polynomial" method by Surrel [35], which is closely related to the Z-transform of a quadrature linear filter. However, we believe that working explicitly in the Fourier domain is a much more natural approach both for the design and the assessment of linear PSAs.

4.4.6 Design of FTFs by Means of First-Order Building Blocks

In the previous section we showed that the figures of merit of quadrature linear filters for phase-demodulation depend on the frequency transfer function $H(\omega)$, which in turn is dominated by the stop-band behavior or the location of its spectral zeroes. Now, consider the following function, to which we will refer as "first-order building block" since its FTF has a single spectral zero within the range $\omega \in [o, \pi)$

$$H_1(\omega + \omega_0) = F\{h_1(t)\} = F\{\delta(t) - \exp(-i\omega_0)\delta(t-1)\} = 1 - \exp[-i(\omega + \omega_0)], \qquad (4.47)$$

where $F\{\cdot\}$ stands for the Fourier transform of its argument and $h_1(t) = \delta(t) - \exp(-i\omega_0)\delta(t-1)$ is the impulse response of the first-order digital system. Using this first-order building block, we are able

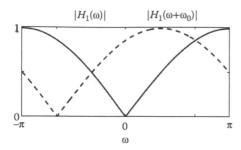 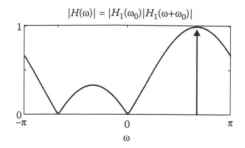

FIGURE 4.10 Building of the FTF for the simplest quadrature linear filter with first-order spectral zeroes at $\omega = \{0, -\omega_0\}$, where $\omega_0 = 2\pi/3$ without loss of generality.

to design the frequency transfer function of N-step phase-demodulation algorithm by locating N-2 spectral zeroes at the sampling frequencies ω_n:

$$H(\omega) = \prod_{n=0}^{N-2} \{1 - \exp[-i(\omega - \omega_n)]\}. \tag{4.48}$$

We need at least two first-order spectral zeroes to fulfill the quadrature conditions (Equation 4.23): $H(0) = H(-\omega_0) = 0$, and $H(\omega_0) \neq 0$. As Figure 4.10 shows, this can be achieved by taking the product which defines the minimum $H_1(\omega)H_1(\omega + \omega_0)$, three-step PSA up to an irrelevant global phase piston (see Equation 4.35). Any extra spectral zero can be used to improve the robustness of the custom PSA against detuning-error, distorting harmonics, additive white-Gaussian noise, and so on. Once the location of the spectral zeroes is set, we expand the product of binomials in Equation 4.48 to express it as a polynomial $\exp(-i\omega)$ of from where the searched coefficients for the PSA quadrature linear filter are directly readable (from Equation 4.22 through 4.25):

$$H(\omega) = \sum_{n=0}^{N-1} c_n \exp(-i\omega n); \quad A_0(x,y) \exp[i\phi(x,y)] = \sum_{n=0}^{N-1} c_n I_n(x,y). \tag{4.49}$$

Finally, for completeness, we can compute the arctangent of the modulating phase:

$$\phi(x,y) \bmod 2\pi = \arctan\left\{ \frac{\mathrm{Im}\left[\sum c_n I_n(x,y)\right]}{\mathrm{Re}\left[\sum c_n I_n(x,y)\right]} \right\} = \arctan\left\{ \frac{\sum b_n I_n(x,y)}{\sum a_n I_n(x,y)} \right\}, \tag{4.50}$$

where $c_n = a_n + ib_n$. We want to emphasize that this last step is actually unnecessary since any modern computer language manages complex algebra without difficulty. In other words, in practice we only need to compute the angle of $\sum c_n I_n(x,y)$ to find $\phi(x,y) \bmod 2\pi$.

Next we will illustrate how this simple approach of collocating the FTF's spectral zeros (and therefore fine shaping the detuning robustness, the signal-to-noise ratio and harmonic rejections) allows us to design robust quadrature linear filters according to the following simple rules:

- Introducing N-1 spectral zeroes in a FTF (considering multiplicity) requires N phase-stepped interferograms.

- The FTF has to fulfill the quadrature conditions $H(0) = H(-\omega_0) = 0$, while $H(\omega_0) \neq 0$. Consequently, $\omega_0 = 0$ and $\omega_0 = \pi$ are not allowed.
- Locating two or more spectral zeroes at the same frequency produces a second-order smoother zero. This allows us to design robust PSAs against detuning error (at expense of reducing the *S/N* power-ratio and the harmonics rejection capabilities respect to LS-PSAs using the same number of phase-steps).
- The FTF has to fulfill $H(-n\omega_0) = H(n\omega_0) = 0$ to reject both components of the *n*th order distorting harmonic. Fortunately, several of these components are automatically rejected whenever $\pm n\omega_0$ is an alias of $\{0, -\omega_0\}$; this may be easily assessed with a normalized-frequency plot $|H(\omega)|$ versus the normalized frequency ω/ω_0.
- Special care must be paid to avoid introducing a spectral zero near (or precisely at) an alias frequency of $\omega = \omega_0$. Since this is not always easy to predict, we strongly suggest to always plot the meta-frequency response $G_{S/N}(\omega_0)$ for $\omega_0 \in (0, \pi)$.

In our opinion, Schwider-Hariharan's PSA has the most convenient spectral response among 5-step PSAs (apart from Bruning's LS-PSA). Therefore, for the following examples we will assume six or more phase-shifted interferograms.

4.4.7 Six-Step PSA with Robustness against Background Rejection and Detuning Error

When designing a robust PSA we are restricted to the number of phase-shifted interferograms available and sometimes to a given phase-step value ω_0. For instance, let us assume that for a given experiment we are able to obtain up to six phase-shifted interferograms with a relative phase-step $\omega_0 \approx \pi/2$, but due to technical constraints we also recorded a slowly varying background signal. Given these conditions we would like to place two spectral zeroes at $\omega = 0$ to cope with the background signal, and additionally two zeroes at $\omega = -\pi/2$ for robustness against detuning; finally a single zero at $\omega = \pi$ to improve the *S/N* power ratio gain and to reject both components of the high-energy second-order distorting harmonic:

$$H(\omega) = H_1^2(0)H_1^2(\omega + \pi/2)H_1(\omega + \pi),$$

$$= [1 - \exp(-i\omega)]^2 \{1 - \exp[-i(\omega + \pi/2)]\}^2 \{1 - \exp[-i(\omega + \pi)]\}, \quad (4.51)$$

$$= 1 - (1 - 2i)e^{-i\omega} - (2 + 2i)e^{-i2\omega} + (2 - 2i)e^{-i3\omega} + (1 + 2i)e^{-i4\omega} - e^{-i5\omega}.$$

From where we obtain the following phase demodulation formulas:

$$A_0(x,y)\exp[i\phi(x,y)] = I_0 - (1 - 2i)I_1 - (2 + 2i)I_2 + (2 - 2i)I_3 + (1 + 2i)I_4 - I_5,$$

$$\phi(x,y) \bmod 2\pi = \arctan\left(\frac{2(I_1 - I_2 - I_3 + I_4)}{I_0 - I_1 - 2I_2 + 2I_3 + I_4 - I_5}\right) \quad (4.52)$$

Figure 4.11 shows the spectral plots for this six-step PSA. Since the robustness against detuning error is guaranteed by design, here we show the meta-frequency response plot $G_{S/N}(\omega_0)$. The S/N power ratio gain equals 4.6 for $\omega_0 = \pi/2$, which is acceptable for six phase-steps. However, its harmonic rejection capability is similar to that of the four-step LS-PSA: failing to reject the $\{\ldots, -7, -3, +5, \ldots\}$ components within the range shown.

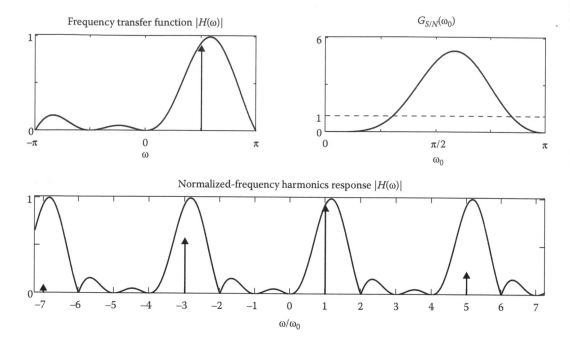

FIGURE 4.11 Spectral response of the custom six-step PSA with second-order spectral zeroes at $\omega = \{0, -\pi/2\}$, and a first-order spectral zero at $\omega = \pi$.

4.4.8 Seven-Step PSA with a Second-Order Spectral Zero at $\omega = -\omega_0$

Since N-step least-squares phase-shifting algorithms (LS-PSAs, $\omega_0 = 2\pi/N$) have the highest S/N power ratio gain $G_{S/N}(\omega_0)$ and the best harmonics rejection capabilities, they are a logical starting point for the design or robust quadrature linear filters.

As we know from Equation 4.46 and Figure 4.9, the FTFs of an N-step LS-PSA contains exactly $N-1$ first-order spectral zeroes equally spaced within the frequency range $\omega \in (-\pi, \pi]$, therefore making LS-PSAs very sensitive to detuning-error. Assuming a total of $N + 1$ phase-shifted interferograms with a phase-step $\omega_0 = 2\pi/N$, we can construct a FTF with robustness against detuning-error by taking the FTF of an N-step LS-PSA (Equation 14.46) and introducing an extra spectral zero at $\omega = -\omega_0$:

$$H(\omega) = \left\{1 - \exp[-i(\omega + \omega_0)]\right\} \times \prod_{n=0}^{N-2}\left\{1 - \exp[-i(\omega + n\omega_0)]\right\}; \qquad (\omega_0 = 2\pi/N)$$

$$= \sum_{n=0}^{N-1}\left\{\exp[-in(\omega - \omega_0)] - \exp[-i(n+1)\omega + i(n-1)\omega_0]\right\}. \tag{4.53}$$

From where we obtain the following phase-demodulation algorithm in analytic formulation:

$$A_0(x,y)\exp[i\phi(x,y)] = \sum_{n=0}^{N-1}\left\{e^{in\omega_0}I_n(x,y) - e^{i(n-1)\omega_0}I_{n+1}(x,y)\right\}, \quad \omega_0 = \frac{2\pi}{N}. \tag{4.54}$$

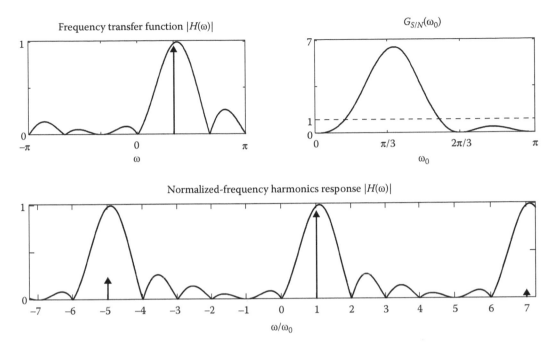

FIGURE 4.12 Spectral response of the custom seven-step (6+1) PSA with a second-order spectral zero at $\omega = -\omega_0$, with $\omega_0 = 2\pi/6$.

For a particular value, say $N = 6$ (thus having seven phase-shifted interferograms), we have

$$A_0(x,y)\exp[i\phi(x,y)] = 2I_0 - i\sqrt{3}I_1 - \left(3+i\sqrt{3}\right)I_2 + i\sqrt{3}I_4 + \left(3+i\sqrt{3}\right)I_5 + \left(1-i\sqrt{3}\right)I_6,$$

$$\phi(x,y)\bmod 2\pi = \arctan\left[\frac{\sqrt{3}\left(-2I_1 - I_2 + I_3 + 2I_4 + I_5 - I_6\right)}{2I_0 - 3I_2 - 3I_3 + 3I_5 + I_6}\right]. \tag{4.55}$$

The corresponding spectral plots are shown in Figure 4.12. Once again, the robustness against detuning-error of this PSA is guaranteed by design. Note that the S/N power ratio gain is between those of the six-step and seven-step LS-PSA, $G_{S/N}(2\pi/6) = 6.4$, and its harmonics rejection capability is identical to that of the six-step LS-PSA: failing to reject only the $\{-5,+7\}$ components within the range shown.

4.4.9 Eight-Step Broadband PSA Tuned at $\omega_0 = \pi/2$

As last example of custom PSAs consider the following FTF with seven spectral zeroes (considering multiplicity), three of which are located at $\omega = -\pi/2$:

$$H(\omega) = \{1 - \exp(i\omega)\}\{1 - \exp[-i(\omega + \pi/2)]\}^3\{1 - \exp[-i(\omega + \pi)]\}$$

$$\times\{1 - \exp[-i(\omega + \pi/6)]\}\{1 - \exp[-i(\omega + 5\pi/6)]\}. \tag{4.56}$$

Proceeding as before it is easy to obtain (omitting an irrelevant amplitude factor):

$$A_0(x,y)\exp\left[i\phi(x,y)\right] = I_0 - i4I_1 - 8I_2 + i11I_3 + 11I_4 - i8I_5 - 4I_6 + iI_7,$$

$$\phi(x,y) \bmod 2\pi = \arctan\left(\frac{-4I_1 + 11I_3 - 8I_5 + I_7}{I_0 - 8I_2 + 11I_4 - 4I_6}\right); \qquad \omega_0 \in \left(\frac{\pi}{4}, \frac{3\pi}{4}\right). \tag{4.57}$$

At this point, the reader should be able to deduce rejection of both components of the second-order distorting harmonic (thanks to the spectral zero at $\omega = \pi$) and a very robust behavior against detuning-error (due to the third-order spectral zero at $\omega = -\pi/2$). Moreover, as Figure 4.13 shows, the additional spectral zeroes at $\omega = \{-\pi/6, -5\pi/6\}$ contribute to obtain a broad stop-band for this FTF. From the meta-frequency response plot we found that $G_{S/N}(\omega_0) > 1$ for $\omega_0 \in (\pi/4, 3\pi/4)$. This means that we can apply this PSA and obtain a reliable estimation of $\phi(x,y)$ even without knowing the actual phase-step value. As for the harmonics rejection capability, the highest harmonic rejection is found for $\omega_0 = \pi/2$, where it behaves like the four-step LS-PSA. Finally, we must remark that the arctangent formula in Equation 4.57 is not new: it was introduced by Schmit and Creath [45] following a completely different approach, but they restricted its applicability to its optimum value $\omega_0 = \pi/2$ instead of a continuous range.

Here ends our discussion about synthesis of robust phase-shifting algorithms. Equations 4.51 through 4.58 illustrate some custom FTFs that we found interesting, but clearly the possibilities are virtually unlimited. One thing that we need to take into account when designing a custom PSA is that optimizing the FTF for a specific feature involves a compromise of another figure of merit

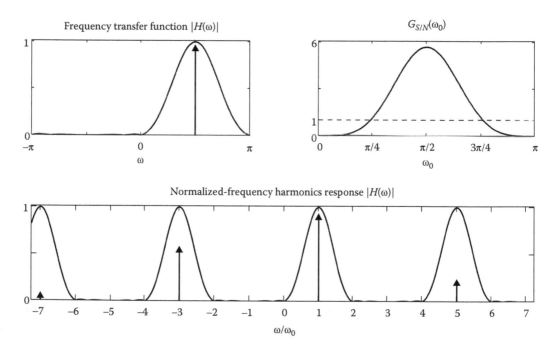

FIGURE 4.13 Spectral response of the broadband eight-step PSA by Schmit and Creath with an applicability range $G_{S/N}(\omega_0) > 1$, $\forall \omega_0 \in (\pi/4, 3\pi/4)$. Its best S/N ratio is reached for $\omega_0 = \pi/2$, where $G_{S/N}(\pi/2) = 5.7$ and its harmonics rejection capability is identical to those of the four-step LS-PSA: failing to reject the $\{-7, -3, 5\}$ components within the range shown.

(for instance, detuning robustness decreases the *S/N* power ratio gain). Regularizing methods can also be used when several phase-shifting interferograms are available [47,48].

4.5 Spatial Phase-Measuring Methods

As seen before, PSI uses a sequence of phase-shifted interferograms to find the searched phase. The phase-estimating system is a one-dimensional quadrature filter tuned at the fundamental time frequency of the signal. Sometimes we cannot have several phase-shifted interferograms. In such cases one needs to deal with only one interferogram with either closed or open fringes [17,49–52]. Open fringes may be obtained by introducing a large tilt in the reference beam of a two-path interferometer or by projecting a linear ruling in profilometry. Closed-fringes interferograms are difficult to demodulate due to the non-monotonic variation of the phase field within the fringe pattern. An interferogram with open-fringes (having a spatial linear carrier) is much easier to demodulate than an interferogram with closed-fringes.

An open-fringes interferogram can always be written as

$$I(x,y) = a(x,y) + b(x,y)\cos[\phi(x,y) + \omega_0 x].\tag{4.58}$$

The carrier frequency must be higher than the maximum frequency content of the phase in the carrier direction, that is,

$$\omega_0 > \left[\frac{\partial\phi(x,y)}{\partial x}\right]_{max}.\tag{4.59}$$

This condition ensures that the total phase of the interferogram will grow (or decrease) monotonically, so the slope of the total phase will always be positive (or negative). Figure 4.14 shows an open-fringes interferogram phase-modulated by spherical aberration and defocusing.

4.5.1 Fourier-Transform Method for Open-Fringes Interferograms

To estimate the modulating phase of open-fringes interferograms one may use frequency domain techniques [14]. The so-called Fourier method is based in a band-pass quadrature filtering in the frequency domain and it works as follows: rewrite the carrier frequency interferogram given above as

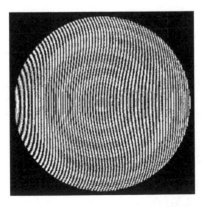

FIGURE 4.14 Computer simulation of a typical open-fringes interferogram.

$$I(x, y) = a(x, y) + c(x, y)\exp(i\omega_0 x) + c^*(x, y)\exp(-i\omega_0 x), \tag{4.60}$$

where

$$c(x, y) = \frac{1}{2}b(x, y)\exp\left[i\phi(x, y)\right], \tag{4.61}$$

and * denotes the complex conjugate. Then the Fourier transform of $I(x, y)$ is given by

$$F\{I(x, y)\} = A(\omega_x, \omega_y) + C(\omega_x + \omega_0, \omega_y) + C^*(\omega_x - \omega_0, \omega_y). \tag{4.62}$$

Using a quadrature (one-sided) band-pass filter, one keeps only one of the two $C(.)$ terms of the frequency spectrum; therefore,

$$C(\omega_x + \omega_0, \omega_y) = H(\omega_x + \omega_0, \omega_y)F\{I(x, y)\} \tag{4.63}$$

where $H(\omega_x + \omega_0, \omega_y)$ represents the FTF of a quadrature filter centered at $-\omega_0$ and with a bandwidth large enough to contain the spectrum of $C(\omega_x + \omega_0, \omega_y)$. Then, the following step is either to translate the information peak towards the origin (to remove the carrier frequency ω_0) to obtain $C(\omega_x, \omega_y)$, or to take directly the inverse Fourier transform of the filtered signal; so that one gets alternatively:

$$F^{-1}\{C(\omega_x, \omega_y)\} = \frac{1}{2}b(x, y)\exp\left[i\phi(x, y)\right],$$

$$F^{-1}\{C(\omega_x - \omega_0, \omega_y)\} = \frac{1}{2}b(x, y)\exp\left[i(\phi(x, y) + \omega_0 x)\right]. \tag{4.64}$$

And their respective wrapped phases are given by

$$\phi(x, y)\bmod 2\pi = \arctan\left(\frac{\mathrm{Im}[c(x, y)]}{\mathrm{Re}[c(x, y)]}\right),$$

$$[\phi(x, y) + \omega_0 x]\bmod 2\pi = \arctan\left(\frac{\mathrm{Im}[c(x, y)]\exp(-i\omega_0 x)}{\mathrm{Re}[c(x, y)]\exp(-i\omega_0 x)}\right). \tag{4.65}$$

In the second case, the estimated phase $\phi(x, y)$ is usually obtained after a plane-fitting procedure performed on the total phase function. Of course, the last step in either case is to process the estimated phase, wrapped because of the arctangent function involved, to obtain a continuous *unwrapped* phase. Figure 4.15a shows the Fourier spectrum of the open-fringes interferogram in Figure 4.14. The estimated phase resulting from keeping only one of the two spectral lobes is shown in Figure 4.15b. We can see how the recovered phase has some distortion at the boundary of the pupil. This phase distortion at the edge of the fringe pattern may be reduced by apodizing the fringe pattern intensity using, for example, a Hamming window [53].

While using the Fourier-transform method, which involves the global but point-wise operation on the interferogram spectrum, one has to be aware of the main sources of errors [13,54–57]:

- The errors associated with the use of fast Fourier transform (FFT): aliasing if the sampling frequency is too low; the picket fence effect, if the analyzed frequency includes a frequency that is not one of the discrete frequencies; and, the most significant error, leakage of the energy

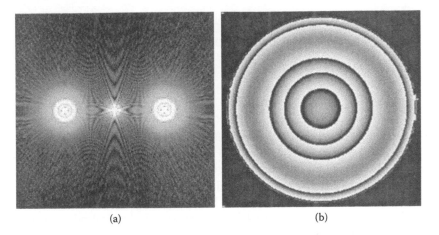

(a)   (b)

FIGURE 4.15   Phase demodulation using Takeda's Fourier technique. Panel (a) shows the Fourier spectrum of the interferogram shown in Figure 4.14. Panel (b) shows the wrapped phase estimated with quadrature filtering and translating the information towards the origin.

from one frequency into adjacent ones, due to fringe discontinuity or inappropriate truncation of the data.

- The errors due to incorrect filtering in the Fourier space, especially if a nonlinear recording of the fringe pattern has occurred.
- The influence of random noise and spurious fringes in the interferogram.

Here we refer to the most significant errors. The leakage of the energy can be reduced significantly by apodizing the fringe pattern intensity using Haming, Hanning, bell, or $\cos^4$ windows. The errors due to incorrect filtering are sometimes difficult to avoid, especially if the assumption of $[\partial\phi/\partial r]_{max} \ll \omega_0$ is not fulfilled and minimizing the filtering window is required due to noise in the image. Bone et al.[54] have shown that with an optimum filter window, the errors due to noise are approximately equal to the errors from information components lost from the filter window.

The Fourier transform method can also be modified by a technique proposed by Kreis [58]. He transformed an interferogram without a spatial frequency carrier added (but with a certain linear phase term intrinsic to the data) and obtained a complex analytic signal by applying a filter function that covers nearly a half plane of the Fourier space, which gives us the possibility of evaluating more complex fringe patterns.

4.5.2 Spatial Carrier Phase-Shifting Method

The spatial carrier phase-shifting method (SCPI) is based on the use of the same phase-stepping quadrature filters used in temporal PSI but in the space domain. So the most simple quadrature filter to use is the three-step filter [18]. This filter along the x-direction looks like

$$h(x) = \sin\omega_0\left[2\delta(x)-\delta(x-1)-\delta(x+1)\right]+i(1-\cos\omega_0)\left[\delta(x-1)-\delta(x+1)\right]. \tag{4.66}$$

When this filter is convolved with an open-fringes interferogram, given by

$$I(x,y) = a(x,y)+b(x,y)\cos[\phi(x,y)+\omega_0 x], \tag{4.67}$$

one obtains the searched analytic signal:

$$A_0(x,y)\exp[i\phi(x,y)] = I(x,y)*h(x). \tag{4.68}$$

From where we solve for the phase of interest as

$$\phi(x,y)\bmod 2\pi = \arctan\left[\frac{1-\cos(\omega_0)}{\sin(\omega_0)}\frac{I(x-1,y)-I(x+1,y)}{2I(x,y)-I(x-1,y)-I(x+1,y)}\right]. \tag{4.69}$$

But, as we mentioned before, the three-step PSI filter has the disadvantage of being too sensitive to detuning. The detuning weakness of the three-step algorithm is evident when dealing with wide-band carrier-frequency interferograms. For that reason, one may need to use a more robust PSI algorithm (such as the 5-step Schwider-Hariharan's). The main inconvenience of using a larger-size convolution filter is that the first two or three pixels inside the boundary of the fringe pattern are not going to be phase estimated. One possible solution is to stick with three samples but, instead of assuming a constant phase over three consecutive pixels, we will make a correction due to the instantaneous frequency variation within the three sample window. This was made by [59] and later, independently, by Servin and Cuevas [16]. The first step is to filter the interferogram $I(x,y)$ with a high-pass filter in order to eliminate the DC term $a(x,y)$. Now consider the following three consecutive pixels of the high-pass filtered fringe pattern $I'(x,y)$:

$$I'(x-1,y) = b(x-1,y)\cos\big[\omega_0(x-1)+\phi(x-1,y)\big],$$

$$I'(x,y) = b(x,y)\cos\big[\omega_0 x+\phi(x,y)\big],$$

$$I'(x+1,y) = b(x-1,y)\cos\big[\omega_0(x+1)+\phi(x+1,y)\big]$$

$$I'(x,y) = b(x,y)\cos\big[\omega_0 x+\phi(x,y)\big] \tag{4.70}$$

$$I'(x+1,y) = b(x-1,y)\cos\big[\omega_0(x+1)+\phi(x+1,y)\big]$$

Assuming that the modulating function $b(x,y)$ remains constant over three consecutive pixels, and using a first-order approximation of $\phi(x,y)$ around the pixel at (x,y) along the x-axis, we obtain

$$I'(x-1,y) = b(x,y)\cos\left[\omega_0+\phi(x,y)-\omega_0-\frac{\partial\phi(x,y)}{\partial x}\right],$$

$$I'(x,y) = b(x,y)\cos[\omega_0 x+\phi(x,y)], \tag{4.71}$$

$$I'(x+1,y) = b(x,y)\cos\left[\omega_0 x+\phi(x,y)+\omega_0+\frac{\partial\phi(x,y)}{\partial x}\right].$$

In these equations we have three unknowns—namely, $b(x,y)$, $\phi(x,y)$, and $\partial\phi(x,y)/\partial x$—so we may solve for $\phi(x,y)$ as

$$\tan[\omega_0 x+\phi(x,y)] = \left(\frac{I'(x-1,y)-I'(x+1,y)}{\mathrm{sgn}[I'(x,y)]\sqrt{[2I'(x,y)]^2-[I'(x-1,y)]+I'[(x+1,y)]^2}}\right), \tag{4.72}$$

where the function sgn[.] takes the sign of its argument. For a more detailed discussion of this method, see Servin and Cuevas [16] or, in general, the n-point phase-shifting technique, as explained by Schmit and Creath [60] and Küchel [61]. A similar approach, with the assumption of constancy of the first derivative of phase within the convolution filter, is also used to the five-point algorithm [62], together with the concept of the two-directional spatial-carrier phase-shifting method [63]. This last approach allows us to analyze multiplexed information coded into a fringe pattern.

4.5.3 Synchronous Demodulation of Open-Fringes Interferograms

This synchronous demodulation method for open-fringes interferograms was introduced in digital form by Womack [64]. This method is equivalent to the Fourier method but in this case all operations are performed in the spatial domain: consider as usual an open-fringes interferogram given by

$$I(x,y)=a+b\cos\left[\phi(x,y)+\omega_0 x\right] \tag{4.73}$$

The dependence on the spatial coordinates of $a(x, y)$ and $b(x, y)$ will be omitted in this section for notation clavarity. This fringe pattern is now multiplied by the sine and cosine of the carrier phase as follows:

$$
\begin{aligned}
g_r(x,y) &= I(x,y)\ \cos(\omega_0 x) = a\cos(\omega_0 x)\ + b\cos(\omega_0 x)\ \cos\left[\omega_0 x+\phi(x,y)\right], \\
g_i(x,y) &= I(x,y)\ \sin(\omega_0 x) = a\sin(\omega_0 x)\ + b\sin(\omega_0 x)\ \cos\left[\omega_0 x+\phi(x,y)\right]
\end{aligned}
\tag{4.74}
$$

This may be rewritten as

$$
\begin{aligned}
g_r(x,y) &= a\cos(\omega_0 x)+\frac{b}{2}\cos[2\omega_0 x+\phi(x,y)]+\frac{b}{2}\cos\left[\phi(x,y)\right], \\
g_i(x,y) &= a\cos(\omega_0 x)+\frac{b}{2}\cos[2\omega_0 x+\phi(x,y)]-\frac{b}{2}\sin\left[\phi(x,y)\right].
\end{aligned}
\tag{4.75}
$$

To obtain the searched phase $\phi(x, y)$, we have to low-pass filter the signals $g_r(x,y)$ and $g_i(x,y)$ to eliminate the two first high-frequency terms. Finally to find the searched phase, we need to compute their ratio as

$$-\phi(x,y)\bmod 2\pi = \arctan\left(\frac{g_i(x,y)**h(x,y)}{g_r(x,y)**h(x,y)}\right)=\arctan\left(\frac{-(b/2)\sin[\phi(x,y)]}{(b/2)\cos[\phi(x,y)]}\right), \tag{4.76}$$

where $h(x, y)$ is an averaging mask for low-pass convolution filtering.

4.5.4 Robust Quadrature Filters

A robust quadrature bandpass filter may be obtained by simply shifting in the frequency domain the regularizing potentials seen in Section 4.3.3 to the carrier frequency ω_{0x} of the fringe pattern [25]. That is

$$U[f(x,y)]=\sum_{(x,y)\in S}\sum\{[f(x,y)-2g(x,y)]^2+\eta R_1[f(x,y)]\}, \tag{4.77}$$

in which the first-order regularizer is now

$$R_1\left[\phi(x,y)\right] = \left[\phi(x,y) - \phi(x-1,y)exp(-\omega_{0x}x)\right]^2 + \left[\phi(x,y) - \phi(x,y-1)\right]^2, \tag{4.78}$$

where we have shifted the first-order regularizer in the x-direction. The minimizer of this cost function given the observation field $2g(x, y)$ is a quadrature bandpass filter. To see this, let us find the frequency response of the filter that minimizes the above cost function; consider an infinite two-dimensional lattice. Setting the gradient of $U[f(x, y)]$ to zero, one obtains the following set of linear equations:

$$0 = f(x,y) - 2g(x,y) \, \eta\left[-f(x-1,y)\exp(i\omega_{0x}) + 2f(x,y) - f(x+1,y)\exp(i\omega_{0x})\right]$$
$$+\eta\left[-f(x,y-1) + 2f(x,y) - f(x,y+1)\right], \tag{4.79}$$

and taking the discrete Fourier transform of this equation, one obtains

$$\eta\left[1 + 2\eta\left[2 - cos\left(\omega_x - \omega_{0x}\right) - cos\left(\omega_y\right)\right]F(\omega) = G(\omega), \tag{4.80}$$

which leads to the following transfer function $H(\omega)$:

$$H(\omega_x,\omega_y) = \frac{F(\omega_x,\omega_y)}{G(\omega_x,\omega_y)} = \frac{1}{1+2\eta[2-\cos(\omega_x-\omega_{0x})-\cos(\omega_y)]}, \tag{4.81}$$

which is a bandpass quadrature filter centered at the frequency ($\omega_x = \omega_{0x}$, $\omega_y = 0$) with a bandwidth controlled by the parameter η. As this frequency response shows the form of the filter, it is exactly the same as the membrane low-pass filter studied in Section 4.3.3 but moved in the frequency domain to the coordinates (ω_{0x}, 0). So this filter may be used for estimating the phase of a carrier frequency interferogram.

An even better signal-to-noise ratio and edge-effect immunity may be obtained if one lets the tuning frequency vary in the two-dimensional space ($\omega_{0x} = \omega_x(x, y)$); that is,

$$R_f\left[f(x,y)\right] = \left[f(x,y) - f(x-1,y)exp(-i\omega_x(x,y))\right]^2 + \left[f(x,y) - f(x,y-1)\right]^2 \tag{4.82}$$

Using this regularizer, one obtains an adaptive quadrature filter [26]. In this case, one must optimize not only for the filtered field $f(x, y)$ but also for the two-dimensional frequency field $\omega_x(x, y)$. Additionally, if we want the estimated frequency field $\omega_x(x, y)$ to be smooth, one needs also to use a regularizer for this field: for example, a first-order regularizer,

$$R_{\omega x}\left[\omega_x(x,y)\right] = \left[\omega_x(x,y) - \omega_x(x-1,y)\right]^2 + \left[\omega_x(x,y) - \omega_x(x,y-1)\right]^2 \tag{4.83}$$

The final cost function will have the following form:

$$U[f(x,y)] = \sum_{(x,y)\in S}\sum \{[f(x,y) - 2g(x,y)]^2 + \eta_1 R_f[f(x,y)] + \eta_2 R_{\omega x}[\omega_x(x,y)]\}. \tag{4.84}$$

Unfortunately, this cost function contains a nonlinear quadrature term (the $R_f[\omega_x(x, y)]$ term), so the use of fast convergence techniques, such as conjugate gradient or transformed methods, are precluded. One needs then to optimize this cost function following simple gradient search or Newtonian descent [26]. Finally, we may also optimize for the frequency in the y-direction $\omega_y(x, y)$. By estimating $\omega_x(x, y)$ and $\omega_y(x, y)$ altogether, it is possible to demodulate a single fringe pattern containing closed fringes [26].

4.5.5 The Regularized Phase-Tracking Technique

The regularized phase-tracking (RPT) technique may be applied to almost every aspect of the fringe pattern processing. The RPT evolved from an early phase-locked loop (PLL) technique that was applied for the first time to fringe processing by Servin et al. [15,65]. In the RPT [22] technique, one assumes that locally the phase of the fringe pattern may be considered as spatially monochromatic, so its irradiance may be modeled as a sinusoidal function phase modulated by a plane $p(.)$. Additionally, this phase plane $p(.)$ located at (x, y) must be close to the phase values $\phi_0(\xi, \eta)$ already detected in the neighborhood of the site (x, y).

Specifically, the proposed cost function to be minimized by the estimated phase $\phi_0(x, y)$ at each site (x, y), is

$$U(x,y) = \sum_{(\xi,\eta) \in (\Omega \cap S)} \{ I'(\xi, \eta) - \cos p(x,y,\xi,\eta)]^2 + \tau [\phi_0(\xi,\eta) - p(x,y,\xi,\eta)]^2 m(\xi,\eta) \} \tag{4.85}$$

and

$$p(x,y,\xi,\eta) = \phi_0(x,y) + (x - \xi)\omega_x(x,y) + (y - \eta)\omega_y(x,y), \tag{4.86}$$

where S is a two-dimensional lattice having valid fringe data (good amplitude modulation); Ω is a neighborhood region around the coordinate (x, y) where the phase is being estimated $m(x, y)$; is an indicator field, which equals one if the site (x, y) has already been phase estimated, and zero otherwise. The fringe pattern $I'(\in, \eta)$ is the high-pass filtered and amplitude normalized version of $I(x, y)$; this operation is performed in order to eliminate the low-frequency background $a(x, y)$ and to approximate $b(x, y) \approx 1.0$. The functions $\omega_x(x, y)$ and $\omega_y(x, y)$ are the estimated local frequencies along the x- and y-directions, respectively. Finally, τ is the regularizing parameter that controls (along with the size of Ω) the smoothness of the detected phase.

The first term in Equation 4.85 attempts to keep the local fringe model close to the observed irradiance in a least-squares sense within the neighborhood Ω. The second term enforces the assumption of smoothness and continuity using only previously detected pixels $\phi_0(x, y)$ marked by $m(x, y)$. To demodulate a given fringe pattern we need to find the minimum of the cost function $U(x, y)$ with respect to the fields $\phi_0(x, y)$, $\omega_x(x, y)$, and $\omega_y(x, y)$. This may be achieved using the algorithm described in the next paragraph.

The first phase estimation on S is performed as follows: to start, the indicator function $m(x, y)$ is set to zero ($m(x, y) = 0$ in S). Then, one chooses a seed or starting point (x_0, y_0) inside S to begin the demodulation of the fringe pattern. The function $U(x_0, y_0)$ is then optimized with respect to $\phi_0(x_0, y_0)$, $\omega_x(x_0, y_0)$, $\omega_y(x_0, y_0)$; the visited site is marked as detected, that is, we set $m(x_0, y_0) = 1$. Once the seed pixel is demodulated, the sequential phase demodulation proceeds as follows:

1. Choose the (x, y) pixel inside S (randomly or with a prescribed scanning order).
2. If $m(x, y) = 1$, return to the first statement.
3. If $m(x, y) = 0$, then test if $m(x', y') = 1$ for any adjacent pixel (x', y').
4. If no adjacent pixel has already been estimated, return to the first statement.
5. If $m(x', y') = 1$ for an adjacent pixel, take $[\phi_0(x', y'), \omega_x(x', y'), \omega_y(x', y')]$ as initial condition to minimize $U(x, y)$ with respect to $[\phi_0(x, y), \omega_x(x, y), \omega_y(x, y)]$.
6. Set $m(x, y) = 1$.
7. Return to the first statement until all the pixels in S are estimated.

An intuitive way of considering the first iteration just presented is as a crystal growing (CG) process where new molecules are added to the bulk in that particular orientation, which minimizes the local crystal energy given the geometrical orientation of the adjacent and previously positioned molecules. This demodulating strategy is capable of estimating the phase within any fringe pattern's boundary.

To optimize $U(x, y)$ at site (x, y) with respect to $(\phi_0, \omega_x, \omega_y)$, we may use simple gradient descent:

$$\phi_0^{k+1}(x, y) = \phi_0^k(x, y) - \mu \frac{\partial U(x, y)}{\partial \phi_0(x, y)},$$

$$\omega_x^{k+1}(x, y) = \omega_x^k(x, y) - \mu \frac{\partial U(x, y)}{\partial \omega_x(x, y)}, \qquad (4.87)$$

$$\omega_y^{k+1}(x, y) = \omega_y^k(x, y) - \mu \frac{\partial U(x, y)}{\partial \omega_y(x, y)},$$

where μ is the step size and k is the iteration number. Only one or two iterations are normally needed (except for the demodulation of the starting seed point, which may take about 20 iterations); this is because the initial conditions for the gradient search are taken from a neighborhood pixel already estimated. In this way the initial conditions are already very close to the stable point of the gradient search. It is important to remark that the two-dimensional RPT technique gives the estimated phase $\phi_0(x, y)$ already unwrapped, so no additional phase-unwrapping process is required.

The first global phase estimation in S (using the gradient search along with the CG algorithm) is usually very close to the actual modulating phase; if needed, one may perform additional global iterations to improve the phase-estimation process. Additional iterations may be performed using, again, Equation 4.87, but now taking as initial conditions the last estimated values at the same site (x, y) (not the ones at a neighborhood site as done in the first global CG iteration). Note that for the additional iterations, the indicator function $m(x, y)$ in $U(x, y)$ is now everywhere equal to one; therefore, one may scan S in any desired order whenever all the sites are visited at each global iteration. In practice, only three or four additional global iterations are needed to reach a stable minimum of $U(x, y)$ at each site (x, y) in S.

Figure 4.16 shows the result of applying the RPT technique to the fringe pattern shown in Figure 4.14. From Figure 4.16 we can see that the estimated phase at the borders is given accurately.

As mentioned previously, the RPT technique may also be used to demodulate closed-fringes interferograms. Figure 4.17a shows a closed-fringes interferogram and Figure 4.17b shows its estimated phase using the RPT technique. Some additional modifications [22] are needed to the RPT to make it more robust to noise when dealing with closed-fringes interferograms.

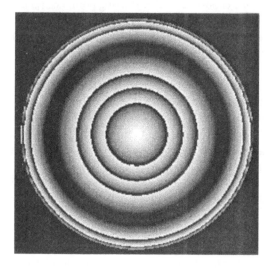

FIGURE 4.16 Estimated phase using the phase-tracking technique (RPT) applied to the carrier frequency interferogram shown in Figure 4.14. As we can see, the interferogram's edge was properly estimated.

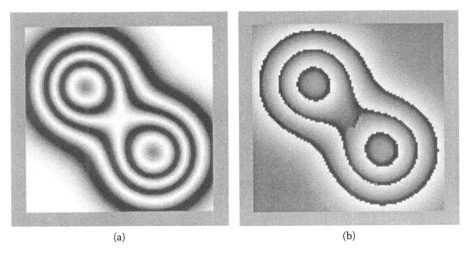

(a) (b)

FIGURE 4.17 Illustrative example closed-fringes demodulation from a single interferogram using the RPT technique. Panel (a) shows the closed-fringes and panel (b) shows the estimated phase.

4.6 Phase Unwrapping

Except for the RPT technique, all other interferometric methods discussed in this chapter so far produce a wrapped phase estimation (the principal value or modulo 2π of the actual phase) due to the arctangent function involved in the phase-estimation process. The relationship between the wrapped phase and the unwrapped phase may be stated as

$$\phi_w(x,y) = \phi(x,y) + 2\pi k(x,y), \tag{4.88}$$

where $\phi_w(x, y)$ is the wrapped phase, $\phi(x, y)$ is the unwrapped phase, and $k(x, y)$ is an integer-valued correcting field. The unwrapping problem is trivial for phase maps calculated from good-quality fringe data; in such phase maps, the absolute phase difference between consecutive phase samples in both the horizontal and vertical directions is less than π, except for the expected 2π discontinuities. Unwrapping is therefore a simple matter of adding or subtracting 2π offsets at each discontinuity (greater than π radians) encountered in the phase data or integrating wrapped phase differences [19,20,66–70].

Unwrapping becomes more difficult when the absolute phase difference between adjacent pixels at points other than discontinuities in the arctan function is greater than π. These unexpected discontinuities may be introduced, for example, by high-frequency, high-amplitude noise, discontinuous phase jumps and regional undersampling in the fringe pattern, or a real physical discontinuity of the domain [38,71].

4.6.1 Unwrapping Using Least-Squares Integration of Gradient Phase

The least-squares technique was first introduced by Ghiglia et al. [72] to unwrap inconsistent phase maps. To apply this method, start by estimating the wrapped phase gradient along the x- and y-direction, that is,

$$\phi_y(x,y) = W\big[\phi_w(x,y) - \phi_w(x,y-1)\big]$$
$$\phi_x(x,y) = W\big[\phi_w(x,y) - \phi_w(x-1,y)\big] \tag{4.89}$$

having an oversampled phase map with moderately low noise, the phase differences in Equation 4.89 will be everywhere in the range $(-\pi, \pi)$. In other words, the estimated gradient will be unwrapped.

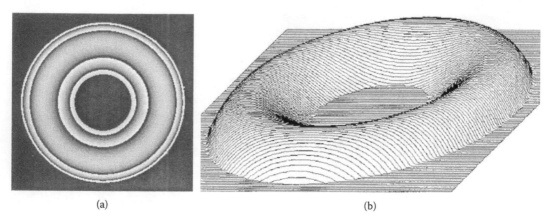

(a)   (b)

FIGURE 4.18   Phase unwrapping. (a) Wrapped phase. (b) Unwrapped phase estimated using path-independent least-squares integration.

Now we may integrate the phase gradient in a consistent way by means of least-squares integration. The integrated or searched continuous phase will be the one that minimizes the following cost function:

$$U[\phi(x,y)] = \sum_{(x,y)\in S} \left\{ \left[\phi(x,y) - \phi(x-1,y) - \phi_x(x,y)\right]^2 + \left[\phi(x,y) - \phi(x,y-1) - \phi_y(x,y)\right]^2 \right\}. \quad (4.90)$$

The estimated unwrapped phase $\phi(x, y)$ may be found, for example, using simple gradient descent, as

$$\phi^{k+1}(x,y) = \phi^k(x,y) - \varepsilon\, \frac{\partial U}{\partial \phi(x,y)}, \quad (4.91)$$

where k is the iteration number and ε is the convergence rate of the gradient search system. There are faster algorithms of obtaining the searched unwrapped phase among the techniques of conjugate gradient or the transform methods [72]. Figure 4.18 shows the resulting unwrapped phase applying the least-squares technique to a noiseless phase map.

We may also include regularizing potentials to the least-squares unwrapper in order to smooth out some phase noise and possibly interpolate over regions of missing data with a predefined behavior [73]. When the phase map is too noisy, the fundamental basis of the least-squares integration technique may be broken; that is, the wrapped phase difference may no longer be a good estimator of the gradient field due to a high amplitude noise. In this severe noise situation, the wrapped phase difference among neighborhood pixels is no longer less than π. As a consequence, a reduction of the dynamic range of the resulting unwrapped phase is obtained [38,73].

4.6.2 Unwrapping Using the Regularized Phase-Tracking Technique

The main motivation to apply the RPT method to phase unwrapping [73] is its superior robustness to noise with respect to the least-squares integration technique. The RPT technique as the least-squares integration of phase gradient is also robust to the edge effect at the boundary of the phase map.

The first step to unwrap a given phase map using the RPT technique is to put the wrapped phase into two-phase orthogonal fringe patterns. These fringe patterns may be obtained using the cosine and the sine of the map phase being unwrapped; that is,

$$I_C(x,y) = \cos[\phi_w(x,y)],$$
$$I_S(x,y) = \sin[\phi_w(x,y)],$$

(4.92)

where $\phi_w(x,y)$ is the phase map being unwrapped. Now the problem of phase unwrapping may be treated as a demodulation of two phase-shifted fringe patterns using the RPT technique [73]. Therefore, the cost function to be minimized by the unwrapped phase $\phi_0(x,y)$ at each site (x,y) is

$$U(x,y) = \sum_{(\varepsilon,\eta)\varepsilon(N_{x,y}\cap S)} \{[I_C(\varepsilon,\eta) - \cos p(x,y,\varepsilon,\eta)]^2 + [I_S(\varepsilon,\eta) - \sin p(x,y,\varepsilon,\eta)]^2$$

$$+\lambda[\phi_0(\varepsilon,\eta) - p(x,y,\varepsilon,\eta)]^2 m(\varepsilon,\eta)\}$$

(4.93)

and

$$p(x,y,\varepsilon,\eta) = \phi_0(x,y) + \omega_x(x,y)(x-\epsilon) + \omega_y(x,y)(y-\eta),$$

(4.94)

where S is a two-dimensional lattice having valid fringe data (good amplitude modulation); $N_{x,y}$ is a neighborhood region around the coordinate (x,y) where the phase is being unwrapped; $m(x,y)$ is an indicator field that equals 1 if the site (x,y) has already been unwrapped, and 0 otherwise. The functions $\omega_x(x,y)$ and $\omega_y(x,y)$ are the estimated local frequencies along the x- and y-directions respectively. Finally, λ is the regularizing parameter that controls (along with the size of $N_{x,y}$) the smoothness of the detected-unwrapped phase.

The algorithm to optimize this cost function is the same as the one described in the RPT Section 4.5.5, so we are not going into the details here.

4.6.3 Temporal Phase Unwrapping

This phase unwrapping technique was introduced by Huntley and Saldner in 1993 [74] and it has been applied in optical metrology to measure deformation. The basic idea of this technique is to take several interferograms as the object is deformed; therefore, the number of deformation fringes within that object will grow due to the increasing applied force. If one wants to analyze each fringe pattern using the PSI technique, one needs to take at least three phase-shifted interferograms for each object's deformation to obtain its corresponding phase map. The sampling theorem must be fulfilled in the temporal space for every pixel in the fringe pattern; that is, consecutive pixels should have a phase difference less than π in the time domain. The main advantage of this method is that the unwrapping of each pixel is an independent process from the unwrapping process of any other pixel in the temporal sequence of phase maps at hand.

As mentioned before, for each object's deformation the modulating phase of the object is estimated using the PSI technique. This gives us a temporal sequence of wrapped phases. The sequence of phase maps may be represented by

$$\phi_w(1,x,y), \phi_w(2,x,y), \phi_w(3,x,y), \ldots, \phi_w(N,x,y), (x,y) \in S$$

(4.95)

where S is the region of valid phase data and N is the total number of intermediate-phase maps. In order to fulfill the sampling theorems, the following condition must be fulfilled:

$$\left| W\left[\phi_w(i+1,x,y) - \phi_w(i,x,y)\right]\right| < \pi, \quad (1 \leq i \leq N-1), \quad (x,y) \in S \tag{4.96}$$

and $W[.]$ is the wrapping operator. The unwrapping process then proceeds according to

$$\phi(x,y) = \sum_{i=1}^{N-1} W\left[\phi_w(i+1,x,y) - \phi_w(i,x,y)\right]. \tag{4.97}$$

The distribution given by $\phi(x,y)$ gives the unwrapped phase difference between the initial state $\phi_w(1,x,y)$ and the final state $\phi_w(N,x,y)$.

Figure 4.19 shows a sequence of phase maps that may be unwrapped using the temporal phase unwrapping technique herein described. Each wrapped phase has less than half a fringe among them, as required

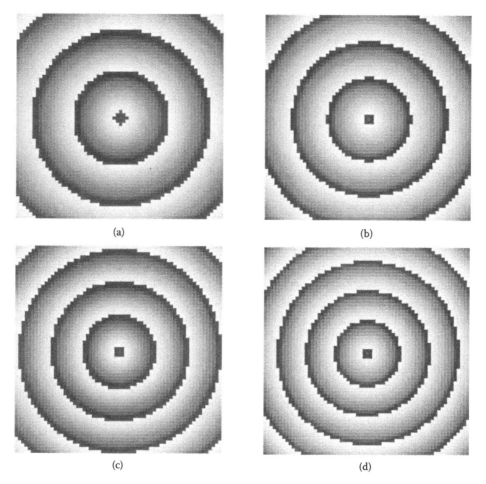

(a)

(b)

(c)

(d)

FIGURE 4.19 A sequence of wrapped phases suitable for being unwrapped by the method of temporal phase unwrapping. The unwrapped phase will be the difference between the first wrapped phase in panel (a) and the one in panel (d).

by this technique. It must be pointed out that the unwrapped phase is going to be equal to the phase difference between the phase map shown in Figure 4.19a and the phase map shown in Figure 4.19d.

4.7 Extended Range Fringe Pattern Analysis

Extended range interferometry allows us to measure larger numbers of aspheric wavefronts than standard two-arms interferometers. The reason for this extended range is that using these techniques enables one to directly measure the gradient or curvature of the wavefront instead of the wavefront itself. If the wavefront being measured is smooth, then one needs fewer image pixels to represent its spatial variations. As we will see later, an exception to this is the sub-Nyquist interferometry, where the wavefront is taken in a direct way, with the disadvantage of needing a special-purpose CCD video array.

4.7.1 Phase Retrieval from Gradient Measurement Using Screen-Testing Methods

The screen-testing methods are used to detect the gradient of the wavefront under analysis. One normally uses a screen with holes or strips lying perpendicular to the propagation direction of the testing wavefront. Then, one collects the irradiance pattern produced by the shadow or the self-image (whenever possible) of the testing screen at some distance from it. If the testing wavefront is aberrated, then the shadow or self-image of the screen will be distorted with respect to the original screen. The phase difference between the screen's shadow and the screen is related to the gradient of the aberrated wavefront at the screen plane. Thus, to obtain the shape of the testing wavefront, one must use an integration procedure. The integration procedure that we employ is the least-squares solution, which has the advantage of being path-independent and robust to noise.

The two most used screens to test the wavefront aberration are the Ronchi ruling [75] and the Hartmann testing plate [76]. The Ronchi ruling, which is a linear grating, is inserted in a place perpendicular to the average direction of propagation of the wavefront being tested. The other frequently used screen is the Hartmann plate, which is a screen with a two-dimensional array of small circular holes evenly spaced in the screen. A linear grating, such as a Ronchi ruling, is only sensitive to the ray aberration perpendicular to the ruling strips. This means that, in general, we will need two shadow Ronchi images to fully determine the aberration of the wavefront. In contrast, only one Hartmann testing screen is needed to collect all the data regarding the wavefront aberration. Unfortunately, a Hartmanngram (the irradiance shadow of the Hartmann screen at the testing plane) is more difficult to analyze than a Ronchigram. That is because a Ronchigram may be analyzed using robust and well-known carrier frequency interferometry.

The main advantage of using screen tests along with a CCD camera is to increase the measuring dynamic range of the tested wavefront; that is, sensing the gradient of the testing wavefront instead of the wavefront itself allows us to increase the number of aberration waves that can be tested. This is why screen tests are the most popular way of testing large optics (such as telescopes' primary mirrors). Thus, for a given number of pixels of a CCD camera, it is possible to measure more waves of aberration using screen tests than using a standard interferometer, which measures the aberration waves directly.

4.7.2 Wavefront Slope Analysis with Linear Gratings (Ronchi Test)

As mentioned earlier, a linear grating is easier to analyze using standard carrier fringe detecting procedures, such as the Fourier method, the synchronous method, or the spatial phase shifting (SPSI) method. These techniques have already been discussed above. The Ronchi test has been a widely used technique and has been reported by several researchers in metrology [77–80].

We may start with a simplified mathematical model for the transmittance of a linear grating (Ronchi rulings are normally made of binary transmittance):

$$T_x(x,y) = \frac{[1+\cos(\omega_0 x)]}{2}, \tag{4.98}$$

where ω_0 is the angular spatial frequency of the Ronchi ruling. The linear ruling is then placed at the plane where the aberrated wavefront is being estimated. If a light detector is placed at a distance d from the Ronchi plate then, as a result of wavefront aberrations, we obtain a distorted irradiance pattern that will be given, approximately, by

$$I_x(x,y) = \frac{1}{2} + \frac{1}{2}\cos\left(\omega_0 x + \omega_0 d \frac{\partial W(x,y)}{\partial y}\right), \tag{4.99}$$

where $I_x(x,y)$ is the distorted shadow of the transmittance $T_x(x,y)$, and $W(x,y)$ represents the wavefront under test. As the above equation shows, it is necessary to detect two orthogonal shadow patterns to completely describe the gradient field of the wavefront under test. The other linear ruling, located at the same testing plane but with its strip lines oriented in the y-direction, is

$$T_y(x,y) = \frac{[1+\cos(\omega_0 y)]}{2}. \tag{4.100}$$

Thus, the distorted image of the Ronchi ruling at the collecting data plane is given by

$$I_y(x,y) = \frac{1}{2} + \frac{1}{2}\cos\left(\omega_0 y + \omega_0 d \frac{\partial W(x,y)}{\partial x}\right). \tag{4.101}$$

We may use any of the carrier fringe methods described in this chapter to demodulate these two Ronchigrams. Figure 4.20a shows a Ronchi ruling and Figure 4.20b shows the same Ronchi ruling modulated by a wavefront containing aspheric aberration.

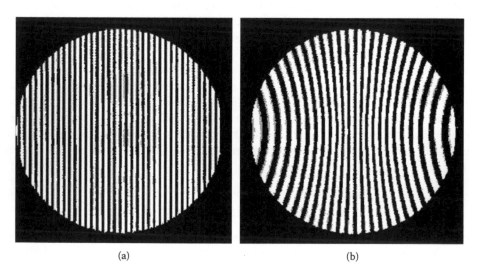

(a)　　　　　　　　　　　　(b)

FIGURE 4.20　(a) Unmodulated Ronchi ruling. (b) Modulated Ronchi ruling.

Once the detected and unwrapped phase of the ruling's shadows has been obtained, one needs to integrate the resulting gradient field. To integrate this phase gradient, one may use path-independent integration, such as least-squares integration. The least-squares integration of the gradient field may be stated as the function that minimizes the following quadratic cost function:

$$U(\hat{W}) = \sum_{(x,y)\in S} \left[\check{W}(x+1,y) - \hat{W}(x,y) - \frac{\partial W(x,y)}{\partial x} \right]^2 + \sum_{(x,y)\in S} \left[\hat{W}(x,y+1) - \hat{W}(x,y) - \frac{\partial W(x,y)}{\partial y} \right]^2, \qquad (4.102)$$

where the hat function $\hat{W}(x,y)$ is the estimated wavefront, and we have approximated the derivative of the searched phase along the x- and y-axis as first-order differences of the estimated wavefront. The least-squares estimator may then be obtained from $U(x,y)$ by simple gradient descent as

$$\hat{W}^{k+1}(x,y) = \hat{W}^{k}(x,y) - \tau \frac{\partial U(\hat{W})}{\partial \hat{W}(x,y)}, \qquad (4.103)$$

or using a faster algorithm, such as conjugate gradient or transform methods [72,82–88].

4.7.3 Moiré Deflectometry

We may increase the sensitivity of the Ronchi test by placing the collecting data plane at the first self-image of the linear ruling. The first Talbot self-image for a collimated light beam appears at the so-called Rayleigh distance L_R, given by

$$L_R = \frac{2d^2}{\lambda}. \qquad (4.104)$$

The resulting deflectograms may be analyzed in the same way as the one described for the Ronchigrams.

4.7.4 Wavefront Analysis with Lateral Shearing Interferometry

Lateral shearing interferometry consists in obtaining a fringe by constructing an interfering pattern using two lateral displaced copies of the wavefront under analysis [89–94]. The mathematical form of the irradiance of a lateral sheared fringe pattern may be written as

$$I_x(x,y) = \frac{1}{2} + \frac{1}{2}\cos\left\{ \frac{2\pi}{\lambda}[W(x-\delta x,y) - W(x+\delta x,y)] \right\},$$

$$I_x(x,y) = \frac{1}{2} + \frac{1}{2}\cos\left[\frac{2\pi}{\lambda}\Delta_x W(x,y) \right], \qquad (4.105)$$

where δx is half of the total lateral displacement. As the Equation 4.105 shows, one also needs the orthogonally displaced shearogram to describe the wavefront under analysis completely. The orthogonal shearogram in the y-direction may be written as

$$I_x(x,y) = \frac{1}{2} + \frac{1}{2}\cos\left\{ \frac{2\pi}{\lambda}[W(x,y-\delta y) - W(x,y+\delta y)] \right\},$$

$$I_x(x,y) = \frac{1}{2} + \frac{1}{2}\cos\left[\frac{2\pi}{\lambda}\Delta_y W(x,y) \right]. \qquad (4.106)$$

These fringe patterns may be transformed into carrier frequency interferograms by introducing a large and known amount of defocusing to the testing wavefront [95]. Having linear carrier fringe patterns, one may proceed to their demodulation using standard techniques of fringe carrier analysis, as seen in this chapter. A shearing interferogram is shown in Figure 4.21. This shearing interferogram corresponds to a defocused wavefront having a circular pupil and, as we can see from Figure 4.21, interference fringes are only present in the common area of the two copies of the laterally displaced wavefront.

We may analyze in the frequency domain the modulating phase of a sheared interferogram (in the x-direction, for example) as the output of a linear filter:

$$F\big[\Delta_x W(x)\big] = F\big[W(x-\delta x) - W(x+\delta x)\big]$$
$$= 2i\sin(\delta x \omega_x) F\big[W(x)\big]. \qquad (4.107)$$

As can be seen, the transfer function of the shearing operator in the frequency domain is a sinusoidal-shaped filter. As a consequence, the inverse filter (the one needed to obtain the searched wavefront) has poles in the frequency domain, so its use is not straightforward. Instead of using a transformed method to recover the wavefront from the sheared data, we feel it is easier to use a regularization approach in the space domain.

Assume that we have already estimated and unwrapped the interesting phase differences $\Delta_x W(x, y)$ and $\Delta_y W(x, y)$. Using this information, the least-squares wavefront reconstruction may be stated as the minimizer of the following cost function:

$$U(\hat{W}) = \sum_{(x,y)\in S_x} \left\{ \hat{W}(x-\delta x, y) - \hat{W}(x+\delta x, y) - \Delta_x W(x, y) \right\}^2$$
$$+ \sum_{(x,y)\in S_y} \left\{ \hat{W}(x, y-\delta y) - \hat{W}(x, y+\delta y) - \Delta_y W(x, y) \right\}^2 \qquad (4.108)$$

$$U(\hat{W}) = \sum_{(x,y)\in S_x} U_x(x, y)^2 + \sum_{(x,y)\in S_y} U_y(x, y)^2,$$

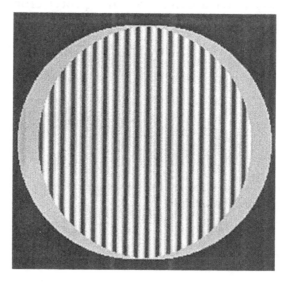

FIGURE 4.21 Lateral shearing of a circular pupil containing a defocused wavefront. Interference fringes form only at the superposition of both sheared pupils.

where the "hat" function represents the estimated wavefront and S_x and S_y are two-dimensional lattices containing valid phase data in the x- and y-shearing directions (the common area of the two laterally displaced pupils). Unfortunately, the least-squares cost function stated above is not well posed, because the matrix that results from setting the gradient of U equal to 0 is not invertible (as seen previously, the inverse filter may contain poles in the frequency range of interest). Fortunately, we may regularize this inverse problem and find the expected smooth solution of the problem [96]. As seen before, the regularizer may consist of a linear combination of squared magnitude of differences of the estimated wavefront within the domain of interest. In particular, one may use a second-order or thin-plate regularizer:

$$R_x(x,y) = \hat{W}(x-1,y) - 2\hat{W}(x,y) + \hat{W}(x+1,y),$$
$$R_y(x,y) = \hat{W}(x,y-1) - 2\hat{W}(x,y) + \hat{W}(x,y+1)$$

(4.109)

Therefore, the regularized cost function becomes

$$U(\hat{W}) = \sum_{(x,y)\in S_x} U_x(x,y)^2 + \sum_{(x,y)\in S_y} U_y(x,y)^2 + \eta \sum_{(x,y)\in \text{Pupil}} [R_x(x,y)^2 + R_y(x,y)^2],$$

(4.110)

where Pupil refers to the two-dimensional lattice inside the pupil of the wavefront being tested. The regularizing potentials discourage large changes in the estimated wavefront among neighboring pixels. As a consequence, the searched solution will be relatively smooth. The parameter η controls the amount of smoothness of the estimated wavefront. It should be remarked that the use of regularizing potentials in this case is a must (even for noise-free observations) to yield a stable solution of the least-squares integration for lateral displacements greater than two pixels, as analyzed by Servin et al. [96].

The estimated wavefront may be calculated using simple gradient descent as

$$\hat{W}^{k+1}(x,y) = \hat{W}^{k+1}(x,y) - \tau \frac{\partial U(\hat{W})}{\partial \hat{W}(x,y)},$$

(4.111)

where τ is the convergence rate. This optimizing method is not very fast. One normally uses faster algorithms such as conjugate gradient.

4.7.5 Wavefront Analysis with Hartmann Screens

The Hartmann test is a well-known technique for testing large optical components [76,94]. The Hartmann technique samples the wavefront under analysis using a screen of uniformly spaced holes situated at the pupil plane. The Hartmann screen may be expressed as

$$\text{HS}(x,y) = \sum_{n=-N/2}^{N/2} \sum_{m=-N/2}^{N/2} h(x-pn, y-pm),$$

(4.112)

where $\text{HS}(x,y)$ is the Hartmann screen and $h(x,y)$ are the small holes that are uniformly spaced in the Hartmann screen. Finally, p is the space among the holes of the screen. A typical Hartmann screen may be seen in Figure 4.22, where the two-dimensional arrangement of holes is shown. The measuring wavefront must pass through these holes, and their shadow is recorded at a distance d from it. If we have wavefront aberrations higher than defocusing, the Hartmann screen's shadow will be geometrically distorted.

The collimated rays of light that pass through the screen holes are then captured by a photographic plate at some distance d from it. The uniformly spaced array of holes at the instrument's pupil is then

FIGURE 4.22   A typical Hartmann screen used in the Hartmann test.

distorted at the photographic plane by the aspherical aberrations of the wavefront under test. The screen deformations are then proportional to the slope of the aspheric wavefront; that is,

$$H(x,y) = \left[\sum_{(n,m)=-N/2}^{N/2} h'\left(x - pn - d\frac{\partial W(x,y)}{\partial x}, y - pm - d\frac{\partial W(x,y)}{\partial x} \right) \right] P(x,y), \qquad (4.113)$$

where $H(x, y)$ is the Hartmanngram (the irradiance of the screen's shadow) obtained at a distance of d from the Hartmann screen. The function $h'(x, y)$ is the image of the screen's hole $h(x, y)$ as projected at the Hartmanngram plane. Finally, $P(x, y)$ is the pupil of the wavefront being tested. As seen in Equation 4.113, only one Hartmanngram is needed to fully estimate the wavefront's gradient. The frequency content of the estimated wavefront will be limited by the sampling theorem to the hole's period p of the screen. A typical Hartmanngram may be seen in Figure 4.23. This Hartmanngram corresponds to an aspheric wavefront having a strong spherical aberration component.

Traditionally, these Hartmanngrams (the distorted image of the screen at the observation plate's plane) are analyzed by measuring the centroid of the spots' images $h'(x, y)$ generated by the screen holes $h(x, y)$. The deviations of these centroids from their uniformly spaced positions (unaberrated positions) are recorded. These deviations are proportional to the aberration's slope. The centroids' coordinates give a two-dimensional discrete field of the wavefront gradient that needs integration and interpolation over regions without data. Integration of the wavefront's gradient field is normally done by using the trapezoidal rule [76]. The trapezoidal rule is carried out following several independent integration paths and their outcomes averaged. In this way, one may approach a path-independent integration. Using this integration procedure, the wavefront is only known at the hole's position. Finally, a polynomial or spline wavefront fitting is necessary to estimate the wavefront's values at places other than the discrete points where the gradient data is collected. A two-dimensional polynomial for the wavefront's gradient may be proposed. This polynomial is then fitted by least squares to the slope data; it must contain every possible type of wavefront aberration, otherwise some unexpected features (especially at the edges) of the wavefront may be filtered out. On the other hand, if one uses a high-degree of polynomial (in order to ensure not to filter out any wavefront aberration), the estimated

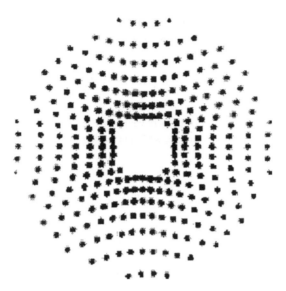

FIGURE 4.23 A distorted image of the Hartmann screen as seen inside the paraxial focus of a paraboloid under test.

continuous wavefront may wildly oscillate in regions where no data are collected. Robust quadratic filters have been used to demodulate Hartmanngrams [97]. Also the regularized phase tracker (RPT) has been used to demodulate the gradient information of the Hartmanngram. Using the RPT technique one is able to estimate the gradient field not only at the hole's positions but continuously over the whole pupil of the Hartmanngram [98].

4.7.6 Wavefront Analysis by Curvature Sensing

Teague [99], Streibl [100], and Roddier [100] have analyzed and demonstrated phase retrieval using the irradiance transport equation. Assuming a paraxial beam propagating along the z-axis, we may obtain the irradiance transport equation as [99–100]:

$$\frac{\partial I(x,y,z)}{\partial z} = -\nabla I(x,y,z) \cdot \nabla W(x,y,z) - I(x,y,z)\nabla^2 W(x,y,z) \tag{4.114}$$

where $I(x, y, z)$ is the distribution of the illumination along the propagating beam, $W(x, y, z)$ is the wavefront surface at distance z from the origin, and ∇ is the $(\partial/\partial x, \partial/\partial y)$ operator. In the analysis of wavefronts using the transport equation, there is no need for a codifying screen pupil, as in the case of the Ronchi or Hartmann test.

Following an interesting interpretation of the irradiance transport equation given by Ichikawa et al. [101], one may note in the transport equation the following interpretation for each term:

- The first term $\nabla I \cdot \nabla W$ may be seen as the irradiance variation caused by a transverse shift of the inhomogeneous ($\nabla I \neq 0$) beam due to the local tilt of the wavefront whose normal ray direction is given by ∇W; this may be called a prism term.
- The second term $I\nabla^2 W$ may be interpreted as the irradiance variation caused by convergence or divergence of the beam whose local focal length is inversely proportional to $\nabla^2 W$; this may be called a "lens term."

Thus, the sum expresses the variation of the beam irradiance caused by the prism and lens effect as it propagates along the z-axis. Rewriting the transport equation as

$$-\frac{\partial I(x,y,z)}{\partial z} = \nabla \cdot \left[I(x,y,z)\nabla W(x,y,z)\right]$$

(4.115)

and remarking that ∇W is the direction of the ray vector, we can easily see that the transport equation represents the law of light energy conservation, which is analogous to the law of mass or charge conservation, frequently expressed by

$$\frac{\partial \rho}{\partial t} = \text{div}(\rho \upsilon),$$

(4.116)

with ρ and v being the mass or charge density and the flow velocity, respectively.

The technique proposed by Roddier [102] to use the transport equation in wavefront estimation is as follows. Let $P(x, y)$ be the transmittance of the pupil. That is, $P(x, y)$ equals 1 inside the pupil and 0 outside. Furthermore we may assume that the illumination at the pupil's plane is uniform and equal to I_0 inside $P(x, y)$. Hence ∇I equals zero everywhere except at the pupil's edge, where it has the value

$$\nabla I = -In\delta_c$$

(4.117)

Here δ_c is a Dirac distribution around the pupil's edge and $\mathbf{n}$ is the unit vector perpendicular to the edge and pointing outward. Substituting the irradiance Laplacian in $P(x, y)$ into the irradiance transport equation, one obtains

$$\frac{\partial I(x,y,z)}{\partial z} = -I_0 \cdot -\frac{\partial W(x,y,z)}{\partial n}\delta_c - I_0 P(x,y)\nabla^2 W(x,y,z)$$

(4.118)

where $\partial W/\partial n = \mathbf{n}\cdot\nabla W$ is the wavefront derivative in the outward direction perpendicular to the pupil's edge. Curvature sensing consists in taking the difference between the illumination observed in two close planes separated a distance $\pm\Delta z$ from the reference plane where the pupil $P(x, y)$ is located. Then we obtain the following two measurements as

$$I_1 = I_0 + \frac{\partial I}{\partial z}\Delta z$$

$$I_2 = I_0 - \frac{\partial I}{\partial z}\Delta z.$$

(4.119)

Having these data, one may form the so-called sensor signal as

$$S = \frac{I_1 - I_2}{I + I_2} = \frac{1}{I_0}\frac{\partial I}{\partial z}\Delta z.$$

(4.120)

Substituting this into Equation 4.112 yields

$$S = \left\{ \frac{\partial W(x,y)}{\partial n} \delta_c - P(x,y)\nabla^2 W(x,y) \right\} \Delta z. \tag{4.121}$$

Solving this differential equation, one is able to estimate the wavefront inside the pupil $P(x,y)$, knowing both the Laplacian of $W(x,y)$ inside $P(x,y)$ and $\partial W/\partial n$ along the pupil's edge as Neumann boundary conditions.

4.7.7 Sub-Nyquist Analysis

Testing of aspheric wavefronts is nowadays routinely achieved in the optical shop by the use of commercial interferometers. The testing of deep aspheres is limited by the aberrations of the interferometer's imaging optics as well as the spatial resolution of the digital camera used to gather the interferometric data. The maximum recordable frequency over the image sensor of the recording camera is π radians per pixel; this is called the Nyquist limit of the sampling system. The detected phase map of an interferogram having frequencies higher than the Nyquist limit is said to be aliased and cannot be unwrapped using standard techniques, such as the ones presented so far.

The main prior knowledge that is going to be used by us is that the expected wavefront is smoothness [33,103]. Then, one may introduce this prior knowledge into the unwrapping process. The main requirement to apply sub-Nyquist techniques is to have a CCD camera with detectors much smaller than the spatial separation among them. Another alternative is to use a mask with small holes over the CCD array to reduce the light-sensitive area of the CCD pixels. This requirement allows us to have a strong signal even for thin interferogram fringes. To obtain the undersampled phase map, one may use any well-known PSI techniques using phase-shifted undersampled interferograms.

The undersampled interferogram may be imaged directly over the CCD video array with the aid of an optical interferometer, as seen in Chapter 1. If the CCD sampling rate is Δx over the x-direction, and Δy over the y-direction, and the diameter of the light-sensitive area of the CCD is d, we may write the mathematical expression for the sampling operation over the interferograms irradiance as

$$S[I(x,y)] = \left[I(x,y) ** \text{circ}\left(\frac{\rho}{d}\right) \right] \text{comb}\left(\frac{x}{\Delta x}, \frac{y}{\Delta y}\right), \quad \rho = (x^2 + y^2)^{1/2}, \tag{4.122}$$

where the function $S[I(x,y)]$ is the sampling operator over the interferogram's irradiance. The symbol ($**$) indicates a two-dimensional convolution. The $\text{circ}(\rho/d)$ function represents the circular size of the CCD detector. The comb function is an array of delta functions with the same spacing as the CCD pixels. The phase map of the subsampled interferogram may be obtained using, for example, three phase-shifted interferograms, as

$$I(x,y,t) = \sum_{n=-1}^{1} a_n(x,y) + b_n(x,y) \cos\left[\frac{2\pi}{\lambda} \phi(x,y) + t\right] \delta(t - n\alpha), \tag{4.123}$$

where the variable α is the amount of phase shift. Using well-known formulae we can find the subsampled wrapped phase by

$$\phi_{w(x,y)} = \tan^{-1}\left(\frac{1 - \cos(\alpha)}{\sin(\alpha)} \frac{S[I_1(x,y)] - S[I_3(x,y)]}{2S[I_1(x,y)] - S[I_2(x,y)] - S[I_3(x,y)]} \right). \tag{4.124}$$

As Equation 4.124 shows, the obtained phase is a modulo-2π of the true undersampled phase due to the arc tangent function involved in the phase-detection process.

Now we may treat the problem of unwrapping undersampled phase maps due to smooth wavefronts: i.e., the only prior knowledge about the wavefront being analyzed is smoothness. This is far less restrictive than the null testing technique presented in the last section. Analysis of interferometric data beyond the Nyquist frequency was first proposed by Greivenkamp [103], who assumed that the wavefront being tested is smooth up to the first or second derivative. Greivenkamp's approach to unwrap subsampled phase maps consists of adding multiples of 2π each time a discontinuity in the phase maps is found. The number of 2π values added is determined by the smoothness condition imposed on the wavefront in its first or second derivative along the unwrapping direction. Although Greivenkamp's approach is robust against noise, its weakness is that it is a path-dependent phase unwrapper.

In this section we present a method [104] that overcomes the path dependency of the Greivenkamp approach while preserving its noise robustness. In this case, an estimation of the local wrapped curvature (or wrapped Laplacian) of the subsampled phase map $\phi_w(x, y)$ is used to unwrap the interesting deep aspheric wavefront. Once having the local wrapped curvature along the x- and y-directions, one may use least-squares integration to obtain the unwrapped continuous wavefront. The local wrapped curvature is obtained as

$$L_x(x,y) = W\left[\phi_w(x-1,y) - 2\phi_w(x,y) + \phi_w(x+1,y)\right],$$

$$L_y(x,y) = W\left[\phi_w(x,y-1) - 2\phi_w(x,y) + \phi_w(x,y+1)\right]$$

(4.125)

If the absolute value of the discrete wrapped Laplacian is less than π, its value will be non-wrapped. Then we may obtain the unwrapped phase $\phi(x, y)$ as the function which minimizes the following quadratic cost function (least squares):

$$U[\phi(x,y)] = \sum_{(x,y)\in S} U_x(x,y)^2 + U_y(x,y)^2,$$

(4.126)

where S is a subset of a two-dimensional regular lattice of nodes having good amplitude modulation. The functions $U_x(x, y)$ and $U_x(x, y)$ are given by

$$U_x(x,y) = L_x(x,y) - \left[\phi(x-1,y) - 2\phi(x,y) + \phi(x+1,y)\right],$$

$$U_y(x,y) = L_y(x,y) - \left[\phi(x,y-1) - 2\phi(x,y) + \phi(x,y+1)\right]$$

(4.127)

The minimum of the cost function is obtained when its partial with respect to $\phi(x, y)$ equals zero. Therefore, the set of linear equations that must be solved is

$$\frac{\partial U[\phi(x,y)]}{\partial \phi(x,y)} = U_x(x-1y) - 2U_x(x,y) + U_x(x+1,y) + U_y(x,y-1)$$

$$- 2U_y(x,y) + U_y(x,y+1).$$

(4.128)

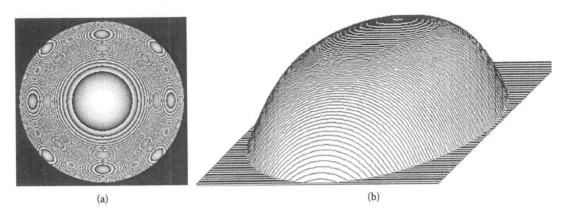

(a)   (b)

FIGURE 4.24 Undersampled phase unwrapping. (a) Undersampled phase map. (b) Unwrapped phase.

Several methods may be used to solve this system of linear equations; among others, there is the simple gradient descent:

$$\phi^{k+1}(x,y) = \phi^k(x,y) - \eta \frac{\partial U}{\partial \phi(x,y)}, \tag{4.129}$$

where the parameter η is the rate of convergence of the gradient search. The simple gradient descent is quite slow for this application; instead, we may conjugate gradient or transformed techniques to speed up the computing time. Figure 4.24 shows a subsampled phase map along with its unwrapped version using the technique herein described.

Finally, if one has a good knowledge of the wavefront being tested up to a few wavelengths, one may use this information to obtain an oversampled phase map [105]. This oversampled phase map is then the estimation error between what is the expected wavefront and the actual wavefront being tested. One also may reduce the number of aspherical aberration wavelengths by introducing a compensating hologram [106,107].

4.8 Applicability of Fringe Analysis Methods

The success in implementation of optical full-field measuring methods into industrial, medical, and commercial areas depends on proper retrieval of a measurand coded in an arbitrary fringe pattern. This is the reason why such variety of techniques exists. Table 4.2 provides a comparison of the fringe pattern analysis methods, and indicates which techniques are most commonly used in commercial systems.

In order to fulfill the conditions of fast, automatic, accurate, and reliable analysis of fringe data, the new solutions of phase-measuring methods focus on the following issues:

- Active approach to fringe-pattern forming.
- Active approach to design of phase analysis algorithms (phase shifting).
- Improving the methods.
- Given that many problems involved with fringe analysis are ill-posed, it is convenient to search for regularizers for the solution according to prior information available.

TABLE 4.2 Comparison of Features of Fringe Pattern (FPs) Analysis Methods[a]

Method	Number of FPs per Frame	Detector Resolution Requirements	Real Time Method for mod (21)	Inherent Image Enhancement	Inherent Phase Interpolation	Automatic Sign Detection	Achievable Accuracy	Experimental Requirements	Complexity of Processing	Dynamic Events Analysis	Commercial Systems
Fringe extreme localization	1/1	R_0[b]	No	No	No	No	Low	Low	Low	No	**
Regularization	1/1	R_0	No	Yes	Yes	Yes	Medium	Low	High	No	*
Temporal heterodyning	+	Single[c] detector and scanning	No	Yes	Yes	Yes	Very high	Very high	Hardware	No	*
Phase shifting:											
• temporal	min 3/3	R_0	Partly	No	No	Yes	High	High	Low	No	****
• spatial	3/1 3/3	$3R_0$	Yes	No	No	Yes	Medium	Medium	Low	Yes	
• carrier frequency	1/1	min $2R_0$	Yes	No	No	Yes	Medium	Low	Low	Yes	***
Fourier transform	1/1	min $2R_0$	Partly	Yes	Yes	Yes	High inside domain low at edges	Low	High	Yes	..
PLL	1/1	min $2R_0$	No	Yes	Yes	Yes	Medium	Low	Medium	No	
Space-domain processing	1/1	min $2R_0$	Yes	Yes	Yes	Yes	High	High	Hardware	Yes	

[a] The choice of the methods most often used is given.

[b] R_0 is detector resolution for the infinite fringe detection mode.

[c] The possibility of the use of an image dissector camera is not considered.

Acknowledgments

Dr. Servin and Dr. Padilla want to acknowledge the financial support of the CONACYT.

References

1. Robinson DW, Reid GT eds., *Interferogram Analysis: Digital Fringe Pattern Measurement Techniques*, Bristol: Institute of Physics Publishing House, 1993.
2. Schwider J. Automated evaluation techniques in interferometry, in *Progress in Optics*, Wolf, E., ed., Vol. 28, Amsterdam, the Netherlands: Elsevier Science Publishers, 1990.
3. Creath K. Phase measuring interferometry: Beware of these errors, *Proc. SPIE*, **1559**, 313–220; 1991.
4. Kujawinska M, Kosinski C. Adaptability: Problem or Solution, in Jüptner W and Osten W, eds., Proceedings of the 3rd International Workshop on Automatic Processing of Fringe Patterns, 419–431 (Akademie-Verlag, Berlin, 1997).
5. Kozlowska A, Kujawinska M. Grating interferometry with a semiconductor light source, *Appl. Opt.*, **36**, 8116–8120; 1997.
6. Efron U. et al. Special issue on the technology and application of SLMs, *Appl. Opt.*, **28**, 4708–4954; 1989.
7. Jones JD. Engineering applications of optical fiber interferometers, *Proc. SPIE*, **2341**, 222–238; 1994.
8. Yamaguchi I, Liu J, Kato J Active phase shifting interferometers for shape and deformation measurements, *Opt. Eng.*, **35**, 2930–2937; 1996.
9. Olszak A, Patorski K Modified electronic speckle interferometer with reduced number of elements for vibration analysis, *Opt. Commun.*, **138**, 265–269; 1997.
10. Joenathan C, Khorana BM. Phase measuring fiber optic electronic speckle pattern interferometer: Phase steps calibration and phase drift minimization, *Opt. Eng.*, **31**, 315–321; 1992.
11. Van der Heijden F. *Image-Based Measurement Systems, Object Recognition and Parameter Estimation*, Chichester: John Wiley & Sons, 1994.
12. Takeda M, Kitoh M. Spatio-temporal frequency-multiplex heterodyne interferometry, *J. Opt. Soc. Am.*, **9**, 1607–1614; 1992.
13. Towers DP, Judge TR, Bryanston-Cross PJ. Automatic interferogram analysis techniques applied to quasy-heterodyne holography and ESPI, *Opt. Lasers Eng.*, **14**, 239–282; 1991.
14. Takeda M, Ina H, Kobayashi S. Fourier transform methods of fringe-pattern analysis for computer-based topography and interferometry, *J. Opt. Soc Am.*, **72**, 156–160; 1982.
15. Servin M, Rodriguez-Vera R. Two dimensional phase locked loop demodulation of carrier frequency interferograms, *J. Mod. Opt.*, **40**, 2087–2094; 1993.
16. Servin M, Cuevas FJ. A novel technique for spatial-phase-shifting interferometry, *J. Mod. Opt.*, **42**, 1853–1862; 1995.
17. Bruning JH, Herriott DR, Gallagher JE, Rosenfel DP, White AD, Brangaccio DJ. Digital wavefront measuring interferometer for testing optical surfaces and lenses, *Appl. Opt.*, **13**, 2693–2703; 1974.
18. Shough DH, Kwon OY, Leavy DF. High-speed interferometric measurements of aerodynamic phenomena, *Proc. SPIE*, **1221**, 394–403; 1990.
19. Huntley JM. Phase Unwrapping—Problems & Approaches, in Proc. FASIG, *Fringe Analysis '94, York University*, 391–393; 1994a.
20. Takeda M. Recent progress in phase-unwrapping techniques, *Proc. SPIE*, **2782**, 334–343; 1996.
21. Osten W, Nadeburn PA. General hierarchical approach in absolute phase measurement, *Proc. SPIE*, **2860**, 2–13; 1996.
22. Servin M, Marroquin JL, Cuevas FJ. Demodulation of a single interferogram by use of a two-dimensional regularized phase-tracking technique, *Appl. Opt.*, **36**, 4540–4548; 1997b.
23. Schreiber W, Notni G, Kühmstedt P, Gerber J, Kowarschik R. Optical 3D measurements of objects with technical surfaces, in *Academic Verlag Series in Optical Metrology*, Füzessy Z, Juptner W, and Osten W, eds., Vol. 2, pp. 46–51; 1996.

24. Marroquin JL. Deterministic interactive particle models for image processing and computer graphics, *Computer and Vision, Graphics and Image Processing*, **55**, 408–417; 1993.

25. Marroquin JL, Servin M, Figueroa JE. Robust quadrature filters, *J. Opt. Soc. Am. A*, **14**, 779–791; 1997a.

26. Marroquin JL, Servin M, Rodriguez-Vera R. Adaptive quadrature filters and the recovery of phase from fringe pattern images, *J. Opt. Soc. Am. A*, **14**, 1742–1753; 1997b.

27. Shagam RN, Wyant JC. Optical frequency shifter for heterodyne interferometers, *Appl. Opt.*, **17**, 3034–3035; 1978.

28. Morgan CJ. Least squares estimation in phase measurement interferometry, *Opt. Lett.*, **7**, 368–370; 1982.

29. Schwider J, Burow R, Elssner KE, Grzanna J, Spolaczyk R, Merkel K. Digital wavefront interferometry: Some systematic error sources, *Appl. Opt.*, **22**, 3421–3432; 1983.

30. Greivenkamp JE. Generalized data reduction for heterodyne interferometry, *Opt. Eng.*, **23**, 350–352; 1984.

31. Ai C, Wyant JC. Effect of piezoelectric transducer non-linearity on phase shift interferometry, *Appl. Opt.*, **26**, 1112–1116; 1987.

32. Asundi A, Yung KH. Phase shifting and local Moiré, *J. Opt. Soc. Am. A*, **8**, 1591–1600; 1991.

33. Greivenkamp JE, Bruning JH. Phase shifting interferometry, in Optical Shop Testing, Malacara, D., ed., New York, NY: John Wiley & Sons, 1992, pp. 501–598.

34. Freischlad K, Koliopoulos CL. Fourier description of digital phase measuring interferometry, *J. Opt. Soc. Am. A*, **7**, 542–551; 1990.

35. Surrel Y. Design of algorithms for phase measurement by use of phase stepping, *Opt. Lett.*, **35**, 51–60; 1996.

36. Servin M, Malacara D, Marroquin JL, Cuevas FJ. Complex linear filters for phase shifting with very low detuning sensitivity, *J. Mod. Opt.*, **44**, 1269–1278; 1997a.

37. Servin M, Estrada JC, Quiroga JA. The general theory of phase shifting algorithms, *Opt. Express*, **17**, 21867–21881; 2009.

38. Servin M, Quiroga JA, Padilla JM. *Fringe Pattern Analysis for Optical Metrology: Theory, Algorithms, and Applications*, Weinheim: Wiley-VCH, 2014.

39. Creath K. Phase-measurement interferometry techniques, in *Progress in Optics XXVI*, Wolf, E., ed., Amsterdam, the Netherlands: Elsevier Science Publishers, 1988.

40. Wyant JC. Interferometric optical metrology: Basic principles and new systems, *Laser Focus*, **5**, 65–71, May 1982.

41. Hariharan P, Areb BF, Eyui T. Digital phase-stepping interferometry: A simple error-compensating phase calculation algorithm, *Appl. Opt.*, **26**, 3899; 1987.

42. Groot PD. Derivation of algorithms for phase shifting interferometry using the concept of a data-sampling windows, *Appl. Opt.*, **34**, 4723–4730; 1995.

43. Hibino K, Oreb F, Farrant DI. Phase shifting for nonsinusoidal waveforms with phase-shifting errors, *J. Opt. Soc. Am. A*, **12**, 761–768; 1995.

44. Schmit J, Creath K. Extended averaging technique for derivation of error compensating algorithms in phase shifting interferometry, *Appl. Opt.*, **34**, 3610–3619; 1995a.

45. Schmit J, Creath K. Window function influence on phase error in phase-shifting algorithms, *Appl. Opt.*, **35**, 5642–5649; 1996.

46. Hibino K, Yamaguchi M. Phase determination algorithms compensating for spatial nonuniform phase modulation in phase shifting interferometry, *Proc. SPIE*, **3478**, 110–120; 1998.

47. Marroquin JL, Servin M, Rodriguez-Vera R. Adaptive quadrature filters for multiphase stepping images, *Opt. Lett.*, **24**, 238–240; 1998.

48. Servin M, Rodriguez-Vera R, Marroquin JL, Malacara D. Phase shifting interferometry using a two dimensional regularized phase-tracking technique, *J. Mod. Opt.*, **45**, 1809–1820; 1998b.

49. Ichioka Y, Inuiya M. Direct phase detecting system, *Appl. Opt.*, **11**, 1507–1514; 1972.

50. Macy W Jr. Two-dimensional fringe pattern analysis, *Appl. Opt.*, **22**, 3898–3901; 1983.

51. Mertz L. Real time fringe pattern analysis, *Appl. Opt.*, **22**, 1535–1539; 1983.

52. Patorski K. *Handbook of the Moiré Fringe Technique*, Amsterdam, the Netherlands: Elsevier, 1993.

53. Malacara D, Servin M, Malacara Z. *Optical Testing: Analysis of Interferograms*, New York, NY: Marcel Dekker, 1998.

54. Bone DJ, Bachor H-A, Sandeman RJ. Fringe pattern analysis using a 2D Fourier transform, *Appl. Opt.*, **25**, 1653–1660; 1986.

55. Roddier C, Roddier F. Interferogram analysis using Fourier transform techniques, *Appl. Opt.*, **26**, 1668–1673; 1987.

56. Kujawinska M, Wojciak J. High accuracy Fourier transfer fringe pattern analysis, *Opt. Lasers Eng.*, **14**, 325–329; 1991a.

57. Kujawinska M, Wojciak J. Spatial carrier phase shifting technique of fringe pattern analysis, *Proc. SPIE*, **1508**, 61–67; 1991b.

58. Kreis T. Digital holographic interference-phase measurement using the Fourier transform method, *J. Opt. Soc. Am. A*, **3**, 847–855; 1986.

59. Ransom PL, Kokal JB. Interferogram analysis by a modified sinusoidal fitting technique, *Appl. Opt.*, **25**, 4199–4205; 1986.

60. Schmit J, Creath K. Fast calculation of phase in spatial n-point phase shifting technique, *Proc. SPIE*, **2544**, 102–111; 1995b.

61. Küchel, MF. Some Progress in Phase Measurement Techniques, in W. Jüptner and W. Osten (eds.), Fringe' 97, Automatic Processing of Fringe Patterns, Bremen, Germany, Sept. 15–17, 1997, Akademie Verlag, Berlin pp. 27–44.

62. Pirga M, Kujawinksa M. Errors in two-directional spatial-carrier phase shifting for closed fringe pattern analysis, *Proc. SPIE*, **2860**, 72–83; 1996.

63. Pirga M, Kujawinksa M. Two dimensional spatial-carrier phase-shifting method for analysis of crossed and closed fringe patterns, *Opt. Eng.*, **34**, 2459–2466; 1995.

64. Womack KH. Interferometric phase measurement using spatial synchronous detection, *Opt. Eng.*, **23**, 391–395; 1984.

65. Servin M, Malacara D, Cuevas FJ. Direct phase detection of modulated Ronchi rulings using a phase locked loop, *Opt. Eng.*, **33**, 1193–1199; 1994.

66. Itoh K. Analysis of the phase unwrapping algorithm, *Appl. Opt.*, **21**, 2470; 1982.

67. Ghiglia DC, Mastin GA, Romero LA. Cellular automata method for phase unwrapping, *J. Opt. Soc. Am.*, **4**, 267–280; 1987.

68. Bone DJ. Fourier fringe analysis: The two dimensional phase unwrapping problem, *Appl. Opt.*, **30**, 3627–3632; 1991.

69. Owner-Petersen M. Phase unwrapping: A comparison of some traditional methods and a presentation of a new approach, *Proc. SPIE*, **1508**, 73–82; 1991.

70. Huntley JM. New methods for unwrapping noisy phase maps, *Proc. SPIE*, **2340**, 110123; 1994b.

71. Ghiglia DC, Pritt MD. *Two-Dimensional Phase Unwrapping: Theory, Algorithms, and Software*, New York, NY: John Wiley & Sons, 1998.

72. Ghiglia DC, Romero LA. Robust two dimensional weighted and unweighted phase unwrapping that uses fast transforms and iterative methods, *J. Opt. Soc. Am. A*, **11**, 107–117; 1994.

73. Marroquin JL, Rivera M. Quadratic regularization functionals for phase unwrapping, *J. Opt. Soc. Am. A*, **12**, 2393–2400, 1995.

74. Huntley JM, Saldner H. Temporal phase unwrapping algorithm for automated interferogram analysis, *Appl. Opt.*, **32**, 3047–3052; 1993.

75. Servin M, Cuevas FJ, Malacara D, Marroquin JL. Phase unwrapping through demodulation using the RPT technique, *Appl. Opt.*, **37**, 1917–1923; 1998a.

76. Cornejo A. The Ronchi test, in *Optical Shop Testing*, Malacara, D., ed., New York, NY: John Wiley & Sons, 1992.

77. Ghozeil I. Hartmann and other screen tests, in *Optical Shop Testing*, Malacara, D., ed., New York, NY: John Wiley & Sons, 1992.

78. Yatagai T. Fringe scanning Ronchi test for aspherical surfaces, *Appl. Opt.*, **23**, 3676–3679; 1984.

79. Omura K, Yatagai T. Phase measuring Ronchi test, *Appl. Opt.*, **27**, 523–528; 1988.

80. Wan D-S, Lin D-T. Ronchi test and a new phase reduction algorithm, *Appl. Opt.*, **29**, 3255–3265; 1990.

81. Fischer DJ. Vector formulation for Ronchi Shear surface fitting, *Proc. SPIE*, **1755**, 228–238; 1992.

82. Fried DL. Least-squares fitting of a wave-front distortion estimate to an array of phase-difference measurements, *J. Opt. Soc. Am.*, **67**, 370–375; 1977.

83. Hudgin RH. Wave-front reconstruction for compensated imaging, *J. Opt. Soc. Am.*, **67**, 375–378; 1977.

84. Noll RJ. Phase estimates from slope-type wave-front sensors, *J. Opt. Soc. Am.*, **68**, 139–140; 1978.

85. Hunt BR. Matrix formulation of the reconstruction of phase values from phase differences, *J. Opt. Soc. Am.*, **69**, 393–399; 1979.

86. Freischlad K, Koliopoulos CL. Wavefront reconstruction from noisy slope or difference data using the discrete Fourier transform, *Proc. SPIE*, **551**, 74–80; 1985.

87. Freischlad K. Wavefront integration from difference data, *Proc. SPIE*, **1755**, 212–218; 1992.

88. Takajo H, Takahashi T. Least squares phase estimation from phase differences, *J. Opt. Soc. Am. A*, **5**, 416–425; 1988.

89. Rimmer MP, Wyant JC. Evaluation of large aberrations using a lateral shear interferometer having variable shear, *Appl. Opt.*, **14**, 142–150; 1975.

90. Hung YY. Shearography: A new optical method for strain measurement and nondestructive testing, *Opt. Eng.*, **21**, 391–395; 1982.

91. Yatagai T, Kanou T. Aspherical surface testing with shearing interferometry using fringe scanning detection method, *Opt. Eng.*, **23**, 357–360; 1984.

92. Gasvik KJ. *Optical Metrology*, New York, NY: John Wiley & Sons, 1987.

93. Hardy JW, MacGovern AJ. Shearing interferometry: A flexible technique for wavefront measuring, *Proc. SPIE*, **816**, 180–195; 1987.

94. Welsh BM, Ellerbroek BL, Roggemann MC, Pennington TL. Fundamental performance comparison of a Hartmann and a shearing interferometer wave-front sensor, *Appl. Opt.*, **34**, 4186–4195; 1995.

95. Mantravadi MV. Lateral shearing interferometers, in *Optical Shop Testing*, Malacara, D., ed., New York, NY: John Wiley & Sons, 1992.

96. Servin M, Malacara D, Marroquin JL. Wave-front recovery from two orthogonal sheared interferograms, *Appl. Opt.*, **35**, 4343–4348; 1996b.

97. Servin M, Malacara D, Cuevas FJ. New technique for ray aberration detection in Hartmanngrams based on regularized band pass filters, *Opt. Eng.*, **35**, 1677–1683; 1996d.

98. Servin M, Cuevas F, Malacara D, Marroquin JL. Direct ray aberration estimation in Hartmanngrams using a regularized phase tracking system, *Appl. Opt.*, **38**, 2862–2869; 1999.

99. Teague MR. Deterministic phase retrieval: A green's function solution, *J. Opt. Soc. Am.*, **73**, 1434–1441; 1983.

100. Streibl N. Phase imaging by the transport equation of intensity, *Opt. Commun.*, **49**, 6–10; 1984.

101. Ichikawa K, Lohmann AW, Takeda M. Phase retrieval based on the irradiance transport equation and the Fourier transport method: Experiments, *Appl. Opt.*, **27**, 3433–3436; 1988.

102. Roddier F. Wavefront sensing and the irradiance transport equation, *Appl. Opt.*, **29**, 1402–1403; 1990.

103. Greivenkamp JE. Sub-Nyquist interferometry, *Appl. Opt.*, **26**, 5245–5258; 1987.

104. Servin M, Malacara D. Path-independent phase unwrapping of subsampled phase maps, *Appl. Opt.*, **35**, 1643–1649; 1996a.

105. Servin M, Malacara D. Sub-nyquist interferometry using a computer-stored reference, *J. Mod. Opt.*, **43**, 1723–1729; 1996c.

106. Horman MH. An application of wavefront reconstruction to interferometry, *Appl. Opt.*, **4**, 333–336; 1965.

107. Dörband B, Tiziani HJ. Testing aspheric surfaces with computer generated holograms: Analysis of adjustment and shape errors, *Appl. Opt.*, **24**, 2604–2611; 1985.

108. Frankowski G. The ODS 800—a new projection unit for optical metrology, in Proc. Fringe 97, 532–539; 1997.

Optical Methods in Metrology: Point Methods

5.1	Introduction 153	
5.2	Linear Distance Measurements.. 153	
	Large-Distance Optical Measurements Moiré Techniques in Medium Distances • Interferometric Methods in Small Distance Measurement	
5.3	Angular Measurements 171	
	Non-Interferometric Methods • Moiré Methods in Level, Angle, Parallelism, and Perpendicularity Measurements • Interferometric Methods	
5.4	Velocity and Vibration Measurement 182	
	Velocity Measurement Using Stroboscopic Lamps • The Laser Interferometer • Laser Speckle Photography and Particle Image Velocimetry • Laser Doppler Velocimetry • Optical Vibrometers • Types of Laser Doppler Vibrometers • Vibration Analysis by Moiré Techniques	
	References..205	

Zacarías Malacara-
Hernández
and Ramón
Rodriguez-Vera

5.1 Introduction

Light and optics have been used as the ultimate tools for metrology since olden days. An example of this is the way a ray beam is used as a reference for straightness, or in modern times, the definition of a given wavelength as a distance standard [1]. In this chapter, we provide an overview of some optical measuring methods and their applications to optical technology. We do not intend to cover all the optical methods in metrology. Several metrology techniques are described more extensively in other chapters in this book.

During the measuring process we need to adopt a common measuring standard from which all the references are made. The SI measuring system has a worldwide acceptance, and it is the one used in this chapter. Among the main characteristics of the SI system is the definition of a primary standard for every defined fundamental physical unit. A set of derived units is also defined from the primary units. The primary standard in the SI system is the meter. After several revised definitions [2], the meter is now defined as the distance traveled by light in $1/299,792,458$ of a second [1]. Under this new definition, the meter is a derived unit from the time standard. To avoid the meter being a derived unit, it has been proposed to define the meter as "the length equal to $9,192,631,770/299,792,458$ wavelengths in the vacuum of the radiation corresponding to the transition between the two hyperfine levels of the ground state of the cesium 133 atom" [3,4].

5.2 Linear Distance Measurements

Linear distance measurements are made through a direct comparison to a scale or a secondary standard. In other cases, an indirect measurement to the standard is done with known precision. Optics gives us the flexibility and simplicity of both methods. We have an assortment of optical methods for distance measurements, depending on the scale of distances to be measured. Some representative methods are described next.

5.2.1 Large-Distance Optical Measurements

Large distances (about a human body and larger) can be measured by optical means to a high degree of precision. Direct distance comparison, distance and angle measurement, and time-of-flight methods are used.

5.2.1.1 Range Finders and Optical Radar

Range finders are devices used to measure distances in several ways. The simplest case of a range finder is called stadia. A stadia is made from a telescope and a precision rotating mirror (Figure 5.1). A bar with a known distance w is placed at the range to be measured R. A beamsplitting prism superimposes two images from the bar in the telescope. First, both images are brought into coincidence; then, opposite ends of the bar are put together by rotating the mirror at an angle θ from the coincidence point. This gives the angle subtense for the reference bar. The range R is then

$$R = \frac{W}{\theta} \tag{5.1}$$

where θ is small and expressed in radians. Another stadia technique is used by some theodolites that have a reticle for comparison against a graduated bar. If a bar with known length W is seen through a telescope with focal length f, the image of the bar on the reticle has a size i; then, the distance is

$$R = \left(\frac{f}{i}\right)W. \tag{5.2}$$

A range finder uses a baseline instead of a reference bar as the standard length. A basic range finder schematic is shown in Figure 5.2. Two pentaprisms separated by a distance B each have two identical telescope objectives. Both telescope images are brought to the same eyepiece by means of a beamsplitter or a coincidence prism. Originally, the instrument is built so that images from an infinite-distance located object are overlapped. A range compensator is a device inserted in one of the instrument branches to displace one image laterally on the focal plane (angularly on the object space), and bring into

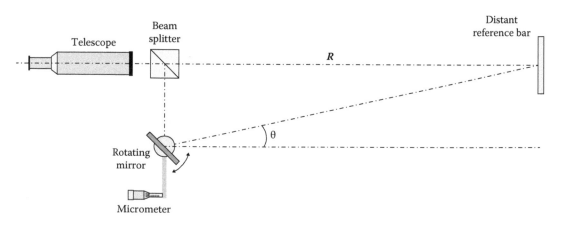

FIGURE 5.1   A stadia range finder.

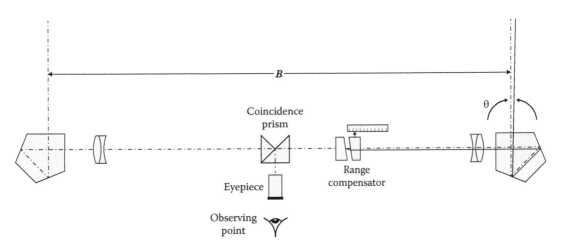

FIGURE 5.2 Range finder schematics.

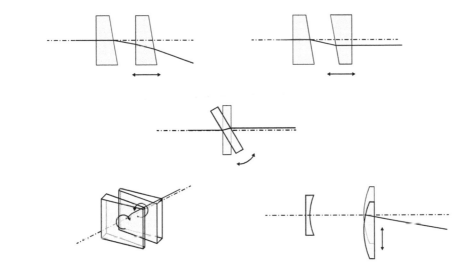

FIGURE 5.3 Range compensators for rangefinders.

coincidence objects that are not at infinity. Figure 5.3 shows some devices used as range compensators. The baseline B viewed from the distant point subtends an angle θ and the range is then

$$R = \frac{B}{\theta}.$$ (5.3)

For a given error in the angle measurement, the range error is

$$\Delta R = -b\theta^{-2}\Delta\theta$$ (5.4)

also, from Equation 5.3,

$$\Delta R = -\frac{R^2}{B}\Delta\theta$$ (5.5)

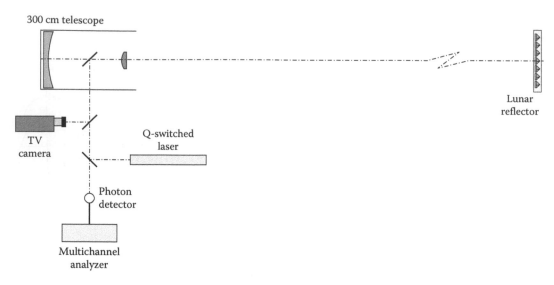

FIGURE 5.4 Moon ranging experiment.

To keep the range finder error low, the baseline must be as large as possible, but the error increases with the square of the distance. Since the eye has an angular acuity of about 10 arc seconds, the angular error is about 0.00005 radians [5]. The image displacement can be measured using two CCD (charge-coupled device) cameras, one for each branch, and measuring the parallax automatically with a computer [6]. Another method correlates the images from both telescopes electronically [7].

Another method for distance measurement is the time of flight method, tested shortly after the laser invention [8–10]. For a light beam traveling at a known speed, the distance can be measured by the time a light beam takes to go and return from the measuring point. For very large distance scales, optical radar is used, as in the moon distance determination [11]; but it has also been used in surveying instruments [12]. While in RF radar, the main problem was to obtain a fast rising pulse; the problem is well solved with Q-switched lasers. A block diagram for the moon ranging experiment is shown in Figure 5.4.

A precise measurement of the time-of-flight method is done by measuring the phase of an amplitude-modulated beam. Several systems that use this method are described in the literature [13,14]. The light beam is amplitude modulated at a frequency ω, the output light beam has an amplitude $s_o = A_o \sin \omega t$, and the returning beam has an amplitude $S_r = A_r \sin \omega(t + \Delta t)$, where Δt is the time of flight for the light beam. The resulting phase difference between the returning and a local reference beam will be $\Delta\varphi = \omega\Delta t$. Since the modulating signal is periodical, the returning and reference signal will be in phase for distances that are multiples of the modulating wavelength, or $\Delta\varphi = n\pi$, n being an integer. The phase difference is equivalent to a distance of $x = c\Delta t = c\Delta\varphi/\omega$; c is the speed of light in the medium.

For a returning beam in phase with the sending beam, the distance can be found from the modulating wavelength:

$$x = \frac{c\Delta\phi}{2\pi f} = n\lambda. \tag{5.6}$$

For any given distance, the distance in terms of the modulating wavelength is

$$D = n_1\lambda_1 + \Delta\lambda. \tag{5.7}$$

Since the number of full wavelengths in Equation 5.7 is unknown, we can use a wavelength longer than the range to be measured with the consequent loss in precision. An alternative is to use at least three close frequencies:

$$D = n_1\lambda_1 + \Delta\lambda_1,$$
$$D = n_2\lambda_2 + \Delta\lambda_2, \tag{5.8}$$
$$D = n_3\lambda_3 + \Delta\lambda_3.$$

To solve Equations 5.8 we use three close frequencies, and assume the same n for the three measurements. Another solution could be made making λ_1 and λ_2 an exact multiple (about 1000) from one another, and solving for D and n. In another method, a system that resembles a phase-locked loop (PLL) is used [15]. The system sweeps in frequency until it locks, repeats for the next frequency ($n + 1$), and then two more frequencies. When the oscillator is locked, the phase difference is zero ($\Delta\lambda = 0$); we only need to determine D and n_1, n_2, and n_3.

Absolute distance laser ranging: For a laser distance measuring system, we assume a given refractive index n. The limiting factor will always be any change in the refractive index that is a function of the temperature, pressure, and moisture content. By measuring at two or more wavelengths, all the sources of uncertainty are removed. Systems have been designed for two wavelengths [16,17]. A non-ambiguity measuring range of 0.2 parts per million has been achieved.

5.2.1.2 Curvature and Focal Length

Optical manufacturing has several ways to measure a radius of curvature, including templates, spherometers, and test plates. Templates are rigid metal sheets with both concave and convex curvature cuts at opposite faces (Figure 5.5). Templates are brought into direct contact with the sample. A minimal light

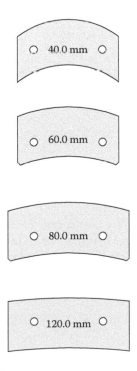

FIGURE 5.5 Curvature measuring templates.

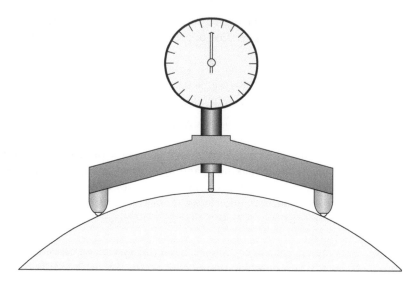

FIGURE 5.6   Bar spherometer.

space must be observed at the contact point. If the space between surfaces is small, the light between surfaces turns blue, due to diffraction. A template has the simplicity that it can be made in a mechanical shop with appropriate measuring tools, but it is also commercially available. For a more precise curvature measurement, a spherometer is used. Essentially, a spherometer measures the sagitta in a curved surface. Assume a bar spherometer, as shown in Figure 5.6 [18]. For a leg separation y and a ball radius r, the radius of curvature R can be obtained from the sagitta z:

$$R = \frac{z}{2} + \frac{y^2}{2z} \pm r, \tag{5.9}$$

where the plus sign is used for concave surfaces and minus for convex surfaces. The uncertainty in the measurement is found by differentiating the previous equation:

$$\Delta R = \frac{\Delta z}{2}\left(1 - \frac{y^2}{z^2}\right). \tag{5.10}$$

Assuming no error in the determination of the parameters for the instrument, the leg's separation is perfectly known.

   Several variants of the basic spherometer include the ring spherometer (Figure 5.7), which is commonly used where a spherical surface is assumed. An interchangeable set of rings extends its use to improve the accuracy for a larger lens diameter. The Geneva gauge is a portable bar spherometer that reads the diopter power directly, assuming a nominal refractive index for ophthalmic lenses. Modern spherometers are now digital, and include a microprocessor to convert the sagitta distance to a radius of curvature, diopter power, or curvature. Spherometer precision and accuracy are analyzed by Jurek [19].

   Test plates [20] are glass plates polished to a curvature opposite to the one we want to check. Curvature is tested by direct contact. A test plate is made for each curvature to be tested. Test plates are appropriate for online curvature testing. Plane testing plates are common on the optical shop.

   Several means have been devised for optical curvature measuring. For concave curvature measurements, probably the simplest method is the Foucault test. Analysis and applications for this test are

FIGURE 5.7 Ring spherometer.

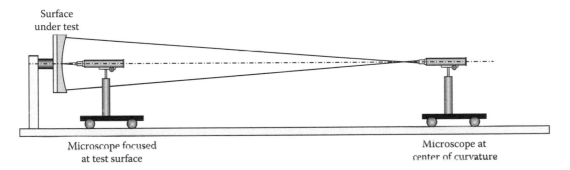

FIGURE 5.8 Traveling microscope to measure radius of curvature.

covered in another chapter in this book. Another precise curvature radii's measurement device is the traveling microscope [21], shown in Figure 5.8. A point source is produced at the front focus of a traveling microscope. Also, an illuminated reticle eyepiece could be used. In both cases, a sharp image of the point source or the reticle is sought when the surface coincides with the front focus. Then, the microscope is moved until a new image is found. The curvature radius is then the distance between these positions. Carnell and Welford [22] describe a method using only one measurement. After focusing the microscope at the curvature center, a bar micrometer is inserted with one end touching the vertex of the surface. This method is also suitable for convex surfaces by inserting a well-corrected lens such that the conjugate focus is larger than the curvature radius under test. Convex curvatures can be measured using an autocollimator and an auxiliary lens [23].

For a large curvature radius, an autocollimator and a pentaprism are used, as shown in Figure 5.9. A shift in the reticle image is measured in the autocollimator. This is an indirect measure of the surface's slope. By scanning the surface with the prism, samples of the slope are obtained and by integration, the curvature is calculated. This system is appropriate for large curvature measurements and can be used both for concave and convex surfaces [24].

Another commonly used method for curvature measurements is the confocal cavity method. The so-called optical cavity technique, described by Gerchman and Hunter [25,26], interferometrically

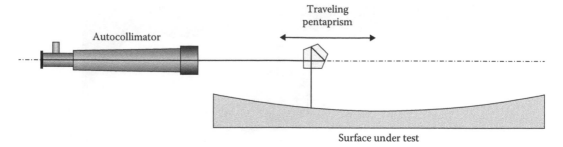

FIGURE 5.9 Autocollimator and pentaprism to measure curvature.

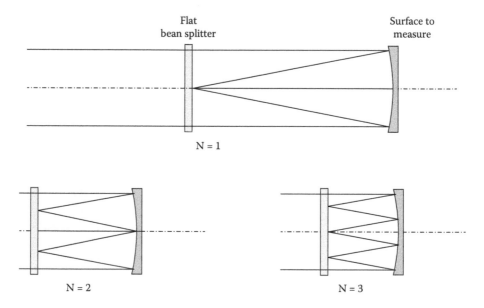

FIGURE 5.10 Confocal cavity curvature measurement.

measures the radii of curvature for long curvature concave surfaces. A Fizeau interferometer is formed, as shown in Figure 5.10. An nth order confocal cavity is obtained where n is the number of times the optical path is folded. The radius of curvature is equal to $2\,n$ times the cavity length Z. The accuracy is about 0.1%.

5.2.2 Moiré Techniques in Medium Distances

When two periodic objects are superimposed, a well-known effect takes place; this is the *moiré effect* [27–29]. This effect produces a low-frequency pattern of secondary fringes known as *moiré fringes*. In this manner, for example in Figure 5.11, we can see several overlapping periodic objects, showing moiré fringes. For measuring purposes, the periodic objects are usually constituted by gratings of alternating clear and dark lines. Figure 5.12a is a sample of a linear grid and Figure 5.12b is the moiré pattern taking place by the overlapping of two such gratings. Ronchi gratings are a particular case of linear grids with a quadratic profile that can easily be reproduced.

FIGURE 5.11 Formation of moiré fringes with different periodic objects.

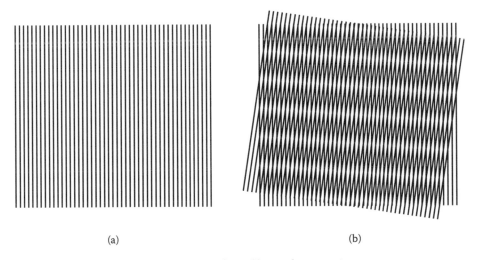

(a)   (b)

FIGURE 5.12 (a) Linear grating; (b) Moiré pattern formed by two linear gratings.

From the geometric point of view [30], the moiré fringes are defined as a locus of points of two overlapping periodical objects. It is possible to determine the period p' and the angle φ of the moiré fringes knowing the periods p_1, p_2, and the angle θ among the lines of the gratings (Figure 5.13). Then,

$$p' = \frac{p_1 p_2}{\sqrt{p_1^2 + p_2^2 - 2p_1 p_2 \cos\theta}} \tag{5.11}$$

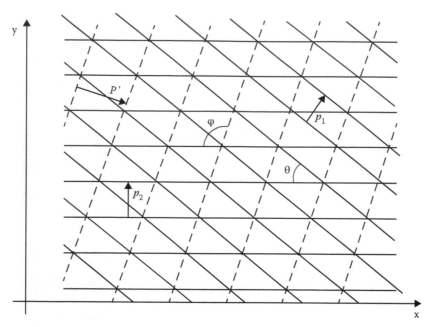

FIGURE 5.13 Diagram showing the intersection points of two superimposed linear gratings of periods p_1 and p_2; θ is the angle among the lines of the gratings. Dotted lines correspond to moiré fringes, which make an angle φ with the lines of the grating of period p_2.

and

$$\sin \varphi = \frac{p_2 \sin \theta}{\sqrt{p_1^2 + p_2^2 - 2 p_1 p_2 \, \cos \theta}}. \tag{5.12}$$

When $\theta = 0$, then $\varphi = 0$, and Equation 5.11 becomes

$$p' = \frac{p_1 p_2}{|p_1 - p_2|}. \tag{5.13}$$

Now, if $p_1 = p_2 = p$, then Equation 5.11 transforms into

$$p' = \frac{p}{2 \, \sin \theta/2}. \tag{5.14}$$

If $\theta \approx 0$, then $p' = p/\theta$ and $\varphi \approx 90°$. A quick analysis of Equation 5.14 shows that when the angle between the gratings is large, the moiré pattern frequency increases (the period is reduced); otherwise, the frequency diminishes until the moiré fringes disappear.

One of the fundamental characteristics of a moiré pattern is that if one of the gratings is deformed and the other remains fixed, the moiré pattern is also deformed, as shown in Figure 5.14. Deformation of one of the gratings can arise because of the large size to be measured. For this reason, it is possible to call the deformed grating a grating object, while the one not deformed is the reference grating. A simple way to obtain a couple of linear gratings is to photocopy in a transparency the image of Figure 5.12a. When superimposing these gratings, the moiré effect is observed, similar to that of Figure 5.12b. Another characteristic of the moiré effect is that if we maintain a fixed grating and the other displaced a small distance in the direction of its lines, the moiré pattern has a large displacement.

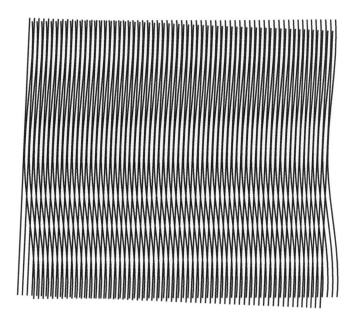

FIGURE 5.14 Moiré patterns formed by two linear gratings; one of them is lightly deformed.

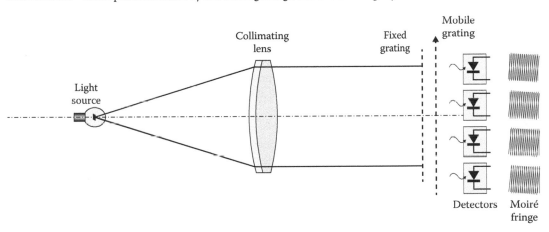

FIGURE 5.15 Opto-electronic arrangement for distance measurement based on the moiré effect.

There are several ways of overlapping two gratings. The simplest, as already mentioned, is by contact. Another manner is to project the image of a grating by means of an optical system over another grating. A third form is by means of some logic or arithmetic operation in a digital way [31,32]. This last method can be carried out when the gratings are stored in a computer.

Several metrology techniques are based on the moiré effect. These techniques have been used in several applications of science and engineering. Some of the techniques will now be explained along with different measuring tasks.

5.2.2.1 Photoelectric Fringe Counting

Typical examples of distance meters based on the moiré effect are verniers and digital micrometers, and coordinate measuring machines (CMM). Their operation consists of the photoelectric detection of the moiré fringes' movement [33]. The basic elements of such a system are shown in the Figure 5.15. This system consists of a light source, a collimating lens, two linear gratings, and a set of four photocells.

The displacement is measured by counting moiré fringes when one of the gratings moves in the normal direction to its lines. The displacement of the moiré fringes will be analogous to the lateral displacement of the lines of the gratings. Using a photocell, the moiré fringes are detected during the movement [34]. In practice, to determine the displacement sense, four photocells are required. These detectors are positioned at four points in a moiré fringe pattern, and they are spaced to a quarter of their period. An alternate count of phase steps and signs combine to be fed to an amplifier, in a pair of symmetrical signals in quadrature. The sensibility of the instruments based on this moiré technique can be up to 0.000250 inches [35].

5.2.2.2 Talbot Effect

When a periodic object is illuminated with a spherical or plane wavefront, replicas of the object are obtained in precisely defined planes along the light propagation direction [28]. This phenomenon is known as the Talbot effect or self-imaging. In spherical illumination, the replication of the object is amplified and the distances between planes are not a constant. On the other hand, when the illumination of the periodic object is made by means of a collimated wave, the self-image planes are equally spaced and well-defined (Figure 5.16). If the periodic object is a linear grating with a period p, the self-images are formed in planes, given by the equation [36]:

$$\Delta_n = \frac{n\,p^2}{\lambda},$$
(5.15)

where n is the n-order plane and λ is the illumination wavelength.

This effect is very useful for measuring the distance Δ between a grating and its Talbot image, forming a moiré pattern. In this manner, for example, Jutamulia et al. [37] and Rodriguez-Vera et al. [38] used the Talbot effect to identify depths or separate planes in color. The color appears naturally by illuminating the grating with a white light source and by superimposing the self-image reflected from different planes of the scene into a second grating. Moiré fringes look at different colors depending on the plane position.

5.2.2.3 Liquid Level Measurement

As an extension of moiré techniques, the Talbot effect is used for liquid level and volume determination in containers [39]. This technique uses the reflected image from the liquid-air interface inside a container. The incident light to this interface comes from the image of a linear grating illuminated from a collimated monochromatic light source. By means of an appropriate optical system, the reflected image is formed onto a second grating, forming a sharp moiré pattern, when the second grating is at a distance equal to the Talbot plane. For a given longitudinal displacement in the liquid-air interface that corresponds to a level change, the moiré pattern becomes unsharp. Consequently, it is necessary to adjust the second grating mechanically to observe, again, a sharp moiré pattern. This linear mechanical adjustment reflects the level change in the container.

5.2.2.4 Focal Length Measurement

Moiré fringes and the Talbot effect have also been useful in the focal length measurement of an optical system. Two categories are used for these techniques: the first is based on the measurement of the moiré pattern's rotation [40–42]; the second is based on the beating of a moiré pattern [43,44]. In the first case, the focal length f is obtained by measuring the rotation angle α_n; for the moiré fringes due to beam divergence changes to non-collimation, the incident beam suffers at the first grating, as shown in Figure 5.16. The moiré fringes are observed through a diffuse screen. The focal length will be calculated using the following equation [40]:

$$f = \frac{1}{\sin\vartheta\tan\alpha_n + \cos\vartheta - 1}\,\frac{np^2}{\lambda},$$
(5.16)

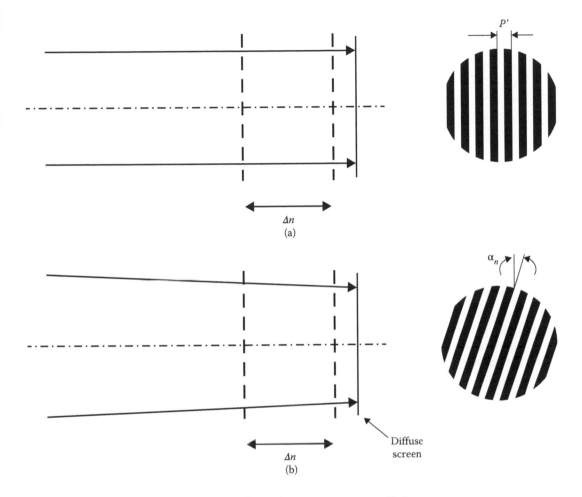

FIGURE 5.16 Moiré fringes formed by (a) collimated and (b) noncollimated light.

where θ is the angle between the lines of the gratings and p is the period of the gratings. A typical experimental setup used to make these arrangements is shown in Figure 5.17.

The second case uses the beating of the moiré pattern produced when the lines of the grills are parallel ($\theta = 0$). In this case, the moiré fringes are caused by different periods on both grills, and the focal length is calculated by means of the equation

$$f = \frac{npp'}{\lambda}. \tag{5.17}$$

This last case is more difficult to obtain in practice, since the setup is more difficult to implement, and measuring the moiré pattern period p' is more time consuming than the angle α_n.

An additional method that does not use collimated illumination has been reported [45]. However, this method requires additional mechanical displacements to the gratings while the measurement is being performed.

Based also on moiré techniques and Talbot interferometry, systems have been built that not only measure the focal length of a lens but also its refractive power [46,47]. The strength of these techniques resides in the use of automatic digital processing of the moiré fringes.

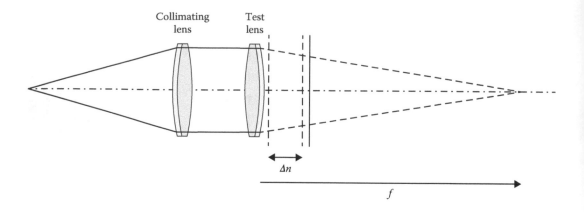

Collimating
lens
Test
lens

Δn

f

FIGURE 5.17   Experimental setup to determine the focal length of a lens.

5.2.2.5 Thickness Measurement

The thickness measurement is based on the very well-known method of projection moiré contouring [48]. Basically, it consists in projecting a grating over the object under study. The projected grating is deformed according to the topography of the object. By means of an optical system, the deformed grating is imaged on a similar reference grating, in such a way that a moiré pattern is obtained on the overlapping plane. Under this outline, the moiré fringes represent contours or level curves of the surface object. This technique has been broadly used for measuring tasks in ways using a Talbot image as projection grating [49], or determining form and deformation of engineering structures by means of digital grating superposition [50].

A simple way to interpret the moiré fringes in this outline is through the contour interval [51]. For collimated illumination and far away observation, the contour interval [49] is given by

$$\Delta z = \frac{p}{\sin\beta},\tag{5.18}$$

where p is the projected grating period and β is the angle between illumination and detection optical axes, as shown in Figure 5.18. Equation 5.18 displays height differences between a contour and the next one.

This projection outline can be simplified by analyzing the projected grating or a light line. Figures 5.19a and b show schematically a grating and a light line on a cube, respectively. Note that both projected grating and light line are deformed according to the topography of the surface. In this case we do not deal with closed contours or level lines, but with "grid contours" [52]. These grid contours are similar to those of carrier frequency introduced in an interferometer. For the case of a step like the one shown in Figure 5.20a, the fringes are displaced. This displacement is a change of phase of the grid contour and, therefore, is related to the difference of height between the two planes. Figure 5.20b shows the three-dimensional plot of phase shift when applying the technique of phase-locked loop detection to determine the height between the planes that form the step [50].

A limitation of this technique is in the ambiguity that results when the differences of height are large enough so that projected fringes surpass the phase change by 2π To solve this problem, the possibility of using a narrow light sheet to illuminate the object has been investigated [53], as shown in Figure 5.19b. The object to be analyzed is placed on a servomechanism to be moved along an axis. During the movement, the object is illuminated with a sheet light beam (flat-shaped beam); the images are captured on a computer and processed. Figure 5.21 shows the object with the projected deformed light line formed by the incident light sheet and its topographical reconstruction.

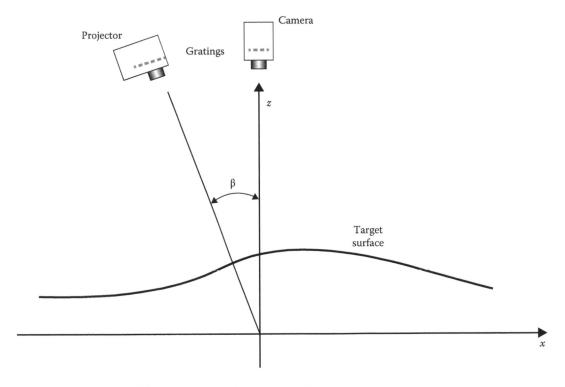

FIGURE 5.18 Scheme of the projection moiré contouring technique.

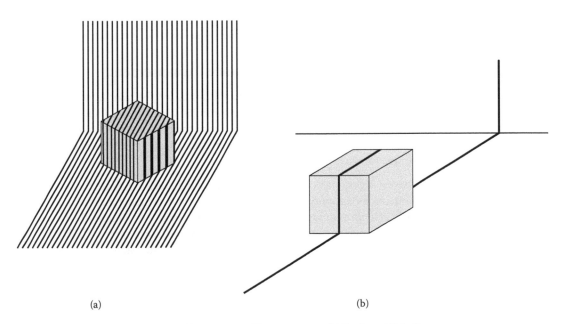

(a)   (b)

FIGURE 5.19 (a) Fringes produced by a projected linear grating. (b) Projected light line.

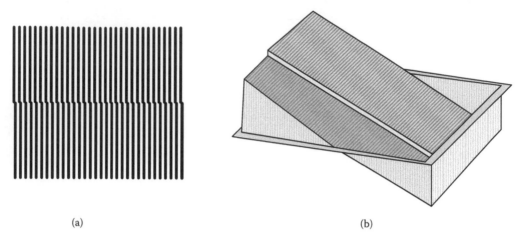

(a)   (b)

FIGURE 5.20 Projected linear grating on a step produced by two planes. (a) Grid contours. (b) Difference of height between planes.

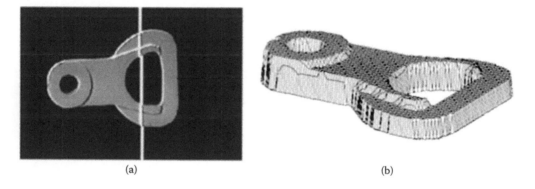

(a)   (b)

FIGURE 5.21 (a) Sheet of light projected on a metallic surface. (b) Three-dimensional reconstruction of the object (a).

5.2.3 Interferometric Methods in Small Distance Measurement

Fringe counting in an interferometer suggests an obvious application for the laser in short distance measurement. Interferometric methods are reviewed by Hariharan [54]. The high radiance and monochromaticity are the main useable characteristics in the laser. From the conceptually simple outline of a Michelson interferometer, several improvements make it a more convenient and practical instrument. Some problems from the simple form are

1. A light source must have a highly stable frequency to improve the precision. Since the light wavelength is used as a standard, as the more stable the source remains, the more precise the measurement will be. The original length standard was chosen as a discharge ^{86}Kr source. The frequency of this source is stable up to one part in 10^8, even better than some laser sources [2]. To improve the laser light stability, several means have been developed for frequency stabilization [55]. For a precise measurement device, frequency stabilization is essential.

2. A light beam reflected back to the laser makes it unstable and the radiance will fluctuate with the moving-mirror displacement. For example, a Michelson interferometer forms two complementary interference patterns, one reflected back to the source. This reflected pattern eventually makes the laser unstable and the intensity will fluctuate as the optical path in one branch is changed. To avoid this problem, a non-reacting configuration must be used.

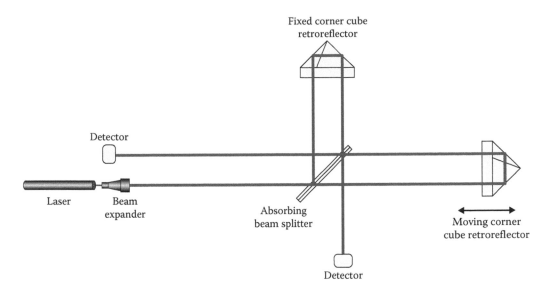

FIGURE 5.22   DC interferometer for distance measurement.

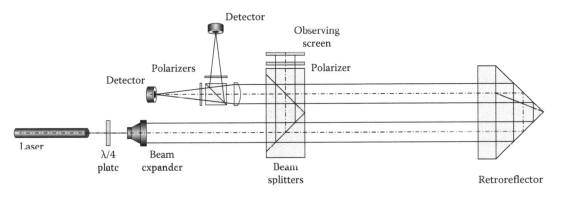

FIGURE 5.23   Minkowitz nonreacting interferometer.

3. Fringe counting in the interferometer is usually made electronically, which means the electronic counters should discriminate the fringe movement direction. To discriminate the direction movement, two detectors at a 90° phase difference are used.

4. The fringe count across the aperture should be kept low and constant as the mirror is moved. This problem is easily solved by using retroreflectors instead of plane mirrors. Now, the phase shift across the full aperture is kept almost constant.

The so-called DC interferometer is shown in Figure 5.22. The use of two corner cube retroreflectors is twofold: to have a non-reacting configuration and to have a constant phase shift across the full aperture. For a non-absorbing beamsplitter, the phase shift between both interference patterns will be 180°, but for an absorbing one, it can be adjusted to a 90° phase shift, making it possible to discriminate for the movement direction.

Another non-reacting interferometer for distance measuring is described by Minkowitz et al. [56]. A circularly polarized beam of light (Figure 5.23) is obtained through a linearly polarized laser and a λ/4 phase plate. A first beamsplitter separates the reference and measuring beam. Upon reflection, the

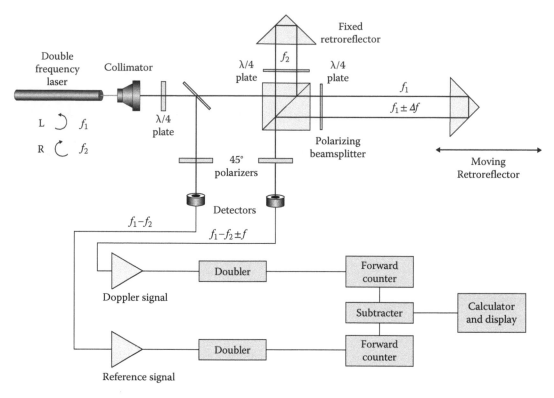

FIGURE 5.24   AC interferometer for distance measuring.

reference beam changes its polarization while the measuring beam changes twice its polarization at both reflections on the moving corner cube prism reflector. Since both beams have opposite polarizations, the resulting beams are linearly polarized, with a polarization angle determined by the optical path difference. The polarization angle rotates 360° for every $\lambda/2$ path difference. Two polarizers are placed at each complementary output of the beamsplitter. Since the polarizers are rotated 90° each from the other, the irradiance varies sinusoidally with the corner cube prism displacement with a quadrature phase shift at the detectors.

A different approach is taken in an ac interferometer (Figure 5.24) described by Burgwald and Kruger [57] and Dukes and Gordon [58] and commercially produced by Hewlett-Packard. A frequency stabilized He-Ne laser is Zeeman-split by a magnetic field and the optical beam now has two frequencies about 2 MHz apart. The two signals with frequencies f_1 and f_2 are circularly polarized and with opposite handedness. A $\lambda/4$ phase plate changes the signals f_1 and f_2 into two orthogonal linearly polarized beams, one in the vertical and the other in the horizontal plane. A beamsplitter takes a sample of the beam. Since both signals are mixed, the resulting beam is linearly polarized, rotating at a frequency f_1-f_2. Later, the polarizer makes a sinusoidally modulated beam at this frequency. Another portion of the beam is sent to a polarizing beamsplitter. One branch of the interferometer receives the f_1 component while the other receives the f_2 part of the signal. On each branch, a $\lambda/4$ phase plate converts the polarization to circularly polarized, but while one branch is left-handed, the other is right-handed. After two reflections at the comer cube reflectors, the handedness is preserved but the $\lambda/4$ phase plate changes its polarization to a perpendicular one. Assuming a moving corner cube prism, the signal will be Doppler-shifted to a frequency $f_1 + \Delta f$. Both beams meet again at the polarizing beamsplitter and the outgoing beams are opposite-handed circularly polarized beams that add to form a slowly varying circularly polarized beam at a frequency f_1 -f_2 + Δf. The signals from the detectors are fed to a couple of digital counters, one increasing at a frequency f_1–f_2 while the other at a frequency f_1–f_2 + Δf. A digital subtracter obtains the accumulated pulse difference. For a stationary reflector, each counter increases at about two million counts per second, but the difference remains constant. Some advantages of this system lie in its relative insensibility to radiance variations from the laser.

5.2.3.1 Multiple-Wavelength Interferometry

At the best, all the measuring instruments can measure incremental fringes from an initial point. Absolute distance measurements can be made from a multiple-wavelength laser. The technique is similar to that described in Section 5.2.1 and is reviewed by Dändliker et al. [17]. For a double-pass interferometer, the distance is expressed as

$$D = N\left(\frac{\lambda}{2}\right) + \epsilon, \tag{5.19}$$

where N is an integer number and ϵ is a fractional excess less than half the wavelength. For a multiple wavelength laser, the distance is

$$D = N_i\left[(\lambda_i)/2\right] + \epsilon_i, \tag{5.20}$$

Since all N_i are integer constants, it is possible to know D from several wavelengths (λ_i's). Bourdet and Orszag [59] use six wavelengths from a CO_2 laser for an absolute distance determination.

By using two wavelengths, a beating is obtained at the detector. This beating has a synthetic wavelength:

$$\Lambda = \frac{\lambda_1\lambda_2}{\lambda_1 - \lambda_2}. \tag{5.21}$$

In practice, each wavelength has to be optically filtered for measurement. The absolute distance accuracy depends on the properties of the source. Both wavelengths must be known very accurately. To increase the non-ambiguity range, multiple wavelengths are used by dispersive comb spectrum interferometry [60]. If one laser is continuously tuned, a variable synthetic wavelength is used [61].

5.3 Angular Measurements

Angle measurements are done with traditional and interferometrical methods. For traditional methods, templates are used in the lower end for low-precision work, while goniometers and autocollimators are used for high precision. The best precision is obtained with interferometric angle-measuring methods. Interferometric angle measurements have the additional advantage that they can be interfaced to automatic electronic systems.

5.3.1 Non-Interferometric Methods

5.3.1.1 Divided Circles and Goniometers

For a rough scale, protractors and divided circles are used for angle determination. Although these devices are limited to a precision of about 30 min, modern electronic digital protractors and inclinometers can measure precise angles up to 0.5 min. At the optical shop, angle measurements can be made by means of a sine plate (Figure 5.25). The sine plate can both support a piece of glass and measure its angle by itself or with a collimator. The angle is defined from the base plate's length and a calibrated plate

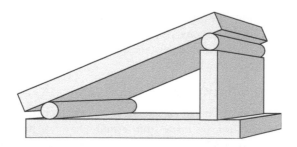

FIGURE 5.25   Sine plate.

FIGURE 5.26   Goniometer.

inserted in one point. The angle is then calculated from the plate length and the calibrated plate length inserted in one leg to form a triangle. The angle is calculated assuming a rectangular triangle: hence its name of sine plate. With a good sine plate, an accuracy of 30 min can be achieved. A serrated table for angle measurement is described by Horne [21] with an accuracy of 0.1 s.

Angle blocks are also used for comparison. These are available commercially to an accuracy of ±20 s. By reversing and combining the blocks, any angle between 1° and 90° can be obtained. Since angle blocks have precision flat-polished faces, together with a goniometer, they can be used as an angle standard. A glass polygon used with an autocollimator can precisely calibrate goniometers or divided circles. A goniometer (Figure 5.26) is a precision spectrometer table. Its divided circles are used for precise angle measurement. For angle measurement, the telescope reticle is illuminated and the reflected beam is observed. Prism or polygon angles are measured in a goniometer. The divided circle sets the precision for the instrument. The polygon under test is set at the table while the telescope is turned around until the reflected reticle is centered.

5.3.1.2 Autocollimators and Theodolites

An autocollimator is essentially a telescope with an illuminated reticle: for example, a Gauss or Abbe eyepiece (Figure 5.27) at the focal plane of the objective. By placing a mirror at a distance from the autocollimator, the reflected beam puts an image from the reticle displaced from its original position at a distance

$$d = 2\alpha f, \tag{5.22}$$

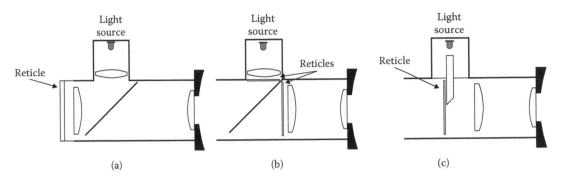

FIGURE 5.27 Illuminated eyepieces: (a) Gauss, (b) Bright line, and (c) Abbe.

where f is the collimator's focal distance and α is the mirror tilt angle from perpendicularity. Since the angle varies with the focal length, the objective is critically corrected to maintain its accuracy. Precise focus adjustment can be achieved through Talbot interferometry [62]. Autocollimators are manufactured with corrected doublets, although some include a negative lens to form a telephoto system to get a more compact system. A complete description of autocollimators and applications can be found in Hume [63].

Some autocollimators have a drum micrometer to measure the image displacement precisely, while others have an electronic position sensor to obtain the image centroid. Electronic autocollimators can go beyond diffraction limit, about an order of magnitude from visual systems. Micro-optic autocollimators use a microscope to observe the image position.

Autocollimators, besides measuring angles of a reflecting surface, can be used for parallelism in glass plates, or divided circles manufacturing [64]. By slope integration, from an autocollimator, flatness measurements can be obtained for a machine tool bed or an optical surface [65]. In an autocollimator measurement, the reflecting surface must have high reflectivity and be very flat. When a curved surface is measured, this is equivalent of introducing another lens in the system with a change in the effective focal length [65].

Theodolites are surveying instruments made from a telescope mounted in a precise altitude-azimuth mounting with a spirit level and a tree screw base for leveling. Besides giving a precise means of measuring both the elevation and the azimuth angles, a reticle with stadia markings permits distance measurements. Old instruments with engraved circles had an accuracy of 20 arcmin. Modern electronic theodolites contain digital encoders for angle measurements and electronic range finders. Angle accuracy is better than 20 arcsec. Errors derived from eccentricity and perpendicularity are removed by rotating both axes 180° and repeating the measurement.

Besides surveying, theodolites are used for angle measurement in the optical shop from reference points visually taken through the telescope for large baseline angles.

5.3.1.3 Angle Measurement in Prisms

Angle measurement in prisms is very important, since prisms are frequently used as angle standards. Optical prisms are made with 30°, 40°, and 90° angles. Optical means of producing these angles are easily obtained without the need for a standard.

A very important issue in prism measurement is the assumption of no pyramidal error. For a prism to be free of pyramidal error, surface normals for all faces must lie in a single plane (Figure 5.28). Pyramidal error can be visually checked [66–68] for a set of three faces by examining both the reflected and refracted image from a straight line (Figure 5.29). Under pyramidal error, the straight line appears broken. A far target can be used as a reference to measure the angle error.

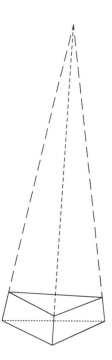

FIGURE 5.28   Pyramidal error.

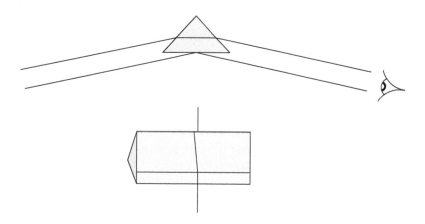

FIGURE 5.29   Pyramidal error check by reflected and refracted reference line observation.

Precise angle replication can be made by mounting the blank glass pieces in the same axis as the master prism [69,70]. An autocollimator is directed to see the reflected beam from each face on the master prism (Figure 5.30).

Precise 90° prisms can be tested either visually or with auxiliary instruments. By looking from the hypotenuse side, a retroreflected image [67] from the pupil is seen that depends on the departure from the 90° angle, as shown in Figure 5.31. This test can be improved, as shown by Malacara and Flores [71] by using a target with a hole and a cross in the path, as shown in Figure 5.32.

An autocollimator can be used to increase the sensitivity of the test [72,68]. Two overlapped images are observed, with a separation $2 N\alpha$, where α is the magnitude of the prism angle error and its sign is unknown. To determine the error of the sign, DeVany [73] suggests defocusing the autocollimator

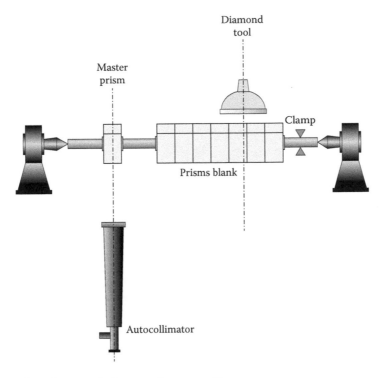

FIGURE 5.30 Angle replication of polygons with an autocollimator.

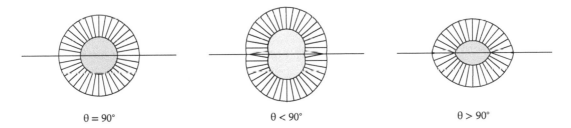

$\theta = 90°$ $\theta < 90°$ $\theta > 90°$

FIGURE 5.31 Pupil's image in a rectangular prism with and without error.

inward. If the images tend to separate, then the angle in the prism is larger than 90°. Another means of determining the angle error is by introducing, between the autocollimator and the prism, a glass plate with a small wedge with a known orientation. The wedge is introduced to cover half the aperture. Polarized light can also be used as suggested by Ratajczyk and Bodner [74].

5.3.1.4 Level Measurement

Levels are optical instruments that define a horizontal line of sight. The traditional form is a telescope with a spirit level. Once the spirit level is adjusted, the telescope is aimed at a constant level line. For an original adjustment of the spirit level [75], a pair of surveying staves are set some distance apart (Figure 5.33). The level is directed to a fixed point in one of the staves from two opposite directions and the corresponding point in the other staff is compared. Any difference in the reading is compensated by moving the telescope to a midpoint in the second staff; then, the spirit level is fixed for the horizontal position.

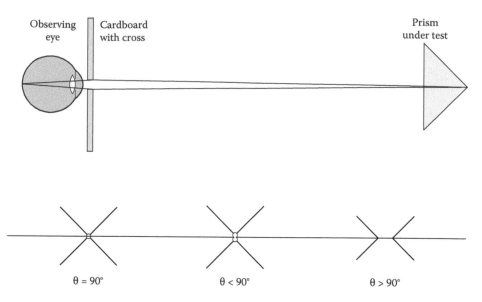

FIGURE 5.32   Prism angle error observation.

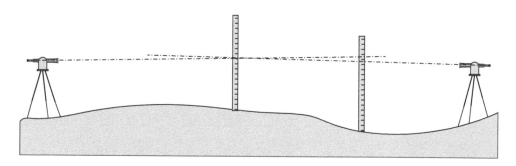

FIGURE 5.33   Level adjustment with two staves.

An autoset level (Figure 5.34) relies on a pendulum-loaded prism inside the telescope tube. A small tilt is compensated by the pendulum movement, although other mechanisms are also used [76]. A typical precision for this autoset level lies within 1 arc second and works properly for a telescope angle within 15 arc min [65].

5.3.1.5 Parallelism Measurement

Rough parallelism measurements can be done with micrometers and thickness gauges, but they need a physical contact with a damage risk; besides, several measurements are necessary to make a reliable test. Optically, parallelism is measured with the versatility provided by an autocollimator. The reticle from the autocollimator is reflected back by both surfaces simultaneously in a transparent optical plane. Any departure from parallelism is seen as two reticles. Under this setup, the reflected image from the first surface is adjusted for perpendicularity and the angle for the second surface is simply found from Snell's law. Parallelism in opaque surfaces can be measured after Murty and Shukla [77] by using a low-power laser incident in an optical wedged plate as shown in Figure 5.35. The plate under test is placed in a three-point support base, and then the reflected point is noted on a distant screen. Next the plate is

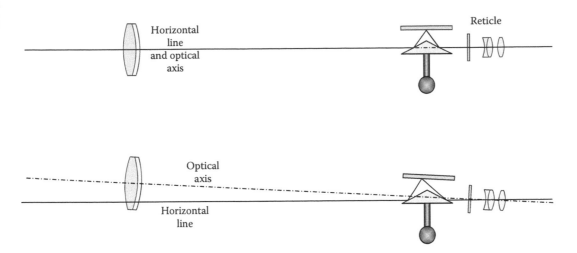

FIGURE 5.34 The autoset level.

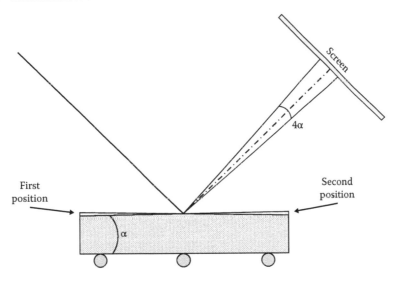

FIGURE 5.35 Parallelism measurements in an opaque plate.

rotated 180° and the new position for the beam is recorded. Let the separation for both points be d, and the screen distance from the plate is D; then the wedge angle α is

$$\alpha = \frac{d}{4D}. \tag{5.23}$$

By using two opposed collimators in a single optical bench, they are adjusted to center the other collimator image. The plate to be checked for parallelism is inserted between both collimators and first adjusted for perpendicularity in one of the collimators while the other reads the amount of departure for the second surface [78]. The latter technique could be used for nontransparent surfaces.

5.3.2 Moiré Methods in Level, Angle, Parallelism, and Perpendicularity Measurements

The different optical techniques that follow have their foundation in the formation of moiré fringes (see Section 5.2.2). These techniques, in principle, will give the shadowy characteristics of angular measurement with applications toward the same angular measurement, parallelism or collimation, perpendicularity, alignment, slope, and curvature.

If a grating of period P_p is projected on a flat object, the size of the period changes if it is observed to a different angle from that of projection, as shown in Figure 5.36. In this figure, it is supposed that the projection and observation systems are far from the surface, in such a way that the illumination and reflected beams are plane wavefronts. An angular change of the surface also produces a change in the period of the observed grating, as shown in Figure 5.37. If the observed gratings are superimposed before and after the surface is tilted, a moiré pattern of period p', given by Equation 5.14, will be observed. Then, we can find the relationship between the observed pattern's period and the angular displacement of the surface, knowing the period of the projected grating. The observed period p_1, before the surface is moved, is given by (see Figure 5.37):

$$p_1 = p_P \frac{\cos\delta}{\cos\beta}, \qquad (5.24)$$

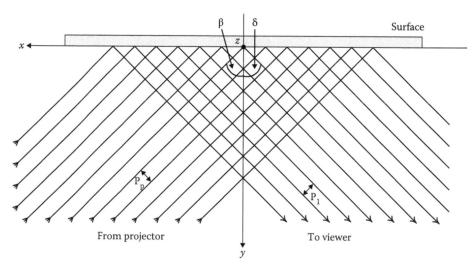

FIGURE 5.36   Diagram of a linear grating projected on a flat surface. The grating lines are perpendicular to the *x-y* plane.

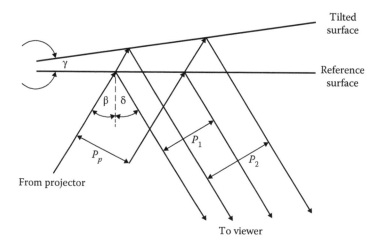

FIGURE 5.37   Parameters for calculating the moiré fringe period when the test surface is tilted.

where P_p is the period of the projected grating and δ and β are the angles to the z-axis of the optical axes of observation and projection systems, respectively. Figure 5.37 shows all these parameters. The observed period P_2 after the surface is tilted is given by

$$p_2 = p_P \frac{\cos(\delta - \gamma)}{\cos(\beta + \gamma)}, \tag{5.25}$$

where γ is the angular displacement that suffered the surface. Substituting Equations 5.24 and 5.25 in 5.13, one has

$$p' = p_P \left| \frac{\cos\delta \, \cos(\delta - \gamma)}{\cos(\beta + \gamma)\cos\delta - \cos\beta\cos(\delta - \gamma)} \right|. \tag{5.26}$$

For the particular case of a perpendicular observation to the surface, where $\delta = 0$, Equation 5.26 is transformed into

$$\tan\gamma = \frac{p_P}{p' \sin\beta}. \tag{5.27}$$

This last equation allows us to calculate the inclination that suffers a surface by means of the moiré method. The tilt angle γ that moves the surface from its original position is determined from the moiré pattern's period p', the projected grating period P_p, and the projection angle β.

5.3.2.1 Tilt Measurement

Moiré fringes are used in three methods to measure angular variations of a surface. One is based on the use of linear grating [79,80]; another is the use of nonlinear gratings [81–83]; and, the third, in the projection of interference fringes [84,85].

The interference fringe projection method is based on the detection of phase changes of the projected fringes when the object is tilted at a small angle [84]. The projected interference fringes on the object are reflected and they are detected in two points. The change in the phase difference between the two detected phase points is a function of the object rotation angle. The sensibility of the technique depends on the position of the two detection reference points in the fringe pattern. With this technique, angular variations up to 17 mrad/arc sec can be detected. The technique suffers from an ambiguity: the maximum detected phase change is 2π. In order to increase the sensibility of the method, projection of two interference fringe patterns with different periods were employed [85].

Techniques of angular measurement with nonlinear gratings can use two [81,83] or a single circular grating [82]. The basic technique consists of detecting the moiré pattern when the circular gratings are superimposed. For the case of a single circular grating, the shadow moiré technique is used [86]. The grating is placed in front of a reference plane and illuminated. A projected shadow of the grating is formed on the reference plane and is superimposed with the physical grating, producing in this way the moiré pattern. The observed moiré pattern, with a camera located perpendicularly to the reference plane, is similar to that of the Figure 5.11c.

Nakano [81] uses the Talbot effect to measure small angular variations of a surface. Figure 5.38 is an experimental diagram to carry out tilt measurements. By means of a collimated monochromatic beam, a grating g_1 is illuminated. The wavefront passing through the grating is reflected on a surface M that makes an angle γ with respect to the optical axis. This reflected wavefront impinges on a second grating g_2. The grating g_2 is placed at a Talbot distance, with respect to g_1. So, in the observation plane, a moiré pattern is formed. If the angle between the lines of the gratings is θ, the moiré fringes appear inclined a quantity α_1 to the x'-axis. If the surface object is tilted an angle Δy from its original position, the Talbot

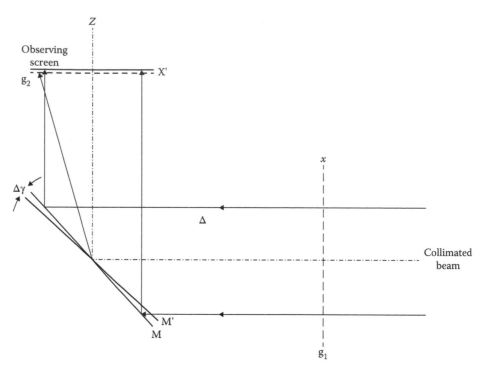

FIGURE 5.38 Experimental outline to measure the tilt angle $\Delta\gamma$ of the surface M.

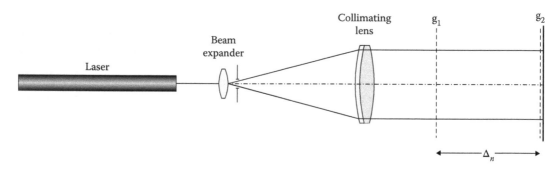

FIGURE 5.39 Experimental arrangement for optical collimating.

self-image suffers a modification. This modification means that the moiré fringes lean at an angle α_2, to the same x'-axis. Then, the small angular variation Δy will be given by Nakano [80].

The inclination angles of the moiré fringes are measured in a direct way or automatically, taking as reference the first position of the surface M. The accuracy of this technique depends on the method used to measure the inclination angles of the moiré fringes α. Other error sources affecting the sensitivity of the method are the measurement of the Talbot distance and the angle between the grating lines θ.

5.3.2.2 Optical Collimation

A very simple technique for beam collimating is also based on the moiré and Talbot effects [87–89]. The experimental arrangement consists of a couple of identical linear gratings, a laser, a beam expander, and the collimating lens, as shown in Figure 5.39. The principle consists of adjusting the collimating lens mechanically to coincide its focal points with that of the expander lens. A way to corroborate this coincidence of points is by means of the moiré pattern. It is known, by Equation 5.14, that when two identical linear gratings are superimposed, the moiré pattern's period becomes infinitely big when the angle among the lines of the gratings goes to zero. On the other hand, if the beam illuminating the first grating, in the outline of Figure 5.39, is not collimated, the self-image is amplified and the moiré pattern appears; even the angle among the grating lines becomes zero (see Equation 5.13). So, the mechanical movement of the collimating lens (along the optical axis) and of the second grating must be adjusted until the moiré pattern disappears when θ = 0. A similar outline to that previously described, has been reported by Kothiyal et al. [90] and Kothiyal and Sirohi [61]. Each one of the gratings is built with two different frequencies and dispositions of the lines, giving a greater sensitivity to the technique.

5.3.2.3 Optical Level and Optical Alignment

Under certain conditions, an optical collimator, as the one described in previous sections, can serve as an optical level. This way, for example, it is possible to place one or several objects in a straight line. This line can be parallel or perpendicular to the optical axis of the system. In consequence, this system can serve as help for alignment, and can center and measure perpendicular deviation. This idea of using the moiré as an optical measurement tool has resulted in instruments for precise optical levels and "aligners" [91]. Some of these use circular gratings [92] or combinations of linear and circular gratings [93].

5.3.2.4 Slope and Curvature Measurements

Measurement of a surface by means of the moiré effect is very well known [86,48]. The fundamental characteristic of this technique is the formation of contours or level curves, as mentioned in Section 5.2.2, Equation 5.18. However, for some applications, it is useful to know contours of slopes or local surface curvature, mainly when it is required to know the field of mechanical strains on this surface. A useful technique to measure the slope and curvature of the surface is the reflection moiré [94–96]. This technique is based on projecting a linear grating on a mirror-like test surface. The projected grating is recorded photographically or electronically twice before and after subjecting the object to a flection. As a result, a moiré pattern is formed and interpreted as a slope map (derived) in the perpendicular direction to the lines of the projected grating.

Another possibility is to obtain a local curvature map on the plate [97]. To obtain this map, two slope moiré maps are superimposed, giving a second moiré pattern (moiré of moiré). The slope maps that will be superimposed to give the moiré of moiré are displaced a small quantity. This curvature map (second derived) is also perpendicular to the lines of the projected grating.

5.3.3 Interferometric Methods

Interferometric angle measurement seems an obvious task for a simple Michelson interferometer. A small tilt in one mirror produces a fringe pattern. Unfortunately, since the interferometer sensitivity is very high, this device can only practically measure very small angles. Interferometric angle measurement with both high precision and large range is a desired device. Angular measurements can be done [98] with a distance-measuring interferometer, built as shown in Figure 5.40. By tilting the retroreflectors assembly, the distance difference changes and the angle can be obtained from the equation:

$$\vartheta = \arcsin \frac{\Delta x}{L}, \tag{5.28}$$

where L is the mirror separation and Δx is the distance difference between retroreflectors. From this equation, it is evident that the angle precision depends on the angle, so the measurement range is also limited.

A laser Doppler displacement interferometer is used for a large angle precise measurement in a telescope. The setup is the same as the previously described system. With a single interferometer, the

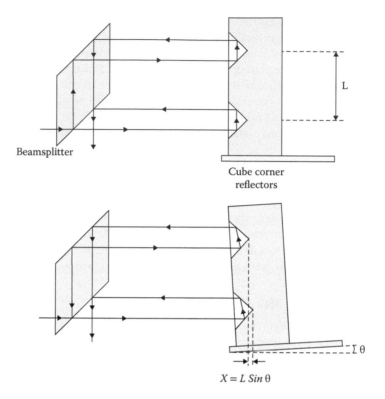

Beamsplitter

Cube corner
reflectors

$X = L \, Sin \, \theta$

FIGURE 5.40 Interferometric angular measurement.

resolution is 0.01 arc sec. For a mirror separation of 28 mm, the maximum angle is 4°. To cover the full 360° circle, a polygon and two measuring systems can be used [99].

A Murty's shearing interferometer has been proposed by Malacara and Harris [100] that is accurate within tenths of arc seconds. The basic setup is shown in Figure 5.41. A collimated beam of light is reflected from the two faces of a plane parallel faces glass. The angle can be obtained from the fringe count as the plate rotates. This method can be used for any angle but only for a limited angle span.

Laser speckle interferometry has been suggested for angle measurements. The objected is illuminated with a laser and a defocused image is formed at H (Figure 5.42). A double exposure of the object will form a fringe pattern that reflects the amount of rotation [101,102].

5.4 Velocity and Vibration Measurement

Velocity measurement u involves two physical magnitudes to be determined: displacement Δx and time Δt. The measurement of time is essentially a process of counting. Any phenomenon that repeats periodically can be used as a time measurement; this measure consists on counting the repetitions. Of the many phenomenon of this type that occur in nature, the rotation of the Earth around its axis is adopted. This movement, when reflected in the apparent movement of the stars and the sun, is a basic unit that one has easily to reach.

Another way of measuring time is artificially, by means of two apparatuses: one that generates periodic events and another to count these events. Today, optical clocks operate by means of atomic transition measurements between energy levels of the cesium atom [103,104]. Also, electronics can be used to implement instrumentation of time measurement. Perhaps, the most convenient and widely utilized instruments for accurate measurement of time interval and frequency are based on piezoelectric crystal oscillators, which generate a voltage whose frequency is very stable [105].

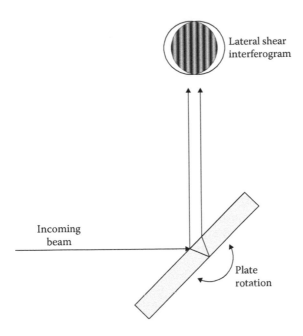

FIGURE 5.41 Lateral shear interferometer for angle measurement.

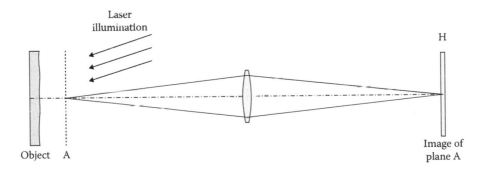

FIGURE 5.42 Speckle angle measurement.

Distance measurement also has its historical ingredient, and has been based on the observation of natural phenomena. For example, Eratosthenes (273 BC) tried to calculate the diameter of the Earth. This was based on his observations in Assuan that, in the summer solstice, the sun's rays fell perpendicularly and, therefore, a shadow was not cast, while at the same time, in Alexandria, there was a ray inclination of 7.2° in the shadows. Performing the appropriate calculations, this Greek astronomer obtained the Earth's diameter with an error of 400 km from the actual value (Eratosthenes measured 39,690 km; the actual diameter is 40,057 km). Currently, popular techniques for distance measurement use graduated rules, traced to the standard meter [106]. However, as we have seen in the present chapter, there are several optical techniques that can be used to make this measurement with a very high precision.

This section describes some different optical methods for local displacement, velocity, acceleration, and vibration measurements.

5.4.1 Velocity Measurement Using Stroboscopic Lamps

Translation and rotational velocity can conveniently be measured using electronic stroboscopic lamps, which emit light in a controlled and intermittent way. For the case of translation velocity measurement, the movement of the target is recorded on a photographic film when the camera shutter is opened and several shots of the strobe lamp are fired during that time. The measurement is made of the number of lamp shots time, and the serial position of the registered target. In order to measure the target displacement, a fixed rule is placed along the target path.

Sometimes, it is desirable to measure a value of average velocity of an object over a short distance or time interval, and velocity or time is not required in a continuous way. A useful basic optical method is to somehow generate a pulse of light when the object in movement passes through two points whose spacing is exactly known [107]. If the velocity were constant, any spacing could be used: for large spacing, of course, one has a better accuracy. If the velocity is varying, the spacing Δx should be small enough so that the average velocity over Δx is not very different from the velocity at either end of Δx.

Rotational velocity can be measured using a strobe lamp fixing the target and adjusting its shot frequency until the object is observed motionless. At this setting, the lamp frequency and motion frequency are identical, and the numerical value can be read from the lamp's calibrated dial for pulse repetition rate to an accuracy of about ±1% of the reading; or up to 0.01% in some units with crystal-controlled time base [107].

5.4.2 The Laser Interferometer

Although the principle of the interference of light as a measuring tool is very old, it continues to be advantageous. Albert A. Michelson, in the 1890s, was the first to use the interferometer, which bears his name, to measure distance and position in a very precise way.

The first efforts to use interferometry for the study of mechanical vibrations dates back to the 1920s [108,109]. The use of the interferometer replaced stroboscopic techniques in those experiments where measuring oscillation frequencies are higher than those in the range of stroboscopic lamps.

The advent of the laser in the 1960s popularized the interferometer as a useful optical instrument for measuring distances and displacements with high precision [58,54,110,111], as well as accelerometer and vibrometer calibration [112–114] to high measuring velocities.

Figure 5.43 shows a Michelson interferometer. A laser beam is divided in two parts: beam 1 hits the moving (test) mirror directly and beam 2 impinges on the fixed (reference) mirror. Due to the optical path difference between the overlapping beams, light and dark fringes are produced on the observing screen. The motionless test mirror produces a static fringe pattern. Cycles of maxima and minima

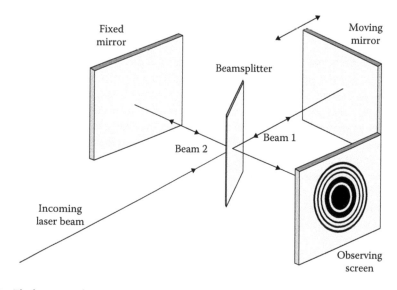

FIGURE 5.43 The laser interferometer.

passing through a fixed point on the observation screen are detected when the mobile mirror moves. Fringe displacement is due to a phase change (and therefore, optical path change) of the beam 1 with respect to the fixed one. If we know the laser light wavelength to be, for example 0.5×10^{-6} m, then each 0.25×10^{-6} m of mirror movement corresponds to one complete cycle (light to dark to light) at the fixed point on the observation screen. It is possible to calculate the displacement of the mobile mirror between two positions by counting the number of cycles. Therefore, the mobile mirror velocity is determined by calculating the number of cycles in the unit of time.

Electronic fringe counting involves two uniform interferograms, one of them with an additional quarter wavelength optical path difference introduced between the interfering beams. Two detectors viewing these fields provide signals in quadrature, which are used to drive a bidirectional counter, which gives the changes in the integral fringe order [115]. Nowadays, the two signals can also be processed in a microcomputer to give an accurate estimate of the fractional fringe order [116,114].

Another technique used to carry out fringe counting uses a laser emitting two frequencies, which avoids low-frequency laser noise [58]. These kinds of interferometers have been widely used for industrial measurements over distances up to 60 m.

Additional techniques, where the phase of moving interference fringe patterns is measured, have been applied. These methods are heterodyne phase measurement and phase lock detection [117–119]. In the heterodyne technique, also known as AC interferometry, the interferometer produces a continuous phase shift by introducing two different optical frequencies between both arms. With this approach, the interferogram intensity is modulated at the difference frequency. Phase lock interferometry involves applying a small sinusoidal oscillation to the reference mirror [120,121,110].

5.4.3 Laser Speckle Photography and Particle Image Velocimetry

Laser speckle photography (LSP) and particle image velocimetry (PIV) are two nearly related optical techniques for the measurement of in-plane two-dimensional displacement, rotation, and velocity [122,123]. LSP is used primarily for the measurement of the movement of solid surfaces, while PIV is used in applications of fluid dynamics. In both cases, the principle of operation is based on photographic recording under light laser illumination.

In LSP, light scattered from a moving object illuminated by coherent laser light is double-exposure photographed with a known time delay between exposures, as shown in Figure 5.44. In this way, locally identical but slightly shifted speckle patterns are recorded, which can be analyzed optically to find local displacement vectors at the surface of the moving object surface.

In PIV, a double-pulsed laser sheet of light is used to illuminate a plane within a seeded flow, which is photographed to produce a double-exposure transparency as shown in Figure 5.45. It is important

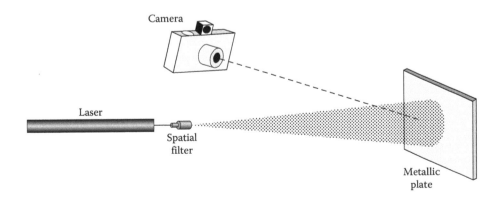

FIGURE 5.44 Experimental recording of laser speckle photography.

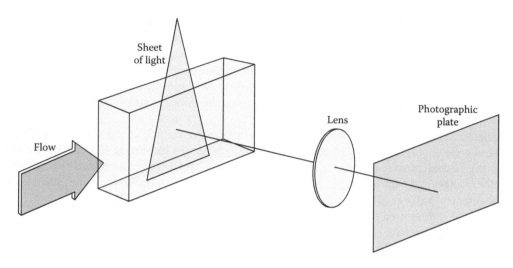

FIGURE 5.45   Particle image velocimetry.

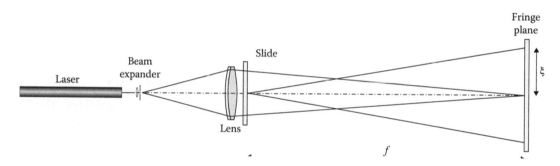

FIGURE 5.46   Fringe formation by the Fourier Method.

to note, however, that for practical seeding densities in PIV, the recorded image consists no longer of two overlapping speckle patterns but of discretely resolved particle-image pairs corresponding to both exposures.

In both cases, the double-exposure transparencies could be analyzed in two ways. The first method, a whole-field approach, consists of optically filtering the recorded transparency in a setup, as schematized in Figure 5.46.

Supposing that the object shift is s along the x-axis, displacement fringe contours appear on the Fourier focal plane described by [124]

$$\left| \mathcal{F}\left(\left| U(x) \right|^2 \right) \right|^2 \cos^2\left(\frac{s\omega}{2} \right),\tag{5.29}$$

where $\mathcal{F}\{U(x)\}$ means the Fourier transform of $U(x)$, the complex amplitude distribution of the object, and $\omega = 2\pi\xi/\lambda f$, with f being the focal length. Then, the object's displacement can be measured by

$$s_0 = \frac{\lambda f}{M\xi},\tag{5.30}$$

where M is a magnification produced by the recorder camera objective, and ξ denotes the distance from zero diffraction order, parallel to the x-axis.

The second method, a point-by-point approach, consists of scanning the transparency by means of a narrow laser beam (Figure 5.47). The laser beam diffracted by the speckles (or seed images in the PIV case) lying within the beam area gives rise to a diffraction halo. The halo is modulated by an equidistant system of fringes arising from the interference of two identical but displaced speckles, like the Young's interferometer. The directions of these fringes are perpendicular to the displacement direction. The magnitude of displacement, inversely proportional to the fringe spacing d, is given by

$$S = \frac{\lambda z}{Md}, \tag{5.31}$$

where z is the separation between the transparency and the observation screen. Figure 5.48 shows examples of such a point-by-point approach obtained for different linear displacements of a metallic plate [125]. Young fringes were formed by displacing the metallic plate 120, 100, 80, 60, 40, and 20 μm, respectively. Note that the number of fringes increases with the displacement.

By using LSP, in-plane rotation of a metallic plate is measured. In this case, in-plane displacements are not uniform point to point, because different speckles move in different directions. Near the rotation axis, speckles will move less than at the ends. The recording step is made in such a way that the rotation and viewing optical axes coincide. By placing the transparency to slide, in a movable stage in both horizontal and vertical directions, measurements on different points on the transparency were carried out. Knowing the distance r from rotation center to the sampling point and the displacement s, it is easy to know the rotation angle by means of the relationship

$$\alpha = \frac{Ms}{y}, \tag{5.32}$$

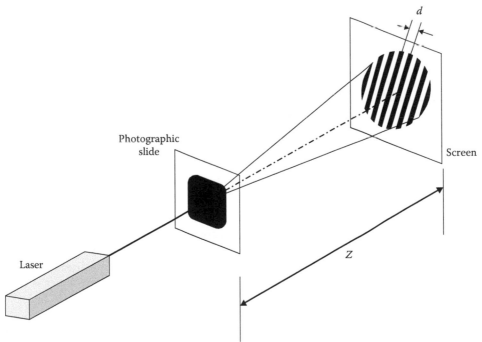

FIGURE 5.47 Young's fringe formation.

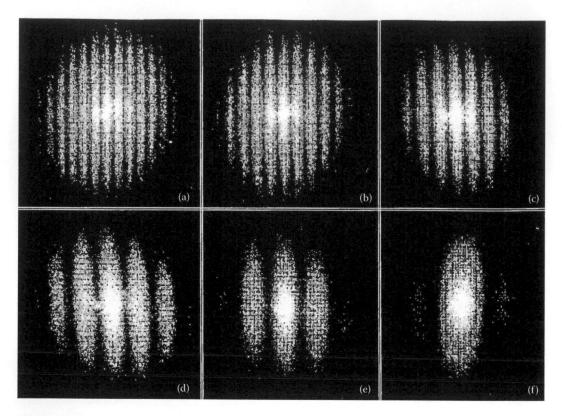

FIGURE 5.48 Young's experimental fringes for different displacements [125]. (With permission from J. Manuel Lopez-Ramirez.)

TABLE 5.1 Displacement and Angle for Rotational Movement Calculation

Zone	d (mm)	r (mm)	s (mm)	α (rad)
A	20.66	9.50	63.40	0.00148
B	21.60	9.00	60.64	0.00150
C	36.00	5.50	36.40	0.00147

by combining Equations. 5.32 and 5.31, we obtain:

$$\alpha = \frac{Mz}{yd}. \tag{5.33}$$

Figure 5.49A shows Young's fringe images for different points on the same plate. These points are localized, as shown in Figure 5.49B. Both direction and fringe separation can be measured directly on the observation screen. In order to check experimentally, three zones on the surface object are analyzed. Figure 5.50 shows such zones. Measurements from the center to each zone are measured directly on the photographic plate. In this case, the distance between transparence and the observation screen is $z = 46$ cm, the wavelength is 632.8 nm, and the optical recording system amplification is 0.2222. Table 5.1 show figures of the measurements obtained by using Equation 5.33.

Presented results have been made by manual evaluation of Young's fringes; therefore, it is a time-consuming, tedious and impractical procedure. For this reason, considerable effort has

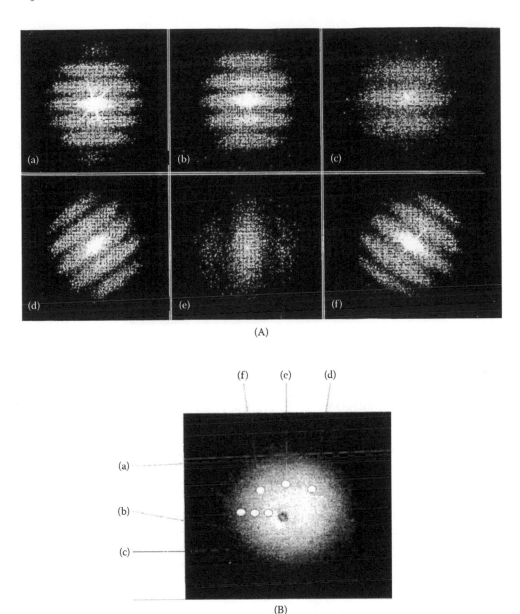

FIGURE 5.49 (A) Young's experimental fringes obtained for a plate under rotation from the points shown in (B) [125]. (With permission.)

been put into developing optical systems with digital image processing to automate the analysis procedure [126,127,119].

Other ways of reducing times of dynamic phenomena by means of PIV and LSP are using streak and CCD cameras instead of making photographic recordings [129–131]. Nowadays, experimental setups have been implemented by using two cameras to make three-dimensional measurements from stereoscopic images. Together with the modern recording instruments, computational techniques have been developed for a reliable and quick interpretation, such as the use of neural networks [128].

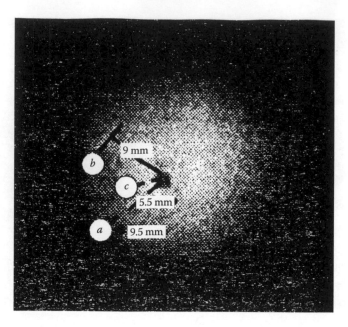

FIGURE 5.50   Zones analyzed in the plate [125]. (With permission.)

5.4.4 Laser Doppler Velocimetry

5.4.4.1 Physical Principle

Laser Doppler Velocimetry (LDV), also known as Laser Doppler Anemometry (LDA), is now a well-established optical nondestructive technique for local measurements of velocity. Although initially this technique, developed at the end of the 1960s, was applied in the field of fluid flows, it has now been extended to solid mechanics [132].

The physical principle of all LDVs lies in the detection of Doppler frequency shift of coherent light scattered from a moving object. The frequency shift is proportional to the component of its velocity along the bisector angle between illuminating and viewing directions. This frequency shift can be detected by beats produced either by the scattered light and a reference beam or by scattered light from two illuminating beams incident at different angles. An initial frequency offset can be used to distinguish between positive and negative movement direction [132].

Figure 5.51a shows a schematic diagram of the effect when a particle moving with a velocity u scatters light in a direction k_2 from a laser beam, fundamentally, a single frequency traveling in a direction k_1 where k_1 and k_2 are wavenumber vectors. The light frequency shift, Δf, produced by the moving point is given by [133]:

$$\Delta f = (k_2 - \mathbf{k}_1) \cdot u = K \cdot u, \tag{5.34}$$

where $K = k_2 - k_1$.

Figure 5.51b shows a geometry that is appropriate for solid surface velocity measurement where the laser beam is directed at a solid surface, which acts like a dense collection of particle scatters. In this situation, k_1 and k_2 are parallel so that the Doppler frequency shift measured corresponds to the surface velocity component in the direction of the incident beam and is given by

$$\Delta f = 2ku. \tag{5.35}$$

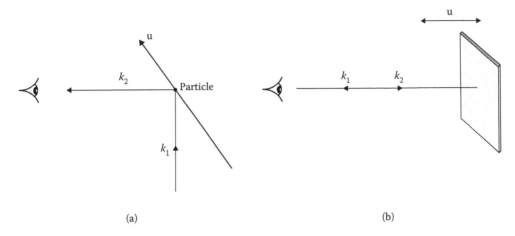

FIGURE 5.51 The Doppler effect: (a) transverse movement and (b) longitudinal movement.

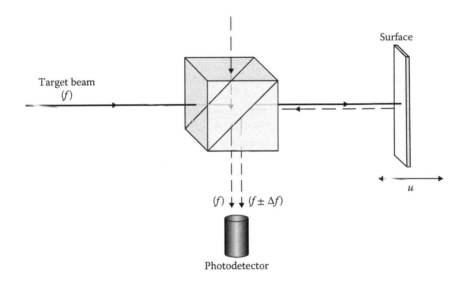

FIGURE 5.52 Reference beam mixing.

Measuring the frequency shift Δf, then the linear relationship with surface velocity to measure u can be used. Furthermore, tracking the changing Doppler frequency, it is possible to have a means of time-resolved measurement. The scattered light, as shown in Figures 5.51a and b has a frequency that is typically 10^{15} Hz; that is too high to demodulate directly. Therefore, Δf should be measured electronically by mixing the scattered light with another frequency shifted reference beam so that the two signals are heterodyned on a photodetector face. This is shown schematically in Figures 5.52 and 5.53, where a beam splitter has been used to mix the two beams. These figures also demonstrate the need to frequency pre-shift the reference beam. If the reference beam is not pre-shifted, as in Figure 5.52, then, when the target surface moves through zero velocity, the Doppler signal disappears and cannot be tracked [134]. Figure 5.53 shows the situation required where the target surface frequency modulates the carrier frequency, which is provided by the constant frequency pre-shift (f_R) in the reference beam. Frequency tracking the changing Doppler frequency then provides a time-resolved voltage analogue to the surface velocity.

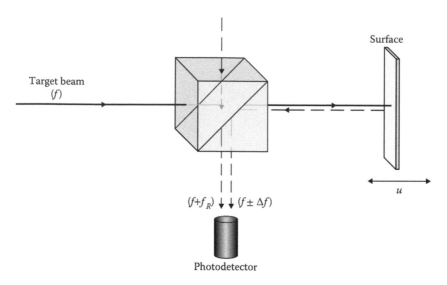

FIGURE 5.53   Pre-shifted reference beam mixing.

5.4.4.2 Frequency Shifting and Optical Setups

All LDVs work on the physical principle described above and differ only in the choice of optical geometry and the type of frequency-shifting device used. Just as frequency shifting is paramount for solid surface velocity measurement, it is also included as a standard item in commercially available LDV systems, which are used for flow measurement. It is obviously necessary for measurements in highly oscillatory flows and, in practice, it is extremely useful to have a carrier frequency corresponding to zero motion for alignment and calibration purposes. For instrumentation purposes, there are six types of signal that can be used to modulate sensors. These are mechanical, thermal, electrical, magnetic, chemical, and radiant [135]. But, the most commonly used form of frequency shifting is the Bragg cell [136]. The incident laser beam passes through a medium in which acoustic waves are traveling and the small-scale density variations diffract the beam into several orders. Water or glass is amongst the media of choice and usually the first-order diffracted beam is used. In this way, the frequency-shifted beam emerges at a slight angle to the incident beam, which requires compensation in some setups. Physical limitations often restrict the frequency shift provided by a Bragg cell to tens of megahertz, which is rather high for immediate frequency tracking demodulation. Consequently, two Bragg cells are often used, which shift both target and reference beams by typically 40 MHz and 41 MHz, respectively. Subsequent heterodyning then provides a carrier frequency that is readily demodulated. An alternative scheme sometimes utilized is to pre-shift with one cell and to electronically downbeat the photodetector output prior to demodulation [137].

Figures 5.54 and 5.55 show two possible optical setups, which can be used for solid surface velocity measurement and which incorporate Bragg cells for frequency shifting. Their compactness, electronic control, and freedom from mechanically moving parts make them popular as pre-shift devices in commercially available LDV systems for laboratory uses. Their presence does, however, add to the expense of the system, which then requires additional electronic and optical components.

Rotating a diffraction grating disk through an incident laser beam provides another common means of frequency shifting. Just as the small density variations in the Bragg cell diffracted the beam, in the diffraction grating case, the small periodic thickness variations perform the same task. Advantages over the Bragg cell are the smaller shifts obtained (~1 MHz) and the easy and close control of the latter through modification of the disk speed. Disadvantages are the mechanically moving

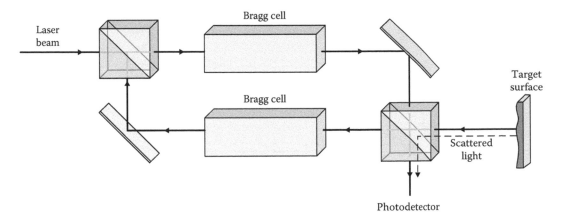

FIGURE 5.54 Geometry on axis.

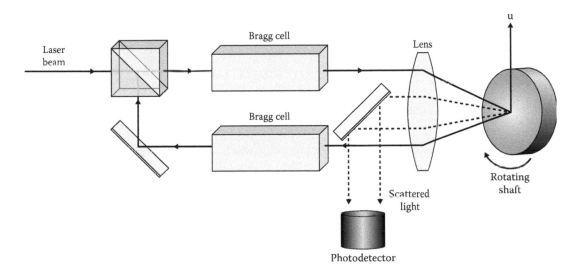

FIGURE 5.55 Torsional geometry.

parts and the inherent fragility of the disk itself, which is expensive to manufacture. Extra optical components are again needed to control the cross-sectional areas of the diffracted orders. A typical optical geometry for the measurement of solid surface vibration using a rotating diffraction grating as a frequency shifter is shown in Figure 5.56. Other frequency shifting devices that have been utilized consist of Kerr cells [138], Pockel's cells, and rotating scatter plates [139,140]. In the case of an LDV using a CO_2 laser, a Liquid-nitrogen-cooled mercury-cadmium-telluride (MCT) detector is used to detect a Doppler-shift signal [141]. The liquid nitrogen should be replenished in the MCT detector at a predetermined time interval. Lately, the optogalvanic effect has been investigated with the purpose of detecting a Doppler-shift signal in the self-mixing-type CO_2 LDV without using an MCT detector [142].

Another important variation of the LDVs uses stabilized double-frequency lasers [107,143]. These designs provide portable measuring systems that are highly precise and easy to use. For example, the one already mentioned in Figure 5.24 is shown a diagram of one such system.

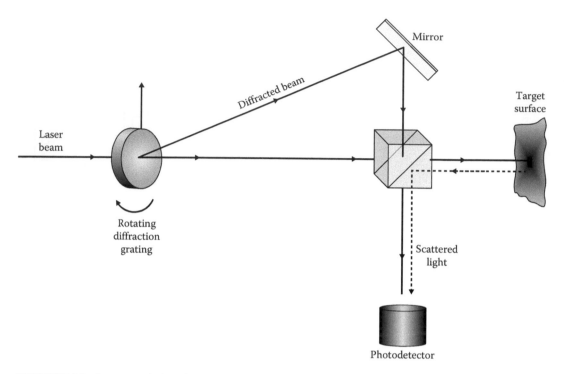

FIGURE 5.56 Frequency shifting by a diffraction grating.

Another particular geometry is shown in Figure 5.57, in which the in-plane surface velocity measurement is made. This experimental geometry is called *Doppler differential mode*, to differentiate it from the previous ones known as *reference beam mode*. In the Doppler differential mode, two symmetrical beams from the same laser are used for illuminating the moving point-object (or particle). The Doppler frequency shift is given by [144]

$$\Delta f = \frac{2u}{\lambda} \sin\frac{\theta}{2} \tag{5.36}$$

where λ is the wavelength of the laser light and θ is the angle between the two illuminating beams.

The operation of this dual-beam system is possible to visualize in terms of fringes applied in fluid dynamics. Where the two beams cross the region, the light waves interfere to form alternate regions of high and low intensity. If one particle traverses the fringe pattern, it will scatter more light when it crosses through regions of high intensity. Thus, the light received by the detector will show a varying electrical signal whose frequency is proportional to the rate at which the particle crosses the interference fringes.

5.4.4.3 Photodetector and Doppler Signal Processing

5.4.4.3.1 Photodetectors

A very important step in the measurement process of velocity is the manner of how the Doppler signal is detected and processed. Normally, a photodetector and its associated electronics for use in LDVs convert incident optical energy into electrical energy. Since early days, *photo sensors* have been used for fringe detection and are based on the photoelectric effect; that is, an incident photon on a given material may remove an electron from its surface (cathode). These removed electrons are accelerated toward, and collected by, an anode. This mechanism involves vacuum and gas-filled phototubes. Electron-ion pairs are also produced as the primary electrons collide and atoms of the gas, so that electron multiplication

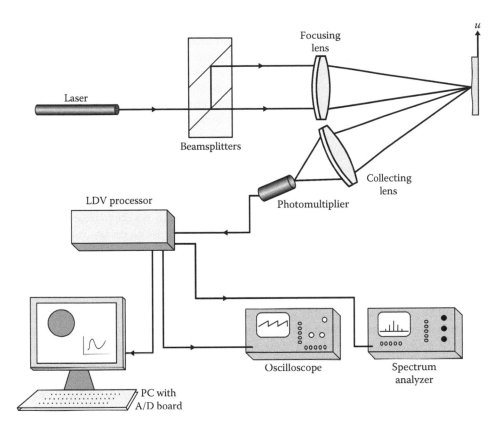

FIGURE 5.57 The differential geometry.

occurs. One of the most sensitive sensors of this family is the photomultiplier. This same effect can be carried out in solid-state sensors in which the photoelectric effect is made internally [34]. Phototubes are primarily sensitive to the energetic photons of the UV and blue regions, while solid-state detectors are sensitive to the red and IR regions of the spectrum.

There is a variety of different configurations and technologies used for the manufacture of solid-state sensors, including frame transfer and interline transfer charge-coupled devices (CCDs), charge injection devices (CIDs), and photodiode arrays. Solid-state sensors can be classified by their geometry as either area arrays or linear arrays. Area arrays are the most commonly used for image detection and permit the measurement of a two-dimensional section of a surface. The system spatial resolution is related to the number of pixels along each dimension of the sensor, and commonly available sensors have resolutions appropriate for television applications, typically about 500 × 500 pixels or less. Some newer sensors designed for machine vision applications or high-definition television, have dimensions of about 1000 × 1000 or even 2000 × 2000 pixels, but these sensors are expensive and are difficult to use. These kinds of sensors make slower frame rates and the amount of required computer memory and processing is quite high. Linear arrays, on the other hand, measure only one-dimensional trace across the part, but make up for this disadvantage by providing a higher spatial resolution along this line. Linear sensors containing over 7,000 pixels are available today, so that measurements with extremely high spatial resolution can be obtained. The amount of data from these large linear arrays is small and easily handled when compared with the 100,000 or more pixels on even a low-resolution area array. Because of advances in semiconductor fabrication, we can expect to see the dimensions of available sensors continuing to grow.

Photoresistors are sensors that consist essentially of a film of material whose resistance is a function of incident illumination. This film can either be an intrinsic or extrinsic semiconductor material, but for the visible part of the spectrum, the chalcogenides, cadmium sulphide and cadmium selenide, are the more common. The size and shape of the active film determines both the dark resistance and the sensitivity. However, for fringe counting, a small strip of material upon which the fringes can be focused is most suitable, and this strip must be connected to a pair of electrodes. The resistance of the cell decreases as the illumination is increased, so that it is common to consider the inverse of this resistance or the conductance, as being the basic parameter, which accounts for the alternative name of the sensor, the photoconductive cell. Electrically, therefore, it is resistance or conductance changes that must be detected. This implies that a current must be passed through the sensor, and either variation in this, or in associated voltage drops, should be measured.

If a single crystal of a semiconductor, such as silicon or germanium, is doped with both donor or acceptor impurities to form P and N regions, then the junction between these regions causes the crystal to exhibit diode properties. The commonly used photodiodes include the simple P-N junction diode, the P-I-N diode, and the avalanche diode [145]. The operation region of a photodiode is limited to that of light polarization change. Incident light to the junction P-N will give an energy transfer as a result of the atomic structure, which originates a bigger inverse current level. The current returns to zero once the polarization of the light changes 90°, and so forth.

5.4.4.3.2 Doppler Signal Processing

The choice of a Doppler signal-processing method is dictated by the characteristics of the Doppler signal itself, which is directly related to the particular measurement problem. In fluid flows (water excepted), for example, it is usually necessary to seed the flow with scattering particles in order to detect sufficient intensities of Doppler-shifted light [146]. Clearly for time-resolved measurements the ideal situation requires a continuous Doppler signal but unfortunately, in practice, the latter is often intermittent due to changes in seeding particle density, which occur naturally. The three most popular methods of signal processing are frequency tracking, frequency counting, and photon correlation (although in the specialized literature others like burst counting, Fabry-Perot interferometry, and filter bank are found). Tracking requires a nearly continuous Doppler signal, while correlation has been developed to deal with situations where seeding is virtually absent. The increased ability to deal with intermittent signals (in what is really a statistical sampling problem) is usually indicative of the expense of the commercial processor concerned, and the relationship is not linear.

Fortunately, in the case of solid-surface vibration measurements, Doppler signals are continuous and frequency tracking demodulation is the treatment of choice. With this form of processing, a voltage-controlled oscillator (VCO) is used to track the incoming Doppler signal and is controlled via a feedback loop [147,148]. Usually, a mixer at the input stage produces an "error" signal between the Doppler and VCO frequencies, which is bandpass filtered and weighted before being integrated and used to control the oscillator to drive the error to a minimum. The feedback loop has an associated "slew rate," which limits the frequency response of the processor. With respect to the Doppler signal, the tracker is really a low-pass filter that outputs the VCO voltage as a time-resolved voltage analogue of the changing frequency. The frequency range of interest for vibration measurements (20 kHz in dc) is well within the range of this form of frequency demodulation. Some trackers carry sophisticated weighting networks that tailor the control of the VCO according to the signal-to-noise ratios of the incoming signal. A simple form of this network will hold the last value of Doppler frequency being tracked if the amplitude of the signal drops below a preset level. In this way the Doppler signal effectively "drops out" and careful consideration must be given to the statistic of what is essentially a sampled output, especially when high-frequency information of the order of a drop-out period is required.

In several outdoors applications, the sensitivity of the LDV suffers along the free space path, from co-channel interference arising from spurious scattered light from rain, moisture, speckle, refractive index change, and dust, among others. New methods of demodulation signals from LDVs have been proposed. In order to minimize spurious scattering, an amplitude-locked loop (ALL) is combined with a phase-locked loop (PLL). The signal from the photodiode is down-converted to an intermediate frequency before being demodulated by a PLL to obtain baseband information, that is, the vibration frequency of the mirror. The ALL is a high-bandwidth servo loop that is able to obtain extra information on the amplitude of the spikes. The incoming corrupted FM signal is directly connected to the ALL. The output of the ALL gives a fixed output FM signal, which is connected to the PLL input [149,150].

Commercially available LDVs were originally designed for use in fluid-flow situations. Consequently, a great deal of research and development work has been directed toward solving the signal drop-out problem and other Doppler uncertainties produced by the finite size of the measurement volume. Since the early work in the middle to late 1970s, manufacturers now appear to prefer frequency counting for the standard laboratory system. This represents a successful compromise between tracking and correlation. Modern electronics will allow very fast processing so that a counter will provide a time-resolved analogue in a continuous signal situation while producing reliable data when seeding is sparse.

Compact LDVs have come to revolutionize the instrumentation of the velocity sensors. Optical systems of gradient index, together with diode lasers and optical fibers, have made the use of portable systems a reality. Fast response of modern photodetectors, as well as rapid computers for Doppler-shifted signal analysis and data reduction, has come to aid performance of such instruments [151–153]. Basic investigation has also continued to be carried out, such as near-field investigations to determine three-dimensional velocity fields by combining differential and reference beam LDVs and the use of evanescent waves for determining flow velocity [154,155].

5.4.5 Optical Vibrometers

Displacement sensors and transducers are used to modulate signals with the purpose of making measurements of dynamic magnitudes. In previous sections, a number of optical transducers have been analyzed, which have provided different forms of signal modulating to measure displacement and velocity. Now, in this section, a survey of punctual optical techniques to determine mechanical vibrations is given.

The measurement of vibration of a solid surface is usually achieved with an accelerometer or some other form of surface-contacting sensor [156]. There are, however, many cases of engineering interest where this approach is either impossible or impractical, such as for lightweight, very hot, or rotating surfaces. Practical examples of these are loudspeakers, engine exhausts, crankshafts, and so on. Since the advent of the laser in the early 1960s, optical metrology has provided a means of obtaining remote measurements of vibration in situations that had been hitherto thought unobtainable. The first demonstration of the use of the laser as a remote velocity sensor was in the measurement of a fluid flow [133]. The physical principles of the optical vibrometers have their roots in velocity and displacement measurement systems, like those described above. These concepts have already been treated in previous sections; however, they will be described again in an electronics context.

A variety of methods are at hand for utilizing the frequency (temporal) coherence, spatial coherence, or modulation capacity of laser light to measure the component's dynamics of a moving object. For general applications, the element under study is moving with displacements of normal and angular deflection, or tilt, with respect to some axis in the plane of the surface. Methods for detection for both types of motion will be described.

In general, the surface motion can be detected by observing its effect on the phase or frequency of a high-frequency subcarrier, which has been amplitude modulated onto the optical carrier, or by observing the phase or frequency changes (Doppler shift) on the reflected optical carrier itself. In addition, the surface acts as a source of reflected light that changes its orientation in space with respect to the optical receiver; hence, the arrival angle of reflected light varies with target position and can be detected. All of the vibration measurement techniques that will be described fit into these general classifications. From the historical point of view, optical vibrometers will be described, starting with the first systems that appeared in scientific literature (based on subcarriers) and their evolution, up to the state of the art (scanning LDV).

When the subcarrier methods are used, the vibration-induced phase shift is proportional to the ratio of the vibration amplitude to the subcarrier wavelength. Obviously, then, it is desirable to use the highest possible subcarrier or modulation frequency at which efficient modulation and detection can be performed.

When the optical carrier is used, a photodetector alone is not adequate, as already mentioned above, to detect frequency changes as small as those produced in this application. Some type of optical interference is required to convert optical phase variations on the reflected signal into intensity variations (interference fringes), which the photosensor can detect. A reference beam from the laser transmitter may be used to produce the interference with the incoming signal beam collected by the receiver optics. In such a system, the reference beam is often called the *local oscillator* beam, using the terminology established for radiofrequency receivers. Demodulation using a reference beam is known as coherent detection. If the frequency of the reference or local oscillator beam is the same as the transmitted signal, the system is a coherent optical phase detector, or *homodyne* system [157,158]. When the frequency of the reference is shifted with respect to the transmitted wavelength, an electrical beat is produced by the square-law photodetector at the difference, or intermediate, frequency between the two beams. Such a system is called a *heterodyne* detector or coherent optical intermediate frequency system [159,160]. Obviously, the coherent phase detector or homodyne is a special case of heterodyne detection with the intermediate frequency equal to zero.

5.4.5.1 Subcarrier Systems

Three different subcarrier systems are analyzed. Each of them uses an electro-optic amplitude modulator to vary the intensity of the transmitted laser beam at a microwave rate [135]. In practical devices, only a fraction M of the light intensity is modulated. These intensity variations represent a subcarrier envelope on the optical carrier; they can be detected by a photodetector, which has good enough high-frequency performance to respond to intensity variations at the sub-carrier frequency. When the modulated light is reflected from the moving surface, the phase of the subcarrier envelope will vary with time, according to the expression:

$$\phi(z) = \frac{4\pi z_0}{\lambda_m} \sin \omega_r t \qquad (5.37)$$

where z_0 is the zero-to-peak vibration amplitude, $\omega_r = 2\pi f_r$ is the vibration frequency, and λ_m is the microwave subcarrier wavelength. For a subcarrier frequency of 3 GHz, λ_m is 10 cm; thus, the peak of phase deviation for small vibrations is much less than 1 rad. The information on the vibration state of the surface appears on the reflected light in the subcarrier sidebands produced by the time-varying phase changes. For small peak-phase deviations, which are always of interest in determining the maximum sensitivity of a given system, only the first two sidebands are significant (the exact expression gives an infinite set of sidebands with Bessel function amplitudes, most of which are negligibly small). In that case, the reflected spectrum of light intensity is given by the approximation:

$$P = \frac{P_0}{2} + \frac{MP_0}{2} \sin \omega_m t + \frac{MP_0}{2\lambda_m}\pi\chi_0 \sin(\omega_m + \omega_r)t + \frac{MP_0\pi\chi_0}{2\lambda_m} \sin(\omega_m - \omega_r)t, \qquad (5.38)$$

where ω_m is the microwave signal, M is the modulation index of the incident light, and P_0 is the reflected power. The first term of Equation 5.38 is the unmodulated light; the second term, the subcarrier; and, the third term shows the sidebands. The necessary receiver bandwidth is simply the maximum range of vibration frequencies to be measured. Thus, any usable microwave system must be able to detect the sideband amplitudes in the presence of receiver noise integrated over the required bandwidth. It should be noted that Equation 5.38 is for optical power; the demodulated power spectral components in the electrical circuits will be proportional to the squares of the individual terms in Equation 5.38.

5.4.5.1.1 *Direct Phase Detection System*

This approach is the simplest subcarrier technique and is shown schematically in Figure 5.58. The subcarrier is recovered after reflection by the photodetector, which might be a tube or solid state diode. The diode has the disadvantages of no multiplication and effective output impedances of only a few hundred ohms at best. However, solid-state quantum efficiencies can approach unity [161].

The low-noise traveling-wave amplifier and tunnel diode limiter are required to suppress low-frequency amplitude fluctuations on the received signal due to surface tilt, laser amplitude fluctuations, and other effects. The balanced mixer further rejects amplitude modulation on either of the microwave input signals.

The voltage-tunable magnetron exciter for the optical modulator is used. This might be tuned by a servo-control system, not shown in Figure 5.58, to follow the drift in modulator resonant frequency during warm-up. Such a servo-system would include a sensor placed in or near the modulator cavity to sample the phase and amplitude of the cavity field. This signal would be processed by narrow-band electronics to derive the tuning voltage needed to make the oscillator track the cavity resonance.

The directional coupler, attenuator, and variable phase shifter provide the reference or local oscillator signal to the balanced mixer at the phase and amplitude for phase quadrature detection.

5.4.5.1.2 *Intermediate Frequency Detection System*

This technique is illustrated schematically in Figure 5.59. The receiver portion of this system differs from the one above because the local oscillator signal to the micro-wave mixer has been shifted in frequency. This is the purpose of the intermediate frequency (IF) generator and sideband filter. The microwave mixer produces a beat frequency when a signal is present. A second mixer, known as a phase detector, is used to demodulate the phase modulation produced on the IF by the surface motion. Thus, the mixing down to audio is done in two steps. This improves the mixer noise figures, and with limiting in the IF amplifier, it might be possible to operate the phototube directly into the first mixer. As pointed out in

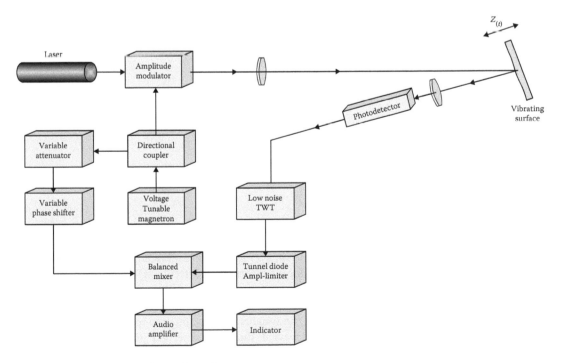

FIGURE 5.58 Microwave direct-phase detection system.

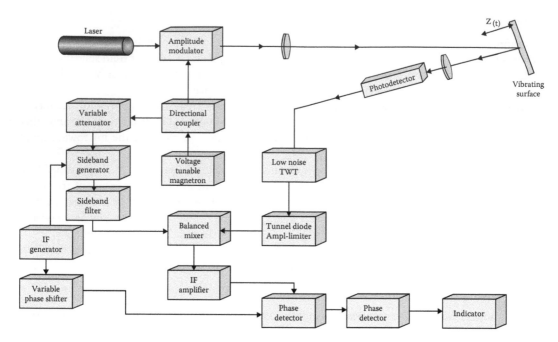

FIGURE 5.59   Microwave intermediate frequency detection system.

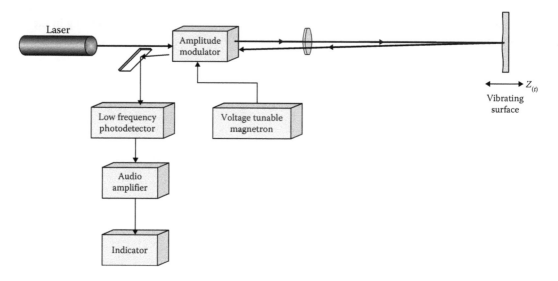

FIGURE 5.60   Microwave double modulation system.

the previous section, the mixer noise is negligible anyway, so the additional complexity of this approach appears unnecessary. Sensitivity is the same as for the direct phase detector analyzed above.

5.4.5.1.3 Double Modulation System

The fundamental limitation in performance of the systems above is imposed by the low quantum efficiency of the microwave phototube. A method that avoids this is illustrated in Figure 5.60. Here, the phase demodulation is done not with a high-frequency detector but with a gated receiver and low-frequency detector. Gating of the microwave rate is accomplished by passing the reflected light back through the microwave optical modulator. A beamsplitter mirror or calcite prism and quarter-wave plate might be used to allow transmission and reception through the same optical system. The re-modulated signal in this case will have audio frequency intensity variations corresponding to the subcarrier phase shifts (Doppler effect) produced by the vibration. Efficient, low-noise phototubes and diodes are now available at these frequencies.

With the external optical path adjusted so that the second modulation occurs 90° out of phase with the first one, the detected optical power is of the form:

$$P = P_r \left[1 + M^2 \frac{2\pi z_0}{\lambda_m} \sin(\omega, t) \right], \tag{5.39}$$

where P_r is the average power reaching the photodetector.

Several limitations were found in this system. It was not possible to detect the phase shift on the subcarrier because of the large spurious amplitude fluctuations due to surface tilt. This was true even when the vibrating mirror was placed at the focal point of a lens, an optical geometry that minimizes the angular sensitivity. Thus, any substantial amplitude change due to laser noise or reflected beam deflection will overcome the desired signal. Another limitation is that the subcarrier demodulation to audio is done optically; thus, it is not possible to use limiting to remove the large amplitude variations before detection. For this reason, it is doubtful that this system could be used to its theoretical limit of performance even if the modulator were improved to make M almost unity. If this problem could somehow be eliminated, the double-modulation system would realize the advantage of requiring no microwave receiver components and could make use of the best possible optical detectors.

5.4.6 Types of Laser Doppler Vibrometers

For overcoming the limitations imposed by subcarrier systems, coherent optical detection methods, based on the Doppler Effect, began to be used. Combining the signal beam with a local oscillator beam that acts as a phase reference can coherently demodulate the vibration-induced phase shifts on the optical carrier. When both beams are properly aligned and are incident on an optical detector, the output current is proportional to the square of the total incident electric field. This current may be written as [160]

$$I(t) = I_r + I_s + 2\sqrt{I_r I_s} \cos(\omega_r - \omega_s)t, \tag{5.40}$$

where I_r is the local oscillator field only, l_s, is the direct field due to signal alone, and ω_r and ω_s are the frequencies of the local oscillator and the signal wave, respectively (see Figure 5.24). The instantaneous frequency shifts from the local oscillator and target surface are, equivalently,

$$\omega_r = 2ku + \phi, \tag{5.41}$$

$$\omega_s = 2kz_0 \,\omega_V \cos(\omega_V t), \tag{5.42}$$

where k is the wavenumber of the laser light and $(z_0 \cos \omega_v t)$ represents the target surface displacement of amplitude z_0 and frequency ω_v.

$$i(t) = A \cos\left\{ 2k\left[u - z_0 \omega_V \cos(\omega_V t) \right] t + \int(t) \right\}, \tag{5.43}$$

The function $\varphi(t)$ represents a pseudo-random phase contribution due to the changing population of particulate scatters in the laser spot. Neglecting constant terms, we can write $A = (I_r l_s)^{1/2}$ and $\epsilon(t) = t\varphi(t)$. The function $\epsilon(t)$ is *pseudo-random* since, when using a reference beam oscillating to frequency shift, the spatial distribution of scatters repeats after each revolution. These cause the frequency spectrum of the noise floor of the instrument to be a period gram since the random amplitude modulation of $i(t)$ due to $\epsilon(t)$ repeats exactly after each oscillation period.

With reference to Equation 5.43, a frequency tracking demodulator follows the frequency modulation of the carrier frequency ($2\,ku$) to produce a voltage output that is an analogue of the changing surface velocity of amplitude ($z_o\omega_v$).

5.4.6.1 Referenced (Out-of-Plane) Vibrometer

This LDV measures the vibrational component $z(t)$ that lies along the laser beam, already analyzed for velocity measuring, which is the most common type of LVD system. The system is a heterodyne interferometer, as shown in Figure 5.61, which means that the signal and reference beams are frequency shifted relative to one another to allow the FM carrier generation. The two outputs of the interferometer provide complementary signals, which, when differentially combined, generate a zero-centered FM carrier signal. The superimposed frequency (or phase) shifts are related to the surface position via the wavelength of the laser used as optical source, given by Equation 5.40. This system has a resolution up to 10^{-15} m $(Hz)^{1/2}$ [148].

Axial measurements can be obtained by approaching the same measurement point from three different directions.

5.4.6.2 Dual Beam (In-Plane) or Differential Vibrometer

The basic differential LDV vibrometer arrangement is shown in Figure 5.57. The output from a laser is split into two beams of about equal intensity. A lens focuses the two beams together in a small spot on the face of a vibrating plate in direction $x(t)$, which is the target surface. Light scattered by the target surface is collected by a second lens and focused into a pinhole in front of a photodetector. The photodetector output is processed by an LDV counter in a similar manner to that of referenced vibrometer [154]. By rotating the probe by 90°, $x(t)$ or $y(t)$ can be measured.

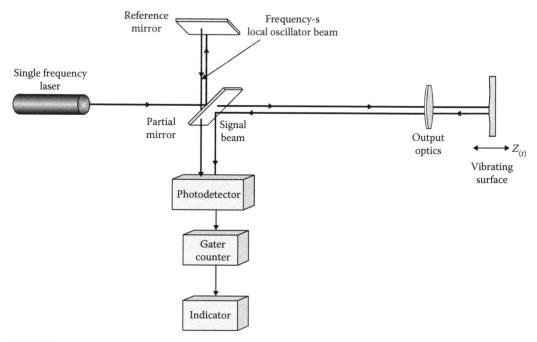

FIGURE 5.61 Out-of-plane system.

5.4.6.3 Scanning Vibrometer

An extension of the standard out-of-plane system, the scanning LDV uses computer-controlled deflection mirrors to direct the laser to a user-selected array of measurement points. The system automatically collects and processes vibration data at each point, scales the data in standard displacement, velocity, or acceleration engineering units, performs fast Fourier transform (FFT) or other operations, and displays full-field vibration pattern images and animated operational deflection shapes. The role of this system is found in its scanning system. Scanning mirrors can be moved by using voltage changes [162] or galvanometer-based [163,164], requiring high-precision mechanical mounts. But once the scanning system is calibrated, the unit gives a haughty standard LDV.

5.4.6.4 Spot Projection Systems

Other not less important optical vibrometers are spot projection-based or triangulation principle methods [107]. If the moving surface is a diffuse reflector, it is possible to obtain information about some components of the motion by projecting one or more spots of laser light onto the surface and measuring the motion-induced effects on reflected light collected by an optical receiver. One method, in which the apparent motion of the spot is measured, does not make use of the spectral coherence of the laser. This system has been called the incoherent spot projection technique or triangulation-based technique. Another approach, in which two spots are projected and the interference between reflected waves from both of them is used, is known as coherent spot projection, because the laser coherence is utilized.

5.4.6.4.1 Incoherent Spot Projection System

This system is illustrated in Figure 5.62. The laser beam is projected to a small spot on the vibrating surface. A rectangular receiver aperture collects some of the reflected light. In the receiver focal plane, motion of the surface produces a lateral motion of the spot image. If a knife edge stop is placed a short distance behind the image, where the beam has expanded to a rectangle, motion of the image affects the fraction of the light that passes the stop. In practice, the stop would cover half the beam on average, and the distance behind focus would be adjusted to accommodate the largest expected image displacements in the linear range. The fraction of power passing the knife edge is measured by a photodetector whose current output is a linear analogue of the surface displacement along the axis of the transmitted beam.

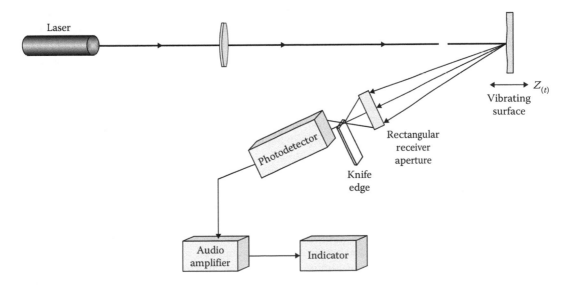

FIGURE 5.62 Incoherent spot projection.

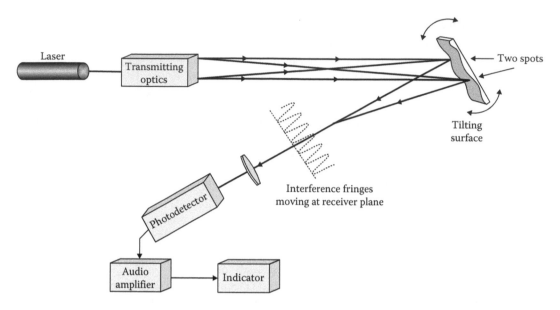

FIGURE 5.63   Coherent spot projection.

It is interesting to notice that the sensitivity can approach that of the Doppler-shift-based systems; if the maximum displacement is no greater than an optical wavelength, good optics are used, and the receiver aperture is large. In such a situation, the spatial coherence of the laser is fully utilized. Such a system is therefore not incoherent in the limit.

For large vibrations the system has many advantages and a few disadvantages. The main disadvantage is the need for careful alignment of the receiver and the knife edge. For scanning, the separate receiver and transmitter are troublesome. However, tolerance on the optical components is not severe for large maximum displacements, and other components are simple and reliable.

5.4.6.4.2 Coherent Spot Projection System

The above *incoherent* system measures normal displacement of the surface. The coherent system to be described measures angular tilt of the surface in the plane determined by the transmitter and receiver axes. Figure 5.63 illustrates the system schematically. Two small spots are projected by high-quality optics. They are separated on the surface by a distance approximately equal to a spot diameter. Then, at the receiver plane there will be interference fringes produced by reflected light from the two spots. If the spot separation is sufficiently small, the fringes can become large enough to fill a receiver aperture of a few inches. A tilt of the surface corresponding to a relative motion between the spots equal to 1/2 wavelength will move the fringe pattern laterally by a full spatial period. Power changes due to motion of the fringe pattern are detected in the receiver. The spot separation corresponding to a 3-inch receiver at 3 feet is about 4×10^{-4} inches or 10^{-2} mm. The tilt associated with a half-wave relative motion is then approximately 30 mrad, or about 1.7°.

Some limitations of this technique are encountered. First of all, the diffuse surface causes self-interference in the reflected light from each spot. Thus, at any point in the receiver plane, the fields from the spots are likely to be far from equal. Consequently, the desired fringes have very low contrast over most of the plane, and must be detected in the presence of a large, strongly modulated background of random interference. The second problem is the inability to produce very small spots on a diffuse surface. Scattering among the rough elements near the illuminated region spreads the effective spot size considerably, thus reducing the maximum usable receiver aperture. Because of these problems, the receiver aperture needs to be carefully positioned in the fringe pattern for linear demodulation (a condition difficult to meet if the surface must be scanned).

5.4.7 Vibration Analysis by Moiré Techniques

The first works on moiré techniques that appeared for vibration measuring were based on the use of two gratings in contact [165]. Maximum and minimum of the moiré pattern in movement, due to vibration, are detected by means of photoelectric cells and transduced through an electronic system. Recently, the use of techniques, such as fringe projection [48,51], reflection moiré [166,96], shadow moiré [167], moiré deflectometry [168], and holographic moiré [169] have been applied to vibration analysis. All these techniques have been used in a similar way to the time-average speckle technique [170]. On vibrating surfaces, antinodal positions will continuously give varying deflections, while nodes will produce zero deflection. A time-averaged photograph of a projected grating on to a vibrating object will produce areas of washed-out grating at the antinodes and a sharp grating image at the nodal positions. If a second grating is superimposed on the photograph to produce dark fringes at the nodes, the antinodes will appear brilliant, and one immediately has a contour map of the nodal positions.

References

1. Swyt DA The international standard of length, in *Coordinate Measuring Machines and Systems.*, Bosch, J. A., ed., New York, NY: Marcel Dekker, 1995.
2. Baird KM, Howlett LE The international length standard, *Appl. Opt.*, **2**, 455–463; 1963.
3. Giacomo P. Metrology and fundamental constants, *Proceedings of the International School of Physics Enrico Fermi, course 68*, North Holland, the Netherlands, 1980.
4. Goldman DT. Proposed new definitions of the meter, *J. Opt. Soc. Am.*, **70**, 1640–1641; 1980.
5. Smith WJ. *Modern Optical Engineering*, New York, NY: McGraw-Hill, 1966.
6. Tocher AJ. US Patent # 5483336, 1992.
7. Schmidt H. US Patent # 4465366, 1984.
8. Buddenhagen DA, Lengyel BA, McClung FJ, Smith GF. *Proceedings of the IRE International Convention*, New York, NY, Part 5 (Institute of Radio Engineers, New York), 1961, p. 285.
9. Stitch ML, Woodbury EJ, Morse JH. Optical ranging system uses laser transmitters, *Electronics*, **34**, 51–53; 1961.
10. Stitch ML. *Laser Ranging, in Laser Handbook*, Vol. **2**, Arecchi, F. T. and E. O. Schulz- Dubois, eds, Elsevier (North-Holland Publishing Co), Amsterdam, Netherlands, 1972, pp. 1745–1804.
11. Faller EF, Wampler EJ. The lunar laser reflector, *Sci. Am.*, **223**(3), 38–47, March, 1970.
12. Rüeger JM, Paseo RW. Performance of a distomat wild DI 3000 distance meter, Tech. Paper No. 19, *Aust. Survey Congress, Hobart*, 1989.
13. Sona A. *Laser Handbook*, Vol. **2**, Arecchi, F., and E. Schulz-Dubois, eds, Elsevier (North-Holland Publishing Co), Amsterdam, Netherlands, 1972.
14. Burnside CD. *Electronic Distance Measurement*, 3rd edn, London, UK: BSP Professional Books, 1991.
15. Takeshima A. US Patent # 5534992, 1996.
16. Shipley G, Bradsell RH. Georan 1, a Compact Two-Color EDM Instrument, *Survey Review, XXIII*, 1976.
17. Dändliker R, Savadé Y, Zimmermann E. Distance measurement by multiple-wavelength interferometry, *J. Opt.*, **29**, 105–114; 1998.
18. Cooke F. The bar spherometer, *Appl. Opt.*, **3**, 87–88; 1964.
19. Jurek B. *Optical Surfaces*, New York, NY: EIsevier Scientific Publ. Co, 1977.
20. Malacara D. (ed.), *Optical Shop Testing*, 2nd edn, New York, NY: Wiley, 1992.
21. Horne DF. *Optical Production Technology*, Adam Hilger, London, and Crane Russak, New York, 1972, Chapter XI.
22. Carnell KH, Welford WT. A method for precision spherometry of concave surfaces, *J. Phys.*, **E4**, 1060–1062; 1971.

23. Boyd WR. 1969, *The use of a collimator for measuring domes*, Appl. Opt., **8**, 792; 1969.
24. Cooke F. Optics activities in industry, *Appl. Opt.*, **2**, 328–329; 1963.
25. Gerchman MC, Hunter GC. Differential technique for accurately measuring the radius of curvature of long radius concave optical surfaces, *Proc. SPIE*, **192**, 75–84; 1979.
26. Gerchman MC, Hunter GC. Differential technique for accurately measuring the radius of curvature of long radius concave optical surfaces, *Opt. Eng.*, **19**, 843–848; 1980.
27. Kafri O, Glatt I. *The Physics of Moiré Metrology*, New York, NY: John Wiley, 1990.
28. Patorski K. *Handbook of the Moiré Fringe Technique*, Amsterdam, the Netherlands: Elsevier, 1993.
29. Post D, Han B, Ifju P. *High Sensitivity Moiré*, New York, NY: Springer-Verlag, 1994.
30. Nishijima Y, Oster G. Moiré patterns: Their application to refractive index and refractive index gradient measurements, *J. Opt. Soc. Am.*, **54**, 1–5; 1964.
31. Asundi A, Yung H. Phase-shifting and logical moiré, *J. Opt. Soc. Am.*, **8**, 1591–1600; 1991.
32. Rodriguez-Vera R. Three-dimensional gauging by electronic moiré contouring, *Rev. Mex. Fis.*, **40**, 447–458; 1994.
33. Luxmoore AR. *Optical Transducers and Techniques in Engineering Measurement*, London, UK: Applied Science Publishers; 1983.
34. Watson J. Photodetectors and electronics, in *Optical Transducers and Techniques in Engineering Measurement*, Luxmoore, A. R., ed., London, UK: Applied Science Publishers, 1983, Chapter 1.
35. Farago FT. *Handbook of Dimensional Measurement*, 2nd edn, New York, NY: Industrial Press, 1982.
36. Malacara D. Some properties of the near field of diffraction gratings, *Opt. Acta.*, **21**, 631–641; 1974.
37. Jutamulia S, Lin TW, Yu FTS. Real-time color-coding of depth using a white-light Talbot interferometer, *Opt. Comm.*, **58**, 78–82; 1986.
38. Rodriguez-Vera R, Kerr D, Mendoza-Santoyo F. Three-dimensional contouring of diffuse objects using Talbot interferometry, *Proc. SPIE*, **1553**, 55–65; 1991a.
39. Silva AA, Rodriguez-Vera R. Design of an optical level using the Talbot effect, *Proc. SPIE*, **2730**, 423–426; 1996.
40. Nakano Y, Murata K. Talbot interferometry for measuring the focal length of a lens, *Appl. Opt.*, **24**, 3162–3166; 1985.
41. Chang Ch-W, Su D-Ch. An improved technique of measuring the focal length of a lens, *Opt. Comm.*, **73**, 257–262; 1989.
42. Su D-Ch, Chang Ch-W. A new technique for measuring the effective focal length of a thick lens or a compound lens, *Opt. Comm.*, **78**, 118–122; 1990.
43. Glatt I, Kafri O Determination of the focal length of non-paraxial lenses by moiré deflectometry, *Appl. Opt.*, **26**, 2507-2508; 1987b.
44. Bernardo LM, Soares ODD. Evaluation of the focal distance of a lens by Talbot interferometry, *Appl. Opt.*, **27**, 296–301; 1988.
45. Sriram KVSr., Kothiyal MP, Sirohi RS. Use of non-collimated beam for determining the focal length of a lens by Talbot interferometry, *J. Optics*, **22**, 61–66; 1993.
46. Nakano Y, Ohmura R, Murata K. Refractive power mapping of progressive power lenses using Talbot interferometry and digital image processing, *Opt. Laser Technol.*, **22**, 195–198; 1990.
47. Malacara Doblado D. *Problems Associated to the Analysis of Interferograms and their Possible Applications*, PhD thesis, CIO-Mexico, 1995.
48. Hovanesian JD, Hung YY. Moiré contour-sum, contour-difference, and vibration analysis of arbitrary objects, *Appl. Opt.*, **10**, 2734; 1971.
49. Rodriguez-Vera R, Kerr D, Mendoza-Santoyo F. 3-D contouring of diffuse objects by Talbot-projected fringes, *J. Mod. Opt.*, **38**, 1935–1945; 1991b.
50. Rodriguez-Vera R, Servin M. Phase locked loop profilometry, *Opt. Laser Technol.*, **26**, 393–398; 1994.
51. Dessus B, Leblanc M. The 'Fringe Method' and its application to the measurement of deformations, vibrations, contour lines and differences, *Opto-Electronics*, **5**, 369–391; 1973.

52. Rodriguez-Vera R, Kerr D, Mendoza-Santoyo F. Electronic speckle contouring, *J. Opt. Soc. Am. A*, **9**, 2000–2008; 1992.

53. Muñoz Rodriguez A, Rodriguez-Vera R, Servin M. Direct object shape detection based on skeleton extraction of a light line, *Opt Eng.*, **39**, 2463–2471; 2000.

54. Hariharan P. Interferometric metrology: Current trends and future prospects, *Proc. SPIE*, **816**, 2–18; 1987.

55. McDuff OP. Techniques of gas lasers, in *Laser Handbook*, Vol. 1, Arecchi, F. T., and E. O. Schulz-Dubois, eds, Elsevier (North-Holland Publishing Co), Amsterdam, Netherlands, 1972, pp. 631–702.

56. Minkowitz S, Smith-Vanir WA. Laser interferometer, *J. Quantum Electronics*, **3**, 237; 1967.

57. Burgwald GM, Kruger WP. An instant-on laser for length measurements, *Hewlett Packard J.*, 21; 1970.

58. Dukes JN, Gordon GB. A two-hundred-foot yardstick with graduations every microinch, *Hewlett-Packard J.*, **21**, 2–8; 1970.

59. Bourdet GL, Orszag AG. Absolute distance measurements by CO_2 laser multiwavelength interferometry, *Appl. Opt.*, **18**, 225–227; 1979.

60. Rovati L, Minoni U, Bonardi M, Docchio F. Absolute distance measurement using combsSpectrum interferometry, *J. Opt.*, **29**, 121–127; 1998.

61. Bechstein K, Fuchs W. Absolute interferometric distance measurements applying a variable synthetic wavelength, *J. Opt.*, **28**, 179–182; 1998.

62. Kothiyal MP, Sirohi RS. Improved collimation testing using Talbot interferometry, *Appl. Opt.*, **26**, 4056–4057; 1987.

63. Hume KJ. *Metrology with Autocollimators*, London, UK: Hilger and Watts, 1965.

64. Horne DF. *Dividing, Ruling and Mask Making*, London, UK: Adam Hilger, 1974, Chapter VII.

65. Young AW. Optical workshop instruments, in *Applied Optics and Optical Engineering*, Kingslake, R., ed., New York, NY: Academic Press, 1967, Chapter 7.

66. Martin LC. *Optical Measuring Instruments*, London, UK: Blackie and Sons Ltd, 1924.

67. Johnson BK. *Optics and Optical Instruments*, New York, NY: Dover, 1947, Chapters II and VIII.

68. Taarev AM. Testing the angles of high-precision prisms by means of an autocollimator and a mirror unit, *Sov. J. Opt. Technol.*, **52**, 50–52; 1985.

69. Twyman F. *Prisms and Lens Making.*, 2nd edn, London, UK: Hilger and Watts, 1957.

70. DeVany AS. Reduplication of a penta-prism angle using master angle prisms and plano interferometer, *Appl. Opt.*, **10**, 1371–1375; 1971.

71. Malacara D, Flores R. A simple test for the 90 degrees angle in prisms, *Proc. SPIE*, **1332**, 678; 1990.

72. DeVany AS. Making and testing right angle and dove prisms, *Appl. Opt.*, **7**, 1085–1087; 1968.

73. DeVany AS. Testing glass reflecting-angles of prisms, *Appl. Opt.*, **17**, 1661–1662; 1978.

74. Ratajczyk F, Bodner Z. An autocollimation measurement of the right angle error with the help of polarized light, *Appl. Opt.*, **5**, 755–758; 1966.

75. Kingslake R. *Optical System Design*, New York, NY: Academic Press, 1983, Chapter 13.

76. Ahrend M. Recent photogrammetric and geodetic instruments, *Appl. Opt.*, **7**, 371–374; 1968.

77. Murty MVRK, Shukla RP. Methods for measurement of parallelism of optically parallel plates, *Opt. Eng.*, **18**, 352–353; 1979.

78. Tew EJ. Measurement techniques used in the optics workshop, *Appl. Opt.*, **5**, 695–700; 1966.

79. Nakano Y, Murata K. Talbot interferometry for measuring the small tilt angle variation of an object surface, *Appl. Opt.*, **25**, 2475–2477; 1986.

80. Nakano Y. Measurements of the small tilt-angle variation of an object surface using moiré interferometry and digital image processing, *Appl. Opt.*, **26**, 3911–3914; 1987.

81. Glatt I, Kafri O. Beam direction determination by moiré deflectometry using circular gratings, *Appl. Opt.*, **26**, 4051–4053; 1987a.

82. Ng TW, Chau FS. Object illumination angle measurement in speckle Interferometry, *Appl. Opt.*, **33**, 5959–5965; 1994.

83. Ng TW. Circular grating moiré deflectometry analysis by zeroth and first radial fringe order angle measurement, *Opt. Comm.*, **129**, 344–346; 1996.

84. Dai X, Sasaki O, Greivenkamp JE, Suzuki T. Measurement of small rotation angles by using a parallel interference pattern, *Appl. Opt.*, **34**, 6380–6388; 1995.

85. Dai X, Sasaki O, Greivenkamp JE, Suzuki T. High accuracy, wide range, rotation angle measurement by the use of two parallel interference patterns, *Appl. Opt.*, **36**, 6190–6195; 1997.

86. Takasaki H. Moiré topography, *Appl. Opt.*, **9**, 1457; 1970.

87. Silva DE. A simple interferometric method of beam collimation, *Appl. Opt.*, **10**, 1980–1982; 1971.

88. Foueré JC, Malacara D. Focusing errors in a collimating lens or mirror: Use of a moiré technique, *Appl. Opt.*, **13**, 1322–1326; 1974.

89. Yokozeki S, Patorski K, Ohnishi K. Collimating method using fourier imaging and moiré technique, *Opt. Comm.*, **14**, 401–405; 1975.

90. Kothiyal MP, Sirohi RS, Rosenbruch KJ. Improved techniques of collimation testing, *Opt. Laser Technol.*, **20**, 139–144; 1988.

91. Palmer CH. Differential angle measurements with moiré fringes, *Opt. Laser Technol.*, **1**(3), 150–152; 1969.

92. Patorski K, Yokozeki S, Suzuki T. Optical alignment using fourier imaging phenomenon and moiré technique, *Opt. Laser Technol.*, **7**, 81–85; 1975.

93. Reid GT. A moiré fringe alignment aid, *Opt. Lasers Eng.*, **4**, 121–126; 1983.

94. Rieder G, Ritter R. Krummungsmessung an Belasteten Platten Nach dem Ligtenbergschen Moiré-Verfahen, *Forsch. Ing.- Wes.*, **31**, 33–44; 1965.

95. Kao TY, Chiang FP. Family of grating techniques of slope and curvature measurements for static and dynamic flexure of plates, *Opt. Eng.*, **21**, 721–742; 1982.

96. Asundi A. Novel techniques in reflection moiré, *Exp. Mech.*, **34**, September, 230–242; 1994.

97. Ritter R, Schettler-Koehler R. Curvature measurement by moiré effect, *Exp. Mech.*, **23**, 165–170; 1983.

98. Sirohi RS, Kothiyal MP. *Optical Components, Systems and Measurement Techniques*, New York, NY: Marcel Dekker, 1990.

99. Ravensbergen M, Merino R, Wang CP. Encoders for the altitude and azimuth axes of the VLT, *Proc. SPIE*, **2479**, 322–328; 1995.

100. Malacara D, Harris O. Interferometric measurement of angles, *Appl. Opt.*, **9**, 1630–1633; 1970.

101. Tiziani HJ. A study of the use of laser speckle to measure small tilts of optically rough surfaces accurately, *Opt. Commun.*, **5**, 271–276; 1972.

102. Francon M. *Laser Speckle and Applications in Optics*, New York, NY: Academic Press, 1979.

103. Lee WD, Shirley JH, Lowe JP, Drullinger RE. The accuracy evaluation of NIST-7, *IEEE Trans. Instrum. Meas.*, **44**, 120–123; 1995.

104. Teles F, Magalhaes DV, Santos MS, Bagnato VS. A cesium-beam atomic clock optically operated, *Proceedings of the International Symposium on Laser Metrology for Precision Measurement and Inspection*, Florianópolis, Brazil, October 13–15, 1999, pp. 3.37–3.45.

105. Bottom VE. *Introduction to Quartz Crystal Unit Design*, New York, NY: Van Nostrand, 1982.

106. Busch T. *Fundamentals of Dimensional Metrology*, 2nd edn,: Delmar Publishers Inc, Albany, New York, 1989.

107. Doebelin EO. *Measurement Systems, Application and Design*, 4th edn, Singapore: McGraw-Hill International Editions, 1990.

108. Osterberg H. An interferometer method for studying the vibration of an oscillating quartz plate, *J. Opt. Soc. Am.*, **22**, 19–35; 1932.

109. Thorton BS, Kelly JC. Multiple-beam interferometric method of studying small vibrations, *J. Opt. Soc. Am.*, **46**, 191–194; 1956.

110. Fischer E, Dalhoff E, Heim S, Hofbauer U, Tiziani HJ. Absolute interferometric distance measurement using a FM-demodulation technique, *Appl. Opt.*, **34**, 5589–5594; 1995.

111. Gouaux F, Servagent N, Bosch T. Absolute distance measurement with an optical feedback interferometer, *Appl. Opt.*, **37**, 6684–6689; 1998.

112. Ruíz Boullosa R, Pérez López A. Interferómetro Láser y Conteo de Franjas Aplicado a la Calibración de Acelerómetros y Calibradores de Vibraciones, *Rev. Mex. Fis.*, **36**, 622–629; 1990.

113. Ueda K, Umeda A. Characterization of shock accelerometers using Davies bar and laser interferometer, *Exp. Mech.*, **35**, 216–223; 1995.

114. Martens von H-J, Taubner A, Wabinsk W1, Link A, Schlaak H-J. Laser interferometry-tool and object in vibration and shock calibrations, *Proc. SPIE*, **3411**, 195–206; 1998.

115. Peck ER, Obetz SW. Wavelength or length measurement by reversible fringe counting, *J. Opt. Soc. Am.*, **43**, 505–509; 1953.

116. Smythe R, Moore R. Instantaneous phase measuring interferometry, *Opt. Eng.*, **23**, 361–364; 1984.

117. Greivenkamp JE, Bruning JH. Phase shifting interferometry, in *Optical Shop Testing*, Malacara, D., ed., 2nd edn, New York, NY: John Wiley & Sons, 1992.

118. Dändliker R, Hug K, Politch J, Zimmermann E. High-accuracy distance measurements with multiple-wavelength interferometry, *Opt. Eng.*, **34**, 2407–2412; 1995.

119. Malacara D, Servin M, Malacara Z. *Interferogram Analysis for Optical Testing*, New York, NY: Marcel Dekker, 1998.

120. Johnson GW, Moore DT. Design and construction of a phase-locked interference microscope, *Proc. SPIE*, **103**, 76–85; 1977.

121. Matthews HJ, Hamilton DK, Sheppard CJR. Surface profiling by phase-locked interferometry, *Appl. Opt.*, **25**, 2372–2374; 1986.

122. Sirohi RS. ed., *Speckle Metrology*, New York, NY: Marcel Dekker, 1993.

123. Raffel M, Willert C, Kompenhans J. *Particle Image Velocimetry, a Practical Guide*, Berlin, Germany: Springer- Verlag, 1998.

124. Gasvik KJ. *Optical Metrology*, New York, NY: John Wiley & Sons, 1987, pp. 150–156.

125. Lopez-Ramirez JM. *Medición de Desplazamientos de Partículas Mediante Holografía y Moteado*, MSc thesis, CIO-Universidad de Guanajuato, Mexico, 1995.

126. Pickering CJD, Halliwell NA. Laser speckle photography and particle image velocimetry: Photographic film noise, *Appl. Opt.*, **23**, 2961–2969; 1984.

127. Kaufmann GH. Automatic fringe analysis procedures in speckle metrology, in *Speckle Metrology*, Sirohi, R. S., ed., New York, NY: Marcel Dekker, 1993.

128. Grant I, Pan X, Romano F, Wang X. Neural-network method applied to the stereo image correspondence problem in three-component particle image velocimetry, *Appl. Opt.*, **37**, 3656–3663; 1998.

129. Fomin N, Laviskaja E, Merzkirch W, Vitkin D. Speckle photography applied to statistical analysis of turbulence, *Opt. Laser Technol.*, **31**, 13–22; 1999.

130. Funes-Gallanzi M. High accuracy measurement of unsteady flows using digital particle image velocimetry, *Opt. Laser Technol.*, **30**, 349–359; 1998.

131. Drain LE. *The Laser Doppler Technique*, Chichester, UK: John Wiley & Sons, 1980.

132. Durst F, Mellin A, Whitelaw JH. *Principles and Practice of Laser-Doppler Interferometry*, London, UK: Academic Press, 1976.

133. Yeh Y, Cummins Z. Localized fluid flow measurements with an He-Ne laser spectrometer, *Appl. Phys. Lett.*, **4**, 176–178; 1964.

134. Halliwell NA. Laser properties and laser doppler velocimetry, in *Vibration Measurement Using Laser Technology, Course notes*, Sira Communications Ltd, Loughborough University of Technology, 3–5 April 1990.

135. Medlock RS. Review of modulating techniques for fibre optic sensors, *J. Opt. Sensors*, **1**, 43–68; 1986.

136. Crosswy FL, Hornkohl JO. *Rev. Sci. Instrum.*, **44**, 1324; 1973.

137. Hurst F, Mellin A, Whitelaw JH. *Principles and Practice of Laser Doppler Anemometry*, 2nd edn, London, UK: Academic Press, 1981.

138. Drain LE, Moss BC. The frequency shifting of laser light by electro-optic techniques, *Opto-Electronics*, **4**, 429; 1972.

139. Rizzo JE, Halliwell NA. Multicomponent frequency shifting self-aligning laser velocimeters, *Rev. Sci. Instrum.*, **49**, 1180–1185; 1978.
140. Halliwell NA. Laser doppler measurement of vibrating surfaces: A portable instrument, *J. Sound and Vib.*, **62**, 312–315; 1979.
141. Churnside JH. Laser doppler velocimetry by modulating a CO_2 laser with backscattered light, *Appl. Opt.*, **23**, 61–66; 1984.
142. Choi J-W, Kim Y-P, Kim Y-M. Optogalvanic laser doppler velocimetry using the self-mixing effect of CO_2 laser, *Rev. Sci. Instrum.*, **68**, 4623–4624; 1997.
143. Müller J, Chour M. Two-frequency laser interferometric path measuring system for extreme velocities and high accuracy's, *Int. J. Optoelectron.*, **8**, 647–654, 1993.
144. Ready JF. *Industrial Applications of Lasers*, New York, NY: Academic Press, 1978.
145. Sirohi RS, Chau FS. *Optical Methods of Measurement, Wholefield Techniques*, New York, NY: Marcel Dekker, 1999, Chapter 4.
146. Hinsch KD. Particle image velocimetry, in *Speckle Metrology*, Sirohi, R. S., ed., New York, NY: Marcel Dekker, 1993.
147. Watson RC, Lewis RD, Watson HJ. Instruments for motion measurements using laser doppler heterodyning techniques, *ISA Transactions*, **8**, 20–28; 1969.
148. Lewin A. The implications of system 'sensitivity' and 'resolution' on an ultrasonic detecting LDV, *Proc. SPIE*, **2358**, 292–304; 1994.
149. Dussarrat OJ, Clark DF, Moir TJ. A new demodulation process to reduce cochannel interference for a laser vibrometer sensing system, *Proc. SPIE*, **3411**, 2–13; 1998.
150. Crickmore Rl, Jack SH, Hann DB, Greated CA. Laser doppler anemometry and the acousto-optic effect, *Opt. Laser Technol.*, **31**, 85–91; 1999.
151. Jentink HW, de Mul FF, Suichies HE, Aarnoudse JG, Greve J. Small laser doppler velocimeter based on the self-mixing effect in a diode laser, *Appl. Opt.*, **27**, 379–385; 1988.
152. D'Emilia G. Evaluation of measurement characteristics of a laser doppler vibrometer with fiber optic components, *Proc. SPIE.*, **2358**, 240–246; 1994.
153. Lewin AC. Compact laser vibrometer for industrial and medical applications, *Proc. SPIE*, **3411**, 61–67; 1998.
154. Ross MM. Combined differential reference beam LDV for 3D velocity measurement, *Opt. Lasers Eng.*, **27**, 587–619; 1997.
155. Yamada J. Evanescent wave doppler velocimetry for a wall's near field, *Appl. Phys. Lett.*, **75**, 1805–1806; 1999.
156. Harris CM. ed., *Shock and Vibration Handbook*, New York, NY: McGraw-Hill, 1996.
157. Deferrari HA, Darby RA, Andrews FA. Vibrational displacement and mode-shape measurement by a laser interferometer, *J. Acoust. Soc. Am.*, **42**, 982–990; 1967.
158. Gilheany JJ. Optical homodyning, *theory and experiments*, Am. J. Phys., **39**, May, 507–512; 1971.
159. Ohtsuka Y. Dynamic measurement of small displacements by laser interferometry, *Trans. Inst. Meas. Control*, **4**, 115–124; 1982.
160. Oshida Y, Iwata K, Nagata R. Optical heterodyne measurement of vibration phase, *Opt. Lasers Eng.*, **4**, 67–69; 1983.
161. Blumenthal RH. Design of a microwave-frequency light modulator, *Proc. IRE*, **50**, April, 452; 1962.
162. Zeng X, Wicks AL, Mitchell LD. The determination of the position and orientation of a scanning laser vibrometer for a laser-based mobility measurement system, *Proc. SPIE*, **2358**, 81–92; 1994.
163. Li WX, Mitchell LD. Error analysis and improvements for using parallel-shift method to test a galvanometer-based laser scanning system, *Proc. SPIE*, **2358**, 13–22; 1994.
164. Stafne MA, Mitchell LD, West RL. Positional calibration of galvanometric scanners used in laser doppler vibrometers, *Proc. SPIE*, **3411**, 210–223; 1998.
165. Aitchison TW, Bruce JW, Winning DS. Vibration amplitude meter using Moiré-Fringe technique, *J. Sci. Instrum.*, **36**, September, 400–402; 1959.

166. Theocaris PS. Flexural vibrations of plates by the moiré method, *Brit. J. Appl. Phys.*, **18**, 513–519; 1967.

167. Dirckx JJJ, Decraemer WF, Janssens JL. Real-time shadow moiré vibration measurement: Method featuring simple setup, high sensitivity, and exact calibration, *Appl. Opt.*, **25**, 3785–3787; 1986.

168. Kafri O, Band YB, Chin T, Heller DF, Walling JC. Real-time moiré vibration analysis of diffusive objects, *Appl. Opt.*, **24**, 240–242; 1985.

169. Sciammarella CA, Ahmadshahi MN. Nondestructive evaluation of turbine blades vibrating in resonant modes, *Proc. SPIE*, **1554B**, 743–753; 1991.

170. Jones R, Wykes C. *Holographic and Speckle Interferometry*, London, UK: Cambridge University Press, 1983.

6

Optical Metrology of Diffuse Objects: Full-Field Methods

6.1   Introduction 213
6.2   Fringe Projection 214
6.3   Holographic Interferometry 218
      Optical Holographic Interferometry • Digital Holographic
      Interferometry
6.4   Electronic Speckle Pattern Interferometry................................224
      Optical Setups for Electronic Speckle Pattern Interferometry • Fringe
      Formation by Video Signal Subtraction or Addition • Real-Time
      Vibration Measurement
6.5   Electronic Speckle Pattern Shearing Interferometry.................232
6.6   In-Plane Grid and Moiré Methods 234
      Basis of Grid Method • Conventional and Photographic Moiré
      Patterns • Moiré Interferometry • Multiple-Channel Grating
      Interferometric Systems
References..241

Marija Strojnik,
Malgorzata
Kujawinska,
and Daniel
Malacara-Hernández

6.1 Introduction

Optical metrology of diffuse objects may be contrasted with metrology of specularly reflecting optical surfaces [1]. Some of the procedures that are employed in its implementation are described in this chapter. Techniques that are presented are actually useful in the engineering to measure state and changes in state of solid bodies or structures that do not possess a specularly reflecting surface. Optical metrology of diffuse objects focuses on gathering information about shape and displacement or deformation vector $\mathbf{r}(u, v, w)$ of three-dimensional solid bodies or structures with nonspecularly reflecting surfaces. In fact, most objects with notable exception of mirrors fall into this category.

The full-field coherent and non-coherent methods that are most frequently applied in engineering are considered in this chapter. In Table 6.1 $w()$ is an out-of-plane displacement in the z-direction, and $u()$ and $v()$ are in-plane displacements along the x- and y-directions, respectively. In these methods, we code the measured quantity into an irradiance distribution (fringe pattern) that is posteriorly processed and analyzed using one of the methods described in detail in the chapter dealing with the fringe interpretation.

TABLE 6.1 Review of the Methods Used in Optical Metrology of Diffuse Objects

Method	Features of Object	Measured Quantity (Range)
Grid projection	Arbitrary shape, diffuse object	Shape, w (μm–cm)
Out-of-plane moiré		
• Projection	Arbitrary shape diffuse object	Shape, w (μm–cm)
• Shadow		
Holographic interferometry		
• Classical	Arbitrary shape, arbitrary surface including	u, v, w (nm–μm)
• Digital	diffuse objects	Shape (μm–cm)
Speckle interferometry		
• ESPI in plane	Flat object, diffuse surface	v, v (nm–μm)
• ESPI out of plane	Arbitrary shape, diffuse surface	w (nm–μm)
		Shape (μm–cm)
Shearography		
• In-plane	Flat object, diffuse surface	Derivatives of u and v
• Out-of-plane	Arbitrary shape, diffuse surface	derivative of w
Speckle photography		
• Conventional	Diffuse object, flat surface often painted white	u, v (μm-mm)
• Digital		
In-plane moiré;	Flat sample with grating attached	u, v
• Conventional	$f < 40$ lines/mm	(μm-mm)
• Photographic (high resolution)	$f < 300$ lines/mm	(μm-cm)
Grating interferometry (moiré)	Flat sample with high frequency grating	u, v
	$f < 3000$ lines/mm	(nm–μm)

6.2 Fringe Projection

Projecting the image of a periodic structure or a ruling over the surface of a three-dimensional (3D) solid body may facilitate determination of its shape [2–6]. Likewise, the shape may be coded in the interferometric pattern generated by two tilted plane or spherical wavefronts [7]. The fringes may be projected on the body by a lens or a slide projector [8–14]. Surface profilometry may also be accomplished by applying a relative displacement between two gratings moving at a constant velocity.

Determination of the surface shape, referred to as surface profilometry, may be accomplished by applying a relative displacement between two gratings. Chang et al. [15] describe two gratings moving relative to each other at a constant velocity, allowing each pixel of a CMOS camera to capture a heterodyne moiré signal. A one-dimension, fast Fourier transform is performed on all pixels of the field simultaneously, yielding the phase distribution. This phase field is easily converted to a height to render surface profile.

Fringe projection profilometry is a simple instrumental technique used to recover the three-dimensional shape of a body. More than 90% of the phase error may be eliminated by four-step equi-phase shifting with an initial offset of 22.5 degrees. Noise arising from system vibration may be reduced by averaging 20 interferometric images, captured continuously [16].

High-resolution moiré patterns may project and image interferometric patterns to be superimposed on a CCD array. With electronic storage of imaged patterns, we perform sampling at a rate of 3 pixels per fringe period using photo-lithographically printed binary pattern with 33% duty cycle. Direct phase integration avoids ambiguity of phase unwrapping [17–20]. Accuracy in the generation of the surface map is increased by 33%. Surface features with a lateral resolution of less than 10 μm are achievable [21].

Depending on the topology of the surface, the fringes are usually distorted. They may be described by the following equation:

$$I(x, y) = a(x, y) + b(x, y)\cos\left[2\pi f_o x + \phi(x, y)\right]. \tag{6.1}$$

Here $a(x, y)$ and $b(x, y)$ are background and local modulation functions, while f_o is the frequency of projected fringes in a reference plane close to the object and $\phi(x, y)$ is the phase related to the height of the object.

The fringes imaged on the observation plane are an optical system, photographic camera, a charged-coupled device (CCD) camera, or a CMOS device. These fringes may be analyzed directly or pre-processed by one of the moiré techniques: projection moiré [2] and shadow moiré [11]. In projection moiré, the distorted fringes are superimposed on a linear ruling with approximately the same frequency as the projected fringes. The reference ruling may be a real, physical object or software generated, with the same experiment-controlling computer that also captures and analyzes the image [22]. In shadow moiré, an obliquely illuminated Ronchi ruling is placed just in front of the object under study. The moiré fringes are formed as a result of beating between the distorted shadow grid and the grid itself. In recent years, improvement in shadow moiré technology benefited from developments in image processing advances in the speed of computation, and flexibility of image acquisition systems. Zhao et al. [23] developed a rapid technique for the measurement of surface topology. They combine principal component analysis and image demodulation to extract the wrapped phase map, using multiple light sources to sequentially illuminate the grating.

Moiré tomography and shadowing may effectively be employed for structural visualization and parameter identification in flow fields. Chen et al. [24] demonstrate that the integration of measured results represents a feasible method to accomplish this objective.

Fringe projection has most often been used in engineering, medical, and multimedia applications. The measurement process may be described as consisting of two steps:

1. A phase $\phi(x, y)$ assigned to each image point of the object is primary measured quantity. The phase is calculated from the fringe pattern (Equation 6.1), often by phase shifting or by the Fourier-transform method [25]. In the case of steep slopes or step-like objects, the methods that allow circumventing the phase process are applied. They might include the coded light, gray code technique or hierarchical absolute phase decoding [26] that rely on using the combination of at least two projected patterns with different spatial frequencies. Recently direct integration of irradiance distributions has been increasingly applied [17].

2. Three-dimensional coordinates are determined from these phase values on the basis of the geometrical model of the image formation process. Certain parameters of measurement system have to be known in advance to make this method perform optimally.

The triangulation principle forms the basis of the height evaluation in most cases. A point light beam is projected onto the surface. The illuminated point is observed under the so-called triangulation angle θ (Figure 6.1). An optical system images this point on a light-sensitive sensor. Consequently, the

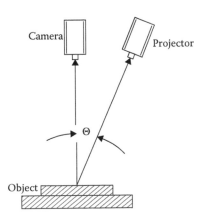

FIGURE 6.1 Triangulation principle used in fringe projection.

measurement of the height h is reduced to the measurement of the lateral displacement Δx on the CCD surface or the CMOS chip. The imaging geometry and the triangulation angle θ is needed for the calculation of height h.

Three basic configurations are used most frequently, as illustrated in Figures 6.2 and 6.3. In the first two cases (presented in Figure 6.2a and b), the optical axes of the projection and the observation systems intersect at an angle θ. Additionally, the telecentric lens projects the fringes. If the observation point is located at a height l from the reference plane, the contour surfaces are not planes (except for the reference one), and the evaluation h of a body is given by (see Figure 6.3a)

$$h = \frac{l\Delta x}{l\tan\theta + x},$$ (6.2)

where the fringe deviation Δx is given by

$$\Delta x = \frac{d}{2\pi}\phi(x, y).$$ (6.3)

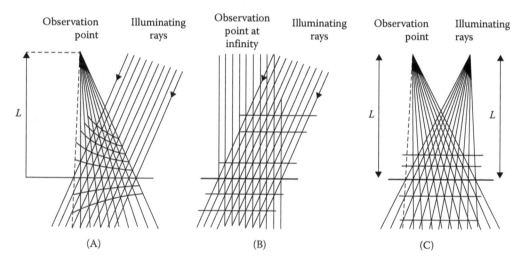

(A) (B) (C)

FIGURE 6.2 Three configurations to project a periodic structure over a solid body to measure its shape: (a) near observation point, (b) observation point and illuminating light source at infinity, (c) observation point and illuminating light source at a finite distance.

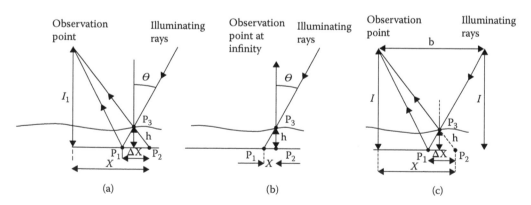

(a) (b) (c)

FIGURE 6.3 Geometries for the three basic configurations used to calculate the object height in the fringe projection.

Here d is the period of the projected fringes in the reference plane and x is measured in a direction perpendicular to the projected fringes, with the origin at the point where the axis of the projection intersects the optical axis of the observation optical system.

The height of an object may be calculated upon substituting Equation 6.3 into 6.2, upon determining the phase values $\phi(x, y)$ from fringe pattern analysis [27],

$$h = \frac{1}{2\pi}\left(\frac{ld}{l\tan\theta + x}\right)\phi(x, y). \tag{6.4}$$

The most elegant scaling of the height distribution of a three-dimensional object is given by a system with telecentric projection and detection (Figures 6.2b and 6.3b). The height h is then given by

$$h = \frac{1}{2\pi}\left(\frac{d}{\tan\theta}\right)\phi(x, y). \tag{6.5}$$

In this configuration, unfortunately, the object size is restricted to the diameter of the telecentric optics.

The third configuration is based on the geometry, with mutually parallel optical axes of projecting and observation systems (Figures 6.2c and 6.3c). In this case, the height of the point on the body is given by

$$h = \frac{1}{2\pi}\left(\frac{ld}{b + \dfrac{d}{2\pi}\phi(x, y)}\right)\phi(x, y). \tag{6.6}$$

Here, b is the separation along the x-axis between the illumination and observation points. If the geometry of the fringe projection system is not known in advance, a calibration of the measurement volume is required [28]. Photogrammetric and triangulation/phase-measuring approaches may also be combined [29]. Shape measurement systems often deliver data about an object's coordinates (x, y, z) in the form of a cloud of points obtained and merged from different directions. This data is extensively used in CAD–CAM environment and rapid prototyping systems, as well as in computer graphics and virtual reality applications.

En et al. [30] describe fiber-optical sinusoidal phase demodulating interferometer for 3D profilometry. The experimental setup incorporates piezoelectric transducer and closed loop feedback system to decrease the signal drift. Perciante et al. [31] proposed a simple and robust technique for phase retrieval, side-stepping the phase unwrapping procedure often used in phase-shifting interferometry (PSI) or spatial-carrier interferometry [17–20]. The proposed approach is based on the direct integration of the spatial derivatives of the irradiance distribution patterns. The principal steps of the technique are illustrated in Figure 6.4. They present practical results with applications in interferometry, 3D-shape profiling, and deflectometry.

These methods have become so well established that miniaturized and portable devices have been implemented. Steckenride and Steckenrider [32] describe a two-dimensional optical profilometry system where fringe projection and image processing are optimally employed to provide micro-scale precision in a portable device.

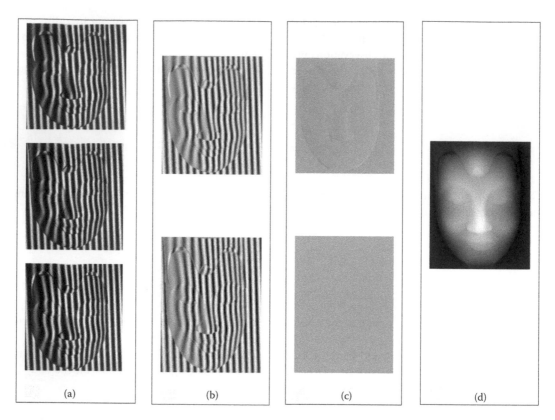

FIGURE 6.4 The principal steps in acquiring profile of a 3D object, side-stepping the unwrapping stage of phase retrieval algorithms: (a) acquired fringe patterns; (b) cosine and sine functions from irradiance distributions; (c) phase derivatives; and (d) phase determined upon integration. (After Perciante et al., *Appl. Opt.*, 54, 3018–3023, 2015.)

6.3 Holographic Interferometry

Holographic interferometry [33] is a highly successful method to investigate a diffuse object. Historically, the bottleneck of holographic interferometry was the recording medium. Silver halides provide the best resolution with high sensitivity and good-quality holographic reconstruction, but their use requires wet chemical processing. Photo-thermoplasts involve special electronics; they are limited in size, resolution, and diffraction efficiency. The current systems that incorporate high-resolution CCD cameras and fast computers are known as electronic digital holographic interferometry. The principles of both optical (conventional) and digital holographic interferometry are described next. Kim et al. [34] offer an application of optical holography on a micro-scale. Optical images at a cellular level, obtained with plane waves incident with different illumination angles, are reconstructed and provide a 3D refractive index tomogram.

6.3.1 Optical Holographic Interferometry

The basis of the holographic interferometry lies in the fact that a reconstructed hologram contains all the information (phase and amplitude) about the recorded object. If this holographic image is superimposed on the object wave originated at the same object, but only slightly changed, the two waves interfere generating a volume hologram [35]. This may be accomplished by taking the first exposure of the object in the reference state with an amplitude $E_1 \exp(i\phi_1(\mathbf{r}))$, and the second one—after change in the

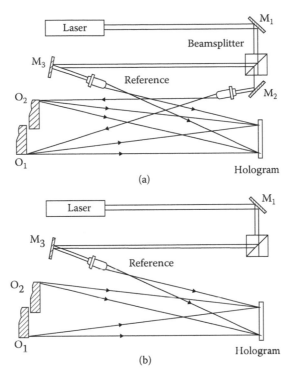

FIGURE 6.5 Optical arrangement in digital holographic interferometry with an off-axis reference beam. (a) Recording of a double-exposure hologram; (b) reconstruction of a double-exposure hologram.

state of the object—with an amplitude $E_2\exp(i\phi_2(\mathbf{r}))$ (for more details see Figure 6.5a). Two exposures result in the so-called double-exposure hologram. After the development of the photographic plate, the hologram is illuminated with the reference wave, leading to the concurrent reconstruction of both states of the object (Figure 6.5b). It is also possible to record only the object reference state. Then, we monitor the interference pattern created from the object wavefront and the wavefront reconstructed from the hologram. This is referred to as real-time holographic interferometry. If the change in the state of the object is sufficiently small, then the two reconstructed waves interfere forming a fringe pattern, given by

$$I = E_1^2 + E_2^2 + E_1 E_2 \exp[i(\phi_1(\mathbf{r}) - \phi_2(\mathbf{r}))] + E_1 E_2 \exp[-(i(\phi_1(\mathbf{r}) - \phi_2(\mathbf{r})))]$$
$$= I_2 + I_2 + 2I_1 I_2 \cos[\phi_1(\mathbf{r}) - \phi_2(\mathbf{r})].$$
(6.7)

The difference in phase terms may be combined in to a single phase $\phi(x, y, z)$ that is calculated from the irradiance distribution using one of the available algorithms for the fringe pattern analysis. The phase difference between two object states is related to the optical path difference (OPD) by

$$\phi(x, y) = \left(\frac{2\pi}{\lambda}\right)\text{OPD}.$$
(6.8)

The optical path difference OPD is a geometrical concept. However, it is projected onto the sensitivity vector $\mathbf{s} = \mathbf{p} - \mathbf{q}$ [36]. This vector is given by the difference of the unit vector $\mathbf{q}$ in direction from the illumination source to the point P with coordinates (x, y, z) located on the object and the unit vector $\mathbf{p}$ direct from point P to the observation point (see Figure 6.6).

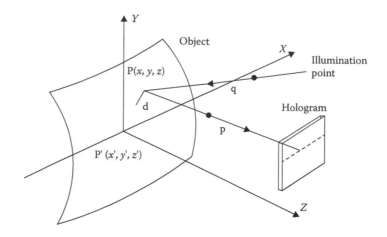

FIGURE 6.6 Optical path difference calculation in holographic interferometry.

$$OPD = \mathbf{d}(\mathbf{p} - \mathbf{q}) \tag{6.9}$$

Here **d** is the displacement vector from the point P with coordinates (x, y, z) to the shifted point with coordinates (x', y', z'). The sensitivity vector **s** is directed along the bisectrix of the angle between the illuminating ray and the ray traveling from the illuminated object to the observing point. Its maximum magnitude is 2 when two light rays coincide.

The phase term is proportional only to the projection of the displacement vector onto the sensitivity vector. However, three scalar sensitivities are required to specify three components of vector **d**. They may be found by introducing different directions of observation and most often by changing the illumination directions, often using multiple illumination angles.

An interesting specific application of displacement measurement is to the analysis of vibrating objects where the displacement of each point is a simple harmonic function

$$\mathbf{d}(P,t) = \mathbf{d}(P)\sin \omega t. \tag{6.10}$$

This task may be performed by stroboscopic holographic interferometry, whereby we record a hologram using a sequence of short pulses that are synchronized with the vibrating object [37]. It may also be implemented in the time-average holographic interferometry where the object is recorded holographically, with a single exposure that is long compared with the period of vibration [38]. The resulting irradiance of the reconstructed image is

$$I(P) = I_0(P) J_0^2 \left[\frac{2\pi}{\lambda} \mathbf{d} \cdot \mathbf{s} \right]. \tag{6.11}$$

Here $\mathbf{s} = [\mathbf{p} - \mathbf{q}]$. J_0 is the Bessel function of zero order of the first kind, with the quantities in the square bracket being its argument. Maximum irradiance is detected at the nodes of the vibration modes. Dark fringes characterize those values of the argument, $(2\pi/\lambda)(\mathbf{d} \cdot \mathbf{s})$, for which the zero-order Bessel function of the first kind achieves zero (0) value. Furthermore, holographic interferometry enables us to measure the object shape in addition to generating the object displacement map. In holographic contouring, two holograms of a static object with two different sensitivity vectors are recorded. The different sensitivity vectors may be introduced by alternatively changing the laser wavelength from λ to λ' [39]:

$$\Delta z = \frac{\lambda\lambda'}{[(\lambda-\lambda')(1+\cos\theta)]}, \tag{6.12}$$

where Δz is the depth difference. Similarly, the direction of the illuminating beams may be changed [40]. The depth modulation probe and lines are most frequently setup with collimated beams that generate equidistant parallel contour surfaces with the change in depth distance.

$$\Delta z = \frac{\lambda}{2\sin(\theta/2)} \tag{6.13}$$

Finally, the index of refraction of the medium surrounding the diffuse object (from n to n') may likewise be modulated [41] to produce the depth modulation.

$$\Delta z = \frac{\lambda}{2(n-n')} \tag{6.14}$$

6.3.2 Digital Holographic Interferometry

The principle used in the digital holographic interferometry is basically the same as that employed in the conventional optical holography. A CCD detector rather than the traditional holographic photographic plate records the image. The typical size of a CCD is about 5 mm on a side, at this writing incorporating up to 50,000 × 50,000 pixels. Unfortunately, the CCD resolution is still considered low compared to that of the holographic plate. Therefore, the experimental setup must generate correspondingly larger fringe separation to maintain the same distance recording resolution. Thus, γ_{max} is the maximum angle between the reference beam and the object point at the largest distance from the light source, as seen from the detector. Then we should have

$$\sin\gamma_{max} - \frac{\lambda}{d}, \tag{6.15}$$

This geometry is illustrated in Figure 6.7. Here d is the center-to-center distance between two adjacent pixels on the CCD. This requirement can be satisfied when the object is small, on the order of a few millimeters, and the light source producing the collimated reference beam is at least at a distance of about 1 m away from the location where fringe pattern is formed. Better resolution distances are being achieved with the development of increasingly larger CCD cameras, incorporating ever-smaller pixels.

In classical holography, the object reconstruction is performed optically. In digital holography [42,43], the object reconstruction is performed numerically in a computer where discrete finite

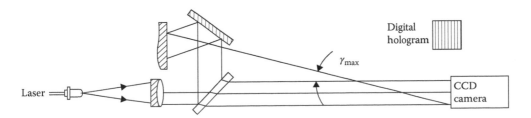

FIGURE 6.7 Optical arrangement in digital holographic interferometry.

Fresnel transformation is carried out. The phase on the object surface is thus computed. When two holograms are recorded, one of them after a small displacement or deformation of the body, the resulting phase difference may be calculated. The basic difference between the digital holographic interferometry and the optical one is that the optical steps are implemented in software in the former. In the early digital holographic experiments, the off-axis setup just described was often used. Three images generate a zero order and two conjugate images or orders. The limitation of off-axis systems was removed upon employing phase-shifting techniques that incorporate a piezoelectric mount to translate one of the mirrors. This allowed the use of the on-axis configurations. The distribution of the complex electric field, at the holographic plane, may be deduced, applying one of the many available phase reconstruction techniques [44].

The hologram is usually imaged on a digital detector, such as for example a CCD detector array. The reconstruction plane is located very close to the measured object. In the theoretical development that follows, complex amplitude at the CCD detector is represented by $h(x, y)$. Then, the complex amplitude $E(\xi, \eta)$ on the reconstruction plane at a distance z from the hologram is given by [45]

$$E(\xi,\eta) = \frac{iU_0}{\lambda z}\exp\left[-\frac{i\pi}{\lambda z}(\xi^2+\eta^2)\right]\iint h(x,y)\exp\left[-\frac{i\pi}{\lambda z}(x^2+y^2)\right]$$
$$\exp\left[-\frac{i\pi}{\lambda z}(x\xi+y\eta)\right]d\xi d\eta; \tag{6.16}$$

The phase on plane near object surface is given by

$$\phi(\xi,\eta) = \arctan\frac{\mathrm{Im}[E(\xi,\eta)]}{\mathrm{Re}[E(\xi,\eta)]}. \tag{6.17}$$

There exist a number of methods to calculate the phase difference between two body states. The general principles involved in two representative procedures are illustrated in the block diagram in Figure 6.8 [35]. More recently, Waghmare et al. [46] describe the use of unscented Kalman filters to estimate phase when reconstructed interference fields are noisy. Researchers perform phase tracking for phase estimation in digital holographic interferometry, particularly applicable to estimating rapidly varying phase fields in presence of noise.

When phase-shifting photorefractive spatial light modulator is incorporated into holographic interferometry setup, simultaneous measurement of optical inhomogeneity and thickness variation may be accomplished. Li et al. [47] report that neither the application of special coatings nor any other modifications to the sample or its surface are needed.

Digital holographic interferometry has recently been implemented also in off-axis configuration. Belashov et al. [48] report adjustable smoothing, decrease in noise, and increased fringe resolution upon employing a simulated wavefront.

With the development of large array, small pixel size bolometers, Georges et al. [49] demonstrated holographic interferometry at carbon dioxide (CO_2) wavelengths. A side benefit of IR measurements is that temperature may be determined concurrently. The digital age has augmented the performance achieved with optical holographic interferometry, hindered only by the lack of availability of suitable cameras and spatial light modulators with the same or corresponding pixel size. Porras-Aguilar et al. [50] evaluate performance of five methods to decrease deleterious effects of such mismatch. Cashmore et al. [51] study the feasibility of using binary reconfigurable holograms. Ferroelectric-liquid-crystal-on-silicon spatial light modular allows easy and rapid recording of optical holograms.

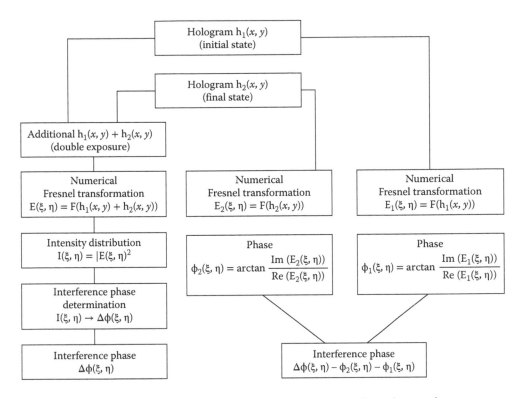

FIGURE 6.8 Procedures used to compute the object phase distribution in holographic interferometry.

Using holographic interferometry, the displacement vector at any point of the body being examined may be detected and measured. Holographic interferometry is sensitive to the in-plane and to the out-of-plane displacement.

Kumar et al. [52] describe the application of digital holographic interferometry to measuring the local convective heat transfer coefficient. Its value may be determined upon using phase shifting procedures. The maturity of the field of endeavor encompassing digital interferomety has been demonstrated by using it successfully aboard the International Space Station [53]. Results of thermo-diffusion experiments confirmed the applicability and efficacy of the implementation of the windowed Fourier transform during the data analysis process.

Kumar and Shalcher [54] employ lens-less Fourier transform in a digital holographic interferometry to study heat transfer mechanisms. Quantitative information about temperature distribution is obtained from reconstructed phase difference maps. Optical path differences between heated and ambient air (OPDs) are subtracted for propagation through heated and ambient air.

Symmetrical configurations may be employed to measure vibratory strain field by transverse digital holography [55]. One beam is phase stepped by quarter wavelength, while the other one is modulated at the frequency of the vibrating object. The unwrapped images are processed to yield average slope and strain.

Zhang et al. [56] describe a dynamic measurement of refractive index distribution with digital holographic interferometer. Double exposure is achieved by phase shifting of the additional light beam generated by the total internal reflection. The phase shift is accrued upon the extra path taken by the reflected radiation refractive to the direct image.

The increase in the signal-to-noise ratio in digital holographic interferometry is desirable, facilitating unambiguous phase extraction from measured irradiance distribution. Kulkarni and Rastogi [57] describe the use of amplitude discriminating algorithms, in tandem with different irradiances of illuminating beams.

One of the popular techniques to determine the diffusion coefficient by digital holographic interferometry employs measuring the distance between two peaks in concentration difference profile at two different times [58]. The measurement considers the importance of the determination of the initial time, when the diffusion is initiated He et al. [59] describe mathematical techniques employed to determine initial time and data processing methods of hologram analysis to determine the diffusion coefficient with increased precisions. Zhang et al. [56] found that the "initial time" actually decreases during the duration of the experiment when the digital holographic interferometry is used to measure diffusion coefficient.

Moisture absorption and desorption induce dimensional changes and deformations in wood that might lead to unanticipated failure of such products. Kumar and Shakher [54] describe how lens-less Fourier transform digital holographic interferometry is used to determine deformation field parameters. The experimental results show that the strain and the coefficient of hygroscopic absorption may actually be minimized when wood is dried under specific, optimal humidity conditions.

6.4 Electronic Speckle Pattern Interferometry

Speckle pattern interferometry has been developed to study vibrations and deformations by many authors, especially by Macovski et al. [60] in the United States and by Butters and Leendertz [61] in Great Britain. Later, other interesting developments followed: for example, by Løkberg [62] and by Jones and Wykes [63,64]. The article by Ennos [65] and the book by Cloud [66] also present excellent reviews on this subject. To fully appreciate the experimental procedure, let us consider an extended, nonspecular, diffusing surface, illuminated with a single, well-defined wavefront. In other words, the illuminating light beam must be spatially coherent. If a second diffusing surface is located in front of the illuminated surface, as illustrated in Figure 6.9a, each bright point on the illuminated surface makes a small

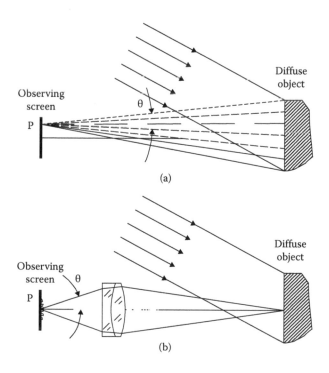

FIGURE 6.9 Speckle formation: (a) subjective speckles and (b) objective speckles.

contribution to illuminate the surface in front of it. The light arriving at point P comes from all points on the illuminated surface, generating coherent illumination. This, in turn, assures that the multiple-beam interference may take place. The second surface appears as if covered by many small and bright speckles [67]. These speckles are real: this may be easily demonstrated by replacing the second surface with a photographic plate. These are called objective speckles. Their average size d depends on the semi-angle θ of the illuminated surface as follows:

$$d = 1.22\lambda \sin\theta. \tag{6.18}$$

If we place a lens in front of the illuminated surface, as in Figure 6.9b, the lens forms the image of the illuminated surface on the plane of another diffusing surface. Each point of the illuminated surface is imaged as a small Airy disk with a diameter given by Equation 6.18. Here θ is the angular diameter of the imaging lens, as seen from the image point P. If F is the focal length of the lens, D its diameter, and m its magnification, then we find the speckle diameter,

$$d = \frac{1.22\lambda D}{2F(m+1)}. \tag{6.19}$$

The interference between neighboring Airy disks produces speckles, called subjective speckles.

The speckle patterns are quite complicated. However, they depend only on two factors, that is, the roughness of the illuminated surface and the phase distribution of the illuminating light beam. We may draw the following empirical observation for a diffusing surface: if the relative phase distribution for all points on the diffusing surface or on its image at the observing screen changes, then the speckle pattern structure is likewise modified. Furthermore, it may be seen that this relative phase distribution changes only when there are two interfering beams present. One of these beams may illuminate the diffusing surface and the other is incident directly onto the observing screen, or both may superimpose at the diffusing surface.

Electronic speckle pattern interferometry has also been successfully used for the dynamic thermal measurements of the printed circuit board [68]. Increased temperature results in local fringe deformation and change in fringe density. Advanced image processing techniques, usually also incorporating the BL-HilbertL2 methods, work to smooth the low-density fringes and contrast-enhance the high-density ones.

6.4.1 Optical Setups for Electronic Speckle Pattern Interferometry

We next consider some typical illumination configurations producing subjective speckle patterns in order to understand how the structure of the speckle pattern may change. Any possible displacement takes place along the coordinate axes x, y, z affixed to the body under study. They are denoted by u, v, w, where the displacement coordinate w is along the line perpendicular to the surface on the body.

6.4.1.1 Coaxial Arrangement

In the coaxial layout, the speckle pattern also moves with the surface under study, but its structure remains unchanged. When the object is illuminated with a normally incident flattop wavefront, the illuminated surface moves in its own plane. When the illuminated surface moves in the direction perpendicular to the incident illumination, the speckle structure also remains unchanged. However, when a reference wavefront traveling directly to the detector is added, out-of-plane movements could additionally be detected. This setup is easily configured when one of the mirrors in a Twyman–Green interferometer is replaced with the diffusing surface whose out-of-plane movement is to be determined.

6.4.1.2 Asymmetrical Arrangement

As the asymmetrical arrangement, neither of the two possible displacements (in-plane or out-of-plane movement) of the illuminated surface changes the speckle structure when the illuminating wavefront has oblique incidence. Similarly, this is also the case with normally incident illumination, unless a reference wavefront is introduced. A possible experimental arrangement with the reference beam incident directly onto the detector with normal incidence is illustrated in Figure 6.10a. If the plane of incidence is the x-z plane, then the phase difference either increases or decreases by an amount $\Delta\phi$ for a small in-plane displacement u in the plane of incidence (along the x-axis) and a small out-of-the-plane displacement w (along the z-axis). It is given by

$$\Delta\phi = ku \sin\theta + kw(1 + \cos\theta), \tag{6.20}$$

where $k = 2\pi/\lambda$ and θ is the angle of incidence for the oblique illuminating beam. A simple expression obtained for a normally illuminating beam (coaxial arrangement) when $\theta = 180°$, as

$$\Delta\phi = 2kw. \tag{6.21}$$

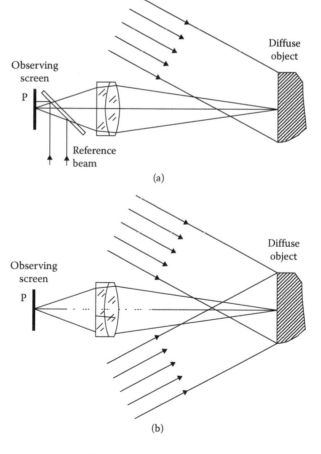

FIGURE 6.10 Schematic arrangements for out-of-plane movement detection in speckle interferometry: (a) out-of-plane and (b) in-plane movements.

The displacement values u and w may be separated upon two measurements employing two different values of θ in Equation 6.20. When the object experiences a combined displacement u and w, the movement cannot be detected where u and w are related as follows:

$$\frac{u}{w} = \frac{1 + \cos\theta}{\sin\theta}. \tag{6.22}$$

Figure 6.10a illustrates an experimental setup to perform electronic speckle pattern interferometry with a single beam with oblique incidence.

6.4.1.3 Symmetrical Arrangement

Let us assume that two illuminating wavefronts in the plane of incidence x-z have the same angle of incidence, and are traveling in the opposite directions, as illustrated in Figure 6.10b. We may consider that the observed speckle pattern is formed by the interference of two speckle patterns; each one produced two illuminating light beams. These two speckle patterns have identical structures when observed independently. However, their average phase changes linearly in opposite directions. This superposition, therefore, results in a structure different from any of its two constituent components.

In-plane v movements in a direction perpendicular to the plane of incidence of the illuminating beams (y-axis) do not change the speckle structure. The phase difference between two illuminating beams remains the same for all points on the surface. Thus, the experimental setup is insensitive for the movement along the direction of the y-axis.

With an in-plane movement u (x-axis direction), the phase upon movement increases for one of the beams and decreases for the other beam. The phase difference between the two beams increases in one direction and decreases in the opposite direction. Therefore, the phase difference changes by an amount $\Delta\phi$, given by

$$\Delta\phi = 2ku\sin\theta. \tag{6.23}$$

Here θ is the angle of incidence for the two illuminating beams. $k = \frac{2\pi}{\lambda}$. Sensitivity increases with the angle θ. It is important to note that the introduced displacement must be smaller than the speckle diameter. Figure 6.11b presents an experimental arrangement to perform electronic speckle pattern interferometry with two symmetrical divergent beams with oblique incidence. When the illuminating beams are not collimated, a variable sensitivity over the illuminated area is obtained for the in-plane displacements. Another configuration incorporates a flat mirror on one side of the object, thus creating two illuminating beams from one, as illustrated in Figure 6.11c. Figure 6.12 shows an arrangement with two collimated symmetrical beams to provide constant sensitivity over the illuminated area. Its disadvantage is that the area being probed is relatively small.

Another possible experimental setup using fiber optics is offered in Figure 6.13. The original beam is divided into two beams of equal intensity by a directional coupler. The light in one of the interferometer arms passes through a phase shifter, which is formed by a PTZ cylinder with fiber loops wrapped around it. A computer may be interfaced to control phase-shifting process, facilitating posterior interferogram interpretation.

In speckle interferometry, two similar speckle patterns are superimposed on each other by an additive or subtractive procedure. One speckle pattern is taken before and the other after a certain object displacement has taken place. The correlation between these two patterns appears as moiré fringes. These fringes represent the phase difference between two positions on, or the change in, the shape of the object on which the speckle pattern is formed. Each of the superimposed speckle patterns is generated by the interference of the speckle pattern of the diffuse surface being measured with a

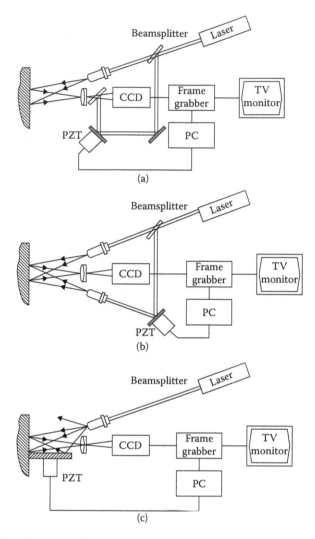

FIGURE 6.11 Optical configurations for movement detection in speckle interferometry: (a) in-plane and out-of-plane movements, (b) in-plane movement, and (c) in-plane movement.

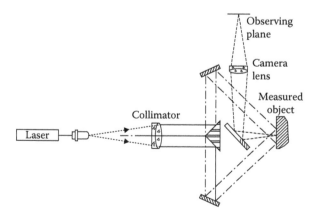

FIGURE 6.12 Optical configuration for in-plane movement detection with uniform sensitivity.

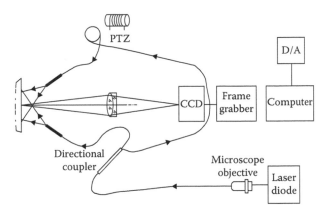

FIGURE 6.13   Configuration for in-plane movement detection using fiber optics.

reference beam. Jones and Wykes [63] point out that the speckle interferometers can be classified into two categories:

1. *Class 1,* where the reference beam is a single wavefront, generally plane or spherical, as in the arrangement just described
2. *Class 2,* where both the object beam and the reference beams are speckle patterns, as in a Michelson interferometer with both mirrors replaced by diffuse objects

In speckle photographic interferometry [61], a photograph of the previous speckle pattern is superimposed on the new speckle pattern, after the diffusing screen has been distorted or bent. In another method, two photographs may be concurrently placed on top of each other. In electronic speckle pattern interferometry (ESPI) [65], two video images are electronically super-imposed. An important operational condition is imposed by the requirement that the speckle size must be greater than the size of each detector element.

Stetson [69] implemented further development of phase-step digital holography by incorporating speckled reference beam. A speckled field that interferes with the object replaces the additional reference beam. The interference between the two random fields results in a recording equivalent to the product of two interfering fields. One field may be thought of as a transmission element through which the other field propagates.

A significant recent improvement in the usage of speckle interferometry includes employment of waveguide phase modulator, achieving phase-shifting rates of up to 100 Hz [70]. One-shot phase shifting further incorporates a CMOS camera to measure time-dependent deformations. The electronic speckle pattern interferometry has also experienced renaissance due to the incorporation of ever more efficient phase extraction methods. F. Zhang et. al. [71,72] find the gradient field (see also Peez and Strojnik [17–20]) and interpolate the whole field with anisotropic partial differential equations. Gu et al. [73] describe a triple-optical path digital speckle pattern interferometer to measure three-dimensional displacement fields. Single laser source and three cameras produce the data for phase-shift demodulation techniques with a fast, discrete, curvelet transform for phase recovery.

6.4.2 Fringe Formation by Video Signal Subtraction or Addition

Next, we consider the interference of two speckle patterns with incidences I_0 and I_R, and with random phases ψ_0 and ψ_R. Both of these irradiances and phases change very rapidly from point to point, due to the nature of the speckle. The sum of these two speckle patterns produces another speckle pattern with irradiance I_1, given by

$$I_1 = I_0 + I_R + 2\sqrt{I_0 I_R}\,\cos\psi, \tag{6.24}$$

where $\psi = \psi_0 - \psi_R$ is a different random phase function. Now, an additional smooth (as opposed to random) phase $\Delta\phi$ may be added onto one of the two constituent speckle patterns by a screen displacement or deformation. The new phase is $\psi + \Delta\phi$. Then, the incidence pattern in the image volume is

$$I_2 = I_0 + I_R + 2\sqrt{I_0 I_R}\,\cos(\psi + \Delta\phi). \tag{6.25}$$

The last two speckle patterns given by Equation 6.24 and 6.25 produced by interference are quite similar, with relatively small differences introduced by the phase $\Delta\phi$. A high-pass spatial filter may be applied to each of two speckle patterns to remove low-frequency noise and low-frequency variation in mean speckle irradiance. When the filtered patterns are combined, a moiré pattern provides evidence for their correlation. Moiré fringes between these two speckle patterns may be generated in one of two distinct ways. The first one is by subtracting these two incidences, obtaining

$$I_1 - I_2 = 2\sqrt{I_0 I_R}\,\sin\left(\psi + \frac{\Delta\phi}{2}\right)\sin\left(\frac{\Delta\phi}{2}\right). \tag{6.26}$$

This irradiance exhibits positive as well as negative values because the DC terms subtract. Thus, its absolute value must be taken

$$I_1 - I_2 = 2\sqrt{I_0 I_R}\,\left|\sin\left(\psi + \frac{\Delta\phi}{2}\right)\right|\left|\sin\left(\frac{\Delta\phi}{2}\right)\right|. \tag{6.27}$$

Then, a low-pass filter is applied to eliminate the high-frequency components produced by the speckle, resulting in a single bright fringe, given by

$$B_- = 2K\sqrt{I_0 I_R}\,\left|\sin\left(\frac{\Delta\phi}{2}\right)\right|. \tag{6.28}$$

The subtraction method is used for the analysis of static distributions. It requires separate storage for two images.

The second method of the fringes formation involves addition of the irradiances of the two speckle patterns, as follows:

$$I_1 + I_2 = 2(I_0 + I_R) + 4\sqrt{I_0 I_R}\,\cos\left(\psi + \frac{\Delta\phi}{2}\right)\cos\left(\frac{\Delta\phi}{2}\right). \tag{6.29}$$

This irradiance has a constant average value given by

$$\langle I_1 + I_2 \rangle = 2\langle I_0 \rangle + 2\langle I_R \rangle. \tag{6.30}$$

However, the added phase $\Delta\phi$ introduces a variable contrast. The variance σ of this irradiance for many points in the vicinity of a given point in the pattern is given by

$$\sigma^2 = \langle I^2 \rangle - \langle I \rangle, \tag{6.31}$$

and the standard deviation σ is

$$\sigma = \sqrt{\langle I^2 \rangle - \langle I \rangle^2}, \tag{6.32}$$

where $\langle I \rangle$ is the mean value of the irradiance values in the vicinity of the point being considered. The standard deviation of the irradiance over many points in the vicinity of a point has the advantage of eliminating the DC bias while, at the same time, it both rectifies the signal and applies a low-pass filter to eliminate the speckle. The square at the standard deviation is given by

$$\sigma^2 = 4\sigma_R^2 + 4\sigma_0^2 + 8\langle I_R I_0 \rangle \cos^2\left(\frac{\Delta\phi}{2}\right). \tag{6.33}$$

Furthermore, we may also assume a Poisson probability distribution for each of the speckle patterns [74]. It is then possible to show that the standard deviation for each speckle pattern is

$$\sigma_R = \left[\langle I_R^2 \rangle - \langle I_R \rangle^2\right]^{1/2} = \langle I_R \rangle \tag{6.34}$$

and

$$\sigma_0 = \left[\langle I_0^2 \rangle - \langle I_0 \rangle^2\right]^{1/2} = \langle I_0 \rangle. \tag{6.35}$$

Thus, the average irradiance B_+ on the screen, is given by

$$B_+ = K\left[\langle I_R \rangle^2 + \langle I_0 \rangle^2 + 2\langle I_R I_0 \rangle \cos^2\left(\frac{\Delta\phi}{2}\right)\right]^{1/2}. \tag{6.36}$$

Two speckle patterns, which are correlated in order to produce the fringes, are added on the sensitive surface of the CCD detector. The generation and formation of two patterns need not be simultaneous, because of the detector integration time of up to about 0.1 s. The observation of dynamic or transient events or modal analysis of membranes is therefore easy to implement, employing two consecutive laser pulses, for illumination and imaging, at two different times.

Points (pixels) in the fringe pattern with a high correlation have a high contrast. A potential disadvantage of the addition scheme is that its contrast is lower than that of the subtraction method. An important advantage, however, is that storage is only needed for a single image, rather than two. Low fringe contrast may usually be increased by judicious application of image processing techniques.

Figure 6.14 shows two speckle images, one before and one after the object was modified. The noise effects of adding and subtracting the second image are also illustrated. Figure 6.15 lays out a block diagram presenting the basic steps involved in electronic speckle interferometry. An important advantage of this method is the relatively high speed of its implementation with correspondingly low environmental requirements. A potentially serious disadvantage is presented by its relatively high noise content.

6.4.3 Real-Time Vibration Measurement

The out-of-plane vibrations of an object, like a membrane or diaphragm of a loudspeaker, may be measured with speckle interferometry. The system is then operated in time average mode. Usually, the frames are added. At any time t, the irradiance in the combined speckle image is given by

$$I(t) = I_0 + I_R + 2\sqrt{I_0 I_R} \cos\left(\Delta\phi + \frac{4\pi}{\lambda} a_0 \sin \omega t\right), \tag{6.37}$$

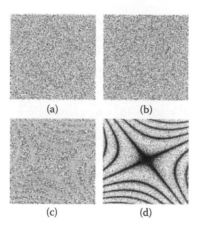

FIGURE 6.14   Speckle images: (a) first frame, (b) second frame, (c) subtraction, and (d) addition.

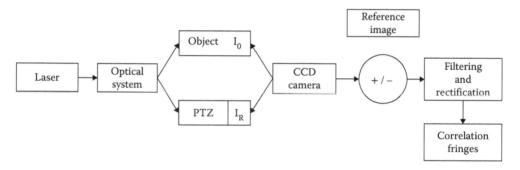

FIGURE 6.15   Speckle interferometry procedure to obtain the fringes.

where the factor $a_0 \sin \omega t$ represents the position of the vibrating membrane at the time t. This irradiance is averaged over a time τ, assuming that $2\pi/\omega = \tau$ obtaining

$$I_\tau = I_0 + I_R + \frac{2}{\tau} \sqrt{I_0 I_R} \int \cos \left(\Delta\phi + \frac{4\pi}{\lambda} a_0 \sin \omega t \right) d\tau. \qquad (6.38)$$

We take the time average over several cycles. Then, it is possible to show that

$$I_\tau = I_0 + I_R + 2\sqrt{I_0 I_R} \; J_0^2 \left(\frac{a_0}{\lambda} \right) \cos \Delta\phi, \qquad (6.39)$$

where J_0 is the zero-order Bessel function.

Figure 6.16 exhibits a photo of an interferogram of the vibrating modes of a square diaphragm. Vibration analysis with an in-plane sensitive arrangement using pulsed phase-stepped electronic speckle pattern interferometry is now routinely being applied.

6.5 Electronic Speckle Pattern Shearing Interferometry

Electronic speckle pattern shearing interferometry [75] also called shearography was introduced by Hung [76] and further developed by many researchers [77–79]. In this method, irradiances at two different points, laterally separated by a small distance often called the shear, are added upon detection by the same detector pixel.

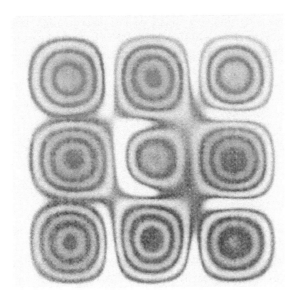

FIGURE 6.16   Speckle interferometry pattern for a square vibrating membrane.

Shearography may provide information about out-of-plane deformations in an indirect manner, by finding the slopes (derivatives) corresponding to these deformations. As in conventional lateral shearing interferometry, the fringes are the locus of the points with the same slope in the direction of the shear [80]. Thus, tilt, for example, cannot introduce any fringes. The derivative properties make such interferograms more difficult to analyze and interpret. The deformations may be obtained only after numerical integration of the slopes in two perpendicular directions. This image processing procedure requires the generation of two shearograms with lateral shear in two orthogonal directions.

According to the theory of small deflections of a thin plate, fortuitously the strains are proportional to the second derivative of the deflection. Thus, to obtain the strains with shearography, only one additional derivative is necessary [81]. As described by Sirohi [78] and Ganesan et al. [82], speckle shear interferometry may also be performed with radial or rotational shear, though the lateral shear is by far the most common incorporation of the technique. It is sometimes convenient, as in conventional lateral shearing interferometry, to introduce a linear carrier by defocusing the beam, as described by Templeton and Hung [83].

The principle of shearography is very similar to that of conventional speckle pattern interferometry. Two images with a relative lateral shear are recorded, one before and the other after the body deformation upon the application of external forces. Then, the irradiances of these two images are added or subtracted. Low-pass filtering removes the speckle noise, resulting in a fringe pattern encoded with the information about the body deformations. Using an oblique illumination beam with an angle of incidence θ, the phase differences with small lateral shear Δx for the sheared speckle patterns before deformation of the sample are

$$\Delta\phi = k(\Delta x)\,\sin\theta \qquad (6.40)$$

and after deformation

$$\Delta\phi = k(\Delta x + \Delta u)\sin\theta + k(\Delta v)(1 + \cos\theta), \qquad (6.41)$$

where Δx is the lateral shear and u and w are small local in-plane (along x-axis) and out-of-plane (along z-axis) displacements. Thus, the change in the phase difference introduced upon such small local displacements may be found upon subtracting Equation 6.41 from 6.40

$$\Delta\phi = k(\Delta u)\sin\theta + k(\Delta v)(1+\cos\theta). \tag{6.42}$$

Equation 6.42 may also be written for small changes $\dfrac{\partial u}{\partial x}\Delta x$ and $\dfrac{\partial w}{\partial x}\Delta x$

$$\Delta\phi = k(\Delta x)\left[\frac{\partial u}{\partial x}\sin\theta + \frac{\partial w}{\partial x}(1+\cos\theta)\right]. \tag{6.43}$$

When both types of displacements u and w are present, but only the slope of w is desired, normal illumination may be used. However, when the slope of u is additionally required, two different measurements with different illumination angles are necessary. Figure 6.17 illustrates three possible experimental layouts to produce the superposition of two laterally sheared speckle images on the detector surface. In the first example (Figure 16a), the lens aperture is divided into two parts, one of them covered with a glass wedge. In the second example (Figure 16b case (b)), a Michelson interferometric configuration is used, with a tilt in one of the mirrors. Finally, the third example (Figure 16c) exhibits a system incorporating a Wollaston prism. Two symmetrical beams are incident on the diffuse object, with opposite angles of incidence. This system is insensitive to out-of-plane displacements w, but responsive to in-plane displacements u. The change in the phase $\Delta\phi$ due to the in-plane displacements is obtained from Equation 6.43 by setting $\partial w/\partial x = 0$ equal to zero

$$\Delta\phi = k\frac{\partial u}{\partial x}(\Delta x)\sin\theta. \tag{6.44}$$

When two illuminating beams are circularly polarized with opposite senses, the analyzer should be oriented at $+45°$ or $-45°$ forming two complementary interferometric patterns.

The advances in computing power and implementation of advanced image processing techniques are combined for the detection of tire defects using laser shearography. Zhang et al. [84] first segment the shearography image. Then, they perform the curvelet transform and Canny edge detection for optimal results. Bai et al. [85] describe the non-uniform out-of-plane displacement measuring system in a modified shearography setup. A variant of Michelson interferometer shears the image in two orthogonal directions. A two-step integration method is applied to the orthogonal displacement derivative field, without requiring the knowledge of the boundary conditions on the object surface [17–20].

Electronic pattern shearing interferometry and shearography methods feature many advantages, including good vibration isolation. Even under rough environmental conditions of in-situ measurements, fringes with adequate contrast are formed.

6.6 In-Plane Grid and Moiré Methods

In grating or moiré interferometry, a specially prepared diffraction grating is attached onto a nearly flat surface of a solid body, as illustrated in Figure 6.18. [86,87] originally developed moiré methods that are often referred to as *moiré* interferometry. Similarly to the techniques of electronic speckle pattern interferometry, small deformations or displacements of a solid body may be measured. Depending on the specimen material and the required grating frequency (that in turn depends also on the specimen itself), the object surface may be prepared according to one among several methods. Printed patterns, photographic prints or stenciled paper patterns may be glued on the object surface (frequencies up to 100 lines/mm).

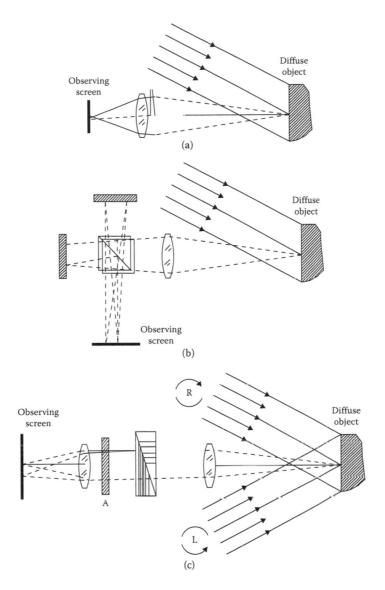

FIGURE 6.17 Configuration for speckle shearing interferometry: (a) With a glass wedge covering half of the lens aperture, (b) with a Michelson interferometer configuration with a tilted mirror, and (c) with a Wollaston prism.

Spatial frequencies up to 4000 lines/mm may be achieved upon replicating a relief-type master grating in epoxy or by exposing interferometric fringes in a photoresist layer at the specimen surface.

The attached grating is deformed and locally displaced when the external forces are applied to the specimen. The grating may be illuminated with two oblique and symmetrically positioned collimated light beams, with an angle of incidence θ given by

$$\sin\theta = \frac{\lambda}{d}, \tag{6.45}$$

Here, d is the period of the grating attached to the solid body and λ is the wavelength.

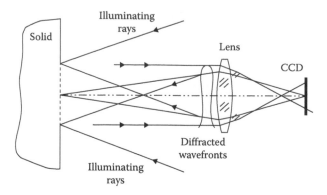

FIGURE 6.18   Grating interferometry configuration.

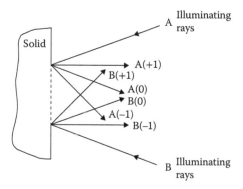

FIGURE 6.19   Diffracted beams in grating interferometry.

Thus, two conjugated diffraction orders (+1 from illuminating beam A and –1 from illuminating beam B, exhibited in Figure 6.19) emerge from the grating along the normal to the grating. These two wavefronts produce an interference pattern. The illuminated surface is then imaged on the sensitive surface of a CCD camera to detect, capture and store the fringe pattern in the computer for further processing, and analysis. Recently, Zhou et al. [88] described an application of moiré interferometry using higher orders (m,-m), and (m,0) to achieve high precision wafer-mask alignment in a proximity imaging lithographic system.

The moiré interferometry is sensitive primarily to the in-plane displacements along the plane of incidence of the illuminating beams. The fringes are formed at the locus of the points on the object undergoing the same displacement. The displacement u of a point on the fringe with order n, relative to the point on the fringe of the order 0 (zero), is given by

$$u = \frac{n}{f} = nd, \tag{6.46}$$

Here, f is the grating frequency and d is the grating period.

When changes are introduced to the spatial frequency of the grating attached onto the solid body due to stresses, expansions, contractions, or bends, both diffracted beams become likewise distorted. The physical model of Moiré interferometry may be formulated as the interference of two wavefronts. This is precisely what we have just done. Additionally, it may also be formulated as a purely geometrical (moiré) effect.

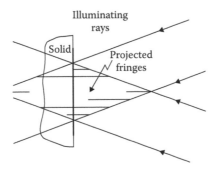

FIGURE 6.20 Projection of a virtual grating in grating interferometry.

6.6.1 Basis of Grid Method

Moiré patterns are frequently analyzed and interpreted from a geometrical point of view. A beating between two illuminating structures is observed in the form of another periodic structure with a significantly lower spatial frequency, called a moiré pattern, when two slightly different periodic structures are superimposed. As sketched in Figure 6.20, two illuminating wavefronts interfere, projecting a fringe pattern on the grating. Thus, the grating may be illuminated with a periodic linear structure, with a period d equal to that of the grating. The observed fringe pattern may then be interpreted as a moiré of two superimposed structured fields. Sciammarella [89], Patorski [90], and many other authors have employed moiré techniques in metrology in diverse configurations and applications.

In moiré interferometry, the image of a grating generated upon the interference of two illuminating beams is projected over the grating attached to the solid body being examined. Two illuminating flattop wavefronts produce parallel interference fringes in space as dark and bright surfaces, perpendicular to the plane of incidence and to the illuminated surface. Thus, a fringe pattern composed of straight lines, called a virtual grating, is being projected onto the body. The superposition of this virtual grating and the real grating attached to the body produces the moiré pattern. This superposition is sometimes interpreted as a product of the irradiance on the virtual grating by the reflectivity of the real grating.

De Oliviera et al. [91] describe the use of photorefractive $Bi_{12}TiU_{20}$ crystal to generate two or more moiré-like fringe patterns, with at least two different variation directions. Such multidimensional fringe patterns may be projected on the object to provide mesh in the Fourier transform profilometry.

6.6.2 Conventional and Photographic Moiré Patterns

The superposition of the two periodic structures to produce the moiré patterns may be accomplished by multiplication, addition, or subtraction.

The multiplication may be implemented, for example, by the superposition of the slides with two images, as in moiré interferometry. The irradiance transmission of the combination is equal to the product of two transmittances in the case of two slides. The contrast in the final moiré image is lower than that in each of two image slides. This is not the case in addition or subtraction schemes where the contrast of the moiré pattern is higher than that of the individual components.

In moiré interferometry, the fringes are formed upon the superposition of the grating attached to the surface to be analyzed and the illuminating projected virtual grating. We consider the fixed grating as the periodic structure that is phase modulated (distorted) and whose relative reflectance $R(x, y)$ may be described by

$$R(x, y) = 1 + \cos\left[\frac{2\pi}{d}(x + u(x, y))\right], \tag{6.47}$$

Here, d is the grating period and $u(x, y)$ represents the local displacement of the grating at the point (x, y). Let us now illuminate this distorted grating to be measured by a projected virtual reference grating with an amplitude $E_r(x, y)$ given by

$$E_r(x, y) = A(x, y)\cos\left[\frac{2\pi}{d}x\right]. \tag{6.48}$$

This virtual grating includes no DC term. Thus, the observed amplitude $E(x, y)$ in the moiré pattern is the product of right-hand-side terms in Equations 6.47 and 6.48

$$E(x, y) = A(x, y)\left[1 + \cos\left(\frac{2\pi}{d}(x + u(x, y))\right)\right] \times \left[\cos\left(\frac{2\pi}{d}x\right)\right], \tag{6.49}$$

The right-hand side may be rewritten as a sum of three terms:

$$E(x, y) = A(x, y)\cos\left(\frac{2\pi}{d}x\right) + \frac{A(x, y)}{2}\cos\left(\frac{2\pi}{d}(2x - u(x, y))\right)$$
$$+ \frac{A(x, y)}{2}\cos\left(\frac{2\pi}{d}u(x, y)\right). \tag{6.50}$$

The first term includes two zero-order beams. The second term represents the (+1) order beam from illuminating beam A and the (–1) order beam from the illuminating beam B. Finally, the last term represents the negative first-order (–1) beam from illuminating beam B and the negative (–1) first-order beam from the illuminating beam A. Figure 6.21 exhibits a grating interferogram of a distorted object obtained with moiré interferometry. Sometimes, instead of a projected virtual grating, an actual reference grating may be placed over the body grating [87].

Similarly to the electronic speckle pattern shearing interferometry, moiré interferometry may be modified to produce fringes with information about the object slopes for strain analysis.

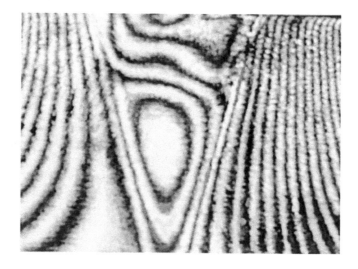

FIGURE 6.21 Grating interferogram of a distorted object.

6.6.3 Moiré Interferometry

A local grating frequency change produces a wavefront distortion of the diffracted beams, as mentioned previously. The orders of two diffracted interfering beams are of opposite sign, that is, they are conjugate. In other words, the wavefront deformations $W(x, y)$ produced by the local displacement $u(x, y)$ have opposite signs for conjugated beams. Consequently, the sensitivity to the in-plane grating deformations is doubled.

Any in-plane lateral displacement $u(x, y)$ of the grating in the direction perpendicular to the grating lines produces a relative phase shift on both diffracted beams, shifting the interference fringes. The amplitude of the (+1) order diffraction beam, produced by the illuminating beam A, with amplitude $A(x, y)$ is

$$E_{+1}^A = A(x,y)\exp\left\{i\left[\frac{2\pi}{d}u(x,y)+\frac{2\pi}{\lambda}w(x,y)\right]\right\}. \tag{6.51}$$

The amplitude of the −1 order diffraction beam, produced by the illuminating beam B, with amplitude $B(x, y)$ is

$$E_{-1}^B = B(x,y)\exp\left\{-i\left[\frac{2\pi}{d}u(x,y)-\frac{2\pi}{\lambda}w(x,y)\right]\right\}. \tag{6.52}$$

Here, $k = 2\pi/\lambda$. Thus, the interferogram irradiance distribution becomes upon adding Equations 6.51 and 6.52,

$$I(x,y)=\left[A(x,y)\right]^2+\left[B(x,y)\right]^2+2\left[A(x,y)\right]\left[B(x,y)\right]\cos\left[\frac{4\pi}{d}u(x,y)\right]. \tag{6.53}$$

The phase changes produced by an out-of-plane displacement w are of the same magnitude and sign for both beams, keeping the interference pattern unchanged.

When there is a local change in the slope of the grating in the direction perpendicular to the grating lines, the diffracted beams tilt by different amounts, as illustrated in Figure 6.22. If the local grating tilt is θ, then the angle $\delta\beta$ between the two diffracted wavefronts becomes

$$\delta\beta = \frac{\lambda\alpha^2}{d} = \alpha^2 \sin\theta, \tag{6.54}$$

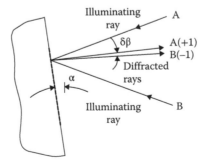

FIGURE 6.22 Grating interferometry with a tilted object.

In the last equality, we made use of Equation 6.5. Here, d is the grating period and a is the inclination of the front surface of the diffuse object with respect to the vertical. This effect provides added sensitivity to slope changes in the object being measured [92]. It must be taken into consideration when considerable out-of-plane displacements are contemplated.

Two independent interferograms may be obtained from two diffracted complementary wavefronts with special experimental arrangements. They permit recovery of the information about out-of-plane displacements in addition to that about in-plane displacements [93].

6.6.4 Multiple-Channel Grating Interferometric Systems

Phase-shifting techniques for the interferogram interpretation require at least three interference patterns, obtained with three different phase-difference values. In the most popular procedures, the three images are recorded sequentially with a CCD camera. This is often not possible in a field environment with vibrations and mechanical instabilities. It is then necessary to generate three phase-shifted irradiance patterns simultaneously in the so-called multichannel interferometer [94]. These types of interferometers are particularly useful when measuring stresses and deformations of diffuse objects with speckle or moiré interferometry. An example of the three-channel phase-stepped system for moiré interferometry is illustrated in Figure 6.23.

This moiré interferometer is similar to the one displayed in Figure 6.9c. A collimated beam of circularly polarized light illuminates the diffuse object. The light reflected at the flat mirror changes the sense of circular polarization. Thus, the body under examination is illuminated by two symmetrically oriented beams of circularly polarized light, as in the speckle interferometer featured in Figure 6.17c. The key of this embodiment is the incorporation of the high-frequency grating that splits the interference beam into three beams. Similarly to the system in Figure 6.17c, an analyzer selects a single polarization plane to generate the interference pattern. Two interfering beams are circularly polarized in opposite directions, so a rotation of the analyzer changes the phase difference between these beams. Thus, using three analyzers with different orientations, three interference patterns with different phase differences are produced.

Several diffracted beams traveling in different directions are created when a light beam passes through a diffraction grating. A lateral displacement of the grating in the direction perpendicular to the grating lines changes the relative phases of the diffracted beams. This effect has been used to construct several multichannel systems, described by Kujawinska [94].

Kulkarni and Rastogi [96] retrieve multiple interference phases corresponding to the in-plane and out-of-plane displacement components, using digital holographic moiré. A single moiré fringe pattern is recorded and analyzed with a pseudo-Wigner-Hough transform upon segmentation of the interference field. The method is robust against object beam irradiance nonuniformity.

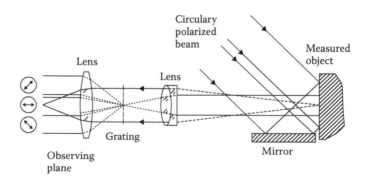

FIGURE 6.23 Three-channel system for moiré interferometry.

References

1. Malacara D. ed., *Optical Shop Testing*, 3nd ed., New York, NY: Marcel Dekker (2007).
2. Idesawa M, Yatagai T, Soma T. Scanning moiré method and automatic measurement of 3-D shapes, *Appl. Opt.*, **16**, 2152–2162; 1977.
3. Takeda M. Fringe formula for projection type moiré topography, *Opt. Las. Eng.*, **3**, 4552; 1982.
4. Doty JL. Projection moiré for remote contour analysis, *J. Opt. Soc. Am.*, **76**, 366–372; 1983.
5. Gåsvik KJ. Moiré technique by means of digital image processing, *Appl. Opt.*, **22**, 3543–3548; 1983.
6. Kowarschik R, Kuhmstedt P, Gerber J, Notni G. Adaptive optical three-dimensional measurement with structured light, *Opt. Eng.*, **39**, 150–158; 2000.
7. Brooks RE, Heflinger LO. Moiré gauging using optical interference fringes, *Appl. Opt.*, **8**, 935–939; 1969.
8. Takasaki H. Moiré topography, *Appl. Opt.*, **9**, 1467–1472; 1970.
9. Takasaki H. Moiré topography, *Appl. Opt.*, **12**, 845–850; 1973.
10. Parker RJ. Surface topography of non-optical surfaces by oblique projection of fringes from diffraction gratings, *Opt. Acta.*, **25**, 793–799; 1978.
11. Pirodda L. Shadow and projection moiré techniques for absolute and relative mapping of surface shapes, *Opt. Eng.*, **21**, 640–649; 1982.
12. Suganuma M, Yoshisawa T. Three-dimensional shape analysis by use of a projected grating image, *Opt. Eng.*, **30**, 1529–1533; 1991.
13. Halioua M, Krishnamurthy RS, Liu H, Chiang FP. Projection moiré with moving gratings for automated 3-D topography, *Appl. Opt.*, **22**, 805–855; 1983.
14. Gåsvik KJ. *Optical Metrology*, 2nd edn, New York, NY, John Wiley & Sons, 1995.
15. Chang W-Y, Hsu F-H, Chen K-H, Chen J-H, Hsu K-Y. Heterodyne moiré surface profilometry, *Opt. Express*, **22**, 2845–2852; 2014.
16. Yao J, Xiong C, Zhou Y, Miao H, Chena J. Phase error elimination considering gamma nonlinearity, system vibration, and noise for fringe projection profilometry, *Opt. Eng.*, **53**, 094102; 2014.
17. Peez G, Strojnik M. Fringe analysis and phase reconstruction from modulated intensity patterns, *Opt. Lett.*, **22**(22), 1669–1971; 1997. doi:10.1364/OL.22.001669.
18. Peez G, Strojnik M. Convergent, recursive phase reconstruction from noisy, modulated intensity patterns using synthetic interferograms, *Opt. Lett.*, **23**(6), 406–408; 1998. doi:10.1364/OL.23.000406.
19. Peez G, Scholl MS. Phase-shifted interferometry without phase unwrapping: Reconstruction of a decentered wavefront, *J. Opt. Soc. Am. A*, **16**, 475–480; 1999. doi:10.1364/JOSAA.
20. Peez G, Strojnik M. Phase reconstruction from underdetected intensity pattern(s), *J. Opt. Soc. Am. A*, **17**(1), 46–52; 2000. DOI: 10.1117/12.372691.
21. Steckenrider JJ, Steckenrider JS. High-speed alternative phase extraction method for imaging array-coupled binary moiré interferometry, *Appl. Opt.*, **54**, 9037–9045; 2015a.
22. Asundi A. Projection moiré using PSALM, *Proc. SPIE*, **1554B**, 254–265; 1991.
23. Zhao H, Du H, Li J, Qin Y. Shadow moiré technology based fast method for the measurement of surface topography, *Appl. Opt.*, **52**, 7874–7881; 2013.
24. Chen Y-Y, Xie A, Zhong X, Zhang Y-Y. Feasibility of integrating moiré tomography and shadowing in flow field's visualization and diagnosis, *Opt. Laser Tech.*, **66**, 125–128; 2015.
25. Patorski K. *Handbook of the Moiré Fringe Technique*, Amsterdam, London, Elsevier, 1993.
26. Osten W, Nadeborn W, Andra T. General approach in absolute phase measurement, *Proc. SPIE*, **2860**, 2–13; 1996.
27. Malacara D, Servin M, Malacara Z. *Interferogram Analysis for Optical Testing*, New York, NY: Marcel Dekker, 1998.
28. Sitnik R, Kujawinska M. Opto-numerical methods of data acquisition for computer graphics and animation system, *Proc. SPIE*, **3958**, 36–45; 2000.

29. Reich C, Ritter R, Thesing J. 3-D shape measurement of complex objects by combining photogram-metry and fringe projection, *Opt. Laser Tech.*, **3**, 224–231; 2000.

30. En B, Fa-jie D, Chang-rong L, Fu-kai Z, Fan F. Sinusoidal phase modulating interferometry system for 3D profile measurement, *Opt. Laser Tech*, **59**, 137–142; 2014.

31. Perciante CD, Strojnik M, Peez G, et al. Wrapping-free phase retrieval with application to interfer-ometry, 3D-shape profiling, and deflectometry, *Appl. Opt.*, **54** (10), 3018–3023; 2015. doi:10.1364/AO.54.003018.

32. Steckenrider JJ, Steckenrider JS. High-resolution moiré interferometry for quantitative low-cost, real-time surface profilometry, *Appl. Opt.*, **54**, 8298–8305; 2015b.

33. Ostrovsky YI, Butusov MM, Ostrovskaya GV.*Interferometry by Holography*, Berlin, Germany: Springer-Verlag, 1980, p. 142.

34. Kim Y, Shim H, Kim K, et al. Common-path diffraction optical tomography for investigation of three-dimensional structures and dynamics of biological cells, *Opt. Express*, **22**, 10398–10407; 2014.

35. Kreis T.*Holographic Interferometry*, Akademie Verlag, Berlin, Germany, 1996.

36. Pryputniewicz RJ.Quantitative determination of displacements and trains from holograms, in Holographic Interferometry, Rastogi, P. K., ed., Springer Series in Optical Sciences, Vol. 68; 1994.

37. Hariharan P, Oreb B, Freund CH. Stroboscopic holographic interferometry measurement of the vec-tor components of a vibration, *Appl. Opt.*, **26**, 3899–3903; 1987.

38. Pryputniewicz RJ. Time-average holography in vibration analysis, *Opt. Eng.*, **24**, 843–848; 1985.

39. Friesem AA, Levy U. Fringe formation in two wavelength contour holography, *Appl. Opt.*, **15**, 3009–3020; 1976.

40. DeMattia P, Fossati-Bellani V. Holographic contouring by displacing the object and the illuminating beams, *Opt. Comm.*, **26**, 17–21; 1978.

41. Tsuruta T, Shiotake N, Tsujiuchi J, Matsuda K. Holographic generation of contour map of diffusely reflecting surface by using immersion method, *Jap. J. Appl. Phys.*, **6**, 66; 1967.

42. Yaraslavskii LP, Merzlyakov NS.*Methods of Digital Holography* New York, NY: Consultants Bureau, 1980.

43. Schnars U. Direct phase determination in hologram interferometry with use of digitally recording holograms, *J. Opt. Soc. Am. A*, **11**, 2011–2015; 1994.

44. Yamaguchi I. Phase-shifting digital holography with applications to microscopy and interferometry, *Proc. SPIE*, **3749**, 434–435; 1999.

45. Goodman JW.*Introduction to Fourier Optics*, New York, NY: McGraw-Hill, 1975a, Chapter 4.

46. Waghmare RG, Mishra D, Subrahmanyam GRKS, Banoth E, Gorthi SS. Signal tracking approach for phase estimation in digital holographic interferometry, *Appl. Opt.*, **53**, 4150–4157; 2014.

47. Li J, Wang YR, Meng XF, Yang XL, Wang QP. Simultaneous measurement of optical inhomogeneity and thickness variation by using dual-wavelength phase-shifting photorefractive holographic inter-ferometry, *Opt. Laser Tech.*, **56**, 241–246; 2014.

48. Belashov AV, Petrov NV, Semenova IV. Digital off-axis holographic interferometry with simulated wavefront, *Opt. Express*, **22**, 28363–28376; 2014.

49. Georges MP, Vandenrijt JF, Thizy C, et al. Combined holography and thermography in a single sensor through image-plane holography at thermal infrared wavelengths, *Opt. Express*, **22**, 25517–25529; 2014.

50. Porras-Aguilar R, Kujawinska M, Zaperty W. Capture and display mismatch compensation for real-time digital holographic interferometry, *Appl. Opt.*, **53**, 2870–2880; 2014.

51. Cashmore MT, Hall SRG, Love GD. Traceable interferometry using binary reconfigurable holo-grams, *App. Opt.*, **53**, 5353–5358; 2014.

52. Kumar V, Kumar M, Shakher C. Measurement of natural convective heat transfer coefficient along the surface of a heated wire using digital holographic interferometry, *Appl. Opt.*, **53**, G74–G83; 2014.

53. Ahadi A, Khoshnevis A, Saghir M. Optical image processing windowed fourier transform as an essential digital interferometry tool to study coupled heat and mass transfer, *Opt. Laser Tech*, **57**, 304–317; 2014.

54. Kumar V, Shakher C. Study of heat dissipation process from heat sink using lensless Fourier transform digital holographic interferometry, *Appl. Opt.*, **54**, 1257–1266; 2015.

55. Stetson KA. Phase-step digital holography with speckled reference beams, *Appl. Opt.*, **54**, 4116–4119; 2015a.

56. Zhang S, He M, Zhang Y, Peng S, He X. Study of the measurement for the diffusion coefficient by digital holographic interferometry, *Appl. Opt.*, **4**, 9127–9135; 2015.

57. Kulkarni RP, Rastogi P. Iterative signal separation based multiple phase estimation in digital holographic interferometry, *Opt. Express*, **23**, 26842–52; 2015.

58. Bochner N, Pipman J. A simple method of determining diffusion constants by holographic interferometry, *J. Phys. D Appl. Phys.*, **9**(13), 1825–1830; 1976.

59. He MG, Zhang S, Zhang Y, Peng SG. Development of measuring diffusion coefficients by digital holographic interferometry in transparent liquid mixtures, *Opt. Express*, **23**, 2015.

60. Macovski A, Ramsey SD, Shaefer LF. Time-lapse interferometry and contouring using television systems, *Appl. Opt.*, **10**, 2722–2727; 1971.

61. Butters JN, Leendertz JA. Speckle pattern and holographic techniques in engineering metrology, *Opt. Laser Tech.*, **3**, 26–30; 1971.

62. Løkberg OJ. The present and future importance of ESPI, *Proc. SPIE*, **746**, 86–97; 1987.

63. Jones R, Wykes C. General parameters for the design and optimization of electronic speckle pattern interferometers, *Optica Acta*, **28**, 949–972; 1981.

64. Jones R, Wykes C.*Holographic and Speckle Interferometry*, Cambridge University Press, Cambridge, 1989.

65. Ennos AE. Speckle interferometry, in Progress in Optics, Vol. 16, Wolf, E., ed., North-Holland, Amsterdam, 1978, pp. 231–288.

66. Cloud G.*Optical Methods of Engineering Analysis*, Cambridge University Press, Cambridge, 1995.

67. Boone PM. Use of close range objective speckles for displacement measurement, *Opt. Eng.*, **21**(3), 213407; 1982.; http://dx.doi.org/10.1117/12.7972923

68. Zhu X, Tang C, Ren H, Sun C, Yan S. Image decomposition model BL-Hilbert-L2 for dynamic thermal measurements of the printed circuit board with a chip by ESPI, *Opt. Laser Tech.*, **63**, 125–131; 2014.

69. Stetson KA. Vibratory strain field measurement by transverse digital holography, *Appl. Opt.*, **54**, 8207–8211; 2015b.

70. Rodríguez-Zurita G, García-Arellano A, Toto-Arellano NI, et al. One-shot phase stepping with a pulsed laser and modulation of polarization:: Application to speckle interferometry, *Opt. Express*, **23**, 23414–23427; 2015.

71. Zhang F, Wang D, Xiao Z, et al. Skeleton extraction and phase interpolation for single ESPI fringe pattern based on the partial differential equations, *Opt. Express*, **23**, 29625–29638; 2015a.

72. Zhang J, Di J, Li Y, Xi T, Zhao J. Dynamical measurement of refractive index distribution using digital holographic interferometry based on total internal reflection, *Opt. Express*, **23**, 27328–27334; 2015b.

73. Gu G, Wang K, Wang Y, She B. Synchronous triple-optical-path digital speckle pattern interferometry with fast discrete curvelet transform for measuring three-dimensional displacements, *Opt. Laser Tech*, **80**, 104–111; 2016.

74. Goodman JW.*Laser Speckle and Related Phenomena*, Dainty, J. C., ed., Berlin, Germany: Springer-Verlag, 1975b, Chapter 2.

75. Sirohi RS.*Speckle Metrology*, New York, NY: Marcel Dekker, 1993.

76. Hung YY. Shearography: A new method for strain measurement and non destructive testing, *Opt. Eng.*, **21**, 391–395; 1982.

77. Sirohi RS. Speckle shear interferometry – A review, *J. Optics (India)*, **13**, 95–113; 1984a.

78. Sirohi RS. Speckle shear interferometry, *Opt. Las. Tech.*, **84**, 251–254; 1984b.

79. Owner-Petersen M. Digital speckle pattern shearing interferometry: Limitations and prospects, *Appl. Opt.*, **30**, 2730–2738; 1991.

80. Strojnik M, Mantravadi M.Lateral shearing interferometry, in Optical Shop Testing, Malacara, D, ed., pp. 649–700, New York, NY: Marcel Dekker, 2007.

81. Toyooka S, Nishida H, Takesaki J. Automatic analysis of holographic and shearographic fringes to measure flexural strains in plates, *Opt. Eng.*, **28**, 55–60; 1989.

82. Ganesan AR, Sharma DK, Kothiyal MP. Universal digital speckle interferometer, *Appl. Opt.*, **27**, 4731–4734; 1988.

83. Templeton DW, Hung YY. Shearographic fringe carrier method for data reduction computerization, *Opt. Eng.*, **28**, 30–34; 1989.

84. Zhang Y, Li T, Li Q. Defect detection for tire laser shearography image using curvelet transform based edge detector, *Opt. Laser Tech*, **47**, 64–71; 2013.

85. Bai P, Zhu F, He X. Out-of-plane displacement field measurement by shearography, *Opt. Laser Tech*, **73**, 29–38; 2015.

86. Post D. The moiré grid-analyzer method for strain analysis, *Exp. Mech.*, **5**, 368–377; 1965.

87. Post D. Developments in moiré interferometry, *Opt. Eng.*, **21**, 458–467; 1982.

88. Zhou S, Hu S, Fu Y, Xu X, Yang J. Moiré interferometry with high alignment resolution in proximity lithographic process, *Appl. Opt.*, **53**, 951–959; 2014.

89. Sciammarella CA. The moiré method. A review, *Exp. Mech.*, **22**, 418–433; 1982.

90. Patorski K. Moiré methods in interferometry, *Opt. Lasers Eng.*, **8**, 147–170; 1988.

91. De Oliviera ME, de Oliviera GN, de Souza JC, Dos Santos PAM. Photorefractive moiré like patterns for the multi fringe projection method in Fourier transform profilometry, *Appl. Opt.*, **55**, 0; 2016.

92. McKelvie J, Patorski K. Influence of the slopes of the specimen grating surface on out-of-plane displacements measured by moiré interferometry, *Appl. Opt.*, **27**, 4603–4606; 1988.

93. Basehore ML, Post D.Displacement field (U, W) obtained simultaneously by moiré interferometry, *Appl. Opt.*, **21**, 2558–2562; 1982.

94. Kujawinska M.Spatial phase measuring methods, in Interferogram Analysis, Robinson DW, Reid GT, eds, Bristol and Philadelphia: Institute of Physics Publishing, 1993.

7

Active and Adaptive Optics

7.1   Introduction and History 245
7.2   Principles of Adaptive and Active Optics 246
7.3   Wavefront Sensing in Adaptive Optics 247
         Shack-Hartmann Wavefront Sensor • Pyramid Wavefront Sensor
7.4   Adaptive Optics in Astronomy 250
7.5   Adaptive Optics in Ophthalmology 253
References 256

Daniel Malacara-
Hernández and
Pablo Artal

7.1 Introduction and History

When observing the stars on a clear night, it is common to notice that their brightness changes fast and at random, producing the phenomenon known as scintillation, and twinkling or, in the language of astronomers, "atmospheric seeing." This is due to the continuous flow of cold and warm layers of air in the upper and low layers of the atmosphere. When observing with a telescope, this atmospheric turbulence changes the shape and size of the images of the stars. Due to their extremely large distances, the angular diameter of the stars is so small that the images in a telescope should ideally be extremely small points. However, due to the atmospheric seeing, the images of the stars have an angular diameter whose magnitude depends on the local weather and the altitude of the observing place.

It is interesting to mention that as early as the eighteenth century, Newton mentions in his book *Opticks* [1] that "telescopes cannot be so formed as to take away that confusion of the rays that arises from the tremors of the atmosphere. The only remedy is a most serene and quiet air, such as may perhaps be found on the tops of the highest mountains above the grosser clouds."

Due to the atmospheric seeing, the wavefront arriving from the star, instead of being a plane, has deformations, changing fast and randomly. With this interpretation, it was a natural consequence to imagine a telescope with a dynamic flexible mirror whose shape adapted to the instantaneous wavefront shape, to reflect an aspherical or flat wavefront. This procedure would eliminate the atmospheric seeing, increasing the angular resolution of the telescope. With time methods to implement this concept, it appeared with the names of adaptive and active optics.

The concepts of adaptive and active optics are not due to a single person. One of the initial proposals was due to Horace Babcock [2], then director of the Mount Wilson and Palomar observatories. Independently, Linnik [3] in the former Soviet Union proposed a similar system. Later, in 1970 it was even considered a possibility in a science fiction book by Poul Anderson titled Tau Zero. However, it was not possible to construct a practical system due to the lack of many necessary technological advances, such as computers, microelectronics, and lasers. The initial developments were made by the military in the United States during the Cold War in the late 1960s. During the 1970s, the United States classified the research on adaptive optics as military work, for missile detection, and communications [4]. The first working system was constructed at the Haleakala Observatory in Maui, Hawaii in order to track Soviet satellites. To the delight of the astronomical community, this research was finally declassified in May 1991.

Several books on the general subject of adaptive optics, at different levels, nontechnical or historical [5,6], introductory [7,8], and advanced [9–12] had been written.

7.2 Principles of Adaptive and Active Optics

The main elements in an adaptive system, as illustrated in Figure 7.1, are the following:

1. The optical observing system, which is a telescope or an ophthalmoscope, to observe the sky or the retina of the eye. Frequently, a collimating system is used after this observing system.
2. A reference luminous point object, beacon or guide star, to produce an identifiable distorted wavefront whose distortions can be measured. In a telescope this can be a bright star in the field of view or an artificial star created with a laser beam.
3. A wavefront sensor to measure the deformations in the wavefront originating at the luminous point object. The most common is a Shack-Hartmann or a pyramid sensing device, as will be described in the next section with more detail.
4. Frequently, a tip-tilt mirror is used to keep the light beam centered as much as possible. In a small telescope where the atmospheric seeing is moderate, a tip-tilt mirror system could be enough to improve the image.
5. A flexible mirror on which the distorted wavefronts from the object reflect and then form the image. This shape of this flexible mirror is adjusted by a computer analyzing the optical signals

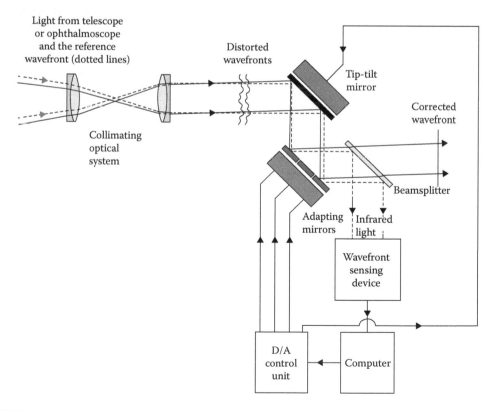

FIGURE 7.1 Main elements in an adaptive optics system.

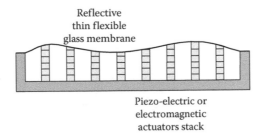

Reflective
thin flexible
glass membrane

Piezo-electric or
electromagnetic
actuators stack

FIGURE 7.2   A flexible mirror.

provided by the wavefront sensor. The flexible mirrors acquire a shape such that the wavefronts deformations are canceled out, producing aberration free (spherical or plane) wavefronts. As described by Roorda and Williams [13], there are several different systems of adaptive mirrors, based on flexible mirrors, as illustrated in Figure 7.2, liquid crystals, etc.

6. The reference wavefront and the wavefronts from the observed objects are separated by means of a dichroic mirror.

Active optics is quite similar to adaptive optics. The only difference is that it works on a much lower time scale. The fluctuations in the wavefront deformations are much slower, due to mechanical deformation of the primary mirror or misalignments due to telescope tube bendings or flexures. Remaining deformations in the figuring of the mirrors can also be compensated.

7.3 Wavefront Sensing in Adaptive Optics

The most common methods to measure the wavefront deformations is with the Shack-Hartmann [14] test described in Chapter 2, the pyramid sensor [16] or the curvature sensor based on the irradiance transport equation [16]

7.3.1 Shack-Hartmann Wavefront Sensor

The Shack-Hartmann test is a modification of the classic Hartmann test, illustrated in Figure 7.2, as proposed by Platt and Shack [17], in order to measure nearly collimated wavefronts. The basic sensor consists of a bi-dimensional array of small lenses molded in plastic or glass, as shown in Figure 7.3.

The distances from the positions of the spots in the image detector to the ideal positions are the ray transverse aberrations. If these transverse aberrations are measured, the real shape of the optical surface under test can be calculated. Let us represent the transverse aberrations in the x direction by $TA_x(x, y)$ and the transverse aberrations in the y direction by $TA_y(x, y)$. These transverse aberrations are related to the wavefront slopes with respect to an ideal spherical wavefront in the same direction by

$$TA_x(x,y) = r\frac{\partial W(x,y)}{\partial x} \tag{7.1}$$

and

$$TA_y(x,y) = r\frac{\partial W(x,y)}{\partial y}, \tag{7.2}$$

Shack-Hartmann
wavefront sensor

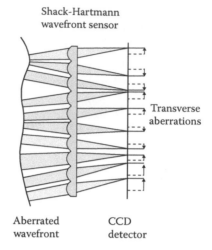

Transverse
aberrations

Aberrated   CCD
wavefront   detector

FIGURE 7.3   Shack-Hartmann sensor.

where r is the radius of curvature of the reference sphere and $W(x, y)$ is the function representing the wavefront deformations with respect to this reference sphere. By integration of these expressions we obtain:

$$W(x,y)=\frac{1}{r}\int_0^x TA_x(x,y)\,dx =\frac{1}{r}\int_0^y TA_y(x,y)\,dy. \tag{7.3}$$

In order to carry on this integration, the one-to-one correspondence between the sampling points on the exit pupil of the system and the Hartmann spots on the observation plane must be known. When the wavefront is highly aspherical, this is quite difficult if the observation plane is inside the caustic zone. For this reason the observation plane must be either inside or outside of focus, outside of the caustic limits.

7.3.2 Pyramid Wavefront Sensor

Another wavefront sensor sometimes used in adaptive optics is the pyramid sensor proposed by Ragazzoni [15] and Ragazzoni et al. [18]. Chew et al. [19] described this sensor and compared it with the Shack-Hartmann sensor. Iglesias et al. [20] and Daly and Dainty [21] used it for ophthalmic purposes. The design and fabrication of this sensor was described by Wang et al. [22]. As illustrated in Figure 7.4, the pyramid glass is located at the image plane.

There is a strong similarity between this sensor and the knife edge in the Foucault test. The difference is that four images are formed, each one of them as if a 90° knife is used instead of the straight knife. If the wavefront has rotational symmetry, the images are like Foucault images with the knife edge at 45°, with the four possible orientations. In this figure the images with primary spherical aberration are illustrated.

The illumination at any point on the images A, B, C and D of the pupil of the system is represented by $I_A(x, y)$ $I_B(x, y)$ $I_C(x, y)$ and $I_D(x, y)$. The total amount of light on each of these images is directly proportional to the light from the image passing through the corresponding quadrant in the pyramid prism.

It is easy to show that these four images A, B, C and D can be added, obtaining four Foucault images as illustrated in Figure 7.5, as if they were produced with the following knife orientations:

The image $I_A(x, y) + I_C(x, y)$ is complementary to $I_B(x, y) + I_D(x, y)$ and the image $I_C(x, y) + I_D(x, y)$ is complementary to $I_A(x, y) + I_B(x, y)$. Thus, the image $I_A(x, y) + I_B(x, y) + I_C(x, y) + I_D(x, y)$ is a constant. For all these images, the brightness of each pixel is between zero and a maximum value.

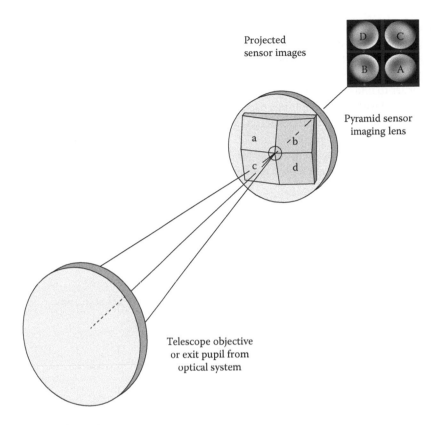

FIGURE 7.4 Pyramid wavefront sensor.

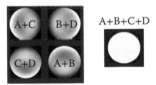

FIGURE 7.5 Images obtained by summing images in the pyramid wavefront sensor.

A + C   Vertical knife, introduced from the left hand side
B + D   Vertical knife, introduced from the right hand side
C + D   Horizontal knife, introduced from below
A + B   Horizontal knife, introduced from the above
A + B + C + D   Uniformly illuminated pupil

A transverse movement of the pyramid sensor is equivalent to a movement of the knife in the Foucault test. For any position (x, y) of the center of the pyramid sensor, we may measure two signals, $S_x(x, y)$ and $S_y(x, y)$, defined as

$$S_x(x,y) = \frac{(I_B(x,y) + I_D(x,y)) - (I_A(x,y) + I_C(x,y))}{I_A(x,y) + I_B(x,y) + I_C(x,y) + I_D(x,y)}$$

$$S_y(x,y) = \frac{(I_C(x,y) + I_D(x,y)) - (I_A(x,y) + I_B(x,y))}{I_A(x,y) + I_B(x,y) + I_C(x,y) + I_D(x,y)}$$

(7.4)

whose values are between zero and one. Their values are directly proportional to the difference between the two complementary illumination values for two knife edge orientations in opposite directions. Thus, these signals have values between -1 and $+1$. Their value is zero when the two complementary illuminations are equal, which happens at the limit of the geometrical shadow that the knife edge projects onto the pupil. In order to find this edge of the geometrical shadow with good accuracy, the glass pyramid is transversally moved with small amplitude oscillations on the image plane; the whole transverse aberrations map can be obtained. Practical details on how these oscillations can be implemented can be found in the references given in this chapter.

If the amplitudes of the components of the oscillation in the focal plane of the pyramid sensor, in the directions x and y respectively, are represented by δV_x and δV_y, the wavefront slopes can be shown to be

$$\frac{\partial W(x,y)}{\partial x} = S_x(x,y)\frac{r}{\delta V_x} \tag{7.5}$$

$$\frac{\partial W(x,y)}{\partial y} = S_y(x,y)\frac{r}{\delta V_y}, \tag{7.6}$$

where r is the radius of curvature of the reference sphere. It can be shown that in analogy to Equations 7.1 and 7.2, we can write for the pyramid test:

$$TA_x(x,y) = r\frac{\partial W(x,y)}{\partial x} \tag{7.7}$$

and

$$TA_y(x,y) = r\frac{\partial W(x,y)}{\partial y}. \tag{7.8}$$

7.4 Adaptive Optics in Astronomy

The most popular application of adaptive optics techniques is in astronomy, to eliminate the image degradation produced by the atmospheric turbulence [23–25]. The state of the art in this field, until 2010, has been described by Hart [1].

The main advantage of adaptive optics is that astronomical telescopes on the surface of the earth can produce images with a quality only limited by diffraction. In space, where the atmospheric turbulence is absent, the only limitation to the angular resolution of a perfect telescope is diffraction. The angular radius of the Airy disc in radians is

$$\theta = 1.22\frac{\lambda}{D}, \tag{7.9}$$

where D and l have the same units. In arc-seconds this expression becomes:

$$\theta = \frac{14}{D}, \tag{7.10}$$

where D is the objective diameter in centimeters (see Figure 7.6).

The image of a star with scintillation has a certain angular size represented by θ_{seeing} defined by the full width at half the maximum value. This broadening of the image is due to the deformations in

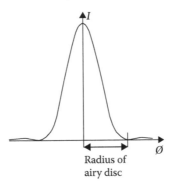

FIGURE 7.6 Airy disc due to diffraction.

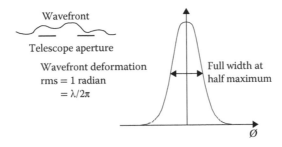

FIGURE 7.7 Image when the aperture has a diameter equal to the Fried parameter.

the incident wavefront due to the atmospheric turbulence. This is not a monochromatic effect since the refractive index is not a constant for all wavelengths.

In an atmospheric model by Kolmogorov to study the turbulence, the atmosphere is considered as a structure of many stratified layers. Local differences in temperature produced by the sun induce local differences in pressure. The result is flow or air and wind in different layer sections. A flat wavefront coming from a star, passing through the turbulent layers of the atmosphere, will suffer wavefront deformations with amplitudes as large as several wavelengths and spatial frequencies of the order of a few centimeters.

A telescope with a small pupil diameter will not be affected by the atmospheric seeing if the period of the lowest spatial frequency component of the wavefront deformations is larger than the aperture diameter. However, its image will move randomly in the field, without changing its size, only limited by diffraction. If the aperture diameter of this small diffraction limited telescope increases, a pupil diameter r_0 will be reached, where the atmospheric seeing will start to be noticed. This happens when the shortest spatial frequency components of the wavefront deformations get smaller than the telescope aperture. Fried [26] found that the aperture diameter r_0 of a telescope that is just in the limit between diffraction limited images and atmospheric seeing broadened images is

$$r_o = \frac{\lambda}{\theta_{seeing}}, \tag{7.11}$$

where this aperture diameter in cm is called the Fried parameter or the Fried coherence length and θ_{seeing} is the half width angular diameter of the seeing image in radians. When the pupil diameter of the telescope is increased, the magnitude of rms value of the phase deformations in the incident wavefront also increases. An interesting property is that this rms value is almost one radian, when the pupil reaches a diameter equal to the Fried parameter (see Figure 7.7).

The Fried parameter ranges from about 5 cm with poor seeing to more than 20 cm with good seeing in the best conditions. Thus, the scintillation image diameter in urban places can be as large as one or even two arc-seconds. In the best astronomical locations, it can be only as small as two or three tenths of an arc-second. An important conclusion is that all telescopes with diameters greater than or equal to about the Fried parameter have the same angular resolution independently of its diameter, due to the atmospheric seeing. Larger telescopes, with apertures greater than about the Fried parameter, will permit the observation of fainter objects, but not with greater detail. The limitation in the resolving power of a telescope due to the atmospheric seeing was one of the main motivations for the placing in orbit of the Hubble telescope in 1990.

If the turbulent layer of air displaces at a velocity v, the whole telescopic aperture D is traveled in a time

$$t = \frac{r_0}{v}. \tag{7.12}$$

On the average, at a good observatory site, this time is of the order of 30 ms. To monitor and remove the turbulence effects, the adaptive optics system has to operate in the order of ten times faster.

One important characteristic in astronomical implementations of adaptive optics is the need for a reference light source whose wavefront has the same deformations as the wavefronts from the observed stellar objects. This reference light source can be a bright star in the field, with a visual magnitude greater than 12, but quite frequently it is not available. A solution independently invented by several researchers is the creation of an artificial star, frequently called a beacon or laser guide star by means of a high power laser aimed in the direction of observation. They create the artificial star by producing a 589.0 nm resonance of mesospheric sodium atoms at an altitude of about 90 km. A problem to be solved with this technique is that the whole path gets illuminated due to backscattering. For this reason, a pulsed laser is used, with a shutter in front, synchronized with the pulse. The laser beacon is not the only solution. Some other techniques are a subject of research.

Even with the laser guide stars, the angle between this star and the observed stellar objects is not zero. Since the light from them travel slightly different paths to the telescope, the two wavefronts are not identical, but have some differences (see Figure 7.8).

The reference luminous image from the guide star has to be inside the field being observed, as close as possible to the stars being observed, so that the wavefront deformations are almost the same for the

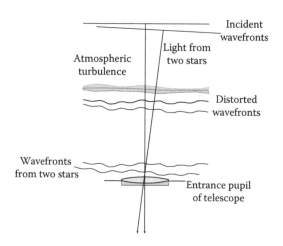

FIGURE 7.8   Different paths for two close stars.

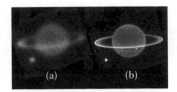

FIGURE 7.9 Images of Uranus without and with adaptive optics, taken with the Keck telescope (Photo courtesy of Heidi B. Hammel and Imke de Pater, Keck Observatory).

reference measured wavefront and for all wavefronts produced by each observed star in the field. In other words, the measured and the observed wavefronts have to be affected by the same atmospheric path. This small field where all the observed images have aberrations within a small range, due to the atmospheric turbulence, is called the isoplanatic region. The angular size of this field is the isoplanatic angle such that the root mean squared (rms) difference between the two wavefronts is one radian ($\lambda/2\pi$).

It is quite important, however, to know that the atmosphere not only creates the atmospheric seeing that degrades the resolving power of the telescope. It is also a window that blocks the light coming from the stars in certain wavelength regions, mainly in the ultraviolet and far infrared. These are the reasons why the Hubble space telescope will not be replaced. Instead, the James Webb space telescope is designed to provide images in the infrared.

The flexible adaptive mirror can be placed in the optical system at several different positions. A possibility is outside the telescope focus, but it has also been proposed at the secondary mirror. Of course, ideally, the best location would be at an image of the pupil of the telescope.

An example of the image difference between an astronomical image without and with adaptive optics is the image of Uranus in Figure 7.9.

7.5 Adaptive Optics in Ophthalmology

Adaptive optics also has applications in biology [27] and in particular in the human eye [28]. It was first used in the late 1990s to correct the eye's higher-order aberrations [29–32]. One application was to obtain high-resolution images of the retina to resolve individual photoreceptors in vivo [29] and to identify the type of photopigment in each cell [33].

To observe the retina of the eye has been a subject of interest for nearly two centuries. The first great advance was due to Helmholtz [34], who revolutionized the field with his invention of the ophthalmoscope. Fifteen years later, Jackman and Webster [35] obtained the first photograph of the retina by attaching a photographic camera to the ophthalmoscope. Retinal or fundus cameras were developed over the next century by applying new technical developments, obtaining even stereoscopic retina images. Real time video images of the retina were obtained thanks to the scanning laser ophthalmoscope [36].

Typical fundus cameras and ophthalmoscopes provide macroscopic images of the human eye retina, as illustrated in Figure 7.10. They do not have the spatial resolution to allow the observation of single cells, like the cones and rods.

Several different alternative procedures were tried to see the photoreceptor mosaic (cones) for several years. An early attempt was the use of speckle interferometry to obtain the inter-center spacing between cones [37]. The first application of a segmented deformable mirror in a scanning laser ophthalmoscope only allowed correction of astigmatism without imaging the cones [38]. Miller et al. [39] used a fundus camera with incoherent light. The pupil was dilated to reduce the size of the diffraction image and all existing ammetropies were very carefully corrected. However, the photoreceptor mosaic was observed only in young people with very good eyes. Soon later, the first static correction of the eye's aberrations using a deformable mirror was performed showing the first direct images of the cone photoreceptors in vivo [29].

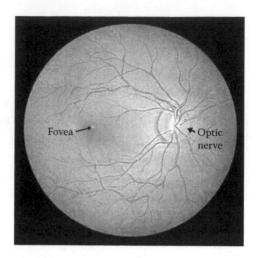

FIGURE 7.10   Image of the retina obtained with an ophthalmoscope.

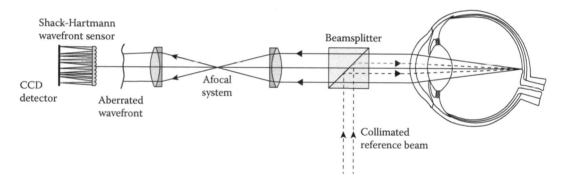

FIGURE 7.11   Measurement of the wavefront aberrations in the human eye by means of the Shack-Hartmann test.

Adaptive optics techniques are necessary in order to see any person's retina with a higher resolution. The reason was that high-order aberrations are very common and vary quite rapidly with time due to many factors and also that there are constant micro-fluctuations in the accommodation [40]. Real time closed-loop adaptive optics was later demonstrated in the eye. As in the astronomical applications, the preferred wavefront sensor for the eye was based in the application of the Shack-Hartmann sensor [40,41], using the arrangement in Figure 7.11. The light from a point light source, or laser, is used to form a spot on the retina. To obtain the smallest point image possible, only a small artificial pupil is used to illuminate the eye. The light is scattered and reflected back as a spherical wavefront with a larger angle, covering the complete eye pupil. When traveling the eye, this wavefront is distorted by aberrations and imaged onto a microlenses array. The wavefront is then reconstructed in real time. The signal processing devices had not been illustrated in this figure.

In adaptive optics imaging of the retina, as illustrated in Figure 7.12, a laser is used to illuminate the eye through and to form a point image at the center of the retina. A flexible mirror, to compensate the aberrations, is located at an image of the pupil of the eye. The Shack-Hartmann sensor as well as the pupil in front of the retinal image camera is also at planes conjugates to the observed eye pupil and the flexible mirror. To illuminate an area of the retina, a krypton flash lamp is used. This is the system used by Roorda and Williams [13]. More recent systems work with basically the same principles, but many interesting modifications have been made in the last few years to improve it, with impressive results.

An image of the retina showing the cones is in Figure 7.13.

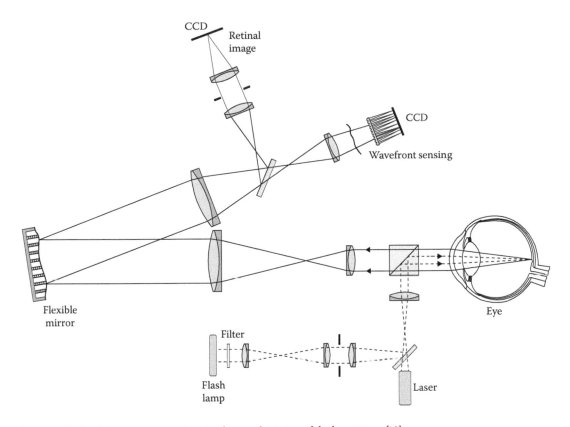

FIGURE 7.12 Adaptive optics system to observe the retina of the human eye [13].

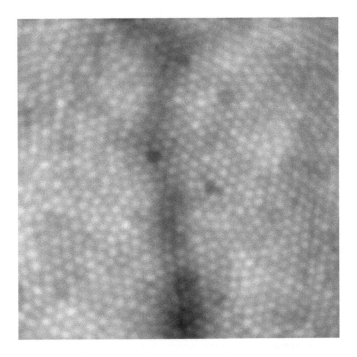

FIGURE 7.13 An image *in vivo* of the cones in a human retina [33].

A confocal adaptive optics laser scanning ophthalmoscope was reported by Roorda et al. [43]. The scanning is performed by means of small oscillating mirrors, with a vertical frequency of thirty Hertz and a horizontal frequency of 16 kHz. With this system, they were able to obtain images of the retina with unprecedented contrast and brightness in real time, increasing both lateral and axial resolution. This permits axial sectioning of the retinal tissue *in vivo*, due to its confocal arrangement, visualizing photoreceptors, nerve fibers, and flow of white blood cells in retinal capillaries.

Dubra et al. [44] have reported another confocal adaptive optics scanning ophthalmoscope capable not only of imaging the cones, but also the rods. To observe the rod is quite important, since rod dysfunction is involved in a large number of devastating sources of blindness, like retinitis pigmentosa, cone-rod dystrophy, congenital night blindness, macular degeneration, and so on. By observing the rods, we may be able to monitor the efficacy of treatments for these diseases. In order to image the rods, this instrument required cyclopegia and a good stabilization of the patient. The light illumination sources were two super-luminescent diodes with peak wavelengths of 680 and 775 nm.

Adaptive optics was also incorporated to other retinal imaging modalities, most notably optical coherence tomography (OCT) providing significant improvements [45].

Another different, but also important application of adaptive optics, is to produce optical aberration patterns in the eye, enabling visual testing with controlled optics. In this configuration, instead of the retinal imaging path, a screen or micro-display presents visual stimuli to the eye. This allows it to perform interesting new visual experiments, for instance to better understand the role of the eye's optics in visual perception [46] and to design new ophthalmic correcting devices [47]. More recently, binocular versions of the adaptive optics instrument for visual testing have been also developed [48].

References

1. Hart M. Recent advances in astronomical adaptive optics, *Appl. Opt.*, **49**, D17–D26, 2010.
2. Babcock HW. The possibility of compensating atmospheric seeing, *Publ. Astron. Soc. Pac.*, **65**, 229–236, 1953.
3. Linnik VP. On the possibility of reducing atmospheric seeing in the image quality of stars, *Opt. Spektrosc.*, **3**, 401–402, 1957.
4. Collins GP. Making stars to see stars: DOD adaptive optics work is declassified, *Phys. Today*, **18**, 17–21, February 1992.
5. Tyson RK. *The Lighter Side of Adaptive Optics, SPIE Monograph*, Vol. PM191. Bellingham, WA: SPIE Press, 2009.
6. Duffner RW. *The Adaptive Optics Revolution: A History by Robert W. Duffner*. Alburquerque, NM: University of New Mexico Press, 2016.
7. Tyson RK. *Introduction to Adaptive Optics, SPIE Tutorial Texts in Optical Engineering*, Vol. TT41. Bellingham, WA: SPIE Press, 2000a.
8. Tyson RK. *Principles of Adaptive Optics*, 4th Edn. Boca Raton, FL: CRC Press, 2016.
9. Tyson RK. Ed., *Adaptive Optics Engineering Handbook*. New York, NY: Marcel Dekker, 2000b.
10. Ames K. *Advanced Concepts in Adaptive Optics*. New York, NY: Research Press, 2015.
11. Zaretsky N. Ed., *Advanced Concepts in Adaptive Optics*. Albany, NY: Delve Publishing, 2015.
12. Tyson RK, Frazler BW. *Field Guide to Adaptive Optics, SPIE Field Guide*, 2nd Edn., Vol. FG24. Bellingham, WA: SPIE Press, 2012.
13. Roorda A, Williams DR. Adaptive optics and retinal imaging, in vision science and its applications, *OSA Tops*, **35**, 151–162, 2000.
14. Platt BC, Shack RV. Lenticular Hartmann screen, *Opt. Sci. Newsl.*, **5**, 15–16, 1971.
15. Ragazzoni R. Pupil plane wavefront sensing with an oscillating prism, *J. Mod. Opt.*, **43**, 289–293, 1996.
16. Roddier, F., Curvature sensing and compensation: A new concept in adaptive optics, *Appl. Opt.*, **27**, 1223–1225 (1988).

17. Platt BC, Shack RV. Lenticular Hartmann Screen, *Opt. Sci. Newsl.*, **5**, 15–16 (1971).

18. Ragazzoni R, Ghedina A, Baruffolo A, et al. Testing the pyramid wavefront sensor on the sky, *Proc. SPIE.*, **4007**, 423–430, 2000.

19. Chew TY, Clare RM, Lane RG. A comparison of the Shack-Hartmann and pyramid wavefront sensors, *Opt. Commun.*, **268**, 189–195, 2006.

20. Iglesias I, Ragazzoni R, Julie Y, Artal P. Extended source pyramid wavefront sensor for the human eye, *Opt. Express*, **10**, 419–428, 2002.

21. Daly EM, Dainty C. Ophthalmic wavefront measurements using a versatile pyramid sensor, *Appl. Opt.*, **49**, G67–G77, 2010.

22. Wang A, Yao J, Cai D, Ren H. Design and fabrication of a pyramid wavefront sensor, *Opt. Eng.*, **49**, 73401–73405, 2010.

23. Lukin VP. *Atmospheric Adaptive Optics*. Bellingham, WA: SPIE Press, 1996.

24. Hardy JW. *Adaptive Optics for Astronomical Telescopes, Oxford Series in Optical and Imaging Sciences*. Oxford: Oxford University Press, 1998.

25. Roddier F. Ed., *Adaptive Optics in Astronomy*. Cambridge: Cambridge University Press, 1999.

26. Fried DL. Optical resolution through a randomly inhomogeneous medium for very long and very short exposures. *J. Opt. Soc. Am.*, **56**, 1372–1379, 1966.

27. Kubby JA. *Adaptive Optics for Biological Imaging*. Boca Raton, FL: CRC Press, 2013.

28. Porter J, Queener H, Lin J, Thorn K, Awwal A. Eds., *Adaptive Optics for Vision Science: Principles, Practices, Design and Applications*. Hoboken, NJ: Wiley-Interscience, 2006.

29. Liang J, Williams DR, Miller DT. Supernormal vision and high-resolution retinal imaging through adaptive optics, *J. Opt. Soc. Am. A*, **14**, 2884–2892, 1997.

30. Vargas-Martín F, Prieto P, Artal P. Correction of the aberrations in the human eye with liquid crystal spatial light modulators: Limits to the performance. *J. Opt. Soc. Am. A*, **15**, 2552–2562, 1998.

31. Fernández EJ, Iglesias I, Artal P. Closed-loop adaptive optics in the human eye, *Opt. Lett.*, **26**, 746–748, 2001.

32. Hofer H, Artal P, Singer B, Aragón JL, Williams DR. Dynamics of the eye's wave aberration, *J. Opt. Soc. Am. A*, **18**, 497–506, 2001a.

33. Roorda A, Williams, D. R., The arrangement of the three cone classes in the living human eye, *Nature*, **397**, 520–522, 1999.

34. Helmholtz HLF. *Beschreibung eines Augen-Spiegels zur Untersuchung der Netzhaunt im lebenden Auge* (Description of the Eye Mirror for the Investigation of the Retina of the Living Eye). Berlin: A Forstner'sche Verlagsbuchhandlung, 1851.

35. Jackman WT, Webster JD. On photographing the retina of the living human eye, *Philadelphia Photographer*, **23**, 340–341, 1886.

36. Webb RH, Hughes GW, Pomerantzeff O. Flying spot TV ophthalmoscope, *Appl. Opt.*, **19**, 2991–2997, 1980.

37. Artal P, Navarro R. High-resolution imaging of the living human fovea: Measurement of the inter-center cone distance by speckle interferometry, *Opt. Lett.*, **14**, 1098–1100, 1989.

38. Dreher A, Bille J, Weinreb R. Active optical depth resolution improvement of the laser tomographic scanner, *Appl. Opt.*, **28**, 804–808, 1989.

39. Miller DT, Williams DR, Morris GM, Liang J. Images of cone photoreceptors in the living human eye, *Vision Res.*, **36**, 1067–1079, 1996.

40. Hofer H, Chen L, Yoon GY, Singer B, Yamauchi Y, Williams DR. Improvement in retinal image quality with dynamic correction of the eye's aberrations. *Opt. Express*, **8**, 631–643, 2001b.

41. Liang J, Grim B, Goelz S, Bille JF. Objective method of measuring aberrations of wave aberrations in the human eye with the use of the Hartmann-Shack wave-front sensor, *J. Opt. Soc. Am. A*, **11**, 1949–1957, 1994.

42. Prieto PM, Vargas-Martín F, Goelz S, Artal P. Analysis of the performance of the Hartmann-Shack sensor in the human eye, *J. Opt. Soc. Am. A*, **17**, 1388–1398, 2000.

43. Roorda A, Romero-Borja F, Donnelly WJ III, Queener H, Hebert TJ, Campbell MCW. Adaptive optics scanning laser ophthalmoscopy, *Opt. Express*, **10**, 405–412, 2002.

44. Dubra A, Sulai Y, Norris JL, et al. Noninvasive imaging of the human rod photoreceptor mosaic using a confocal adaptive optics scanning ophthalmoscope, *Biomed. Opt. Express*, **2**, 1864–1876, 2011.

45. Hermann B, Fernández EJ, Unterhuber A, et al. Adaptive-optics ultrahigh-resolution optical coherence tomography, *Opt. Lett.*, **29**, 2142–2144, 2004.

46. Artal P. Optics of the eye and its impact in vision: A tutorial, *Adv. Opt. Photonics*, **6**, 340–367, 2014.

47. Manzanera S, Prieto P, Ayala DB, Lindacher JM, Artal P. Liquid crystal adaptive optics visual simulator: Application to testing and design of ophthalmic optical elements, *Opt. Express*, **15**, 16177–16188, 2007.

48. Fernández, E. J., Prieto, P., and Artal, P., Binocular adaptive optics visual simulator, *Opt. Lett.*, **34**, 2628–2630 (2009).

Holography

8.1 Introduction 259
8.2 Diffraction Gratings..261
    Waves, Interference, and Grating Criteria • Thick Gratings
    Characteristics • Thin Gratings Characteristics
8.3 Scalar Theory of Diffraction 272
    Kirchhoff Diffraction Integral • Fresnel Diffraction
    Integral • Fraunhofer Diffraction Integral • Diffraction by Simple
    Apertures
8.4 Computer Generated Holograms 278
    Fourier Hologram • Fresnel Hologram • Iterative Computation for
    Hologram • Errors in CGH
8.5 Holographic Setups...282
    Formalism • Inline Transmission Hologram (Gabor) • Inline
    Reflection Hologram (Denisyuk) • Off-Axis Transmission Hologram
    (Leith and Upatnieks) • Transfer Holograms • Rainbow Hologram
    (Benton) • Holographic Stereogram • Holographic Interferometry
8.6 Holographic Recording Materials..290
    Silver Halide • Dichromated Gelatin • Photopolymer • Photoresist
    and Embossed Holograms • Miscellaneous Materials • Electronic
    Devices
Acknowledgment...294
References..294

Pierre-Alexandre
Blanche

8.1 Introduction

There are three ways to alter or change the trajectory of light: reflection, refraction, and diffraction. In our everyday experiences we encounter reflections from mirrors and flat surfaces, and refraction when we look through a glass or water or wear prescription glasses. Scientists have employed reflection and refraction for over 400 years to engineer powerful instruments, such as telescopes and microscopes. Isaac Newton [1] championed the classical theory of light propagation as particles, which accurately described reflection and refraction. Diffraction on the other hand could not be explained by this corpuscular theory and was only understood much later with the concept of wave propagation of light first described by Huygens [2], and extensively developed later by Young [3] and Fresnel [4]. Wave propagation theory predicts that when the light encounters an obstacle such as a slit, the edges do not "cut" a sharp shape in the light beam as the particle theory predicts, but rather there is formation of wavelets that propagate on the side in a new direction. This is the diffraction phenomenon. Eventually, the particle and wave points of view will be reconciled by the quantum theory, and the duality of wave particle developed by Schrödinger [5] and de Broglie [6, 7].

While the light propagation from mirrors and lenses can be explained with the thorough understanding of reflection and refraction, holography can only be explained by recognizing diffraction.

A hologram is nothing but precisely positioned apertures that diffract the light and form a complex wavefront, such as a three-dimensional (3D) images. And since the light can be considered as a wave, both the amplitude and the phase can be modulated to form the hologram. In the later case of phase modulation, the hologram is totally transparent and it is the refractive index that is engineered to provide the discontinuities, which will generate the side wavelets.

Holograms are very well known for the awe inspiring 3D images they can recreate. But they can also be used to generate any desired wavefront, such as a focusing exactly like a lens, or reflecting exactly like a mirror. The difference from the original structure is that, in both cases, diffraction is involved, not reflection or refraction. That type of hologram, frequently called a holographic optical element (HOE), is found in optical setups where for reason of space, weight, size, complexity, it is not possible to use classical optical elements.

There are two very different techniques for manufacturing holograms. One can either compute it, or record it optically. Computing the hologram involves the calculation of the position of the apertures and/or phase shifters, according to the laws of light propagation derived by Maxwell [8]. This calculation can be fairly easy for simple wavefronts, such as a lens, to extremely complicated for high-resolution three-dimensional images. To optically record a hologram, the amplitude and the phase of the wavefront need to be captured. Recording the intensity was first achieved with the invention of photography by Niépce in 1822, but recording the phase eluded scientists until 1948. Although the concept of optical interference was known for ages, it is Gabor [9, 10] who introduced the concept of making an object beam interfere with a reference beam to reveal and record the phase. Indeed, when two coherent beams intersect, constructive and destructive interferences occur according to the phase difference; this transforms the phase information into intensity information that can be recorded the same way photographs are taken. In some sense, the reference beam is used to generate a wave carrier that is modulated by the information provided by the object wave.

Gabor coined the term holographic from the Greek words holos: "whole" and graphe: "drawing," since the technique recorded for the first time the entire light field information: amplitude and phase. Gabor used the technique to increase the resolution in electron microscopy and received the Nobel Prize in physics in 1971 for this discovery.

Due to the very short coherence length of the light sources available to Gabor at the time, the object and reference beams require it to be co-linear. Unfortunately, this configuration yields to very poor imaging quality since the transmitted beam and ± 1 diffracted orders were superimposed, leading to high noise and a "twin image" problem.

Holography will have to wait for the invention of the visible light laser in 1960 by Maiman [11], and for Leith and Upatnieks [12–14] to resolve the twin-image problem. With a long coherence length laser source, one may divide a beam into two parts—one to illuminate the object (the object beam) and the other (the reference beam) is at an angle directly to the hologram recording plane. As a result of the high degree of coherence, the object and reference beams will still interfere to form the complex interference pattern: the hologram. Upon reconstruction, a monochromatic beam is incident to the recorded hologram and the different diffracted waves are angularly separated. This way, the 0, +1 and −1 orders can be observed independently.

In parallel, and independently to Leith and Upatnieks, Denisyuk worked on holograms where the object and reference beams are incident the hologram plane from opposite directions [15–17]. Such holograms are formed by placing the photosensitive medium between the light source and a diffusely reflecting object. In addition to being much simpler and more stable to record, these reflection holograms can be viewed by a white light source, since only a narrow wavelength region is reflected back in the reconstruction process.

Once high-quality imaging and computer generated holograms were demonstrated [18, 19], the research on holography experienced phenomenal growth expanding to encompass a large variety of applications, such as data storage, information processing, interferometry, and dynamic holography to only cite a few. Today, with the widespread access to active LCoS and MEMS devices, there is a rejuvenation of holographic field where a new generation of researchers are applying the discoveries of the past decades to dynamic spatial light modulators. New applications are only limited by the imagination of scientists and engineers, and developments are continuously being reported in the scientific literature.

This chapter will start by developing the theory of thick and thin diffraction gratings. Once these bases have been established, we will move to the scalar theory of diffraction that shows how to calculate the field from a diffractive element and vice versa. We will finish by describing several important holographic recording setups, and discuss materials.

8.2 Diffraction Gratings

8.2.1 Waves, Interference, and Grating Criteria

A lot about holography can be understood without the complication of imaging, and by simply looking at diffraction gratings. Diffraction gratings are particular holograms where the interferences fringes, or Bragg's planes, are parallel. They transform one set of planar wavefronts into another set of planar wavefronts such that the mathematical formalism is simplified. After this analysis, imaging can simply be viewed as the superposition of several planar wavefronts.

Maxwell's equation defines the properties of the electromagnetic field. However, for most holographic applications, the magnetic field can be neglected, and only the Helmholtz equation defining the electric field E remains:

$$\frac{1}{c^2}\left(\frac{\partial^2 E}{\partial t^2}\right) - \nabla^2 E = 0 \tag{8.1}$$

A solution of this differential equation has the form of a plane wave:

$$E(r,t) = A\cos\left(k \cdot r - \omega t + \varphi\right), \tag{8.2}$$

where A is an imaginary vector describing the direction of the electric field oscillation, and contains the polarization information, k is the wave vector pointing in the direction of light propagation which magnitude is related to the wavelength $k = 2\pi/\lambda$. r is the position vector defining the position at which the field is calculated, w is the frequency, and φ is the phase. Two equivalent representations of a plane wave are illustrated in Figure 8.1. It has to be noted that a spherical wavefront is also a solution of the Helmholtz equation.

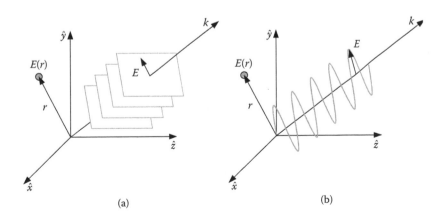

(a)   (b)

FIGURE 8.1 Plane wave representation as (a): plane of equal field intensity, (b): oscillation of the amplitude of the field along the wave vector **k**.

Using the Euler's formula $\exp(ix) = \cos x + i\sin x$, the plane wave solution can be re-written as

$$U(r,t) = A\exp\left[i(k \cdot r - \omega t + \varphi)\right]\hat{a},$$ (8.3)

where the polarization vector $\hat{a}$ has been extracted from the vector $\mathbf{A}$, which is now a scalar A.

Keeping in mind that the actual electric field is the real part of the complex notation:

$$E(r,t) = \Re\left[U(r,t)\right] = \frac{1}{2}U(r,t) + \frac{1}{2}U^*(r,t)$$ (8.4)

when two plane waves of the form of Equation 8.3 cross each other, interference occurs. The total field is

$$U_{total}(r,t) = A_1\exp\left[i(k_1 \cdot r - \omega_1 t)\right]\hat{a}_1 + A_2\exp\left[i(k_2 \cdot r - \omega_2 t)\right]\hat{a}_2.$$ (8.5)

In this formulation, we can see that the pattern is not static but change as a function of time. In the special case of w_1-w_2, Equation 8.5 becomes time invariant and can be expressed in a simpler form where the total intensity is

$$\begin{aligned}
I(r) &= \int U_{total}(r,t)U_{total}^*(r,t)dt \\
&\propto A_1^2 + A_2^2 + 2A_1A_2\left(\hat{a}_1 \cdot \hat{a}_2\left(\cos\left[(k_1-k_2)\cdot r + \arg(\hat{a}_1 \cdot \hat{a}_2)\right]\right)\right. \\
&= I_1^2 + I_2^2 + 2I_1I_2\left(\hat{a}_1 \cdot \hat{a}_2\left(\cos\left[(k_1-k_2)\cdot r + \arg(\hat{a}_1 \cdot \hat{a}_2)\right]\right)\right.
\end{aligned}$$ (8.6)

To maximize the contrast between dark (destructive interference) and bright (constructive interference) regions, the polarization of the wave should be identical: $\hat{a}_1 = \hat{a}_2$, and the equation reduces to the familiar form:

$$I(r) = I_1^2 + I_2^2 + 2I_1I_2\cos\left[(\mathbf{k}_1 - \mathbf{k}_2)\cdot \mathbf{r}\right].$$ (8.7)

This intensity modulation can be recast as a static plane wave with a wave vector defined as

$$K = k_1 - k_2.$$ (8.8)

This intensity modulation can be recorded inside a material as an index modulation or absorption modulation pattern to form a diffraction grating. K is then called the grating vector, and its magnitude is related to the spacing between two planes of equal magnitude Λ, also called the Bragg's plans:

$$K = \frac{2\pi}{\Lambda}.$$ (8.9)

When an incident plane wave defined by the wave vector $\mathbf{k}_i$ encounters the recorded grating K, it is diffracted in the direction $\mathbf{k}_d$ according to the K-vector closing condition (see Figure 8.2):

$$K = k_d - k_i.$$ (8.10)

This condition is identical to the grating equation devised from crystallographic measurements where the modulation planes were actually rows of atoms:

$$\sin(\theta_d) - \sin(-\theta_i) = m\frac{\lambda}{\Lambda},$$ (8.11)

where θ_d is the angle of diffraction, θ_t is the angle of incidence, and m is an integer number that defines the diffraction order.

The angular dispersion as a function of the frequency can be directly derived from the grating equation:

$$\frac{d\theta_d}{d\lambda} = \frac{m}{\Lambda \cos\theta_d}.$$ (8.12)

It can be seen from Equation 8.16 that the lower frequencies (red) are diffracted at a larger angle than the higher frequencies (blue). Conversely, the dispersion from a refractive prism (with normal index dispersion) with higher frequencies exit at a larger angle. This opposition can be used to make an optical system achromatic.

It should be noted that the grating equation expressed in Equations 8.14 through 8.16, does not give any indication on the intensity of the wave being diffracted, only the direction and frequency. The direction of the maximum diffraction efficiency is given by the Bragg's law that can be understood as the condition for constructive interference for the light interacting with two successive diffraction planes:

$$\sin(\theta + \varphi) = m\frac{\lambda_B}{2n\Lambda},$$ (8.13)

where θ is the angle of incidence, φ is the slant angle: the angle between the grating vector and the normal to the grating surface.

The energy distribution around the Bragg's angle and wavelength for thick diffraction gratings was first derived by Kogelnik in his coupled wave theory [20]. Another derivation that also gives very good results is called the parallel stacked mirror model and has been introduced by Brotherton-Ratcliffe [21, 22]. These two models give analytical solutions in the case when the grating satisfies the Bragg's condition for thick grating. This condition is somewhat misnamed since it is not based on the physical thickness of the material, but on the premise that most of the energy is concentrated in the first diffraction order. By contrast, thin gratings are subject to the Raman-Nath regime of diffraction where appreciable energy can be found in higher orders of diffraction.

There is not a clear dividing line between thin and thick gratings, but two criterion have been devised according to the approximations used in solving the coupled wave equation, and according to the results observed experimentally:

The Klein and Cook criteria [23] is

$$Q' = \frac{2\pi\lambda d}{n\Lambda^2\cos\theta}$$ (8.14)

with $Q' < 1$ for thin gratings, and $Q' > 1$ for thick gratings.

The Moharam and Young criteria [24] is

$$\rho = \frac{\lambda^2}{n\Delta n\Lambda^2\cos\theta}$$ (8.15)

with $\rho < 1$ for thin gratings, and $\rho \geq 1$ for thick gratings.

When the diffraction grating does not satisfy the thick grating condition(s), neither the Kogelnik's or Brotherton-Ratcliffe's theories can be applied and one has to solve the more general rigorous coupled wave analysis developed by Moharam and Gaylord that do not have analytical solution [25].

8.2.2 Thick Gratings Characteristics

In order to satisfy the thick grating criterion introduced in Equations 8.18 and 8.19, the Bragg's planes need to extend to a certain volume inside the material (thus the name). Such a diffraction structure cannot be just overlaid on the surface. The advantage of thick grating is that most of the diffracted energy is found in the first order. For that reason, thick grating is of particular interest in holographic imaging and engineering since one does not have to deal with light present in higher diffraction orders. These higher-order modes contain some power and reduce the overall signal to noise of the desired first-order image.

The manufacturing of thick grating generally involves the optical recording with an interferometric setup, techniques that will be detailed in a subsequent section. The reason for optical recording is that the diffractive structures need to be imbedded inside the volume of the material, which is difficult to access otherwise. It is also possible to produce diffractive structures that satisfy the thick grating conditions using multi-layer coating similar to what is used to create dichroic mirrors.

The rigorous derivation of the diffraction efficiency for thick gratings in various conditions can be found in the original publications [20, 21]. We will summarize here the principal results in the special case of unslanted ($\varphi = 0$ or $\pi/2$) phase (Δn) or amplitude ($\Delta \alpha$) sinusoidal modulation. For these conditions, the mathematical expressions simplify dramatically and other modulation formats, such as square can be calculated from the Fourier components.

In addition to both types of modulation (phase and amplitude), two different configurations of the grating will be discussed: transmission and reflection. Illustration of these two geometries can be seen in Figure 8.2 where a slant angle ($\varphi \approx 0$ or $\approx \pi/2$) has been introduced for the sake of generality, and the K-vector closing condition is depicted.

In transmission geometry, the diffracted light exits the grating by *the opposite side* of the incident light: the light goes through the grating. To do so, the Bragg's planes are oriented more or less orthogonal

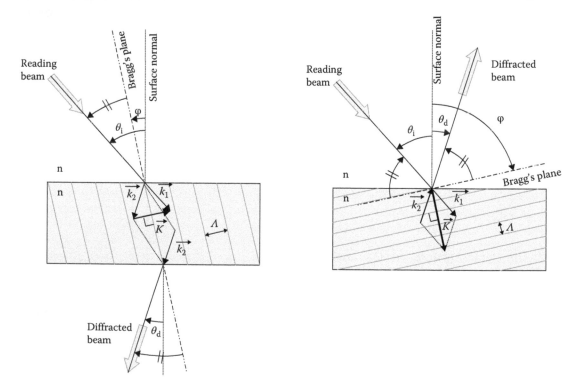

FIGURE 8.2 Geometry of (a) transmission grating, (b) reflection grating, with K-vector closing condition and slant angle.

to the grating surface. The grating frequency in transmission geometry ranges from 300 to 3000 line pairs per mm (lp/mm) for visible light.

In the reflection geometry, the diffracted light exits the grating by *the same side* of the incident light: the light bounces back from the grating. In this geometry the Bragg's planes are oriented more or less parallel to the surface and the frequency is over 4000 lp/mm.

Phase gratings can reach 100% efficiency either in transmission or reflection. The expression for TE (transverse electrical) mode is respectively given by

$$\eta_{TE} = \sin^2\left[\left(\frac{\pi\Delta n}{\lambda}\right)\left(\frac{d}{\cos\theta_i}\right)\right] \tag{8.16}$$

for transmission configuration, and

$$\eta_{TE} = \tanh^2\left[\left(\frac{\pi\Delta n}{\lambda}\right)\left(\frac{d}{\cos\theta_i}\right)\right] \tag{8.17}$$

for reflection configuration.

In transmission, the efficiency is a periodic function (sin) that reaches maximum when the phase modulation equals $\Delta n = (2m + 1)\pi/2$. When the phase modulation extends past the first maximum, the grating is said to be over-modulated and the light starts to be coupled back to the zero order, reducing the efficiency. The minimum is reached when $\Delta n = m\pi$.

In the reflection configuration, the diffraction efficiency monotonically increases and there is no over-modulation. The reason is that the light transferred into the first order exits the media and does not propagate into the volume where it would have had a chance to be coupled back into the zero order.

Plots of the diffraction efficiency according to the modulation Δn for transmission (Equation 8.20) and reflection (Equation 8.21) Bragg's gratings are presented in Figure 8.3.

Amplitude gratings are less efficient than phase gratings because the modulation is based on the absorption of a portion of the incident light. Amplitude gratings cannot reach 100% efficiency, and cannot exceed 7.2% efficiency. The expression for TE (transverse electrical) for a transmission grating is

$$\eta_{TE} = \exp\left(\frac{-2\alpha d}{\cos\theta_i}\right)\sinh^2\left[\left(\frac{\Delta\alpha}{2}\right)\left(\frac{d}{\cos\theta_i}\right)\right], \tag{8.18}$$

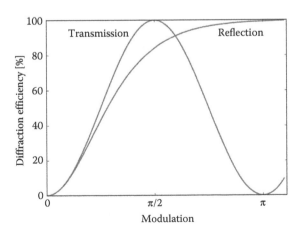

FIGURE 8.3 Diffraction efficiency of a phase Bragg's grating in transmission or reflection according to the modulation amplitude Δn.

266 *Advanced Optical Instruments and Techniques – Volume 2*

where α is the absorption coefficient, and can reach values higher than 1, $\Delta\alpha$ is the modulation of that coefficient.

Equation 8.22 experiences a maximum efficiency of 3.7% when $\alpha = \Delta\alpha = \ln 3$.

For any value of α, the maximum efficiency is achieved when the modulation is maximum: $\alpha = \Delta\alpha$.

For the reflection case, the TE efficiency is given by

$$\eta_{TE} = \frac{-\Delta\alpha^2}{4}\left[A + \sqrt{A^2 - \left(\frac{\Delta\alpha^2}{4}\right)}\coth\sqrt{A^2 - \left(\frac{\Delta\alpha^2}{4}\right)}\right]^{-2}, \tag{8.19}$$

where A denotes the absorption term: $A = \alpha\, d\,/\cos\theta_i$. The efficiency for the reflection amplitude grating increases with the modulation, asymptotically approaching a maximum of 7.2%.

Plots of the diffraction efficiency according to the modulation when $\alpha = \Delta\alpha$ for transmission (Equation 8.22) and reflection (Equation 8.23) Bragg's gratings are presented in Figure 8.4.

The solutions for the TM (transverse magnetic) mode are the same as TE with the coupling factor κ added to the modulation either Δn or $\Delta\alpha$:

$$\kappa = \cos(2\theta_i). \tag{8.20}$$

From Equations 8.20 through 8.23, it is possible to derive the dispersion of each type of gratings according to wavelength and angular incidence. However, the important point to understand is that even with the same characteristics of thickness and modulation amplitude, transmission and reflection gratings have very different behaviors when it comes to dispersion and selectivity. This is due to the very different Bragg's plane frequencies (Λ) generated by the two geometries. In the transmission geometry, the grating frequency ranges from 300 to 3000 line pairs per mm, while in the reflection geometry the grating frequency is over 4000 lp/mm.

Typical angular and spectral dispersion characteristics of reflection Bragg's gratings are illustrated in Figures 8.5 and 8.6. Transmission grating is diffracting a large bandwidth of wavelengths each at a very specific angle; they produce a rainbow. They are wavelength tolerant and angularly selective. On the other hand, reflection gratings diffract a very narrow bandwidth at any angle; they act as a notch filter, reflecting one color. Reflection Bragg's gratings are wavelength selective and angularly tolerant.

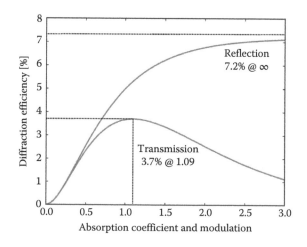

FIGURE 8.4 Diffraction efficiency of an amplitude Bragg's grating in transmission or reflection according to the absorption coefficient, and assuming maximum modulation amplitude $\alpha = \Delta\alpha$.

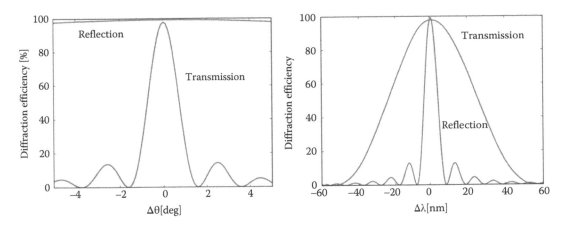

FIGURE 8.5 Typical angular and spectral dispersion of a transmission or reflection Bragg's gratings. Transmission gratings are angularly selective, when reflection gratings are wavelength selective.

FIGURE 8.6 Picture of dispersion by a transmission thick grating (left), and monochromatic diffraction by a reflection thick grating (right). Both holograms are illuminated by a halogen white light.

There exists a third kind of thick grating where the Bragg's planes are oriented more or less at 45° to the material surface. This kind of grating is called "edge lit" because in order to be incident at the correct angle, the grating needs to be illuminated from the thickness side of the material. This type of grating is useful for injecting or extracting the light from a waveguide and has recently gained popularity for solar concentration application [26] and augmented reality see-through displays [27, 28]. The angle and frequency selectivity properties of edge lit holograms are in between those of transmission and reflection gratings.

8.2.3 Thin Gratings Characteristics

Thin gratings operate in the Raman-Nath regime where substantial energy can be coupled in higher diffraction orders ($m > 1$). Equations 8.18 and 8.19 mathematically describe the condition for the thin grating regime. In thin gratings, the light only interacts with the diffractive structure once, where in thick gratings, there are multiple interactions.

Thin gratings are extremely important because they can easily be manufactured by printing the structure obtained by computer calculation. Thin gratings can also be dynamically displayed using electronically controlled spatial light modulators such as LCoS (liquid crystal on silicon) and DLP (digital light processor). Other thin grating manufacturing techniques include: stilet ruling, lithography, and embossing.

The efficiency of thin gratings depends on the modulation format [29]. One can distinguish between amplitude or phase modulation, but also the geometrical shape of the modulation, such as square, sinusoidal, or sawtooth pattern. The rigorous calculation of the efficiency and number of orders is based on Fourier decomposition of the complex amplitude of the transmitted wave function $t(x)$ according to the grating modulation $M(x)$. By finding an expression of the form:

$$t(x) = \sum_{m=-\infty}^{\infty} A_m \exp(imKx), \tag{8.21}$$

the portion of the intensity in the mth order is $\eta_m = (A_m{}^2)$, and the direction of propagation given by the vector $m\mathbf{K}$.

We are going to analyze six cases that are relevant to today's holographic manufacturing and displays. These include the three modulation shapes presented in Figure 8.7: sinusoidal, binary (or square), and sawtooth (or blazed) that will be derived for two modulation formats: amplitude and phase. We will also consider what happens when the sawtooth is digitized into m levels.

For a sinusoidal amplitude grating, which can be fabricated by printing the hologram on a gray scale printer, the modulation is given by

$$\begin{aligned}|t(x)| = M(x) &= M_0 + \frac{\Delta M}{2}\sin(Kx)\\ &= M_0 + \frac{\Delta M}{4}\exp(iKx) + \frac{\Delta M}{4}\exp(-iKx)\end{aligned} \tag{8.22}$$

where $0 < M_0 < 1$ is the average transmittance, $0 < \Delta M < 1$ is the transmittance peak to valley modulation, and $K = 2\pi/\Lambda$ is the wave vector.

The different terms on the right side of Equation 8.26 are associated with the amplitude of the different diffraction orders (0, +1, −1 respectively). There are no higher orders for such a grating; the efficiency $(\eta = |t_{\pm 1}|^2)$ is given by

$$\eta_{\pm 1} = (\Delta M / 4)^2 \le 6.25\%, \tag{8.23}$$

which is maximum when $M_0 = \Delta M/2 = \frac{1}{2}$.

The behavior of the diffraction efficiency as a function of the amplitude modulation ΔM is presented in Figure 8.8.

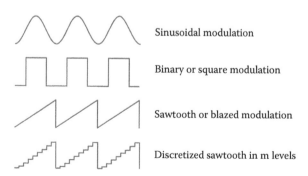

Sinusoidal modulation

Binary or square modulation

Sawtooth or blazed modulation

Discretized sawtooth in m levels

FIGURE 8.7   Modulation for the phase or amplitude of thin gratings that will be analyzed.

For a sinusoidal phase grating, which can be fabricated by recording the hologram with photoresist, there is no absorption: $|t(x)| = 1$, but the complex amplitude transmittance is

$$t(x) = \exp(-iM_0)\exp\left(-i\frac{\pi\Delta M}{2}\cos(Kx)\right),$$ (8.24)

where M_0 is a constant phase shift, and ΔM is the peak-to-valley phase modulation.

Ignoring the constant phase shift, the right hand side of Equation 8.28 can be expanded in Fourier series as

$$t(x) = \sum_{m=-\infty}^{\infty} J_m\left(\frac{\pi\Delta M}{2}\right)\exp(imKx),$$ (8.25)

where J_m is the Bessel function of the first kind and mth order represents the amplitude of the waves, when the exponential terms represents plane waves, that is, the direction of the diffracted orders.

From this decomposition, it can be seen that there is an infinite number of diffracted orders (each term of the sum). The diffraction efficiency in the first order is

$$\eta_{\pm 1} = J_1^2\left(\frac{\pi\Delta M}{2}\right) \le 33.8\%,$$ (8.26)

which is maximum when $\Delta M = 1.18$.

The behavior of the diffraction efficiency according to the peak-to-valley phase modulation ΔM is presented in Figure 8.8.

For a binary amplitude grating, the modulation is a square function, and the Fourier decomposition is:

$$M(x) = M_0 + \frac{2\Delta M}{\pi}\sum_{m=1}^{\infty}\frac{\sin\left[(2m-1)Kx\right]}{2m-1}$$
$$- M_0 + \frac{\Delta M}{\pi}\sum_{m=1}^{\infty}\frac{\exp\left[i(2m-1)Kx\right] + \exp\left[-i(2m-1)Kx\right]}{2m-1}.$$ (8.27)

The terms of the decomposition are all odds due to the $2m$-1 expression in the exponential functions, so there are no even diffraction orders. The diffraction efficiency for the ± 1 orders is

$$\eta_{\pm 1} = (\Delta M/\pi)^2 \le 10.1\%.$$ (8.28)

Maximum efficiency is achieved when $M_0 = \Delta M/2 = 1/2$.

The behavior of the diffraction efficiency according to the peak-to-valley amplitude modulation ΔM is presented in Figure 8.8.

This is the case when a holographic pattern is displayed on a DLP light modulator. DLP pixels are composed of mirrors that can be flipped left or right. For the light, the mirrors act as nearly perfect reflector or absorber depending on the direction they are oriented.

For a binary phase grating, which is obtained using an LCoS light modulator, the complex amplitude transmittance is

$$t(x) = \exp(-iM_0)\exp\left[-i\pi\frac{\Delta M}{2}\sum_{m=1}^{\infty}\frac{\sin\left[(2m-1)Kx\right]}{2m-1}\right].$$ (8.29)

The terms of the decomposition are all odds due to the $2m-1$ expression in the exponential functions, so there are no even diffraction orders as was the case for the amplitude binary grating. But conversely to the amplitude case, the phase modulation term is now contained in the exponential and needs to be expanded to find the value of the efficiency. For the $\pm m$ orders the efficiency is

$$\eta_{\pm m} = \left[sinc\left(\frac{m}{2}\right) \sin\left(\frac{\pi \Delta M}{2}\right) \right]^2 \tag{8.30}$$

with $sincx = (\sin\pi x)/(\pi x)$.

For the first orders ($\pm$) we have:

$$\eta_{\pm 1} = \left[\frac{2}{\pi} \sin\left(\frac{\pi \Delta M}{2}\right) \right]^2 \leq 40.5\%, \tag{8.31}$$

which is maximum for $\Delta M = 1$.

The behavior of the diffraction efficiency according to the peak-to-valley phase modulation ΔM is presented in Figure 8.8.

A counter-intuitive and important result from this decomposition is that the maximum diffraction efficiency is larger for square gratings (40.5% phase, 10.1% amplitude) than for sinusoidal gratings (33.8% phase, 6.25% amplitude).

For a sawtooth phase grating, which is the case of a blazed grating, the complex amplitude transmittance is

$$t(x) = \exp(-iM_0)\exp\left[-i\pi \frac{\Delta M}{2} \sum_{m=1}^{\infty} \frac{(-1)^m}{m} \sin(mKx) \right]. \tag{8.32}$$

The diffraction efficiency for the ± 1 orders is

$$\eta_{\pm 1} = \left[sinc\left(1 - \frac{\Delta M}{2}\right) \right]^2 \leq 100\%. \tag{8.33}$$

To maximize the efficiency the amplitude of the phase modulation should be $\Delta M = 2$ (see Figure 8.8).

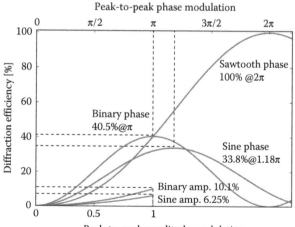

FIGURE 8.8 Diffraction efficiency of thin gratings according to the modulation format and amplitude.

Note that when phase patterns are used in a reflection configuration, the modulation is half of the transmission value because the path length difference is twice as large due to the reflection.

In many applications, it is not possible to reproduce a perfect sawtooth, and the slope is discretized and composed of m steps called a **discretized sawtooth phase grating**. For example, it is possible to expose and etch photo-resist resin several times (m) to approximate a sawtooth, or to dedicate several pixels (m) of an LCoS modulator to have a stepped phase slope. In this case, the diffraction efficiency for the ± 1 orders is [30, 31]:

$$\eta_{\pm 1} = \left[\frac{\sin\left(\pi\left(1-\dfrac{\Delta M}{2}\right)\right)}{\pi} \frac{\sin\left(\dfrac{\pi}{m}\right)}{\sin\left(\dfrac{\pi}{m}\left(1-\dfrac{\Delta M}{2}\right)\right)} \right]^2 \leq 100\%, \tag{8.34}$$

which for the limit as $m \to \infty$ yields the same result as Equation 8.37.

Equation 8.38 is maximum for a modulation $\Delta M = 2$.:

$$\eta_{\pm 1} \approx \mathrm{sinc}^2\left(\frac{1}{m}\right) \leq 100\%. \tag{8.35}$$

Of course, while the function of Equation 8.39 is continuous, in the real world m can only takes discrete values starting at two (see Figure 8.9).

For a binary level grating: $m = 2$, Equation 8.38 logically gives the same value for the efficiency as the one derived for a binary phase grating: 40.5% (Equation 8.35). Even though the modulation is $\Delta M = 2$ for Equation 8.38 and $\Delta M = 1$ for Equation 8.35, the two results are coherent because during the digitization of the modulation, the average level is divided by two (see Figure 8.10).

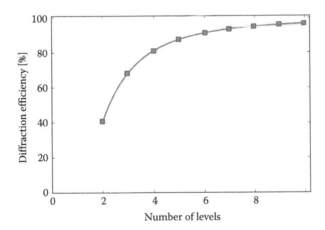

FIGURE 8.9 Diffraction efficiency of a discretized sawtooth grating with a 2π modulated phase, according to the number of levels defining the sawtooth structure.

FIGURE 8.10 Division of the modulation by a factor of 2 during digitization of the sawtooth function.

8.3 Scalar Theory of Diffraction

Now that we have introduced the diffraction by periodic structures, that is, gratings, we are going to generalize the formalism for any aperture. Finding the mathematical formulation for such a transformation between aperture and field will not only allow us to determine the field diffracted by the aperture, but also to calculate the aperture to generate a particular field, the so-called computer generated hologram (CGH).

The field of digital holography encompasses the computation of the diffraction pattern from an object or the retrieval of the object from the diffracted pattern. Digital holography offers many advantages compared to regular digital photography, for example by allowing the calculation of both phase and amplitude of the object, we can retrieve the 3D structure of the object or change the focus after the data is captured (post-focusing) [32, 33].

8.3.1 Kirchhoff Diffraction Integral

To start, we would like to determine the propagation of the field after going through an aperture as presented in Figure 8.11.

The energy carried by the magnetic field is usually much weaker (10^{-4}) than the energy in the electrical field, so we are going to simplify the calculation by limiting ourselves to the E field.

According to the Huygen's principle, the aperture acts as a homogeneous source, and the field is null in the opaque portions of the aperture. So the field at distance z is the summation over all the points of the aperture multiplied by the wave propagation to the distance z:

$$E\left(x_z, y_z\right) = \sum_{\text{aperture}} \left[\text{incident field at } x_0, y_0\right] \times \left[\text{wave propagation to } z{:}r_{z0}\right] \tag{8.36}$$

The wave propagation is solution of the Helmoltz equation introduced in Equation 8.1, and we will choose the spherical wave solution:

$$E(r,t) = \frac{A}{r}\cos\left(k \cdot r - t + \varphi\right). \tag{8.37}$$

Inserting Equation 8.41 into expression 8.40, we obtain the *Kirchhoff diffraction integral*:

$$E\left(x_z, y_z\right) = \frac{1}{i\lambda} \int_{\text{aperture}} E_{\left(x_0, y_0\right)} \frac{\exp\left(ikr_{z0}\right)}{r_{z0}} \cos\theta \, ds \,, \tag{8.38}$$

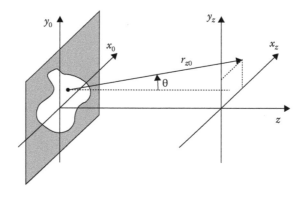

FIGURE 8.11 Field propagation through an aperture, geometry, and coordinate axes.

where, for Cartesian coordinates: $r_{z0} = \sqrt{z^2 + (x_z - x_0)^2 + (y_z - y_0)^2}$.

8.3.2 Fresnel Diffraction Integral

As elegant as Equation 8.42 is, it is very hard to compute and some simplifications are necessary to obtain a manageable expression.

Let us consider the expansion of the z term in Taylor series: $\sqrt{1+\varepsilon} = 1 + \dfrac{\varepsilon}{2} - \dfrac{\varepsilon^2}{8} + \ldots$

$$r_{z0} = z + \frac{1}{2}\left\{ \left(\frac{x_z - x_0}{z}\right)^2 + \left(\frac{y_z - y_0}{z}\right)^2 \right\} + \ldots \tag{8.39}$$

We can neglect the third term inside the complex exponential, and the second term in the denominator. This set of simplifications is referred to as the *paraxial approximation* since it can be applied for a small aperture in regard to the distance z: $z \gg x_z - x_0$ and $z \gg y_z - y_0$. This leads to the following equation:

$$E(x_z, y_z) = \frac{\exp(ikz)}{i\lambda z} \int_{\text{aperture}} E_{(x_0, y_0)} \exp\left[\frac{ik}{2z}(x_z - x_0)^2 + (y_z - y_0)^2 \right] ds, \tag{8.40}$$

known as the *Fresnel diffraction integral*.

The paraxial approximation validity criteria can also be expressed as the Fresnel number F:

$$F = \frac{(D/2)^2}{z} \geq 1, \tag{8.41}$$

where D is the aperture diameter.

Equation 8.46 expresses that the distance z should be larger than the wavelength λ but not necessarily much larger than the aperture D. So, the Fresnel approximation is valid in the so-called "near field."

8.3.3 Fraunhofer Diffraction Integral

Further approximation can be used for an observation plane farther away from the aperture: $z \gg k\left(x_0^2 + y_0^2\right)_{\text{max}}$. If we expend the quadratic terms of Equation 8.45 as $(a-b)^2 = a^2 + b^2 - 2ab$:

$$E(x_z, y_z) = \frac{\exp(ikz)}{i\lambda z} \exp\left[\frac{ik}{2z}(x_z^2 + y_z^2) \right]$$
$$\times \int_{\text{aperture}} E_{(x_0, y_0)} \exp\left[\frac{ik}{2z}(-2x_z x_0 - 2y_z y_0) + (x_0^2 + y_0^2) \right] ds \tag{8.42}$$

then the quadratic phase factor can be set to unity over the entire aperture:

$$\int_{\text{aperture}} E_{(x_0, y_0)} \exp\left[\frac{ik}{2z}(x_0^2 + y_0^2) \right] ds = 1. \tag{8.43}$$

So Equation 8.47 can be written as

$$
E(x_z, y_z) = \frac{\exp(ikz)}{i\lambda z} \exp\left[\frac{ik}{2z}(x_z^2 + y_z^2)\right]
$$
$$
\times \int\limits_{\text{aperture}} E_{(x_0, y_0)} \exp\left[\frac{-ik}{z}(x_z x_0 + y_z y_0)\right] ds,
$$

(8.44)

which is called the *Fraunhofer diffraction integral*.

This result is particularly important once it is recognized that the integration term is simply the Fourier transform of the aperture, and since it is the optical intensity that is relevant for most applications: $I = |E|^2$, the phase factor can be neglected.

Ultimately, this long mathematical development leads to the very convenient formulation:

$$
E(x_z, y_z) \propto \Im\left(\text{aperture}(x_0, y_0)\right).
$$

(8.45)

The criteria for the Fraunhofer diffraction integral to be used is that observation distance z must be much larger than the aperture size and wavelength:

$$
F = \frac{(D/2)^2}{z\lambda} \ll 1.
$$

(8.46)

This condition is known in optics as the "far field."

8.3.4 Diffraction by Simple Apertures

Considering the relative simplicity of the Fraunhofer diffraction integral, it is possible to find analytical solutions for simple apertures illuminated by a plane wave:

$$
U(x, y) = A\exp\left[i(2\pi ct/\lambda)\right]
$$

(8.47)

8.3.4.1 Diffraction by a Slit

The slit is a rectangular function located at $z = 0$ of width W:

$$
f(x) = \text{rect}\left(\frac{x}{W}\right)
$$

(8.48)

The integration of the field over the slit is

$$
U(x, z) = A \int\limits_{-W/2}^{W/2} \exp\left(-i\frac{2\pi}{\lambda z}xx'\right) dx
$$
$$
= \frac{-Az}{ikx}\left|\exp\left(-i\frac{2\pi}{\lambda z}xx'\right)\right|_{-W/2}^{W/2}.
$$

(8.49)

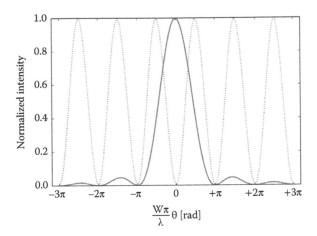

FIGURE 8.12 Interference by a slit of width W. The dotted line is a sin² function for which secondary minima are collocated.

Using Euler's formula, we have

$$U(x,z) = AW \frac{\sin\left(\dfrac{W\pi x}{\lambda z}\right)}{\dfrac{W\pi x}{\lambda z}}$$

$$= AW \operatorname{sinc}\left(\frac{W\pi x}{\lambda z}\right) \tag{8.50}$$

or:

$$U(\theta) = AW\operatorname{sinc}\left(\frac{W\pi}{\lambda}\sin\theta\right). \tag{8.51}$$

Since the intensity can be expressed as $I = U\,U^*$:

$$I(\theta) = I_0 W \operatorname{sinc}^2\left(\frac{W\pi}{\lambda}\sin\theta\right) \tag{8.52}$$

The intensity distribution of Equation 8.60 is presented in Figure 8.12.

8.3.4.2 Diffraction by a Pinhole

The diffraction by a pin hole of diameter D is a two-dimensional generalization of the case we just analyzed for a slit, with a rotational symmetry. The intensity distribution in the far field becomes

$$I(\theta) = I_0 D \left[\frac{2J_1\left(\dfrac{D\pi}{\lambda}\sin\theta\right)}{\dfrac{D\pi}{\lambda}\sin\theta} \right]^2. \tag{8.53}$$

The diffraction pattern formed by the pinhole aperture is called the Airy disk and is presented in Figure 8.13. A cross section of this Airy disk pattern is a *sinc²* function as presented in Figure 8.12.

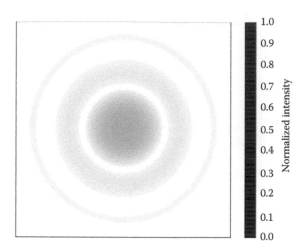

FIGURE 8.13 Diffraction pattern formed by a pinhole: the Airy disk.

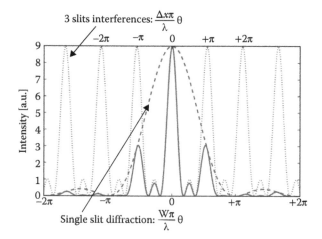

FIGURE 8.14 Diffraction by 3 slits (plain line) obtained by the multiplication of the interference of 3 slits (dotted line) by the diffraction by a single slit (dashed line).

8.3.4.3 Diffraction by Multiple Slits

Under the Fraunhofer condition, the intensity distribution diffracted by m slits of width W each separated by a distance Δx is

$$I(\theta) = I_0 W \operatorname{sinc}^2\left(\frac{W\pi}{\lambda}\sin\theta\right) \times \left[\frac{\sin\left(m\dfrac{\Delta x\pi}{\lambda}\sin\theta\right)}{\dfrac{\Delta x\pi}{\lambda}\sin\theta}\right]^2. \tag{8.54}$$

This equation can be obtained either by the Fourier transform of the aperture as expressed in Equations 8.50 and 8.52, or by multiplying the expressions for the diffraction by a single slit (first term of Equation 8.62) by the interference of m slits (second term of Equation 8.62). Figure 8.14 presents an example obtained for three slits.

8.3.4.4 Fresnel Zone Plate

The Fresnel zone plate is a diffractive structure that acts as a lens: it diffracts an incident plane wave into a point. This type of diffractive structure is particularly interesting for electromagnetic radiation where there is no refractive material (such as for x-rays), or the refractive materials are too expensive.

The pattern of a Fresnel zone plate is composed of alternating opaque and transparent rings that act like slits. The radii R_m of these rings is such that the interference is constructive along the axis at the focal distance f:

$$R_m = \sqrt{m\lambda f + \frac{m^2\lambda^2}{4}} \, . \tag{8.55}$$

Figure 8.15 presents a Fresnel zone plate structure and the condition on the radii of the rings to obtain constructive interference at a distance f: the distance from the radii to the focal point must be a multiple of $\lambda/2$. It has to be noted that the black and transparent rings can be inverted without any alteration in the diffraction properties.

Considering that the Fresnel zone plate is a binary amplitude modulation, we have seen in Section 8.2.3 on thin grating characteristics that such a structure diffracts multiple odd diffraction orders. So, the positive higher orders $(2m + 1)$ will form a focal point at $f/(2m + 1)$, when the negative orders $(-2m + 1)$ will act as negative lenses with focal length $-f/(2m + 1)$.

To reduce the number of higher orders, it is possible to replace the binary modulation pattern by a sinusoidal modulation that will only diffract the ±1 orders. In this case, the structure is called the Gabor zone plate, and is presented in Figure 8.16. The amplitude of the modulation according to the distance to the center is given by

$$M = \frac{1 \pm \cos\left(\dfrac{kr^2}{f}\right)}{2} \tag{8.56}$$

with $k = 2\pi/\lambda$.

The Gabor zone plate modulation is the same as the one obtained by interfering a plane wave with a point source located at a distance f from the plane of observation, itself normal to the propagation direction of the plane wave. This becomes obvious when the hologram recording setup is understood (see Section 8.5).

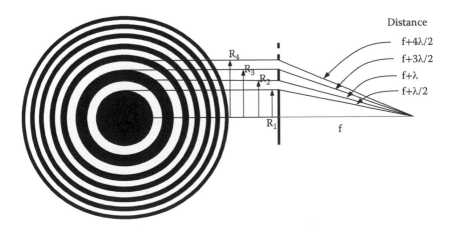

FIGURE 8.15 Fresnel zone plate and radii of the successive rings to obtain constructive interference at the distance f.

FIGURE 8.16   Sinusoidal modulation pattern to focus an incident plane wave: the Gabor zone plate.

To increase the diffraction efficiency of the zone plate, phase modulation can be used instead of amplitude modulation.

8.4 Computer Generated Holograms

Deriving the aperture from the wavefront that we want to achieve involves "inverting" the diffraction integral. We can use the Kirchhoff (21.38), Fresnel (21.40), or Fraunhofer (21.44) diffraction integrals to calculate the specific aperture that will generate the desired wavefront. Since, for the most general case, an exact solution cannot be found, the holographic pattern must be calculated using a computer. The field of computer generated holograms (CGH) started in the late 1960s when scientists gained greater access to computers. The field expanded rapidly with the implementation of the fast Fourier transform (FFT) algorithm that made it possible to compute FFT over two-dimensional images of significant size. [18, 19, 34].

8.4.1 Fourier Hologram

The simplest example of the aperture computation is to use the Fraunhofer diffraction integral (21.45) and take the inverse Fourier transform of each side, which yields to:

$$\text{aperture}(x_0, y_0) \propto f^{-1}\big(E(x_z, y_z)\big). \tag{8.57}$$

The same condition regarding the image distance being much greater than the aperture size and wavelength applies to both this expression and the diffraction integral. The common shorthand for this condition is that the hologram will be formed in the far field: $z \to \infty$. To observe the image one can use a lens to form the diffracted image at the focal length as shown in Figure 8.17.

Considering a Fourier transform of 2D function is always a 2D function and that the solution is independent of the distance z, there is only one image plane for the Fourier hologram and the image will be two-dimensional.

Example of a Fourier binary amplitude hologram is presented in Figure 8.18.

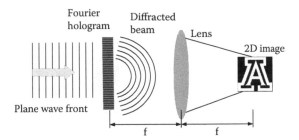

FIGURE 8.17   Image formation with a Fourier computer generated hologram.

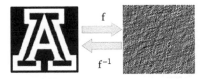

FIGURE 8.18   Fourier holographic pattern.

8.4.2 Fresnel Hologram

Inverting the Fresnel diffraction integral (40) is somewhat more complicated because the E field is a function of the propagation distance z. Equation 8.45 can be simplified by creating a parabolic wavelet function (58) and substituting $h(z)$ into Equation 8.45:

$$h(z) = \exp\left[\frac{ik}{2z}(x_z - x_0)^2 + (y_z - y_0)^2\right]. \tag{8.58}$$

Inverting the $h(z)$ substituted from Equation 8.45 gives

$$\text{Aperture}_{(x_0,y_0)} \propto \frac{f^{-1}\left[\dfrac{f\left(E_{(x_z,y_z)}\right)}{h(z)}\right]}{E_{(x_0,y_0)}}, \tag{8.59}$$

while Equation 8.59 looks more daunting than the simpler Fourier expression 8.57, the Fresnel hologram contains Fresnel-zone plate functions that act as lenses to focus the diffracted light at finite locations. For the Fresnel hologram, there is no need for an additional lens in the setup to bring the image to a focus. The image generated by the Fresnel hologram can be three-dimensional, i.e. composed of several focal planes. The reconstruction setup of a Fresnel hologram is presented in Figure 8.19.

An example of a computer generated Fresnel hologram is presented in Figure 8.20 where the two sections of the image will be formed at different distances. A close observation of the diffractive pattern will reveal some centrosymmetric structures that are due to the Fresnel zone plates.

The Fresnel diffraction integral equation can also be used to reconstruct an object when the hologram (or interferogram) is captured as an image. Today CMOS and CCD sensors have pixels small enough to resolve the interference produced by small objects located on the top of the sensor plane. The advantage of not imaging the object of interest is that: first no lens is needed, and second the 3D information can be reconstructed. This technique has been used to demonstrate very high-resolution holographic microscopes capable of resolving single cells, such as red blood cells and lymphocytes [35].

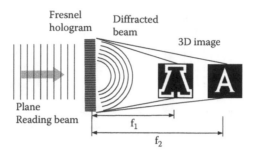

FIGURE 8.19   Image formation with a Fresnel computer generated hologram.

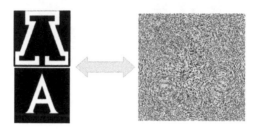

FIGURE 8.20   Fresnel holographic pattern.

8.4.3 Iterative Computation for Hologram

When computing the diffraction pattern from the image, or when reconstructing the image from the diffraction pattern (inverse transformation), the Fourier transform generates two terms: the real part, which is the transmittance (amplitude), and the imaginary part, which is the phase modulation. Quite often one or the other is lost during the measurement or during the reproduction of the hologram, due to the properties of the image sensor (amplitude only) or due to the spatial light modulator (amplitude only for a DLP). If this problem is not addressed during reconstruction, the efficiency of the hologram is reduced, or the noise in the image is increased.

To minimize the degradation, iterative computation, such as the Gerchberg-Saxton algorithm can be used [36]. The principle of this algorithm is that the phase (intensity) of the mth iteration can be used along the source intensity (phase) distribution to calculate the m+1th function via Fourier transforms and its inverse. A schematic diagram of the iteration is presented in Figure 8.21 where a phase hologram is computed from the image intensity distribution.

The Gerchberg-Saxton algorithm converges quite rapidly as it can be seen in Figure 8.22 where only three iterations are necessary for the hologram to eliminate most of the noise in the image it reproduces.

8.4.4 Errors in CGH

The limitation in the resolution of CGH comes from several factors. The first one being the computation of the Fourier transform, which usually uses a fast Fourier transform (FFT) algorithm that samples the function and limits the number of frequencies. So, the result is not continuous but discretized, which generates some high frequency noise.

Noise is also introduced during the physical reproduction of the hologram because the finite pixel size and pitch of the device introduce quantization. Whether it is a printer, lithography, or an electronic spatial light modulator, the technique has a limited space-bandwidth product (SBP), and is not able to reproduce the entire spectrum of frequencies that are contained in the holographic pattern.

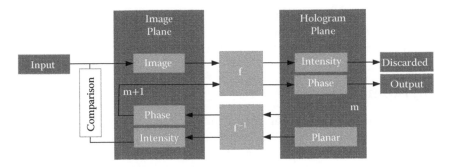

FIGURE 8.21 Diagram of the iterative Gerchberg-Saxton algorithm.

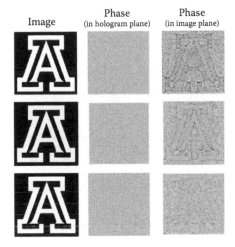

FIGURE 8.22 Convergence of a Gerchberg-Saxton algorithm. The first image is the input, subsequent images are the computed back from the phase hologram. Note the noise in the second image that is dramatically reduced in the third image generated by the second iteration.

More rigorously, the SBP is a measure of the information contained in a signal or the rendering capacity of a device. For an optical system, it is defined as the product of the spatial frequency $\Delta\nu$ by the spatial extent of the image Δx. According to the Nyquist sampling theorem, a signal can only be perfectly reproduced by a system if the area of its SBP fits inside the area of the system SBP. The shape of the SBP itself can be modified by lenses, reducing the spatial extent but increasing the frequency.

$$SBP_{\text{signal}} \le SBP_{\text{system}} \tag{8.60}$$

When computing a hologram, the Fourier transform rotates the SBP by 90 degrees because the role of the space and frequency are inverted. For a two-dimensional image, and an image sensor, the bandwidth and spatial extent are both two-dimensional. However, in holography, the image can be three-dimensional, but the holographic pattern is only two-dimensional, which imposes a very high burden on the system SBP:

$$SBP_{\text{signal}} = \Delta x^3 \cdot \Delta\nu^3 \le SBP_{\text{system}} = \Delta x^2 \cdot \Delta\nu^2 \tag{8.61}$$

To satisfy the Nyquist theorem, we see that the number of "pixels" composing the system (hologram) should be larger by a power of 6/4 to fully reconstruct the 3D image (signal). This is the principal reason why computer generated holograms are still not able to reproduce small details, such as object textures, even with today's high-resolution SLM and computer capacity. New devices, such as leaky mode waveguides, with very high SBP might help in that regard in the near future [37].

8.5 Holographic Setups

In this section, we will describe the optical configurations for recording and replaying of holograms, as well as the respective properties of the reconstructed wavefronts for these particular holograms. From the specific cases developed in this section, there are many variations that will not be detailed. Readers are invited to look at the following references for further developments [22, 38–40].

8.5.1 Formalism

In the most general terms, the intensity modulation pattern created when a reference beam R and an object beam O interfere is

$$|O + R|^2 = |O|^2 + |R|^2 + OR^* + O^*R. \tag{8.62}$$

When the intensity modulation is recorded inside a material (silver halide, dichromated gelatin, photopolymer,...), the transmittance variation includes the material response β:

$$T(x, y) = \beta|O|^2 + \beta|R|^2 + \beta OR^* + \beta O^*R. \tag{8.63}$$

Then, this transmittance pattern is interrogated with a reading beam R identical to the reference beam, so the output field is

$$E_{\text{out}}(x, y) = \beta|O|^2 R + \beta|R|^2 R + \beta O|R|^2 + \beta O^* R^2. \tag{8.64}$$

The different terms of Equation 8.64 can be interpreted as follows:

- $\beta|O|^2 R = E_{\text{scat}}$ is an intermodulation term, also called halo, resulting from the interference of wave coming for the different points of the object. This 'information' is contained in the $|O|^2$ term, and is generally considered as noise.
- $\beta|R|^2 R = E_{\text{scat}}$ is the transmitted beam: the zero order; it does not contain any object information, only R terms.
- $\beta O|R|^2 = E_{+1}$ is the +1 diffraction order; it is the reconstructed object beam since it contains an O term. It will produce a virtual image of the object at the object position. It actually reproduces the exact wavefront as the one scattered by the object.
- $\beta O^*|R|^2 = E_{-1}$ is the –1 diffraction order; it contains O^* term reconstructing a conjugate image of the object. It will produce a real image of the object that will appear pseudoscopic: the relief is inverted as the front part is seen on the back, and the background on the front (like a molding cast seen from the inside out).

A graphical representation of the diffraction and the different terms is presented in Figure 8.23.

In the following sections, we will review the different relationship between the positions of the object and the reference beams. Because of the historical role that the different geometries have

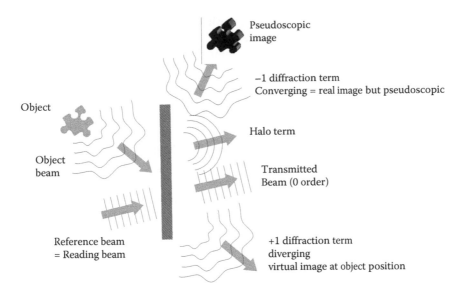

FIGURE 8.23 Recording and diffraction terms produced by a hologram.

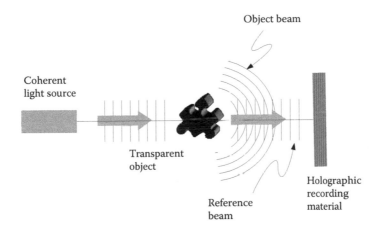

FIGURE 8.24 Inline transmission hologram (Gabor) recording geometry.

played in the development of holography, the different geometries are associated with the name of their inventor.

8.5.2 Inline Transmission Hologram (Gabor)

Introduced by Denis Gabor for improving the resolution of the electron microscope [9, 10], this particular recording geometry is depicted in Figure 8.24. The object is positioned in front of the recording media and the interference occurs between the wavefront transmitted but not perturbed through the object, and the light transmitted and scattered by the object. Obviously, the object must to be transparent for this configuration to work.

The advantages of this configuration are that the coherence of the light source can be reduced since the path difference between the object and reference beam is very small. Minimizing the path length difference was critical in Gabor's original work, which occurred before the invention of the laser.

Also, the hologram can be read with a polychromatic light source since the chromatic dispersion only occurs when moving off-axis. The disadvantage is that all the terms of Equation 8.64: transmission, diffraction orders and halo, superimpose in the same direction, which reduces the visibility of the information.

8.5.3 Inline Reflection Hologram (Denisyuk)

Introduced by Yuri Denisyuk for 3D imaging [15, 17], the object to be recorded is located behind the holographic recording material as presented in Figure 8.25. The interference is produced between the light incident to the material and the light scattered back from the object.

The advantages of this geometry are that the setup is quite stable since the optical path difference between the object and reference beams could be kept to a minimum if the object is close to the recording medium. Also, since the reflection hologram is wavelength selective (see Section 8.2.2), the hologram can be read with a polychromatic light source, reproducing the color at which the hologram was recorded.

The disadvantages of this geometry are that the beam reflected by the front face of the hologram is directed in the same direction as the reconstructed beam and can superimpose, showing an annoying glare. Second is that the color of the hologram is dictated by the wavelength of the recording light source, so in order to produce a color hologram with this geometry, three different light sources centered on red, green, and blue are required. Finally, since the hologram has a very large acceptance angle, it can diffract light from different points of an extended source, smearing the reproduced image especially in planes that are further away from the recording material, which limit the depth of field (see Figure 8.26).

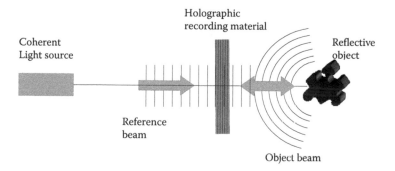

FIGURE 8.25   Inline reflection hologram (Denisyuk) recording geometry.

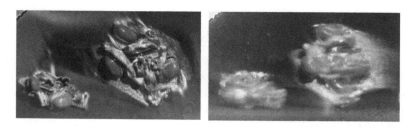

FIGURE 8.26   Pictures of an inline reflection hologram that has been recorded with a green laser on bleached silver halide. The hologram is replayed by, left: a halogen (polychromatic, point-like) light source, right: fluorescent tube (large etendue) light source. (Hologram courtesy of Arkady Bablumian.)

8.5.4 Off-Axis Transmission Hologram (Leith and Upatnieks)

To separate the image (+1 order) from the transmitted (0 order) and halo beams, Emmett Leith and Juris Upatnieks used an off-axis geometry where the object and reference recording beams are incident to the material at different incident angles [12–14]. The geometry is presented in Figure 8.27.

When reading such a hologram, a monochromatic point-like source is needed. Indeed if a polychromatic light source is used, the wavelength dispersion is such that multiple copies of the image superimpose at different colors, and the object cannot be observed (see Figure 8.28). The color of the reproduced image is given by the color of the replay source and not the recording source as is the case for reflection holograms.

While coherence is not necessary, oftentimes a laser is used to read transmission holograms since they are monochromatic. The narrower the bandwidth, the sharper the image appears. However, with very narrow spectral lines, speckle pattern becomes visible, which gives a grainy aspect to the image (see Figure 8.28, left). The speckle is produced by the interference between different points of the image on the observer's retina/detector.

Color can be reproduced with transmission holograms by recording three different holograms at three different angles, and reproducing each one with a different monochromatic light source (red, green, and blue). To make sure the three different diffracted images superimpose in the same direction during the replay, the recording angles should be calculated using the Bragg's law (Equation 8.17) to correct for the wavelength difference between recording and reading sources.

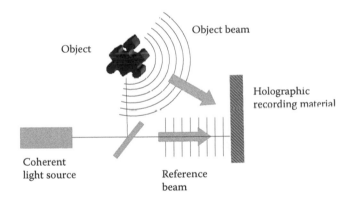

FIGURE 8.27 Off-axis transmission hologram (Leith and Upatnieks) recording geometry.

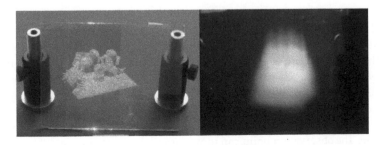

FIGURE 8.28 Picture of an off-axis transmission hologram recorded on bleached silver halide replayed by left: a red monochromatic laser diode, and right: a halogen light source. (Hologram courtesy of Pierre Saint Hilaire.)

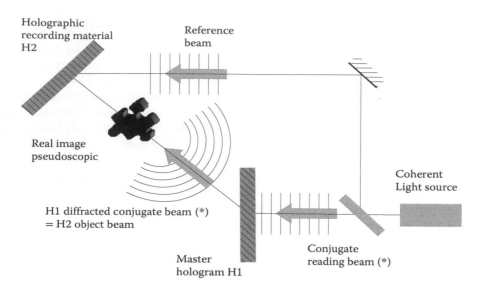

FIGURE 8.29   Recording a transmission transfer hologram for the image to appear in front of the media.

8.5.5 Transfer Holograms

In both the Denisyuk, and Leith and Upatnieks configuration, the reconstructed image appears on the back of the recording medium. For a more dramatic effect, it is often wished that the image appears floating in front of the plate. To do so, a second hologram called transfer hologram (H2) can be recorded from the first one, called the master hologram (H1). The recording geometry is presented in Figure 8.29 for transmission hologram. The master hologram is recorded as described in Figure 8.27. When replaying it with a reading beam that is the conjugate (*) of the initial reference beam, it forms a real image that is pseudoscopic (inside-out relief, see Section 8.5.1). This image is used as the object beam for the transfer hologram, along with another reference beam.

When replaying the transfer hologram, the conjugate of the reference beam is used, which generates a pseudoscopic image of the object, itself pseudoscopic. The double inversion (pseudoscopic of pseudoscopic) restores the original relief.

The parts of the real image that were in front of the material during the recording of the transfer hologram appear in front of the plate, as if there were freely floating in thin air.

8.5.6 Rainbow Hologram (Benton)

To avoid the constraint of using a monochromatic source to read transmission hologram, Stephen Benton invented the rainbow hologram [41]. Rainbow holograms are recorded as a transfer hologram where a horizontal slit is put in front of the master hologram. The slit sacrifices the vertical parallax to be able to read the hologram with white light source. The recording setup is presented in Figure 8.30.

If the rainbow hologram is read with a monochromatic light source, the image reproduces not only the object but also the slit that was used. So the eyes of the viewer must align exactly with the slit to observe the object, and if the spatial extent is too large, the object is cropped (Figure 8.31, top). In the case of a polychromatic light source, the different colors are dispersed and each one reproduces the slit at a different angle. The observation point can move up and down, catching a different slit, viewing the object at different colors (Figure 8.31, bottom and Figure 8.32). While the vertical extent of the object is restored, there is still no vertical parallax, which has been lost during the transfer recording.

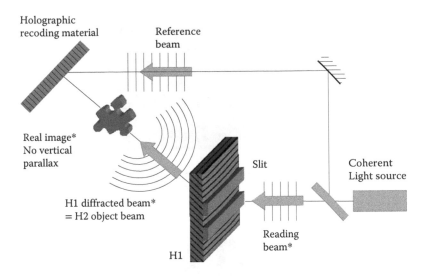

FIGURE 8.30 Recording setup for a transfer rainbow hologram (Benton).

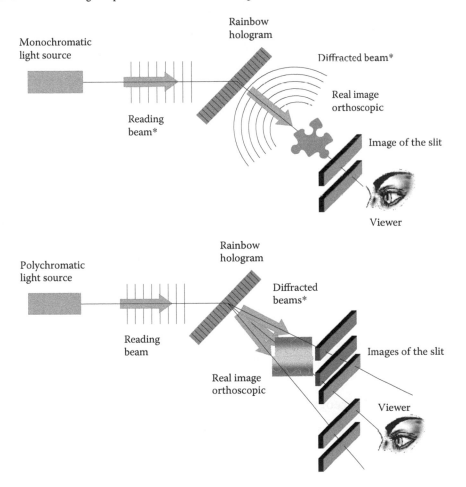

FIGURE 8.31 Replaying a rainbow hologram with, top: monochromatic light source, bottom: polychromatic light source.

FIGURE 8.32 Picture of a rainbow hologram showing the color dispersion. (Coal Molecule hologram from Jody Burns, photographed at the MIT museum by the author.)

FIGURE 8.33 Recording one holographic pixel (left), and replaying a holographic stereogram (right).

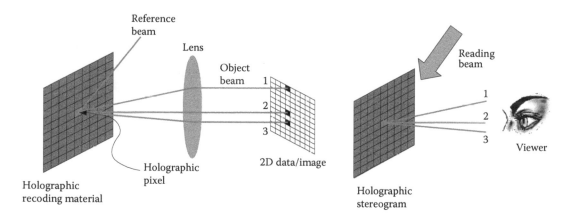

FIGURE 8.34 Left and right view of the holographic sterogram "Tractatus Holographis." Image created by the artist Jacques Desbiens, hologram recorded by RabbitHoles, and pictures taken by the author.

8.5.7 Holographic Stereogram

Holographic stereograms were invented by DeBitetto to overcome the requirement that the other holographic techniques need the actual object to be present in the optical setup to be recorded [42]. The system uses multiple 2D views (pictures) that have been recorded at different angles to reconstruct the parallax. Unfortunately, the wavefront (phase) information is lost with this technique, which cannot reproduce the depth of field. However, in addition to the parallax, movement can also be reproduced when the viewer moves around the image (Figure 8.33).

In holographic stereogram, the entire image is formed by pixels that are themselves holograms. A single 2D image is focused by a lens into an element of the stereogram where it is recorded by interference with a reference beam (Figure 8.34). Once all the holographic pixels have been recorded, the stereogram can be replayed using a reading beam that will recompose the original images. Since the images are now separated angularly, the viewer experiences parallax when moving in front of the display. The same technique can be used for holographic data storage [43], and is related to integral imaging [44].

8.5.8 Holographic Interferometry

In interferometry, the fringe structure produced by the interference of two coherent beams is analyzed to retrieve the optical path difference between these two beams. The fringe pattern can be used to determine the shape of an object to a fraction of a wavelength. However, for objects with large differences in optical path, the fringe structure can be so small that it is not distinguishable. To have a coarse fringe structure, the wavefront of the reference beam should have a similar shape as the wavefront of the object beam. Holography can help in that regard by recording and then replaying the wavefront(s). The

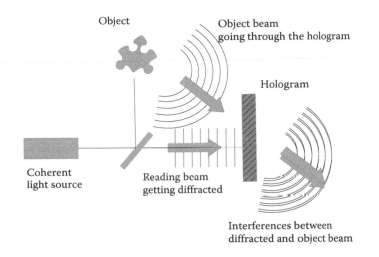

FIGURE 8.35 Holographic interferometry setup.

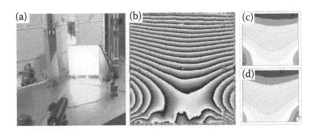

FIGURE 8.36 Holographic interferometric measurement of a composite material structure deformed by heat. (a) Picture of the setup, (b) recorded phase map, (c) retrieved deformation, (d) computed deformation. Images courtesy of Marc Georges. (From Thizy, C. et al., Comparison between finite element calculations and holographic interferometry measurements, of the thermo-mechanical behavior of satellite structures in composite materials. In Nolte, D. D., Ed., *Photorefractive Effects, Materials, and Devices*, Washington, Optical Society of America, 2005.)

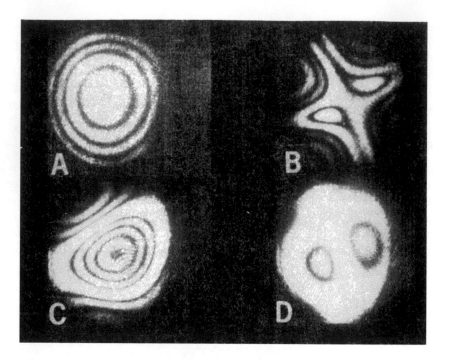

FIGURE 8.37 Mode shapes of a membrane vibrating at different frequencies obtained using Holographic interferometry recorded with a dynamic photorefractive polymer. (From Volodin, B. L. et al., *Opt. Eng.*, 34, 2213, 1995.)

wavefront can also be produced by a computer generated hologram (CGH), so the object is compared to a theoretical shape [33].

More generally, holographic interferometry uses the interference produced between either an object and a hologram, or two holograms. An example of a setup is presented in Figure 8.35. A hologram with a wavefront similar to the object has been recorded. It is then replayed with a reading beam, and the diffraction is superimposed with the object beam going through the hologram. If the object is deformed between recording and replaying, the deformation is visible through the fringe pattern (Figure 8.36, [45]).

Holography can also be used to record vibration modes of an object since the nodes are fixed and will produce a stable interference pattern that will diffract when the hologram is replayed. Conversely, the anti-nodes are moving and will not be recorded and subsequently diffract. The observed diffraction pattern is composed of bright zones where the nodes are, and dark fringes where the anti-nodes are located as shown in Figure 8.37 [46, 47].

The use of dynamic holographic recording materials, such as photorefractives, opened the door to a large variety of techniques in holographic interferometry [48, 49].

8.6 Holographic Recording Materials

8.6.1 Silver Halide

Silver halide refers to an emulsion of gelatin, a natural polymer obtained from collagen, loaded with different silver halide crystals, such as AgBr, AgCl, or AgI, which act as light sensitizers. The emulsion is coated on film or advantageously on a glass plate for holography for better stability.

The same material has been used in photographic and movie cameras for a long time, before being replaced by electronic sensors. The difference between the regular picture and the holographic recording is the higher resolution needed for holograms, which requires a much smaller crystal grain size.

With the size reduction of the grains, the sensitivity of the emulsion decreases, which requires longer exposure or a powerful laser.

To resolve a line density of 5000 lp/mm, the grain size is in the order of 10 nm, and the sensitivity requires an exposure of a few mJ/cm^2.

Silver halide materials necessitate a post-exposure wet processing for the image/hologram to be revealed. A developer is used to convert the latent image into macroscopic particles of metallic silver, then a stop bath halts the reaction, and a fixer dissolves the remaining silver halide to prevent further sensitization. At this point, the modulation pattern is of the absorption type, and the diffraction efficiency can only reach a maximum of 7% (see Equations 8.22 and 8.23).

To increase the diffraction efficiency, eventually up to 100% (see Equations 8.20 and 8.21), this amplitude modulation is converted into a phase modulation by a bleaching process that involves the dissolution of the silver crystals to leave air voids, lowering the index of refraction in the dark regions of the pattern. These steps in silver halide processing are represented in Figure 8.38.

8.6.2 Dichromated Gelatin

In dichromated gelatin (DCG), the photo-sensitizer is ammonium dichromate: $(NH_4)_2Cr_2O_7$. Once exposed, a wet processing is required to amplify the latent image. First, a fixer is used to remove the dichromate, and the material loses its brown-orange coloration. Then the film is soaked in warm water that makes the gelatin swell. Where the material has been exposed, the polymer chain undergoes cross linking. The cross-linked polymer chains are more rigid, and the material does not swell as much in these regions. Following the water bath, several baths with increased concentration of alcohol are used to dehydrate the gelatin and shrink it to its final thickness. In the regions where the material swelled the most, micro-voids decrease the average refractive index.

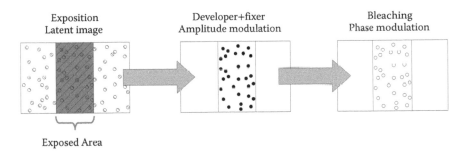

FIGURE 8.38 Silver halide processing to obtain a phase hologram: Exposure creates the latent image, developer and fixer reveal an amplitude modulation pattern, bleaching transforms the amplitude modulation into phase modulation.

FIGURE 8.39 Dispersion by a large volume holographic grating recorded in dichromated gelatin. The fact that the complementary color is seen in the zero order is an indication that the efficiency is very large (close to 100%). (Reprint from Blanche, P.-A. et al., *Opt. Eng.*, **43**, 2603, 2004.)

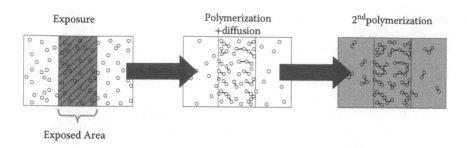

FIGURE 8.40 Schematic of the formation of a phase hologram in photopolymer: Exposure of the monomer to activate the photopolymerization, diffusion of the monomers in the regions that have been exposed, second polymerization with UV light, polymerize the entire material and passivate it.

Once processed, the DCG hologram should be protected from humidity by encapsulation, or it could swell back, which makes the diffraction shift in wavelength and finally disappear.

The index modulation that can be reached using DCG can be as large as $\Delta n = 0.14$ [50]. Such a value allows the use of very thin films to have a very large chromatic dispersion and efficiency near 100%, which is particularly useful to realize disperser for spectrometer instruments (Figure 8.39).

Since there are no crystal grains in the DCG material, the resolution is very large and frequencies as large as 10,000 lp/mm can be recorded.

The drawback of DCG versus silver halide is its sensitivity, which is limited to the blue-green region of the spectrum, and requires exposure as large as several hundred of mJ/cm^2.

8.6.3 Photopolymer

This class of organic materials undergoes a polymerization under illumination. The production of a phase hologram using photopolymer is reproduced in Figure 8.40: During the holographic recording, only the bright regions of the interference pattern polymerize. In the minute or so following the exposure, the monomers diffuse into the polymerized regions by osmosis increasing the density and the refractive index in these locations. Some materials require heat to be applied to accelerate the diffusion. A second exposure with a homogeneous beam polymerizes the remaining monomer and stops further reaction. This second exposure is often done using UV light.

The sensitivity of photopolymers depends on the exact composition, but panchromatic systems have been developed that require exposure as low as 10 mJ/cm^2.

8.6.4 Photoresist and Embossed Holograms

Photoresists such as SU8, which are used in the photo-lithography process, can be used to make diffractive structure. The exposure to light polymerizes the material and the non-exposed parts are then washed out by a solvent, leaving a surface relief pattern.

However, this structure is quite fragile, and is better used as a master to make a metal stamp by electroplating. The metal stamp can then be used to emboss the relief into heat-softened polymer, which are ultimately coated with metal and protected with a cover layer to prevent environmental damage. The final structure is a phase hologram that is reflective due to the metal coating. The whole process is illustrated in Figure 8.41.

The advantage of the embossing technique is that once the stamp has been manufactured, a large quantity of holograms can be replicated in a short period of time by using the stamp on a roller machine much like the way newspapers are printed. Most holograms produced in large quantities, such as for security on bank notes, are manufactured by the embossing technique.

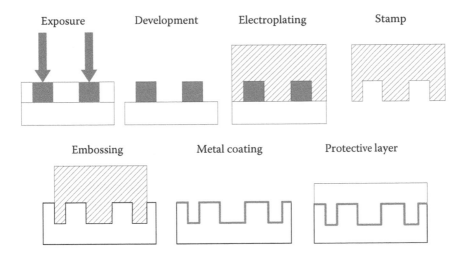

FIGURE 8.41 Use of photoresist to generate a master hologram, and replication of diffractive structures by embossing using a stamp made by electroplating.

8.6.5 Miscellaneous Materials

In addition to the classical materials we just described, there exists a large variety of more exotic holographic recording materials with specific properties.

Photochromic materials change their absorption coefficient upon illumination such that an amplitude hologram can be recorded. There exist both inorganic, such as doped glass [51], and organic molecules, such as bacteriorhodopsin [52]. This latter molecule can undergo transition at different excitation wavelengths, which make it useful for re-writable data storage applications [53].

Azoic colorants experience molecular reorientation due to multiple trans-cis-trans photo-isomerizations, which align their transition axis perpendicularly to the electric field of the incident light. Since the molecular permittivity is different along the transition axis than perpendicular to it (index ellipsoid), the material index of refraction changes with the molecular orientation and organization. This allows it to record polarization holograms where there is no intensity modulation [54, 55].

In photorefractive crystals and polymers, the light induced index of refraction modulation is dynamic and non-local due to charge migration and redistribution in the bulk of the material. The non-locality induces a phase shift between the illumination interference pattern and the index grating that leads to unique phenomena, such as beam coupling, phase conjugation mirror, and wave mixing [56, 57].

Phase holograms can also be written in doped glass, such as germanium-doped silica, and are used to manufacture very efficient reflectors in telecommunication optical fibers called fiber Bragg gratings [58].

8.6.6 Electronic Devices

In a same way electronic devices, such as digital sensors and spatial light modulators (SLM), have replaced materials for image capture and projection, they are nowadays replacing hologram recording materials. Their advantages are their ease of use and their dynamism, allowing them to capture and display holograms at videorate. However, electronic devices are only able to create thin holograms and their resolution is still limited compared to materials that can reproduce the more efficient and versatile volume holograms with a much higher resolution.

Although the technology is essentially the same in both the imaging and holographic field, there exists some specific technical requirement for holography.

8.6.6.1 Digital Sensors

Pixelatex sensors, such as CCD (charge-coupled device) and CMOS (complementary metal-oxyde-semiconductors), can both be used to record holograms. More specifically they record the intensity of an interference pattern. In order to capture the highest possible frequencies, the sensor should have a very small pixels pitch. Microns size pitch is required to be useful in the visible region. Such a system is widely used for digital holographic microscopy [59, 60].

8.6.6.2 Digital Displays

SLM, initially developed for micro-display, can be separated into two technologies: LCoS (liquid crystal on silicon), and mirror MEMS (micro-electro mechanical system).

LCoS is based on the alignment of birefringent liquid crystal molecules under the application of an electrical field. The birefringence can be used to modulate the phase of the incident beam (retardation), or modulate the amplitude by the use of an analyzer. Phase modulation provides highly efficient diffraction, only limited by the reflection loss and the device fill factor. LCoS pixel pitch can be as small as a couple of microns giving a diffraction angle of a few degrees in the visible [61]. The limitation of the LCoS technology is that the maximum refresh speed is several hundred frames per second, and is restricted by the visco-elastic relaxation of the molecules. This speed, which is fast enough for imaging and display application, reduces the space-bandwidth product, and prevents the use of time multiplexing schemes for holography.

Mirror MEMS are electronic devices where the position or orientation of micro-mirrors is controlled by the application of an electric field. The most successful is the Texas Instruments DMD (digital micro-mirror device) where the mirrors can be tilted left or right by about 10 degrees [62]. As such, the DMD can only display binary amplitude diffraction patterns, and its efficiency is relatively modest (10%, see Equation 8.32). Pixel pitch for the DMD can be as small as 5 microns (for the pico™ DLP'), but its biggest advantage is its refresh rate that can be as high as 20 kHz.

The Texas Instruments DMD is the only commercial one example in the class of MEMS, however, it is not perfectly adapted to holography due to the binary amplitude diffraction pattern it displays. A promising approach for the field, is the development of piston MEMS where the up and down movement of the mirror modulate the phase of the incident beam. This phase modulation allows a much higher diffraction efficiency (see Figure 8.9) [63, 64]. With a similar pixel pitch and refresh speed as the actual DMD, a piston MEMS could be extremely useful for holographic application, such as 3D display, beam steering in lidar, maskless lithography, or optical switch [65–67].

Acknowledgment

The author would like to thank Dr. Alexander Miles for his help with the Python script for some of the figures, as well as Dr. Lloyd LaComb for the careful revision of the manuscript.

References

1. Newton, I. *Opticks: Or a treatise of the reflections, refractions, inflexions and colors of light,* Courier Corporation, 1704.
2. Huygens C. *Traité de la lumière: où sont expliquées les causes de ce qui luy arrive dans la reflexion, & dans la refraction, et particulierement dans l'etrange refraction du cistal d'Islande,* Chez Pierre vander Aa, 1690.
3. Young T. The bakerian lecture: Experiments and calculations relative to physical optics, *Philos. Trans. R. Soc. London,* **94**, 1–16; 1804.
4. Fresnel A. *The Wave Theory of Light.* New York, NY: American Book, 1819.
5. Schrodinger E. Quantisierung als Eigenwertproblem, *Ann. Phys.,* **384**, 361–376; 1926.

6. De Broglie L. Waves and quanta, *Nature*, **112**, 2815, 540; 1923.
7. De Broglie L. XXXV. A tentative theory of light quanta, *Philos. Mag. Ser.*, *6*, **47**, 446–458; 1924.
8. Maxwell JC. On physical lines of force. Part I to IV, *Philos. Mag. J. Sci.*, **90**, 11–23; 1861.
9. Gabor D. A new microscopic principle, *Nature*, **161**, 777; 1948.
10. Gabor D. Microscopy by reconstructed wave-fronts, *Proc. R. Soc. A., Math. Phys. Eng. Sci.*, **197**, 454–487; 1949.
11. Maiman TH. Stimulated optical radiation in ruby, *Nature*, **187**, 493–494; 1960.
12. Leith EN, Upatnieks J. Reconstructed wavefronts and communication theory*, *J. Opt. Soc. Am.*, **52**, 1948; 1961.
13. Leith EN, Upatnieks J. Wavefront reconstruction with continuous-tone objects, *J. Opt. Soc. Am.*, **53**, 1377; 1963.
14. Leith EN, Upatnieks J. Wavefront reconstruction with diffused illumination and three-dimensional objects, *J. Opt. Soc. Am.*, **54**, 1295–1301; 1964.
15. Denisyuk YN. Photographic reconstruction of the optical properties of an object in its own scattered radiation field, *Sov. Physics-Doklady*, **7**, 543–545; 1962.
16. Denisyuk YN. On the reproduction of the optical properties of an object by the wave field of its scattered radiation. Part 1, *Opt. Spectrosc.*, **15**, 279–284; 1963.
17. Denisyuk YN. On the reproduction of the optical properties of an object by its scattered radiation II, *Opt. Spectrosc.*, **18**, 152–157; 1965.
18. Lohmann AW, Paris DP. Binary fraunhofer holograms, generated by computer, *Appl. Opt.*, **6**, 1739–1748; 1967.
19. Brown BR, Lohmann AW. Computer-generated binary holograms, *IBM J. Res. Dev.*, **13**, 160–168; 1969.
20. Kogelnik H. Coupled wave theory for thick hologram gratings, *Bell Syst. Tech. J.*, **48**, 2909–2947; 1969.
21. Brotherton-Ratcliffe D. A treatment of the general volume holographic grating as an array of parallel stacked mirrors, *J. Mod. Opt.*, **59**, 1113–1132; 2012.
22. Bjelkhagen H, Brotherton-Ratcliffe D. *Ultra-Realistic Imaging: Advanced Techniques in Analogue and Digital Colour Holography*. CRC Press, 2013.
23. Klein WR, Cook BD. Unified approach to ultrasonic light diffraction, *IEEE Trans. Sonics Ultrason.*, **14**, 123–134; 1967.
24. Moharam M, Young L. Criterion for Bragg and Raman-Nath diffraction regimes, *Appl. Opt.*, **17**, 1757–1759; 1978.
25. Moharam MG, Gaylord TK. Three-dimensional vector coupled-wave analysis of planar-grating diffraction, *J. Opt. Soc. Am.*, **73**, 1105; 1983.
26. Castro JM, Zhang D, Myer B, Kostuk RK. Energy collection efficiency of holographic planar solar concentrators, *Appl. Opt.*, **49**, 858–870; 2010.
27. Han J, Liu J, Yao X, Wang Y. Portable waveguide display system with a large field of view by integrating freeform elements and volume holograms, *Opt. Express*, **23**, 3534; 2015.
28. Cameron A. Optical waveguide technology and its application in head-mounted displays, *Proc. SPIE*, **83830E**, 1–11; 2012.
29. Magnusson R, Gaylord TK. Diffraction efficiencies of thin absorption and transmittance gratings, *Opt. Commun.*, **28**, 1–3; 1979.
30. Wyrowski F. Diffractive optical elements: Iterative calculation of quantized, blazed phase structures, *J. Opt. Soc. Am. A*, **7**, 961; 1990.
31. Swanson GJ. Binary optics technology: Theoretical Limits on the Diffraction Efficiency of Multilevel Diffractive Optical Elements, *Technical Report 914*, Lincoln Laboratory, 1991.
32. Picart P. Ed., *New Techniques in Digital Holography*. Hoboken, NJ: Wiley, 2015.
33. Nehmetallah GT, Aylo R, Williams L. *Analog and Digital Holography with MATLAB*. Bellingham, WA: SPIE Press, 2015.

34. Cooley JW, Tukey JW. An algorithm for the machine computation of the complex Fourier series, *Math. Comput.*, **19**, 297; 1965.

35. Greenbaum A, Luo W, Su T-W, et al. Imaging without lenses: Achievements and remaining challenges of wide-field on-chip microscopy, *Nat. Methods*, **9**, 889–895; 2012.

36. Gerchberg RW, Saxton WO. A practical algorithm for the determination of phase from image and diffraction plane pictures, *Optik*, **35**, 237–246; 1972.

37. Smalley D, Smithwick Q, Bove V. Anisotropic leaky-mode modulator for holographic video displays, *Nature*, **498**, 313–317; 2013.

38. Benton SA, Bove, Jr. VM. *Holographic Imaging*. Hoboken, NJ: Wiley, 2008.

39. Hariharan P. *Optical Holography: Principles, Techniques, and Application*. Cambridge: Cambridge University Press, 2004.

40. Blanche P-A. *Field Guide to Holography*. Bellingham, WA: SPIE Press, 2014.

41. Benton SA. Hologram reconstructions with extended incoherent sources in program of the 1969 Annual Meeting of the Optical Society of America, *J. Opt. Soc. Am.*, **59**, 1545; 1969.

42. Debitetto DJ. Holographic panoramic stereograms synthesized from white light recordings, *Appl. Opt.*, **8**, 1740–1741; 1969.

43. Curtis K, Dhar L, Blanche PA. Holographic data storage technology, in *Optical and Digital Image Processing: Fundamentals and Applications*, pp. 227–250, Cristobal, G. and P. Schelkens, Eds. Wiley-VCH, 2011.

44. Roberts DE, Smith T. The History of Integral Print Methods, pp. 1–21, http://lenticulartechnology.com/files/2014/02/Integral-History.pdf.

45. Thizy C, Lemaire P, Georges M, et al. Comparison between finite element calculations and holographic interferometry measurements, of the thermo-mechanical behavior of satellite structures in composite materials, in *Photorefractive Effects, Materials, and Devices*, p. 700, Nolte, D. D., Ed. Washington, DC: Optical Society of America, 2005.

46. Molin N-E, Stetson KA. Measuring combination mode vibration patterns by hologram interferometry, *J. Phys. E.*, **2**, 609–612; 1969.

47. Volodin BL, Sandalphon, Meerholz K, Kippelen B, Kukhtarev NV, Peyghambarian N. Highly efficient photorefractive polymers for dynamic holography, *Opt. Eng.*, **34**, 2213; 1995.

48. Georges M. Photorefractives for holographic interferometry and nondestructive testing, in *Photorefractive Organic Materials*, Blanche, P.-A., Ed. Switzerland Springer Science, 2016.

49. Lemaire P, Georges M. Dynamic holographic interferometry: Devices and applications, in *Photorefractive Materials and Their Applications 3. Applications*, pp. 223–251, Gunter, P. and Huignard J.P., Eds. New York, NY: Springer, 2007.

50. Blanche P-A, Gailly P, Habraken S, Lemaire P, Jamar C. Volume phase holographic gratings: Large size and high diffraction efficiency, *Opt. Eng.*, **43**, 2603; 2004.

51. Kirk JP. Hologram on photochromic glass, *Appl. Opt.*, **5**, 1684–1685; 1966.

52. Downie JD, Smithey DT. Measurements of holographic properties of bacteriorhodopsin films, *Appl. Opt.*, **35**, 5780–5789; 1996.

53. Hampp N. Bacteriorhodopsin as a photochromic retinal protein for optical memories, *Chem. Rev.*, **100**, 1755–1776; 2000.

54. Blanche P-A, Lemaire P, Maertens C. Polarization holography reveals the nature of the grating in polymers containing azo-dye, *Opt. Commun.*, **185**, 1–12; 2000.

55. Blanche P-A, Lemaire PC, Maertens C, Dubois P, Jérôme R. Photoinduced birefringence and diffraction efficiency in azo dye doped or grafted polymers: Theory versus experiment of the temperature influence, *J. Opt. Soc. Am. B*, **17**, 729; 2000.

56. Günter P, Huignard J-P. *Photorefractive Material and their Applications. 1 Basic Effects*. New York, NY: Springer, 2006.

57. Blanche P-A. Ed., *Photorefractive Organic Materials and Applications*. AG Switzerland: Springer, 2016.

58. Kashyap R. *Fiber Bragg Gratings*. San Diego, CA: Academic Press, 1999.

59. Yu X, Hong J, Liu C, Kim MK. Review of digital holographic microscopy for three-dimensional profiling and tracking, *Opt. Eng.*, **53**, 112306; 2014.
60. Marquet P, Depeursinge C, Magistretti PJ. Review of quantitative phase-digital holographic microscopy: Promising novel imaging technique to resolve neuronal network activity and identify cellular biomarkers of psychiatric disorders, *Neurophotonics*, **1**, 020901; 2014. http://www.ti.com/lit/wp/dlpa051a/dlpa051a.pdf.
61. Bleha WP, Lei LA. Advances in Liquid Crystal on Silicon (LCOS) Spatial Light Modulator Technology, in *SPIE Defense, Security, and Sensing*, Baltimore, Maryland, United States, p. 87360A, 2013.
62. DLP ® Technology for Near Eye Display White paper, Texas Instruments, 2014.
63. Yoo B-W, Megens M, Sun T, et al. A 32 × 32 Optical phased array using polysilicon sub-wavelength high-contrast-grating mirrors, *Opt. Express*, **22**, 19029; 2014.
64. Lin T. Implementation and characterization of a flexure-beam micromechanical spatial light modulator, *Opt. Eng.*, **33**, 3643–3648; 1994.
65. Lopez D, Aksyuk V, Watson G, et al. Two Dimensional MEMS Piston Array for DUV Optical Pattern Generation, in *IEEE/LEOS International Conference on Optical MEMS and their Applications Conference*, pp. 148–149, Big Sky, MT, 2006.
66. Rhoadarmer T. Flexure-beam micromirror spatial light modulator devices for acquisition, tracking, and pointing, *Proc. SPIE*, **2227**, 418–430; 1994.
67. Blanche P-A, Carothers D, Wissinger J, Peyghambarian N. Digital micromirror device as a diffractive reconfigurable optical switch for telecommunication, *J. Micro/Nanolithogr. MEMS, MOEMS*, **13**(1), 011104, December 2013.

9

Fourier Optics and Image Processing

9.1   Fourier Optics and Image Processing 299
     Fourier Transformation by Optics • Coherent and Incoherent
     Processing • Fourier and Spatial Domain Processing • Exploitation
     of Coherence
9.2   Image Processing 312
     Coherent Processing • Processing with Incoherent Light
9.3   Neural Networks... 319
     Optical Neural Net Architectures • Hopfield Neural
     Network • Interpattern Association Neural Network
9.4   Wavelet Transform Processing 325
     Wavelet Transform • Optical Implementations
9.5   Computing with Optics 331
     Logic-Based Computing • Matrix–Vector and Matrix–Matrix
     Processors • Systolic Processor
9.6   Optical Processing for Radar Imaging and Broadband Signal..... 338
     Radar Imaging with Optics • Optical Processing of Broadband
     Signals • Broadband Signal Processing with Area Modulation • Final
     Remarks

Francis T.S. Yu

Suggested Reading.. 350

9.1 Fourier Optics and Image Processing

The discovery of intensive lasers has enabled us to build more efficient optical systems for communication and signal processing. Most of the optical processing architectures to date have confined themselves to the cases of complete coherence or complete incoherence. However, a continuous transition between these two extremes is possible. Added to the recent development of spatial light modulators, this has brought optical signal processing to a new height. Much attention has been focused on high-speed and high-data-rate optical processing and computing.

In this chapter, we shall discuss the basic principles of Fourier optics as applied to image processing and computing.

9.1.1 Fourier Transformation by Optics

To understand the basic concept of Fourier optics, we shall begin our discussion with the Fresnel–Kirchhoff theory. Let us start from the Huygens' principle, in which the complex amplitude observed at the point p′ of a coordinate system $\sigma(\alpha,\beta)$, due to a monochromatic light field located in another coordinate system $\rho(x,y)$, as shown in Figure 9.1, can be calculated by assuming that each point of light

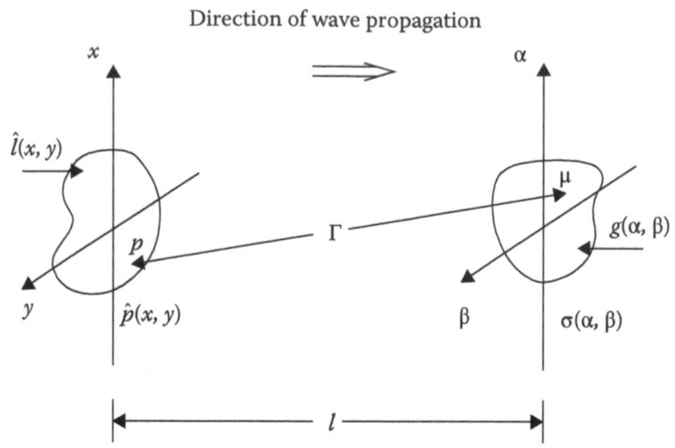

Direction of wave propagation

FIGURE 9.1 Fresnel–Kirchhoff theory.

source is an infinitesimal spherical radiator. Thus, the complex light amplitude $h_l(\alpha,\beta;k)$ contributed by a point p in the (x, y) coordinate system can be considered to be that from an unpolarized monochromatic point source, such that

$$h_l = -\frac{i}{\lambda r}\exp[i(kr - \omega t)],\qquad(9.1)$$

where λ, k, and ω are the wavelengths, wave number, and angular frequency, respectively, of the point source, and r is the distance between the point source and the point of observation.

If the separation l of the two-coordinate systems is assumed to be large compared with the regions of interest in the (x, y) and (α, β) coordinate systems, r can be approximated by

$$r = l + \frac{(\alpha - x)^2}{2l} + \frac{(\beta + y)^2}{2l},\qquad(9.2)$$

which is known as a paraxial approximation. By substituting it into Equation 9.1, we have

$$h_j(\sigma - \rho;k) \cong -\frac{i}{\lambda l}\exp\left\{ik\left[l + \frac{(\alpha - x)^2}{2l} + \frac{(\beta - y)^2}{2l}\right]\right\},\qquad(9.3)$$

which is known as the *spatial impulse response,* where the time-dependent exponent has been dropped for convenience. Thus, we see that the complex amplitude produced at the (α, β) coordinate system by the monochromatic radiating surface $f(x, y)$ can be written as

$$g(\alpha,\beta) = \iint\limits_{x,y} f(x,y)h_l(\sigma - \rho;k)\mathrm{d}x\mathrm{d}y,\qquad(9.4)$$

which is the well-known *Kirchhoff's integral.* In view of the proceeding equation, we see that the Kirchhoff's integral is in fact the convolution integral, which can be written as

$$g(\alpha,\beta) = f(x,y) * h_l(x,y),\qquad(9.5)$$

FIGURE 9.2 Linear system representation.

where the asterisk denotes the convolution operation

$$h_l(x,y) = C \exp\left[1\frac{k}{2l}(x^2 + y^2) \right]$$ (9.6)

and $C = (i/\lambda l)\exp(ikl)$ is a complex constant. Consequently, Equation 9.5 can be represented by a block box system diagram, as shown in Figure 9.2. In other words, the complex wave field distributed over the (α, β) coordinate system plane can be evaluated by the *convolution integral* of Equation 9.4.

It is well known that a two-dimensional Fourier transformation can be obtained with a positive lens. Fourier-transform operations usually require complicated electronic spectrum analyzers or digital computers. However, this complicated transform can be performed extremely simply with a coherent optical system.

To perform Fourier transformations in optics, it is required that a positive lens is inserted in a monochromatic wave field of Figure 9.1. The action of the lens can convert a spherical wavefront into a plane wave. The lens must induce a phase transformation, such as

$$T(x,y) = C \exp\left[-i\frac{\pi}{\lambda f}(x^2 + y^2) \right],$$ (9.7)

where C is an arbitrary complex constant.

Let us now show the Fourier transformation by a lens, as illustrated in Figure 9.3, in which a monochromatic wave field at input plane (ξ, η) is $f(\xi, \eta)$. Then, by applying the Fresnel–Kirchhoff theory of Equation 9.5, the complex light distribution at (α, β) can be written as

$$g(\alpha,\beta) = C\left\{ \left[f(\xi,n) * hl(\xi,n) \right] T(x,y) \right\} * h_f(x \cdot y),$$ (9.8)

where C is a proportionality complex constant, $h_l(\xi,\eta)$ and $h_f(x,y)$ are the corresponding spatial impulse responses, and $T(x, y)$ is the phase transform of the lens, as given in Equation 9.7.

By a straightforward, tedious evaluation, we can show that

$$g(\alpha,\beta) = C_1 \exp\left[-i\frac{k}{2f}\frac{1-v}{v}(\alpha^2 + \beta^2) \right] \iint f(\xi,\eta) \exp\left[-i\frac{k}{f}(\alpha\xi + \beta\eta) \right] d\xi d\eta,$$ (9.9)

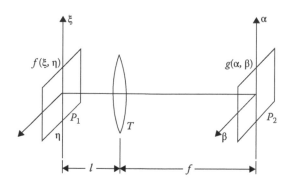

FIGURE 9.3   Fourier transformation by a lens.

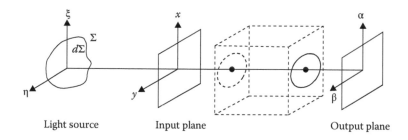

FIGURE 9.4   A hypothetical optical processing system.

which is essentially the Fourier transform of $f(\xi, \eta)$ with a quadratic phase factor. If the signal plane is placed at the front focal plane of the lens, that is $l = f$, the quadratic phase factor vanishes, which leaves an exact Fourier transformation,

$$G(p,q) = C_1 \iint f(\xi,\eta)\exp[-i(p\xi + q\eta)]\mathrm{d}\xi\mathrm{d}\eta, \tag{9.10}$$

where $p = k\alpha/f$ and $q = k\beta/f$ are the angular spatial frequency coordinates.

9.1.2 Coherent and Incoherent Processing

Let a hypothetical optical processing system be as shown in Figure 9.4. Assume that the light emitted by the source Σ is monochromatic, and $u(x, y)$ is the complex light distribution at the input signal plane due to the incremental source $\mathrm{d}\Sigma$. If the complex amplitude transmittance of the input plane is $f(x, y)$, the complex light field immediately behind the signal plane is $u(x,y)f(x,y)$. We assume the optical system in the black box is linearly spatially invariant with a spatial impulse response of $h(x, y)$, the output complex light field, due to $\mathrm{d}\Sigma$, can be calculated by

$$g(\alpha,\beta) = [u(x,y)f(x,y)] * h(x,y), \tag{9.11}$$

which can be written as

$$\mathrm{d}I(\alpha,\beta) = g(\alpha,\beta)g^*(\alpha,\beta)\mathrm{d}\Sigma, \tag{9.12}$$

where the superscript asterisk denotes the complex conjugate. The overall output density distribution is therefore

$$I(\alpha,\beta) = \iint |g(\alpha,\beta)|^2 \, d\Sigma,$$

which can be written in the following convolution integral:

$$I(\alpha,\beta) = \int\int\int\int \Gamma(x,y;x',y')h(\alpha-x,\beta-y)h^*(\alpha-x',\beta-y')$$
$$\cdot f(x,y)f^*(x',y')\,dx\,dy\,dx'\,dy', \tag{9.13}$$

where

$$\Gamma(x,y:x',y') = \iint_\Sigma u(x,y)u^*(x'y')\,d\Sigma$$

is the *spatial coherence function*, also known as the *mutual intensity function*, at input plane (x, y).

By choosing two arbitrary points Q_1 and Q_2 at the input plane, and if r_1 and r_2 are the respective distances from Q_1 and Q_2 to $d\Sigma$, the complex light disturbances at Q_1 and Q_2 due to $d\Sigma$ can be written as

$$u_1(x,y) = \frac{[I(\xi,\eta)]^{-1/2}}{r_1}\exp(ikr_1) \tag{9.14}$$

and

$$u_2(x',y') = \frac{[I(\xi,\eta)]^{-1/2}}{r_1}\exp(ikr_2), \tag{9.15}$$

where $I(\xi,\eta)$ is the intensity distribution of the light source. By substituting Equations 9.14 and 9.15 into Equation 9.13, we have

$$\Gamma(x,y;x',y') = \iint_\Sigma \frac{I(\xi,\eta)}{r_1 r_2}\exp[ik(r_1 - r_2)]\,d\Sigma. \tag{9.16}$$

In the paraxial case, $r_1 - r_2$ may be approximated by

$$r_1 - r_2 \simeq \frac{1}{r}[\xi(x-x')+\eta(y-y')],$$

where r is the separation between the source plane and the signal plane. Then, Equation 9.16 can be reduced to

$$\Gamma(x,y;x',y') = \frac{1}{r^2}\iint_\Sigma I(\xi,\eta)\exp\left\{i\frac{k}{r}[\xi(x-x')+\eta(y-y')]\right\}d\xi d\eta, \tag{9.17}$$

which is known as the *Van Cittert–Zernike theorem*. Notice that Equation 9.17 forms an inverse Fourier transform of the source intensity distribution.

Now one of the two extreme cases is that by letting the light source become infinitely large, for example, $I(\xi,\eta) \approx K$, then Equation 9.17 becomes

$$\Gamma(x, y; x', y') = K_1 \delta(x - x', y - y'), \tag{9.18}$$

which describes a completely *incoherent illumination*, where K_1 is an appropriate constant.

On the other hand, if the light source is vanishingly small, that is, $I(\xi,\eta) \approx K, \delta(\xi,\eta)$, $\delta(\xi,\eta)$, Equation 9.17 becomes

$$\Gamma(x, y; x', y') = K_2, \tag{9.19}$$

which describes a completely *coherent illumination*, where K_2 is a proportionality constant. In other words, a monochromatic point source describes a strictly coherent processing regime, while an extended source describes a strictly incoherent system. Furthermore, an extended monochromatic source is also known as a *spatially incoherent* source.

By referring to the completely incoherent illumination, we have $\Gamma(x, y; x', y') = K_1 \delta(x - x', y - y')$, the intensity distribution at the output plane, can be shown

$$I(\alpha,\beta) = \iint |h(\alpha - x, \beta - y)|^2 |f(x, y)|^2 \, dx \, dy, \tag{9.20}$$

in which we see that the output intensity distribution is the convolution of the input signal intensity with respect to the intensity impulse response. In other words, for the completely incoherent illumination, the optical signal processing system is linear in intensity, that is,

$$I(\alpha,\beta) = |h(x, y)|^2 * |f(x, y)|^2, \tag{9.21}$$

where the asterisk denotes the convolution operation. On the other hand, for the completely coherent illumination, that is, $\Gamma(x, y; x', y') = K_2$, the output intensity distribution can be shown as

$$I(\alpha,\beta) = g(\alpha,\beta)g^*(\alpha,\beta) = \iint h(\alpha - x, \beta - y)f(x, y)dxdy$$

$$\cdot \iint h^*(\alpha - x', \beta - y')f(x', y')dx'dy' \tag{9.22}$$

when

$$g(\alpha,\beta) = \iint h(\alpha - x, \beta - y)f(x, y)dxdy, \tag{9.23}$$

for which we can see that the optical signal processing system is linear in *complex amplitude*. In other words, a coherent optical processor is capable of processing the information in complex amplitudes.

9.1.3 Fourier and Spatial Domain Processing

The roots of optical signal processing can be traced back to Abbe's work in 1873, which led to the discovery of spatial filtering. However, optical pattern recognition was not appreciated until the complex spatial filtering of Van der Lugt in 1964. Since then, techniques, architectures, and algorithms have been developed to construct efficient optical signal processors. The objective of this section is to discuss the optical architectures and techniques as applied to image processing. Basically, there are two frequently used signal-processing architectures in Fourier optics: namely, the Vander Lugt correlator (VLC) and the joint transform correlator (JTC). Nevertheless, image processing can be accomplished either by

Fourier domain filtering or *spatial domain filtering* or both. Processors that use Fourier domain filtering are known as VLCs and the spatial-domain filtering is often used for JTC. The basic distinctions between them are that VLC depends on a Fourier-domain filter, whereas JTC depends on a spatial domain filter. In other words, the complex spatial filtering of Vander Lugt is input scene *independent,* whereas the joint transform filtering is input scene *dependent.* The reason is that once the Fourier domain spatial filter is synthesized, the structure of the filter will not be altered by the input scene (e.g., for multi-target detection or background noise). Thus, the performance of Fourier domain filtering is independent of the input scene. On the other hand, the joint transform power spectrum would be affected by the input noise and multi-target detection, for which the performance in the JTC is input scene dependent.

It is apparent that a pure optical correlator has drawbacks, which make certain tasks difficult or impossible to perform. The first problem is that optical systems are difficult to program, in the sense of a general-purpose digital computer. A second problem is that accuracy is difficult to achieve in Fourier optics. A third problem is that optical systems cannot be easily used to make decisions. Even a simple decision cannot be performed optically without the intervention of electronics. However, many deficiencies of optics happen to be the strong points of the electronic counterpart. For example, accuracy, controllability, and programmability are the obvious traits of digital computers. Thus, by combining an optical system with its electronic counterpart is rather natural to have a better processor, as shown in Figures 9.5 and 9.6, in which spatial light modulators (SLMs) are used for input-object and spatial filter devices. One of the important aspects of these hybrid-optical architectures is that decision-making can be done by the computer.

We shall now consider the VLC (depicted in Figure 9.5), which we assume is illuminated by a coherent plane wave. If an input object $f(x, y)$ is displayed at the input plane, the output complex light can be shown as

$$g(\alpha,\beta) = Kf(x, y) * h(x, y), \tag{9.24}$$

where * denotes the convolution operation, K is a proportionality constant, and $h(x, y)$ is the spatial impulse response of the Fourier domain filter, which can be generated on the SLM. We note that a Fourier domain filter can be described by a complex-amplitude transmittance, such as

$$H(p,q) = |H(p,q)| \exp[i\phi(p,q)], \tag{9.25}$$

for which the physically realizable conditions are imposed by

$$|H(p,q)| \leq 1 \tag{9.26}$$

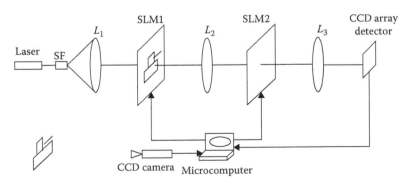

FIGURE 9.5 Hybrid optical Vander Lugt correlator (VLC). SLMs, spatial light modulators; CCDs, charge-coupled detectors; SF, spatial filter; L, lense.

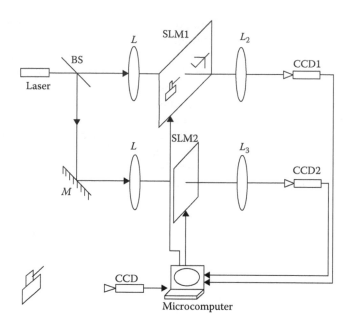

FIGURE 9.6 Hybrid optical joint transform correlator (JTC). SLMs, spatial light modulators; CCDs, charge-coupled detectors; BS, beamsplitter; M, mirror; L, lense.

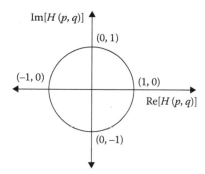

FIGURE 9.7 Complex amplitude transmittance.

and

$$0 \leq \phi(p,q) \leq 2\pi, \tag{9.27}$$

where (p, q) is the angular spatial frequency coordinator system. We stress that the complex transmittance imposed by the physical constraints can be represented by a set of points within or on a unit circle in a complex plane shown in Figure 9.7. In other words, a physically realizable filter can only be synthesized within or on the unit circle of the complex plane.

Let us now illustrate a complex signal detection using the VLC. We assume that a Fourier-domain matched filter is generated at the spatial frequency plane, as given by

$$H(p,q) = K\left\{1 + |F(p,q)|^2 + 2|F(p,\ q)|\cos[\alpha_0 p + \phi(p,q)]\right\}, \tag{9.28}$$

which is a positive real function subject to the physical constraints of Equations 9.26 and 9.27, where α_0 is the spatial carrier frequency. It is straightforward to show that the output complex light distribution is given by

$$g(\alpha,\beta) = K[f(x,y) + f(x,y) * f(x,y) * f^*(-x,-y) + f(x,y) * f(x+x_0,y)$$

$$+ f(x,y) * f^*(-x+\alpha_0,-y)],$$

(9.29)

in which the first and second terms represent the zero-order diffraction; the third and fourth terms are the convolution and cross-correlation terms, which are diffracted in the neighborhood of $\alpha = -\alpha_0$, and $\alpha = \alpha_0$, respectively. The zero-order and convolution terms are of no particular interest here; it is the cross-correlation term that is used in signal detection.

Now, if the input signal is assumed to be embedded in an additive white Gaussian noise n, that is,

$$f'(x,y) = f(x,y) + n(x,y),$$

(9.30)

then the correlation term in Equation 9.29 would be

$$R(\alpha,\beta) = K[f(x,y) + n(x,y)] * f^*(-x+\alpha_0,-y).$$

(9.31)

Since the cross-correlation between $n(x, y)$ and $f^*(-x+\alpha_0, -y)$ can be shown to be approximately equal to zero, Equation 9.31 reduces to

$$R(\alpha,\beta) = f(x,y) * f^*(-x+\alpha_0 - y),$$

(9.32)

which is the autocorrelation distribution of $f(x, y)$.

To ensure that the zero-order and the first-order diffraction terms will not overlap, the spatial carrier frequency α_0 is required that

$$\alpha_0 > l_f + \tfrac{3}{2} l_s,$$

(9.33)

where l_f and l_s are the spatial lengths in the x-direction of the input object transparency and the detecting signal $f(x, y)$, respectively. To show that this is true, we consider the length of the various output terms of $g(\alpha,\beta)$, as illustrated in Figure 9.8.

Since lengths of the first, second, third, and fourth terms of Equation 9.29 are l_f, $2l_s + l_f$, and $l_f + l_s$, respectively, to achieve complete separation, the spatial carrier frequency α_0 must satisfy the inequality of Equation 9.33.

Complex spatial filtering can also be performed by JTC (shown in Figure 9.6) in which the input object displayed on the SLMs are given by

$$f_1(x-\alpha_0,y) + f_2(x+\alpha_0,y),$$

(9.34)

where $2\alpha_0$ is the main separation of the input objects. The corresponding joint transform power spectrum (JTPS), as detected by charge-coupled detector (CCD_1), is given by

$$I(p,q) = |F_1(p,q)|^2 + |F_2(p,q)|^2 + 2|F_1(p,q)|^2 F_2|(p,q)|^2 \cdot$$

$$\cos[2\alpha_0 p - \phi_1(p,q) + \phi_2(p,q)],$$

(9.35)

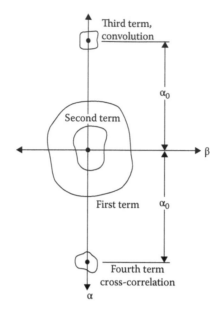

FIGURE 9.8   Sketch of the output diffraction.

where

$$F_1(p,q) = \left| F_1(p,q) \right| e^{i\phi_1(p,q)},$$

and

$$F_2(p,q) = \left| F_2(p,q) \right| e^{i\phi_2(p,q)}.$$

If the JTPS is displayed on SLM2, the output complex light distribution would be

$$g(\alpha,\beta) = f_1(x,y) \otimes f_1^*(x,y) + f_2(x,y) \otimes f_2^*(x,y) + f_1(x,y)$$

$$\otimes f_2^*(x - 2\alpha_0, y) + f_1^*(x,y) \otimes f_2(x + 2\alpha_0, y),$$

(9.36)

where $\otimes$ denotes the correlation operation. The first two terms represent overlapping correlation functions $f_1(x,y)$ *and* $f_2(x,y)$, which are diffracted at the origin of the output plane. The last two terms are the two cross-correlation terms, which are diffracted around $\alpha = 2\alpha_0$ and $\alpha = -2\alpha_0$, respectively.

To avoid the correlation term overlapping with the zero diffraction, it is required that the separation between the input object should be

$$2\alpha_0 > 2\chi,$$

(9.37)

where χ is the width of the input object.

For complex target detection, we assume that the target $f_1(x - \alpha_0, y)$ is embedded in an additive white Gaussian noise, that is, $f_1(x - \alpha_0, y) + n(x - \alpha_0, y)$ and $f_2(x - \alpha_0, y)$ is replaced by $f_1(x - \alpha_0, y)$. Then, it can be shown that the JTPS is given by

$$I(p,q) = 2|F_1|^2 + |N|^2 + F_1 N^* + N F_1^* + (F_1 F_1^* + N F_1^*)e^{-2\alpha_0 p}$$

$$+ (F_1 F_1^* F_1 N^*)e^{-2\alpha_0 p}.$$

(9.38)

Since the noise is assumed to be additive and Gaussian distributed with zero mean, we note that

$$\iint f_1(x,y)n(\alpha+x,\beta+y)\,dxdy = 0.$$

Thus, the output complex light field is

$$g(\alpha,\beta) = 2f_1(x,y)\otimes f_1^*(x,y) + n(x,y)\otimes n^*(x,y) + f_1(x,y)\otimes f_1^*(x-2\alpha_0,y)$$
$$+ f_1(x,y)\otimes f_1^*(x+2\alpha_0,y),$$

(9.39)

in which the autocorrelation terms are diffracted at $\alpha_0 = 2\alpha_0$ and $\alpha_0 = -2\alpha_0$, respectively.

9.1.4 Exploitation of Coherence

The use of a coherent source enables optical processing to be carried out in complex amplitude processing, which offers a myriad of applications. However, coherent optical processing also suffers from coherent artifact noise, which limits its processing capabilities. To alleviate these limitations, we discuss methods to exploit the coherence contents from an incoherent source for complex amplitude processing. Since all physical sources are neither strictly coherent nor strictly incoherent, it is possible to extract their inherent coherence contents for coherent processing.

Let us begin with the exploitation of spatial coherence from an extended incoherent source. By referring to the conventional optical processor shown in Figure 9.9, the spatial coherence function at the input plane can be written as

$$\Gamma(x_2 - x_2') = \iint \gamma(x_1)\exp\left[i2\pi\frac{x_1}{\lambda_f}(x_2 - x_2')\right]dx_1,$$

(9.40)

which is the well-known Van Citter–Zernike theorem, where $\gamma(x_1)$ is the extended source, f is the focal length of the collimated lens, and λ is the wavelength of the extended source. Thus, we see that the spatial coherence at the input plane and the source-encoding intensity transmittance form a Fourier-transform pair, given by

$$\gamma(x_1) = \mathcal{F}[\Gamma(x_2 - x_2')], \text{and } \Gamma(x_2 - x_2') = \mathcal{F}^{-1}[\gamma(x_1)],$$

(9.41)

where $\mathcal{F}$ denotes the Fourier-transform operation. In other words, if a specific spatial coherence requirement is needed for information processing, a source-encoding can be performed. The source-encoding

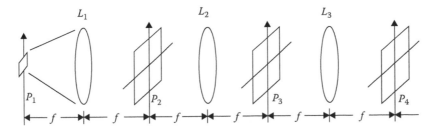

FIGURE 9.9 Incoherent source optical processor: L_1, collimating lens; L_2 and L_3, achromatic transformation lenses; P_1, source encoding mask; P_2, input plane; P_3, Fourier plane, and P_4, output plane.

$\gamma(x_1)$ can consist of apertures of different shapes or slits, but it should be a positive real function that satisfies the following physically realizable constraint:

$$0 \leq \gamma(x_1) \leq 1. \tag{9.42}$$

For the exploitation of *temporal coherence,* we note that the Fourier spectrum is linearly proportional to the wavelength of the light source. It is apparently not capable of (or inefficient at) using a broadband source for complex amplitude processing. To do so, a narrow-spectral-band (i.e., temporally coherent) source is needed. In other words, the spectral spread of the input object should be confined within a small fraction fringe spacing of the spatial filter, which is given by

$$\frac{p_m f \Delta\lambda}{2\pi} \ll d, \tag{9.43}$$

where d is the fringe spacing of the spatial filter, p_m is the upper angular spatial frequency content of the input object, f is the focal length of the transform lens, and $\Delta\lambda$ is the spectral bandwidth of the light source. In order to have a higher temporal coherence requirement, the spectral width of the light source should satisfy the following constraint:

$$\frac{\Delta\lambda}{\lambda} \ll \frac{\pi}{h p_m}, \tag{9.44}$$

where λ is the center wavelength of the light source, $2h$ is the size of the input object transparency, and $2h = (\lambda f) / d$.

There are ways to exploit the temporal coherence content from a broadband source. One of the simplest methods is by dispersing the Fourier spectrum, which can be obtained by placing a spatial sampling grating at the input plane P_2. For example, if the input object is sampled by a phase grating, as given by

$$f(x_2) T(x_2) = f(x_2) \exp(i p_0 x_2), \tag{9.45}$$

then the corresponding Fourier transform would be

$$F(p,q) = F\left(x_3 - \frac{\lambda f}{2\pi} p_0 \right), \tag{9.46}$$

in which we see that $F(p, q)$ is smeared into rainbow colors at the Fourier plane. Thus, a high temporal coherence Fourier spectrum within a narrow-spectral-band filter can be obtained, as given by

$$\frac{\Delta\lambda}{\lambda} \cong \frac{4 p_m}{p_0} \ll 1. \tag{9.47}$$

Since the spectral content of the input object is dispersed in rainbow colors, as illustrated in Figure 9.10a, it is possible to synthesize a set of narrow-spectral-band filters to accommodate the dispersion.

On the other hand, if the spatial filtering is a 1D operation, it is possible to construct a fan-shaped spatial filter to cover the entire smeared Fourier spectrum, as illustrated in Figure 9.10b. Thus, we see that a high degree of temporally coherent filtering can be carried out by a simple white light source. Needless to say, the (broadband) spatial filters can be synthesized by computer-generated techniques.

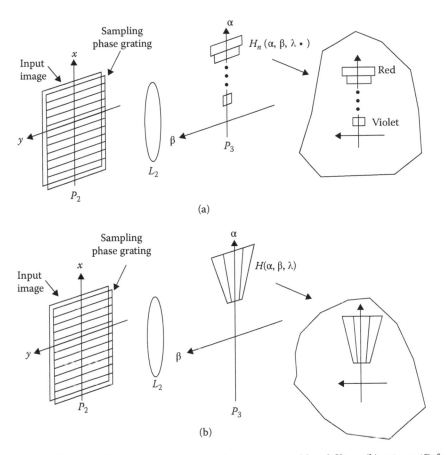

FIGURE 9.10 Broad spectral filtering: (a) using a set of narrow-spectral-band filters, (b) using a 1D fan-shaped broadband filter.

In the preceding, we have shown that spatial and temporal coherence can be exploited by spatial encoding and spectral dispersion of an incoherent source. We have shown that complex amplitude processing can be carried out with either a set of 2D narrow-spectral-band filters or with a 1D fan-shaped broadband filter. Let us first consider that a set of narrow-spectral-band filters is being used, as given by

$$H_n(p_n, q_n; \lambda_n), \quad \text{for } n = 1, 2, ..., N, \tag{9.48}$$

where (p_n, q_n) represents the angular frequency coordinates and λ_n is the center wavelength of the narrow-width filter; then, it can be shown that the output light intensity would be the incoherent superposition of the filtered signals, as given by

$$I(x, y) \cong \sum_{n=1}^{N} \Delta\lambda_n |f(x, y; \lambda_n) * h(x, y; \lambda_n)|^2, \tag{9.49}$$

where * denotes the convolution operation, $f(x, y; \lambda_n)$ represents the input signal illuminated by λ_n, $\Delta\lambda_n$ is the narrow spectral width of the narrow-spectral-band filter, and $h(x, y; \lambda_n)$ is the spatial impulse response of $H_n(p_n, q_n; \lambda_n)$, that is,

$$h_n(p_n, q_n; \lambda_n) = \mathcal{F}^{-1}\left[H(p_n, q_n; \lambda_n) \right]. \tag{9.50}$$

Thus, we see that by exploiting the spatial and temporal coherence, an incoherent source processor can be made to process the information in complex amplitude as a coherent processor. Since the output intensity distribution is the sum of mutually incoherent image irradiances, the annoying coherent artifact noise can be avoided.

On the other hand, if the signal processing is a 1D operation, then the information processing can be carried out with a 1D fan-shaped broadband filter. Then, the output intensity distribution can be shown as

$$I(x,y) = \int_{\Delta\lambda} \left| f(x,y;\lambda) * h(x,y;\lambda_n) \right|^2 d\lambda, \tag{9.51}$$

where the integral is over the entire spectral band of the light source. Again, we see that the output irradiance is essentially obtained by incoherent superposition of the entire spectral band image irradiances, by which the coherent artifact noise can be avoided. Since one can utilize a conventional white light source, the processor can indeed be used to process polychromatic images. The advantages of exploiting the incoherent source for coherent processing are that it enables the information to be processed in complex amplitude as a coherent processor and it is capable of suppressing the coherent artifact noise as an incoherent processor.

9.2 Image Processing

9.2.1 Coherent Processing

In the preceding sections, we have seen that by simple insertion of spatial filters, a wide variety of image processing can be performed by coherent processors (i.e., VLC or JTC). In this section, we shall describe a couple of image processings by coherent light.

One of the interesting applications of coherent optical image processing is the restoration of blurred photographic images. The Fourier spectrum of a blurred (or distorted) image can be written as

$$G(p) = S(p)D(p), \tag{9.52}$$

where $G(p)$ is the distorted image, $S(p)$ is the non-distorted image, $D(p)$ is the distorting function, and p is the angular spatial frequency. Then, the inverse filter transfer function for the restoration is

$$H(p) = \frac{1}{D(p)}. \tag{9.53}$$

However, the inverse filter function is generally not physically realizable. If we would accept some restoration errors, then an approximated restoration filter may be physically realized. For example, let the transmission function of a linear smeared point image be

$$f(\xi) = \begin{cases} 1, & -1/2\Delta\xi \leq \xi \leq 1/2\Delta\xi, \\ 0, & \text{otherwise} \end{cases} \tag{9.54}$$

where $\Delta\xi$ is the smear length. If the preceding smeared transparency is displayed on the input SLM1 of the VLC shown in Figure 9.5, the complex light field at the Fourier plane is given by

$$F(p) = \Delta\xi \frac{\sin(p\Delta\xi/2)}{p\Delta\xi/2}, \tag{9.55}$$

which is plotted in Figure 9.11. In principle, an inverse filter as given by

$$H(p) = \frac{p\Delta\xi/2}{\sin(p\Delta\xi/2)} \qquad (9.56)$$

should be used for the image restoration. However, it is trivial to see that it is not physically realizable, since it has infinite poles. If one is willing to sacrifice some degree of restoration, an approximate filter may be realized as follows:

An approximated inverse filter can be physically constructed by combining an amplitude filter with a phase filter, as shown in Figures 9.12 and 9.13, respectively, by which the combined transfer function is given by

$$H(p) = A(p)e^{i\phi(p)}. \qquad (9.57)$$

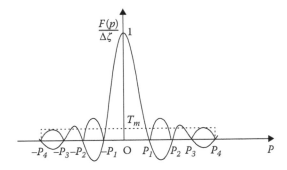

FIGURE 9.11 The solid curve represents the Fourier spectrum of a linear smeared point image. The shaded area represents the restored Fourier spectrum.

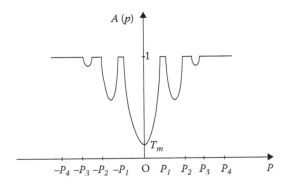

FIGURE 9.12 Amplitude filter.

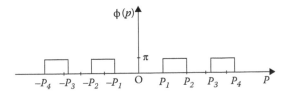

FIGURE 9.13 Phase filter.

Thus, by displaying the preceding inverse filter in SLM2, the restored Fourier spectrum is given as given by

$$F_1(p) = F(p)H(p). \tag{9.58}$$

By denoting T_m as the minimum transmittance of the amplitude filter, then the restored Fourier spectrum is the shaded area of Figure 9.11. Let us define the degree of image restoration as given by

$$\vartheta(T_m)(\%) = \frac{1}{T_m \Delta p} \int_{\Delta p} \frac{F(p)H(p)}{\Delta \xi} dp \times 100, \tag{9.59}$$

where Δp is the spectral bandwidth of restoration. A plot as a function of T_m is shown in Figure 9.14. We see that high degree of restoration can be achieved as T_m approaches zero. However, at the same time, the restored Fourier spectrum is also vanishing small, for which no restored image can be observed. Although the inverse filter in principle can be computer generated, we will use a holographic phase filter for the demonstration, as given by

$$T(p) = \tfrac{1}{2}\{1 + \cos[\phi(p) + \alpha_0 p]\}, \tag{9.60}$$

where α_0 is an arbitrarily chosen constant, and

$$\phi(p) = \begin{cases} \pi, & p_n \le p \le p_{n+1}, \quad n = \pm 1, \pm 3, \pm 5, \dots \\ 0, & \text{otherwise} \end{cases} \tag{9.61}$$

Thus, by combining with the amplitude filter, we have

$$H_1(p) = A(p)T(p) = \tfrac{1}{2}A(p) + \tfrac{1}{4}[H(p)\exp(i\alpha_0 p) + H^*(p)\exp(-i\alpha_0 p)]. \tag{9.62}$$

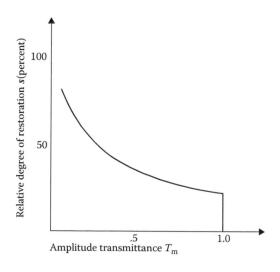

FIGURE 9.14   Relative degree of restoration as a function of T_m.

If this complex filter $H_1(p)$ is inserted in the Fourier domain, then the transmitted Fourier spectrum would be

$$F_2(p) = \tfrac{1}{2}F(p)A(p) + \tfrac{1}{4}[F(p)H(p)\exp(i\alpha_0 p) + F(p)H^*(p)\exp(-i\alpha_0 p)], \tag{9.63}$$

in which we see that the second and third terms are the restored Fourier spectra, which will be diffracted around $\alpha = \alpha_0$ and $\alpha = -\alpha_0$, respectively, at the output plane. It is interesting to show the effects of restoration due to amplitude filter alone, phase filter alone, and the combination of both, as plotted in Figure 9.15. We see that by phase filtering alone, it offers a significant effect of restoration as compared with the one using the amplitude filter. To conclude this section, a restored image using this technique is shown in Figure 9.16. We see that the smeared image can indeed be restored with a coherent processor, but the stored image is also contaminated with coherent noise, as shown in Figure 9.16b.

Another coherent image processing we shall demonstrate is image subtraction. Image subtraction can be achieved by using a diffraction grating technique, as shown in Figure 9.17. By displaying two input images on the input SLM1,

$$f_1(x-h, y) + f_2(x+h, y). \tag{9.64}$$

The corresponding Fourier spectrum can be written as

$$F(p,q) = F_1(p,q)e^{-ihp} + F_2(p,q)e^{ihp}, \tag{9.65}$$

where $2h$ is the main separation between the two input images. If a bipolar grating, given by

$$H(p) = \sin hp, \tag{9.66}$$

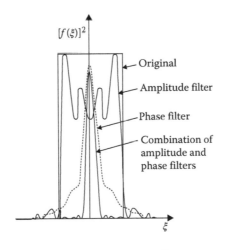

FIGURE 9.15 Restoration due to amplitude, phase, and complex filters.

(a)　　　(b)

FIGURE 9.16 Image restoration: (a) smeared image, (b) restored image with coherent light.

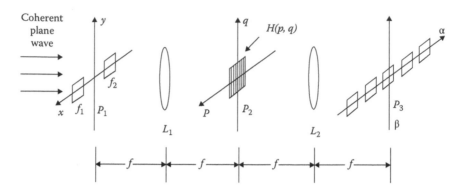

FIGURE 9.17   Optical setup for image subtraction.

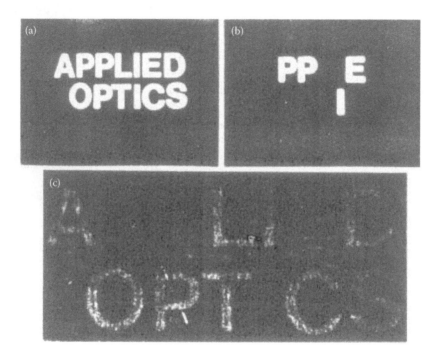

FIGURE 9.18   (a) Input images. (b) Subtracted image with coherent light.

is displayed on SLM2 at the Fourier domain, it can be shown that the output light-intensity distribution is given by

$$I(x,y) = \int f_1(x-2h,y)|^2 + |f_1(x,y) - f_2(x,y)|^2 + |f_1(x+2h,y)|^2, \qquad (9.67)$$

in which we see that the subtracted image (i.e., $f_1 - f_2$) is diffracted around the origin at the output plane. Figure 9.18 shows an experimental result obtained with the coherent processor. Again, we see that the subtracted image is contaminated by severe coherent artifact noise.

Mention must be made that JTC can also be used as an image processor. Instead of using the Fourier domain filter, JTC uses the spatial domain filter. For example, an image $f(x, y)$ and a spatial filter domain $h(x, y)$ are displayed at this input plane of the JTC shown in Figure 9.2, as given by

$$f(x-h,y) + h(-x+h,-y). \qquad (9.68)$$

It can be shown that output complex light distribution is

$$g(\alpha,\beta) = f(x,y) \otimes f(x,y) + h(-x,-y) \otimes (-x,-y) + f(x,y) * h(x-2h,y)$$
$$+ f(x,y) * h(x+2h,y), \tag{9.69}$$

in which two processed images (i.e., the convolution terms) are diffracted around $x = \pm 2h$. Thus, we see that JTC can indeed be used as an image processor, for which it uses a spatial domain filter.

9.2.2 Processing with Incoherent Light

In Section 9.1.2, we have shown that the incoherent processor is only capable of processing the image in terms of intensity and it is, however, not possible to process the information in complex amplitude. It is for this reason that coherent processors are more attractive for optical signal processing. Nevertheless, we have shown that the complex amplitude processing can be exploited from an incoherent source, as described in Section 9.1.4. In this section, we demonstrate a couple of examples that complex amplitude image processing can indeed be exploited from an incoherent light processor.

Let us now consider the image deblurring under the incoherent illumination. Since smeared image deblurring is a 1D processing operation, inverse filtering takes place with respect to the smeared length of the blurred object. Thus, the required spatial coherence depends on the smeared length instead of the entire input plane. If we assume that a spatial coherence function is given by

$$\Gamma(x_2 - x_2') = \sin c \left\{ \frac{\pi}{\Delta x_2}(x_2 - x_2') \right\}, \tag{9.70}$$

then the source-encoding function can be shown as

$$\gamma(x_1) = \text{rect}\left(\frac{x_1}{w} \right), \tag{9.71}$$

where Δx_2 is the smeared length, $w = (f\lambda)/(\Delta x_2)$ is the slit width of the encoding aperture as shown in Figure 9.19a, and

$$\text{rect}\left(\frac{x_1}{w} \right) = \begin{cases} 1, & -\dfrac{w}{3} \leq x_1 \leq \dfrac{w}{2} \\ 0, & \text{otherwise} . \end{cases} \tag{9.72}$$

As for the temporal coherence requirement, a sampling phase grating is used to disperse the Fourier spectrum in the Fourier plane. Let us consider the temporal coherence requirement for a 2D image in the Fourier domain. A high degree of temporal coherence can be achieved by using a higher sampling frequency. We assume that the Fourier spectrum dispersion is along the x-axis. Since the smeared image deblurring is a 1D processing, a *fan-shape* broadband spatial filter to accommodate the smeared Fourier spectrum can be utilized. Therefore, the sampling frequency of the input phase grating can be determined by

$$p_0 \geq \frac{4\lambda p_m}{\Delta \lambda}, \tag{9.73}$$

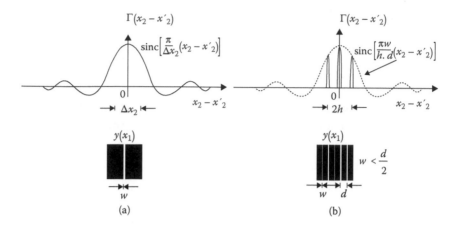

FIGURE 9.19 Source encoding and spatial coherence Γ, spatial coherence function γ, and source-encoding mask (a) for image deblurring and (b) for image subtraction.

(a)   (b)

FIGURE 9.20 (a) Smeared image. (b) Restored image with incoherent light.

in which λ and $\Delta\lambda$ are the center wavelength and the spectral bandwidth of the light source, respectively, and p_m is the x-axis spatial frequency limit of the blurred image.

Figure 9.20a shows a set of blurred letters (OPTICS) due to linear motion. By inserting this blurred transparency in an incoherent source processor of Figure 9.9, a set of deblurred letters is obtained, as shown in Figure 9.20b. Thus, we see that by properly exploiting the coherence contents, complex amplitude processing can indeed be obtained from an incoherent source. Since the deblurred image is obtained by incoherent integration (or superposition) of the broadband source, the coherent artifact can be suppressed. As compared with the one obtained with coherent illumination of Figure 9.16b, we see that the coherence noise has been substantially reduced, as shown in Figure 9.20b.

Let us now consider an image subtraction processing with incoherent light. Since the spatial coherence depends on the corresponding point-pair of the images to be subtracted, a strictly broad spatial coherence function is not required. Instead, a point-pair spatial coherence function is needed. To ensure the physical reliability of the source-encoding function, we let the point-pair spatial coherence function be given by

$$\Gamma\left(x_2 - x_2'\right) = \frac{\sin\left[N\left(\pi/h\right)\left(x_2 - x_2'\right)\right]}{N\sin\left[\left(\pi/h\right)\right]\left(x_2 - x_2'\right)}\,\sin c\left[\frac{\pi w}{hd}\left(x_2 - x_2'\right)\right], \tag{9.74}$$

where $2h$ is the main separation of the two input image transparencies. As $N \gg 1$ and $w \ll d$, Equation 9.74 converges to a sequence of narrow pulses located at $\left(x_2 - x_2'\right) = nh$, where n is a positive integer, as shown in Figure 9.19b. Thus, a high degree of coherence between the corresponding point-pair can be obtained. By Fourier transforming Equation 9.74, the source-encoding function can be shown as

$$\gamma(x_1) = \sum_{n=1}^{N} \text{rect}\left(\frac{x_1 - nd}{w}\right), \tag{9.75}$$

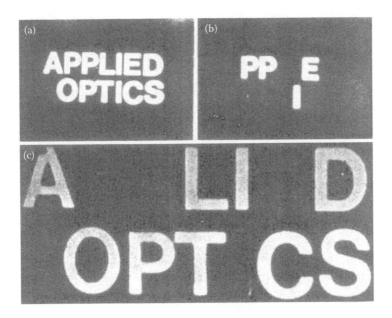

FIGURE 9.21 (a) Input images. (b) Subtracted image with incoherent light.

where w is the slit width, and $d = (\lambda f)/h$ is the separation between the slits. By plotting the preceding equation, the source-encoding mask is represented by N equally spaced narrow slits, as shown in Figure 9.19b.

Since the image subtraction is a 1D processing operation, the spatial filter should be a *fan-shaped* broadband sinusoidal grating, as given by

$$G = \frac{1}{2}\left[1 + \sin\left(\frac{2\pi xh}{\lambda f}\right)\right]. \tag{9.76}$$

Figure 9.21a shows two input image transparencies inserted at the input plane of the incoherent source processor of Figure 9.9. The output subtracted image obtained is shown in Figure 9.21b in which we see that the coherent artifact noise is suppressed.

To conclude this section we note that the broadband (white-light) source contains all the visible wavelengths; the aforementioned incoherent source processor can be utilized to process color (or polychromatic) images.

9.3 Neural Networks

Electronic computers can solve some classes of computational problems thousands of times faster and more accurately than the human brain. However, for cognitive tasks, such as pattern recognition, understanding and speaking a language, and so on, the human brain is much more efficient. In fact, these tasks are still beyond the reach of modern electronic computers.

A neural network consists of a collection of processing elements, called neurons. Each neuron has many input signals, but only one output signal, which is fanned out to many pathways, connected to other neurons. These pathways interconnect with other neurons to form a network. The operation of a neuron is determined by a transfer function that defines the neuron's output as a function of the input signals. Every connection entering a neuron has an adaptive coefficient called a weight assigned to it. The weight determines the interconnection strength between neurons, and they can be changed

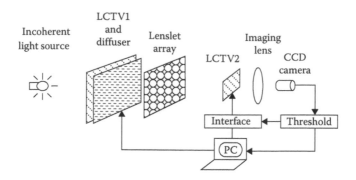

FIGURE 9.22 An optical hybrid neural network. CCD, charge-coupled detectors; LCTVs, liquid-crystal televisions.

through a learning rule that modifies the weights in response to input signals and the transfer function. The learning rule allows the response of the neuron to change with time, depending on the nature of the input signals. This means that the network adapts itself to the environment and organizes the information within itself, which is a type of learning.

9.3.1 Optical Neural Net Architectures

Generally speaking, a one-layer neural network of N neurons has N^2 interconnections. The transfer function of a neuron can be described by a nonlinear relationship, such as a step function, making the output of a neuron either zero or one (binary), or a sigmoid function, which gives rise to analog values. The state of the ith neuron in the network can be represented by a *retrieval equation*, given by

$$u_i = f\left\{ \sum_{j=1}^{N} T_{ij} u_j - \theta_i \right\}, \tag{9.77}$$

where u_i is the activation potential of the ith neuron, T_{ij} is the *interconnection weight* matrix (IWM) (associative memory) between the jth neuron and the jth neuron, θ_i is a phase bias, and f is a nonlinear processing operator. In view of the summation within the retrieval equation, it is essentially a matrix-vector outer-product operation, which can be optically implemented.

Light beams propagating in space will not interfere with each other and optical systems generally have large space-bandwidth products. These are the traits of optics that prompted the optical implementation of neural networks (NNs). An optical NN using a liquid-crystal television (LCTV) SLM is shown in Figure 9.22, in which the lenslet array is used for the interconnection between the IWM and the input pattern. The transmitted light field after LCTV2 is collected by an imaging lens, focusing at the lenslet array and imaging onto a CCD array detector. The array of detected signals is sent to a thresholding circuit and the final pattern can be viewed at the monitor, and it can be sent back for the next iteration. The data flow is primarily controlled by the microcomputer, by which the LCTV-based neural net just described is indeed an *adaptive* optical NN.

9.3.2 Hopfield Neural Network

One of the most frequently used neural network models is the Hopfield model, which allows the desired output pattern to be retrieved from a distorted or partial input pattern. The model utilizes an associative memory retrieval process equivalent to an iterative thresholded matrix–vector outer product expression, as given by

$$V_i \rightarrow 1 \quad \text{if} \qquad \sum_{j=1}^{N} T_{ij}V_j \quad \geq 0,$$
$$V_i \rightarrow 1 \qquad\qquad\qquad\qquad\qquad < 0, \tag{9.78}$$

where V_i and V_j are binary output and binary input patterns, respectively, and the associative memory operation is written as

$$T_{ij} = \begin{cases} \sum_{m=1}^{N} \left(2V_i^m - 1 \right)\left(2V_j^m - 1 \right), & \text{for } i \neq j \\ 0, & \text{for } i = j \end{cases} \tag{9.79}$$

where V_i^m and V_j^m are ith and jth elements of the mth binary vector.

The Hopfield model depends on the outer-product operation for construction of the associated memory, which severely limits storage capacity, and often causes failure in retrieving similar patterns. To overcome these shortcomings, neural network models, such as back propagation, orthogonal projection, multilevel recognition, interpattern association, moment invariants, and others have been used. One of the important aspects of neural computing must be the ability to retrieve distorted and partial inputs. To illustrate partial input retrieval, a set of letters shown in Figure 9.23a were stored in a Hopfield neural network. The positive and negative parts of the memory matrix are given in Figure 9.23b and c, respectively. If a partial image of A is presented to the Hopfield net, a reconstructed image of A converges by iteration, and is shown in Figure 9.23d. Thus, we see that the Hopfield neural network can indeed retrieve partial patterns.

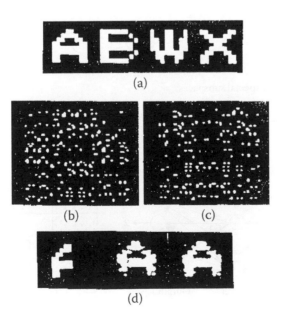

FIGURE 9.23 Results from a Hopfield model: (a) training set; (b and c) positive and negative IWMs, respectively; and (d) retrieved image.

9.3.3 Interpattern Association Neural Network

Although the Hopfield neural network can retrieve erroneous or partial patterns, the construction of the Hopfield neural network is through intrapattern association, which ignores the association among the stored exemplars. In other words, Hopfield would have a limited storage capacity and it is not effective or even incapable of retrieving similar patterns. One of the alternative approaches is called the *interpattern association* (IPA) neural network. By using simple logic operations, an IPA neural network can be constructed. For example, consider three overlapping patterns given in the Venn diagram shown in Figure 9.24, where the common and the special subspaces are defined. If one uses the following logic operations, then an IPA neural net can be constructed:

$$
\begin{aligned}
I &= A \wedge (\overline{B \vee C}), \\
II &= B \wedge (\overline{A \vee C}), & V &= (B \wedge C) \wedge \overline{A}, \\
III &= C \wedge (\overline{A \vee B}), & VI &= (C \wedge A) \wedge \overline{B}, \\
IV &= (A \wedge B) \wedge \overline{C}, & VII &= (A \wedge B \wedge C) \wedge \overline{\phi}.
\end{aligned}
\tag{9.80}
$$

If the interconnection weights are assigned 1, –1, and 0, for excitory, inhibitory, and null interconnections, then a tristate IPA neural net can be constructed. For instance, in Figure 9.25a, pixel 1 is the common pixel among patterns A, B, and C, pixel 2 is the common pixel between A and B, pixel 3 is the common pixel between A and C, whereas pixel 4 is the special pixel, which is also an exclusive pixel with

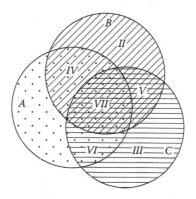

FIGURE 9.24 Common and special subspaces.

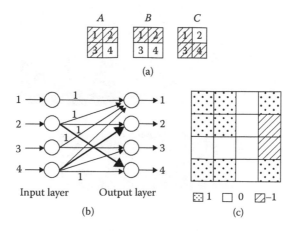

FIGURE 9.25 Construction of IPA neural network: (a) three reference patterns; (b) one-layer neural net; and (c) IWM.

respect to pixel 2. Applying the preceding logic operations, a tri-state *neural network* can be constructed as shown in Figure 9.25b, and the corresponding IPA IWM is shown in Figure 9.25c.

By using the letters B, P, and R as the training set for constructing the IPA IWM shown in Figure 9.26a, the positive and negative parts of the IWM are shown in Figure 9.26b and c. If a noisy pattern of B, (SNR =7dB), is presented to the IPA neural network, a retrieved pattern of B is obtained, as shown in Figure 9.26e. Although the stored examples B, P, and R are very similar, the retrieved pattern can indeed be extracted by using the IPA model.

For comparison of the IPA and the Hopfield models, we have used an 8 ×8 neuron optical NN for the tests. The training set is the 26 letters lined up in sequence based on their similarities. Figure 9.27 shows the error rate as a function of the number of stored letters. In view of this plot, we see that the Hopfield model becomes unstable to about four patterns, whereas the IPA model is quite stable even for 10% input

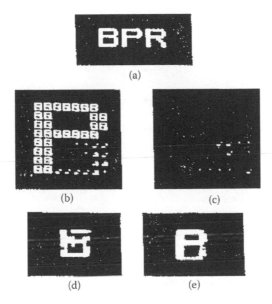

FIGURE 9.26 IPA neural network: (a) input training set; (b and c) positive and negative IWMs; (d) noisy input SNR = 7dB; and (e) retrieved image.

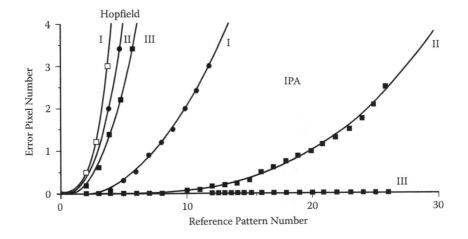

FIGURE 9.27 Comparison of the IPA and the Hopfield models. Note I, 10% noise level; II, 5% noise level; III, no noise.

noise, which can retrieve 12 stored letters effectively. As for the noiseless input, the IPA model can in fact produce correct results for all 26 stored letters.

Pattern translation can be accomplished using the hetero-association IPA. Using similar logic operations among input–output (translation) patterns, a hetero-associative IWM can be constructed. For example, an input–output (translation) training set is given in Figure 9.28a. Using the logic operations, a hetero-association neural net can be constructed, as shown in Figure 9.28b, while Figure 9.28c is its IWM. To illustrate the optical implementation, an input–output training set is shown in Figure 9.29a. The positive and negative parts of the hetero-association IWMs are depicted in Figure 9.29b. If a partial

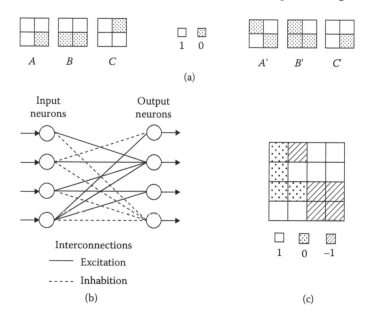

FIGURE 9.28 Construction of a hetero-association IPA neural network: (a) input-output training sets; (b) a hetero-association neural net; and (c) hetero-association IWM.

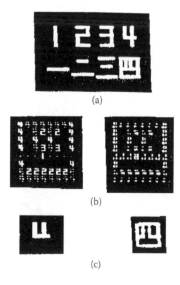

FIGURE 9.29 Pattern translation: (a) Arabic-Chinese training set; (b) hetero-association IWM (positive and negative parts); and (c) partial Arabic numeral to the translated Chinese numeral.

Arabic numeral 4 is presented to the optical neural net, a translated Chinese numeral is obtained, as shown in Figure 9.29c. Thus, the hetero-association neural net can indeed translate Arabic numerals into Chinese numerals.

9.4 Wavelet Transform Processing

A major question concerning this section may be asked: What is a wavelet? Why is it interesting for solving signal-processing problems? These are crucial remarks that a signal analyst would want to know. The answer to these questions may be summarized as: wavelet representation is a versatile technique having, very much, physical and mathematical insights with great potential applications to signal and image processing. In other words, wavelets can be viewed as a new basis signal and images representation, which can be used for signal analysis and image synthesis.

9.4.1 Wavelet Transform

For nonstationary signal processing, the natural way to obtain joint time–frequency resolution of a signal is to take the Fourier transform of the signal within a time window function. This transform is known as *short-time Fourier transform* (STFT), where the size of the window is assumed invariant. However, if the size of the window changes as the analyzing frequency changes, then the transform is known as a *wavelet transform* (WT). The expression of the STFT can be written as

$$\text{STFT}(\tau,\omega) = \int x(t)h^*(t-\tau)\exp(-i\omega t)\,dt, \tag{9.81}$$

where $h(t)$ is an analyzing window function, ω is the analyzing frequency, and τ is an arbitrary time shift. Notice that if $h(t)$ is a Gaussian function, then the transform is also known as the *Gabor transform*. The STFT has been widely used in signal processing, such as time-varying signal analysis and filtering, spectral estimation, signal compression, and others. Usually, the STFT offers very good performances for signals having uniform energy distribution within an analyzing window. Thus, the selection of the analyzing window size is critically important for achieving an optimum joint time–frequency resolution. However, the apparent drawback of STFT must be the invariant size of the analyzing window. To overcome this shortcoming, the WT can be used:

$$\text{WT}(\tau,a) = \frac{1}{\sqrt{a}} \int x(t)\psi^*\left(\frac{t-\tau}{a}\right) dt, \tag{9.82}$$

where a is a scaling factor and $\psi(t)$ is called the *mother wavelet*. We note that the shape of $\psi(t)$ shrinks as the scaling factor a decreases, while it dilates as a increases. The shrunken and dilated wavelets are also known as the *daughter wavelets*. Thus, to have a better time resolution, a narrower WT window should be used for higher frequency content. In principle, the WT suffers from the same time–frequency resolution limitation as the STFT; that is, the time resolution and the frequency resolution cannot be resolved simultaneously, as imposed by the following inequity:

$$\Delta t\Delta\omega \geq 2\pi, \tag{9.83}$$

where Δt and $\Delta\omega$ are defined as

$$\Delta t = \frac{\int |h(t)|\,dt}{|h(0)|}, \text{ and } \Delta\omega = \frac{\int |H(\omega)|\,d\omega}{|H(0)|}. \tag{9.84}$$

Since window functions having a smaller resolution cell are preferred, the Gaussian window function is the best in the sense of meeting the lower bound of the inequality. However, the Gaussian function lacks either the biorthogonality or the orthogonality, which is the constraint of window functions for perfect signal (or image) reconstruction. We note that perfect reconstruction is one of the objectives for using the STFT and the WT: for example, as applied to nonlinear filtering, image compression, and image synthesis.

Let us begin with the basic definitions of the semi-continuous WT, which is given by

$$\mathrm{WT}(\tau,n) = a_0^n \int X(t)\psi^*\left(a_0^n(t-\tau)\right)\mathrm{d}tc, \tag{9.85}$$

and its Fourier domain representation is written as

$$\mathrm{WT}(\tau,n) = \frac{1}{2\pi}\int X(\omega)\psi^*(a_0^{-n}\omega)\exp(i\omega\tau)\mathrm{d}\omega, \tag{9.86}$$

where $a_0 > 1$ is a scaling factor (i.e., $a = a_0^{-n}$), n is an integer, $\psi(t)$ is the mother wavelet, and $\psi(\omega)$ is its Fourier transform. Equation 9.85 is somewhat different from Equation 9.82, where the normalization factor $1/\sqrt{a}$ (i.e., $a_0^{n/2}$) is used instead of a_0^n (i.e., $1/a$). We note that this modification simplifies the optical implementation of the WT, as will be shown later. Similar to the STFT, the WT can be regarded as a multi-filter system in the Fourier domain by which a signal can be decomposed into different spectral bands. Although WT uses narrower and wider band filters to analyze the lower and the higher frequency components, the operation is essentially similar to that of STFT.

To meet the admissibility of WT, $\psi(t)$ has to be a bandpass filter; however, for signals having rich low-frequency contents, a scaling function $\varphi(t)$ to preserve the low-frequency spectrum is needed. The scaling transform of the signal is defined as

$$\mathrm{ST}(\tau) = \int x(t)\phi^*(t-\tau)\mathrm{d}t. \tag{9.87}$$

Thus, the inverse operation of the WT can be written as

$$x(t) = \int \mathrm{ST}(\tau)s_\phi(t-\tau)\mathrm{d}\tau + \sum_{n=-\infty}^{\infty}\int \mathrm{WT}(\tau,n)s_\psi(a_0^n(t-\tau))a_0^n\mathrm{d}\tau, \tag{9.88}$$

and its Fourier domain representation can be shown as

$$x(t) = \frac{1}{2\pi}\int F\{\mathrm{ST}(\tau)\}S_\psi(\omega)\exp(i\omega t)\mathrm{d}\omega + \frac{1}{2\pi}\sum_{n=-\infty}^{\infty}\int F\{\mathrm{WT}(\tau,n)\}$$

$$\tau S_\psi(a_0^{-n}\omega)\exp(i\omega t)\mathrm{d}\omega, \tag{9.89}$$

where $s_\phi(t)$ and $s_\psi(t)$ are the synthesis scaling function and wavelet function, respectively, and $s_\phi(\omega)$ and $s_\psi(\omega)$ are the corresponding Fourier transforms.

If the WT is used for signal or image synthesis, for a perfect reconstruction $\varphi(t)$ and $\psi(t)$ must satisfy the following biorthogonality and orthogonality constraints:

$$\Phi^*(\omega) + \sum_{n=-\infty}^{+\infty}\psi^*(a_0^{-n}\omega) = C, \quad \text{for } s_\phi(t) = \delta(t), \text{ and } s_\psi(t) = \delta(t), \tag{9.90}$$

$$\left|\Phi*(\omega)\right|^2 + \sum_{n=-\infty}^{+\infty}\left|\psi*(a_0^{-n}\omega)\right|^2 = C, \quad \text{for } s_\phi(t)=\phi(t), \text{ and } s_\psi(t)=\psi(t). \tag{9.91}$$

Similar to Short-time Fourier-transform (STFT), that satisfies the biorthogonality constraints is given by

$$\psi(\omega)=\begin{cases} 0, & \omega<\omega_0, \\[2mm] \sin^2\left[\dfrac{\pi}{2}v\dfrac{\omega-\omega_0}{\omega_0(a_0-1)}\right] & \omega_0\leq\omega\leq a_0\omega_0, \\[3mm] \cos^2\left[\dfrac{\pi}{2}v\dfrac{\omega-a_0\omega_0}{\omega_0 a_0(a_0-1)}\right] & a_0\omega_0\leq\omega\leq a_0^2\omega_0, \\[3mm] 0, & \omega>a_0^2\omega_0, \end{cases} \tag{9.92}$$

for which the scaling function F(ω) can be shown as

$$\Phi(\omega)=\cos^2\left[\frac{\pi}{2}v\frac{|\omega|}{(a_0+1)\omega_0/2}\right], \quad |\omega|\leq(a_0+1)\omega_0/2, \tag{9.93}$$

where the function $v(\cdot)$ has the same definition as the STFT shown in Figure 9.30. Thus, we see that $\psi(\omega)$ forms a biorthogonal wavelet, and the squared-root $\sqrt{\psi(\omega)}$ is known as the *Mayer's wavelet*, which is orthogonal in terms of the constraint of Equation 9.92.

Figure 9.30 shows a set of Fourier domain wavelets, given by

$$\psi_1(\omega)=\begin{cases} 1, & 1.5\leq\omega\leq 3, \\ 0, & \text{other} \end{cases} \tag{9.94}$$

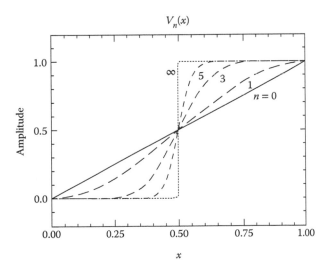

FIGURE 9.30 Function v as n increases.

$$\psi_2(\omega) = \begin{cases} \sin[(\omega-1)\pi/2], & 1 \le \omega \le 2 \\ \cos[(\omega-2)\pi/4], & 2 \le \omega \le 4 \\ 0, & \text{others} \end{cases}$$

$$\psi_3(\omega) = \begin{cases} \sin[(\omega-1)\pi/2], & 1 \le \omega \le 2 \\ \cos[(\omega-2)\pi/4], & 2 \le \omega \le 4 \\ 0, & \text{others} \end{cases}$$

$$\psi_4(\omega) = \begin{cases} \exp[-\pi(\omega-2)^2], & \omega \le 2 \\ \exp[-\pi(\omega-2)^2/4], & \omega \ge 2, \end{cases}$$

where we assume $a_0 = 2$. By plotting the real parts, as shown in Figure 9.31, we see that the biorthogonal window $\psi_3(t)$ is an excellent approximation to the Gaussian function $\psi_4(t)$, both in the Fourier and the time domains. Therefore, the wavelet $\psi_3(t)$ has the advantages of having the smaller joint time-frequency resolution and biorthogonality, which simplifies the inversion of the WT. The function $\psi_2(t)$ can be used as an orthogonal wavelet, which has a relatively good joint time-frequency resolution. Nevertheless,

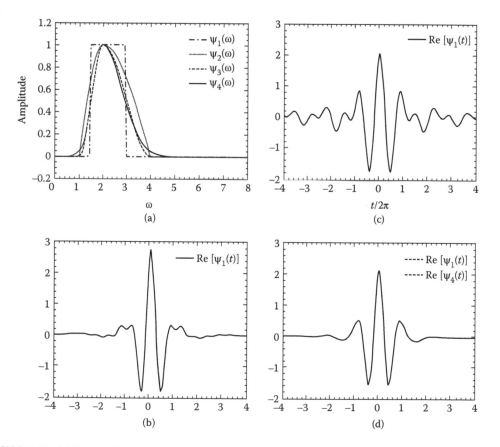

FIGURE 9.31 (a) Fourier domain representations, $\psi_1(\omega)$, $\psi_2(\omega)$, $\psi_3(\omega)$, $\psi_4(\omega)$, of the complex-valued wavelets (b) $\psi_1(t)$, (c) $\psi_2(t)$, (d) $\psi_3(t)$, and $\psi_4(t)$

wavelet $\psi_1(t)$ is often used, since its Fourier transform is a rectangular form, which is rather convenient for the application in Fourier domain processing.

Although our discussions are limited to the wavelets, which have similar forms to the window functions, in practice, the WT offers more solutions than the STFT: namely, one can select the wavelets for specific applications.

9.4.2 Optical Implementations

A couple of possible optical implementations for 1D WT processing are shown in Figure 9.32. The architecture of Figure 9.32a is used for biorthogonal WT, in which we assume that the synthesis function $s(t)$ = $d(t)$. For example, let an input signal be displayed on a spatial light modulator (SLM) at P1 and a set of filter banks are placed at the Fourier plane P2. Then, WT signals can be obtained in the back focal plane P3. Thus, the reconstructed signal can be obtained at P4, by summing all the WT signals diffracted from the filter banks. We notice that real-time processing can be realized by simply inserting an SLM filter at P3. Let us assume that the desired filter is $F(t, n)$; then, the reconstructed signal would be

$$x'(t) = ST(t) + SWT(t,n)F(t,n). \tag{9.95}$$

Figure 9.32b shows the implementation of the orthogonal WT, in which we assume that the orthogonal wavelets $s(t) = \psi(t)$. Notice that the optical configuration is rather similar to that of Figure 9.32a except the inverse operation. By virtue of the reciprocity principle, the inverse operation can be accomplished by placing a phase conjugate mirror (PCM) behind plane P3. The major advantage of the PCM must be the self-alignment, for which the filter alignment can be avoided. As the return phase conjugate signal, $WT^*(t, n)$, arrives at plane P2, it is subjected to the Fourier transformation. By inserting a desired filter at plane P3, we see that real-time processing can indeed be obtained. We note that the filter at the

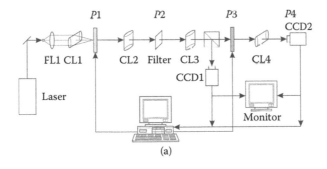

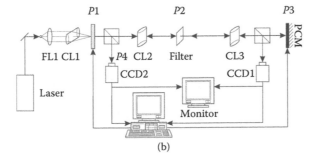

FIGURE 9.32 Optical implementations for WT: (a) for biorthogonal windows, i.e., $s(t) - \delta(t)$; (b) for orthogonal windows, i.e., $S(t) = \psi(t)$.

plane P3 would be proportional to $F(t, n)^{1/2}$, since the signal has gone through the filter twice. Thus, the reconstructed signal at plane P4 would be

$$x_r^*(t) = \int F\{ST^*(\tau)F(\tau)\}\Phi^*(\omega)\exp(i\omega t)\,d\omega$$

$$+ \sum_n \int F\{WT^*(\tau,n)\}_\tau \, \psi^*(a_0^n \omega \exp(i\omega t)\,d\omega), \qquad (9.96)$$

which can be observed with a CCD camera. Although the preceding discussion is limited to 1D signal processing, the basic concepts can be easily extended to 2D image processing.

For comparison, a set of Fourier domain filter banks, for complex-valued STFT and WT, are shown in Figure 9.33a and b, respectively. Note that the filter banks for WT are constant Q-filters, by which the bandwidth varies. We have omitted the filters for negative frequency components since the scalegram from the negative frequency components is the same as that from the positive components. A test signal that includes a chirp and a transient shown at the bottom of Figure 9.33c and d is given by

$$x(t) = \sin\left[\frac{\pi}{256}(t-127)^2\right] + \cos\left[\frac{3\pi}{4}(t-27)\right]\exp\left[\frac{|(t-127)|}{2}\right], \quad 0 \le t < 256. \qquad (9.97)$$

The STFT spectrogram and the WT scalegram are shown at the upper portion of these figures. Although STFT offers a relatively good time-frequency resolution for the chirp signal, it gives rise to a

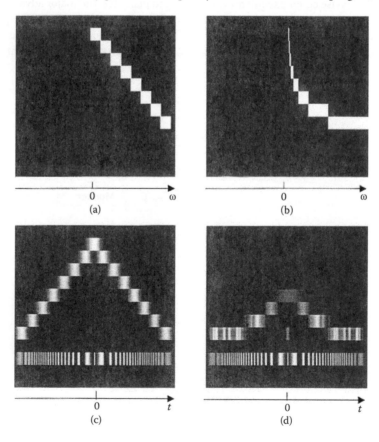

FIGURE 9.33 Computer simulations, for a chirp and transient signal: (a) filter banks for STFT; (b) filter banks for WT; (c) spectrogram; and (d) scalegram.

weaker response for the transient, which is located at the center. On the other hand, the WT provides a higher time resolution for the transient signal, but offers relatively poorer performance for the uniformly distributed chirp signal.

9.5 Computing with Optics

Reaffirmation of optics parallelism and the development of picosecond and femtosecond optical switches have thrust optics into the nontraditional area of digital computing. The motivation primarily arises from the need for higher performance general-purpose computers. However, computers with parallelism require complicated interconnections, which are difficult to achieve by using wires or microcircuits. Since both parallelism and space interconnection are the inherent traits of optics, it is reasonable to look into the development of a general-purpose optical computer. We shall, in this section, restrict ourselves to discussing a few topics where computing can be performed conveniently by optics.

9.5.1 Logic-Based Computing

All optical detectors are sensitive to light intensity: they can be used to represent binary numbers 0 and 1, with dark and bright states. Since 1 (bright) cannot be physically generated from 0s (dark), there are some difficulties that would occur when a logical 1 has to be the output from 0s (e.g., NOR, XNOR, NAND). Nevertheless, the shadow-casting method can solve these problems, by simply initially encoding 1 and 0 in a dual-rail form.

The shadow-casting logic essentially performs all sixteen Boolean logic functions based on the combination of the NOT, AND, and OR. For example, 1 and 0 are encoded with four cells, as shown in Figure 9.34, in which the spatially encoded formats A and B are placed in contact, which is equivalent to an AND operation, as shown in Figure 9.34b. On the other hand, if the uncoded 1 and 0 are represented by transparent and opaque cells, they provide four AND operations, that is, AB, $A\bar{B}$, $\bar{A}B$, and $\overline{AB}$, as shown in Figure 9.34b. This superposed format is the input-coded image of the optical logic array processor, and is set in the input plane. Four spatially distributed light-emitting diodes (LEDs) are employed to illuminate the encoded input. The shadows from each LED will be cast onto an output screen, as shown in Figure 9.35. A decoding mask is needed to extract only the true output. The shadow casting is essentially a selective OR operation among AB, $A\bar{B}$, $\bar{A}B$, and $\overline{AB}$. If the on–off states of the LEDs are denoted by α, β, γ, δ (where on is 1, and off is 0), the shadow-casting output can be expressed as follows:

$$G = \alpha(AB) + \beta(A\bar{B}) + \gamma(\bar{A}B) + \delta(\overline{AB}) \tag{9.98}$$

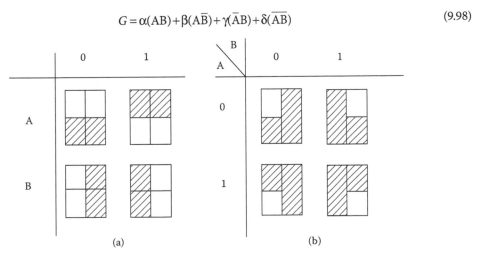

(a) (b)

FIGURE 9.34 (a) Encoded input patterns. (b) Product of the input patterns for the shadow-casting logic array processor.

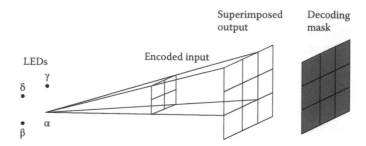

FIGURE 9.35   A shadow-casting logic array processor. The shadows of the encoded input generated by four laser-emitting diodes (LEDs) are superimposed.

TABLE 9.1   Generation of Sixteen Boolean Functions

	α	β	Y	δ
F0	0	0	0	0
F1	1	0	0	0
F2	0	1	0	0
F3	1	1	0	0
F4	0	0	1	0
F5	1	0	1	0
F6	0	1	1	0
F7	1	1	1	0
F8	0	0	0	1
F9	1	0	0	1
F10	0	1	0	1
F11	1	1	0	1
F12	0	0	1	1
F13	1	0	1	1
F14	0	1	1	1
F15	1	1	1	1

which is the intensity at the overlapping cell. The complete combination for generating the 9 Boolean functions is given in Table 9.1. A schematic diagram of a hybrid-optical logic array processor is depicted in Figure 9.36, in which the endoded inputs A and B are displayed on LCTV SLM1 and SLM2, respectively, and the SLM3 is employed to determine the values of α, β, γ, and δ. However, the OR operation is performed electronically rather than optically. This has the advantage that no space is needed for shadow-casting NOR. The problems with this system are (1) the encoding operation would slow down the whole process, (2) if the OR operation is performed in parallel, a large number of electronic OR gates and wire intercommunications are required, and if it is performed sequentially, a longer processing time is required. Since all the optic logic processors use encoded inputs, if the coding is done by electronics, the overall processing speed will be substantially reduced. On the other hand, the optical output is also an encoded pattern that requires a decoding mask to obtain only the true output. Although the decoding process is parallel, and thus takes no significant processing time, the decoding mask does change the format of the output. A non-coded shadow-casting logic array that is free from these difficulties is shown in Figure 9.37. An electronically addressed SLM, such as an LCTV, can be used to write an image format. The negation can be done by rotating the LCTV's analyzer (by 90°) without altering the addressing electronics or software processing. The electronic signal of input A is split into four individual SLMs. Two of them display input A, and the other two display $\overline{A}$. Input B follows a similar procedure to

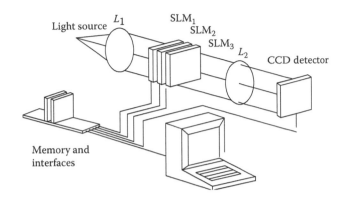

FIGURE 9.36 Shadow-casting logic processor using cascaded spatial light modulators (SLMs).

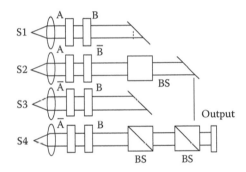

FIGURE 9.37 A noncoded shadow-casting logic array processor: S, light source; BS, beamsplitter; M, mirror.

generate two Bs and two $\bar{B}$s. The products of AB, $A\bar{B}$, $\bar{A}B$, and $\overline{AB}$ are straightforwardly obtained by putting two corresponding SLMs up to each other. Finally, beamsplitters combine AB, $A\bar{B}$, $\bar{A}B$ and $\overline{AB}$. The logic functions are AB, $A\bar{B}$, $\bar{A}B$, and $\overline{AB}$. The logic functions are controlled by the on–off state (α, β, γ, and δ) of the illuminating light (S1, S2, S3, and S4), as illustrated in Table 9.1.

The implementation of these logic functions using LCTVs is straightforward, since no electronic signal modification or software data manipulation is required to encode inputs A and B. Although it seems that more SLMs are needed (four times as many), there is no increase in the space–bandwidth product of the system. In the original shadow-casting logic array, four pixels are used to represent a binary number 1 or 0, while a binary number can be represented by only one pixel in the non-coded system. The use of four sets of individual SLMs is to fully utilize the capability of the LCTV to form a negative image format by simply rotating the analyzer by 90°. This method can eliminate the bottleneck of the shadow-casting optical parallel logic array, which is introduced by the coding process.

9.5.2 Matrix–Vector and Matrix–Matrix Processors

The optical matrix–vector multiplier can be implemented as shown in Figure 9.38. The elements of the vector are entered in parallel by controlling the intensities of N light-emitting diodes (LEDs). Spherical lens L_1 and cylindrical lens L_2 combine to image the LED array horizontally onto the matrix mask M, which consists of $N \times N$ elements. The combination of a cylindrical lens L_3 and a spherical lens L_4 is to collect all the light from a given row and bring it into focus on one detector element that measures the

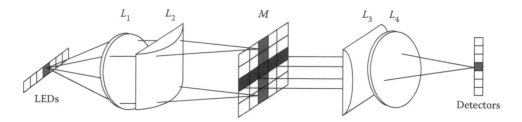

FIGURE 9.38   Schematic diagram of an optical matrix–vector multiplier.

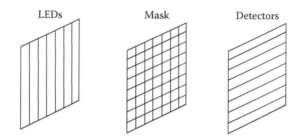

FIGURE 9.39   Optical matrix–vector multiplier using line-shape LEDs and detectors.

value of one output vector element. Thus, we see that it is essentially the matrix–vector multiplication operation. Additionally, this configuration can be further simplified by fabricating line-shape LEDs and a line-shape detector array, as depicted in Figure 9.39. Note that the LED array can be replaced by an SLM with a uniform illumination.

Many important problems can be solved by the iterative multiplication of a vector by a matrix, which includes finding eigenvalues and eigenvectors, solving simultaneous linear equations, computing the discrete Fourier transform, and implementation of neural networks. For example, a neural network consists of several layers of neurons in which two successive neuron layers are connected by an interconnection net. If the neuron structure in a layer is represented by a vector, then the interconnect can be represented by a matrix.

For example, the Hopfield neural network (see Section 9.3.2) uses an associative memory retrieval process, which is essentially a matrix–vector multiplier, as given by

$$
\begin{aligned}
V_i \to 1 \;\; \text{if} \;\; \sum T_{ij} V_i \geq 0, \\
V_i \to 1 \;\; \text{if} \;\; \sum T_{ij} V_i < 0,
\end{aligned}
\tag{9.99}
$$

where V_i and V_j are the output and the input binary vectors, respectively, and T_{ij} is the interconnect matrix.

If the matrix is binary, the matrix–vector multiplier becomes a *crossbar switch*. The crossbar switch is a general switching device that can connect any N inputs to any N outputs; this is called *global interconnect*. Crossbar switches are usually not implemented in electronic computers because they would require N^2 individual switches; however, they are used in telephone exchanges. On the other hand, an optical crossbar interconnected signal processor would be very useful for performing fast-Fourier transforms (FFTs), convolution and correlation operations, by taking advantage of the reconfigurability and parallel processing of crossbar interconnect. Also, the optical crossbar switch can be employed to implement a programmable logic array (PLA). The electronic PLA contains a two-level, AND–OR circuit on a single chip. The number of AND and OR gates and their inputs is fixed for a given PLA. A PLA can be used as a read-only memory (ROM) for the implementation of combinational logic.

The matrix–matrix multiplier is a mathematical extension of the matrix–vector multiplier. In contrast, the implementation of the matrix–matrix multiplier requires a more complex optical arrangement. Matrix–matrix multipliers may be needed to change or process matrices that will eventually be used in a matrix–vector multiplier. Matrix–matrix multiplication can be computed by successive outer-product operations as follows:

$$
\begin{bmatrix} a_{11} & a_{12} & a_{13} \\ a_{21} & a_{22} & a_{23} \\ a_{31} & a_{32} & a_{33} \end{bmatrix} \begin{bmatrix} b_{11} & b_{12} & b_{13} \\ b_{21} & b_{22} & b_{23} \\ b_{31} & b_{32} & b_{33} \end{bmatrix} = \begin{bmatrix} a_{11} \\ a_{21} \\ a_{31} \end{bmatrix} [b_{11}b_{12}b_{13}] + \begin{bmatrix} a_{12} \\ a_{22} \\ a_{32} \end{bmatrix} [b_{21}b_{22}b_{23}]
$$

$$
+ \begin{bmatrix} a_{13} \\ a_{23} \\ a_{33} \end{bmatrix} [b_{31}b_{32}b_{33}].
$$

(9.100)

Since the outer product can be expressed as

$$
\begin{bmatrix} a_{11} \\ a_{21} \\ a_{31} \end{bmatrix} [b_{11}b_{12}b_{13}] = \begin{bmatrix} a_{11}b_{11} & a_{11}b_{12} & a_{11}b_{13} \\ a_{21}b_{11} & a_{21}b_{12} & a_{21}b_{13} \\ a_{31}b_{11} & a_{31}b_{12} & a_{31}b_{13} \end{bmatrix},
$$

(9.101)

it can be obtained by simply butting two SLMs against each other. Equation 9.101 can be realized optically, as shown in Figure 9.40. Each pair of the SLMs performs the multiplication of the outer products, while beamsplitters perform the addition of the outer products.

Thus, we see that the basic operations of matrix–matrix multiplication can be performed by using a pair of SLMs, while the addition is performed by using a beamsplitter. The whole combinational operation can be completed in one cycle. If the addition is performed electronically in a sequential operation, then only a pair of SLMs is required; if the multiplication is performed sequentially, then more than one cycle of operation is needed, and the approach is known as *systolic processing*, and is discussed in Section 9.5.3. The obvious advantage of systolic processing is that fewer SLMs are needed. The trade-off is the increase in processing time.

Figure 9.41 illustrates an example of a systolic array matrix operation. Two transmission-type SLMs are placed close together and in registration at the input plane. By successively shifting the A and B systolic array matrix formats into two SLMs, one can obtain the product of matrices A and B with a time-integrating CCD at the output plane. Although only two SLMs are required, the computation time needed for performing a matrix-matrix multiplication with b bit numbers would be $(2xb-1)+(x-1)$ times that needed by an outer-product processor.

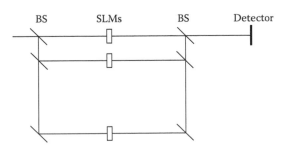

FIGURE 9.40 Matrix–matrix multiplier based on outer product. BS, beamsplitter; SLMs, spatial light modulators.

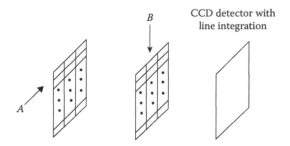

FIGURE 9.41 Matrix–matrix multiplier based on systolic array.

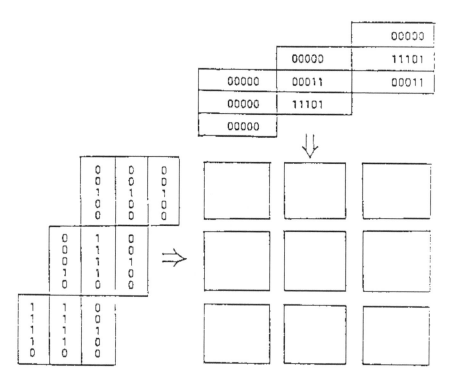

FIGURE 9.42 Matrix–matrix multiplication based on the systolic outer-product method.

Figure 9.42 shows an example of a systolic outer-product processing. The optical processor consists of $n \times n = 3 \times 3 = 9$ pieces of SLM, $b \times b = 5 \times 5$ pixels each. Note that the systolic array representation of matrices A and B differs from the systolic formats described previously. By sequentially shifting the row and column elements of A and B into the SLMs, we can implement the $a_{ij}b_{kl}$ multiplication at each step, with an outer-product operation that has been performed in the 5×5 SLM. The result can be integrated in relation to time with a CCD detector at the output plane. Since more SLMs are employed in parallel, the matrix–matrix multiplication can be completed in fewer steps.

9.5.3 Systolic Processor

The engagement matrix–vector multiplier, in fact a variation of a systolic processor, is illustrated in Figure 9.43. The components of vector B are shifted into multiplier-added modules starting at time t_0.

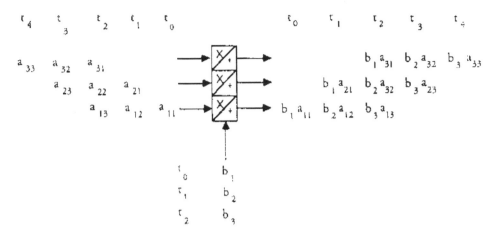

FIGURE 9.43 Conceptual diagram of an engagement systolic array processor.

Subsequent vector components are clocked in contiguously at t_1 for b_2, t_2 for b_3, and so on. At time t_0, b_1 is multiplied with a_{11} in module 1. The resultant $b_1 a_{11}$ is retained within the module to be added to the next product. At time t_1, b_1 is shifted to module 2 to multiply with a_{21}. At the same time, b_2 enters module 1 to multiply with a_{12}, which forms the second product of the output vector component. Consequently, module 1 now contains the sum $b_1 a_{11} + b_2 a_{12}$. This process continues until all the output vector components have been formed. In all, $(2N - 1)$ clock cycles that employ N multiplier-adder modules are required. The main advantage of a systolic processor is that optical matrix–vector multiplication can be performed in the high-accuracy digital mode.

A discrete linear transformation (DLT) system can be characterized by an impulse response. The input–output relationship of such a system can be summarized by the following equation:

$$g_m = \sum f_n h_{mn}. \tag{9.102}$$

Since the output g_m and the input h_n can be considered vectors, the preceding equation can be represented by a matrix–vector multiplication, where h_{mn} is known as a transform matrix. Thus, the different DLTs would have different matrices. The discrete Fourier transform (DFT) is one of the typical examples for DLT, as given by

$$F_m = (1/N) \sum f_n \exp[-i2\pi mn/N], \tag{9.103}$$

where $n = 0, 1, \ldots, N$, and

$$h_{mn} = \exp[-i2\,pmn\,/\,N] \tag{9.104}$$

is also known as the *transform kernel*. To implement the DFT transformation in an optical processor, we present the complex transform matrix with real values, for which the real transform matrices can be written

$$\mathrm{Re}[h_{mn}] = \cos\frac{2\pi mn}{N}, \ \ \mathrm{Im}[h_{mn}] = \sin\frac{2\pi mn}{N}, \tag{9.105}$$

which are the well-known discrete cosine transform (DCT) and the discrete sine transform (DST).

The relationship between the real and imaginary parts of an analytic signal can be described by the Hilbert transform. The discrete Hilbert transform (DHT) matrix can be written as

$$
h_{mn} = \begin{cases} \dfrac{2\sin^2[\pi(m-n)/2]}{\pi(m-n)} & m-n \neq 0, \\[2mm] 0, & m-n = 0. \end{cases} \tag{9.106}
$$

Another frequently used linear transform is the chirp-Z transform, which can be used to compute the DFT coefficients. The discrete chirp-Z transform (DCZT) matrix can be written as

$$
h_{mn} = \exp\frac{i\pi(m-n)^2}{N}. \tag{9.107}
$$

Since the DLT can be viewed as the result of a digital matrix–vector multiplication, systolic processing can be used to implement it. By combining the systolic array processing technique and the *two's complement representation*, a DLT can be performed with a digital optical processor. As compared with the analog optical processor, the technique has high accuracy and a low error rate. Also, it is compatible with other digital processors.

Two's complement representation can be applied to improving the accuracy of matrix multiplication. Two's complement numbers provide a binary representation of both positive and negative values, and facilitate subtraction by the same logic hardware that is used for addition. In a *b-bit* two's complement number, the most significant bit is the sign. The remaining $b - 1$ bits represent its magnitude. An example for DCT is given in Figure 9.44 in which we see that the digital transform matrix is encoded in two's complement representation and arranged in the systolic engagement format.

To conclude this chapter, we remark that it was not our intention to cover the vast domain of Fourier optics. Instead, we have provided some basic principles and concepts. For further reading, we refer the readers to the list of references as cited at the end of this chapter. Nevertheless, several basic applications of Fourier optics are discussed, including coherent and incoherent light image processing, optical neural networks, wavelet transform, and computing with optics.

9.6 Optical Processing for Radar Imaging and Broadband Signal

9.6.1 Radar Imaging with Optics

One of the interesting applications of optical processing must be to the synthetic-aperture antenna data or side-looking antenna imaging. Let us consider a side-looking radar system carried by an aircraft in level flight shown in Figure 9.45, in which we assume a sequence of pulsed radar signals is directed at the terrain and the return signals along the flight path are received. Let the cross-track coordinate of the radar image as ground range coordinate and the along-track coordinate as azimuth coordinate, then the coordinate joining the radar trajectory of the plane of any target under consideration as can be defined as slant range as shown in Figure 9.46. In words, the azimuth resolution will be in the order of $\lambda r_1/D$, where λ is the wavelength of the radar signals, r_1 is the slant range, and D is the along-track dimension of the antenna aperture. Since the radar wavelength is several orders of magnitude larger than the optical wavelength, a very large antenna aperture D is required to have an angular resolution comparable to that of a photoreconnaissance system. The required antenna length may be hundreds, or even thousands of feet and it is impractical to realize on an aircraft. Nevertheless, this difficulty can be resolved by means of the synthetic-aperture technique, as will be shown in the following:

Discrete Cosine Transform (DCT)

$$[h_{m,n}] = \begin{pmatrix} \cos[(2\pi/3)*0*0] & \cos[(2\pi/3)*0*1] & \cos[(2\pi/3)*0*2] \\ \cos[(2\pi/3)*1*0] & \cos[(2\pi/3)*1*1] & \cos[(2\pi/3)*1*2] \\ \cos[(2\pi/3)*2*0] & \cos[(2\pi/3)*2*1] & \cos[(2\pi/3)*2*2] \end{pmatrix}$$

$$= \begin{pmatrix} 1 & 1 & 1 \\ 1 & -1/2 & 1/2 \\ 1 & -1/2 & -1/2 \end{pmatrix} = \begin{pmatrix} 00100 & 00100 & 00100 \\ 00100 & 11110 & 00010 \\ 00100 & 11110 & 11110 \end{pmatrix}$$

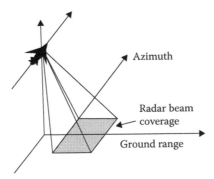

FIGURE 9.44 Transformation matrix for a discrete cosine transform using two-complement representation.

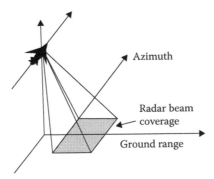

FIGURE 9.45 Side-looking radar.

Let us assume that the aircraft carries a small side-looking antenna and a relatively broad beam radar signal is used to scan the terrain, by virtue of the aircraft motion. The radar pulses are emitted in a sequence of positions along the flight path, which can be treated as if they were the positions occupied by the elements of a linear antenna array. The return radar signal at each of the respective positions can be recorded coherently as a function of time with a local reference signal for which the amplitude and

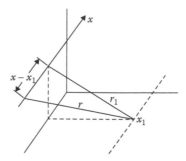

FIGURE 9.46   Geometry of side-looking radar.

the phase quantities are simultaneously encoded. The complex waveforms are then properly combined to simulate a long-aperture element-antenna.

Foe simplicity, we first consider a point-target problem, which then can be quickly extended to multi-target situations owning to spatial invariance condition. Since the radar signal is produced by periodic rectangular pulses modulated with a sinusoidal signal, the pulse periodic gives the range information and the azimuth resolution is provided by the distance traveled by the aircraft between pulses, which is smaller than $\pi/\Delta p$, where Δp is the spatial bandwidth of the terrain reflections. Thus, the returned radar signal can be written as

$$S_1(t) = A_1(x_1, r_1) \exp\left\{ i\left[\omega t - 2k r_1 - \frac{k}{r_1}(x - x_1)^2 \right] \right\}, \tag{9.108}$$

where A_1 is an appropriate complex constant.

Now, if we consider that the terrain at range r_1 consists of a collection of N points targets, then by superposing the returned radar signals, we have

$$
\begin{aligned}
S(t) &= \sum_{n=1}^{N} S_n(t) \\
&= \sum_{n=1}^{N} A_n(x_n, r_1) \exp\left\{ i\left[\omega t - 2k r_1 - \frac{k}{r_1}(vt - x_n)^2 \right] \right\}
\end{aligned}
\tag{9.109}
$$

where v is the velocity of the aircraft. If the returned radar signal is synchronously demodulated, the demodulated signal can be written as

$$S(t) = \sum_{n=1}^{N} |A_n(x_n, r_1)| \cos\left[\omega_c t - 2k r_1 - \frac{k}{r_1}(vt - x_n)^2 + \varphi_n \right], \tag{9.110}$$

where ω_c is the arbitrary carrier frequency and φ_n is the arbitrary phase angle. In order to display the return radar signal of Equation 9.110 on a cathode-ray tube, the demodulated signal is used to modulate the intensity of the electron beam, which is swept vertically across the cathode-ray tube in synchronism with return radar pulses (Figure 9.47). If this modulated cathode-ray display is imaged on a strip of photographic film, which is drawn at a constant horizontal velocity, then the successive range traces will be recorded side by side, producing a two-dimensional format (Figure 9.48).

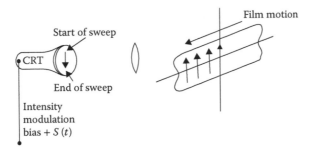

FIGURE 9.47 Recording of the returned radar signal for subsequent optical processing.

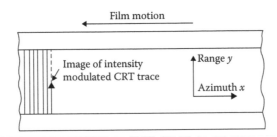

FIGURE 9.48 Radar signal recording format.

The vertical lines represent the successive range sweeps and the horizontal dimension represents the azimuth position, which is the sample version of $s(t)$. This sampling is carried out in such a way that, by the time the samples have been recorded on the film, it is essentially indistinguishable from the un-sampled version. In this recording, the time variable is converted to a space variable as defined in terms of distance along the recorded film. With the proper reading exposure, the transparency of the recorded film represents the azimuth history of the returned radar signals. Thus, considering only the data recorded along the line y-y_1, on the film, the transmittance can be written as

$$T(x_1, y_1) = K_1 + K_2 \sum_{n=1}^{N} |A_n(x_n, r_1)| \cos\left[\omega_x x - 2k r_1 - \frac{k}{r_1}\left(\frac{v}{v_f} x - x_n\right)^2 + \varphi_n\right], \tag{9.111}$$

where K_S are bias constants and $x = v_f t$, v_f is the speed of film motion, and $\omega_f = \omega_x / v_f$.

For simplicity in illustration, we restrict ourselves to the single target problem (i.e., $n = 1$); for which the first exponential term of the cosine function in Equation 9.111 can be written as

$$T_1(x_1, y_1) = C \exp(i\omega_x x) \exp\left[-i\frac{k}{r_1}\left(\frac{v}{v_f}\right)^2\left(x - \frac{v_f}{v} x_j\right)^2\right], \tag{9.112}$$

in which we see that it is essentially a one-dimensional Fresnel lens (or 1D hologram) equation. The first linear phase factor $(\omega_x x)$ represents a prism with an oblique angle of

$$\sin\theta = \frac{\omega_x}{k_1}, \tag{9.113}$$

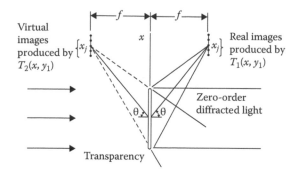

FIGURE 9.49 Image reconstruction produced by the film transparency.

where $k_1 = 2\pi/\lambda_1$, with λ_1 wavelength of the illuminating light source. The second term represents a cylindrical lens centered at

$$x = \frac{v_f}{v} x_j \tag{9.114}$$

with a focal length of

$$f = \frac{1}{2}\frac{\lambda}{\lambda_1}\left(\frac{v_f}{v}\right)^2 r_1. \tag{9.115}$$

Needless to say that for N targets, the corresponding cylindrical lenses would be centered at

$$x = \frac{v_f}{v} x_n, \quad n = 1, 2, \cdots, N. \tag{9.116}$$

In order to reconstruct the radar imagery, the scanned radar transparency is illuminating by a monochromatic plane wave, as depicted in Figure 9.49 in which it is shown that the real and virtual images can be reconstructed relative to the positions of the point images along the foci of the lens-like structure of the recorded transparency, which are determined by the positions of the point targets.

However, the reconstructed images will be spread in the y direction; this is because the transparency is realizing a one-dimensional function along $y = y_i$, and hence exerts no focal power in this direction.

Since it is our aim to reconstruct an image not only in the azimuth direction but also in the range direction, it is necessary to image the y coordinate directly onto the focal plane of the azimuth image. Thus, to construct the terrain map, we must image the y coordinate of the transmitted signal onto a tilted plane determined by the focal distances of the azimuth direction. This imaging procedure can be carried out by inserting a positive conical lens immediately behind the scanned radar image transparency shown in Figure 9.50.

It is clear that if the transmittance of the conical lens is given by

$$T_1(x_1, y_1) = \exp\left(-i\frac{k_1}{2f}x^2\right), \tag{9.117}$$

then it is possible to remove the entire tilted plane of all the virtual diffraction to the point at infinity, while leaving the transmittance in the y direction unchanged. Thus, if the cylindrical lens is placed at

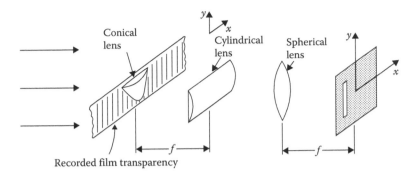

FIGURE 9.50 An optical processing system for imaging synthetic-antenna data.

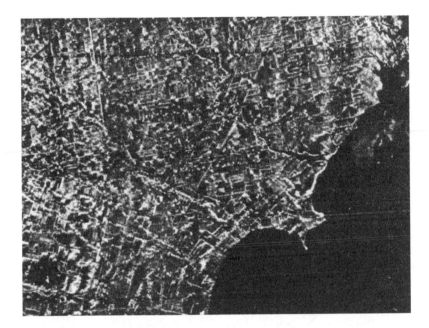

FIGURE 9.51 Synthetic-aperture radar image of the Monroe, Michigan area. (From Cutrona, L. J. et al., *Proc. IEEE*, 54, 1026, 1966. With permission.)

the focal distance from the scanned radar transparency, it will create a virtual image of the *y* direction at infinity. Now, the azimuth and the range images (i.e., x and y direction) are at the point of infinity for which they can be brought back to a finite distance with a spherical lens. Thus, we see that radar imagery can be constructed at the output plane, through a slit, of the optical processing system. An example of the radar imagery is shown in Figure 9.51, in which we see a variety of scatters, including city streets, wooded areas, and farmlands. Lake Erie, with some ice floes, can be seen on the right.

9.6.2 Optical Processing of Broadband Signals

Another interesting application of optical processing is to broadband signals analysis. One of the basic limitations of the multichannel optical analyzer is that the resolution is limited by the width of the input aperture and this limitation can be alleviated by using a two-dimensional raster scan input transparency shown in Figure 9.52.

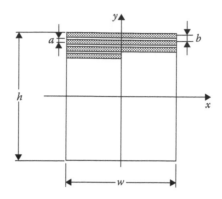

FIGURE 9.52 Raster scan input signal transparency.

Thus, we see that if scanned format transparency is presented to input plane of an optical processor, then the complex light distributed at the output plane would be the Fourier transform of the input scanned format. Thus, a large space-bandwidth signal analysis can be obtained with the scanned format.

Let us assume that a broadband signal is raster scanned on a spatial light modulator (or on a strip of transparency), in which we assumed that the scan rate has been adjusted to accommodate the maximum frequency content of the broadband signal, as represented by

$$v \geq \frac{f_m}{R},$$ (9.118)

where v is the scan velocity, f_m is the maximum frequency content of the broad band signal, and R is the resolution limit of the optical system.

If this format is presented at the input plane of the optical analyzer, then the complex light field distributed at the output plane can be shown by

$$F(p,q) = C_1 \sum_{n=1}^{N} \operatorname{sinc}\left(\frac{qa}{2}\right) \left\{ \operatorname{sinc}\left(\frac{pw}{2}\right) * F(p) \exp\left[i\frac{pw}{2}(2n-1)\right] \right\}$$
$$\times \exp\left\{ -i\frac{q}{2}\left[h - 2(n-1)b - a\right] \right\}$$ (9.119)

where C_1 is a complex constant, and

$$F(p) = \int f(x') \exp(-ipx')dx'$$ (9.120)

$$x' = x + (2n-1)\frac{w}{2}, \text{ and } \operatorname{sinc} X \overset{\Delta}{=} \frac{\sin X}{X}$$

For simplicity, let us assume that the broadband signal is a simple sinusoid, that is,

$$f(x') = \sin p_0 x',$$ (9.121)

then we have

$$F(p,q) = C_1 \text{sinc}\left(\frac{qa}{2}\right) \sum_{n=1}^{N} \left\{ \text{sinc}\left(\frac{pw}{2}\right) * \frac{1}{2}\left[\delta(p-p_0)\right] + \delta(p+p_0) \right.$$

$$\left. \times \exp\left[i\frac{w\,p_0}{2}(2n-1)\right] \right\} \exp\left\{-i\frac{q}{2}\left[w-2(n-1)b\right]\right\} . \tag{9.122}$$

To further simplify the analysis, we consider only one of the components, we have

$$F_1(p,q) = C_1 \text{sinc}\left[\frac{w}{2}(p-p_0)\right] \text{sinc}\left(\frac{qa}{2}\right) \exp\left(-i\frac{qw}{2}\right)$$

$$\times \sum_{n=1}^{N} \exp\left[i\frac{1}{2}(2n-1)(wp_0+bq)\right] . \tag{9.123}$$

The corresponding intensity distribution is given by

$$I_1(p,q) = |F_1(p,q)|^2 = C_1^2 \text{sinc}^2\left[\frac{w}{2}(p-p_0)\right] \text{sinc}^2\left(\frac{qa}{2}\right)$$

$$\times \sum_{n=1}^{N} \exp[i2n\theta] \sum_{n=1}^{N} \exp[-i2n\theta] , \tag{9.124}$$

where

$$\theta = \frac{1}{2}(w\,p_0 + bq) , \tag{9.125}$$

But,

$$\sum_{n=1}^{N} e^{i2n\theta} \sum_{n=1}^{N} e^{-i2n\theta} = \left(\frac{\sin N\theta}{\sin \theta}\right)^2 , \tag{9.126}$$

thus, we have

$$I_1(p,q) = C_1^2 \text{sinc}^2\left[\frac{w}{2}(p-p_0)\right] \cdot \text{sinc}^2\left(\frac{qa}{2}\right) \cdot \left(\frac{\sin N\theta}{\sin \theta}\right)^2 . \tag{9.127}$$

In view of this equation, we see that that the first sinc factor represents a relatively narrow spectral line in the direction p, located at $p = p_0$, which is derived from the Fourier transform of a pure sinusoid truncated within the width of the input transparency. The second sinc factor represents a relatively broad spectral band in the q direction, which is derived from the Fourier transform of a rectangular pulse width (i.e., the channel width). And the last factor deserves special mention for large N (i.e., scanned lines), the factor approaches a sequence of narrow pulses located at

$$q = \frac{1}{b}(2\pi n - w\,p_0), \quad n = 1, 2, \ldots \tag{9.128}$$

Notice that this factor yields the fine spectral resolution in the q direction.

Let us confine within a relatively narrow region in p direction that is the half-width of the spectral spread as given by

$$\Delta p = \frac{2\pi}{w},\qquad(9.129)$$

which is the resolution limit of an ideal transform lens. In the q direction, the irradiance is first confined within a relatively broad spectral band, (primarily depends on the channel width) centered at $q=0$, and it is modulated by a sequence of narrow periodic pulses. The half-width of the broad spectral band is given by

$$\Delta q = \frac{2\pi}{a}.\qquad(9.130)$$

The separation of the narrow pulses is obtained from a similar equation,

$$\Delta q_1 = \frac{2\pi}{b}.\qquad(9.131)$$

It may be seen from the preceding equation that there will be only a few pulses located within the spread of the broad spectral band for each p_0, as shown in Figure 9.53.

Notice that the actual location of any of the pulses is determined by the signal frequency. Thus, if the signal frequency changes, the position of the pulses also changes, in accordance with

$$dq = \frac{w}{b}dp_0.\qquad(9.132)$$

We further note that the displacement in the q direction is proportional to the displacement in the p direction. The output spectrum includes the required frequency discrimination, which is equivalent to analysis of the entire signal in one continuous channel. In order to avoid the ambiguity in reading the output plane, all but one of the periodic pulses should be ignored. This may be accomplished by masking the entire output plane, except the region, shown in Figure 9.54. Thus, we see that, as the input signal frequency advances, one pulse leaves the open region and another pulse enters the region. As a result, a single bright spot would be scanned out diagonally on the output plane. The frequency locus of the input signal can be determined in cycles per second. To remove the non-uniformity of the second sinc factor, a weighting transparency may be placed at the output plane of the processor.

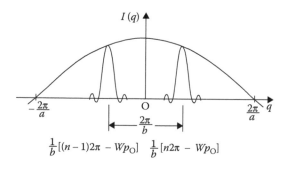

FIGURE 9.53 A broad spectral band modulated by a sequence of narrow pulses.

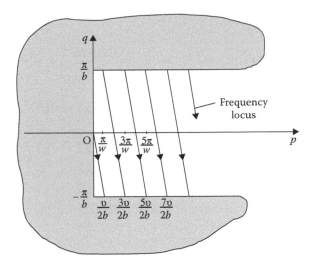

FIGURE 9.54 Frequency loci at the output spectral plane.

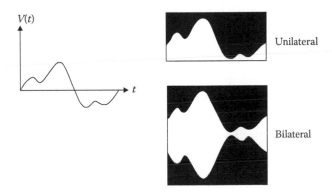

FIGURE 9.55 Unilateral and bilateral area modulation formats.

Although our analysis was carried out on a one-frame basis, the operation can be performed on a continuously running manner, as the input scanned raster format is continuously moving in the q direction.

9.6.3 Broadband Signal Processing with Area Modulation

Area maculation for movie sound track recording had been used for years. And the formats can also be used for spectrum analysis. There are, however, two formats that have been used, namely the unilateral and bilateral shown in Figure 9.55.

Notice that area modulation has the advantages of simplicity implementation and the nonlinearity of the density modulation can be avoided, but there is a trade-off for input spatial limited resolution. Let us now describe the area modulation transmittance functions for unilateral and bilateral respectively as follows:

$$T_1(x, y) = rect\left(\frac{x}{L}\right) rect\left\{\frac{y}{2\left[B + f(x)\right]}\right\} \tag{9.133}$$

and

$$T_2(x,y) = rect\left(\frac{x}{L}\right) rect\left\{\frac{y-[f(x)-B]/2}{[B+f(x)]}\right\},$$

(9.134)

where L is the width of the transmittance, B is the appropriate bias, $f(x)$ is the input function, and

$$rect\left(\frac{x}{L}\right) \overset{\Delta}{=} \begin{cases} 1, & |x| \le L/2 \\ 0, & otherwise \end{cases}.$$

(9.135)

If an area modulated transmittance is presented to an optical processor, the correspondent output light distributions can respectively be shown as

$$\begin{aligned} G_1(p,0) &= 2C\int rect\left(\frac{x}{L}\right)[B+f(x)]\exp(-ipx)\,dx \\ &= 2C\int_{-L/2}^{L/2} B\exp(-ipx)\,dx + 2C\int_{-L/2}^{L/2} f(x)\exp(-ipx)\,dx \end{aligned}$$

(9.136)

and

$$\begin{aligned} G_2(p,0) &= \int T_2(x,y)\exp(-ipx)\,dx \\ &= C\left[\int_{-L/2}^{L/2} B\exp(-ipx)\,dx + \int_{-L/2}^{L/2} f(x)\exp(-ipx)\,dx\right]. \\ &= \frac{1}{2}G_1(p,0) \end{aligned}$$

(9.137)

It is now evident that, if a bilateral sinusoidal area function is implemented as given by

$$f(x) = A\sin(p_0 x + \theta),$$

(9.138)

then the output light distribution can be shown as

$$G_1(p,0) = K1\,\delta(p) + K_2\,\delta(p-p_0) + K_3\,\delta(p+p_0),$$

(9.139)

in which we see that two spectral points located at $p = p_0$ and $p = -p_0$ can be found. In view of this technique, a larger class of recording material can be used, since the recording has the advantage of not restricting to the linear region of the input device. To sum up this section we provide a section of a speech spectrogram generated by the area modulation technique, shown in Figure 9.56.

Needless to say that the area modulation technique can also be applied to broad-band spectrum analysis, using the raster scanned format, as shown in Figure 9.57.

The corresponding input transmittance to the optical processor can be written as

$$f(x,y) = \sum_{n=1}^{N} rect\left(\frac{x}{L}\right) rect\left\{\frac{y-nD}{2[B+f_n(x)]}\right\},$$

(9.140)

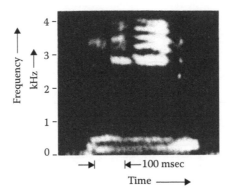

FIGURE 9.56 Speech spectrogram obtained by optical processing with area modulation. (Courtesy of C. C. Aleksoff.)

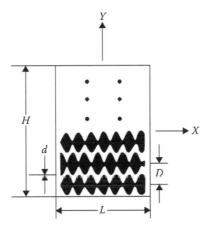

FIGURE 9.57 Geometry of area modulated input for wide-band processing.

where N is the number of scanned lines. Then, by optically performing the Fourier transformation, the output light distribution is given by

$$F(p,q) = \sum_{n=-N}^{n=N} C \int_{-L/2}^{L/2} [B + f_n(x)] \text{sinc}\{q[B + f_n(x)]\} \exp[-i(px - nDq)] dx. \qquad (9.141)$$

If we restrict the observation for a single sinusoid, the corresponding output intensity distribution can be shown as

$$I_1(p,q) = \left[K_1 \text{sinc}\left(\frac{LP}{2}\right) \right]^2 + K_2 \left\{ \text{sinc}\left[\frac{L}{2}(p - p_0) \right] \right\}^2 \cdot \left(\frac{\sin N\theta}{\sin\theta} \right)^2. \qquad (9.142)$$

We see that the result is actually similar to the density modulating case as we have described before. For large value of N (i.e., scanned lines), the last factor would converge to a sequence of narrow periodic pulses located at

$$q = (1/D)(2n\pi - L p_0) \quad n = 1, 2, \ldots \qquad (9.143)$$

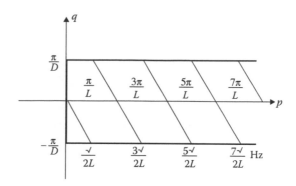

FIGURE 9.58　Output spectrum frequency locus.

Thus we see that as the input frequency changes, the location of the pulses change accordingly as given by

$$dq = (L/D)dp_0. \tag{9.144}$$

The displacement in the q direction is proportional to the displacement in the p direction. Since the pulse width decreases as the number of scan lines N increases, the output spectrum would yield a frequency discrimination equivalent to that obtained with a continuous signal of NL long.

We further note that as the input frequency advances, one of the pulses leaves the open aperture at $q = -\pi/D$ and a new pulse enters at $q = \pi/D$, which is resulting in a diagonal frequency locus as shown in Figure 9.58.

We emphasize again that area modulation offers the advantage of simple implementation for which the nonlinearity of density modulation can be avoided. However, the available space bandwidth product for the processor is smaller. Aside from this disadvantage, by efficiently utilizing the two-dimensional format, its space bandwidth product is still warranted for many interesting applications.

9.7　Final Remarks

The major advantages of optical information processing must be the massive interconnectivity, very high density, parallel processing capability and others. These are the trademarks for using optical processing. We have shown that synthetic aperture radar imaging can be processed with optics, for which was the first time that the naked human eye could actually see radar images, with high resolution. Since optical processing is generally a two-dimensional spatial processor, we have shown by exploiting the two-dimensional processing capability that a very large space-bandwidth signal can actually be processed with a conventional optical processor.

To conclude this chapter we remark that it was not our intention to cover the vast domain of Fourier optics. Instead, we have provided some basic principles and concepts. For further reading, we refer the readers to the list of references as cited at the end of this chapter. Nevertheless, several basic applications of Fourier optics are discussed, including coherent and incoherent light image processing, optical neural networks, wavelet transform, computing with optics, radar imaging with optics and broadband signal processing with optics.

Suggested Reading

1. Yu FTS, Jutamulia S. *Optical Signal Processing, Computing, and Neural Networks*, New York, NY: Wiley-Interscience, 1992.

2. Yu FTS, Yang X. *Introduction to Optical Engineering*, Cambridge: Cambridge University Press, 1997.
3. Reynolds GO, Develis JB, Parrent GB Jr., Thompson BJ. *Physical Optics Notebook: Tutorials in Fourier Optics*, Bellingham, WA: SPIE Optical Engineering Press, 1989.
4. Saleh BEA, Teich MC. *Fundamentals of Photonics*, New York, NY: Wiley-Interscience, 1991.
5. Yu FTS, Jutamulia S. (eds.), *Optical Pattern Recognition*, Cambridge: Cambridge University Press, 1998.
6. Yu FTS. *Entropy and Information Optics*, New York, NY: Marcel Dekker, 2000.
7. Cutrona LJ, Leith EN, Porciello LJ. On the application of coherent optical processing techniques to synthetic-aperture radar. *Proc. IEEE*, **54**, 1026; 1966.
8. Yu FTS. *Introduction to Diffraction, Information Processing, and Holography*, Cambridge, MA: MIT Press, 1973.
9. Thomas CE. Optical spectrum analysis of large space-bandwidth signals. *Appl. Opt.*, **5**, 1782; 1966.
10. Tai A, Yu FTS. Wide-band spectrum analysis with area modulation. *Appl. Opt.*, **18**, 460; 1979.
11. Yu FTS, Jutamulia S, Yin S. *Introduction to Information Optics*, San Diego: Academic Press, 2001.

10

Light-Sensitive Materials: Silver Halide Emulsions, Photoresist, and Photopolymers

10.1 Introduction 353
10.2 Commercial Silver Halide Emulsions..354
Black-and-White Films • Color Films • Sensitometry • Image
Structure Characteristics
10.3 Silver Halide Emulsions for Holography.......................................363
Introduction • Characterizing Methods for Holographic
Emulsions • Reciprocity Failure • Processing: Developing and
Bleaching
10.4 Photoresist..365
Types of Photoresist • Application and Processing • Applications in
Optics
10.5 Recent Applications of Photographic Emulsions 366
10.6 Photopolymer Holographic Recording Materials.......................368
Introduction • Fundamentals • The Early History of Photopolymer
Recording Materials • The State of the Art • Doped Polymer
Materials • Spectral Hole Burning
10.7 Recent Advances in Photopolymer Materials Research.............379
Optimization of the Monomer • Optimization of the
Crosslinker • Optimization of the Sensitizer/Initiator • Binder
Optimization • Photopolymer Nanocomposites • Surface Relief in
Photopolymers • Theoretical Modeling of the Holographic Recording
in Photopolymers • Polymer Dispersed Liquid Crystals • Polymer
Shrinkage • New Applications of Photopolymers
10.8 Conclusion 393
Acknowledgments..394
References 394

Sergio Calixto,
Daniel J. Lougnot,
and Izabela
Naydenova

10.1 Introduction

The term "light sensitive" comprises a variety of materials, such as silver compounds [1], non-silver compounds [2], ferroelectric crystals, photochromic materials, and thermoplastics [3], and photopolymers and biopolymers [4]. Due to the scope of the handbook and limited space, we review only a fundamental description of the three most widely used light-sensitive materials. Undoubtedly, the light-sensitive medium that is mostly used is photographic emulsion; this is because of its high sensitivity, resolution, availability, low cost, and familiarity with handling and processing. Two other popular emulsions are photoresist and photopolymers.

Photographic emulsion has come a long way since its introduction in the nineteenth century. Extensive research, basic and private, had led to its development. In contrast, photoresist began to be applied in the 1950s, when Kodak commercialized the KPR solution. Since then, due to its application in the microelectronics industry, a fair amount of research has been done. Finally, holographic photopolymers began to be developed more recently (1969) and, with a few exceptions, are still in the development stage.

Information presented in this chapter intends to provide a basic understanding to people who are not familiar with light-sensitive materials. With this foundation, it will be possible to read and understand more detailed information. To be well acquainted with stock products and their characteristics, the reader should contact the manufacturers directly. We give a list of their addresses in Section 10.2.1.6.

10.2 Commercial Silver Halide Emulsions

10.2.1 Black-and-White Films

The earliest observations of the influence of sunlight on matter were made on plants and on the coloring of the human skin [5]. In addition, in ancient times, people who lived on the Mediterraean coast dyed their clothes with a yellow substance secreted by the glands of snails. This substance (tyrian purple) develops under the influence of sunlight into a purple-red or violet dye. This purple dyeing process deserves much consideration in the history of photochemistry. The coloring matter of this dye was identified as being a 6-6^{00} dibromo indigo.

10.2.1.1 The Silver Halide Emulsion

The silver halide photographic emulsion [6] consists of several materials, such as the silver halide crystals, a protective colloid, and a small amount of compounds, such as sensitizers and stabilizers. Usually, this emulsion is coated on some suitable support that can be a sheet of transparent plastic (e.g., acetate) or over glass plates, depending on the application. In the photographic emulsion, the silver halide crystals could consist of any one of silver chloride (AgCl), silver bromide (AgBr), or silver iodide (AgI). Silver halide emulsions can contain one class of crystals or mixed crystals, such as AgCl + AgBr or AgBr + AgI. Sensitivity of emulsions depends on the mixing ratio. The protective colloid is the second most important component in the emulsion. It supports and interacts with the silver halide grains. During the developing process, it allows processing chemicals to penetrate, and eases the selective development of exposed grains. Different colloids have been used for the fabrication of photographic films; however, gelatin seems the most favorable material for emulsion making.

10.2.1.2 Photographic Sensitivity

Sensitometry is a branch of physical science that comprises methods of finding out how photographic emulsions respond to exposure and processing [6,7]. Photographic sensitivity is the responsiveness of a sensitized material to electromagnetic radiation. Measurements of sensitivity are photographic speed, contrast, and spectral response [8].

10.2.1.2.1 Characteristic Curve, Gamma, and Contrast Index

After the silver halides in the emulsion have absorbed radiation, a latent image is formed. In the development step, certain agents react with the exposed silver halide in preference to the unexposed silver halide. In this form, the exposed silver is reduced to tiny particles of metallic silver. The unreduced silver halide is dissolved out in the fixing bath. The developed image can be evaluated in terms of its ability to block the passage of light; that is, measuring its transmittance (T) [7]. As the amount of silver in the negative goes up, the transmittance goes down. Bearing this fact in mind, the reciprocal of transmittance (opacity) is more directly related to the amount of silver produced in the developing step. Although opacity increases as silver increases, it does not do so proportionally: the quantity that does, is called the density (density log $1=1/T$), which is used as a measure of the responsiveness of the emulsion [8–14].

To quantify the amount of light that falls on the emulsion two quantities should be considered: the irradiance (E) of the light and the time (t) that the light beam illuminates the emulsion, which are related by the exposure (H), defined by $H = Et$.

The responsiveness of an emulsion is represented in a graphic form that illustrates the relationship between photographic densities and the logarithm of the exposures used to produce them. This graph is called the characteristic curve and varies for different emulsions, radiation sources, kinds of developers, processing times, and temperatures and agitation method. Published curves can be reliably useful only if the processing conditions agree in all essentials with those under which the data were obtained.

A typical characteristic curve for a negative photographic material [8] contains three main sections (Figure 10.1): the toe, the straight-line portion, and the shoulder. Before the toe, the curve is parallel to the abscissa axis. This part of the curve represents the film where two densities are present even if illumination has not reached the film; they are the base density (density of the support) and the fog density (density in the unexposed but processed emulsion), that is, the base-plus-fog density region. The toe represents the minimum exposure that will produce a density just greater than the base-plus-fog density. The straight-line portion of the curve density has a linear relationship with the logarithm of the exposure. The slope, gamma (γ), of this straight-line portion is defined as $\gamma = \Delta D/\Delta \log H$ and indicates the inherent contrast of the photographic material. Numerically, the value of γ is equal to the tangent of the angle α, the angle that makes the straight-line portion of the curve with the abscissa.

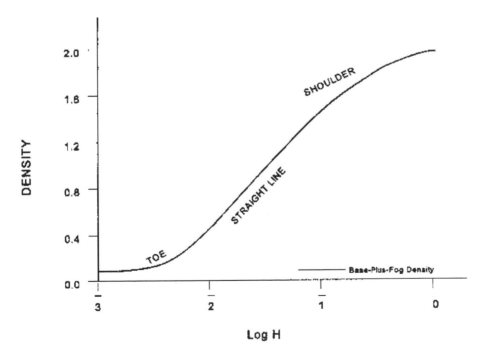

FIGURE 10.1 Characteristic curve.

It is usual that manufacturers present not only one characeristic curve for a given material, but a family of them. Each of these curves is obtained with a different developing time and presents a different value of γ. With these data, a new graph can be made by plotting γ as a function of developing time (Figure 10.2).

Another quantity that can be inferred from the characteristic curve is the contrast index [15], which is defined as the slope of the straight line drawn through three points on the characteristic curve. It is an average slope. The straight line makes an angle with the abscissa, and the tangent of this angle is the contrast index. This contrast also changes with development time (Figure 10.3).

The slope of the characteristic curve changes with the wavelength of the light used to expose the emulsion. This behavior is evident if we plot the contrast index as a function of recording wavelength. This dependency of contrast varies considerably from one emulsion to another (Figure 10.4).

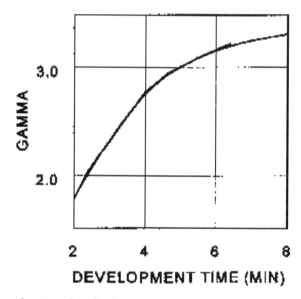

FIGURE 10.2   Slope γ as a function of the development time.

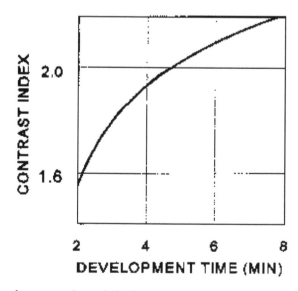

FIGURE 10.3   Behavior of contrast index with the developing time.

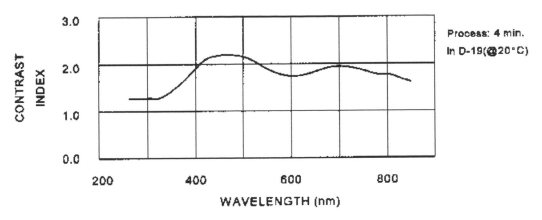

FIGURE 10.4 Behavior of contrast index with wavelength of the recording light.

10.2.1.2.2 Photographic Speed and Meter Setting

The speed of an emulsion [16–18] can be derived from the characteristic curve. If we denote the speed by S, the following relation with exposure will follow: $S = K/H$, where K is a constant and H is the exposure required to produce a certain density above base-plus-fog density. Because the characteristic curve depends on the spectral characteristics of the source used to illuminate the emulsion, the speed will depend also on this source. Therefore, indicating the exposure source when quoting speed values is essential. Speeds applied to pictorial photography (ISO speed) are obtained by using the formula $S = 0.8/H$, where H is the exposure (to daylight illumination) that produces a density of 0.10 above the base-plus-fog density with a specified development. Several factors affect speed adversely, such as aging, high temperature, high humidity, and the spectral quality of the illumination.

For scientific photography, a special kind of speed value called meter setting is used [8]. The published meter setting value is calculated using the relation $M = k/H$, where k is a constant equal to 8 and H is the exposure that produces a reference density of 1.6 above base-plus-fog. This value is chosen because scientific materials are normally used in applications that require higher contrast levels than those used in pictorial photography. The reference density of 0.6 above base-plus-fog density is chosen for certain spectroscopic plates.

10.2.1.2.3 Exposure Determination

Parameters that are important when a camera, or a related instrument, is used, are time of exposure, lens opening, average scene luminance, and speed of the photographic material [19]. The four parameters are related by the equation $t = Kf/Ls$, where t is the exposure time (seconds), f is the f_{number}, L is the average scene luminance, s is the ISO speed, and K is related with the spectral response and transmission losses in the optical system [16,17]. Other equations can be used to calculate the time of exposure [19].

10.2.1.2.4 Spectral Sensitivities of Silver Halide Emulsions

The response of silver halide emulsions to different wavelengths can vary. This response is shown in spectral sensitivity plots. These curves relate the logarithm of sensitivity as a function of the wavelength. Sensitivity is a form of radiometric speed and is the reciprocal of the exposure required to produce a fixed density above a base-plus-fog density [8] (Figure 10.5).

10.2.1.3 Image Structure Characteristics

To select the best photographic emulsion for a specific application, in addition to the sensitometric data, image structure properties should also be considered.

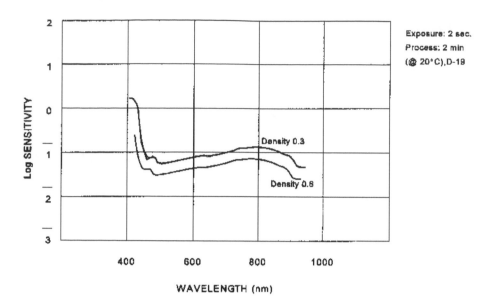

FIGURE 10.5   Spectral sensitivity as a function of the recording wavelength. Density is a parameter.

FIGURE 10.6   Profiles of a slit image given by a microdenistometer for different exposure time.

10.2.1.3.1 Granularity

When a developed photographic image is examined with a microscope, it can be seen that the image is composed of discrete grains formed of filaments of silver [6,8,20–22].

The subjective sensation of the granular pattern is called the graininess. When a uniformly exposed and processed emulsion is scanned with a microdensitometer having a small aperture, a variation in density is found as a function of distance, resulting from the discrete granular structure of the developed image. The number of grains in a given area varies and causes density fluctuations that are called granularity. The microdensitometer shows directly the rms (root mean square) granularity [8], which has values ranging between 5 and 50; the lower numbers indicate finer grain. Granularity has been studied extensively. For a more in-depth knowledge we suggest consulting the references.

10.2.1.3.2 Photographic Turbidity and Resolving Power

The turbidity of a photographic emulsion results from light scattered by the silver halide grains and light absorption by the emulsion [23,24]. This causes a gradual widening of the recorded image as the exposure is increased [8] (Figure 10.6).

Resolving power is the ability of a photographic material to maintain in its developed image the separate identity of parallel bars when their relative displacement is small. Resolving power values specify the number of lines per millimeter that can be resolved in the photographic image of a test object, commonly named the test chart. The test object contrast (TOC) has a direct relation to the resolving power values that are often reported for test objects with the low contrast of 1.6:1 and test objects with the high contrast of 1000:1. Resolving power can be affected by factors, such as turbidity, spectral quality of the exposed radiation, developer, processing conditions, exposure, and grain sizes.

10.2.1.3.3 Modulation Transfer Function (MTF)

For light-sensitive emulsions, the MTF [6,8,25] is the function that relates the distribution of incident exposure in the image formed by the camera lens to the effective exposure of the silver halide grains within the body of the emulsion layer. To obtain these data, patterns having a sinusoidal variation in illuminance in one direction are exposed to the film. The "modulation" M_o for each pattern can be expressed by the formula $M_o = (H_{max} - H_{min})/(H_{max} + H_{min})$, where H is the exposure. After development, the photographic image is scanned in a microdensitometer in terms of density. These densities of the trace are interpreted in terms of exposure, by means of the characteristic curve, and the effective modulation of the image M_i is calculated. The MTF (response) is the ratio M_i/M_o plotted (on a logarithmic scale) as a function of the spatial frequency of the patterns (cycles/mm). Parameters that should be mentioned when specifying an MTF are spatial frequency range, mean exposure level, color of exposing light, developer type, conditions of processing and, sometimes, the f_{number} of the lenses used in making the exposures, Figure 10.7.

10.2.1.3.4 Sharpness

Sometimes the size of the recorded images is larger than the inverse of the highest spatial frequency that can be recorded in the film. However, the recorded image shows edges that are not sharp. The subjective impression of this phenomenon is called sharpness and the measurement of this property is the acutance. Several methods to measure acutance have been proposed [26].

10.2.1.4 Image Effects

10.2.1.4.1 Reciprocity

The law of reciprocity establishes that the product of a photochemical reaction is determined by the total exposure *(H)* despite the range of values assumed by either intensity or time. However, most photographic materials show some loss of sensitivity (decreased image density) when exposed to very low or very high illuminance levels, even if the total exposure is held constant by adjusting the exposure time. This loss in sensitivity (known as reciprocity-law failure) means more exposure time (than normal calculations indicate) is needed at extreme levels of irradiance to produce a given density. Reciprocity effects can be shown graphically by plotting the log *H* vs. log intensity for a given density. Lines at 45° represent constant time (Figure 10.8). Other image effects, present at exposure time, that affect the

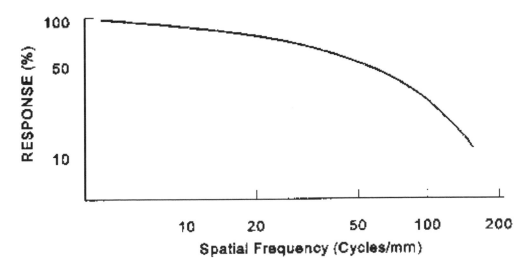

FIGURE 10.7 Modulation transfer function (MTF) of a film.

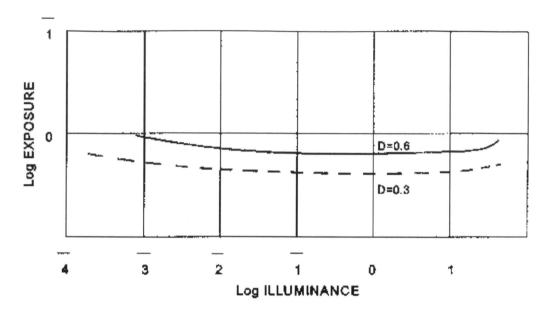

FIGURE 10.8   Plots showing the reciprocity-law failure.

normal behaviors of emulsions are the following: intermittency effect, Clayden effect, Villard effect, solar-ization, Herschel effect, and Sabattier effect [8,27].

10.2.1.4.2 Adjacency Effects in Processing

These effects are due to the lateral diffusion of developer chemicals, reaction products, such as bromide and iodide ions, and exhausted developer within the emulsion layer in the border of lightly exposed and heavily exposed areas [6,19,21]. This phenomenon presents itself as a faint dark line just within the high-density (high-exposure) side of the border. This is the border effect. Related to this effect is the Eberhard effect, Kostsinky effect, and the MTF effect.

10.2.1.5 Development

The result of the interaction of light with the emulsion is the formation of a latent image. This image is not the cause of the reduction of the silver halide grains to silver by means of the developer but it causes the exposed grain to develop in less time than an underexposed grain. To form a latent image, only about 5–50 photons are needed.

The development process [6,28,29] will amplify these phenomena by a factor of 107.

This operation is done by the deposition of silver from the developer on the grain and is proportional, to some extent, to the amount of light action on the grain. The role of the developer on the emulsion has been studied widely and through many years and yet it seems it is not completely clear what happens in the process. A practical developing solution contains the developing agent, an alkali, a preservative, an antifoggant, and other substances, such as wetting agents. For more detailed information, the reader should consult the references.

10.2.1.6 Commercial Black-and-White, Color, and Holographic Films

10.2.1.6.1 Kodak

Kodak manufactures a multitude of photographic products, such as emulsions, books, guides, pamphlets, photographic chemicals, and processing equipment.

A partial list of Kodak photographic emulsions is given in the Kodak catalog L-9 [30]. There are listed about 63 black-and-white films and plates, 60 color films (color negative and reversal films). Also, about 29 black-and-white and color papers are listed.

For the scientific and technological community publication, Kodak P-315 [8] is highly recommended because it describes with detail the application of scientific films and plates. (Table IV on page 6d of this reference should be consulted because it exhibits different emulsion characteristics.) This publication is a good general reference; however, because it is old (1987), some films and plates mentioned in it are not fabricated anymore. The reason for these discontinuities is the wide use of CCDs. Another publication that is also useful for scientists and technicians is the Kodak Scientific Imaging products catalog. For more information write to Eastman Kodak Company, Information Center, 343 State Street, Rochester, NY 14650–0811.

10.2.1.6.2 Agfa

Agfa fabricate black-and-white, color negative, and reversal films; they do not fabricate holographic films and plates anymore [31]. To request information write to Agfa Corporation, 100 Challenge Road, Ridge Field Park, NJ 07660, USA. Tel. (201) 440-00. In Europe, contact Agfa-Gevaert NV, Septstraat 27, B-2510 Mortsel, Antwerp, Belgium.

10.2.1.6.3 Fuji Film

Fuji Film fabricates black-and-white, color negative, and reversal films. Inquiries can be addressed to Fuji Film USA Inc., 555 Taxter Rd, Elsmford, NY 10523. (914) 789-8100, (800) 326-0800, ext 4223 (Western USA).

Other manufacturers of black-and-white, color emulsion, and holographic emulsions are Ilford Limited, Mobberly, Knutsford, Cheshire WA16 7HA, UK and Polaroid Corp., 2 Osborn Street, Cambridge, MA 02139. *Note*: in 1992, Ilford stopped the production of holographic plates.

The following company fabricates the Omnidex photopolymer holographic film: du Pont de Nemours and Co., Imaging Systems Department, Experimental Station Laboratory, Wilmington, Delaware 19880-0352.

A list of commercial color and black-and-white films containing not only the main manufacturers listed above is presented in the magazine *Amateur Photographer*, February 15, 1992.

Copies of ANSI/ISO standards are available from the American National Standards Institute, 1430 Broadway, New York, NY 10018.

10.2.2 Color Films

Silver halide emulsions are primarily sensitive to ultraviolet and blue light. In 1873, W. Vogel [5], working with colloidon dry plates (fabricated by S. Wortley), noted that these plates presented a greatly increased sensitivity to the green part of the spectrum. These plates contained a yellowish-red dye (corallin) to prevent halation.

Vogel studied the role of several dyes in the emulsion and made the discovery of color sensitizers that extend the sensitivity of the silver halide emulsions into the red and near-infrared part of the spectrum [32].

10.2.2.1 Subtraction Process

Color reproduction processes can be divided into two: the direct process, in which each point of the recorded image will show the same spectral composition as the original image-forming light at that point; and the indirect process, in which the original forming light is matched with a suitable combination of colors. The direct process was invented by Gabriel Lippman in 1891 and is based on the interference of light. However, it has not been developed sufficiently to be practical. The indirect method to reproduce color comprises additive and subtractive processes [28]. The former process presents difficulties and is not used now. The subtractive process is the basis for most of the commercial products nowadays [21]. Next, we describe briefly the structure of a simple color film, hoping this will clarify the color process recording [28].

The emulsion will be described in the sense that the light follows. Basically, color films consist of five layers, three of which are made sensitive by dyes to a primary color. Besides these dyes, layers contain color formers or color couplers. The first layer protects the emulsion from mechanical damage. The second layer is sensitive to blue light. A third layer is a yellow filter that lets red and green light pass that react with the following layers that are red and green sensitive. Finally, a base supports the integral tri-pack, as the structure containing the three light-sensitive emulsions, the yellow filter and the protecting layer is called. Green-sensitive and red-sensitive layers are also sensitive to blue light because no matter what additional spectral sensitization is used emulsions are still blue sensitive.

The most common procedure in the color development process is to form images by dye forming, a process called chromogenic development. This process can be summarized in two steps:

Developing agent + Silver halide → Developer oxidation products

+ Silver metal + Halide ions

Then a second reaction follows:

Developer oxidation products + Color couplers → Dye

The color of the dye depends mainly on the nature of the color coupler. At this step, a silver image reinforced by a dye image coexist. Then, a further step removes the unwanted silver image and residual silver halide. In this process, there are dyes formed of just three colors: cyan, magenta, and yellow. With combinations of these colors, the original colors in the recorded scene can be replicated.

A variety of other methods exist for the production of subtractive dye images, such as Polacolor and Cibachrome.

10.2.3 Sensitometry

10.2.3.1 Characteristic Curves

Color films can be characterized in a similar way to black-and-white films. Sensitometric characteriza-tion of color films should consider that color films do not contain metallic silver, as black-and-white films do, but instead, they modulate the light by means of the dyes contained in each of the three layers. Because the reproduction of a given color is made by the addition of three wavelengths, $D \log H$ plots should show three curves, one for each primary color (Figure 10.9) [19,33].

Speed for color films is defined by the relation $S = K/H$ as for black-and-white films (see Section 10.2.1.2); however, this time, $K = 10$ and H is the exposure to reach a specified position on the D vs. $\log H$ curve.

10.2.3.2 Spectral Sensitivity Curves

As described above, color films comprise three emulsion layers. Each layer shows a response to certain wavelengths, or spectral sensitivity, expressed as the reciprocal of the exposure (in ergs/cm2) required to produce a specified density (Figure 10.10) [6].

10.2.4 Image Structure Characteristics

10.2.4.1 Granularity, Resolving Power, and MTF

In the conventional color process, the oxidized developing agent diffuses away from the developing silver grain until color coupling takes place at some distance. In fact, each grain gives place to a roughly spherical dye cloud centered on the crystal and, after all unwanted silver has been removed, only the dye colors remain. Because these dye clouds are bigger than the developed grains, the process of dye development yields a more grainy image. The diffuse character of color images is also responsible for sharpness that is lower than that yielded by black-and-white development of the same emulsion [19].

In Section 10.2.1.3, we mentioned the terms granularity graininess, resolving power, and MTF for black-and-white films. These terms also apply to color films [19].

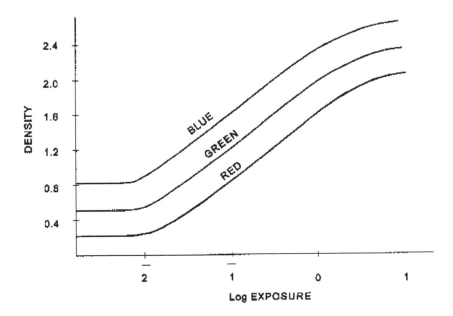

FIGURE 10.9 Characteristic curve for a color film.

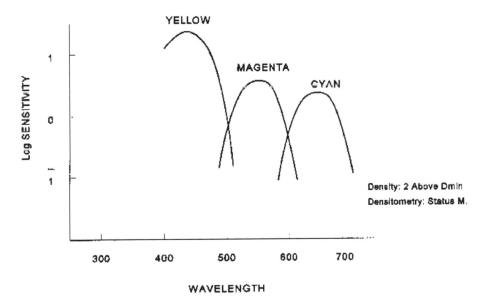

FIGURE 10.10 Spectral sensitivity as a function of recording wavelength.

10.3 Silver Halide Emulsions for Holography

10.3.1 Introduction

The objective of a photosensitive material to be used in holography is the recording of an interference pattern with the highest fidelity. The result of the action of light onto the light-sensitive material can be exposed in any of the three forms: local variations of refractive index, absorption, or thickness. At present, a variety of materials [3] are used to record the interference patterns; however, the most popular material is the holographic silver halide plate. This emulsion can record the high spatial frequencies found in the holographic setups. Interference patterns for transmission holograms can present spatial frequencies between a few lines/mm to about 4000 lines/mm. For reflection holograms, spatial frequencies presented by interference patterns will range between 4500 lines/mm and 6000 lines/mm [34]. Some silver halide holographic emulsions present maximum resolutions of 5000 lines/mm [35].

Full characterization of holographic emulsions cannot be carried out with the D vs. log H curves (Section 10.2). In holographic emulsions, to reconstruct the object wave linearly, it is necessary that differences in amplitude transmittance should be proportional to the differences in exposure. Next, we describe briefly two characterizing methods based on the recoding of sinusoidal interference patterns.

10.3.2 Characterizing Methods for Holographic Emulsions

Several methods have been mentioned in the literature to characterize holographic emulsions. Some are based on the transmittance–exposure curves [36]. These methods emphasize experimental recording parameters, such as exposure, intensity ratio of the reference and object beam, angle between recording beams, and recording wavelength. Parameters measured are transmittance and diffraction efficiency. Through them the optimum bias level, beam ratio, and spatial frequency of the interference pattern can be inferred. Also, the transfer function of the film can be inferred.

A more general method of characterizing holographic materials, which is applicable to all types of holograms was proposed by Lin [37]. This method supposes that a series of sinusoidal interference patterns are recorded in the medium and later diffraction efficiencies (η) are measured. The material can be characterized by the relation $\sqrt{\eta} = SHV$, where S is a constant (holographic sensitivity), H is the average exposure, and V is the interference fringes visibility. For the ideal material, curves of the $\sqrt{\eta}$ as a function of H with constant V as parameter or curves of $\sqrt{\eta}$ as a function of V with H as parameter are families of straight lines. However, real materials only show curves with a limited linear section. Through them, it is possible to know maximum attainable η, optimum exposure value, and holographic sensitivity. Also, these curves allow direct comparison of one material with another or with the ideal one. This cannot be done with the transmittance–exposure characteristic curves mentioned above. Two drawbacks of the $\sqrt{\eta}$ vs. H or $\sqrt{\eta}$ vs. V curves are that they cannot show the spatial frequency response and the noise in reconstruction.

10.3.3 Reciprocity Failure

Failure to follow the law of reciprocity (see Section 10.2.1.4) of holographic materials appears when lasers giving short pulses are employed, such as the Q-switched lasers. These lasers release energy during a very short time, producing a very high output peak power [38]. Contrary to this phenomenon, sometimes when continuous wave (CW) lasers are used and show an intensity-weak output, problems are created for interference pattern recording because of the very long exposure times needed.

10.3.4 Processing: Developing and Bleaching

Developing and bleaching processes recommended by the manufacturers have been optimized and should be followed strictly when a general work on holography is carried out. Much research in developing and bleaching has been done to obtain special effects, such as high-diffraction efficiency, reduced developing time, increased emulsion speed and low noise in reconstruction, for example. It is advisable to consult the references and find the method that best suits the experimental needs [31]. In this last reference, a list of commercial holographic emulsions and some of their characteristics is also exhibited (p. 114).

10.4 Photoresist

Methods of coloring cloth in a pattern by pretreating designed areas to resist penetration by the dye are known as "resist printing" techniques. Batik is an example of this. Photoresists are organic light-sensitive resins suitable for the production of surface-relief patterns and selective stencils that protect the underlying surface against physical, electrical, or chemical attack.

10.4.1 Types of Photoresist

Many kinds of photoresists are commercially available today. Depending on their action, they may be divided into positive- and negative-working photoresists. In a negative-type photoresist, the exposed areas become insoluble as they absorb light so that upon development only the unexposed are dissolved away. In a positive-type photoresist, the exposed areas become soluble and dissolve away during the development process.

The first negative photoresists were produced in the 1950s, and were made possible by the evolution of the electronics industry. Kodak photoresist (KPR), produced by the Eastman Kodak Company (Rochester, NY), was the first of a range of high-resolution materials that polymerize under light and become insoluble in organic developer compounds. Polymerized photoresist is resistant to ammonium persulphate, which can be used for etching copper printed circuit boards. A higher-resolution thin-film resist (KTFR) was developed in the 1960s. These resists are members of two distinct families of negative photoresists: those based on an ester solvent (KPR) and those based on a hydrocarbon solvent (KTFR). Processing solutions, such as developers and thinners are not interchangeable between the two families.

Positive photoresists do not polymerize upon exposure to light, but their structure releases nitrogen and forms carboxyl groups. The solubility acquired through the exposure enables the positive stencil to be washed out in aqueous alkali. Positive photoresists are basically composed of three components: a photoactive compound (inhibitor), a base resin, and a solvent. The base resin is soluble in aqueous alkaline developers and the presence of the photoactive compound strongly inhibits its dissolution. The light neutralizes the photosensitive compound and increases the solubility of the film. After development in such solutions, an intensity light pattern is converted into a relief structure. The Shipley Co. Inc. (Newton, MA 02162) produces good positive photoresists consisting of Novolak resins with naphthoquinone diazodes functioning as sensitizers (AZ formulations).

Most of the available photoresists are sensitive to light in the blue and ultra-violet regions (300–450 nm) of the spectrum, although deep UV (200–270 nm) photoresists are also obtainable.

The choice of photoresist for a particular application is based on a tradeoff among sensitivity, linearity, film thickness, resolution capability, and ease of application and processing. For instance, in holography and other optical applications, the resolution must be better than 1000 lines/mm and a linear response is usually desired. In microlithography, one needs to fabricate line structures of micron (μm) and even submicron dimensions and it is preferable to have a highly nonlinear response of the material.

10.4.2 Application and Processing

Photoresist is provided in the form of a resin suitable for application on to a clean surface. The procedure of deposition, exposure, and development of a photoresist film can vary, depending on the particular kind of photoresist and the application. The preferred method of depositing a photoresist film on the substrate seems to be spin-coating. Fairly even films of photoresist, ranging in thickness from fractions of a micron, to more than 5 mm, can be produced with such a method. For the production of thin films, the substrate must be spun at speeds between 4000 and 7000 rpm. For thicker films, speeds as low as 1000 rpm can be used, at the expense of film uniformity. The quality and adhesion of the coated films to the substrate depends critically on its cleanliness. With positive photoresist, the coatings should be allowed to dry, typically, for a few hours and be subsequently baked to allow for the evaporation of the remaining solvents. After exposure, the sample is developed in a solution of developer and water.

The time and temperature of baking, the dilution ratio of the developer, the time and temperature of development, as well as the wavelength of the exposing light must be chosen for the particular kind of photoresist employed and determine its sensitivity and the linearity of its response.

10.4.3 Applications in Optics

Photoresists find their main application in the electronics industry, where they are used in the production of microcircuits [39]. However, they also have many important applications in optics. Among others, we can mention the fabrication of holographic gratings [40–42], holograms [43,44], diffractive optical elements [45,46], and randomly rough surfaces [47,48]. Perhaps the most common and illustrative application of photoresist in optics is the production of the master plates for the fabrication of embossed holograms, which have become commonplace in modern life.

There is also a trend to create miniature opto-electro-mechanical instruments and subsystems. The existing technology to fabricate integrated circuits is being adapted to fabricate such devices and, since their manufacture is based on micro-lithography, the use of photoresist is a must. For example, refractive microlenses have been made by producing "islands" of photoresist on a substrate and baking them in a convection oven to reshape them. Other examples include microsensors, microactuators, and micromirrors. For a more complete picture of the development of this area of technological research, the interested reader is referred to the November 1994 issue of *Optical Engineering* (vol. 33, pp. 3504–3669).

10.5 Recent Applications of Photographic Emulsions

With the arrival of digital photography, it would seem that photographic emulsions would disappear quickly. However, photographic emulsions (films and papers) are still (year 2017) used in different fields such as ordinary landscape and portrait photography, motion pictures, aerial photography, microfilm, traffic and surveillance, nuclear physics, and more. Companies like Kodak [49], Agfa–Gevaert [50], Fujifilm [51] and Ilford [52] still produce photographic emulsions. Now we will describe some applications of photographic emulsions in scientific fields.

The Alicante group (Spain) has developed research on silver halide materials. They have tested methods to improve and characterize holographic plates like the Slavich PFG–01 and the BB–640 plates. They have studied the role of the developer [53,54], fixation, fixation–free [55,56], bleach [57–59], and reversal bleach methods [60]. Also the rehalogenation process has been studied [54,55].

Other studies have explored the possibility of using photographic emulsion with the Silver Halide Sensitized Gelatin (SHSG) process. Holographic optical elements are recorded in dichromated gelatin because they present high diffraction efficiency and low noise. However, dichromated gelatin presents low exposure sensitivity and limited spectral response. Thus, silver halide emulsions can be processed in such a way that the final hologram will present the properties of dichromated gelatin holograms and besides show high sensitivity. References [61–64] describe the SHSG methods.

Regarding other applications of silver halide emulsions, they are used in nuclear physics. The nuclear emulsion is used to record the tracks of charged particles passing through. Nuclear emulsions have very uniform grain size and grain sensitivity, low chemical fog and the silver to gelatin ratio is higher than in conventional emulsions. When the emulsion is exposed to ionizing radiation or light, clusters of atoms are produced. In the development step, crystals are reduced to metallic silver. Chemical development is dependent on temperature and pH. Development is followed by fixation. This removes the residual silver halide, leaving the metallic silver that forms the image.

At present (2017), Ilford has three nuclear emulsions named: G5, K and L4 with respective crystal diameters of 0.27 µm, 0.20 µm and 0.11 µm. K emulsions can record protons with energies up to 5 MeV and thorium alpha particles. K2 emulsions are used in autoradiography and K5 emulsions for autoradiography with an isotope. L4 emulsions are recommended for electron microscopy. More information can be obtained at the Internet address.

In references 65 and 66, the use of photographic emulsions in nuclear physics is described. Reference 65 explains the use of emulsions in the OPERA experiments (Oscillation Project with Emulsion tRacking Apparatus). This is a long base line appearance search for $\sqrt{\mu}$ à $\sqrt{\tau}$ oscillation. A pure $\sqrt{\mu}$ beam from CERN is sent 730 km far away to the LNGS underground laboratory in Italy where the OPERA detector is located. The charged current $\sqrt{\tau}$ interaction will produce a Tau lepton, which can then be identified according to its decay topology. The whole detector in the OPERA experiment is made of 150,000 modules called bricks. Each brick consists of 57 emulsions and 56 lead plates arranged like a sandwich. Each emulsion film has dimensions of 125 mm x 99 mm with a thickness of 293 μm. About 10 million emulsion films are used in the experiment. Each brick weighs 8.3 kg. Emulsions were developed and fabricated by Fujifilm. For a detailed description of the emulsion used in the OPERA project, the reader may consult reference 66.

An interesting theoretical study that comprises the use of nuclear photographic emulsions for investigation of dark matter is presented in reference 19. Dark matter particles are expected to be Weakly Interacting Massive Particles (WIMP). To detect these WIMP particles with nuclear emulsions, a collision between the WIMP particles with Ag or Br atoms should take place. This article [67] shows details of emulsions containing different AgBr crystals. A good list of references is given there.

Another application of photographic emulsions is given in the fabrication of microstructures. In a series of papers [68–70], Ganzherli proposes that holographic diffusers and microlens arrays can be fabricated using PFG–01 and VR-L emulsions. Methods to obtain these structures using silver halide emulsions are based on the transformation of intensity recording into relief recording. To do this, a tanning bleach is used. The relief is associated with the volume redistribution of the gelatin in the emulsion. The fabricated diffusers [68] could be used to evenly illuminate liquid crystals displays in portable computers or in mobile phones. The fabrication of microlens arrays is proposed in references 69 and 70. After exposure and chemical processing (including development and fixing), an array of positive lenses is obtained. Microlens diameters ranged between 35 μm and 90 μm with a focal length of 30 μm to 500 μm. Researcher Ganzherli gives more references referring to the use of silver halide emulsions.

One more method to make surface relief zone plates and microlenses with silver halide emulsions was described in references 71–73. To make zone plates, halftone masks with 256 gray levels were designed. Then, information in the masks was transferred into the emulsions by projecting an image of the mask over the photographic emulsion. After developing and fixing, relief zone plates were obtained. A zone plate working in reflection at 45o with a diameter of 8.4 mm and a diffraction efficiency of 86% and focal length of 50 cm was fabricated. Regarding the microlenses, binary and halftone masks were also used in their fabrication. Positive and negative microlenses were made. Besides, some lenses showed in the central part a convex concavity and in the periphery a negative concavity. This characteristic permitted the formation of virtual and real images. Microlenses were investigated with a surface analyzer and an interference microscope. Microlens image forming capability was evaluated using a USAF-1951 resolution test target. Lenses made with halftone masks showed better optical performance than those fabricated with binary masks. Also, microlens arrays with 45×45 elements were made. Besides the lenses, relief sinusoidal gratings were fabricated. Diffraction efficiencies of 26%, for transmission gratings, and 29% for reflection gratings were obtained.

We know that silver halide emulsions and photopolymers suffer shrinkage after development. A material, also based on silver halide, but using nanoporous glasses, instead of gelatin, was suggested in reference 74. This material shows no shrinkage because it is made of a rigid skeleton of porous glass. The selected glass was DV–1M (sodium borosilicate), Type I (pore size 7 nm) and Type II (pore size 17 nm). The photosensitive silver–halide composite is synthesized inside the porous skeleton. Solutions containing gelatin, KI, KBr and AgNO3 were used. The region of spectral sensitivity of unsensitized samples is determined by the intrinsic sensitivity of silver halide ($\lambda < 510$ nm). However, dyes can be added and have sensitivities in the whole visible region. The post-exposure processing of the samples is done with the conventional photographic materials. Unfortunately, the development step lasts 20 h. Recorded gratings showed diffraction efficiencies of 50%. The group at S.I. Vavilov State Optical Institute has more articles on this subject.

10.6 Photopolymer Holographic Recording Materials

10.6.1 Introduction

The principle of holography was described by Dennis Gabor some 70 years ago [75]. However, because of the unavailability of monochromatic light sources exhibiting sufficient coherence length, this principle, which was expected to open new vistas in the field of information recording, remained dormant for some 15 years.

The laser effect obtained by Maiman in 1960 [76] from A. Kastler's pioneering theoretical work [77] triggered off a series of experiments that form the early references in practical holography. Among them, the off-axis recording technique described by Leith and Upatnieks [78] must be regarded as a key to the revival of interest in the field of holography. About the same time, Denisyuk's research work [79] on volume holography opened up another fascinating way that resuscitated Lippman's old proposals.

The actual possibility of recording and reconstructing three-dimensional images through a hologram brought about a recrudescence of activity in the scientific community all around the world. Finding out and optimizing new recording materials with a great variety of characteristics became the purpose of a great number of research programs. High-energy sensitivity and resolution, broad spectral response but also simple recording and processing procedure are considered to be indispensable. Furthermore, the demand for self-processing or erasability makes the design of an ideal recording material a challenge for the chemists specializing in photosensitive materials.

The most widely used recording material that has long been known before the emergence of holography is silver halide emulsion. It has been thoroughly investigated for the purpose of holographic recording. In spite of the multifarious improvements introduced during the last 30 years, the typical limitation of these media—the high grain noise resulting in poor signal-to-noise ratio—has prompted investigators to imagine new recording media. Their search has been focused on materials not suffering from this limitation and that would in essence exhibit a grainless structure.

The wide range of applications of holography is a powerful incentive for researchers concerned with this challenge. Indeed, holographic optical elements, holographic displays, videodiscs and scanners, integrated optic devices, optical computing functions, or large-capacity optical memories using volume holography techniques are, as many fields of optics, bursting with activity.

10.6.2 Fundamentals

Basically, photopolymer holographic recording systems are compositions consisting of, at least, a polymerizable substance and a photoiniator, the fate of which is to absorb the holographic photons and generate active species, radicals, or ions capable of triggering the polymerization itself. This composition can also contain a great variety of other structures that are not essential to the polymerization process [80–84]. They are generally incorporated in the formulation with the object of improving the recording properties. In this respect, the following should be remembered:

- The sensitizer is used to extend the spectral range of sensitivity of the sensitive material.
- The binder introduces some flexibility in the viscosity of the formulation and may exercise some influence on the diffusive movements that take place during the recording process.

A large number of chemical structures containing either unsaturations, that is, multiple bonds or cycles, are known to undergo polymerization when subjected to illumination [85]. This kind of process should, theoretically, achieve fairly high efficiency because a single initiating event can trigger a chain polymerization involving a large number of reactive monomer molecules. When the process is photochemically initiated, for example, in holographic recording, the number of monomer units converted per initiating species is commonly referred to as the polymerization quantum yield. It stands to reason that the energy sensitivity of any photopolymer recording system depends to a great extent on this quantum yield.

It is, thus, important to be fully aware of the typical values of this parameter, which never exceeds 104 at the very best. Accordingly, as long as the process used for recording holograms with polymerizable material involves the growth of a polymer chain through a bimolecular-like process requiring the diffusion of a monomer to a living macroradical, the practical values of the energy sensitivity remain far lower than those of silver halide gelatins.

10.6.2.1 Photopolymers and Holographic Recording

Without getting to the roots of holography, it is nevertheless important to examine the adaptation of photopolymer materials to the different characteristics of the recording process. In this respect, a classification of holograms based on the geometry of recording, record thickness, and modulation type allows many specificities of polymer materials to be considered in the perspective of their application to holography.

10.6.2.1.1 Geometry

Basically, two different geometries are used to record holograms [86,87]. When a record is made by interfering two beams arriving at the same side of the sensitive layer from substantially different directions, the pattern is called an off-axis hologram. The planes of interference are parallel to the bisector of the reference-to-object beam angle. Since the volume shrinkage inherent in the polymerization appears predominantly as a contraction of the layer thickness, it does not substantially perturb the periodic structure of the pattern of chemical composition of the polymer as the recording process proceeds to completion. The polymer materials are thus fully suitable for recording transmission holograms.

On the contrary, when a hologram is recorded by interfering two beams travel-ing in opposite direction, that is, the Lippman or Denisyuk configuration, the interfer-ence planes are essentially parallel to the recording layer. Therefore, the shrinkage may cause their spacing to decrease, thus impairing the hologram quality. Consequently, the problem of volume shrinkage must be considered with great care in the formulation of photopolymerizable systems used to record reflection holograms.

10.6.2.1.2 Thickness

Holograms can be recorded either as thin or thick holograms, depending on the thickness of the recording medium and the recorded fringe width. In a simplified approach, these two types of holograms can be distinguished by the Q-factor defined as [88]:

$$Q = 2\pi\lambda d/n\Lambda^2$$

where λ is the holographic wavelength, n is the refractive index, d is the thickness of the medium, and Λ is the fringe spacing. A hologram is generally considered thick when $Q > 10$, and considered thin otherwise.

The suitability of photopolymer materials for recording both thin and thick holograms must be examined with respect to the absorbancy of the corresponding layer at the holographic wavelength and the scattering that depends on the signal-to- noise ratio. In regards to the absorbance parameter, it is related to the extinction coefficient of the initiator or the sensitizer. With a value of this coefficient in the order of 10^4 $M^{-1}.cm^{-1}$, a concentration of 0.5 m, and an absorbance of at least 0.2 (a set of experimental parameters that corresponds to extreme conditions), the minimum attainable value of the film thickness is about 1 μm. Ultimately, the maximum thickness is limited by the inverse exponential character of the absorbed amplitude profile in the recording material: the thicker the sample is, the more the absorbancy has to be reduced. Although there is no limitation to this homographic interdependence between the thickness and the absorbance parameters, the maximum values are determined by the overall absorptivity that determines the photonic yield (absorbed to incident intensity ratio) and, correspondingly, by the energetic sensitivity. In addition, the impurities, the microheterogeneities as well as the submicroscopic grain structure of the glassy photopolymer material, ultimately introduce a practical limitation of a few millimeters.

10.6.2.1.3 Type of Modulation

Holograms are also classified according to the type of modulation they impose on a reconstructing wavefront depending on the change incurred by the recording med-ium during the holographic illu-mination. If the incident pattern is recorded by virtue of a density variation of the sensitive layer that modulates the amplitude of the reconstruction wave, the record is known as an amplitude hologram. If the information is stored as a density and/or thickness variation, the phase of the reading beam is modulated and the record is termed a phase hologram [86].

As a general rule, the photoconversion of a monomer to the corresponding polymer does not go along with any spectral change in the wavelength range used to play back holograms. Thus, polymeriz-able systems apart from some exotic formulations do not lend themselves to the recording of amplitude holograms. On the other hand, the differences in polarizability caused by different degrees of polymer-ization in the dark and the bright regions of recorded fringe patterns and differences in refractive index resulting therefrom can be used for the storage of optical information.

10.6.2.2 The Different Models of Polymer Recording Materials

Without indulging in the details of macromolecular chemistry, it is important to clarify what is meant by a *polymerizable substance* in the composition of a recording medium. Depending on the mechanism of the polymerization process, three different meanings that support a classification in three broad categories have to be distinguished: (1) single-monomer systems; (2) two-mono-mer systems; and (3) polymer systems. In addition, depending on the number of reactive func-tions carried by the various monomers, the different systems undergo either linear or crosslinking polymerization.

10.6.2.2.1 The Single-Monomer Linear Polymerization

Only one polymerizable monofunctional monomer is present in the sensitive formulation and gives rise to a linear polymerization. In this case, the recording process involves a spatially inhomogeneous degree of conversion of this monomer that images the amplitude distribution in the holographic field. A refractive index modulation (Δn) parallels, then, this distribution, the amplitude of which depends both on the exposure, the contrast of the fringes in the incident pattern, and the index change between the monomer and the corresponding polymer. If the recording layer has a free surface, a thickness modula-tion may also be created as the result of a spatially inhomogeneous degree of shrinkage. In spite of the large Δn that are attainable (~ 0.1), these systems are of little value, since they are essentially unstable and cannot lend themselves to a fixing procedure.

If, however, a neutral substance, that is, non-chemically reactive, is added to such a monofunctional monomer, photopolymerization may interfere constructively with diffusion. Due to the gradient of the conversion rate resulting from the spatially inhomogeneous illumination, the photoconverted mono-mers accumulate in the bright fringes. Simultaneously, the neutral substance is driven off to the regions corresponding to the dark fringes. This coupled reaction–diffusion process can be maintained up to complete conversion of the monomer molecules. The segregation induced by the diffusion of the neu-tral substance results in a modulation of refractive index, the amplitude of which depends only on the difference in refractive index between this substance and the polymer. Values exceeding 0.2 are theo-retically possible. In practice, the essential limitation is due to the solubility of the lost molecule in the monomer–polymer system.

Reactive formulations in which the neutral substance is a polymer binder have also been described. Such systems present the advantage of a highly viscous consis-tency that simplifies the coating proce-dure and facilitates the storage. In addition, the polymer structure of this binder often increases the compatibility with the recording monomer–polymer system. The price to pay for these improvements is a slowing down of the two-way diffusion process and, consequently, a decrease in energy sensitivity of the recording formulation.

10.6.2.2.2 The Two-Monomer Linear Polymerization

In contrast with the previously described single-monomer polymerizable systems, it was surmised that using a mixture of monomers with different reactivity parameters and refractive indexes and, occasionally, diffusivity, shrinkage, and solubility parameters, could be advantageous. The mechanism of a hologram formation in the two-monomer systems (or by extension multicomponent systems) involves a linear photocopolymerization process leading to a different composition of the macromolecule, depending on the local rate of initiation that is, itself, controlled by the incident amplitude. The higher rate of polymerization in light-struck areas compared with dark regions associated with very different copolymerization ratios of the two monomers causes the more reactive one to be preferentially converted in the bright fringes, hence a gradual change in the average composition of the copolymer over the interfringe.

Compared with single-monomer/polymer binder systems, the multicomponent systems turn the coupling between diffusion and photochemical reactions to advantage, with an increased compatibility of the components, which is related to the copolymer structure and without impairing the sensitivity.

10.6.2.2.3 The Single-Monomer Crosslinking Polymerization

The simplest system that works according to this model contains a single multi-functional monomer. When subjected to a normal exposure with a pattern of interference fringes, the monomer polymerizes, with the local rate of initiation and polymerization being a function of the illumination.

Consequently, a modulation of refractive index parallels the spatial distribu-tion of the degree of conversion, which allows optical information to be recorded. This system suffers, however, from the same drawbacks as the single monomer system. When the conversion is carried out until the system vitrifies, both the light-struck and the non-light-struck regions of the record should finally reach the same maximum conversion. A careful examination of the polymer structure leads, however, to a less clear-cut conclusion. Indeed, the number of living macroradicals depends on the rate of initiation; hence, the indirect influence on the local value of the average length of the crosslinks and the architecture of the tridimensional network. Unfortunately, the amplitude of the refractive index resulting from such structural heterogeneities is too small ($< 10\text{-}4$) to open up real vistas for practical use.

10.6.2.2.4 The Multicomponent Crosslinking Polymerization

Basically, the principle that governs the building up of the index modulation in these systems shows great similarity to that described in the two-monomer linear polymerization. The only difference is the self-processing character that arises from the use of crosslinking monomers. In fact, substituting monofunctional monomers for a multicomponent mixture containing at least one polyfunctional monomer leads to the fast building up of a tridimensional network that converts the initial liquid system into a gel. This gel behaves, then, like a spongy structure in the holes and channels of which the less-reactive monomers are able to diffuse almost freely—hence a more pronounced effect of the transfer phenomena through diffusion processes. In addition to that very advantageous feature, the presence of crosslinking agents results in an entanglement of the polymer structure that finally annihilates the diffusion of unreacted species. The living macroradicals are, then, occluded in the structure; the record becomes insensitive to the holographic wavelength and can then be played back *in situ* without any degeneration of the hologram quality. Chemical development of a latent image, fixing, and repositioning are, therefore, not required. This can be most beneficial for real-time applications, such as interferometry or information storage.

10.6.2.2.5 Photocrosslinking Polymers

These systems are another major class of recording materials that has been thor-oughly studied by many research groups. On exposure to an appropriate pattern of light distribution, active species are created that trigger, in the end, the geminate attachment of two linear polymer chains, thus generating a crosslinked material. Since they basically involve a starting material that is a linear polymer, most of the systems developed to date from this concept are used in the form of dry films.

Several variations of chemical structure and composition have been described. The chemical function used to crosslink the linear polymer chain can be part of this chain (residual unsaturation or reactive pending group). In such a case, the coupling of the chain is more or less a static process that does not involve important changes of configuration or long-range migration, a feature that goes hand in hand with high-energy sensitivity. In return, since it is impossible to fix the record, the hologram has to be played back under inactinic conditions.

In a different approach, the crosslinking process may involve at least one component that is independent of the polymer chains and is, thus, able to diffuse freely in the matrix. In this case, the record can be made completely insensitive to the recording wavelength by chemical fixing. Systems can also be designed so as to obtain self-processing properties.

10.6.3 The Early History of Photopolymer Recording Materials

In 1969, Close et al. [89,90] were the very first to report on the use of photopoly-merizable materials for hologram recording. Their pioneering experiment involved a mixture of metal acrylates, the polymerization of which was initiated by the methylene blue/sodium *p*-toluene sulfinate system. The holographic source was a ruby laser ($\lambda = 694.3$nm), and energy sensitivities of several hundreds of mJ/cm^2 with a diffraction efficiency of about 40% were reported. The holograms required a post-exposure to a mercury arc lamp to consume the unreacted sensitizer up to complete bleaching and, thus, prevent them from interacting with the reconstructing beam that could result in degeneration of the record.

Jenney et al. [91–94] carried out detailed experiments on the same material and investigated the influence of many optical photonic and chemical parameters. Energy sensitivities lower than 1 mJ/cm2 were achieved on materials sensitized to both the red and green spectral range with various sensitizers. Many fixing methods were pro-posed, such as the flash lamp, thermal treatment or long-term storage in the dark.

The original materials devloped by E. I. Du Pont de Nemours & Co. also date from the early 1970s [95–97]. They were based on acrylate monomers associated with a cellulose binder and a broadband sensitizing system (sensitive both to UV and blue-green light). These materials were reported to be able to record transmission gratings with 90% efficiency; compared with Jenney's formulation, they exhibited a relatively flat MTF up to 3000 lines/mm and 6–8 months shelf life when stored in a cool atmosphere.

Several authors investigated the mechanism of hologram formation [98,99]. Although not completely understood, a complicated process involving both polymerization and monomer diffusion was assumed to account for the density changes and, thus, the refractive index variations that gradually convert the polymerizable layer into a hologram.

As possible substitutes for acrylate monomers, Van Renesse [100], Sukagawara et al. [101,102], and Martin et al. [103] used mixtures of acrylamides and bisacrylamides sensitized to red wavelengths by methylene blue. These systems, which exhibited fairly high energy sensitivity (5mJ/cm^2), were observed to suffer a pronounced reciprocity failure. From these new formulations, Sadlej and Smolinksa [104] introduced a series of variations intended to improve the shelf life of the recording layers and the dark stability of the hologram. Polymer binders, such as polyvinylacetate and derivatives, polyvinyl alcohol, methyl cellulose, and gelatin, were tested with various results. Besides an improved stability, the polymer binders were reported to induce transmission and sensitivity inhomogeneities. If one restricts the review of the early ages of photopolymer recording materials to the years previous to 1975, one cannot afford to ignore Tomlinson's research work on polymethylmethacrylate (PMMA) [105,106]. Three-dimensional gratings were, thus, recorded in very thick PMMA samples (about 2 mm) by interfering two UV laser beams at 325 nm. Spatial frequencies up to 5000 lines/mm with diffraction efficiencies as large as 96% were reported.

The mechanism of hologram formation in PMMA was assigned to the cross-linking of homopolymer fragments by free radicals produced through photocleavage of peroxide or hydroperoxide groups present in the polymer structure as the result of its auto-oxidation during the polymerization of the starting material itself. It was also postulated that the local increase in density may be due to polymerization of small amounts of unreacted monomer trapped in the PMMA structure. No evidence, however, was offered in support of these statements. The most obvious disadvantage of this material was the very poor reproducibility of the holographic characteristics and the great difficulty of extenting its sensitivity to the visible range.

In conclusion, it is worthy to note that the foundations used nowadays to formulate polymerizable systems and to describe the mechanism of hologram formation have been laid from the mid-1970s. The improvements introduced since then are essentially concerned with the adaptability of the systems, their stability, the amplitude and the shape of the modulation of the refractive index and/or the thickness, and the signal-to-noise ratio.

10.6.4 The State of the Art

A great number of state-of-the-art reviews [107–117] on recording materials have been documented in detail over the past decades. Most of these reviews classify photopolymers in categories based either on the mechanism of the photochemical process taking place when holographic irradiation takes place or on the composition of the sensitive layer. Most frequently, they are divided into (i) photopolymerizable systems, (ii) photocrosslinking systems, and (iii) doped polymer systems. Another distinctive feature that is also taken into account to arrange them in categories is the state of the polymer formulation: that is, (1) liquid composition and (2) dry films. As will be discussed later, such classifications do not fully account for the specificities of all the physicochemical mechanisms used to record optical information in polymer materials.

Indeed, all the investigators concur in the belief that hologram formation is a complicated process involving both photopolymerization and mass transport by diffusion. Consequently, the distinction introduced between linear and crosslinking polymerization or copolymerization is hardly justifiable since it does not reflect basically different behaviors of the materials in terms of photoconversion or diffusion rates. Likewise, the subtle distinction between liquid and dry formulation insofar as this refers to the flowing or nonsticky character of a material is quite inappropriate, since it does not give any indication as regards the possibility of undergoing mass transport through diffusive movements. In fact, a dry system consisting of a crosslinked copolymer that contains a substantial percentage of residual monomers may well be the seat of faster liquid–liquid diffusive processes than a homogeneous liquid formulation with very high viscosity.

In the background of the statements developed in the second paragraph of the present review, the various recording systems that are in the process of development in the university research groups and those few ones that are commercially available can be arranged in three classes: (1) systems requiring development, (2) self-processing materials, and (3) dichromated gelatin mimetic systems. This classification does not take into account the physical state of the material before exposition. It puts forward the recording mechanism and its interrelation with the development and the processing step.

10.6.4.1 Systems Requiring Post-Treatment

As described earlier, the recording mechanism in photopolymers involves, with only a few exceptions, the coupling between a photoconversion process and a diffusion mechanism. On initiation of the polymerization in the region illuminated by a bright fringe, the monomer converts to polymer. As a result of several kinds of forcing functions (e.g., gradient of concentration or gradient of solubility), an additional monomer diffuses to these regions from non-light-struck areas while the large size of the living polymer chains or elements of polymer network inhibits their diffusion. When the reactive mixture contains a binder, its solubility in the monomer may also decrease as the conversion proceeds. Consequently, there

TABLE 10.1　Characteristics of the Recording Systems Requiring Development

Reference	Conditioning	Thickness (mm)	Recording Wavelength (nm)	Sensitivity (mJ/cm^2)	Resolution (lines/mm)	Diffraction Efficiency (%)
Dupont Omnidex	Coated on plastic film	5–100	488–633	10–100	6000	>99
		–	–	–	–	–
Polaroid DMP-128	Coated on plastic film	1–30	442–647	5–30	5000	80–95
		–	–	–	–	–
PMMA/titanocene	PMMA block	500–3000	514	4000	—	"" 100
Acrylamide	Coated on glass	100	633	100	3000	80

may be some net migration of binder away from the bright areas. This process continues until no monomer capable of migrating is left and/or until its mobility in the polymer network becomes negligible. At this stage, the information is recorded in the polymer layer as a modulation of the refractive index. The diffusive process may, yet, continue until any remaining monomer is present. In addition, the index modulation resulting from the partial segregation of the component of the mixture is often modest.

Several post-treatments aimed at increasing this modulation and/or at canceling any remaining photosensitivity were described. The following systems exemplify these different development, fixing, or enhancement treatments (Table 10.1).

10.6.4.1.1 du Pont's Omnidex System

This material, which is endowed with several outstanding features, is one of the most attractive photopolymers for holographic recording. Various formulations based on mixtures of aromatic aliphatic acrylates and polymer binders, such as cellulose acetate–butyrate, are available. These materials that can be sensitized either in the red or the blue-green wavelength range can record holograms of almost 100% efficiency with exposure energies down to 50–100 mJ/cm2 [118–124].

When imaging such polymerizable formulations, the hologram builds up in real time, the film vitrifies, and finally the conversion stops. The resulting hologram is, then, stable on further illumination, and modulations of the refractive index in the order of 0.01 are typically attained.

A subsequent thermal processing by stoving at 80–120°C for 1–3 hours induces an important increase of Δn (a maximum of 0.073 has been achieved) while the tuning wavelength changes very little (a few nanometers). Polymer holograms can also be processed by immersing them in organic liquids that swell the coating or extract some elements of it. This treatment results in a shift of the playback wave-length and a bandwidth increase, so that reflection holograms reconstructed with white light appear brighter to the eye.

10.6.4.1.2 Polaroid's DMP-128 System

Another example of a polymer recording material with specially attractive features was developed at the Polaroid Corporation. This formulation was based, among others, on a mixture of acrylates, difunctional acrylamide, and a polyvinylpyrrolidone binder. The sequential processing steps include: an incubation in a humid environment prior to exposure; a holographic illumination with a 5–30 mJ/cm2 exposure, depending on the type of hologram recorded; a uniform white light illumination for a few minutes; an incubation in a liquid developer/fixer; a rinse removing the processing chemicals; and, finally, a careful drying [125–133].

One of the outstanding characteristics of the holograms recorded with these materials is their insensitivity under high humidity environments (no significant alteration of the diffractive properties after nine months incubation at 95% relative humidity at room temperature). In DMP-128, before chemical processing, the holograms formed show a diffraction efficiency lower than 0.1%. The immersion in a developer/fixer bath removes soluble components and produces differential swelling and shrinkage that greatly amplify the modulation of the refractive index and cause very bright high diffraction efficiency holograms to appear.

10.6.4.1.3 Systems with Thermal or Photochemical Repolymerization

An original polymer material based on an advanced formulation that contains a variety of monomers (e.g., acrylate, methacrylate, vinylcarbazole), a binder, radical precursors, solvents, and a sensitizer was developed by Zhang et al. [134]. The fixing procedure of the holograms recorded with this material involves the overcoating of the record with a polymerizable liquid formulation, then its annealing and, finally, the curing of the layer by either a thermal or a photochemical treatment. Since the peak reflection wavelength depends on the condition of the post-treatment, these materials were used to fabricate trichromatic filters or pseudocolor reflection holograms.

10.6.4.1.4 PMMA/Titanocene Dichloride Systems

Thick PMMA samples containing various amounts of residual MMA and titanocene dichloride were also investigated for holographic recording in the blue-green wave-length range [135]. The recording mechanism was assumed to be due to a complex photoprocess in which photodegradation of the homopolymer chain, photopolymer-ization of residual monomers, as well as crosslinking of photogenerated fragments cannot be discriminated. This material presents the outstanding advantage of allowing the superposition of a large number of volume holograms without significant intertalk.

10.6.4.1.5 The Latent Imaging Polymers

A latent imaging polymer was described a few years ago [106]. It was based on a microporous glass substrate with a nanometric pore diameter on the surface of which a photosensitive molecule generating radicals was chemisorbed. The recording process involves: (1) illumination by a pattern of interference fringes, (2) filling the pores with a polymerizable formulation and, then, (3) uniform overall exposure.

10.6.4.2 Self-Processing Materials

In spite of their very attractive features, the recording systems described above do not permit immediate and in-situ reconstruction. Since this property is a prerequisite to the use of a recording material for applications, such as real-time holographic interferometry, a great deal of work has been devoted to the development of materials, a step that could dispense with any further processing after holographic illumination. The basic concept underlying the common approach developed by all the scientists involved in that field is the use of a formulation undergoing both cross-linking copolymerization and segregation of the polymerizable monomer units. The full self-processing and fixing character is achieved when the polymerization process terminates (due to the overall vitrification of the recording sample) just as the amplitude of the refractive index modulation resulting from microsyneresis and differences in the microscopic composition of the crosslinked polymer chains passes through its maximum.

A large number of materials developed according to this principle are described in the literature. They are generally liquid compositions that specially favor the diffusive motions in the early stages of the recording process. Since the initial viscosity of the formulation is a key factor, the major differences are in the choice of polymerizable structures that allows this parameter to be adjusted. The different systems are arranged in two categories: (1) the diluent + oligomer systems and (2) the prepolymerized multicomponent systems (Table 10.2).

10.6.4.2.1 Diluent and Oligomer Systems

These formulations, very popular in the former Soviet Union, were based mainly on oligourethane-, oligocarbonate-, and oligoether-acrylates and multiacrylates [136]. In some cases, inert components with high refractive indexes were also incorporated. These compositions are sensitive over the 300–500 nm spectral range, with a maximum resolution in the order of 2500 lines/mm. Similar materials sensitive to the He–Ne wavelength were also developed using acrylamide derivatives as the reactive monomers, several additives, and the methylene blue/amine system as a sensitizer. An efficiency of 60% was reported at 633 nm with an energy sensitivity of about 50 mJ/cm2 [137–141].

TABLE 10.2 Characteristics for the Systems with Self-Processing Properties

Reference	Conditioning	Thickness (mm)	Recording Wavelength (nm)	Sensitivity (mJ/cm²)	Resolution (lines/mm)	Diffraction Efficiency (%)
Diluent +	Liquid between	20	300–500	20	1500–6000	80
oligomers	glass plates	–	–	–	–	–
(FPK-488)	–	–	–	–	–	–
Diluent +	Liquid between	20	633	50	—	60
oligomers	glass plates	–	–	–	–	–
(FPK-488)	–	–	–	–	–	–
Prepolymerized	Liquid between	20–100	450–800	100–500	=3000	80
multicomponents	glass plates	–	–	–	–	–
(PHG###)	–	–	–	–	–	–

A dry acrylamide–polyvinylalcohol formulation based on the same approach was also used for optical information storage. This polymeric material exhibits a self-processing and fixing character, the completion of which requires a fairly long exposure. The records can, however, be played back with an attenuated reading beam power (200 µW/cm²) before reaching complete inertness. It can also be fixed by illumination with the 365-nm line of a mercury arc, which terminates the polymerization [142,143].

10.6.4.2.2 The Prepolymerized Multicomponent Systems

The mechanism of hologram formation in these multicomponent systems involves a differential cross-linking copolymerization between the areas of higher and lower intensity. Because of the different reactivity parameters, the rates of incorporation of the monomer structures are different, and a modulation of the chemical composition of the final fully polymerized material is created. The novelty of these materials consists in the use of a formulation containing, among others, a highly multifunctional acrylate monomer with such a high reactivity that on preilluminating the layer with a homogeneously distributed beam, it forms almost instantaneously a sponge-like structure in the channels and cavities of which the various remaining monomers diffuse freely during the subsequent holographic illumination. The presence of this network since the very beginning of the recording process allows the coupling between photochemical conversion and transport to be more efficient in the sense that it simultaneously presents the advantage of a liquid formulation over diffusion taking place and that of solid layers over the shrinkage and the resulting hydrodynamical strain therefrom. Finally, the refractive index modulation arising from the modulation of the microscopic chemical composition is much larger than the one paralleling a modulation of segment density or crosslinking average length.

A series of materials (called PHG-###), formulated according to these general lines and having sensitivities from 450 to 800 nm, were proposed [144–155]. The sponge-forming monomer was typically a multifunctional acrylate monomer (pentaerythritol trior tetra-acrylate, dipentaerythritol pentacrylate). It was associated with various low functionality acrylates, a xanthene or polymethine dye sensitizer, an amine cosynergist and, occasionally, specific additives or transfer agents that improved the recording characteristics. Diffraction efficiencies of about 90% were achieved with an almost flat frequency response from 500 to 2500 lines/mm. The application fields, such as holographic interferometry (both real-time and time-average) [147,152], holographic images [149], computer-generated holograms [150], multiple holograms, and recording in chopped mode [145], can be quoted in illustration.

10.6.4.3 Photocrosslinkable Polymers

Photocrosslinkable materials are another important class of polymer recording materials that include two categories of systems: (1) polymers undergoing crosslinking through intercalation of monomer units or by simple coupling of the linear polymer chain and (2) polymers undergoing crosslinking through a complexation process involving a metal ion (Table 10.3).

TABLE 10.3 Characteristics of the Recording Systems Involving the Crosslinking of a Polymer Structure

Reference	Conditioning	Thickness (mm)	Recording Wavelength (nm)	Sensitivity (mJ/ cm²)	Resolution (lines/mm)	Diffraction Efficiency (%)
p-Vinylcarbazole	Solid film on glass	2.5–7	488	50–500	800–2500	80
PMMA	Solid film on glass	100–200	488	7000	2000	HH 100
DCPVA	Solid film on glass	30–60	488	500	3000	"" 70
DCPAA	Solid film on glass	60	488	200	3000	"" 65
FePVA	Solid film on glass	60	488	=15 000	3000	80

10.6.4.3.1 The Monomer/Polymer System

An attractive example of such systems uses poly N-vinylcarbazole as the base polymer, a carbonyl initiator, and a sensitizer. The mechanism of recording involves, first, a holographic exposure that induces photocrosslinking to the polymer base in proportion to the illumination received by every part of the recording layer. A modulation of density is thus created, and a hologram is recorded. This record is then pretreated with a good solvent of the initiator–sensitizer system, to remove these components. The next step consists in swelling the record in a good solvent of the polymer base; the final treatment involves the deswelling, by dipping in a bad solvent of the polymer base. This swelling–deswelling treatment causes the crosslinked regions to reconfigurate, shrink, and finally partially crystallize. Consequently, the attainable modulation of refractive index exceeds by far the values reported in other systems (up to 0.32). Several practical applications were developed with this class of materials, the main feature of which is an almost infinite durability under extreme environmental conditions (at 70°C and 95% RH) [154].

Another system that held the attention of several groups used the photoinitiated coupling of linear PMMA. A radical initiator capable of abstracting hydrogen atoms creates macroradical sites on the polymer backbone, which decay by geminate recombination. The pattern of interference fringes is thus transferred as a pattern of crosslinking density. Typically, such systems exhibit a poor energy sensitivity (several J/cm2 but their angular discrimination capability is excellent over a spatial frequeny range extending from 100 to 2000 lines/mm [155].

10.6.4.3.2 The Metal Ion/Polymer Systems

All the systems categorized under this title were developed on the model of dichro-mated gelatine (DCG). Although the base component of this type of material that has been used as a recording medium ever since 1968 [156] is a biopolymer, it is generally listed and described under a specific heading in the reviews dealing with holographic recording. The recording mechanism prevailing in DCG implies the crosslinking of the gelatin backbone by Cr(III) ions photogenerated through the reduction of dichromate [Cr(VI)] centers. As the result of a chemical post-treatment under carefully controlled conditions, a large refractive index modulation can be achieved (0.08), thus making it possible to record both transmission and reflection volume phase holograms with near-ideal diffraction efficiencies. DCG is, besides, endowed with specially desirable properties, such as uniform spatial frequency response over a broad range of frequencies (from 100 to about 6000 lines/mm) and reprocessing capacity [157–160].

In spite of these outstanding advantages, DCG suffers from some typical draw-backs that may temper the enthusiasm of the holographers. Among others, the fluctuation of many characteristic parameters from batch to batch, the need for a fine-tuning of the prehardening treatment, the complex procedure of development and fixing that requries a specially accomplished know-how, and the sensitivity of the record to environmental factors are not of lesser importance. In addition to these, DCG needs to be sensitized by incorporation of a suitable blue dye to allow extension of its sensitivity to the red. A fair amount of research has been carried out with regard to the optimization and characterization of these sensitized systems as well as to their use in various applications, such as holographic elements, head-up displays, laser scanning systems, fiber-optic couplers, optical interconnects, and dichoric filters [161–166].

Since the basic mechanism of the phototransformation of gelatin is still a controversial question and owing to its typical shortcomings, several DCG-mimetic systems likely to bring about significant improvements were studied.

10.6.4.3.3 Polyvinylalcohol/Cr(III) Systems (DCPVA)

One of the most popular systems uses polyvinylalcohol (PVA) as the polymer base and Cr(III) as the crosslinking agent [167]. It has received continuous attention from many research groups who studied various aspects of its holographic recording performance: exposure energy, beam ratio, polarization, MTF, angular selectivity, acidity of the medium, effect of electron donors, effect of plasticizing substances, and molecular weight of the polymer base. Even though this material exhibits a high real-time diffraction efficiency (~ 65% for an exposure of about 1 J/cm^2), several chemical developing methods allow the final efficiency to be improved (up to 70% for an exposure level of 200 mJ/cm2) [168,169]. In a continuing effort to penetrate the secrets of the recording mechanism of the PVA/Cr(III) systems, many physicochemical studies were carried out. They suggest, convergingly, the intermediacy of Cr(V) in the photoreduction of Cr(VI) to Cr(III). Such systems were also reported to lend themselves to sensitization in the red by incorporation of a blue dye (e.g., methylene blue) [170–180].

10.6.4.3.4 Polyacrylic Acid/Cr(III) (DCPAA)

Polyacrylic acid (PAA) was also used as a possible substitute for gelatin to record volume holograms. Although much less work has been devoted to the optimization of this material, a great similarity with DCPVA was found. Efficiencies exceeding 65% at an exposure of 200 J/cm2 were obtained [178,180].

10.6.4.3.5 Polyvinylalcohol/Fe(II) (FePVA)

As a continuation of the ongoing search for DCG-mimetic systems, the feasability of substituting Fe(II) for Cr(III) in PVA-based materials was examined. After optimi-zation of the initial Fe(III) content, the electron donor structure, and the recording conditions, a diffraction efficiency of about 80% at an exposure of 25 J/cm2 was achieved [173].

10.6.5 Doped Polymer Materials

During the recording step, all of the materials described in this review are the seat of transformations involving the creation of chemical bonds between monomer or polymer structures. In essence, the recording process is thus a nonreversible process. The possibility of obtaining erasable and reusable polymer materials capable of recording a large number of write/read/erase (WRE) cycles was examined by several research groups. Several approaches were proposed that employ matrixes of homopolymers or copolymers (linear or weakly crosslinked) and dye dopants. The selection of the dye dopant is essentially dependent on its photochromic character and on its photostability. Materials doped with azo dyes or spiropyran derivatives were thus reported to be suitable for real-time reversible amplitude or phase hologram recording. Some of the corresponding systems are capable of more than 104 cycles without noticeable fatigue, with thermal self-erasing times of a few seconds. Efforts were also made at developing similar materials where the photochromic active molecule is chemically intercalated in the polymer structure or bound to a side group. Since linearly polarized light induces optical anisotropy in the sensitive film due to reorientation and alignment of the absorbing molecules, these doped materials were mainly used for polarization holographic recording [181–184].

A similar approach was used by several authors who formulated recording materials in doping PVA with bacteriohodopsin (Table 10.4). This structure is a protein transducer involved in the mechanism of mammalian vision. Upon absorption of light energy, this purple ground state sensitizer passes through several intermediate states to a blue light absorbing photoproduct that under standard biological conditions thermally reverts to the ground state material within a decay time of about 10 ms. This

TABLE 10.4 Holographic Characteristics of the Dye-Doped Recording Systems

Reference	Conditioning	Thickness (mm)	Recording Wavelength (nm)	Sensitivity (mJ/cm^2)	Resolution (lines/mm)	Diffraction Efficiency (%)
Methyl orange/ PVA	Solid film on glass	15–30	442–488	=300	500–4000	35
Bacteriorhodopsin	Membrane	150–500	360–600	—	=5000	11

back-process can be drastically slowed down by addition of chemical agents. This product was used as a dopant in holographic recording media that can be read out at wavelengths between 620 and 700 nm in some interesting applications, such as dynamic filtering with a spatial light modulator or optical pattern recognition [185–188].

10.6.6 Spectral Hole Burning

Hole burning is a phenomenon that takes places when molecules undergoing a photoreaction generating photoproducts absorbing at different spectral positions are excited by a sharply monochromatic light. With such assumptions, a "hole" is left in the inhomogeneously broadened absorption band. If this operation is repeated at different wavelengths within the absorption band, every hole can be associated with a bit of information and a high storage capacity becomes available. The feasability of this concept was demonstrated using an oxazine dye in a polyvinylbutyral matrix at 4.1 K [189,190].

10.7 Recent Advances in Photopolymer Materials Research

In the last fifteen years (2000-2015), different approaches have been suggested for improvement of the recording characteristics of photopolymer materials for applications in optical engineering. Among these are optimisation of the monomer [191–225], the crosslinker [226], the sensitizer [227–237], binder permeability and binder/monomer relative refractive indices [238–248], and use of inorganic or organic dopants [249–280]. The characteristics that have been improved are the spatial frequency response, shrinkage of the layers as a result of photopolymerization, dynamic range of the layer (related to the maximum achievable refractive index modulation), stability of the optically recorded patterns and environmental stability of the layers. In addition to the above features, directly related to the photopolymer materials performance in holographic recording, a significant effort has been dedicated to the functionalization of these materials, which allows for the fabrication of interactive photonic structures. The incorporation of liquid crystal molecules in photopolymers has been attracting interest due to the possibility to create electro-optical devices with properties controlled by application of an electric field. The functionalization of photopolymers with materials sensitive to specific chemicals has been used for the development of holographic sensors. Research papers studying different types of functionalized materials are described in greater detail later in this chapter.

10.7.1 Optimization of the Monomer

There are two main photopolymer systems that have been developed in recent years: systems utilizing free radical polymerisation [191–204,208–225] and systems utilizing ring opening polymerisation [205–207]. Different monomers have been proposed for use in these systems. The choice of monomer determines the reactivity of the photosensitive system and thus contributes to the rate of polymerization. The size of the monomer molecule determines its ability to move within a host polymer matrix with defined permeability and thus determines the photopolymer spatial frequency response. The refractive index of the monomer and the produced polymer chains determines the achievable refractive index modulation in holographic recording. Acrylamide [191–196], methyl methacrylate [200,201], dendritic polyisophthalate [204], a mixture of polymerizable acrylates and epoxides [205,206], high refractive index monomers, such as dendritic polyisophthalate endcapped with naphthyl groups [207] and 9-(2,3-bis(allyloxy)propyl)-9H-carbazole (BAPC) [208] are some of the monomers successfully used in development of photopolymer systems. Significant effort has also been directed in development of systems based on less toxic monomers with a view to their application in commercial products [209-213].

The process of photopolymer optimization by selection of suitable monomers has been informed by theoretical [214-217] and experimental [192,218–225] studies of the monomer/short polymer chain diffusion and the polymerization rate during holographic recording.

10.7.2 Optimization of the Crosslinker

Another very important element of the photopolymer system is the crosslinker. It was first introduced by in order to increase the shelf life of the recorded structures. Crosslinking leads to a change of the polymer matrix permeability and thus is also important for the dynamics of the photoinduced processes and the achievable refractive index modulation [227].

10.7.3 Optimization of the Sensitizer/Initiator

Significant effort in photopolymer research has been devoted to optimization of the sensitizer/initiator system. Different dye sensitizers and initiators have been studied with the aim of improving the photopolymer sensitivity [227–231]. Dyes derived from pyrromethene have been successfully used in photopolymers where the dye acts as sensitizer and initiator at the same time [232]. They have been introduced in a hydrophilic binder photopolymer to change the redox initiation system in photopolymers with xanthene dyes. In this way, the use of triethanolamine as the redox initiator, typical in hydrophilic binder photopolymers is avoided [233–235]. Photopolymers sensitive to two colors as well as panchromatic photopolymers have been successfully developed [236–239]. A novel approach to photopolymer sensitization has been proposed in [240] where the sensitizer is added to a unsensitized photopolymer layer at a later stage. This allows for a precise control of the spatial location of the photoinduced refractive index changes that can be utilized for holography, lithography, and photonic device fabrication.

10.7.4 Binder Optimization

The introduction of a binder in the photopolymer system allows for the fabrication of dry layers, which makes their handling significantly easier. The binder permeability and refractive index are the two most important parameters that have to be considered when the photopolymer solution is optimized. Polyvinyl alcohol [241,242], poly(bisphenol-A-carbonate) [244], epoxy resins [245,246], hyperbranched polyisophthalesters [247] and hybrid sol-gel matrices [191,247,248] are some of the binders that have been successfully used. Due to the presence of high optical quality binders in the photopolymer system, it has been possible to fabricate photopolymer layers of thickness approaching the mm scale with relatively low scattering [249–251]. Such thick layers are particularly interesting for applications in holographic data storage.

10.7.5 Photopolymer Nanocomposites

One approach for improvement of the properties of photopolymers, which has been attracting significant scientific interest, is based on development of hybrid organic-inorganic photosensitive nanocomposites. Nanoparticle doped photopolymers have been the subject of intensive research due to their improved optical, mechanical and thermal properties [252-271]. The main advantage of using nanoparticles as dopants is that due to their hybrid nature, the resulting polymer nanocomposites have tuneable characteristics. The small size of the nanoparticles reduces the amount of scattering and facilitates the fabrication of optical layers of very high quality. Another advantage is that compared to traditional additives the loading requirements for nanosize dopants are lower. It has been demonstrated that the incorporation of nanodopants in a host photopolymerizable matrix can lead to a substantial improvement of their recording sensitivity [253,254], dynamic range [252–266], reduction of shrinkage during photopolymerization [253,260,261,264,268] and improved thermal stability [269].

The idea of incorporating inorganic nanoparticles into holographic photopolymers was first introduced in 1996 by Oliveira et al. [253]. The development of photopolymerizable nanocomposites for holographic recording was inspired by the phenomenon of light induced mass transport in photopolymerizable materials, which can lead to a redistribution of the nanoparticles and thus to improved holographic properties. The introduction of nanodopants with significantly different refractive indices from that of the host material and their redistribution during holographic recording improves the material's dynamic range [253,266]. It has been reported that during holographic recording an increased refractive index modulation and thus dynamic range can be achieved for optimum values of recording intensity and volume fractions of incorporated solid nanoparticles, such as SiO_2 [258,259,268], TiO_2 [257,260,261,265], ZrO_2 [253,262,263,277], gold [265] and silver [270,271], and a range of porous nanoparticles, such as zeolites [260,264,276–278]. Quantum dot nanoparticles, such as CdSe [272,273], have been incorporated into photopolymer formulations due to their high fluorescence at certain wavelengths. This property is particularly interesting for holographic applications, such as sensing and product authentication devices.

The incorporation of various types of nanoparticles in a polymer matrix has been a significant challenge due to the poor compatibility between the polymer host and the dopants. Furthermore, the nanodopants produce optical losses [257] due to increased scattering with increasing size and concentration of the nanoparticles and with increasing difference between their refractive index and that of the photopolymer matrix. In order to address this challenge, nanoparticles with functionalized surfaces [259,262,263], controlled chemical composition, hydrophilicity, hydrophobicity and overall particle size and morphology have been used.

The redistribution of the incorporated nanoparticles during holographic recording has been observed by transmission electron microscopy [274] electron probe microanalysis [275], by real time phase shift measurements [276] and by confocal scanning Raman microscopy [277–280]. In most cases, the nanoparticles are excluded from the bright fringes and are moved to the dark fringes. This process is accompanied with a diffusion of the monomer molecules in the opposite direction. The best nanoparticle redistribution typically has been observed at reduced laser power intensity, smaller interferogram periodicity and decreased nanoparticle size, indicating that particle segregation is dominated by diffusion-limited nanoparticle transport directed by a matrix experiencing a gradient of polymerization kinetics [281–283].

10.7.6 Surface Relief in Photopolymers

It has been observed that with particular geometries of holographic recording in photopolymers in addition to the volume holograms, surface relief modulation is inscribed [284–292].

Photoinduced single-step inscription of surface relief modulation in self-processing photopolymer systems opens attractive perspectives for applications, such as diffractive optical elements and biosensors. In addition, this effect can be used to create channels and chambers in microfluidic systems for fabrication of lab-on-a chip devices. Important characteristics of the material in relation to these applications are its spatial resolution, the amplitude of photoinduced surface relief, the required recording time for achievement of maximum surface modulation and its long-term stability.

The models describing surface relief formation in photopolymers can be divided into two main groups. The first group explains the relief formation by local shrinkage of the photopolymer determined by the intensity of recording light. This mechanism is applicable to the systems where the peaks of the surface relief appear in the dark areas [284]. The models of the second group are based on the assumption that diffusion driven mass transport is responsible for the relief formation. These models are consistent with experimental data demonstrating that the surface relief peaks appear in the illuminated areas [285–288].

A common feature of the process of optically inscribed surface relief modulation in self-processing photopolymerizable materials is its limited spatial frequency resolution. It is observed that the amplitude of the surface relief profile decreases with the increase of the spatial frequency and the upper resolution limit in these materials typically does not exceed a few hundred lines/mm [287–290]. For example, the authors of [290] present a detailed study of the surface relief height on the spatial frequency of recording for two photopolymer compositions. Depending on the number of components participating in the formation of the surface relief, the two photopolymers are named binary and mono-compositions. The binary composition was found to perform significantly better. It is demonstrated that in the range 20–70 lines/mm, the surface relief height is at its maximum and in order of 1.6 µm. At spatial frequency of 1100 lines/mm, the surface relief height drops to 80–100 µm.

The limited surface relief height at high spatial frequency in self-processing photopolymers is usually ascribed to surface tension, which prevents the surface deformation at high spatial frequencies. For many practical applications, such as fabrication of switchable electro-optical devices and design of optical sensors, sub-micron resolution is required. Improvement of the surface relief modulation spatial frequency response and surface patterns with sub-micron resolution in acrylamide–based photopolymer by optimization of monomer and initiator (also playing the role of a plasticiser) concentrations and by post recording thermal treatment has been reported in [292]. Surface relief features varying in height from 4 µm at 16 lines/mm to 15 µm height at 1550 l/mm have been achieved in this study.

10.7.7 Theoretical Modeling of the Holographic Recording in Photopolymers

Grating evolution in photopolymer systems has been studied theoretically and experimentally by several authors [192,218,292,300]. Early diffusion models describing holographic recording in photopolymers, based on free radical polymerization, predict that a key factor determining the dynamics, and the refractive index spatial profile and modulation of the recorded holographic grating, is the ratio between the rate of polymerization and the monomer diffusion rate. Experimental evaluation of these two rates is important for determination of the optimal conditions for holographic recording in a particular photopolymer system. The diffusion rate can be controlled by the matrix permeability and the size of the monomer molecules. The polymerization rate can be controlled by the recording intensity. Two extreme regimes of holographic recording at a given spatial frequency can be distinguished with respect to the ratio of the characteristic diffusion and polymerization rates. At recording conditions for which the diffusion rate is faster than the polymerization rate, the grating profile closely resembles that of the recording interference pattern and the saturation value of the refractive index modulation is high due to higher density modulation. When the monomer diffusion rate is slower than the polymerization rate, deviation from the cosinusoidal profile of the grating is observed and the diffraction efficiency at saturation is lower.

The common aim of the theoretical models reported in the literature in recent years has been the analysis of the poor high-spatial-frequency response of photopolymers. The low diffraction efficiency at high-spatial-frequency has been explained using two approaches both referring to a non-local response of the material. A non-local response refers to the case in which the response of the material at given point and time depends on what happens at other points and times in the medium. The non-local photopolymerization driven diffusion model (NPDD) [298,300,301] assumes that the polymer chains grow away from their initiation point and that this is the main factor contributing to the "spreading" of the polymer and smearing of the refractive index contrast. This model predicts that improvement at high spatial frequencies can be achieved if during the holographic recording shorter polymer chains are created. It has been suggested that the incorporation of a chain transfer agent in the photopolymer composition will improve the diffraction efficiency at high spatial frequency. The reported experimental results from recording in materials containing chain transfer agents demonstrate only marginal improvement of the response at spatial frequency around 3000 lines/mm and higher [302]. For example in [302], an improvement of the refractive index modulation from 4.2×10^{-4} to 5×10^{-4} at 2750 l/mm has been reported. Up to 5% diffraction efficiency has been reported at 4600 l/mm in an acrylamide-based photopolymer containing a chain transfer agent in [303]. Alternative models (also based on non-local response of the material) take into account the diffusion of short polymer chains, such as oligomers [304,308,318], away from the bright fringes thus reducing the refractive index modulation. Such process could be responsible for the decrease of diffraction efficiency at high spatial frequencies. The two-way diffusion model [192,304,305] and the immobilization-diffusion model [305] account for both monomer and polymer diffusion and moreover distinguish between short polymer chains capable of diffusing and long polymer chains that are immobile. It has been demonstrated that, similar to the NPDD models, these models can satisfactorily predict the poor high spatial frequency response in highly permeable photopolymers. The two-way diffusion model also predicts that the improvement of the high spatial frequency response of photopolymers must be directed toward decreasing the permeability of the photopolymer matrix and avoiding the creation of diffusing short polymer chains. The model has enabled the optimization of the high spatial frequency response of acrylamide-based photopolymers, to the extent that full color reflection holography [307] is now well established, along with a range of color changing holographic sensors [308,309].

Both diffusion of mobile polymer chains and non-local polymerization processes contribute to the limited spatial resolution in photopolymers. One or the other process has higher significance depending on the permeability of the recording material, the exposure conditions and the spatial frequency of recording.

Another model considering the role of the diffusion of mobile polymer chains has been reported recently [311]. This model is used to analyze the reaction/diffusion kinetics of two-chemistry diffusive photopolymers. It offers a general strategy for characterizing the reaction/diffusion kinetics of photopolymer media in which key processes are decoupled and independently measured.

The NPDD model has also been used to describe the recording in phenanthrenequinone (PQ) doped poly(methyl methacrylate) PMMA photopolymer material. The non-local response of this system is attributed to the semiquinone free radicals HPQ and a radical located at the end of the PMMA chain [311]. The same model has been applied to epoxy resin photopolymers [312].

10.7.8 Polymer Dispersed Liquid Crystals

The introduction of liquid crystals in photopolymerizable materials opened the possibility to develop optical devices with a new functionality—electrically switchable devices. There have been numerous studies of polymer dispersed liquid crystal materials in the last 15 years. The aim of the optimization of these types of devices is to achieve lower switching fields, higher diffraction efficiencies, better optical properties (less scattering), fast response times and higher thermal stabilities.

In the following, a brief review of some of the main achievements in this field will be presented. For more details the reader should consult the literature.

In 2001, Kyu et al. published a theoretical simulation of the emergence of nematic domains during patterned photopolymerization-induced phase separation in holographic polymer-dispersed liquid crystals [313]. The simulated morphological patterns in the concentration and orientation order parameter fields showed discrete layers of liquid-crystal droplets alternating periodically with polymer-network-rich layers. This structure had higher diffraction efficiency than the undoped polymer.

Different polymers were studied as hosts for the liquid crystals. In [314], holographic reflection gratings in polymer-dispersed liquid crystals (H-PDLCs) were formed by thiol–ene photopolymerization. Results indicated that thiol–ene polymers function as better hosts for H-PDLC than multifunctional acrylate. These differences were explained by a different temporal structure development caused by fundamental differences in the polymerization propagation mechanism: a step-growth addition mechanism for the thiol–ene system compared to a chain-growth addition mechanism in multifunctional acrylates. Morphology studies by TEM supported these conclusions, as significant differences in droplet shape and uniformity were observed. Discrete nematic droplets with a nearly spherical shape were observed in the thiol-ene polymer material. Thiol–ene polymers had lower switching fields, higher diffraction efficiencies, better optical properties, and higher thermal stability, but their response times were five times slower than those of acrylates.

The effect of monomer functionality on the morphology and performance of the holographic transmission gratings recorded in polymer dispersed liquid crystals was studied in [315]. Commercially available LTV-curable acrylate monomer was used with the nematic liquid crystal TL203. The morphology and surface topology of the transmission gratings observed through scanning electron microscopy and atomic force microscopy varied profoundly with the different monomer mixtures characterized by different average functionalities. The switching properties of the HPDLC gratings were correlated with the morphological changes.

A detailed characterization of the diffraction efficiency of holographic gratings with a nematic film-polymer-slice sequence structure has been reported in [316]. The dependence of the diffraction efficiency on temperature revealed a non-monotonic behavior, with several maxima and minima. The shapes of curves were observed to be dependent on slight changes in the initial concentration of the nematic component of the mixture; the number of extrema increased with an increase of this concentration. The dependence of the diffraction efficiency on an applied external voltage also appeared to be non-monotonic and depended on the temperature of the sample. The behavior of the gratings was explained in the framework of the conventional Kogelnik theory for the diffraction efficiency of Bragg gratings.

A significant challenge in optimization of the performance of switchable electro-optical devices based on HPDLC structures was their high scattering caused by the size of the liquid crystal droplets. This was addressed by development of novel type of structure—polymer liquid-crystal polymer slices (POLICRYPS) grating proposed in [317]. By preventing the appearance of the nematic phase during the curing process, it was possible to avoid the formation of liquid-crystal droplets and obtain a sharp and uniform morphology, which reduced scattering losses and increased diffraction efficiency. The structure is obtained by irradiating a homogeneous syrup of NLC, monomer and curing agent molecules with an interference pattern of UV/visible light under suitable experimental and geometrical conditions; the spatial periodicity can be easily varied from an almost nanometric (200 μm) to a micrometric (15 μm) scale. Where the effect on an impinging reading light beam is concerned, the POLICRYPS can be utilized either in a transmission or a reflection configuration (depending on the geometry and substrate used) with negligible scattering losses, while the effect of spatial modulation of the refractive index (from polymer to NLC values) can be switched on and off by applying an external electric field of the order of few V μm^{-1}. Possible applications of devices recorded in POLICRYPS, such as a switchable diffraction grating and a switchable optical phase modulator, an array of mirrorless optical micro-resonators for lasing effects and a tuneable Bragg filter, are described in [318].

Holographic polymer-dispersed liquid-crystal (H-PDLC) gratings have been used to fabricate electrically switchable, one-dimensional polymeric resonators [319]. Stimulated emission is demonstrated by optically pumping chromophores within the liquid-crystal domains of 1D bandgap structures. Electrically switchable laser resonance can be achieved in this structure, since applying an electric field across the grating aligns the directors of the liquid crystal, thus diminishing the refractive index profile and, consequently, the lasing action.

10.7.9 Polymer Shrinkage

Reduction of the polymerization induced shrinkage is one of the main challenges in photopolymer materials design. Low shrinkage is particularly important for certain holographic applications, such as holographic data storage and design of holographic optical elements.

For that reason, a significant effort has been dedicated to studies of shrinkage occurring in photopolymer layers during holographic recording. The shrinkage in photopolymers has been successfully characterized using Bragg detuning of slanted holographic gratings [257,320–325].

It has been reported that a significant reduction of shrinkage can be achieved by incorporation of inorganic nanoparticles in photopolymers. The addition of nanoparticles increases the mechanical stability of the layers and reduces the shrinkage during holographic recording from 1%–2% to less than 0.5% [257]. Another successful approach to reducing the shrinkage is the use of a "two chemistry" photopolymer system as suggested by researchers from Bell Laboratories, Lucent Technologies and later used in the development of the photopolymer material commercialized as a holographic data storage medium by InPhase Technologies. Shrinkage below 0.1% has been reported in these types of materials. Negligible shrinkage is observed in photopolymers based on cationic ring opening polymerization, such as the one developed by Polaroid and commercialized for holographic data storage by Aprilis. For example, the Polaroid ULSH500 family of photopolymers exhibit very low shrinkage (<0.1%–0.5%) over much of their useful dynamic range [326].

In most of the studies, it has been assumed that the lateral shrinkage of the material is negligible and only the shrinkage in a direction perpendicular to the photopolymer surface is significant. The importance of lateral shrinkage has been emphasized in [327] where researchers from Aprilis Inc. have characterized its effect by quantitative measurements of the shift and magnification errors of reconstructions from holograms [327].

The disadvantage of using Bragg detuning is that shrinkage measurements have to be carried out after holographic recording. A study utilizing holographic interferometry to measure shrinkage in real time during holographic recording has been reported recently [328]. Through analysis of the real-time shrinkage curves, it was possible to distinguish two processes that determine the value of shrinkage in the photopolymer layer. These processes are ascribed to monomer polymerization and crosslinking of polymer chains. The dependence of shrinkage of the layers on the conditions of recording, such as recording intensity, single or double beam exposure, and the physical properties of the layers, such as thickness, were studied. Higher shrinkage was observed with recordings at lower intensities and in thinner layers. Increased shrinkage was also observed in the case of single beam polymerization in comparison to the case of double beam holographic exposure.

10.7.10 New Applications of Photopolymers

10.7.10.1 Holographic Data Storage (HDS)

The field of Holographic Data Storage has been developed since 1963 when Van Heerden [329] mentioned the possibility of using alkali halides crystals to store information in their volume and not just on the surface of recording materials. Through decades, research groups at universities and companies have made efforts to design and build HDS systems. Great efforts have been made at Stanford University (Prof. Lambertus Hesselink), California Institute of Technology (Prof. Demetri Psaltis) and companies such as IBM, Polaroid (Aprilis), Lucent Technologies (Including InPhase Technologies), and more recently at General Electric. In Japan and Korea, companies such as LG, Sony (Optware), Daewoo and others have also collaborated.

The first attempts to build an HDS system are described in the book by Collier, et al. [330]. The pioneer work developed by Anderson [332] is also worth mention. Efforts developed in HDS until the year 2003 have been well described by Hesselink el al. [332]. More recently researchers from InPhase Technologies presented a book [333] describing their work and that of other collaborators. Unfortunately, this company stopped operating in 2010. Recently, the InPhase facilities were bought to found a new company named Alconia.

Next, we briefly describe the basic elements in a HDS system to take them as a basis to compare the work developed from about 2001 to 2014 in the development of new HDS photopolymeric recording materials. However, the HDS field is very extensive because it has been developed over about 50 years. Thus, there is a vast amount of information. We suggest the reader consult the literature.

HDS comprises mainly two steps: recording and read out of information. Recording media used in HDS are divided mainly in two classes. One class comprises the photorefractive inorganic crystals, such as Lithium Niobate, $LiNbO_3$: Fe, $BaTiO_3$:Rh, Ce and $KNbO_3$. On the other hand, organic media used to record the interference patterns are mainly photopolymers, such as the Dupont Holographic Recording Film (HRF), Polaroid (CROP) ULSH and the one made by Lucent Technologies, Tapestry.

Some of the ideal characteristics of the recording medium to be used in HDS are: fast response time (in the order of microseconds), sensitivity comparable to that of the photographic film, thermal stability and capability of retaining the information for more than 10 years, sensitivity to different light wavelengths, mass producibility (in quantity and large format), present good optical quality, should not be expensive, show dimensional stability (no shrinkage), low scatter, should involve a manufacturing process free of heat and solvents and preferably its manufacture time should be short.

Digital Versatile Disks (DVD) and Blue Ray Disks (BD) present a Disk Capacity of about 100 GB and a Data Transfer Rates of about 100 Mb/s. Photopolymer disks in HDS systems have shown capacities exceeding 250 GB and Data Transfer Rates of 120 Gb/s with access times of 50 μs. Usually photopolymers are a write-once-read-many (WORM) media. The photopolymer mixture is sandwiched between two parallel, flat glass or plastic plates. Photopolymer thickness ranges from 0.5 mm to 1 mm.

Two parameters of the medium used in HDS are its Dynamic Range and the Sample Sensitivity. The Dynamic Range is defined as

$$M_\# = \sum_{i=1}^{N} \sqrt{\eta_i} \qquad (10.1)$$

where N is the total number of holograms and η_i is the diffraction efficiency of the ith hologram. The Sample Sensitivity = s' is given by the following formula:

$$s' = \frac{\frac{\partial\left(\sqrt{\eta}\right)}{\partial t}}{I_0 t} \qquad (10.2)$$

where t is the exposure time. s' is expressed in cm2/Joule. Photorefractive materials show a Sample Sensitivity ranging from about 0.01 to 0.1 cm2/J and M# of about 1–10 for 1 cm thickness crystals. On the other hand, photopolymers show Sample Sensitivities in the range of about 20 cm2/J and M# as high as 10 for samples with a thickness of about 0.5 mm.

Photopolymers are usually mixtures that could comprise monomers/oligomers, sensitizers, a binder and an internal polymer scaffold. When light is absorbed by the sensitizer, it triggers the reaction that converts the monomer into polymers. The binder presents a refractive index different than the monomer. In the recording process, there is amplification because polymerization reactions are chain reactions.

HDS systems comprise high technology components like wavelength, power and image sensors, lasers, spatial light modulators, beam scanners (galvanometers, MEMS, acousto-optics), isometric lenses, micropositioners and more. There are several optical configurations to perform HDS. Here, we present the page composer setup used for recording and reading when photopolymeric discs are used, Figure 10.11.

At recording time, a beamsplitter is used to obtain two beams. The transmitted or object beam is modulated by a Spatial Light Modulator (SLM) or page composer. Data pages could contain one million bits per page. After passing through the SLM, the light is focused by a lens. At near the focal distance, a photosensitive medium is placed. It should be noted that the lens performs the Fourier Transform of the spatial distribution of light that passes through the SLM. The other light beam, after being reflected by the beamsplitter, is reflected by a mirror and then a second mirror sends it to the recording medium where it arrives at an angle. These two beams interfere and the interference pattern is recorded in the volume or bulk of the photosensitive medium. Different interference patterns can be recorded by varying, for example, the interference beam angle or the wavelength of the recording light. This process is called multiplexing. Other multiplexing methods include phase code, peristrophic, shift, correlation,

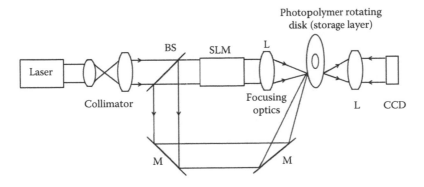

FIGURE 10.11 Basic page composer HDS system used for the recording and reading of information. Interference pattern is recorded in the volume of the photosensitive medium.

polytopic and more. At reading time just the reference beam is sent to the medium where it is diffracted and will reconstruct the object beam. A lens after the medium will perform the Fourier Transform and will send the beam to an array of detectors (CCD).

Another method to store information in photopolymer disks was proposed by the group of Prof. Eichler [335]. This time storage was not recorded in a page-based format but in a bit format by writing micro-holograms of about 1 μm diameter. The configuration used in this recording scheme is depicted in Figure 10.12. Modulation of the refractive index is done when light illuminates the photopolymeric disk and a Bragg grating is formed. To retrieve the stored data the reading beam is sent to the disk and the Bragg grating reflects the light, which is collected by a detector. To improve the system, it was suggested [335] to record the data through several layers. Aprilis photopolymer was used in this experiment. Data densities of about 5 bits/μm2 were recorded when a laser with a wavelength of 532 μm was used. Attempts to use a laser operating at a shorter wavelength (violet light) were proposed and intended to store 10 bit/μm2.

A slight modification of the method described in the last paragraph was shown in reference [341]. The authors also used photopolymer disks. The system layout consists of a standard disk drive but a reflecting unit below the disk has been added. This unit comprises a telecentric reflecting head. With this equipment, micro-holograms were recorded in multilayers. The photopolymer disks had a speed of 3600 rpm with exposure times of 50 ns. The photopolymer was Aprilis in a disk format with 120 mm diameter and 125 μm layer thickness. 12 depth layers were recorded. By using the RZ code, a transfer rate of 19 Mbits/s and a storage density of 9 bits/μm2 were attained.

Later, in 2008, a novel HDS architecture was described in reference [337]. Now, the photopolymer disk had a preformatted reflection grating recorded all over the disk providing a homogenous reflective volume. Later, the disk was patterned by locally modifying the uniform reflection grating. This modification can be done in the X–Y directions and also through the layer depth or Z

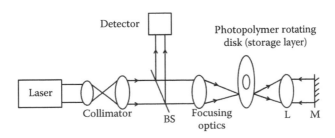

FIGURE 10.12 HDS recording and reading configuration. Bit recording.

direction. In the writing step, the polymer grating is locally destroyed with an ultrafast laser pulse. It has been possible to record 8 data layers separated by 12 microns. Aprilis photopolymer was used, which is sensitive to a wavelength of 532 μm, and presented a post-exposure shrinkage of 0.2%. The energy per bit stored was 12 nJ.

Materials used in HDS are those based on Cationic Ring Opening Polymerization (CROP) from Polaroid and free radical media (Lucent). In this last medium, there are two independently polymerizable systems. One system forms a three-dimensional crosslinked polymer matrix, which creates a mechanically stable and robust medium. The second system will react during the holographic recording or exposure. In the CROP system, a low shrinkage is present. It is composed of a photo-acid generator, a sensitizing dye, some CROP monomers and a binder. This CROP material can be tuned for high photosensitivity.

Efforts have been made to suppress shrinkage in the photopolymers used in HDS systems. One of these efforts is mentioned in reference [338]. A mixture was developed that consists of a two chemistry system. Materials used are Poly (propylene glycoldiglycidylether, PPGDGE) and pentaethylene hexamine (PHA). It is mentioned that this multi-crosslinkable hardener enhances the mechanical strength of the matrix and suppresses shrinkage. This mixture is sensitized for blue light (λ= 405 μm). It was tested recording the Fourier Transform of a SLM with 961 bits/page. At reconstruction, good images were obtained. System stability was verified for a period of roughly two months.

Several research groups have intended to solve the problems associated with the use of photopolymers in HDS. Among these groups, besides those mentioned before like Dupont, Aprilis and InPhase to mention but a few, are the groups at Alicante University, University College Dublin, Dublin Institute of Technology and others. The group at Alicante has worked mainly on polymeric mixtures based on acrylamide. In a paper in 2003 [339], they described the use of a 1 mm thick photopolymer. Later, they outlined a method to optimize the mixture [340] and presented the holographic characteristics of an acrylamide film with thicknesses around 40–1000 micron.

The PVA–acrylamide photopolymer was used for data storage using angular and peristrophic holographic multiplexing [341]. The authors used a PVA–Acrylamide mixture in a HDS system with a configuration that reduced the zero frequency of the Fourier transform of the object [342]. They called it Hybrid Ternary Modulation (HTM). With this method, the photopolymer will not be saturated by the high intensity of the zero frequency. More recently, using the same mixture, they stored information in a HDS system using two different modulations in the LCD [343], Binary Intensity Modulation (BIM) and Hybrid Ternary Modulation (HTM). To compare the two methods the authors used the Bit Error Rate (BER) parameter and quantified the image quality. BER is defined as the probability of having erroneous bits in the image. The lower the BER of the stored images the greater their quality. A BER of 0.2 is taken as the threshold value below which the images obtained are considered to be of good quality. The result of the study was that both methods presented BER values below or about 0.2. The configuration with which the greatest number of holograms has been obtained was with HTM with a given object. However, if it is desired to multiplex, it is preferable to use the BIM method.

Regarding the characterization of PVA/AA mixtures by the Alicante group, it is possible to find more than 50 papers. Characterization includes subjects such as improving performance, linearity in response, stability, diffraction efficiency improvement, temporal evolution, optimization, angular response and more.

Besides the Alicante group, other efforts to improve the photopolymer mixture PVA/AA have been done with this material. For example, researchers at the Dublin Institute of Technology have carried out experiments to find the characteristics of the material by recording transmissive and reflective gratings. For the transmissive gratings [344], they studied the dynamic range, temporal stability and the response of the material when multiple gratings were multiplexed in the same volume. Layers with 160 μm thickness were used. Gratings with 1500 lines/mm were recorded. An M/# of 3.6 was obtained when 18–30 gratings were multiplexed. This value compares to the one presented by InPhase material of 6.7 for samples with a 200 μm thickness. Reflection gratings with spatial frequency of 5460 lines/mm were recorded with 1 μJ of energy. This shows the high spatial resolution of the photopolymer. By optimizing the exposure process, they produced a series of gratings having equal diffraction efficiencies of 0.28%.

Another group that has studied acrylamide films is at the Shangai Institute of Optics and Fine Mechanics. They proposed a method called Angular and Spatial Multiplexing (ASM) [345]. This scheme can achieve a recording capacity and density about 3 to 6 times higher than the spatio-angular multiplexing (SAM) method. The authors attained a recording density of about 30 bits/μm2. The article reports the recording of 10 holograms of a SLM . The average diffraction efficiency of each hologram was 1.6% and the average bit error rate (BER) of the holograms was 6.3×10^{-5}.

Besides the acrylamide films, other materials have been tested for HDS recording. In reference [346], a mixture with multifunctional photoreactive inorganic cages was mentioned. The mixture has a polyhedral oligomeric silsesquioxane-methacrylate (POSS) as a co-monomer plus an organic monomer. This reduced the shrinkage without sacrificing monomer diffusion and photosensitivity. Films with 120–130 μm thickness were used. Holographic gratings were recorded. Refractive index modulation of about 7.9 $\times 10^{-4}$ was obtained. Layer dimensional stability was investigated by measuring the angular selectivity of the diffraction efficiency. Estimated volume shrinkage of about 2% was obtained. By angular multiplexing, a high value of M/# was obtained; 39 holograms were recorded.

In reference [347], the use of bezyether-based dendronized macromonomers for HDS is discussed. This permits the development of a low shrinkage photopolymer without sacrificing the sensitivity or the storage capacity of the recording media. Discs with thickness of 0.5 mm were used and 60 holograms were recorded. M/# values of 6, 8 and 10 were found for different mixtures. Volume shrinkage of 0.23% was observed.

Other materials that have been used in recording HDS systems are nanoparticle–polymer composite films. The group in Japan, lead by Yasuo Tomita, has proposed the use of step growth thiol–ene photopolymerization [348]. Through the recording of 180 page-oriented holograms, they found diffraction efficiencies of about 10^{-2}, symbol-error rates (SERs) lower than 10^{-3} and SNR of about 2. Later [349], they reported the results from recording in another mixture. This time the nanoparticle-polymer composites were based on radical-mediated thiol-ene mixtures. They have made a comprehensive study of the material. They recorded 80 holographic digital data pages and found low values of shrinkage and small changes due to changes in temperature. The SERs values were below 1×10^{-2} and the SNR was 4. The group lead by Y. Tomita has made more studies of nanoparticle mixtures. For an in-depth study the reader should consult more references from this group.

Other material that has been tested for HDS is a biophotopolymer [350]. Usually the PVA/AA mixtures have the following components: Acrylamide, TEA, Yellowish eosin, PVA and MBA (Methylene-bis-acrylamide). Among these components, AA, MBA and YE are toxic components. These components were replaced. Instead of AA, sodium acrylate was used, YE was substituted by sodium salt riboflavinmonophosphate PRF and BMA by DHEBA. Transmissive and reflective elements were recorded in 900 μm thickness layers. Studies of phase shift and shrinkage, as a function of exposure time, and other parameters were carried out. Diffraction efficiencies of about 77% with exposure of 200 mJ/cm2 were obtained. With these data they concluded that the biophotopolymer is an alternative to traditional acrylamide-based polymers for use as holographic recording material and in data storage applications.

At present (2015), the group at Utsunomiya University led by Prof. Toyohiko Yatagai is making efforts to apply polarization holography to Holographic Data Storage. They have developed a new photopolymer named "AK" based on ketones to record the polarized light pattern. When polarized light reaches the recording medium, it suffers a modulation of anisotropy, such as birefringence, dichroism and optical activity. The image or information is read out by the reconstruction wave similarly to an intensity hologram.

Researchers trying to find more information about characterization of materials for HDS can consult reference [351].

10.7.10.2 Holographic Optical Elements

Besides the use of photopolymers in HDS systems, photopolymers have been used in the fabrication of Holographic Optical Elements, such as volume and surface gratings, diffractive lenses and more. An example of Volume Holographic Optical Element (VHOE) is described in reference [352]. The authors report a time-sequential autostereoscopic 3D display with a directional backlight system based on VHOE. The photopolymer used was Bayfol. This VHOE is fabricated by interfering a pair of collimated reference beams and diverging object beams for each of the left and right eyes on a photopolymer film. By time sequentially illuminating the VHOE, similar to the reference beams, two object beams are reconstructed by diffraction. These object beams provide the left and right images alternately shown in a LCD. These images are formed in front of the observer's left and right eyes, from which s/he can finally perceive the 3D images.

Another example of a HOE prepared in photopolymers is mentioned in reference [353]. This article describes the fabrication of Holographic lenses (HL). These lenses are made by recording interference fields and they are lightweight and can be multiplexed allowing various elements to be recorded in the same plate. HLs can be used in optical networks, optical interconnections and Fourier transform holograms. HLs were made with a PVA/AA photopolymer. Layer thickness was about 70 μm. They recorded lenses with focal lengths of 100 mm, 150 mm, and 250 mm. The cut-off frequency for each lens was 61 l/mm, 45 l/mm, and 33 l/mm, respectively. These lenses were used to form images in a 2f configuration. The results indicate that HLs have higher resolution than do the corresponding refractive lenses.

One more example of Holographic Diffractive Optical Element (HDOE) in PVA/AA photopolymer is described in reference [354]. The authors fabricated a HDOE for enhancing light collection from fluorescence–based biochips. They fabricated a HDOE with 50% diffraction efficiency.

10.7.10.3 Photopolymers in Waveguide Fabrication

Photopolymers have been used to make waveguides that link, for example, two optical fibers [355] or waveguides in the bulk of a photopolymer [356]. In the first example [355], waveguides have been made by aligning two opposite ends of optical fibers and placing a drop of photopolymer between them. Then, light is sent through the fibers and induces free-radical polymerization. The liquid formulation then turns to a solid polymer network. At the end of the laser irradiation, the polymer waveguide was developed in ethanol to remove the non-reacted formulation. The result is a polymer guide highly bonded to the fiber ends that appears as an extension of the cores of the fibers.

Regarding the waveguides made in bulk photopolymer, a PVA/AA has been used [356]. To make the waveguide, the photopolymeric mixture is placed in a cuvette. Then, by means of a lens, light from a He-Ne laser is focused and sent into the cuvette. As the beam propagates through the medium, it modifies the refractive index of the material in its path. Permanent waveguides of 10 mm have been formed.

Another method to make waveguides with a direct write lithography system is described in reference [357].

10.7.10.4 Holographic Filters

Another application of photopolymer holography is in the fabrication of holographic filters. One example is given in reference [358]. Dupont film was used in these experiments. The authors recorded two volume Bragg gratings. One has its fringe planes at 45° with respect to the polymer surface. The other volume grating has its planes perpendicular to the surface. Light beam enters through the first grating and then it travels through the emulsion and reaches the second grating. This grating reflects back the light thus acting as a wavelength filter. The spectral bandwidth at the FWHM is about 0.45 μm and reflectance is about 9.7%.

Other examples of a holographic filter are described in reference [359]. This time a PQ-PMMA material was used. Gratings were made with grating vector parallel to the sample surface. Different nano SiO2 concentrations in the mixture were tried. Best results were achieved with 0.1 weight ratios concentrations. Diffraction efficiencies of 80% and 93% were obtained. For the first mixture, a bandwidth of 0.72 μm FWHM was obtained while, for the second mixture, the FWHM was 0.34 μm. Multiplexed gratings were also made.

Another type of holographic filter was described in reference [360]. This time, a sol-gel glass was used, which incorporated monomers and photoinitiator species. Samples with thickness from 50 to

500 μm were used. Samples under investigation presented high optical quality with low scattering. Notch filters were fabricated with a grating spatial frequency of 2800 l/mm. A bandwidth of 0.3 μm was obtained.

Another reference that deals with volume holographic filters is cited in [361].

10.7.10.5 Sensors with Polymers

Photopolymers can be used in the fabrication of sensors in the form of volume holographic reflection and transmission gratings. Different sensors can be fabricated to detect chemicals, glucose, humidity or other parameters. Here, we give some examples. In reference [362], a glucose sensor made of DMPA, acrylamide, MBA and other chemicals is described. Volume reflection gratings were recorded with this mixture. Later, it was found that by immersing the volume reflection gratings, in solutions containing different glucose concentrations, the color reflected by the volume grating changed from green through orange to red as glucose concentration was increased. Glucose diffuses into the gel phase and binds to pendant oronic acid groups. The presence of these charged groups within the polymer generates a potential resulting in osmotic pressure that causes the polymer to absorb more water and swell. This swelling manifests as a red shift in diffraction wavelength as the periodicity of the embedded holographic fringes change. This phenomenon is reversible.

Another use of volume holographic reflection holograms is made in sensing chemicals like testosterone. In reference [363], the use of methacrylate acid (MAA) and ethylene glycol dimethacrylate (EGDMA) to make volume reflection holograms is discussed. Volume gratings can be immersed in the mixtures of testosterone/acetonitrile. The spectral position of the holographic reflection peak was monitored over time. It showed a wavelength shift of 3.65 μm. Another analyte that was sensed was antagonist 5 propanolol. Similar results to those for testosterone were obtained. This study shows that this type of sensor can be used for the detection of analytes including biomolecules, such as peptides and proteins.

Another parameter that has been monitored with volume holographic gratings made with photopolymers has been humidity or the amount of water vapor in the atmosphere. The research group at the Dublin Institute of Technology has proposed a volume holographic grating made with PVA/AA as a humidity holographic sensor. The work has been described in references [225,312,313,364].

In reference [361], the design, fabrication and testing of a volume reflection holographic grating made with PVA/AA is described. Layers with thickness of 30–180 μm were used. These elements are selective with respect to the wavelength. They presented a FWHM of 2.8 μm. The change of the spectral position was determined for different values of relative humidity. Humidity was varied between 5% and 80% and this caused the peak of the spectral response to shift by about 130 μm. This response is predominantly due to the swelling of the material. Reversibility of the humidity response was studied and it was found that the position of the spectral peak was repeatable within 1 μm. Besides, it was found that thin layers responded more quickly than thick layers to changes in relative humidity.

A study of how volume holographic gratings made with PVA/AA depend on humidity and temperature is reported in reference [362]. A dependence was found of the shift of the spectral peak wavelength as a function of relative humidity and on the change in water content in the photopolymer layer. This study was made with a gravimetric method. It revealed that the main contributor to the change in the spectral response originates from the dimensional changes in the layer. Regarding the temperature, it was found that if RH is kept constant at a level below 45%, a temperature change ranging from 15°C to 35°C causes no shift in the spectral response of the reflection grating. However, if the relative humidity increases above that value, the temperature dependence becomes pronounced.

Reference [364] reports a study of transmission volume gratings as a function of humidity. It has been found that on exposure to humidity there is a change in diffraction efficiency. But these changes are reversible if temperatures are kept low. However, irreversible changes in diffraction efficiency, thickness, refractive index modulation and Bragg angle were observed when the temperature during humidity exposure was higher than 16°C.

Another volume reflection holographic sensor was built with porphirin acting as the crosslinker, light absorbing cation and chelating agent [365]. A monomer solution consisting of HEMA and TACPP were mixed in THF. A PMMA substrate was treated with O2 plasma under vacuum to render the PMMA surface hydrophilic and treated with the HEMA–TACPP solution. The final polymer matrix had a thickness of about 10 μm. Volume reflection holographic gratings were recorded. These gratings were used to make pH measurements, and to sense ethanol, methanol, propanol and dimethylsulfide mixtures with water. Cu2 and Fe2+ in salt concentration were also tested.

An interesting application of photopolymers is reported in reference [366]. The authors describe the use of PQ–PMMA phtopolymer to make volume gratings integrated in a rain sensor that is composed of a light source, a teflon coated waveguide, holographic gratings and a detector. This rain detector is used in the windscreen of automobiles to automatically control the wiper system. The whole system is transparent and is embedded in the windscreen. Gratings were recorded with an Argon laser at 514 μm with diffraction efficiencies of about 90%. The proposed system detects humidity, weak rain, and heavy rain. It also detects the internal moisture inside the car.

An effective approach to functionalization of photopolymer materials for sensing applications by incorporation of zeolite nanoparticles has been recently reported [367]. Inkjet printing and patterning strategies have been developed for fabrication of hybrid holographic sensors using zeolite nanocrystals on glass supported photopolymers. The flexibility of the proposed techniques was demonstrated by fabrication and characterization of two types of holographic sensors. The first type, a reversible sensor, is based on a transmission hologram recorded in a hydrophobic MFI-type zeolite doped layer with high sensitivity toward alcohols. In this type of sensor, the patterning of the zeolite nanocrystals in the volume of the polymer layer is achieved by holographic recording; the pattern periodicity is in the sub-micrometer range. The second type of sensor is based on a reflection hologram and it is produced by inkjet printing of zeolite nanocrystals on photopolymer layer before holographic recording. The resulting localized presence of zeolite nanocrystals in the layer is a key factor for the performance of the sensor. Irreversible humidity sensors based on photopolymer layers doped with hydrophilic EMT-type zeolite were fabricated using the second approach and characterized in a controlled humidity environment.

10.7.10.6 Microstructure Fabrication

Photopolymers are used in the fabrication of three-dimensional (3D) micro- and nano-structures like microoptical components, artificial scaffolds, microfluidic devices and more. Different methods have been used to make the structures, for example Direct Laser Writing (DLW), two-photon photopolymerization (TPP) and structuring to mention but a few. An example of Direct-Write Lithography (DWL) has been mentioned before in the section "Waveguides" [357]. Besides this work, reference [368] describes three methods to make micro-tube arrays. These are Direct Laser Writing (DLW), Optical Vortices (OVs) and Holographic Lithography. A femtosecond laser was used generating 300 fs pulses at 1030 μm (fundamental) and 515 μm (second harmonic). Three different configurations were used, one for each of the writing methods. The material used was hybrid organic–inorganic Zr - containing sol–gel phototoresist SZ 2080. Two types of initiators were used. The photo–polymerization reaction was induced via nonlinear absorption. With these methods, micro-tubes were fabricated with different sizes. Diameter of the micro-tubes was about 10 μm with heights of about several tenths of microns.

An example of the two-photon polymerization method is mentioned in reference [369].

Another group involved in the fabrication and copy of microstructures is at the Institute of Physics in Belgrade, Serbia. They have used dichromated pullulan and dental photopolymers [370,371]. To test the photopolymers, first they fabricated diffraction gratings with dichromated pullulan. The gratings had different spatial frequencies and profiles, sinusoidal or saw–tooth. Then, they were copied in photopolymer dental composites. Average grating relief depth was 167 μm for gratings with about 300 l/mm. This depth decreases as spatial frequency increases. Diffraction efficiencies reached about 45%. Stability of the dental composite with temperature was tested at 250°C. No change of optical properties of the dental embossing tool was observed.

An example of the use of photopolymers in microfluidics is given in reference [372]. A composite photopolymerizable material was used to fabricate an optically pumped, organic distributed feedback laser. Characterization of the lasing properties of the device has been carried out.

Besides the application of photopolymers in the fabrication of microstructures mentioned in the above paragraph they also have been used to make three-dimensional photonic crystals. Several methods have been demonstrated. In one of them [373], the authors used a phase mask and, in the other [381], a multiplex-time exposure method. In the first reference, photopolymers have been used to fabricate a circular mask. This was then used to make the photonic crystals with a single beam exposure. The circular mask shows three gratings placed in a disk. Each grating is made using the interference of two beams. This disk is used to make 2D hexagonal surface relief structures having about 600 μm height and 2.7 μm spacing. The material for the 2D–3D structures was PMMA with thicknesses between 50 and 70 μm containing 2% mol PQ. An Ar–Ion laser (λ = 514 μm) was used as a light source. After exposure, the samples were annealed. Other materials used were photocurable monomers and inorganic ZrO_2 or nematic crystals SCB. Ingacure was used as initiator. This time radiation with a wavelength of 364 μm was used. These materials were characterized by shrinkage of 5%.

The method to make photonic crystals shown in reference [374] does not use a single step exposure as described in the above paragraph. It uses three-step exposure to a two-beam interference pattern. This method is called multiple-exposure and refers to the holographic multiplexing as superposition of multiple gratings in the same volume of a recording material. Arbitrary 3D crystals structures can then be created by overlapping three 1D gratings. In the experiments, a six-beam configuration was used. Two photopolymers were tested. One was developed by InTech Technologies, sensitized to 405 μm and the other, based on cationic ring-opening, was developed by Aprilis. Photopolymer layers were 300 μm thick. The distance between the inscribed structures was about 1.7 μm.

More information about the fabrication of photonic crystals can be found in references [375,376].

The application of photopolymers in optics is wide. Others fields where they are used are: Optical Correlators, Couplers, demultiplexers, ESPI, Holographic displays, Holographic interferometry, distributed feed back lasers, Bioimaging, metrology and more.

Some books related to photopolymers can be found in references [377,378].

10.8 Conclusion

The present review takes stock of all the different approaches reported in the specialized literature to formulate polymer or polymerizable classes of materials for holographic recording. The time to achieve an ideal holographic recording material is, no doubt, still a long way off but, in some degree, several existing formulations, nevertheless, meet some of the required properties needed for application.

Even though polymer materials will never excel silver halides in terms of energy sensitivity owing to intrinsic limitations, their high storage capability, rapid access, excellent signal-to-noise ratio and, occasionally, self-processing character, open up new vistas.

Whatever direction in which attempts at perfecting these materials is carried out by multidisciplinary research groups, it must be kept in mind that the key question in designing new polymerizable systems for holographic recording is not concerned with the tailoring of more or less exotic initiator or monomer structures likely to undergo faster curing. It is of paramount importance to realize that the main issue to be dealt with is definitely to gain a fuller insight into the coupling between photochemically induced monomer conversion, shrinkage, and mass transfer. In this respect, the ongoing activity devoted to their improvement from a thorough investigation of the recording mechanism is a cheerful omen.

Acknowledgments

We would like to thank Eugenio R. Mendez (CICESE, Ensenada, c.p. 22860, Mexico) for writing the section devoted to photoresist.

S. Calixto would like to acknowledge Z. Malacara and M. Scholl for fruitful discussions. Thanks are given to Raymundo Mendoza for the drawings.

References

1. Mees CEK, James TH (eds.). The Theory of the Photographic Process, 3rd Ed., New York, NY: Macmillan, 1966.
2. Kosar J. Light Sensitive Systems: Chemistry and Applications of Non-Silver Halide Photographic Processes, New York, NY: John Wiley, 1965.
3. Smith HM. Holographic Recording Materials, in Topics in Applied Physics, Vol. **20**, New York, NY: Springer-Verlag, 1977.
4. Bazhenov V, Yu Marat S, Taranenko VB, Vasnetsov MV. Biopolymers for real time optical processing, in Optical Processing and Computing, Arsenault HH, Zoplik T, Macukow B. eds., San Deigo, CA: Academic Press, 1989, pp. 103–144.
5. Eder JM. History of Photography, Chap. 1, 2 and 64, New York, NY: Dover, 1978.
6. Carrol BH, Higgins GC, James TH. Introduction to Photographic Theory. The Silver Halide Process, New York, NY: John Wiley & Sons, 1980.
7. Todd HN. Photographic Sensitometry a Self Teaching Text, New York, NY: John Wiley, 1976.
8. Kodak. Scientific Imaging with Kodak Films and Plates, Publication P-315, Rochester, NY: Eastman Kodak, 1987.
9. ANSI PH 2.16–1984, Photographic Density: Terms, Symbols and Notation.
10. ANSI PH 2.19–1986, Photographic Density Part 2: Geometric Conditions for Transmission Density Measurements.
11. ISO 5/1–1984, Photographic Density: Terms, Symbols and Notation.
12. ISO 5/2–1984, Photographic Density Part 2: Geometric Conditions for Transmission Density Measurements.
13. ISO 5/3–1984, Photographic Density, Part 3: Spectral Conditions for Transmission Density Measurements.
14. Swing RE. Microdensitometry, SPIE Press, Bellingham, WA: MS 111; 1995.
15. Niederpruem CJ, Nelson CN, Yule JAC. Contrast index, Phot. Sci. Eng., **10**, 35–41; 1966.
16. ISO 6–1975, "Determination of Speed Monochrome, Continuous-Tone Photographic Negative Materials for Still Photography."
17. ANSI PH 2.5–1979, "Determination of Speed Monochrome, Continuous-Tone Photographic Negative Materials for Still Photography."
18. ANSI PH 3.49–1971, Automatic optimization of photographic exposure. R1987.
19. Thomas W. SPSE Handbook of Photographic Science and Engineering, New York, NY: Wiley, 1973.
20. ANSI PH 2.40–1985, Root mean square granularity film. R1991.
21. Altman JH. Photographic films, in Handbook of Optics, M. Bass, ed., New York, NY: McGraw-Hill, 1994.
22. Dainty JC, Shaw R. Image Science, London: Academic Press, 1974.
23. ISO PH 6328–1982, "Method for Determining the Resolving Power of Photographic Materials."
24. ANSI Ph 2.33–1983, "Method for Determining the Resolving Power of Photographic Materials."
25. ANSI PH 2.39–1984, "Method of Measuring the Modulation Transfer Function of Continuous-Tone Photographic Films."
26. Crane EM. Acutance and Graunlance, *Proc. Soc. Photo Opt. Instrum. Eng*, **310**, 125–130; 1981.
27. Bachman PL. Silver halide photography, in Handbook of Optical Holography, J. Caulfield, ed., London: Academic Press, 1979, pp. 89–125.
28. Walls H, Attridge GG. Basic Photoscience, How Photography Works, London: Focal Press, 1977.

29. Grant Haist. Modern Photographic Processing, Chap. 6, New York, NY: Wiley Interscience, 1979, pp. 284, 324.
30. 1994 Kodak Professional Catalog, Publication L-9, Rochester, NY: Eastman Kodak, 1994.
31. Bjelkhagen HI. Silver-Halide Recording Materials for Holography and Their Processing, New York, NY: Springer-Verlag, 1993.
32. Lou GH. Photographic film, in The Infrared and Electrooptical Systems Handbook, Acetta JS, Shumaker DL. eds., Washington, DC: Spie Optical Engineering Press, 1992, pp. 519–539.
33. Kodak Color Films, Publication E-77, Rochester, NY: Eastman Kodak, 1977.
34. Collier RJ, Burckhardt C, Lin LH. Optical Holography, Chap. 10, New York, NY: Academic Press, 1971, p. 271.
35. Agfa-Gevaert NV. B-2510, Mortsel, Belgique, 1983. Holographic Materials. Technical Bulletin.
36. Friesem AA, Kozma A, Adams GF. Recording parameters of spatially modulated coherent wave-fronts, *Appl. Opt.*, **6**, 851–856; 1967.
37. Lin LH. Method of characterizing hologram–Recording materials, *J. Opt. Soc. Am.*, **61**, 203–208; 1971.
38. Nassenstein H, Metz HJ, Rieck HE, Schultze D. Physical properties of holographic materials, *Phot. Sci. Eng.*, **13**, 194–199; 1969.
39. Horne DF. Microcircuit Production Technology, Bristol: Adam Hilger, 1986, pp. 13–19.
40. Hutley MC. Diffraction Gratings, Chap. 4, London: Academic Press, 1982.
41. Popov EK, Tsonev LV, Sabeva ML. Technological problems in holographic recording of plane gratings, Opt. Eng., **31**, 2168–2173; 1992.
42. Mello BA, da Costa IF, Lima CRA, Cescato L. Developed profile of holographically exposed photoresist gratings, *Appl. Opt.*, **34**, 597–603; 1995.
43. Bartolini RA. Characteristics of relief phase holograms recorded in photoresists, *Appl. Opt.*, **13**, 129–139; 1974.
44. Bartolini RA. Photoresists, in Holographic Recording Materials, H. M. Smith, ed., New York, NY: Springer-Verlag, 1977, pp. 209–227.
45. Haidner H, Kipfer P, Sheridan JT, et al. Polarizing reflection grating beamsplitter for the 10.6-μm wavelength, *Opt. Eng.*, **32**, 1860–1865; 1993.
46. Habraken S, Michaux O, Renotte Y, Lion Y. Polarizing holographic beamsplitter on a photoresist, *Opt. Lett.*, **20**, 2348–2350; 1995.
47. O'Donnell KA, Méndez ER. Experimental study of scattering ffrom characterized random surfaces, *J. Opt. Soc. Am.*, **A4**, 1194–1205; 1987.
48. Méndez ER, Ponce MA, Ruiz-Cortés V, Gu Z-H. Photofabrication of one-dimensional rough surfaces for light scattering experiments, *Appl. Opt.*, **30**, 4103–4112; 1991.
49. http://www.kodak.com.
50. http://www.agfa–gevaert.com.
51. http://www.fujifilm.com.
52. http://www.ilfordphoto.com/products.
53. Neipp C, Pascual I, Belendez A. Bleached silver halide volume holograms recorded on Slavich PFG–01 emulsion: The influence of the developer, *J. Mod. Opt.*, **48**, 1479; 2001.
54. Neipp C, Pascual I, Belendez A. Silver volume holograms on BB–640 plates: The influence of the developer in rehalogenating bleach techniques, *Optik*, **22**, 349; 2001.
55. Neipp C, Pascual I, Belendez A. Effects of overmodulation in fixation–free rehalogenating bleached hologram, *Appl. Opt.*, **40**, 3402; 2001.
56. Neipp C, Beronich E, Pascual I, Belendez A. Fixation free bleached silver halide transmission holograms recorded in Slavich PFG–01 red sensitive plates, *J. Mod. Opt.*, **48**, 1643; 2001.
57. Neipp C, Pascual I, Belendez A. Mixed phase amplitude materials, *J. Phys. D*, **35**, 957; 2002.
58. Neipp C, Belendez A, Pascual I. The influence of the procedure on the dynamic range of bleached silver halide emulsions, *J. Mod. Opt.*, **50**, 1773; 2003.
59. Neipp C, Condeza S, Campos AM, Pascual I, Belendez A. Comparative study of bleaches applied to BB–640 plates, *J. Opt. A*, **6**, 71; 2004.

60. Alvarez ML, Camacho N, Neipp C, Márquez A, Belendez A, Pascual I. Holographic gratings with different spatial frequencies recorded on BB–640 bleached silver halide emulsions using reversal bleaches, Mater. Sci. Forum, 480–481, 543; 2005.

61. Kim J-M, Choi B-S, Choi Y-S, et al. Holographic optical elements recorded in silver halide sensitized gelatin emulsions. Part I. Transmission holographic optical elements, *Appl. Opt.*, **40**, 622; 2001.

62. Kim S-L, Choi Y-S, Ham Y-N, Park C-Y, Kim J-M. Holographic optical elements recorded in silver halide sensitized gelatin emulsions part 2. Reflection holographic optical elements, *Appl. Opt.*, **42**, 2482; 2003.

63. Kim S-L, Choi Y-S, Ham Y-N, Park C-Y, Kim J-M. Holographic diffuser by use of a silver halide sensitized gelatin process, *Appl. Opt.*, **42**, 2482; 2003.

64. Neipp C, Pascual I, Belendez A. Silver halide sensitized gelatin derived from BB - 640 holographic emulsion, *Appl. Opt.*, **38**, 12384; 1999.

65. Knevesel J. The photographic emulsion technology of the OPERA experiment on its way to find the $\sqrt{\mu} \to \sqrt{\tau}$ oscillation, *Nucl. Phys. B*, **215**, 66; 2011.

66. Tadeaki T. Characterization of nuclear emulsions in overview of photographic emulsions, *Radiat. Meas.*, **44**, 733; 2009.

67. Ditlov VA. Track theory and nuclear photographic emulsions for dark matter searches, *Radiat. Meas.*, **50**, 7; 2013.

68. Ganzherli NM. Holographic diffusers based on silver halide photoemulsion layers, *J. Opt. Technol.*, **74**, 622; 2007.

69. Ganzherli NM, Maurer IA, Chernykh DF, Gulyaev SN. Microlens rasters and holographic diffusers based on PFG–01 silver halide photographic material, *J. Opt. Technol.*, **76**, 384; 2009.

70. Ganzherli NM, Gurin AS, Kramuschenko DD, Maurer IA, Chernykh DF. Two-dimensional holographic gratings based on silver halide photographic emulsions for the formation of raster images, *J. Opt. Technol.*, **76**, 388; 2009.

71. Navarrete-Garcia E, Calixto S. Surface relief zone plates fabricated with photographic emulsions, *Appl. Opt.*, **37**, 739; 1998.

72. Navarrete-Garcia E, Calixto S. Continuous surface relief micro–optical elements fabricated on photographic emulsions by the use of binary and halftone masks, *Opt. Mater.*, **23**, 501; 2003.

73. Navarrete-Garcia E, Calixto S. Mathematical model for the surface relief formation of photographic emulsions, *J. Microlith. Microfab. Microsyst.*, **4**, 023010–023015; 2005.

74. Andreeva OV, Abyknovennaya IE, Garriluyk ER, Paramov AA, Kushnarenko AP. Silver-halide photographic materials base on nanoporous glasses, *J. Opt. Technol.*, **72**, 916; 2005.

75. Gabor D. *Nature*, **161**, 777; 1948.

76. Maiman TH. *Nature*, **187**, 493; 1960.

77. Kastler A. *J. Phys. Rad.*, **11**, 255; 1950.

78. Leith EN, Upatnieks J. *J. Opt. Soc. Am.*, **52**, 1123; 1962.

79. Denisyuk YN. *Soviet Phys-Doklady*, **7**, 543; 1963.

80. Allen NS. (ed.), Photopolymerization and Photoimaging Science and Technology, New York, NY: Elsevier, 1989.

81. Fouassier JP. Makormol. *Chem. Makromol. Symp.*, **18**, 157; 1988.

82. Decker C. *Coat. Tech.*, **59**(751), 97; 1987.

83. Crivello JV, Lam JWH. *Macromolecules*, **10**, 1307; 1977.

84. Lougnot DJ. in Techniques d'utilisation des photons, Paris: DOPEE, 1992, pp. 245–334.

85. Decker, C, Fouassier, JP, Rabek, JF. Radiation Curing in Polymer Science and Technology, Elsevier Applied Science, 1973.

86. Hariharan P. Optical Holography: Principle, Technology and Applications, Cambridge: Cambridge University Press, 1984, pp. 88–115.

87. Francon M. Holographie, Paris: Masson, 1969.

88. Kogelnik H. *Bell. Syst. Tech. J.,* **48**, 2909; 1969.
89. Margerum JD. Polymer Preprints for the 160th ACS Meeting, 1970, p. 634.
90. Close DH, Jacobson AD, Margerum JD, Brault RG, McClung FJ. *Appl. Phys. Lett.,* **14**, 159; 1969.
91. Jenney JA. *J. Opt. Soc. Am.,* **60**, 1155; 1970.
92. Jenney JA. *Appl. Opt.,* **11**, 1371; 1972.
93. Jenney JA. *J. Opt. Soc. Am.,* **61**, 116; 1971.
94. Jenney JA. Recent developments in photopolymer holography, *Proc. SPIE,* **25**, 105; 1971.
95. Booth BL. *Appl. Opt.,* **11**, 2994; 1972.
96. Haugh EF. US Patent 3 658 526; 1972, assigned to E. I. Dupont de Nemours and Co.
97. Baum MD, Henry CP. US Patent 3 652 275; 1972, assigned to E. I. Dupont de Nemours and Co.
98. Colburn WS, Haines KA. *Appl. Opt.,* **10**, 1636; 1971.
99. Wopschall RH, Pampalone TR. *Appl. Opt.,* **11**, 2096; 1972.
100. Van Renesse RL. *Opt. Laser Tech.,* **4**, 24; 1972.
101. Sukegawa K, Sugawara S, Murase K. *Electron. Commun. Jap.,* **58**, 132; 1975.
102. Sukegawa K, Sugawara S, Murase K. *Rev. Electr. Commn. Labs.,* **25**, 580; 1977.
103. Sukegawara S, Murase K. *Appl. Opt.,* **14**, 378; 1975.
104. Sadlej N, Smolinksa B. *Opt. Laser Tech.,* **7**, 175; 1975.
105. Tomlinson WJ, Kaminow IP, Chandross EA, Fork RL, Silfvast WT. *Appl. Phys. Lett.,* **16**, 486; 1970.
106. Chandross EA, Tomlinson J, Aumiller GD. *Appl. Opt.,* **17**, 566; 1978.
107. Colburn WS, Zech RG, Ralston LM. Holographic Optical Elements, Tech. Report AFAL, TYR-72-409; 1973.
108. Verber CM, Schwerzel RE, Perry PJ, Craig RA. Holographic Recording Materials Development, N.T.I.S. Report N76-23544; 1976.
109. Collier RJ, Burckhardt CB, Lin LH. Optical Holography, New York, NY: Academic Press, 1971, pp. 265–336.
110. Peredereeva SI, Kozenkov VM, Kisilitsa PP. Photopolymers Holography, Moscow, 1978, p. 51.
111. Smith HM. Holographic Recording Materials, Berlin: Springer-Verlag, 1977.
112. Gladden W, Leighty RD. Recording media, in Handbook of Optical Holography, Caulfield HJ, ed., New York, NY: Academic Press, 1979, pp. 277–298.
113. Solymar L, Cooke DJ. Volume Holography and Volume Gratings, New York, NY: Academic Press, 1981, pp. 254–304.
114. Delzenne GA. Organic photochemical imaging systems, in Advances in Photochemistry, J. N. Pitts, Jr., Hammond GS, Gollnick K, eds., New York, NY: Wiley- Interscience, 1980, Vol. 11, pp. 1–103.
115. Tomlinson WJ, Chandross EA. Organic photochemical refractive-index image recording systems, in Advances in Photochemistry, Pitts JN. Jr., Hammond GS, Gollnick K, eds., New York, NY: Wiley-Interscience, 1980, Vol. 12, pp. 201–281.
116. Monroe BM, Smothers WK. Photopolymers for holography and wave- guide applications, in Polymers for Lightwave and Integrated Optics: Technology and Applications, Hornak LA, ed., New York, NY: Marcel Dekker, 1992, pp. 145–169.
117. Monroe BM. Photopolymers, radiation curable imaging systems, in Radiation Curing: Science and Technology, Pappas SP, ed., New York, NY: Plenum Press, 1992, pp. 399–434.
118. Lougnot DJ. Photopolymer and holography, in Radiation Curing Polymer Science and Technology, Polymerization Mechanism, Fousassier JP, Rabek JF, eds., Andover: Chapman and Hall, 1993, Vol. **3**, pp. 65–100.
119. Monroe BM, Smothers WK, Keys DE, Weber AM. *J. Imag. Sci.,* **35**, 19; 1991.
120. Monroe BM. Eur. Patent 0 324 480; 1989 assigned to E. I. Dupont de Nemours and Co.
121. Keys DE. Eur. Patent 0 324 482; 1989 assigned to E. I. Dupont de Nemours and Co. 96.
122. Monroe BM. *J. Imag. Sci.,* **35**, 25; 1991.
123. Heat is on with New Dupont photopolymers, Holographic International, Winter, **26**; 1989.
124. Monroe BM. US Patent 4,917,977; 1990, assigned to E. I. Dupont de Nemours and Co.

125. Fielding HL, Ingwall RT. US Patent 4,588,664; 1986, assigned to Polaroid Corp.
126. Ingwall RT, Troll M. *Opt. Eng.,* **28**, 86; 1989.
127. Ingwall RT, Troll M. Holographic optics: Design and applications, *Proc. SPIE,* **883**, 94; 1998.
128. Ingwall RT, Troll M, Vetterling WT. Practical holography II, *Proc. SPIE,* **747**, 67; 1987.
129. Fielding HL, Ingwall RT. US patent 4 588 664; 1986, assigned to Polaroid Corp.
130. Fielding HL, Ingwall RT. US patent 4 535 041; 1985, assigned to Polaroid Corp.
131. Ingwall RT, Troll MA. US patent 4,970,129; 1990, assigned to Polaroid Corp.
132. Withney DH, Ingwall RT. Photopolymer device physics, chemistry, and applications, *Proc. SPIE,* **1213**(1), 8–26; 1990.
133. Ingwall RT, Fielding HL. *Opt. Eng.,* **24**, 808 1985.
134. Zhang C, Yu M, Yang Y, Feng S. J. *Photopolymer Sci. Tech.,* **4**, 139; 1991.
135. Luckemeyer T, Franke H. *Appl. Phys. Lett.,* **B46**, 181; 1988.
136. Mikaelian AL, Barachevsky VA. Photopolymer device physics, chemistry and applications II, *Proc. SPIE,* **1559**, 246; 1991.
137. Boiko Y, Tikhonov EA. *Sov. J. Quantum Electron.,* **11**, 492; 1981.
138. Boiko Y, Granchak VM, Dilung II, Solojev VS, Sisakian IN, Sojfer VA. *Proc. SPIE,* **1238**, 253; 1990.
139. Boiko Y, Granchak VM, Dilung II, Mironchenko VY. *Proc. SPIE,* **69**, 109; 1990.
140. Gyulnazarov ES, Obukhovskii VV, Smirnova TN. *Opt. Spectrosc.,* **69**, 109; 1990.
141. Gyulnazarov ES, Obukhovskii VV, Smirnova TN. *Opt. Spectrosc.,* **67**, 99; 1990.
142. Calixto S. *Appl. Opt.,* **26**, 3904; 1987.
143. Boiko Y, Solojev VS, Calixto S, Lougnot DJ. *Appl. Opt.,* **33**, 797; 1994.
144. Lougnot DJ, Turck C. *Pure Appl. Opt.,* **1**, 251; 1992.
145. Lougnot DJ, Turck C. *Pure Appl. Opt.,* **1**, 269; 1992.
146. Carre C, Lougnot DJ. *J. Optics,* **21**, 147; 1990.
147. Carre C, Lougnot DJ, Renotte Y, Leclere P, Lion Y. *J. Optics,* **23**, 73; 1992.
148. Lougnot DJ. Proc. OPTO'92, **99**; 1992.
149. Carre C, Lougnot DJ. Proc. OPTO'91, 317; 1991.
150. Carre C, Maissiat C, Ambs P. Proc. OPTO'92, **165**; 1992.
151. Carre C, Lougnot DJ. Proc. OPTO'90, **541**; 1990.
152. Noiret-Roumier N, Lougnot DJ, Petitbon I. Proc. OPTO'92, **104**; 1992.
153. Baniasz I, Loungot DJ, Turck C. Holographic recording material, *Proc. SPIE,* **2405**; 1995.
154. Yamagishi Y, Ishizuka T, Yagashita T, Ikegami K, Okuyama H. Progress in holographic applications, *Proc. SPIE,* **600**, 14; 1985.
155. Matsumoto K, Kuwayama T, Matsumoto M, Taniguchi N. Progress in holographic applications, *Proc. SPIE,* **600**, 9; 1985.
156. Shankoff TA. *Appl. Opt.,* **7**, 2101; 1968.
157. Chang BJ, Leonard CD. *Appl. Opt.,* **18**, 2407; 1979.
158. Meyerhofer D. *RCA Review,* **33**, 270; 1976.
159. Chang BJ. *Opt. Commun.,* **17**, 270; 1976.
160. Changkakoti R, Pappu SV. *Appl. Opt.,* **28**, 340; 1989.
161. Pappu SV, Changkakoti R. Photopolymer device physics, chemistry and applications, *Proc. SPIE,* **1223**, 39; 1990.
162. Cappola N, Lessard RA. Microelectronic interconnects and packages: Optical and electrical technologies, *Proc. SPIE,* **1389**, 612; 1990.
163. Cappola N, Lessard RA, Carre C, Lougnot DJ. *Appl. Phys.,* **B52**, 326; 1991.
164. Calixto S, Lessrad RA. *Appl. Opt.,* **23**, 1989; 1984.
165. Horner JL, Ludman JE. Recent advances in holography, *Proc. SPIE,* **215**, 46; 1980.
166. Wang MR, Sonek GJ, Chen RT, Jannson T. *Appl. Opt.,* **31**, 236; 1992.
167. Zipping F, Juging Z, Dahsiung H. Optica Acta Sinica, **4**, 1101; 1984.
168. Lelievre S, Couture JJA. *Appl. Opt.,* **29**, 4384; 1990.

169. Lelievre S. Holographie de Polarisation au Moyen de Films de PVA Bichromate, MS dissertation, Université Laval, Quebec, Canada, 1989.
170. Manivannan G, Changkakoti R, Lessard RA, Mailhot G, Bolte M. Nonconducting photopolymers and applications, *Proc. SPIE*, **1774**, 24; 1992.
171. Manivannan G, Changkakoti R, Lessard RA, Mailhot G, Bolte M. *J. Phys. Chem.*, **71**, 97; 1993.
172. Trepanier F, Manivannan G, Changkakoti R, Lessard RA. *Can. J. Phys.*, **71**, 423; 1993.
173. Changkakoti R, Manivannan G, Singh A, Lessard RA. *Opt. Eng.*, **32**, 2240; 1993.
174. Couture JJA, Lessard RA. *Can. J. Phys.*, **64**, 553; 1986.
175. Solano C, Lessard RA, Roberge PC. *Appl. Opt.*, **24**, 1189; 1985.
176. Lessard RA, Changkakoti R, Manivannan G. Photopolymer device physics, chemistry and applications II, *Proc. SPIE*, **1559**, 438; 1991.
177. Lessard RA, Changkakoti R, Manivannan G. *Opt. Mem. Neural Networks*, **1**, 75; 1992.
178. Manivannan G, Changkakoti R, Lessard RA. *Opt. Eng.*, **32**, 665; 1993.
179. Lessard RA, Malouin C, Changkakoti R, Manivannan G. *Opt. Eng.*, **32**, 665; 1993.
180. Manivannan G, Changkakoti R, Lessard RA. *Polym. Adv. Technologies*, **4**, 569; 1993.
181. Kakichashvili Sh D. *Opt. Spektrosk.*, **33**, 324; 1972.
182. Todorov T, Tomova N, Niklova L. *Appl. Opt.*, **23**, 4309; 1984.
183. Todorov T, Tomova N, Nikolova L. *Appl. Opt.*, **23**, 4588; 1984.
184. Couture JJA, Lessard RA. *Appl. Opt.*, **27**, 3368; 1988.
185. Gross RB, Can Izgi K, Birge RR. Image storage and retrieval systems, *Proc. SPIE*, **1662**, 1; 1992.
186. Birge RR, Zhang CF, Lawrence AL. Proceedings of the Fine Particle Society, Santa Clara, CA, 1988.
187. Brauchle Ch, Hampp N. *Makromol. Chem., Macromol. Symp.*, **50**, 97; 1991.
188. Hampp N. in Photochemistry and Polymeric Materials, Kelly JJ, McArdle CB, Maudner MJ. de F, eds., Cambridge: Royal Society of Chemistry, 1993.
189. Renn A, Wild UP. *Appl. Opt.*, **26**, 4040; 1987.
190. Wild UP, Renn A, De Caro C, Bernet S. *Appl. Opt.*, **29**, 4329; 1987.
191. Blaya S, Murciano A, Acebal P. Diffraction gratings and diffusion coefficient determination of acrylamide and polyacrylamide in sol-gel glass, *Appl. Phys. Lett.*, **84**, 4765–4767; 2004.
192. Naydenova I, Jallapuram R, Howard R, Martin S, Toal V. Investigation of the diffusion processes in a self-processing acrylamide-based photopolymer system, *Appl. Opt.*, **43**, 2900–2905; 2004.
193. Neipp C, Belendez A, Gallego S. Angular responses of the first and second diffracted orders in transmission diffraction grating recorded on photopolymer material, *Opt. Express*, **11**, 1835–1843; 2003.
194. Lawrence JR, O'Neill F, Sheridan FT. Photopolymer holographic recording material, *Optik*, **112**, 449–463; 2001.
195. Jianhua Z, Guixi W, Yi H. Highly sensitive and spatially resolved polyvinyl alcohol/acrylamide photopolymer for real-time holographic applications, *Opt. Express*, **18**, 18106–18112; 2010.
196. Garcia C, Fimia A, Pascual I. Holographic behavior of a photopolymer at high thicknesses and high monomer concentrations: Mechanism of photopolymerization, *Appl. Phys. B; LasersOpt.*, **72**, 311–316; 2001.
197. Pinto-Iguanero B, Olivares-Pérez A, Fuentes-Tapia I. Acrylates holographic material film composed by Norland Noa 65 (R) adhesive, *Opt. Mater.*, **20**, 225–232; 2002.
198. Duarte-Quiroga RA, Calixto S, Lougnot DJ. Optical characterization and applications of a dual-cure photopolymerizable system, *Appl. Opt.*, **42**, 1417–1425; 2003.
199. Dhar L. High-performance polymer recording materials for holographic data storage, *Bulletin*, **31**, 324–328; 2006.
200. Hsiao YN, Whang WT, Lin SH. Analyses on physical mechanism of holographic recording in phenanthrenequinone-doped poly (methyl methacrylate) hybrid materials, *Opt. Eng.*, **43**, 1993–2002; 2004.

201. Lin SH, Chen P.L, Hsiao Y-N. Fabrication and characterization of poly(methyl methacrylate) photopolymer doped with 9,10-phenanthrenequinone (PQ) based derivatives for volume holographic data storage, *Opt. Commun.*, **281**, 559–566; 2008.

202. Tolstik E, Kashin O, Matusevich A. Non-local response in glass-like polymer storage materials based on poly (methylmethacrylate) with distributed phenanthrenequinone, *Opt. Express*, **16**, 11253–11258; 2008.

203. Liu H, Yu D, Li X. Diffusional enhancement of volume gratings as an optimized strategy for holographic memory in PQ-PMMA photopolymer, *Opt. Express*, **18**, 6447–6454; 2010.

204. Kou HG, Shi WF, Lougnot DJ. Dendritic polyisophthalate endcapped with naphthyl groups for holographic recording. *Polym. Adv. Technol.*, **15**, 508–513; 2004.

205. Cho YH, He M, Kim BK. Improvement of holographic performance by novel photopolymer systems with siloxane-containing epoxides, *Sci. Technol. Adv. Mat.*, **5**, 319–323; 2004.

206. Cho YH, Shin CW, Kim N. High-performance transmission holographic gratings via different polymerization rates of dipentaerythritol acrylates and siloxane-containing epoxides, *Chem. Mater.*, **17**, 6263–6271; 2005.

207. Choi K, Chon J, James WM, Gu M. Ring opening low-distortion holographic data storage media using free-radical ring-opening polymerization, *Adv. Funct. Mater.*, **19**, 3560–3566; 2009.

208. Peng H, Nair D, Kowalski B. High performance graded rainbow holograms via two-stage sequential orthogonal thiol-click chemistry, *Macromolecules*, **47**, 2306–2315; 2014.

209. Ponce-Lee EL, Olivares-Pérez A, Fuentes-Tapia I. Sugar (sucrose) holograms, *Opt. Mater.*, **26**, 5–10; 2004.

210. Ortuno M, Fernández E, Gallego S. New photopolymer holographic recording material with sustainable design, *Opt. Express*, **15**, 12425–12435; 2007.

211. Cody D, Naydenova I, Mihaylova M. New non-toxic holographic photopolymer material, *J. Opt.*, **14**, 1; 2012.

212. Ortuno M, Gallego S, Márquez A. Biophotopol: A sustainable photopolymer for holographic data storage applications, *Materials*, **5**, 5; 2012.

213. Olivares-Pérez A, Toxqui-López S, Padilla-Velasco A. Nopal Cactus (Opuntia Ficus-Indica) as a holographic material, *Materials*, **5**, 2383–2402; 2012.

214. Gallego S, Neipp C, Ortuno M. Analysis of monomer diffusion in depth in photopolymer materials, *Opt. Commun.*, **274**, 43–49; 2007.

215. Yin D, Pu D, Haihui G, Bin G. Analytical rates determinations and simulations on diffusion and reaction processes in holographic photopolymerization, *Appl. Phys. Lett.*, **94**, 21; 2009.

216. Neipp C, Belendez A, Sheridan JT. Non-local polymerization driven diffusion based model: General dependence of the polymerization rate to the exposure intensity, *Opt. Express*, **11**, 16; 2003.

217. Lawrence JR, O'Neill FT, Sheridan JT. Adjusted intensity nonlocal diffusion model of photopolymer grating formation, *J. Opt. Soc. Am. B*, **19**, 621–629; 2002.

218. Moreau V, Renotte Y, Lion Y. Characterization of dupont photopolymer: Determination of kinetic parameters in a diffusion model, *Appl. Opt.*, **41**, 3427–3435; 2002.

219. Martinez-Matos O, Calvo M, Rodrigo JA. Diffusion study in tailored gratings recorded in photopolymer glass with high refractive index species, *Appl. Phys. Lett.*, **91**, 14; 2007.

220. Havranek A, Kveton M, Havrankova J. Polymer holography II - The theory of hologram growth—Polymer growth detected by holographic method y, *Polym. Bull.*, **58**, 261–269; 2007.

221. Babeva T, Naydenova I, Martin S. Method for characterization of diffusion properties of photopolymerisable systems, *Opt. Express*, **16**, 8487–8497; 2008.

222. Jallapuram R, Naydenova I, Byrne HJ. Raman spectroscopy for the characterization of the polymerization rate in an acrylamide-based photopolymer, *Appl. Opt.*, **47**, 206–212; 2008.

223. Gallego S, Márquez A, Mendez D. Direct analysis of monomer diffusion times in polyvinyl/acrylamide materials, *Appl. Phys. Lett.*, **92**, 7; 2008.

224. Close CE, Gleeson MR, Sheridan JT. Monomer diffusion rates in photopolymer material. Part I. Low spatial frequency holographic gratings, *J. Opt. Soc. Am. B*, **28**, 658–666; 2011.
225. Gallego S, Márquez A, Marini S. In dark analysis of PVA/AA materials at very low spatial frequencies: Phase modulation evolution and diffusion estimation, *Opt. Express*, **17**, 18279–18291; 2009.
226. Martinez-Ponce G, Solano-Sosa C. Photocrosslinking using linear polyols in xanthene dye-doped polyvinyl alcohol plates, *Opt. Express*, **14**, 3776–3784; 2006.
227. Liu S, Gleeson MR, Sheridan JT. Analysis of the photoabsorptive behavior of two different photosensitizers in a photopolymer material, *J. Opt. Soc. Am. B*, **26**, 528–536; 2009.
228. Fouassier JP, Lalevee J. Three-component photoinitiating systems: Towards innovative tailor made high performance combinations, *Advances*, **2**, 2621–2629; 2012.
229. Ibrahim A, Ley C, Allonas X. Optimization of a photopolymerizable material based on a photocyclic initiating system using holographic recording, *Photochem. Photobiol. Sci.*, **11**, 1682–1690; 2012.
230. Ibrahim A, Di Stefano.L, Tarzi O. High-performance photoinitiating systems for free radical photopolymerization. Application to holographic recording, *Photochem. Photobiol.*, **89**, 1283–1290; 2013.
231. Lin H, Oliveira PW, Veith M. Ionic liquid as additive to increase sensitivity, resolution, and diffraction efficiency of photopolymerizable hologram material, *Opt. Lett.*, **93**, 14; 2008.
232. Ortuno M, Márquez A, Gallego S. Pyrromethene dye and non-redox initiator system in a hydrophilic binder photopolymer, *Opt. Mater.*, **30**, 227–230; 2007.
233. Gong Q, Wang S, Huang M, Gan F. A humidity-resistant highly sensitive holographic photopolymerizable dry film, *Mater. Lett.*, **59**, 2969–2972; 2005.
234. Gong Q, Wang S, Huang M, Dong Y, Gan F. Effects of dyes and initiators on the holographic data storage properties of photopolymer, *Proc. SPIE*, **5966**, 59660P; 2005.
235. Mikulchyk T, Martin S, Naydenova I. Investigation of the sensitivity to humidity of an acrylamide-based photopolymer containing N-phenylglycine as a photoinitiator, *Opt. Mater.*, **37**, 810–815; 2014.
236. Hirabayashi K, Kanbara H, Mori Y. Multilayer holographic recording using a two-color-absorption photopolymer, *Appl. Opt.*, **46**, 8402–8410; 2007.
237. Bruder FK, Deuber F, Faecke T. Full-color self-processing holographic photopolymers with high sensitivity in red - The first class of instant holographic photopolymers, *J. Photopolym. Sci. Technol.*, **22**, 257–260; 2009.
238. Meka C, Jallapuram R, Naydenova I. Development of a panchromatic acrylamide-based photopolymer for multicolor reflection holography, *Appl. Opt.*, **49**, 1400–1405; 2010.
239. Qi Y, Li H, Guo J. Material response of photopolymer containing four different photosensitizers, *Opt. Commun.*, **320**, 114–124; 2014.
240. Naydenova I, Martin S, Toal V. Photopolymers: Beyond the standard approach to photosensitisation, *J. Eur. Opt. Soc. Rapid Publ.*, **4**, 09042–09046; 2009.
241. Choi DH, Feng DJ, Yoon H. Diffraction gratings of photopolymers composed of polyvinylalcohol or polyvinylacetate binder, *Macromol. Res.*, **11**, 36–41; 2003.
242. Ortuno M, Fernández E, Fuentes R. Improving the performance of PVA/AA photopolymers for holographic recording, *Opt. Mater.*, **35**, 668–673; 2013.
243. Veniaminov A, Bartsch E. Diffusional enhancement of holograms: Phenanthrenequinone in polycarbonate. *J. Opt. A*, **4**, 387–392; 2002.
244. Liu GD, He QS, Luo SJ. Epoxy resin-photopolymer composite with none-shrinkage for volume holography, *Chinese Phys. Lett.*, **20**, 1733–1735; 2003.
245. Jeong Y-C, Lee S, Park JK. Holographic diffraction gratings with enhanced sensitivity based on epoxy-resin photopolymers, *Opt. Express*, **15**, 1497–1504; 2007.
246. Kou HG, Shi WF, Tang L. Recording performance of holographic diffraction gratings in dry films containing hyperbranched polyisophthalesters as polymeric binders, *Appl. Opt.*, **42**, 3944–3949; 2003.
247. Shelkovnikov VV, Russkikh VV, Vasil'ev EV. Production and properties of holographic photopolymeric material in a hybrid sol-gel matrix. *J. Opt. Technol.*, **73**, 480–483; 2006.

248. Cheben P, Calvo M. A photopolymerizable glass with diffraction efficiency near 100% for holographic storage, *Appl. Phys. Lett.,* **78**, 1490–1492; 2001.

249. Neipp C, Sheridan JT, Gallego S. Effect of a depth attenuated refractive index profile in the angular responses of the efficiency of higher orders in volume gratings recorded in a PVA/acrylamide photopolymer, *Opt. Commun.,* **233**, 311–322; 2004.

250. Gallego S, Ortuno M, Neipp C. Physical and effective optical thickness of holographic diffraction gratings recorded in photopolymers, *Opt. Express,* **13**, 1939–1947; 2005.

251. Mahmud MH, Naydenova I, Pandey N. Holographic recording in acrylamide photopolymers: Thickness limitations, *Appl. Opt.,* **48**, 2642–2648; 2009.

252. Vaia R, Maguire J. Polymer nanocomposites with prescribed morphology: Going beyond nanoparticles–filled polymers, *Chem. Mat.,* **19**, 2736–2751; 2007.

253. Oliveira PW, Krug H, Muller P, Schmidt H. Fabrication of GRIN-materials by photopolymerisation of diffusion-controlled organic-inorganic nanocomposite materials, *MRS Proc.* 435, 234–238; 1996.

254. Vaia R, Dennis C, Natarajan L, Tondiglia V, Toml D, Bunning T. One-step, micrometer-scale organization of nano-and mesoparticles using holographic photopolymerization: A generic technique, *Adv. Mat.,* **13**, 1570; 2001.

255. Suzuki N, Tomita Y, Kojima T. Holographic recording in TiO2 nanoparticle-dispersed methacrylate photopolymer films, *Appl. Phys. Lett.,* **81**, 4121–4123; 2002.

256. Tomita Y, Nishibiraki H. Improvement of holographic recording sensitivities in the green in SiO2 nanoparticle-dispersed methacrylate photopolymers doped with pyrromethene dyes, *Appl. Phys. Lett.,* **83**, 410–412; 2003.

257. Suzuki N, Tomita Y. Silica-nanoparticle-dispersed methacrylate photopolymers with net diffraction efficiency near 100%, *Appl. Opt.,* **43**, 2125–2129; 2004.

258. Sanchez C, Escuti M, Heesh C, et al. TiO2 nanoparticle-photopolymer holographic recording, *Adv. Funct. Mater.,* **15**, 1623–1629; 2005.

259. Smirnova T, Sakhno O, Bezrodnyj V, Stumpe J. Nonlinear diffraction in gratings based on polymer-dispersed TiO2 nanoparticles, *Appl. Phys. B,* **80**, 947–951; 2005.

260. Naydenova I, Sherif H, Mintova S, Martin S, Toal V. Holographic recording in nanoparticle-doped photopolymer, *Proc. SPIE,* **6252**, 45; 2006.

261. Kim WS, Jeong YC, Park K. Nanoparticle-induced refractive index modulation of organic-inorganic hybrid photopolymer, *Opt. Express,* **14**, 8967–8973; 2006.

262. Garnweitner G, Goldenberg L, Sakhno O, Antonietti M, Niederberger M, Stumpe J. Large-scale synthesis of organophilic zirconia nanoparticles and their application in organic-inorganic nanocomposites for efficient volume holography, *Small,* **3**, 1626–1632; 2007.

263. Sakhno O, Goldenberg L, Stumpe J, Smirnova T. Surface modified ZrO2 and TiO2 nanoparticles embedded in organic photopolymers for highly effective and UV-stable volume holograms, *Nanotechnology,* **18**, 105704–105706; 2007.

264. Naydenova I, Toal V. Nanoparticle doped photopolymers for holographic applications, *Ordered Porous Solids: Recent Advances and Prospects,* Valtchev V, Mintova S, Tsapatsis M, eds., Amsterdam: Elsevier; 2008.

265. Goldenberg L, Sakhno O, Smirnova T, Helliwell P, Chechik V, Stumpe J. Holographic composites with gold nanoparticles: Nanoparticles promote polymer segregation, *Chem. Mater.,* **20**, 4619; 2008.

266. Han S, Lee M, Kim BK. Effective holographic recordings in the photopolymer nanocomposites with functionalized silica nanoparticle and polyurethane matrix, *Opt. Mat.,* **34**, 131–137; 2011.

267. Goourey GG, Claire B, Israëli Y. Acrylate photopolymer doped with ZnO nanoparticles: An interesting candidate for photo-patterning applications, *J. Mat. Chem. C,* **1**, 3430–3438; 2013.

268. Hata E, Tomita Y. Order of magnitude polymerisation-shrinkage suppression of volume gratings recorded in nanoparticles-polymer composites, *Opt. Lett.,* **35**, 396–398; 2010.

269. Tomita Y, Nakamura T, Tago A. Improved thermal stability of volume holograms in nanoparticles–polymer composite films, *Opt. Lett.,* **33**, 1750–1752; 2008.

270. Balan L, Turck C, Soppera O, Vidal L, Lougnot DJ. Holographic recording with polymer nanocomposites containing silver nanoparticles photogenerated in situ by the interference pattern, *Chem. Mater.*, **21**, 5711–5718; 2009.

271. Pramitha V, Nimmi KP, Subramanyan NV, Joseph R, Sreekumar K, Kartha C. Silver-doped photopolymer media for holographic recording, *Appl. Opt.*, **48**, 2255–2261; 2009.

272. Liu X, Tomita Y, Oshima J, et al. Holographic assembly of semiconductor CdSe quantum dots in polymer for volume Bragg grating structures with diffraction efficiency near 100%, *Appl. Phys. Lett.*, **95**, 2611091–2611093; 2009.

273. Fuentez-Hernandez C, Suh DJ, Kippelen B. High performance photorefractive polymers sensitized by cadium selenide nanoparticles, *Appl. Phys. Lett.*, **85**, 534–536; 2004.

274. Tomita Y, Suzuki N, Chikama K. Holographic manipulation of nanoparticles distribution morphology in nanoparticles-dispersed photopolymers, *Opt. Lett.*, **30**, 839–841; 2005.

275. Tomita Y, Chikama K, Nohara Y, Suzuki N, Furushima K, Endoh Y. Two-dimensional imaging of atomic distribution morphology created by holographically induced mass transfer of monomer molecules and nanoparticles in a silica-nanoparticle-dispersed photopolymer film, *Opt. Lett.*, **31**, 1402–1404; 2006.

276. Suzuki N, Tomita Y. Real-time phase-shift measurement during formation of a volume holographic grating in nanoparticle-dispersed photopolymers, *Appl. Phys. Lett.*, **88**, 011105–011116; 2006.

277. Chikama K, Mastubara K, Oyama S, Tomita Y. Three-dimensional confocal Raman imaging of volume holograms formed in ZrO2 nanoparticle-photopolymer composite materials, *J. Appl. Phys.*, **103**, 113108–113114; 2008.

278. Leite E, Naydenova I, Pandey N, et al. Investigation of the light induced redistribution of zeolite beta nanoparticles in an acrylamide-based photopolymer, *J. Opt. A*, **11**, 024016–024018; 2009.

279. Ostrowski A, Naydenova I, Toal V. Light induced redistribution of Si–MFI zeolite nanoparticles in acrylamide-based photopolymer holographic gratings, *J. Opt. A*, **11**, 034004–034007; 2019.

280. Naydenova I, Leite E, Babeva T, et al. Optical properties of photopolymerisable nanocomposites containing nanosized molecular sieves, *J. Opt.*, 044019–044027; 2011.

281. Juhl A, Busbee J, Koval J, et al. Holographically directed assembly of polymer nanocomposites, *ACS NANO*, **4**, 5953–5961; 2010.

282. Yu D, Liu H, Jiang Y, Sun X. Mutual diffusion dynamics with nonlocal response in SiO2 nanoparticles dispersed PQ-PMMA bulk photopolymer, *Opt. Express*, **19**, 13787–13792; 2011.

283. Liu H, Yu D, Wang W, Geng Y, Yang L. Mutual diffusion dynamics as matter transfer mechanism in inorganic nanoparticles dispersed photopolymer, *Opt. Commun.*, **330**, 77–84; 2014.

284. Jenney J. Holographic recording with photopolymers, *J. Opt. Soc. Am.*, **60**, 1155–1161; 1970.

285. Boiko Y, Slovjev V, Calixto S, Lougnot D. Dry photopolymer films for computer-generated infrared radiation focusing elements, *Appl. Opt.*, **33**, 787–793; 1994.

286. Croutxe-Barghorn C, Lougnot DJ. Use of self-processing dry photopolymers for the generation of relief optical elements: A photochemical study, *Pure Appl. Opt.*, **5**, 811–825; 1996.

287. Smirnova T, Sakhno O. A mechanism of the relief-phase structure formation in self-developing photopolymers, *Opt. Spectrosc.*, **3**, 126–131; 2002.

288. Sakhno O, Smirnova T. Relief structures in the self-developing photopolymer materials, *Optik*, **113**, 130–134; 2002.

289. Naydenova I, Mihaylova E, Martin S, Toal V. Holographic patterning of acrylamide–based photopolymer surface, *Opt. Express*, **13**, 4878; 2005.

290. Kojima T, Tomita Y. Characterization of index and surface-relief gratings formed in methacrylate photopolymers, *Opt. Rev.*, **9**, 222–226; 2002.

291. Shen XX, Yu X, Yang XL. Fabrication of periodic microstructures by holographic photopolymerization with a low-power continuous-wave laser of 532 μm, *J. Opt. A*, **8**, 672–676; 2006.

292. Trainer K, Wearen K, Nazarova D, Naydenova I, Toal V. Optimisation of an acrylamide-based photopolymer system for holographic inscription of surface patterns with sub-micron resolution, *J. Opt.*, **12**, 124012–124015; 2010.

293. Zhao G, Mouroulis P. Diffusion model of hologram formation in dry photopolymer materials, *J. Mod. Opt.*, **41**, 1929–1939; 1994.

294. Piazzola S, Jenkins B. First-harmonic diffusion model for holographic grating formation in photopolymers, *J. Opt. Soc. Am. B*, **17**, 1147–1157; 2000.

295. Colvin VL, Larson RG, Harris AL, Schilling ML. Quantitative model of volume hologram formation in photopolymers, *J. Appl. Phys.*, **81**, 5913–5923; 1997.

296. Blaya S, Carretero L, Mallavia R, Fimia A, Madrigal RF. Holography as a technique for the study of photopolymerization kinetics in dry polymeric films with a nonlinear response, *Appl. Opt.*, **38**, 955–962; 1999.

297. Kwon JH, Hwang HC, Woo KC. Analysis of temporal behavior of beams diffracted by volume gratings formed in photopolymers, *J. Opt. Soc. Am. B*, **16**, 1651–1657; 1999.

298. Neipp C, Gallego S, Ortuno M, et al. First-harmonic diffusion-based model applied to a polyvinyl-alcohol–acrylamide-based photopolymer, *J. Opt. Soc. Am. B*, **20**, 2052–2060; 2003.

299. Lawrence J, O'Neill F, Sheridan JT. Adjusted intensity nonlocal diffusion model of photopolymer grating formation, *J. Opt. Soc. Am. B*, **19**, 621–629; 2002.

300. Wu SD, Glytsis EN. Holographic grating formation in photopolymers: Analysis and experimental results based on a nonlocal diffusion model and rigorous coupled-wave analysis, *J. Opt. Soc. Am. B*, **20**, 1177–1188; 2003.

301. Liu S, Gleeson MR, Guo J. High intensity response of photopolymer materials for holographic grating formation, *Macromolecules*, **43**, 9462–9472; 2010.

302. Gleeson MR, Sabol D, Liu S, Close CE, Kelly JV, Sheridan JT. Improvement of the spatial frequency response of photopolymer materials by modifying polymer chain length, *J. Opt. Soc. Am. B*, **25**, 396–406; 2008.

303. Fernández E, Fuentes R, Ortuño M, Beléndez A, Pascual I. Holographic grating stability: Influence of 4,4'-azobis (4-cyanopentanoic acid) on various spatial frequencies, *Appl. Opt.*, **52**, 6322–6331; 2013.

304. Martin S, Naydenova I, Toal V, Jallapuram R, Howard RG. Two way diffusion model for the recording mechanism in a self developing dry acrylamide photopolymer, *Proc. SPIE*, **6252**, 37–44; 2006.

305. Květoň M, Fiala F. Havránek A. *Polymer Holography in Acrylamide-Based Recording Material*, Holography, Research and Technologies, J. Rosen (Ed.), InTech, 2011.

306. Babeva T, Naydenova I, Mackey D, Martin S, Toal V. Two-way diffusion model for short-exposure holographic grating formation in acrylamide based photopolymer, *J. Opt. Soc. Am. B*, **27**, 197–203; 2010.

307. Meka C, Jallapuram R, Naydenova I, Martin S, Toal V. Development of a panchromatic acrylamide-based photopolymer for multicolor reflection holography, *Appl. Opt.*, **49**, 1400–1405; 2010.

308. Naydenova I, Jallapuram R, Toal V, Martin S. A visual indication of environmental humidity using a colour changing hologram recorded in a self-developing photopolymer, *Appl. Phys. Lett.*, **92**, 031109–031114; 2008.

309. Naydenova I, Jallapuram R, Toal V, Martin S. Characterization of the humidity and temperature responses of a reflection hologram recorded in acrylamide-based photopolymer, *Sens. Actuators B Chem.*, **139**, 35–38; 2009.

310. Benjamin A, Kowalski A, Urness C, Baylor ME, Cole MC, Wilson WL, McLeod RR. Quantitative modeling of the reaction/diffusion kinetics of two-chemistry diffusive photopolymers, *Opt. Mat. Express*, **4**, 1669; 2014.

311. Liu S, Gleeson MR, Guo J. Modeling the photochemical kinetics induced by holographic exposures in PQ/PMMA photopolymer material, *J. Opt. Soc. Am. B*, **28**, 11, 2833–2843; 2011.

312. Sabol D, Gleeson MR, Liu S, Sheridan JT. Photoinitiation study of Irgacure 784 in an epoxy resin photopolymer, *J. Appl. Phys.*, **107**, 053113–053114; 2010.

313. Kyu T, Nwabunma D, Chiu HW. Theoretical simulation of holographic polymer-dispersed liquid-crystal films via pattern photopolymerization-induced phase separation, *Phys. Rev. E*, **63**, 061802–061804; 2006.

314. Natarajan LV, Shepherd CK, Brandelik DM, et al. Switchable holographic polymer-dispersed liquid crystal reflection gratings based on thiol-ene photopolymerization, *Chem. Mater.*, **15**, 2477–2484; 2003.

315. De S, Gill NL, Whitehead JB, Crawford GP. Effect of monomer functionality on the morphology and performance of the holographic transmission gratings recorded on polymer dispersed liquid crystals, *Macromolecules*, **36**, 630–638; 2003.

316. Caputo R, Veltri A, Umeton CP, Sukhov AV. Characterization of the diffraction efficiency of new holographic gratings with a nematic film-polymer-slice sequence structure, *J. Opt. Soc. Am. B*, **21**, 1939–1947; 2004.

317. Caputo R, Sio LC, Veltri A, Umeton C, Sukhov AV. Development of a new kind of switchable holographic grating made of liquid-crystal films separated by slices of polymeric material, *Opt. Lett.*, **29**, 1261–1263; 2004.

318. Caputo R, De Luca A, De Sio L, et al. POLICRYPS: A liquid crystal composed nano/microstructure with a wide range of optical and electro-optical applications, *J. Opt. A: Pure Appl. Opt.*, **11**, 2; 2009.

319. Jakubiak R, Bunning TJ, Vaia RA, Natarajan LV, Tondiglia VP. Electrically switchable, one-dimensional polymeric resonators from holographic photopolymerization: A new approach for active photonic bandgap materials, *Adv. Mater.*, **15**, 241; 2003.

320. Zhao C, Liu J, Fu Z, Chen RT. Shrinkage correction of volume phase holograms for optical interconnects, *Proc. SPIE*, Vol. **3005**; 1997.

321. Trentler TJ, Boyd JE, Colvin VL. Epoxy resin-photopolymer composites for volume holography, *Chem. Mater.*, **12**, 1431–1438; 2000.

322. Tomita Y, Furushima K, Ochi K. Organic nanoparticle (hyperbranched polymer)-dispersed photopolymers for volume holographic storage, *Appl. Phys. Lett.*, **88**, 2006.

323. Lee S, Jeong YC, Heo Y, Kim SI, Choi YS, Park JK. Holographic photopolymers of organic/inorganic hybrid interpenetrating networks for reduced volume shrinkage, *J. Mater. Chem.*, **19**, 1105–1114; 2009.

324. Moothanchery M, Mintova S, Naydenova I, Toal V. Si-MFI Zeolite nanoparticle doped photopolymer with reduced shrinkage, *Opt. Express*, **19**, 25786–25791; 2011.

325. Moothanchery M, Naydenova I, Toal V. Study of the shrinkage caused by holographic grating formation in acrylamide based photopolymer film, *Opt. Express*, **19**, 13395–13404; 2011.

326. Waldman D, Li HYS, Cetin E. Holographic recording properties in thick films of ULSH-500 photopolymer, *Proc. SPIE*, **3291**, 89; 1998.

327. Shelby RM, Waldman DA, Ingwall RT. Distortions in pixel-matched holographic data storage due to lateral dimensional change of photopolymer storage media, *Opt. Lett.*, **25**, 713–715; 2000.

328. Moothanchery M, Bavigadda V, Toal V, Naydenova I. Shrinkage during holographic recording in photopolymer films determined by holographic interferometry, *Appl. Opt.*, **52**, 8519–8527; 2013.

329. Van Heerden PJ. Theory of optical information storage in solids, *Appl. Opt.*, **2**, 393; 1963.

330. Collier RJ, Burckhardt CB, Lin LH. *Optical Holography*, Chap. 16, New York, NY: Academic Press, 1971.

331. Anderson LK. Holographic optical memory for bulk data storage, *Bell Lab. Rec.* **46**, 318; 1968.

332. Hesselink L, Orlov SS, Bashaw MC. Holographic data storage systems, *Proc. IEEE*, **92**, 1231; 2004.

333. Curtis KL, Dhar L, Hiull AJ, Wilson WL, Ayres MR. *Holographic Data Storage*, West Sussex, United Kingdom: Wiley, 2010.

334. Orlic S, Ulm S, Eichler HJ. 3D bit-oriented optical storage in photopolymers, *J. Opt. A: Pure Appl. Opt.*, **3**, 72; 2001.

335. Orlic S, Dietz E, Frohmann S, Gartner J, Mueller L. Microholographic multilayer recording at DVD density, *Proceedings of Optical Data Storage Conference*, Portland, Oregon: paper mb., 2007.

336. McLeod RR, Daiber AJ, McDonald ME, et al. Microholographic multilayer optical disk data storage, *Proceedings of Optical Data Storage Conference*, Honolulu, Hawaii: Paper–MB3, 2005.

337. Macleod RR, Daiber AJ, Honda T, et al. Three-dimensional optical disk data storage via the localized alteration of a format hologram, *Appl. Opt.*, **47**, 2696; 2008.

338. Jeong Y-C, Jung B, Ahn D, Park J-K. Blue laser–Sensitized photopolymer for a holographic high density data storage system, *Opt. Express*, **18**, 25008; 2010.

339. Ortuño M, Gallego S, Garcia C, Neipp C, Pascual I. Holographic characteristics of a 1 mm—Thick photopolymer to be used in holographic memories, *Appl. Opt.*, **42**, 7008; 2003.

340. Ortuño M, Gallego S, Garcia C, Pascual I, Neipp C, Belendez A. Holographic characteristics of an acrylamide/bisacrylamide photopolymer layer 40–1000 μm thick, *Physica Scripta*, **T118**, 66; 2005.

341. Fernández E, Garcia C, Pascual I, Ortuño M, Gallego S, Belendez A. Optimization of a thick poly-vinyl alcohol–acrylamide photopolymer for data storage using a combination of angular and peris-trophic holographic multiplexing, *Appl. Opt.*, **43**, 7661; 2006.

342. Fernández E, Márquez A, Gallego S, Fuentes R, Garcia C, Pascual I. Hybrid ternary modulation applied to multiplexing holograms in photopolymers for data page storage, *J. Lightwave Technol.*, **28**, 776; 2010.

343. Fernández E, Fuentes R, Márquez A, Belendez A, Pascual I. Binary intensity modulation and hybrid ternary modulation applied to multiplexing objects using holographic data storage in a PVA/AA photopolymer, *Int. J. Polym. Sci.*, **2014**, 366534–366535; 2014.

344. Sherif H, Naydenova I, Martin S, McGinn C, Toal V. Characterization of an acrylamide based photopoly-mer for data storage utilizing holographic angular multiplexing, *J. Opt. A: Pure Appl. Opt.*, **1**, 255; 2005.

345. Huang M, Chen Z. Optical setup and analysis of disk–type photopolymer high–density holographic storage, *Opt. Eng.*, **41**, 2315; 2002.

346. Lee S, Jeong Y.C, Lee J, Park JK. Multifunctional photoreactive inorganic cages for three-dimensional holographic data storage, *Opt. Lett.*, **34**, 3015; 2009.

347. Khan A, Daugard AE, Bayles A, et al. Dendronized macromonomer for three–dimensional storage, *Chem. Comm.*, **425**, 340; 2009.

348. Momose K, Takayama S, Jata E, Tomita Y. Shift multiplexed holographic digital data page storage in a nano-particle- (Thiol-ene) polymer composite, *Opt. Lett.*, **37**, 2250; 2012.

349. Mitsube K, Nishimure Y, Nagaya K, Takayama S, Tomita Y. Holographic nanoparticle–polymer composite based on radical mediated thiol–ene photopolymerization characterization and shift multiplexed holographic digital data page storage, *Opt. Mat. Express*, **4**, 982; 2014.

350. Ortuño M, Gallego S, Márquez A, Neipp C, Pascual I, Belendez A. Biophotopol: A sustainable pho-topolymer for data storage applications, *Materials*, **5**, 772; 2012.

351. Orlic S, Dietz E, Feid T, Frohman S. Optical investigation of photopolymer systems for microholo-graphic storage, *J. Opt A: Pure Appl. Opt.*, **11**, 024014–024015; 2009.

352. Hwang Y-S, Bruder F-K, Facke T, et al. Time sequential autostereoscopic 3-D display with novel directional backlight system based on volume holographic optical elements, *Opt. Express*, **22**, 9820; 2014.

353. Garcia C, Rodriguez JD, Fernández E, Campos V, Fuentes R, Pascual I. Holographic lens recorded on photopolymers: Fabrication and study of the image quality, *J. Mod. Opt.*, **56**, 1288; 2009.

354. Macko P, Whelan MP. Fabrication of holographic diffractive optical elements for enhancing light collection from fluorescent—Based chips, *Opt. Lett.*, **33**, 2614; 2008.

355. Jradi S, Soppera O, Lougnot DJ. Fabrication of polymer waveguide between two optical fibers using spatially controlled light induced polymerization, *Appl. Opt.*, **47**, 3987; 2008.

356. Jisha, Kishore CPVC, John BM, Kuriakose VC, Porsezian K, Kartha CS. Self–written waveguide in methylene blue sensitized poly(vinyl alcohol/ acrylamide photopolymer material, *Appl. Opt.*, **35**, 6502; 2008.

357. Sullivan AC, Grtabowski MW, McLeod RR. Three-dimensional direct-write lithography into poly-mer, *Appl. Opt.*, **46**, 295; 2007.

358. Lee K-W. Holographic filter with cascaded volume Bragg gratings in photopolymer waveguide film, *Opt. Express*, **18**, 25649; 2010.

359. Luo Y, Russo JM, Kostuk RK, Stathis GB. Silicon oxide nanoparticles doped PQ-PMMA for volume holographic imaging filters, *Opt. Lett.*, **35**, 1269; 2010.

360. Velasco AV. Photopolymerizable organically modified holographic glass with enhanced thickness for spectral filters, *J. Appl. Phys.*, **113**, 033101–033108; 2013.

361. Zheng GQ, Montemezzini G, Gunter P. Nano-bandwidth holographic reflection filters with photopolymer films, *Appl. Opt.*, **40**, 2423; 2001.

362. Kabilan S, Marshal AJ, Sartian FK, et al. Holographic glucose sensors, *Biosens. Bioelectron.*, **20**, 1602; 2005.

363. Fuchs Y, Kunath S, Soppera O, Haupt K, Mayes AG. Molecularly imprinted silver-halide reflection holograms for label–free optochemical sensing, *Adv. Funct. Mater.*, **24**, 688; 2014.

364. Mikulchyk T, Martin S, Naydenova I. Humidity and temperature effect on properties of transmission gratings recorded in PVA/AA-based photopolymers, *J. Opt.*, **15**, 105301–105304; 2013.

365. Yetisen AK, Qasin M, Nosheen S, Wilkinson TD, Lowe CR. Pulsed laser writing of holographic nanosensors, *J. Mater. Chem. C*, **2**, 3569; 2014.

366. Matusevich V, Wolf F, Tolstik E, Kowarschik R. A transparent optical sensor for moisture detection integrated in PQ-PMMA medium, *IEEE Photon. Technol. Lett.*, **25**, 969; 2013.

367. Naydenova I, Grand J, Mikulchyk T, et al. Hybrid Sensors Fabricated by Inkjet Printing and Holographic Patterning, *Chem. Mater.*, (2015), 27 (17), 6097–6101.

368. Stankevicius E, Gertus T, Rutkauskas M, et al. Fabrication of micro-tube arrays in photopolymer SZ2080 by using three different methods of a direct laser polymerization technique, *J. Micromech, Microeng.*, **22**, 065022–065026; 2010.

369. Bautista G, Romero MJ, Tapong G, Daria VR. Parallel two-photon photopolymerization of micro-gear patterns, *Opt. Commun.*, **282**, 3746; 2009.

370. Pantelic D, Savic S. Dichromated pullulan as a novel photosensitive holographic material, *Opt. Lett.*, **23**, 807–809; 1998.

371. Savic S, Pantelic D. Relief hologram replication using a dental composite as an embossing tool, *Opt. Express*, **13**, 2747–2754; 2005.

372. Luchetta DE, Castagna R, Vita F. Microfluidic transport of photopolymerizable species for laser source integration in Lab-on a-chip photonic devices, *Photonic. Nanostruct.*, **10**, 575; 2012.

373. Goldenberg LM, Gritsai Y, Sakhno O, Kulikovska O, Stumpe J. All optical fabrication of 2D and 3D photonics structures using a single polymer phase mask, *J. Opt.*, **12**, 015103–015108; 2010.

374. Orlic S, Muller C, Schlosser A. All optical fabrication of the three–dimensional photonic crystals in photopolymers by multiplex–exposure holographic recording, *Appl. Phys. Lett.*, **99**, 131105–131114; 2011.

375. Zhang Y, Jianying Z, Wong K-S. Two–photon fabrication of photonic crystals by single beam laser holographic lithography, *J. Appl. Phys.*, **107**, 07431–07433; 2010.

376. Chen J, Jiang W, Chen X, Wang L, Zhang S, Chen RT. Holographic three–dimensional polymeric photonic crystals operating at 1550 nm window, *Appl. Phys. Lett.*, **90**, 093102–093113; 2007.

377. Nakamura K. *Photopolymers: Photoresist Materials, Processes, and Applications,* CRC Press, Taylor and Francis Group, Boca Raton, London, New York; 2014.

378. Munk P. *Introduction to Macromolecular Science,* New York, NY: Wiley, 2001.

11

Electro-Optical and Acousto-Optical Devices

11.1	Introduction	409
11.2	Photoconductors	409
11.3	Photodiodes	411
11.4	Charge-Coupled Imagers	422
11.5	CRT and Imaging Tubes	429
11.6	Electro-Optic Modulation	433
11.7	Electro-Optic Devices	436
11.8	Acousto-Optic Modulation and Devices	453
11.9	Summary	457
	Suggested Reading	457

Mohammad
A. Karim

11.1 Introduction

In this chapter, we introduce and discuss the basic role of photoconduction, photodetection, and electro-optic and acousto-optic modulation. These concepts are vital in the design and understanding of various detection, imaging, amplification, modulation, and signal processing systems, many of which are either electronic or hybrid in nature.

Section 11.2 introduces the concepts of photoconduction. It is followed by Section 11.3, where we discuss design, characteristics, and applications of *p–n* and *p–i–n* photodiodes, avalanche photodiodes, vacuum photodiodes, and photomultipliers. The concept of metal oxide semiconductor (MOS) capacitor and its application in the design of charge-coupled device (CCD) structure, MOS read-out scanner, and CCD imager are introduced in Section 11.4. Next, in Section 11.5, we describe cathode-ray tube (CRT) technology and various imaging tube technologies, such as vidicon, plumbicon, and image intensifier. Section 11.6 introduces the physics of electro-optic (EO) modulation. Section 11.7 discusses the working of EO amplitude modulator, EO phase modulator, Pockels read-out optical modulator, Kerr modulator, liquid-crystal light valve, spatial light modulator, and liquid-crystal display devices. Finally, in Section 11.8, the concept of acousto-optical modulation and its application to a few hybrid systems are elaborated.

11.2 Photoconductors

Almost all semiconductors exhibit a certain degree of photoconductivity. A photoconductor is a simple photodetection device built exclusively of only one type of semiconductor that has a large surface area and two ohmic contact points. In the presence of an energized incident photon, the excited valence band electron of the photoconductor leaves behind a hole in the valence band. Often an extrinsic semiconductor is better suited for the purpose of photoconduction. For example, a far-infrared (IR) sensitive

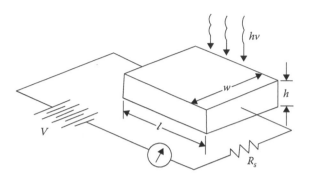

FIGURE 11.1 A photoconductor circuit.

photoconductor can be designed by introducing an acceptor level very close to the valence band or by introducing a donor level very close to the conduction band. Consequently, photoconduction may have two causes. It is caused either by the absorption of photons at the impurity levels in an extrinsic semiconductor or due to band-gap transition in an intrinsic semiconductor. Typically, photo-conductors are cooled in order to avoid excessive thermal excitation of carriers.

Figure 11.1 shows a typical photoconductor circuit where R_s is the series resistance. Assume further that the resistance of the photoconductor is larger than R_s, so that most of the bias voltage appears across the photoconductor surface. To guarantee that in the absence of incoming light the number of carriers is a minimum, the operating temperature is maintained sufficiently low. Incident light continues to affect both generation and recombination of carriers until the photoconductor has reached a new equilibrium at higher carrier concentration. A change in carrier density causes a reduction in the photoconductor's resistance. In fact, there are a great many commercial applications of photoconductors where the fractional change in resistance value is significant. In the presence of an electrical field, the generated excess majority carriers drift away from the appropriate terminals.

The absorbed portion of the incoming monochromatic light that falls normally onto the photoconductor is determined in terms of the absorption coefficient α. In the case of extrinsic semiconductors, α is typically very small (1/cm to 10/cm) since the number of available impurity levels is rather small. But in the case of an intrinsic photoconductor, α is large ($\cong 10^4$/cm) in comparison, as the number of available electron states is very large. The absorbed optical power $P_{abs}(y)$ is given by

$$P_{abs}\left(y\right)= P_{in}(1-R)e^{-\alpha y},\tag{11.1}$$

where P_{in} represents the incoming optical power and R is the surface reflectance of the photoconductor. At a steady state, the generation and recombination rates are equal to each other. Consequently,

$$\begin{aligned}\alpha P_{abs}(y)/h\nu\lambda\omega &= \{\alpha P_{in}(1-R)e^{-\alpha y}\}/h\nu\lambda\omega\\ &= n\left(y\right)/\tau_n\\ &= p\left(y\right)/\tau p\end{aligned}\tag{11.2}$$

where τ_n and τ_p are the mean lifetimes, respectively, of electrons and holes; $n(y)$ and $p(y)$ are the carrier densities, respectively, of electrons and holes; and the product lw represents the surface area of the photoconductor.

The total drift current passing through the intrinsic photoconductor is determined using Equation 11.2:

$$i_s =\left(\eta_{pc}eP_{in}/h\nu l\right)E\left(\mu_n\tau_n +\mu_p\tau_p\right),\tag{11.3}$$

where the quantum efficiency η_{pc} is defined as

$$\eta_{pc} = \alpha\pi(1-R)\left(1-e^{-\alpha h}\right)/\alpha \tag{11.4}$$

and E is the electric field. In the case of an extrinsic photoconductor, the signal current of Equation 11.3 reduces to

$$i_s = \begin{cases} \left(\eta_{pc}eP_{in}/hv\right)\left[\left(E\mu_n\tau_n/l\right)\right], & n-\text{type} \\ \left(\eta_{pc}eP_{in}/hv\right)\left[\left(E\mu_p\tau_p/l\right)\right], & p-\text{type.} \end{cases} \tag{11.5}$$

In either case, the quantity within the square bracket is generally referred to as the photoconductive gain G, as given by

$$G = \begin{cases} \tau_n/\tau_d, & n\text{-type} \\ \tau_p/\tau_d, & p\text{-type,} \end{cases} \tag{11.6}$$

where τ_d is the average carrier drift time or transit time between the two metal contacts, since drift velocity is given by the product of electrical field and carrier mobility. The photoconductive gain generally measures the effective charge transferred through the external circuit due to each of the photo-induced carriers. A high gain is attained by reducing τ_d. It is accomplished by increasing the volume of the photoconductor and decreasing the separation between the metal contacts. Accordingly, photoconductive ribbons are often prepared in the shape of a long ribbon with metal contacts along its edges. However, it should be noted that the carrier lifetime will affect the device response. The current diminishes at a faster rate if light is withdrawn at any instant. Consequently, the device is not sufficiently effective unless the duration of exposure exceeds the carrier lifetime.

Photoconductors are relatively easy to construct, but they are relatively slow in their operation. They require external voltage sources and in most cases are cryogenically cooled to minimize the effect of thermally generated charge carriers. Thus what appears to be a less-expensive detector in the beginning becomes quite expensive when all the peripherals are taken into account. Some of the common photoconductor materials are PbS, CdS, CdSe, InSb, and $Hg_xCd_{1-x}Te$. While InSb has a good response ($\cong 50$ ns), CdS and CdSe have poor responses ($\cong 50$ ms). CdS and CdSe are used for detecting visible light, and both have very high photoconductive gain ($\cong 10^4$).

11.3 Photodiodes

In general, a photovoltaic detector consists of a semiconductor junction so that the equilibrium energy bands on the two sides of the junction are shifted relative to one another. If a sufficiently energized photon strikes a junction, it will result in the generation of an electron–hole pair, which in turn results in a current flowing through the wire that connects the two components of the junction. Such a mode of operation, which requires no external bias, is said to be photovoltaic. Photovoltaic detectors have large surface areas so that they can generate a large photocurrent in the absence of a bias. However, they are nonlinear in their responses. Light-powered meters and solar cells are common examples of this type of detector. Interestingly, we may use photovoltaic detectors in the so-called photoconductive mode by applying a reverse bias. When used in this mode, the detector has a rather linear response.

Photodiodes are examples of bipolar semiconductor junctions that are operated in reverse bias. These photodetectors are generally sensitive, easily biased, small in area, and compatible with integrated optics components. Consequently, they are suitable for use in systems like those of fiber communication links. Beyond a certain bias voltage, the detector response is generally improved at higher bias values. The frequency response is often limited by two factors: carrier diffusion time across the depletion layer and junction capacitance of the diode. The carrier diffusion time is generally reduced by increasing the bias voltage but without exceeding the value of the breakdown voltage, whereas the junction capacitance is improved by incorporating an intrinsic layer between the p and n regions as in the p–i–n photodiode. Our attention later in this section is geared toward the details of such semiconductor devices.

As soon as a semiconductor junction is established, electrons start flowing from the n-region to the p-region, leaving behind donor ions, and holes start flowing from the p-region to the n-region, leaving behind acceptor ions. This flow of electrons and holes builds up a depletion layer at the junction. In the absence of any bias, however, the drift and diffusion components of the total current balance each other out. A reverse bias, on the other hand, greatly reduces the diffusion current across the junction but leaves the drift component relatively unaltered.

The photodiode is reverse-biased such that a current (generated by incoming photons) proportional to the number of absorbed photons can be generated. With an optical energy in excess of the band-gap energy, electron and hole pairs are generated in photodiodes. Those pairs that are generated in the depletion region are driven by the electric field through the junction, thus contributing to the reverse current. In addition, those pairs that are generated in the bulk regions, but within the diffusion length of the depletion region, diffuse into the depletion region and also contribute to the reverse current. If we neglect the amount of recombination loss in the depletion region, we can estimate photocurrent by

$$I_\lambda = e\eta_{pn} P_{abs} / h\nu, \tag{11.7}$$

where P_{abs} is the absorbed optical power and η_{pn} is the conversion efficiency. The effective conversion efficiency is reduced by the fact that some of the electron-hole pairs of the bulk areas diffuse into the depletion region.

The number of minority holes generated in the n-side but within the diffusion length of the depletion region is $AL_p g$, where g is the generation rate and A is the cross-sectional area of the junction. Similarly, the number of minority electrons generated in the p-side but within the diffusion length of the depletion region is $AL_n g$. The net photocurrent in the reverse-biased photodiode is thus given by

$$I = eA\left[\left(L_p p_{n0} / \tau_p\right) + \left(L_n p_{p0} / \tau_n\right)\right]\left[\exp\left(eV_A / kT\right) - 1\right] - eAg\left(L_p + L_n\right), \tag{11.8}$$

where the first term refers to photodiode dark current i_d and the second term accounts for the oppositely directed diffusion photocurrent.

When a photodiode is short-circuited (i.e., when $V_A = 0$), the photocurrent is not any more absent since the current caused solely by the collection of optically generated carriers in the depletion region is nonzero. The equivalent circuit of the photodiode and the corresponding V–I characteristics are shown in Figure 11.2. If the photodiode is open-circuited (i.e., when $I = 0$) in the presence of illumination, an open-circuit photovoltage $V_A = V_{oc}$ appears across the photodiode terminals.

The magnitude of the open-circuit photovoltage V_{oc} is found from Equation 11.8:

$$\begin{aligned}
V_{oc} &= (kT / e) \ln\left[g\left(L_p + L_n\right) / \left\{\left(L_p p_{n0} / \tau_p + L_n p_{p0} / \tau_n\right)\right\} + 1\right] \\
&= (kT / e) \ln\left[\left(I_\lambda / I_0\right) + 1\right],
\end{aligned} \tag{11.9}$$

where $-I_0$ is the peak reverse dark current. The open-circuit voltage is thus a logarithmic function of the incidental optical power P_{abs}. In a symmetrical p–n photo-diode, I_λ / I_0 approaches the value g/g_{th}, where $g_{th} = p_{n0}/\tau_p$ is the equilibrium thermal generation-recombination rate. Thus, as the minority carrier

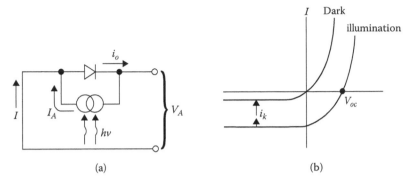

FIGURE 11.2 (a) Equivalent photodiode circuit and (b) its *V–I* characteristics.

concentration is increased, g_{th} increases due to the decrease in the carrier lifetime. Consequently, the increase of the minority carrier concentration does not allow V_{oc} to grow indefinitely. In fact V_{oc} is limited by the value of the equilibrium junction potential.

The power delivered to the load is given by

$$PL = IV_a = I_0 V_A \left[\exp(eV_A / kT) - 1 \right] - I_\lambda V_a. \tag{11.10}$$

Thus, the particular voltage V_{Am}, corresponding to the maximum power transfer, is found by setting the derivative of P_L to zero. Consequently, we obtain

$$[1 + (eV_{Am} / kT) \exp(eV_{Am} / kT) = 1 + (I_\lambda / I_0). \tag{11.11}$$

Since a *p–n* photodiode is also used as a solar cell for converting the sunlight to electrical energy, we may increase the value of V_{Am} as well as the corresponding photocurrent I_m. Note that the photodiode can achieve a maximum current of I_λ and a maximum voltage of V_{oc}. Often, therefore, the efficiency of the photodiode is measured in terms of the ratio, $(V_{Am} I_m / V_{oc} I_\lambda)$, also known as the fill factor. The present thrust of solar-cell research is thus directed toward increasing this ratio. By cascading thousands of individual solar cells, we can generate an enormous amount of power that is sufficient for energizing orbiting satellites.

The mode of operation where the photodiode circuit of Figure 11.2a is applied across a simple load is photovoltaic. The voltage across the load R_L can be used to evaluate the current flowing through it. However, if the photodiode in conjunction with a load is subjected to a relatively large external bias, the operation will be referred to as photoconductive. The latter mode is preferred over the photovoltaic because the current flowing through the load is generally large enough and, therefore, approaches I_λ. Thus, while the current-to-optical power relationship in the photovoltaic mode is logarithmic, it is linear in the photoconductive mode. Since the depletion-layer junction-capacitance C_j in an abrupt junction is proportional to $A[(N_d N_a) / \{V_a (N_d + N_a)\}]^{1/2}$, the photovoltaic mode contributes to a larger capacitance and, therefore, to a slower operation. In comparison, the photoconductive photodiode has a faster response.

A cut-off frequency f_c is generally defined as the frequency when the capacitive impedance of the photodiode equals the value of the load resistance. Therefore,

$$f_c = \frac{1}{2\pi R_L C_j}. \tag{11.12}$$

Thus, the junction capacitance has to be decreased to increase the frequency response. This is achieved by decreasing the junction area, by reducing the doping, or by increasing the bias voltage. While there

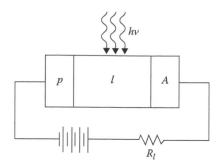

FIGURE 11.3   Reverse-biased *p–i–n* photodiode.

is a physical limit to the smallest junction area, the other two requirements in effect tend to increase depletion width, drift transit time, and bulk resistance, none of which is truly desirable.

The *p–n* photodiode discussed so far has one weakness: the incident optical power is not fully utilized in the optical-to-electrical conversion process because the depletion width of a *p–n* junction is extremely small. Because of this physical limitation, the *p–n* photodiodes do not have a desirable response time. This obstacle is overcome by introducing a semi-insulating thick intrinsic (lightly doped) semiconductor layer between its *p*-layer and its *n*-layer, as shown in Figure 11.3. Such especially organized photodiodes are referred to as *p–i–n* photodiodes.

In *p–i–n* photodiodes, the separating electric field occupies a large fraction of the device. The wider the thickness of the intrinsic layers, the higher the quantum efficiency. High field strength in the intrinsic layer allows the electron-hole pairs to be driven rapidly toward the respective extrinsic regions. However, the carrier transit time is generally proportional to the thickness of the intrinsic layer. Accordingly, there is a design compromise between the expected quantum efficiency and the desirable response time. For a typical *p–n* photodiode, the response time is in the order of 10^{-11} s, whereas for a *p–i–n* photodiode it is about 10^{-9} s. The quantum efficiency of a *p–i–n* photodiode can be anywhere in the range of 50% through 90%. Usually, indirect band-gap semiconductors are preferred over direct band-gap semiconductors as photodiode materials because otherwise there is a significant conversion loss due to surface recombination. Indirect band-gap materials engage photons to conserve momentum during the transfer. A *p–i–n* configuration eliminates part of this problem because of its longer absorption length.

Si photodiodes (having a maximum quantum efficiency of 90% at 0.9 μm) are used mostly in the wavelength region below 1 μm, whereas Ge photodiodes (having a maximum quantum efficiency of 50% at 1.35 μm) are preferred in the ranges above 1 μm. In addition to the single-element semiconductor photodiodes, there are many ternary (e.g., InGaAs, HgCdTe, and AlGaSb) photodiodes that are commercially produced.

Photodiodes have proven to be very successful in their applications in the background-limited photodetection. However, photodiodes lack internal gain and in many cases require an amplifier to provide noticeable signal currents. The avalanche photodiode (APD) is a specific photodiode that makes use of the avalanche phenomenon. By adjusting the bias voltage to a level where it is on the verge of breakdown, we can accelerate the photogenerated carriers. The accelerated carriers, in turn, produce additional carriers by collision ionization.

Avalanche gain is generally dependent on impact ionization encountered in the regions having sufficiently high electric field. This gain is achieved by subjecting the reverse-biased semiconductor junction to voltage below its breakdown field (= 10^5 V/cm). Electrons and holes thereby acquire sufficient kinetic energy to collide inelastically with a bound electron and ionize it generating an extra electron-hole pair. These extra carriers, in turn, may have sufficient energy to cause further ionization until an avalanche of carriers has resulted. Such a cumulative avalanche process is normally represented by a multiplication factor M that turns out to be an exponential function of the bias. Gains of up to 1000 can be realized in this way. This makes an APD that competes strongly with another high-gain photodetector device, known as a photomultiplier tube, in the red and near-infrared.

The probability that carrier ionization will occur depends primarily on the electric field in the depletion layer. Again, since the electric field in the depletion layers is a function of position, the ionization coefficients, α and β, respectively, for the electrons and the holes, turn out also to be functions of position. The ionization coefficients are particularly low at lower values of electric field, as shown in Figure 11.4, for the case of silicon.

Consider the reverse-biased p–n junction of depletion width shown in Figure 11.5. The entering hole current $I_p(0)$ increases as it travels toward the p-side, and the entering electron current $I_n(W)$ increases as it travels toward the n-side. In addition, hole and electron currents due to generation in the depletion layer also move in their respective directions. Thus, for the total hole and electron currents we can write

$$\begin{aligned}
\frac{dI_p(x)}{dx} &= \alpha(x)I_n(x) + \beta(x)I_p(x) + g(x), \\
-\frac{dI_n(x)}{dx} &= \alpha(x)I_n(x) + \beta(x)I_p(x) + g(x),
\end{aligned} \tag{11.13}$$

where $g(x)$ is the rate per unit length with which the pairs are generated thermally and/or optically. With the first of Equation 11.13 integrated from $x = 0$ to $x = x$ and similarly, the second of Equation 11.13 integrated from $x = x$ to $x = W$ we obtain

$$I_p(x) - I_n(x) = \int_0^x [\alpha(x) - \beta(x)]I_n(x)\,dx + I\int_0^x \beta(x)\,dx + \int_0^x g(x)\,dx \tag{11.14}$$

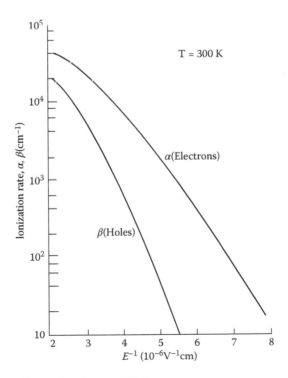

FIGURE 11.4   Ionization coefficients for silicon at 300°K.

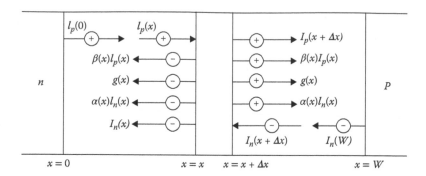

FIGURE 11.5   Avalanche in a reverse-biased *p–n* junction.

and

$$-I_{\mathrm{n}}(W)+I_{\mathrm{n}}(x)=\int_{x}^{w}\left[\alpha(x)-\beta(x)\right]I_{\mathrm{n}}(x)\,\mathrm{d}x+I\int_{x}^{w}\beta(x)\,\mathrm{d}x+\int_{x}^{w}g(x)\,\mathrm{d}x,\tag{11.15}$$

where the sum $I \cong I_{\mathrm{n}}(x) + I_{\mathrm{p}}(x)$ is independent of position. Note, however, that I is equivalent to the saturation current I_0. Adding Equations 11.14 and 11.15, we obtain

$$I=\left[\left(I_0+I_g\right)+\mathrm{Int}\left(\{\alpha(x)-\beta(x)\}I_n(x)\right)\right]/\left[1-\mathrm{Int}\left(\beta(x)\right)\right],\tag{11.16}$$

where I_g represents the total generation current and $\mathrm{Int}(\xi)$ is the integral of ξ with respect to x when evaluated from $x = 0$ to $x = W$.

For a very special case when $\alpha(x)=\beta(x)$, the total current is given by

$$I = M\left(I_0 + I_g\right),\tag{11.17}$$

where M is the avalanche multiplication factor as defined by

$$M =1/\left[1-\mathit{Int}\left(\alpha(x)\right)\right]=1/(1-\delta).\tag{11.18}$$

Ideally speaking, the avalanche condition is thus given by

$$\mathrm{Int}\left(\alpha(x)\right)=1,\tag{11.19}$$

when M becomes infinite. In most practical cases, electron and hole coefficients are not equal and these coefficients vary with the electric field. A practical avalanche photodiode is thus referred to by its ionization rate ratio, $k(\equiv\beta/\alpha)$. We can then arrive at an expression for M after going through some extremely cumbersome mathematics and multiple assumptions:

$$M = \frac{k-1}{k-e^{(k-1)\delta}}.\tag{11.20}$$

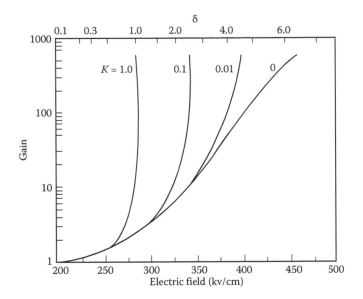

FIGURE 11.6   Gain versus electric field in an APD.

We may note from Figure 11.6 that for most electrical fields of interest, k is negligible. Thus, the avalanche multiplication factor reduces to

$$M \approx e^{\delta}. \tag{11.21}$$

When $k = 0$, the gain increases exponentially with δ, but it does not necessarily become infinite. As shown in Figure 11.6, with k approaching unity, the gain approaches infinity at a still smaller value of the field.

It is appropriate to note that changing the level of doping can easily alter the electric field. The ionization coefficient α is often given by

$$\alpha = Ae^{-B/|\varepsilon|}, \tag{11.22}$$

where A and B are material constants and ε is the electric field in terms of doping level. For silicon, A and B are, respectively, 9×10^{5}/cm and 1.8×10^{6} V/cm. At a gain of 100 when $k = 0.01$, for example, a 0.5% alteration in the doping changes, the gain would have changed by almost 320% if $\alpha = \beta$ The choices of k and doping are, therefore, critical in the design of an APD. APDs are meant for use with small signals, and they require special power supplies to maintain them in their avalanche mode.

The phototransistor, like APD, is a detector that exhibits current gain. It can be regarded as a combination of a simple photodiode and a transistor. Phototransistors are photoconductive bipolar transistors that may or may not have base lead. Light is generally absorbed in the base region. A p–n–p phototransistor is shown in Figure 11.7. When there is no light, no current flows because there is no base control current. Upon illumination, holes that are excited in the base diffuse out leaving behind an overall negative charge of the excess base electrons that are neutralized by recombination. In the photodiode, a much larger current flows through the device. Thus, while the phototransistor works very much like the photodiode, it amplifies the photogenerated current. Also, a longer recombination time for the excess base electrons contributes to higher gain.

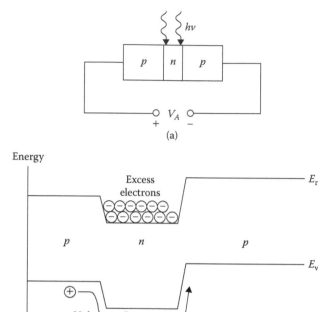

FIGURE 11.7   (a) A biased phototransistor and (b) its energy diagram.

For the phototransistors,

$$I_e = I_c + I_b,$$

(11.23)

where I_e, I_c, and I_b are emitter, collector, and base currents, respectively. In the presence of illumination, the base current is given by $\eta I_{abs} A e \lambda / hc$, where A is the junction area, η is the internal quantum efficiency, and I_{abs} is the intensity of the absorbed light. The collector current I_C has two components: (a) the standard diodes reverse saturation current, I_{CBO} and (b) the portion of the emitter current αI_E that crosses into the collector where $\alpha < 1$. The leakage current I_{CBO} corresponds to the collector current at the edge of the cutoff when $I_E = 0$. Thus,

$$I_E = \left(I_B + I_{CBO}\right)\left[1 + \frac{\alpha}{1-\alpha}\right].$$

(11.24)

The ratio $\alpha/1 - \alpha$ is an active region performance parameter of a phototransistor. This ratio is usually in the order of ~10^2. In the absence of light, the current flowing in a phototransistor is $I_{CBO}[1 + \{\alpha(1 - \alpha)\}]$, which is much larger than that in a photodiode under similar (dark) conditions. When illuminated, phototransistor current approaches $I_B[1 + \{\alpha(1 - \alpha)\}]$, thus contributing to significant gain like that of an APD. The only limitation of the phototransistor happens to be its response time, which is about 5 μs, whereas that in a photodiode is on the order of 0.01 μs.

Electrons may be emitted when light of an appropriate frequency ν strikes the surface of solids. Such light-emitted solids are called photocathodes. The minimum energy necessary for the emission of an electron is referred to as the work function φ of the solid. In the specific case of semiconductors,

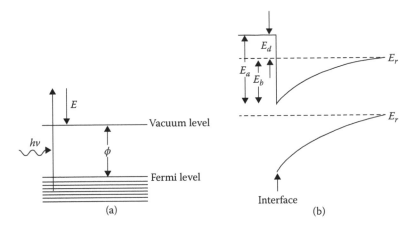

FIGURE 11.8 Energy level diagram in (a) a solid–vacuum interface and (b) a band-bended semiconductor–vaccuum interface.

electron affinity (energy difference between the vacuum level and E_c) plays the role of work function. The behavior of an electron in solids is like that of an electron in a finite potential well, where the difference between the highest occupied (bound) level and vacuum (free) level is φ, as shown in Figure 11.8a. The Fermi level is equivalent to the highest possible bound energy. Thus, the kinetic energy E of an emitted electron is given by

$$E = h\nu - f. \tag{11.25}$$

Since electrons reside at or below the Fermi level, E corresponds to the maximum possible kinetic energy. The emission of an electron thus requires a minimum of $f (= h\nu)$ photoenergy. However, in case of a semiconductor, this minimum energy is equivalent to $E_g + E_a$, where E_a is the electron affinity energy. Often it may become necessary to reduce the value of E_a, which is accomplished by making the semiconductor surface p-type. The band bending at the surface results in a downward shift of the conduction band by an amount E_b, as shown in Figure 11.8b. Consequently, the effective electron affinity becomes

$$E_a' = E_a - E_b. \tag{11.26}$$

In certain semiconductors – those referred to as having negative electron affinity –E_b exceeds E_a. Semiconductors such as these are used for infrared photocathodes.

The vacuum photodiode shown in Figure 11.9 is a quantum detector designed by placing a photocathode along with another electrode (referred to as an anode) within a vacuum tube. In practice, the photocathode consists of a semicylindrical surface while the anode is placed along the cylindrical axis. When an optical energy in excess of the work function illuminates the photocathode, current begins to flow in the circuit. When the bias voltage V_A is large enough ($\cong 100\ V$), the emitted electrons are collected at the anode. When optical energy falls below the work function level, current ceases to exist, irrespective of the bias voltage. For efficient collection of electrons, the distance between the anode and the photodiode is kept to a minimum by making sure that the associated capacitance value remains reasonable. Often the anode is made of highly grid-like wires so as not to impede the incoming optical energy. In comparison, solid-state photodetectors are not only smaller, faster, and less power consuming but also more sensitive. Consequently, vacuum photodiodes are used only when the incoming optical energy is more than a certain maximum that may otherwise damage the solid-state photodetectors.

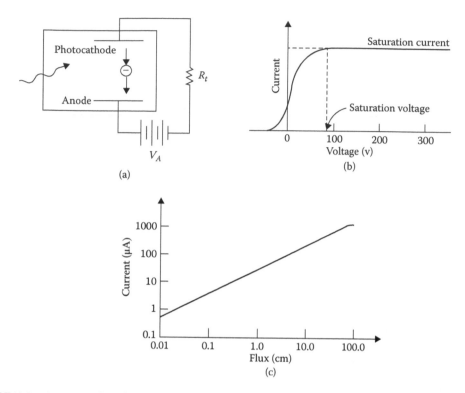

FIGURE 11.9 A vacuum photodiode: (a) circuit; (b) characteristic curve; and (c) current versus light flux.

The characteristic curve of a vacuum phototube shows that the photocurrent for a given illumination is invariant above the saturation voltage. The saturation voltage is mildly wavelength- and illumination-sensitive. Since the operating voltage of a phototube is usually larger than the saturation voltage, minor fluctuations in the supply voltage do not cause any discrepancy in the phototube's performance. An important feature of a phototube is that the photocurrent varies linearly with light flux. A slight departure from linearity occurs at high enough flux values and is caused by the space-charge effects. This nonlinearity is avoided by using a large anode-to-photocathode voltage. In practice, the flux level sets a lower limit on the value of the load resistance. The load, used to produce a usable signal voltage, in turn, sets a lower limit on the time constant.

Gas-filled phototubes are identical to vacuum phototubes except that they contain approximately 0.1 mm of an inert gas. The inert gas provides a noise-free amplification ($\cong$ 5–10) by means of ionization of the gas molecules. However, inert gases have poor frequency responses. Again the response, which is basically nonlinear, is a function of the applied voltage. The phototubes are, therefore, used in applications where the frequency response is not critical.

Photoemissive tube technology is used to develop an alternative but quite popular high-gain device known as a photomultiplier. In a photomultiplier tube (PMT), the photoelectrons are accelerated through a series of anodes (referred to as dynodes) housed in the same envelope; these dynodes are maintained at successively higher potentials. A photoelectron emitted from the photocathode is attracted to the first dynode because of the potential difference. Depending on the energy of the incident electron and the nature of the dynode surface, secondary electrons are emitted upon impact at the first dynode. Each of these secondary electrons produces more electrons at the next dynode, and so on, until the electrons from the last dynode are collected at the anode. The dynodes are constructed from materials that, on average, emit $\delta(>1)$ electrons for each of the incident electrons. One such photomultiplier is shown in Figure 11.10a, where δ is a function of the interdynode

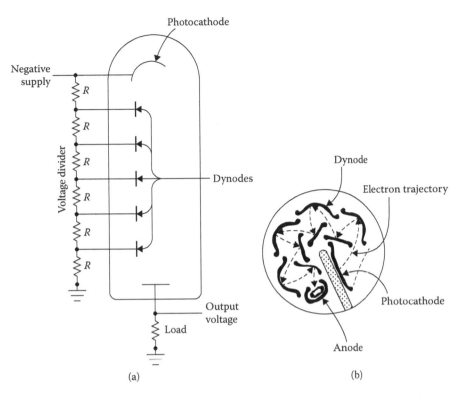

FIGURE 11.10 A photomultiplier tube: (a) schematic of a five-stage PMT and (b) focusing dynode structure.

voltage. When the PMT is provided with N such dynodes, the total current amplification factor is given by

$$G = \left(i_{out} / i_{in} \right) \delta^n. \tag{11.27}$$

Thus, with fewer than 10 dynodes and $\delta (> 1)$, the gain can easily approach 10^6.

The problems of a PMT are quite the same as those of a vacuum photodiode. However, a PMT is undoubtedly more sensitive. The response of a PMT is comparatively slower since electrons have to move through a longer distance. In addition, there is a finite spread in the transit time because all of the electrons may not have identical velocities and trajectories. This transit-time spread is often reduced, not by reducing the number of dynodes but by increasing the value of δ. However, it must be noted that for the most photocathode materials, the maximum wavelength of incoming light is permitted to be about 1200 nm. Thus, for the detection of longer wavelength radiation, a solid-state detector is preferred. PMTs are commonly operated with $\approx 10^2 V$ between the dynodes, which are advantageous because the overall gain of the tube may be varied over a wide range by means of a relatively small voltage adjustment. But, at the same time, it is also disadvantageous because the voltage supply for the PMT must be extremely stable for the calibration to be reliable. We can show that in an N-stage PMT operating at an overall voltage V, a fluctuation ΔV in the voltage produces a change ΔG in the gain G such that

$$\Delta g = GN \frac{\Delta V}{V}. \tag{11.28}$$

Consequently, a 1% fluctuation in a 10-stage PMT will cause a 10% change in the gain.

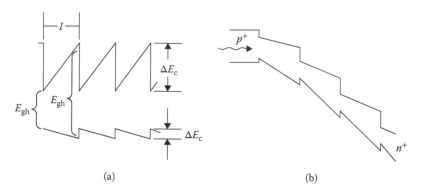

FIGURE 11.11   Staircase APD: (a) unbiased and (b) biased.

Different types of PMTs are distinguishable by their geometrical arrangement of dynodes. In particular, the focusing-type PMT, as shown in Figure 11.10b, employs electrostatic focusing between the adjacent dynodes and thereby reduces the spread in the transit time. These PMTs are, however, somewhat more noisy and unstable than the unfocused types. Like phototubes, PMTs have an exceptionally linear response.

It is appropriate to introduce a solid-state equivalent of PMT, known as the staircase avalanche photodiode (SAPD) that has been added lately to the list of photodetectors. The noise is an APD, which increases with the increase in the ratio of the ionization coefficient $k (= \beta / \alpha)$. On the other hand, a high k is required for a higher gain. An SAPD provides a suitable solution to this apparent anomaly by incorporating PMT-like stages in the solid-state APDs. An unbiased SAPD consists of a graded-gap multi-layer material (almost intrinsic) as shown in Figure 11.11a. Each dynode-like stage is linearly graded in composition from a low band-gap value E_{gl} to a high band-gap value E_{gh}. The materials are chosen so that the conduction band drops ΔE_c at the end of each stage equals or just exceeds the electron ionization energy. Note, however, that ΔE_c is much larger than the valence-band rise ΔE_v. Consequently, only electrons contribute to the impact ionization provided the SAPD is biased, as shown in Figure 11.11b.

A photoelectron generated next to p^+-contact drifts toward the first conduction band under the influence of the bias field and the grading field (given by $\Delta E_c/l$, where l is the width of each step). But this field value is not large enough for the electrons to impact ionize. In this device, only the bias field is responsible for the hole-initiated ionization. The actual impact ionization process in each stage occurs at the very end of the step when the conduction-band discontinuity undergoes a ΔE_c change. The total SAPD gain becomes $(2 - f)^N$, where N is the number of stages and f is the average fraction of electrons that do not impact ionize in each of the stages. The critical bias SAPD field just exceeds $\Delta E_c/l$ so as to provide the electrons with necessary drift through l but not impact ionize.

11.4 Charge-Coupled Imagers

An important solid-state photodetecting device is the charge-coupled imager, which is composed of a closely spaced array of charge-coupled devices (CCDs) arranged in the form of a register. Each of the CCD units is provided with a storage potential well and is, therefore, able to collect photogenerated minority carriers. The collected charges are shifted down the array and converted into equivalent current or voltage at the output terminal. To understand the overall function of such repetitive storage and transfer of charge packets, consider the metal oxide semiconductor (MOS) structure of Figure 11.12, where the metal electrode and the p-type semiconductor are separated by a thin SiO_2 layer of width x_0 and dielectric constant K_0. Silicon nitride, Si_3N_4, is also used for the insulating layer. The capacitance of such a structure depends on the voltage between the metal plate and the semiconductor.

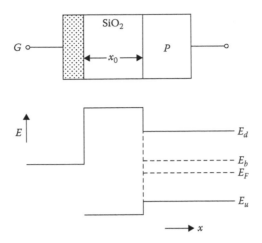

FIGURE 11.12 MOS capacitor and its unbiased energy-band diagram.

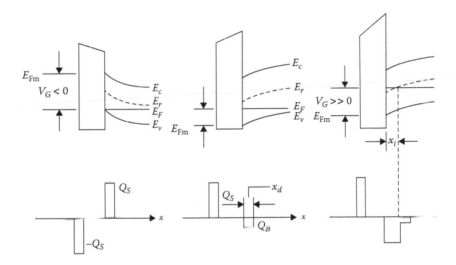

FIGURE 11.13 Energy-band diagram and charge distribution of an unbiased MOS capacitor: (a) $V_G < 0$, (b) $V_G > 0$, and (c) V_G @ 0.

In thermal equilibrium, the Fermi level is constant all throughout the device. For simplicity, we may assume (a) that the work function difference between the metal and the semiconductor is zero and (b) that there is no charge accumulated in the insulator or at the junction between the insulator and the semiconductor. Consequently, the device may be considered to have no built-in potential.

A biased MOS capacitor results in two space-charge regions by the displacement of mobile carriers, as shown in Figure 11.13. The total bias voltage V_G applied at the gate input G is shared between the oxide layer and the semiconductor surface, whereas there is only a negligible voltage across the metal plate. Under reverse bias, the surface potential gives rise to an upward bend in the energy diagram, as shown in Figure 11.13a. At the edges, $E_i - E_F$ becomes comparatively larger, thus resulting in a higher hole density at the surface than that within the bulk region. This condition generally increases the surface conductivity. Figure 11.13b shows the forward-biased case, where a decrease of $E_i - E_F$ at the edges causes a depletion of holes at the semiconductor surface. The total charge per unit area in the bulk semiconductor is given by

$$Q_B = -eN_A x_d, \tag{11.29}$$

where x_d is the width of the depletion layer. Using the depletion approximation in Poisson's equation, we can arrive at the potential within the semiconductor as

$$V_s(x) = V_s(0)\left[1 - \frac{x}{x_d}\right]^2,$$ (11.30)

where

$$V_s(0) = \frac{eN_A}{2K_s\varepsilon_0}x_d^2$$ (11.31)

is often referred to as the surface potential. The voltage characteristic is similar to that of a step junction having a highly doped p-side. Note, however, that as the bias voltage V_G is increased further, the band bending could result in a crossover of E_i and E_F within the semiconductor, as shown in Figure 11.13c. Consequently, the carrier depletion gives rise to an extreme case of carrier inversion whereby electrons are generated at the junction and holes are generated inside the semiconductor with two regions being separated by the crossover point. Therefore, a p–n junction is induced under the metal electrode. The effect of the gate voltage is to remove the majority carriers from the semiconductor region that is closest to the gate and introduce a potential well. Absorbed photons contribute to the freeing of minority carriers that are collected in the well. The resulting output signal corresponds to the photoinduced charge.

If the semiconductor were approximated as a borderline conductor, the metal–semiconductor structure could be envisioned as a parallel-plate capacitor with the oxide layer working as its dielectric material. However, in forward bias, the MOS structure is modeled by incorporating an additional capacitor in series with the oxide capacitor to accommodate the presence of a surface space-charge layer in the semiconductor. The overall MOS capacitance c is thus given by

$$\frac{1}{c} = \frac{1}{c_0} + \frac{1}{c_s},$$ (11.32)

where

$$c_0 = \frac{K_0\varepsilon_0}{x_0}$$ (11.33)

and

$$c_s = \frac{K_s\varepsilon_0}{x_d}.$$ (11.34)

Neglecting the voltage drop in the metal plate, the forward bias can be expressed as

$$V_G = V_s(0) - \frac{Q_s}{c_0},$$ (11.35)

where Q_s is the density of induced charge in the semiconductor surface region and $V_s(0)$ is the surface potential. The gradient of the surface potential generally determines the minority carrier movements. The depth of the potential well is often decreased either by decreasing the oxide capacitance—that is, by increasing the oxide thickness—or by increasing the doping level of the p-type material.

Sufficient forward bias may eventually induce an inversion layer. With the passing of time, electrons accumulate at the oxide–semiconductor junction, and a saturation condition is reached when the electron drift current arriving at the junction counterbalances the electron diffusion current leaving the junction. The time required to reach this saturation condition is referred to as the thermal relaxation time. The net flow of electrons is directed toward the junction prior to from the thermal-relaxation time, whereas the net flow of electrons is directed away from the junction after the thermal-relaxation time has elapsed. Since there was no inversion layer prior to the saturation, the induced charge Q_s is obtained by summing A_B and the externally introduced charge Q_e. Equations 11.29 and 11.30 can be incorporated into Equation 11.35 to give surface potential as

$$V_s(0) = V_G - \frac{Q_e}{c_0} + \frac{eK_s\varepsilon_0}{c_0^2}\left[1 - \left\{1 + \frac{2c_0^2\left(V_G - \frac{Q_e}{c_0}\right)}{eK_s\varepsilon_0 N_A}\right\}^{1/2}\right]. \tag{11.36}$$

The depth of the potential well x_d is often evaluated using Equations 11.25 and 11.30. The value of x_d is used in turn to evaluate c_s using Equation 11.34 and, consequently, we can determine the overall MOS capacitance as

$$c = \frac{c_0}{\left[1 + \frac{2c_0^2}{eN_A K_s\varepsilon_0}V_G\right]^{1/2}}. \tag{11.37}$$

The MOS capacitor in effect serves as a storage element for some period of time prior to reaching the saturation point.

A CCD structure formed by cascading an array of MOS capacitors, as shown in Figure 11.14a, is often referred to as a surface channel charge-coupled device (SCCD). Basically, the voltage pulses are supplied in three lines, each connected to every third gate input (and consequently this CCD is called a three-phase CCD). In the beginning, G_1 gates are turned on, resulting in an accumulation and storage of charge under the gates. This step is followed by a step whereby G_2 is turned on, thus resulting in a charge equalization step across two-thirds of each cell. Subsequently G_1 is turned off, resulting in a complete transfer of all charges to the middle one-third of each cell. This process is repeated to transfer charge to the last one-third of the CCD cell. Consequently, after a full cycle of clock voltages has been completed, the charge packets shift to the right by one cell, as illuminated in Figure 11.14b and c. When the CCD structure is formed using an array of photosensors, charge packets proportional to light intensity are formed and these packets are shifted to a detector for the final readout.

The CCD signal readout is also accomplished by using either two-phase or four-phase clocking schemes. In each of the cases, however, transfer of charges is accomplished by means of the sequentially applied clock pulses. There are three phenomena that enhance the transfer of charges in the SCCD: (a) self-induced drift; (b) thermal diffusion; and (c) fringe-field drift. The self-induced drift, responsible for most of the transfer, is essentially a repulsion effect between the like charges. The thermal diffusion component makes up for most of the remaining signal charge. It can be shown that for most materials the thermal time constant is longer than the self-induced time constant. The upper frequency limit for switching operations is thus determined by the thermal time constant. For the SCCDs, this upper limit can be in the order of 10 MHz. The fringe-field drift is determined by the spacing of the electrodes and results in a smoothing out of the transitional potential fields. This third effect is responsible for the transfer of the final few signal electrons.

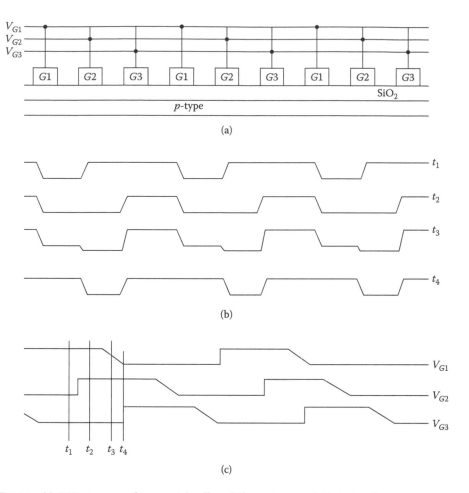

FIGURE 11.14   (a) CCD structure; (b) potential wells at different times; and (c) timing diagram.

Figure 11.15a shows a system of MOS transistors along with a photodiode array, both of which can be embedded under the same monolithic structure. The system is able to perform sequential readout. A voltage pattern can be generated from the shift register so as to turn on only one transistor at a time. The switching voltage is shifted serially around all diodes. This scheme can also be extended to two dimensions, as shown in Figure 11.15b, where one row is switched on and all columns are then scanned serially. The process is repeated for the remaining rows until all the photodiodes have experienced the scanning.

The primary item that hinders transfer of charge is surface-state trapping, which occurs along the semiconductor–oxide interface. These trapping energy levels are introduced by nonuniformities in the interface. These energy levels tend to trap and re-emit electrons at a rate that is a function of clocking frequency and their positions relative to the Fermi level. Transferring a constant amount of charge in each of the CCD wells reduces this hindrance of the transfer of charge. This charge fills most of the trapping states, as a result of which interaction of trapping levels and charge is minimized.

A popular method used to control the problem of surface trapping is accomplished by having what is known as a buried channel CCD (BCCD). It involves implementing a thin secondary layer, which is opposite polarity to that of the substrate material, along the oxide surface. The fringe fields are much smoother in a BCCD than in an SCCD, but the well depth is smaller. Hence, a BCCD can switch information at a faster rate, but it cannot hold as much signal. Switching speeds of up to 350 MHz is not uncommon for BCCDs.

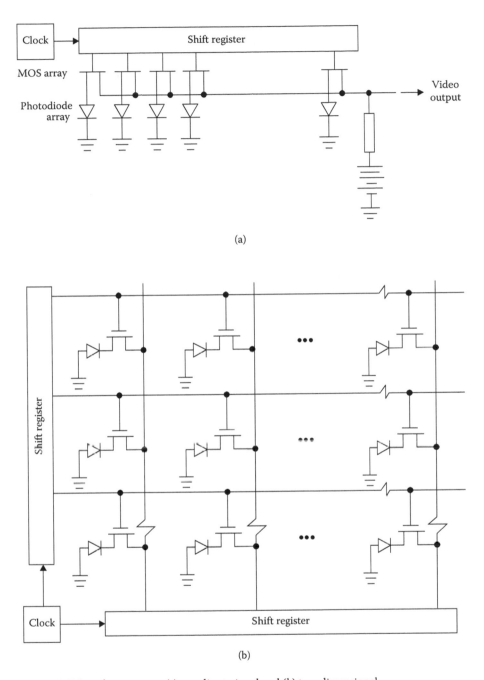

FIGURE 11.15   MOS readout scanner: (a) one-dimensional and (b) two-dimensional.

In applications where a semiconductor depletion region is formed, electron–hole pairs are generated due to the thermal vibration of the crystal lattice at any temperature above 0K. This generation of carriers constitutes a dark current level and determines the minimum frequency with which the transfer mechanism can occur. The time taken by a potential well to fill up with dark electrons can be quite long in some CCDs. There are two basic types of CCD imagers: the line imager and the area imager. In the line imager, charge packets are accumulated and shifted in one direction via one or two parallel CCD shift registers, as shown in Figure 11.16a, where the CCD register is indicated by the shaded regions. The two basic types of CCD area imagers are shown in Figure 11.16b and c.

In the interline transfer CCD (ITCCD), photocells are introduced between the vertical CCD shift registers. Polysilicon strips are placed vertically over each line of the photocells to provide shielding. During an integration period (referred to as one frame), all of the cells are switched with a positive voltage. The ITCCD is obtained by extending the line imager to a two-dimensional matrix of line imagers in parallel. The outputs of the line imagers in parallel are fed into a single-output CCD register. For the case of the frame transfer CCD (FTCCD), the sensor is divided into two halves: a photosensitive section and a temporary storage section. The charge packets of the photosensing array are transferred over to the temporary storage array as frame of picture. Subsequently, the information is shifted down one-by-one to the output register and is then shifted horizontally.

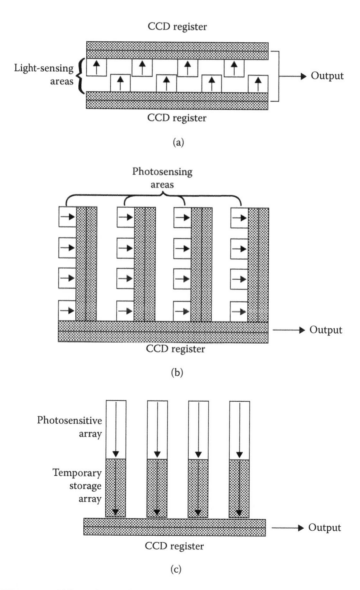

FIGURE 11.16 CCD imager: (a) line, (b) interline transfer, and (c) frame transfer.

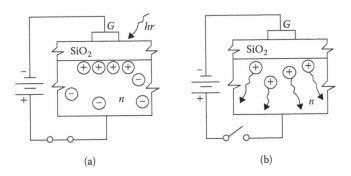

FIGURE 11.17 CID charge: (a) accumulation and (b) injection.

Besides CCDs, a different family of MOS charge transfer devices, referred to as charge-injection devices (CIDs), can also be an integral part of the focal plane array. CIDs involve exactly the same mechanism for detecting photons as CCDs. They differ only in the methods used for reading out the photoinduced charges. The mechanism of charge injection is illustrated in Figure 11.17, where the CID consists of an n-substrate, for example. Application of a negative gate voltage causes the collection of photoinduced minority carriers in the potential well adjacent to the semiconductor–oxide interface. This accumulation of charge is directly proportional to the incident optical irradiance. Once the gate voltage is withdrawn, the potential well dissipates and the minority carriers are injected promptly into the substrate, resulting in current flow in the circuit.

Because of the serial nature of the CCDs, optical input signals cause the resulting charge to spill over into adjacent cells. This effect, referred to as blooming, causes the image to appear larger than its actual magnitude. In comparison, CIDs are basically x–y addressable such that any one of their pixels can be randomly accessed with little or no blooming. However, the CID capacitance (sum of the capacitance of a row and a column) is much larger than the CCD capacitance and, therefore, the CID images tend to be noisier.

11.5 CRT and Imaging Tubes

The cathode-ray tube (CRT) is a non-solid-state display that is deeply entrenched in our day-to-day video world. Even though there are other competing technologies, such as flat-panel display or making serious inroads, it is unlikely that CRTs will be totally replaced. In spite of its large power consumption and bulky size, it is by far the most common display device found in both general- and special-purpose usage, aside from displaying small alphanumeric. The cost factor and the trend toward using more and more high-resolution color displays are the key factors that guarantee the CRT's longevity. CRTs have satisfactory response speed, resolution, design, and life. In addition, there are very few electrical connections, and CRTs can present more information per unit time at a lower cost than any other display technology. A CRT display is generally subdivided into categories, such as having electrostatic or magnetic deflection, monochromatic or color video, and single or multiple beams.

Figure 11.18a shows the schematic of a CRT display where the cathode luminescent phosphors are used at the output screen. Cathodoluminescent refers to the emission of radiation from a solid when it is bombarded by a beam of electrons. The electrons in this case are generated by thermionic emission of a cathode and are directed onto the screen by means of a series of deflection plates held at varying potentials. The electron beam is sequentially scanned across the screen in a series of lines by means of electrostatic or electromagnetic fields (introduced by deflection plates) orthogonal to the direction of electron trajectory. The bulb enclosing the electron gun, the deflectors, and the screen are made air-free for the purpose of having an electron beam and a display area. The video signals to be displayed are applied to both the electron gun and deflectors in synchronization with the scanning signals. The display is

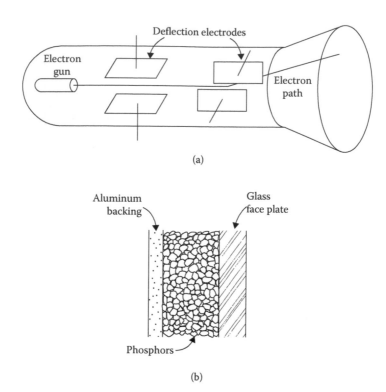

(a)

(b)

FIGURE 11.18   (a) A CRT schematic and (b) a CRT screen.

usually refreshed 60 times a second to avoid having a flickering image. While in the United States CRT displays consist of 525-scan line, the number is about 625 overseas. The phosphor screen is often treated as being split into two interlaced halves. Thus, if a complete refreshing cycle takes t_r time, only odd-numbered lines are scanned during the first $t_r/2$ period, and the even-numbered lines are scanned during the remaining half. Consequently, our eyes treat the refreshing rate as if it were $2/t_r$ Hz instead of only $1/t_r$ Hz.

The deflected electron strikes the CRT phosphor screen causing the phosphors of that CRT location to emit light. It is interesting to note that both absorption and emission distributions of phosphors are bell shaped, but the distribution peaks are relatively displaced in wavelength. When compared with the absorption distribution, the emission distribution peaks at a higher wavelength. This shift toward the red end of the spectrum is referred to as the Stokes' shift. It is utilized to convert ultraviolet radiation to useful visible radiation. It is used in fluorescent lamps to increase their luminous efficiencies. In particular, the CRT illumination, caused by the cathodoluminescent phosphors, is a strong function of both current and accelerating voltage and is given by

$$L_e = Kf(i)V^n, \tag{11.38}$$

where K is a constant, $f(i)$ is a function of current and n ranges between 1.5 and 2. With larger accelerating voltages, electrons penetrate further into the phosphor layer, causing more phosphor cells to irradiate.

The factors that are taken into consideration in selecting a particular phosphor are, namely, decay time, color, and luminous efficiency. Note that even though a phosphor may have lower radiant efficiency in the green, the latter may have a more desirable luminous efficiency curve. Usually, the phosphor screen consists of a thin layer (~5 μm) of phosphor powder placed between the external glass

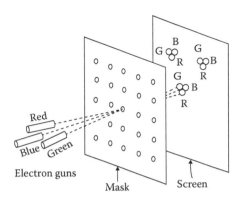

FIGURE 11.19 Shadow masking in a color CRT.

faceplate and a very thin (~0.1 μm) aluminum backing as shown in Figure 11.18b. The aluminum backing prevents charge build-up and helps in redirecting light back toward the glass plate. The aluminum backing is thin enough so that most of the electron beam energy can get through it. A substantial amount of light that reaches the glass at normal (beyond the critical angle of incidence) may be totally internally reflected at the glass-air interface, some of which may again get totally internally reflected at the phosphor-glass interface. Such physical circumstances produce a series of concentric circles of reduced brightness instead of one bright display spot. The combination of diffused display spots results in a display spot that has a Gaussian-distribution profile.

Of the many available methods, the most common one for introducing color in a CRT display involves the use of a metal mask and three electron guns, each corresponding to a primary phosphor granule (red, blue, and green), as shown in Figure 11.19. The three electron guns are positioned at different angles, so that while each of the electron beams is passing through a particular mask-hole, a particular primary phosphor dot strikes. All three beams are deflected simultaneously. In addition, the focus elements for the three guns are connected in parallel so that a single focus control is sufficient to manipulate all beams. The three primary dots are closely packed in the screen so that proper color can be generated for each signal. Misalignment of the three beams causes a loss of purity for the colors. In any event, when compared with the monochrome display, the CRT color reproduction process involves a loss of resolution to a certain degree because the primary phosphor cells are physically disjointed.

Imaging tubes convert a visual image into equivalent electrical signals that are used thereafter for viewing the image on display devices, such as CRTs. They are used as in a television camera tube in which a single multilayer structure serves both as an image sensor and as a charge storage device. A beam of low-velocity electrons to produce a video signal scans the single multilayer structure. In particular, when the characteristics of the photosensor layer during the optical-to-electrical conversion depend on the photosensor's photoconductive property, the imaging tube is referred as a vidicon. Figure 11.20 shows a typical vidicon whose thin target material consists of a photoconductive layer, such as selenium or antimony trisulphide, placed immediately behind a transparent conductive film of tin oxide. Its charge-retention quality is very good since the resistivity of the photoconductor material is very high (~ 10^{12} Ω-cm). The conducting layer is connected to a positive potential V_B via a load resistor. The other side of the target is scanned with an electron beam almost in the same way as CRT scanning. The output video signal is generally capacitively coupled, and the majority of vidicons employ magnetic focusing and deflection schemes.

The vidicon target is often modeled as a set of leaking capacitors, each of which corresponds to a minute area of the target. One side of these capacitors is tied together by means of a transparent conductive layer and is connected to a bias voltage V_B. The low-velocity scanning electron beam makes momentary contact with each of the miniature areas, charging them negatively. The target has high resistance in the

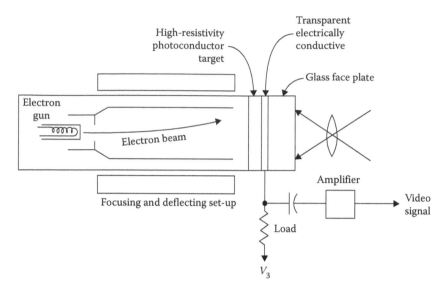

FIGURE 11.20   A vidicon structure.

dark, but when it is photoirradiated, its resistance drops significantly. In the absence of illumination, the scanning beam drives the target to a potential value close to that of the cathode, which allows for a small amount of dark current to flow when the beam is removed. The decrease in resistivity due to photoirradiance causes the capacitor-like target to discharge itself in the absence of the electron beam. However, when the electron beam scans this discharged area, it will recharge the target. More current is being taken away from the illuminated area than from the unilluminated area, thus generating a video signal at the output. The video signal is found to be proportional to nth power of illumination, where n is a positive number less than unity.

The dark current is rather large in vidicons. Again, the spectral response of antimony trisulphide is very poor at wavelengths greater than 0.6 μm. An imaging tube referred to as a plumbicon is often used to overcome the aforementioned shortcomings of a vidicon. The plumbicon is essentially identical to the vidicon, except that the photoconductive target is replaced by a layer of lead oxide that behaves like p–i–n diode and not like a photoconductor.

The band-gap energy of lead oxide is 2eV and, therefore, it is not too sensitive to red. However, the introduction of a thin layer of lead sulfide (with a band gap of about 0.4 eV), along with lead oxide, eliminates the problem of red insensitivity in the plumbicon. The transparent conductive film acts like an n-type region, and the PbO layer acts like a p-type region, whereas the region between the two behaves like an intrinsic semiconductor. Photoirradiance of this p–i–n structure generates carriers in the plumbicon. But the flow of carriers in the opposite direction generally reduces the amount of stored charge. However, in the absence of photoirradiance, the reverse bias gives rise to a dark current that is negligible in comparison to that encountered in the vidicon. There are two serious disadvantages to using plumbicons because their resolution (fewer than 100 lines) is octane limited by the thick lead oxide layer and because the change in target voltage cannot be used to control their sensitivity. In spite of these demerits, they are used widely in color TV studios. However, the lead oxide layer can be replaced by an array formed by thousands of silicon diodes to increase the sensitivity of plumbicons.

Another important imaging tube, referred to as an image intensifier, is of significant importance in the transmittal of images. In principle, it is vacuum photodiode equipped with a photocathode on the input window and a phosphor layer on the output window. Image intensifiers are devices in which the primary optical image is formed on a photocathode surface (with an S20 phosphor layer backing), and the resulting photocurrent from each of the image points is intensified by increasing the energy of the

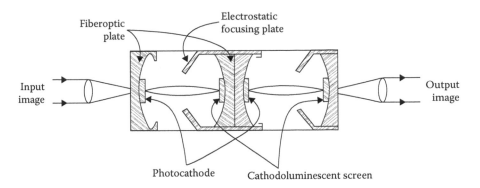

FIGURE 11.21 Two cascaded image intensifiers.

electrons, as shown in Figure 11.21. The windows are made of the fiber-optic plates, so that the plane image surface of the input can be transformed to the curve object and the image surfaces of a simple electrostatic lens. The electrons strike a luminescent screen and the intensified image is produced by cathodoluminescent. It is possible to cascade more than one such intensifier with fiber-optic coupling between them, making sure that an accelerating potential is applied between the photocathode and the screen. Such a cascade device, along with an objective lens and an eyepiece, is used in the direct-view image intensifier. In any event, it is possible to achieve a luminance gain of up to 1000 with each image intensifier. An image intensifier, such as this, can also be designed by increasing the number of electrons (as in a photomultiplier).

11.6 Electro-Optic Modulation

Electro-optic (EO) modulators consist of a dielectric medium and a means of applying an electric field to it. Application of an electric field causes the refractive index of the dielectric medium to be modified. The mechanism involved in this phenomenon is referred to as the electro-optic effect of the dielectric medium. The mechanism is often used for realizing both amplitude as well as the phase modulation of optical signals. The application of such a modulation scheme exists in optical communications Q-switching of lasers, and beam deflections. In principle, electro-optic modulators cause a change in phase shift, which in turn is either a linear or a quadratic function of the applied electric field. This change in phase shift implies a change in optical length or index of refraction.

In an electro-optic crystal, such as those mentioned in the last section, the change in the index of refraction n along a crystal axis may be expressed in series form as

$$\Delta\left(\frac{1}{n^2}\right) = pE + kE^2 + \ldots, \tag{11.39}$$

where E is the electric field, p is the linear electro-optic coefficient, and k is the quadratic electro-optic coefficient. In useful crystal, either the linear electro-optic effect (referred to as the Pockels effect) or the quadratic electro-optic effect (referred to as the Kerr effect) is predominant. In either case, the index of refraction will change at the modulation rate of the electric field. The effect allows a means of controlling the intensity or phase of the propagating beam of light. A Pockels cell uses the linear effect in crystals, whereas a Kerr cell uses the second-order electro-optic effect in various liquids and ferroelectrics; however, the former requires far less power than the latter to achieve the same amount of rotation and thus is used more widely. The Pockels effect, in particular, depends on the polarity of the applied electric field.

In the absence of an external electric field, the indices of refraction along the rectangular coordinate axes of a crystal are related by the index ellipsoid:

$$(x/n_x)^2 + (y/n_y)^2 + (z/n_z)^2 = 1. \tag{11.40}$$

In the presence of an arbitrary electric field, however, the linear change in the coefficients of the index ellipsoid can be represented by

$$\Delta \left(\frac{1}{n^2} \right)_i = \sum_j p_{ij} E_j, \tag{11.41}$$

where p_{ij}'s are referred to as Pockels constants, $i = 1, 2, 3, ..., 6$; and $j = x, y, z$. The 6×3 electro-optic matrix, having p_{ij} as its elements, is often called the electro-optic tensor. In centro-symmetric crystals, all 18 elements of the tensor are zero, whereas in the triclinic crystals, all elements are nonzero. But in the great majority of crystals, while some of the elements are zero many of the elements have identical values. Table 11.1 lists the characteristics of the electro-optic tensors for some of the more important non-centrosymmetric crystals.

Determining the electro-optic effect in a particular modulator thus involves using the characteristics of the crystal in question and finding the allowed polarization directions for a given direction of propagation. Knowledge of refractive indices along the allowed directions can be used to decompose the incident optical wave along those allowable polarization directions. Thereafter, we can determine the characteristics of the emergent wave. Consider, for example, the case of KDP, whose nonzero electro-optic tensor components are p_{41}, p_{52}, and p_{63}. Using Equations 11.40 and 11.41, we can write an equation of the effective index ellipsoid for KDP as

$$\frac{x^2 + y^2}{n_0^2} + \left(\frac{z}{n_E} \right)^2 + 2p_{41} yzE_x + 2p_{63} xyE_z = 1, \tag{11.42}$$

TABLE 11.1 Some of the Important Electro-Optic Constants

Crystal	Nonzero Elements (in 10^{12}m/V)	Refractive Index
BaTiO$_3$	$p_{13} = p_{23} = 8.0$	$N_x = n_y = n_0 = 2.437$
	$p_{33} = 23.0$	$N_E = n_z = 2.365$
	$p_{42} = p_{51} = 820.0$	
KDP	$p_{41} = p_{52} = 8.6$	$N_x = n_y = n_0 = 1.51$
	$p_{63} = 10.6$	$NE = n_z = 1.47$
ADP	$p_{41} = p_{52} = 28.0$	$N_x = n_y \equiv n_0 = 1.52$
	$p_{63} = 8.5$	$N_E = n_z = 1.48$
GaAs	$p_{41} = p_{52} = p_{63} = 1.6$	$N_x = n + y = n_0 = 3.34$
CdTe	$p_{41} = p_{52} = p_{63} = 6.8$	$N_x = n_y \equiv n_0 = 2.6$
Quartz	$p_{11} = 0.29; p_{21} = p_{62} = -0.29$	$N_x = n_y \equiv n_0 = 1.546$
	$p_{41} = 0.2; p_{52} = -0.52$	$Nz = n_E = 1.555$
LiNbO$_3$	$p_{13} = p_{23} = 8.6; p_{33} = 30.8$	$N_x = n_y = n_0 = 2.286$
	$p_{22} = 3.4; p_{12} = p_{61} = -3.4$	$N_z = n_E = 2.200$
	$p_{42} = p_{51} = 28$	
CdS	$p_{13} = p_{23} = 1.1$	$N_x = n_y = n_0 = 2.46$
	$p_{33} = 2.4; p_{42} = p_{51} = 3.7$	$N_z = n_E = 2.48$

with $n_x = n_y = n_0$ and $n_z = n_E$ for this uniaxial crystal where $p_{41} = p_{52}$. To be specific, let us restrict ourselves to the case in which the external field is directed along only the z-direction. Equation 11.42, for such a case, reduces to

$$\frac{x^2 + y^2}{n_0^2} + \left(\frac{z}{n_E}\right)^2 + 2 p_{63} xy E_z = 1. \tag{11.43}$$

Equation 11.43 can be transformed to have a form of the type

$$\left(\frac{x'}{n_{x'}}\right)^2 + \left(\frac{y'}{n_{y'}}\right)^2 \left(\frac{z'}{n_{z'}}\right)^2 = 1, \tag{11.44}$$

which has no mixed terms. The parameters x', y', and z' of Equation 11.44 denote the directions of the major axes of the index ellipsoid in the presence of the external field; $2n'_x$, $2n'_y$, and $2n'_z$ give the lengths of the major axes of the index ellipsoid, respectively.

By comparing Equations 11.43 and 11.44, it is obvious that z and z' are parallel to each other. Again, the symmetry of x and y in Equation 11.43 suggests that x' and y' are related to x and y by a rotation of 45°. The transformation between the coordinates (x, y) and (x', y') is given by

$$\begin{bmatrix} x \\ y \end{bmatrix} = \begin{bmatrix} \cos\left(\frac{1}{4}\pi\right) - \sin\left(\frac{1}{4}\pi\right) \\ \sin\left(\frac{1}{4}\pi\right) \cos\left(\frac{1}{4}\pi\right) \end{bmatrix} \begin{bmatrix} x' \\ y' \end{bmatrix}, \tag{11.45}$$

which, when substituted in Equation 11.43, results in

$$\left[\frac{1}{n^2} + p_{63}E_z\right]x'^2 + \left[\frac{1}{n_0^2} - p_{63}E_z\right]y'^2 + \left(\frac{z'}{nE}\right) = 1. \tag{11.46}$$

Comparing Equations 11.44 and 11.46 and using the differential relation $dn = \frac{1}{2}n^3 d(1/n^2)$, we find that

$$n_{x'} = n_0 - \tfrac{1}{2}n_0^3 p_{63}E_z \tag{11.47}$$

$$n_{y'} = n_0 - \tfrac{1}{2}n_0^3 p_{63}E_z \tag{11.48}$$

$$n_{z'} = n_E, \tag{11.49}$$

when $(1/n_0^2) = p_{63}E_z$. The velocity of propagation of an emerging wave polarized along the x' axis differs from that of an emerging wave polarized along y' axis. The corresponding phase shift difference between the two waves (referred to as electro-optic retardation) after having traversed a thickness W of the crystal is given by

$$\Delta\phi = \frac{2\pi W}{\lambda}|n_{x'} - n_{y'}| - \frac{2\pi W}{\lambda}n_0^3 p_{63}E_z. \tag{11.50}$$

Provided that V is the voltage applied across the crystal, retardation is then given by

$$\Delta\phi = \frac{2\pi}{\lambda}n_0^3 p_{63} V. \tag{11.51}$$

The emergent light is in general elliptically polarized. It becomes circularly polarized only when $\Delta\phi = 1/2\pi$ and linearly polarized when $\Delta\phi = \pi$. Often the retardation is also given by $\pi V / V_\pi$, where $V_\pi (\equiv \lambda / 2n_0^3 p_{63})$ is the voltage necessary to produce a retardation of π.

11.7 Electro-Optic Devices

In the last section, we showed that it is possible to control optical retardation and, thus, flow of optical energy in non-centrosymmetric cyrstals by means of voltage. This capability of retardation serves as the basis of modulation of light. Figure 11.22 shows a schematic of a system showing how a KDP modulator can be used to achieve amplitude modulation. This particular setup is also known as Pockels electro-optic (EO) amplitude modulator. The crossed polarizer-analyzer combination of the setup is necessary for the conversion of phase-modulated light into amplitude-modulated light. Further, in this setup, the induced electro-optic axes of the crystal make an angle of 45° with the analyzer–polarizer axes.

Elliptically polarized light emerges from the EO crystal since, upon modulation, the two mutually orthogonal components of the polarized beams travel inside the crystal with different velocities. To feed an electric field into the system, the end faces of the EO crystal are coated with a thin conducting layer, which is transparent to optical radiation. As the modulating voltage is changed, the eccentricity of the ellipse changes. The analyzer allows a varying amount of outgoing light in accordance with the modulating voltage applied across the EO crystal.

The electric field components associated with the optical wave, immediately upon emerging from the EO crystal, are respectively

$$E'_x = A, \tag{11.52}$$

$$E'_y = Ae^{-j\Delta\phi}. \tag{11.53}$$

Thus, the total field that emerges out of the analyzer is evaluated by summing the E'_x and E'_y components (along the analyzer axis), which gives us

$$\begin{aligned} E_o &= \left[Ae^{-j\Delta\phi} - A \right]\cos(1/4\pi) \\ &= \frac{A}{\sqrt{2}}e^{-j\Delta\phi} - 1. \end{aligned} \tag{11.54}$$

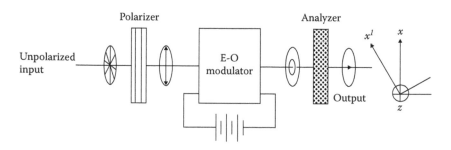

FIGURE 11.22 Pockels EO amplitude modulator: (a) system and (b) orientation of crystal axes.

The resulting irradiance of the transmitted beam is therefore given by

$$I_0 = \text{const}(E_0)(E_0^*) = \text{const } 2A^2 \sin^2\left(\frac{\Delta\phi}{2}\right) = I_i \sin^2\left(\frac{\Delta\phi}{2}\right),$$ (11.55)

where I_i is the irradiance of the light incident on the input side of the EO crystal. One can rewrite Equation 11.55 as

$$\frac{I_0}{I_i} = \sin^2\left[\frac{1}{2}\pi\frac{V}{V_\pi}\right],$$ (11.56)

where $V_\pi = |\lambda/2p_{63}n_0^3|$ is the voltage required for having the maximum transmission. Often V_π is also referred to as the half-wave voltage because it corresponds to a relative spatial displacement of $\lambda/2$ or to an equivalent phase difference of π.

Figure 11.23 shows the transmission characteristics of the cross-polarized EO modulator as a function of the applied voltage. It can be seen that the modulation is nonlinear. In fact, for small voltages, the transmission is proportional to V^2. The effectiveness of an EO modulator is often enhanced by biasing it with a fixed retardation of $\pi/2$. A small sinusoidal voltage will then result in a nearly sinusoidal modulation of the transmitted intensity. This is achieved by introducing a quarter-wave plate between the polarizer and the EO crystal. The quarter-wave plate shifts the EO characteristics to the 50% transmission point. With the quarter-wave plate in place, the net phase difference between the two emerging waves becomes $\Delta\phi = \Delta\phi + (\pi/2)$. Thus, the output transmission is then given by

$$\frac{I_0}{I_i} = \sin^2\left(\frac{\pi}{4} + \frac{1}{2}\pi\frac{V}{V_\pi}\right) \approx \frac{1}{2}\left(1 + \sin\left(\pi\frac{V}{V_\pi}\right)\right).$$ (11.57)

Note that, with no modulating voltage, the modulator output intensity transmission reduces to 0.5 gain. Again, for small V, the transmission factor varies linearly as the crystal voltage.

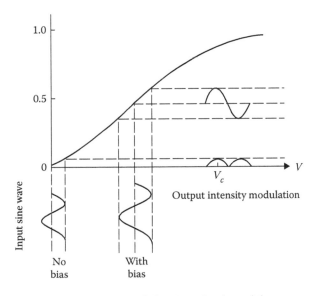

FIGURE 11.23 Transmission versus voltage in a Pockels EO amplitude modulator.

For an input sine wave modulating voltage, the transmission can be expressed as

$$\frac{I_0}{I_i} = \frac{1}{2}\left[1 + \sin\left\{mV_p \sin\left(\omega_p t\right)\right\}\right],$$

(11.58)

where m is a constant of proportionality, V_p is the peak modulating voltage, and ω_m is the modulation angular frequency. When $mV_p \ll 1$, the intensity modulation becomes a replica of the modulating voltage. The irradiance of the transmitted beam begins to vary with the same frequency as the sinusoid voltage. If, however, the condition $mV_p \ll 1$ is not satisfied, the intensity variation becomes distorted. Note that Equation 11.58 can be expanded in terms of Bessel functions of the first kind to give

$$\frac{I_0}{I_i} = \frac{1}{2} + J_1\left(mV_p\right)\sin\left(\omega_m t\right) + J_3\left(mV_p\right)\sin\left(3\omega_m t\right) + J_5\left(mV_p\right)\sin\left(5\omega_m t\right) + \ldots$$

(11.59)

Since

$$\sin\left(x \sin y\right) = 2\left[J_1\left(x\right)\sin\left(y\right) + J_3\left(x\right)\sin\left(3y\right) + J_5\sin\left(5y\right) + \ldots\right]$$

(11.60)

when $J_0(0) = 1$ and, for nonzero n, $J_n(x) = 0$. The ratio between the square root of the sum of harmonic amplitude squares and the fundamental term amplitude often characterizes the amount of distortion in the modulation process. Therefore,

$$\text{Distortion (\%)} = \frac{\left\{\left[J_3\left(mV_p\right)\right]^2 + \left[J_5\left(mV_p\right)\right]^2 + \ldots\right\}^{1/2}}{J_1\left(mV_p\right)} \times 100.$$

(11.61)

Consider the setup of Figure 11.24 where the EO crystal is oriented in such a way that the incident beam is polarized along one of the birefringence axes, say x'. In this specific case, the state of polarization is not changed by the applied electric field. However, the applied field changes the output phase by an amount

$$\Delta\phi_{x'} = \frac{\omega W}{c}\left|\Delta n_{x'}\right|$$

(11.62)

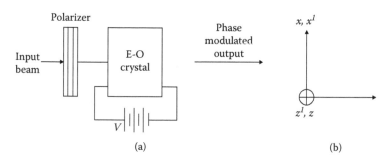

FIGURE 11.24   A phase modulator using an EO crystal: (a) the system and (b) the orientation of crystal axes.

where W is the length of the EO crystal. For a sinusoidal bias field $E_z = E_{z\,p} \sin(\omega_p t)$ and an incident $E_{\text{in}} = E_{\text{in},p}\cos(\omega t)$, the transmitted field is given by

$$E_{\text{out}} = E_{\text{in,p}} \cos\left[\omega t - \frac{\omega W}{c}\{n_0 - \Delta\phi_{x'}\}\right]$$

$$= E_{\text{in,p}} \cos\left[\omega t - \frac{\omega W}{c}\left\{n_0 - \frac{1}{2}n_0^3 p_{63} E_{z,p} \sin(\omega_m t)\right\}\right]. \tag{11.63}$$

If the constant phase factor $\omega W n_0/c$ is neglected, the transmitted electric field can be rewritten as

$$E_{\text{out}} = E_{\text{in,p}} \cos\left[\omega t + \delta \sin(\omega_m t)\right], \tag{11.64}$$

where $\delta \equiv 1/2(\omega W n_0^3 p_{63} E_{z,p}/c)$ is the phase-modulation index. Note that this phase-modulation index is one-half of the retardation $\Delta\varphi$ (as given by Equation 11.51). Using Equation 11.60 and the relationship

$$\cos(x \sin y) = J_0(x) + 2\left[J_2(x)\cos(2y) + J_4(x)\cos(4y) + \ldots\right], \tag{11.65}$$

we can rewrite Equation 11.64 as

$$E_{\text{out}} = E_{\text{in,p}}\Big[J_0(\delta)\cos(\omega t) + J_1(\delta)\big[\cos\{(\omega + \omega_m)t\} - \cos\{(\omega - \omega_m)t\}\big]$$
$$+ J_2(\delta)\big[\cos\{(\omega + 2\omega_m)t\} - \cos\{(\omega - 2\omega_m)t\}\big]$$
$$+ J_3(\delta)\big[\cos\{(\omega + 3\omega_m)t\} - \cos\{(\omega - 3\omega_m)t\}\big] + \ldots\Big]. \tag{11.66}$$

Accordingly, in this case, the optical field is seen to be phase modulated with energy distribution in the side bands varying as a function of the modulation index δ. We observe that, while the EO crystal orientation of Figure 11.23 provides an amplitude modulation of light, the setup of Figure 11.24 can provide a phase modulation of light. Both of these modulators are called longitudinal effect devices because in both cases the electric field is applied in the same direction as that of propagation. In both of these cases, the electric field is applied either by electrodes with small apertures in them or by making use of semitransparent conducting films on either side of the crystal. This arrangement, however, is not too reliable because the field electrodes tend to interfere with the optical beam.

Alternatively, transverse electro-optic modulators can be used, as shown by the system in Figure 11.25. The polarization of the light lies in the x'–z plane at a 45° angle to the x'-axis while light propagates along

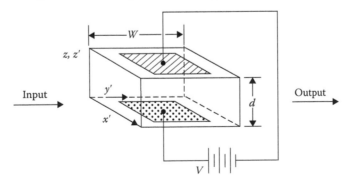

FIGURE 11.25 A transverse EO amplitude modulator.

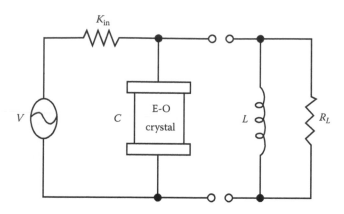

FIGURE 11.26   A circuit equivalent to an EO modulator.

the y'-axis and the field is applied along z. With such an arrangement, the electrodes do not block the incident optical beam and, moreover, the retardation can be increased by introducing longer crystals. In this longitudinal case, the amount of retardation is proportional to V and is independent of the crystal length W according to Equation 11.51. Using Equations 11.47 and 11.49, the retardation caused by the transverse EO amplitude modulator is given by

$$\Delta\phi_t = \phi_{x'} - \phi_{z'} = \frac{2\pi W}{\lambda}\left[(n_0 - n_E) - n_0^3 p_{63}\frac{V}{d}\right], \tag{11.67}$$

where n_0 and n_E are refractive indices, respectively, along ordinary and extraordinary axes. Note that $\Delta\phi_t$ has a voltage-independent term that can be used to bias the irradiance transmission curve. Using a long and thin EO crystal can reduce the halfwave voltage. Such an EO crystal allows the transverse EO modulators to have a better frequency response but at the cost of having small apertures.

There are occasions when we might be interested in driving the modulating signals to have large bandwidths at high frequencies. This can happen when we decide to use the wide-frequency spectrum of a laser. To meet the demand of such a scenario, the modulator capacitance that is caused by the parallel-plate electrodes and the finite optical transit time of the modulator limits both bandwidth and the maximum modulation frequency. Consider the equivalent circuit of a high-frequency, electro-optic modulator as shown in Figure 11.26. Let R_{in} be the total internal resistance of the modulating source V, while C represents the parallel-plate capacitance of the EO crystal. When R_{in} is greater than the capacitance impedance, a significant portion of the modulating voltage is engaged across the internal resistance, thus making the generation of electro-optic retardation relatively insignificant. In order to increase the proportion of the modulating voltage that is engaged across the EO crystal, it is necessary to connect a parallel resistance-inductance circuit in parallel with the modulator. The load R_L is chosen to be very large when compared with R_{in}. The choice guarantees that most of the modulating voltage is employed across the EO crystal. At the resonant frequency $v_0 = \left[1/\{2\pi(LC)\}^{1/2}\right]$, the circuit impedance is equivalent to load resistance. However, this system imposes a restriction on the bandwidth and makes it finite. The bandwidth is given by $\left[1/\{2\pi R_L C\}\right]$ and is centered at the resonant frequency v_0. Beyond this bandwidth, the modulating voltage is generally wasted across R_{in}. Consequently, for the modulated signal to be an exact replica of the modulating signal, the maximum modulation bandwidth must not be allowed to exceed Δv, where Δv is usually dictated by the specific application.

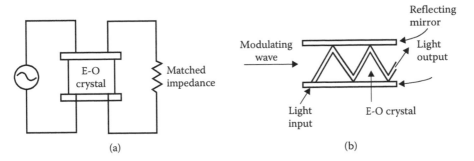

FIGURE 11.27 (a) Ideal traveling wave EO modulator and (b) zig-zag traveling wave modulator.

The power P needed to drive the EO crystal is given by $1/2(V_{max}^2/R_L)$, where $V_{max}\ (=E_{z,max}W)$ is the peak modulating voltage, which produces the peak retardation $\Delta\phi_{max}$. Using Equation 11.51 we can relate the driving power and the modulation bandwidth by

$$P = \frac{(\Delta\phi_{max})^2 \lambda^2 C \Delta v}{4\pi p_{63}^2 n_0^6},$$

$$= \frac{(\Delta\phi_{max})^2 \lambda^2 A K_s \varepsilon_0 \Delta v}{4\pi p_{63}^2 n_0^6 W},$$

(11.68)

since at the modulation frequency v_0, the parallel-plate capacitance C is given by $AK_s\varepsilon_0/W$, where A is the cross-sectional area of the crystal, K_s is the relative dielectric constant for the material, and W is the plate separation between the two electrodes.

As long as the modulating frequency is relatively low, the modulating voltage remains appreciably constant across the crystal. If the above condition is not fulfilled, however, the maximum allowable modulation frequency is restricted substantially by the transit time of the crystal. To overcome this restriction, the modulating signal is applied transversely to the crystal in the form of a traveling wave with a velocity equal to the phase velocity to the optical signal propagating through the crystal. The transmission line electrodes are provided with matched impedance at the termination point, as shown in Figure 11.27a. The optical field is then subjected to a constant refractive index as it passes through the modulator, thus making it possible to have higher modulation frequencies. The traveling modulation field at a time t will have the form

$$E(t, z(t)) = E_p \exp\left[j\omega_m \left\{ t - \frac{z}{v_p} \right\} \right]$$

$$= E_p \exp\left[j\omega_m \left\{ t - \frac{c}{nv_p}(t-t_0) \right\} \right],$$

(11.69)

where v_p is the phase velocity of the modulation field, ω_m is the modulating angular frequency, and t_0 is defined as a reference to account for the time when the optical wavefront enters the EO modulator. The electro-optic retardation due to the field can be written in accordance with Equation 11.51 as

$$\Delta\phi = \Phi\frac{c}{n} \int_{t_0}^{t_0+t_1} E(t, z(t)) dt,$$

(11.70)

where $\Phi = 2\pi n_0^3 p_{63} / \lambda$ and $t_t = nW/c$ is the total transit time (i.e., time taken by light to travel through the crystal). Equation 11.70 can be evaluated to give the traveling wave retardation as

$$\Delta\phi_{\text{travel}} = (\Delta\phi)_0 \frac{e^{j\omega_m t_0}[e^{j\omega_m t_t[1-c/n\upsilon_p]} - 1]}{j\omega_m t_t \left(1 - \dfrac{c}{n\upsilon_p}\right)}, \tag{11.71}$$

where $(\Delta\phi)_0 = \left(\varphi c t_t E_p / n\right)$ is the peak retardation. The reduction factor

$$F_{\text{travel}} = \frac{e^{j\omega_m t_t[1-c/n\upsilon_p]} - 1}{j\omega_m t_t \left(1 - \dfrac{c}{n\upsilon_p}\right)} \tag{11.72}$$

provides the amount of reduction in the maximum retardation owing to transit time limitation. If instead we had begun to calculate the retardation for a sinusoidal modulation field that has the same value throughout the modulator, the retardation would have been

$$\Delta\phi_{\text{travel}} = (\Delta\phi)_0 \exp\left(j\omega_m t_0\right)\left[\frac{e^{j\omega_m t} - 1}{j\omega_m t_t}\right]. \tag{11.73}$$

By comparing Equations 11.71 and 11.73, we find that the two expressions are identical except that, in the traveling wave, the EO modulator t_t is replaced by $t_t\left\{1 - \left(c / n\upsilon_p\right)\right\}$. The reduction factor in this latter case is given by

$$F_{\text{nontravel}} = \frac{e^{j\omega_m t} - 1}{j\omega_m t_t}. \tag{11.74}$$

In the case of a non-traveling system, the reduction factor is unity only when $\omega_m t_t \ll 1$, that is, when the transit time is smaller than the smallest modulation period. But in the case of the traveling system, F approaches unity whenever the two-phase velocities are equal – that is, when $c = n\upsilon_p$. Thus, in spite of the limitation of transit time, the maximum retardation is realized using the traveling wave modulator.

In practice, it might become very difficult to synchronize both electrical and optical waves. For a perfect synchronization, we expect to use an EO crystal for which $n = K_s^{1/2}$. But for most naturally occurring materials, n is less than $K_s^{1/2}$. Thus, synchronization is achieved either by including air gaps in the electrical waveguide cross-section or by slowing down the optical wave by means of a zigzag modulator, as shown in Figure 11.27b.

A useful amplitude modulator involving waveguides, known as the Mach–Zehnder modulator, consists of neither polarizer nor analyzer but only EO material. The system, as shown in Figure 11.28, splits the incoming optical beam into a "waveguide Y" and then recombines them. If the phase shift present in both of the arms is identical, all of the input power minus the waveguide loss reappears at the output. One arm of the Mach–Zehnder system is provided with an electric field placed across it such that the amplitude of the field can be varied. Changing the voltage across the waveguides modulates the output power. When a sufficiently high voltage is applied, the net phase shift difference between the arms can become 180°, thus canceling the power output altogether. Because of the small dimensions involved in electrode separations, relatively small switching voltages are used in such a modulation scheme.

Figure 11.29a shows a prism deflector that uses a bulk electro-optic effect so that we can deflect an optical beam by means of an externally applied electric voltage. The electric voltage induces regions

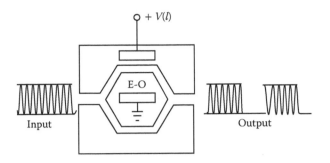

FIGURE 11.28 Mach–Zehnder waveguide EO modulator.

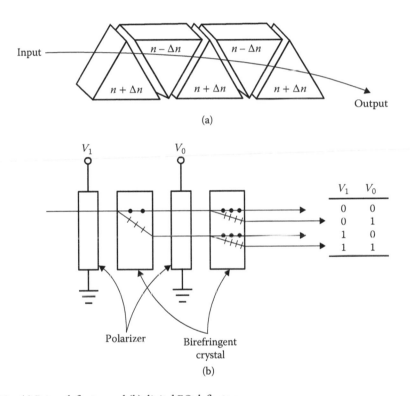

FIGURE 11.29 (a) Prism deflector and (b) digital EO deflector.

having different refractive indices, as a result of which the prism steers the refracted beam of light. Stacks of prisms can be used to provide a larger angle of deflection. Figure 11.29b shows an interesting application where birefringent crystals are combined with a Pockels cell modulator to form a digital EO deflector. By applying an electric field, we can rotate the direction of polarization by $\pi/2$. Accordingly, by manipulating voltage (V_1 and V_0). we can shift the input optical beam to any one of the four spatial locations at the output. Similarly, by using n Pockels cell modulators and n birefringent crystals, we can shift light to a total of 2^n spatial locations. A system such as that of Figure 11.29b can be considered for various optical computing applications.

A device by the name of Pockels read-out optical modulator (PROM) can be fabricated by having an EO crystal, such as Bi_2SiO_{20} or ZnSe, sandwiched between two transparent electrodes. There is also an insulating layer between the EO crystal and the electrode. A dc voltage can create mobile carriers, which in turn causes the voltage in the active crystal to decay. The device is normally exposed with the illumination pattern of a blue light. The voltage in the active area that corresponds to the brightest zones of the input pattern decays, while that corresponding to the comparatively darker area either does not change or changes very little. On the other hand, the read-out beam usually uses a red light. Note that the sensitivity of an EO crystal is much higher in the blue region than in the red region. Such a choice of read-out beam ensures that the read-out beam may not cause a change in the stored voltage pattern. In the read-out mode, the regions having the least amount of voltage act like a half-wave retarder. The reflected light, whose polarization is thus a function of the voltage pattern, is then passed through a polarizer to reproduce the output. For an eraser light E, the amplitude of the output is found to be

$$A = A_0 \sin\left(\frac{\pi V_0}{V_{1/2}}\right) e^{-KE}, \tag{11.75}$$

where A_0 is the amplitude of the input read-out beam, V_0 is the voltage applied across the EO crystal, $V_{1/2}$ is the halfwave voltage, and K is a positive constant. In the reflection read-out mode, $V_{1/2} = 2V_0$. The amplitude of the reflectance of the PROM when plotted against the exposure is surprisingly found to be similar to that of a photographic film with a nearly linear region between $E = 0$ and $E = 2/K$, as shown in Figure 11.30.

There are many isotropic media available that behave like uniaxial crystals when subjected to an electric field E. The change in refractive index Δn of those isotropic media varies as the square of the electric field. Placing one of these media between crossed polarizers produces a Kerr modulator. Modulation at frequencies up to 10^{10} Hz has been realized using a Kerr modulator. The difference between the two refractive indices that correspond respectively to light polarized perpendicular to the induced optic axis and light polarized perpendicular to the induced optic axis is provided by

$$\Delta n = |n_\Pi - n_\perp| = k\lambda E^2, \tag{11.76}$$

where k is the Kerr constant of the material. At room temperature, the Kerr constant has a typical value of 0.7×10^{-12}, 3.5×10^{-10}, and 4.5×10^{-10} cm/V², respectively, for benzene, carbon disulphide, and nitrobenzene.

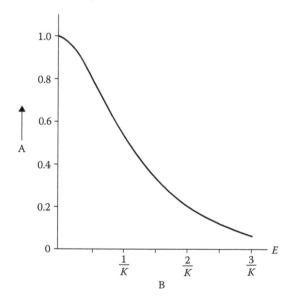

FIGURE 11.30   The characteristics of a PROM device.

The applied electric field induces an electric moment, which in turn reorients the molecules in a manner so that the medium becomes anisotropic. The delay between the application of the field and the appearance of the effect is, though not negligible, on the order of 10^{-12} s. A liquid Kerr cell containing nitrobenzene, as shown in Figure 11.31, has been used for many years, but it has the disadvantage of requiring a large driving power. This problem is often overcome by using, instead, mixed ferroelectric crystals at a temperature near the Curie point, where ferroelectric materials start exhibiting optoelectric properties. Potassium tantalate niobate (KTN) is an example of such a mixture of two crystals, where one has a high Curie point and the other has a low Curie point, but the Curie point of the compound lies very close to room temperature. The transmittance characteristics of a Pockels cell and a Kerr cell are shown, for comparison, in Figure 11.32.

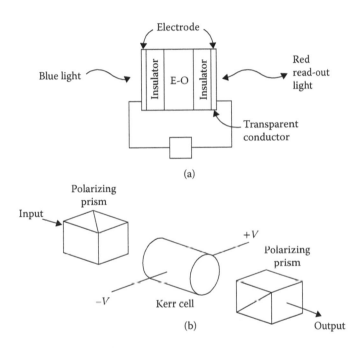

FIGURE 11.31 A Kerr cell light modulator.

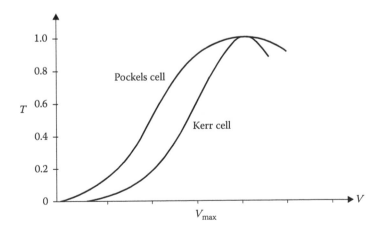

FIGURE 11.32 Transmittance versus voltage curve for the EO modulators.

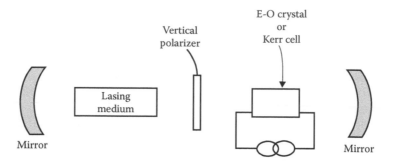

FIGURE 11.33   A Q-switched laser system.

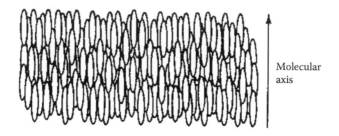

FIGURE 11.34   An aggregate of nematic liquid crystals.

One of the ways one can realize a Q-switched laser involves subjecting either an EO crystal or a liquid Kerr cell to an electric field. Such a non-mechanical system is shown in Figure 11.33. When there is no electric field, there is no rotation. But in the presence of the field, an EO device can introduce a rotation $\pi/2$. Thus, when the EO device is assembled along with a polarizer, the combination works as a shutter. Because of the vertical polarizer, the light coming out of the lasing medium is plane-polarized. Because the polarized light has to traverse through the EO device twice before coming back to the polarizer, only half of the voltage required to produce a rotation of $\pi/2$ is applied to the system. Accordingly, the polarizer of the EO system blocks the light from coming to the main chamber of the laser. This is equivalent to causing a loss in the laser resonator. Such a loss, when suitably timed, can be made to produce a pulsed laser output with each pulse having an extremely high intensity. The voltages required to introduce appropriate fields are usually in the order of kilovolts, and thus it becomes possible to Q-switch at a rate of only nanoseconds or less.

The liquid-crystal light valve (LCLV) is a specific spatial light modulator (SLM) with which one can imprint a pattern on a beam of light in nearly real time. The two aspects of this device that are particularly important are the modulation of light and the mechanism for addressing the device. Besides LCLVs, there are many SLMs that are currently being considered for electro-optic applications. However, most of these devices follow only variations of the same physical principle.

The term liquid crystal (LC) refers to a particular class of materials whose rheological behavior is similar to that of liquids, but whose optical behavior is similar to that of crystalline solids over a given range of temperature. In particular, a type of LC, referred to as the nematic LC, is commonly used in LCLV as well as in most other LC devices. Lately, ferroelectric LCs are also being used in real-time display devices. In comparison, ferroelectric LC-based devices respond at a faster rate than nematic LC-based devices. Nematic LCs generally consist of thin, long molecules, all of which tend to be aligned in the same direction, as shown in Figure 11.34. When an electric field is applied to an LC layer, the molecules are reoriented in space due to both field and ionic conduction effects. The fact that LCs have a rodlike cylindrical molecular shape and that they tend to be aligned are prime reasons for yielding two

EO effects: electric field effect and birefringence. In the "off" state, an LCLV utilizes the properties of a nematic cell, while in the "on" state, it utilizes the birefringence properties.

The shape of the LC molecule introduces a polarization-dependent variation of the index of refraction that contributes to its birefringence characteristics. The difference between the two indices of refraction given by $\Delta n = |n_\Pi - n_\perp|$ is a measure of the anisotropy of the material, where $n_\parallel$ represents the index of refraction for the component of light parallel to the molecular axis, and $n_\wedge$ represents the index of refraction for the light component having an orthogonal polarization. Since LC molecules tend to be aligned, a bulk LC sample exhibits the same anisotropy as that exhibited by an individual LC molecule. In fact, a birefringent LC cell is normally formed by stacking LC layers parallel to the cell wall. Accordingly, all of the LC molecules are aligned along only one direction. When linearly polarized (at 45° to the alignment axis) light enters an LC cell of thickness D, the parallel polarization component lags behind the orthogonal component (due to positive anisotropy) by $\Delta\phi = \left[2\pi D(\Delta n) \right] / \lambda$. In general, the transmitted light turns out to be elliptically polarized. With a suitable choice of D, we can force the LC cell to behave like a half-wave or a quarter-wave plate. When compared with other traditional materials, LCs are preferable for making such retardation plates. In a typical LC material, Δn is non-negligible quantity. For example, for a typical LC, Δn is about 0.2 in the infrared, whereas Δn is only about 0.0135 in CdS in the infrared; in the visible, say, for quartz, Δn can be as low as 0.009. Thinner LC cells produce a comparatively large value of $\Delta\phi$. Note that a thinner LC cell allows for a larger acceptance angle for the incoming light, whereas a thicker, solid crystal like quartz forms a cell that is extremely sensitive to the angle of incidence. Further, LC cells can be grown to have reasonably large aperture sizes, suitable for handling higher laser power; however, the size will be limited by the tolerance for optical flatness.

For positive anisotropy, there is a region of applied voltage over which the LC molecules may gradually rotate, introducing a variable phase delay in the output light. This feature, as illustrated in Figure 11.35, can be used to create a voltage-control phase shifter. As stated earlier, a typical LC device combines the

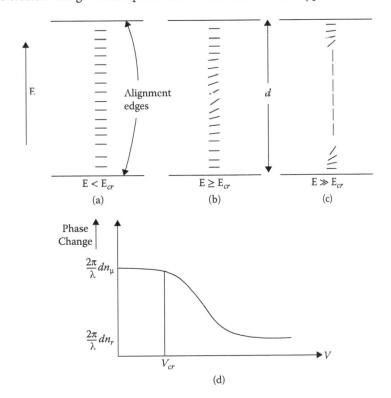

FIGURE 11.35 Field effects in liquid crystals.

characteristics of both birefringence and field effects. A typical SLM structure is shown in Figure 11.36, where the cell is organized in the form of a quarter-wave plate. Consider an incoming light that is linearly polarized at an angle of 45° to the direction of alignment. The transmitted light is then found to be circularly polarized, but the reflected light that passes back through the quarter-wave cell becomes polarized in a direction perpendicular to that of the incident light. The first polarizer acts as an analyzer and thus blocks the light. In the presence of an external voltage, however, the birefringence can be reduced. The polarization characteristics of the resultant reflected light are changed so that the analyzer cannot block all of the reflected light. The external voltage can thus be used to control the transmission of light. In particular, transmission is zero when voltage is zero, but transmission reaches a maximum at high enough voltage. It is possible for the transmission to have a nonbinary value when voltage is set between the two extremes. Devices made using this SLM configuration are generally very sensitive to variations in cell thickness and light wavelength.

Interestingly, the twisted nematic LCs are also used in the common liquid-crystal display (LCD), as shown in Figure 11.37. Instead of having parallel layers of LC stacks, the alignment layers at the opposite faces of the cell are maintained at 90° to each other. The remaining layers, depending on their positions, are oriented in a manner such that there is a gradual change in orientation from one end of the cell to the other. The molecules of a stack tend to line up in the same common direction, and they also tend to

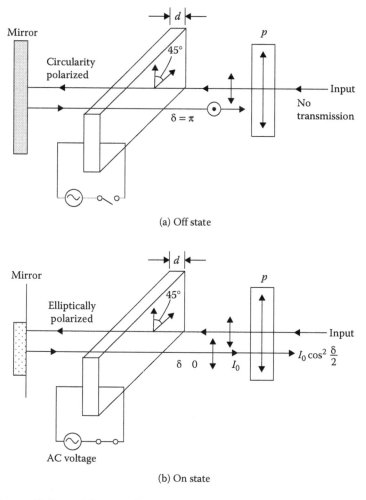

FIGURE 11.36 A spatial light modulator configuration.

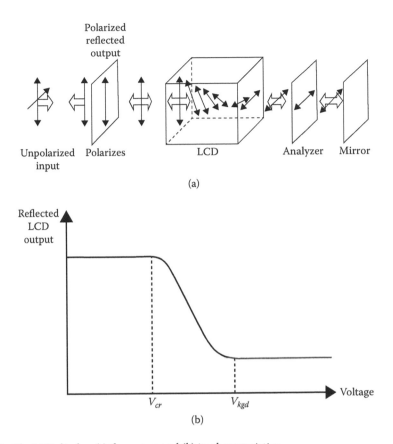

FIGURE 11.37 The LCD display: (a) the system and (b) its characteristics.

align themselves with those of the neighboring stacks. The tilt of the molecular axes changes gradually between the two edges. An externally applied voltage can generally overcome the effects of alignment force and, with sufficient voltage, the molecules can fall back to their isotropic states. When voltages are withdrawn, the light going through the cell undergoes a rotation of 90° and thus passes through the analyzer. But in the presence of voltage, light falls short of 90° and, as a result, the analyzer can block most of the light.

When the voltage is low or completely withdrawn, LC molecules remain gradually twisted across the cell from the alignment direction of one wall to that of the other wall. Such smooth transition is referred to as adiabatic. The light polarization is able to follow this slowly varying twist primarily because the cell width is larger than the light wavelength. As voltage is increased, LC molecules tend to be reorganized along the direction of the applied field. But in order for this to happen, the LC molecules have to overcome the alignment forces of the cell walls. The molecules located at the very center are farthest away from the walls and are, therefore, more likely to be reorganized.

As the tilt angle approaches $\pi/2$, molecules fail to align themselves with their immediate neighbors. In the extreme case of a $\pi/2$ tilt, the cell splits into two distinct halves. While the molecules in one half are aligned with one wall, the molecules in the other half are aligned with the other wall. Such a nonadiabatic system is modeled simply as a birefringent cell with an elliptically polarized transmission. Consequently, this state corresponds to maximum transmission. Other than two extremes, slowly varying and abrupt, there exists transmitted light with an intermediate degree of elliptical polarization. The amount of polarization in the transmission is thus controllable by an externally applied voltage, which allows the LCLV to operate with gray levels. Note that the maximum birefringence occurs when

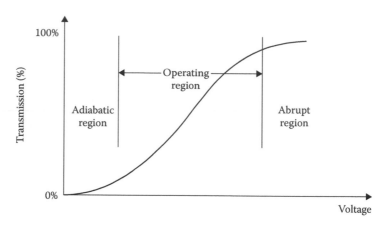

FIGURE 11.38   The transmission characteristics of an LCLV.

the incident light forms an angle of π/4 with the molecular axis. Thus, nematic LCs having a 45° twist provide the maximum LCLV transmission.

The LCLV functions quite similarly to an LCD with only slight differences. The transmission characteristics of the analyzer in the case of an LCLV are the inverse of those in an LCD. This is indicated in Figure 11.38. The LCLV requires only a little energy to produce an output, but it requires a large energy to yield its maximum output. This allows the device to produce a positive image when it is addressed optically. Again, while an LCD deals with binary transmission, an LCLV utilizes gray levels (corresponding to intermediate levels of transmission) to accurately represent an image. Thus, an LCLV is often operated in the transition region that exists between the minimum and the maximum transmission. To produce such an operating characteristic, therefore, an LCLV is organized differently from an LCD.

The light incident on a cell typically strikes its surface along the normal. The incident beam of light becomes linearly polarized in the direction of the molecular alignment of the first layer. In the absence of an applied voltage, the polarization direction rotates through an angle of π/4 along the helical twist of the LCs. The returning reflected light from the mirror undergoes a further twist of π/4, amounting to a total of π/2 rotation, and is thus blocked by the analyzer. With external voltage, however, light transmission increases because light can no longer follow the twist. Thus, by a combination of both birefringence and field effects, the twisted nematic cell can produce EO modulation.

Typical LCDs are addressed via electrode leads; each is connected to only one display segment. Such an addressing technique poses a serious problem when the number of leads begins to increase. LCLVs are, however, addressed differently—by means of optics. Optical addressing allows the image information to be fed into an LCLV in parallel. Thus, in the case of an LCLV, the frame time is the same as the response time of only one pixel. By comparison, in a scanning display, a frame time may equal the response time of one pixel multiplied by the total number of pixels. For example, a typical 20 inch × 20 inch flat panel display may consist of up to 1000 × 1000 pixels, with 2000–3000 pixels per dimension possible. The simplicity of optical addressing is thus obvious. Optical addressing is also preferable because it provides better resolution.

A typical optically addressed LCLV is shown in Figure 11.39. A photoconductive material is used to transfer the optical input to an equivalent voltage on the LC layer. The photoconductor is highly resistive, and it thus utilizes most of the voltage when there is no incident light. Very little or no voltage can be applied across the LC layer to limit the transmission of the read-out light. In the presence of incident light, the resistivity of the photoconductor decreases, as a result of which more voltage appears across the LC layer. The intensity of the incident light thus engages a proportional amount of voltage across the LC layer. Accordingly, an input intensity variation will manifest itself as a voltage variation across

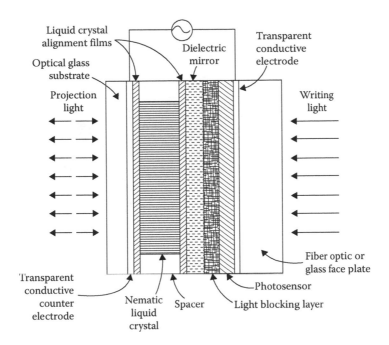

FIGURE 11.39 A schematic of the LCLV.

the LC layer. Again, the coherent read-out beam illuminating the back of the LCLV is reflected back but modulated by the birefringent LC layer. Thus, the input optical image is transferred as a spatial modulation of the read-out beam. The dielectric mirror present in the device provides optical isolation between the input in coherent beam and the coherent read-out beam. In practice, however, the LCLV is driven by the ac voltage. For frequencies with periods less than the molecular response time, the LCLV responds to the rms value of the voltage. An ac-driven LCLV allows flexibility in choosing both the type of photoconductor as well as the arrangement of the intermediate layers.

For the ac-driven LCLV, CdS is generally chosen as the photoconductor, while CdTe is used as the light-blocking material. The CdTe layer isolates the photoconductor from any read-out light that gets through the mirror. It is possible to feed the optical input data by means of a CRT, fixed masks, or even an actual real-time imagery. A typical LCLV has a 25 mm diameter aperture, a 15 ms response time, 60 lines/mm resolution at 50% modulation transfer function, 100:1 average contrast ratio, and a lifetime of several thousand operating hours.

A more recent innovation is the CCD-addressed LCLV, as shown in Figure 11.40. The CCD structure is introduced at the input surface of the semiconductor wafer in the LCLV. The first CCD register is fed with charge information until it is full. The content of this CCD register is then clocked into the next CCD register. The serial input register is again filled and then emptied in parallel. This process is repeated until the total CCD array is loaded with a complete frame of charge information. An applied voltage can then cause the charge to migrate across the silicon to mirror surfaces. The CCD-addressed LCLV generally requires a positive pulse to transfer the charge. This particular display device is very attractive because of its high speed, high sensitivity, high resolution, and low noise distortion.

The LCLV is able to provide image intensification because the read-out beam may be as much as five orders of magnitude brighter than the write beam. The efficient isolation provided by both the mirror and the light-blocking layer is responsible for such intensification as well as for the wavelength conversion between two beams. An LCLV is thus ideally suited to process infrared imagery. Infrared images typically consist of weak signals that require amplification before being processed further. In addition, it is easier to perform optical processing in the visible domain than in the infrared. LCLVs have already

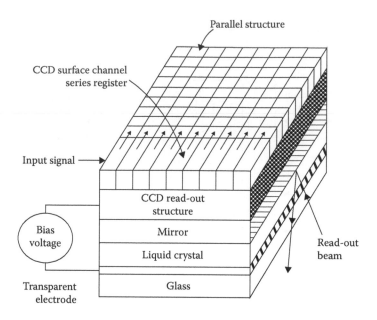

FIGURE 11.40 CCD-addressed LCLV.

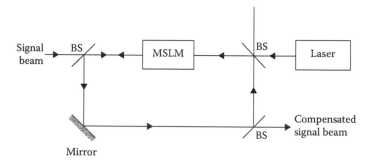

FIGURE 11.41 An adaptive compensating system.

been applied to radar-signal processing, digital-optical computation, optical correlation, and optical image processing; in fact, they have many more applications. However, the response time of an LCLV (determined by the finite time it takes to rotate the LC molecules) is questionable for many of the operations. New LC materials that may improve the response time are currently being developed. The LCLV can be operated with all-coherent, all-incoherent, or mixed read-and-write beams. But the real limitation of this device happens to be its inflated cost.

Often in an adaptive system it becomes necessary to measure the difference between a signal and an estimate of the signal. This measured difference is often used to improve future estimates of the signal. Figure 11.41 shows such a feedback system, which uses a special spatial light modulator, known as a microchannel spatial light modulator, MSLM. The MSLM consists of an EO crystal, a dielectric mirror, a microchannel plate, and a photocathode. The signal wavefront passes through the beamsplitter (BS) and is then reflected by an optically controlled MSLM, which serves as a phase shifter. After further reflections, a part of the beam passes through a second beamsplitter to generate a compensated signal beam. The remaining portion of the beam is mixed with a local oscillator to produce an error signal, which in turn is used to control further phase shifting as well as to maintain phase compensation.

11.8 Acousto-Optic Modulation and Devices

The terms acousto-optic (AO) or elasto-optic effects are used interchangeably to indicate that the refractive index of a medium is being changed either by a mechanically applied strain or by ultrasonic waves. Accordingly, an acousto-optic modulator consists of a medium whose refractive index undergoes a sinusoidal variation in the presence of an externally applied ultrasonic signal, as shown in Figure 11.42. The solid lines indicate the regions of maximum stress, and the dashed lines indicate the regions of minimum stress. There are many materials, such as water, lithium niobate, lucite, cadmium sulphide, quartz, and rutile, that exhibit changes in the refractive index once they are subjected to strain.

As the light enters an AO medium, it experiences a higher value of refractive index at the region of maximum stress, and thus advances with a relatively lower velocity than those wavefronts that encounter the regions of minimum stress. The resultant light wavefront thus inherits a sinusoidal form. The variation in the acoustic wave velocity is generally negligible, and so we may safely assume that the variation of the refractive index in the medium is stationary as far as the optical wavefront is concerned. A narrow collimated beam of light incident upon such a medium is thus scattered into primary diffraction orders. In most practical cases, higher-diffraction orders have negligible intensities associated with them. The zero-order beam generally has the same frequency as that of the incident beam, while the frequencies of +1 and –1 orders undergo a frequency modulation.

In order to appreciate the basics of acousto-optic effects, we can consider the collisions of photons and phonons. Light consists of photons that are characterized by their respective momentum, $hk_l / 2\pi$ and $hk_a / 2\pi$, where k_a and k_l are the respective wave vectors. Likewise, photon and phonon energies are given respectively by $h\nu_l$ and $h\nu_a$, where ν_l and ν_a are the respective frequencies. Consider the scenario

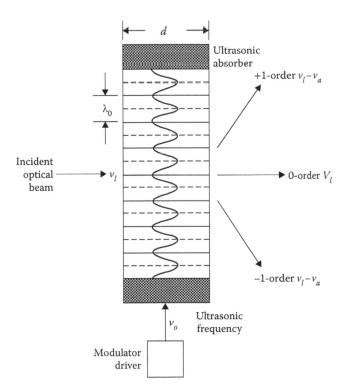

FIGURE 11.42 An acousto-optic modulator.

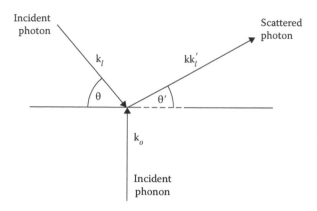

FIGURE 11.43   Photon–phonon collison resulting in the annihilation of a phonon.

where a wave vector is given by k_1', as illustrated in Figure 11.43. The condition for the conversation of momentum, when applied to this collision, yields

$$k_1 \cos\theta = k_1' \cos\theta' \tag{11.77}$$

and

$$K_a = k_1 \sin\theta + k_1' \sin\theta', \tag{11.78}$$

where θ and θ' are the angles formed by the incident and the scattered photons, respectively. Consequently, the angle of the scattered photon is evaluated to give

$$\theta' = \arctan\left[\frac{k_a}{k_1}\sec\theta - \tan\theta\right]. \tag{11.79}$$

It is reasonable to assume that $k_a \ll k_l$ and, thus, for small values of θ, Equation 11.79 reduces to

$$\theta' = \left[\frac{c}{v_a v_1}\right]v_a - \theta, \tag{11.80}$$

where v_a is the acoustic velocity. Equation 11.80 explicitly shows that the angle formed by the scattered photon is proportional to the acoustic frequency. By measuring the deflection angle, we can estimate the acoustic frequency. Further, Equation 11.79 reveals that for an incident angle $\theta = \theta_B = \sin^{-1}(k_a/2k_l)$ in an isotropic medium, $\theta = \theta'$ and $k_1 = k_1'$. At this particular angle of incidence, referred to as the Bragg angle, photon momentum is conserved and the diffraction efficiency reaches a maximum. Note that the power in the scattered beam varies with θ and reaches a maximum when θ is equal to the Bragg angle.

It is important to realize that the acousto-optical effect is produced by multiple collisions of photons and phonons. In any event, the scope of Equation 11.80 is somewhat valid in most practical devices. The condition for conservation of energy is only approximately valid in photon–phonon collision. However, in practice, the frequency of the scattered photon $v_1 = v_1'$, since $v_a \ll .v_1$. In anisotropic materials, k_1'

approaches rk_1, where r is the ratio of the refractive indices corresponding to the diffracted and incident waves, respectively. Equations 11.77 and 11.78 can be modified to give

$$\theta = \sin^{-1}\left[\frac{k_a}{2k_1}\left\{1+\left(\frac{k_1}{k_a}\right)^2(1-r^2)\right\}\right] \tag{11.81}$$

And

$$\theta' = \sin^{-1}\left[\frac{k_a}{2rk_1}\left\{1-\left(\frac{k_1}{k_a}\right)^2(1-r^2)\right\}\right]. \tag{11.82}$$

But to have valid solutions, the condition

$$1-\frac{k_a}{k_1}=r=1+\frac{k_a}{k_1} \tag{11.83}$$

must be satisfied. It is obvious that θ and θ' are equal only when $r=1$ because it is not possible to have $r=(k_a/k_1)-1$ when $k_a \ll k_1$. Thus, the phenomenon $\theta = \theta'$ is associated only with the Bragg angle of incidence and the condition $r=1$. In general, for an incident wave vector, there are two values of k_a (and thus k_1') that satisfy the condition of conservation of momentum. Note that in anisotropic media, the conservation of momentum is satisfied over a wider range of acoustic frequencies or incident light beam directions than is normally realizable in isotropic materials. Consequently, in acousto-optical devices, birefringent diffraction plays a dominant role in determining modulation.

The diffraction of the light beam in AO modulators is justifiably associated with a diffraction grating set up by the acoustic waves. The exact characteristics of this diffraction are indicated by the parameter $Q = k_a^2 d / k_1$, where d is the width of the acoustic-optic device. When $Q < 1$, the diffraction is said to operate in the Raman–Nath regime, and when $Q < 1$, the diffraction is said to operate in the Bragg regime. In the region where $1 \ll Q \ll 10$, the diffraction has a characteristic that is a mixture of the two extremes. Since Q is directly proportional to d, a higher Q requires lesser drive power for any given interaction efficiency. In the Raman–Nath regime, the acoustic-optic grating can be treated as a simple grating, such that

$$m\lambda, = \lambda_a \sin\theta_m, \tag{11.84}$$

where λ_a is the acoustic wavelength, m is an integer, and θ_m is the corresponding angle of diffraction. By comparison, in the Bragg regime, the acoustic field acts very much like a "thick" diffraction grating, requiring that

$$\theta = \theta' = \sin^{-1}\left(\frac{m\lambda}{2\lambda_a}\right). \tag{11.85}$$

Bragg diffraction is identical to that of plane grating when the angle of incidence equals the diffracting angle. Reflected waves, except those for which $\theta = \theta'$, interfere constructively, producing a very strong first-order component.

The fraction of the light diffracted is often characterized by the diffraction efficiency η, defined as $(I_0 - I)/I_0$, where I is the output irradiance in the diffraction orders and I_0 is the output irradiance in the absence of the acoustic waves. While the diffraction efficiency of the Raman–Nath grating is only about 0.35, it approaches 1.00 for the Bragg case. At the Bragg angle, the diffraction efficiency is given by $\sin^2\left[(\pi \Delta n d)/(\lambda \cos\theta_B)\right]$, where Δn is the amplitude of the refractive index fluctuation.

A Bragg cell can be used to switch light beam directions by turning on and off the acoustic source. The intensity of the diffracted light, however, depends on the amplitude of the acoustic wave. An amplitude modulation of the acoustic wave will, therefore, produce amplitude-modulated light beams. But again the movement of the acoustic waves produces a moving diffraction grating, as a result of which the frequencies of the diffracted beams are Doppler-shifted by an amount $+/-mv_a$. This frequency shifting can be effectively manipulated to design frequency modulators. AO modulator transfer function is sinusoidally dependent on the input voltage; however, this presents no difficulty in on-and-off modulation. For analog modulation, it is necessary to bias only the modulator at a carrier frequency such that the operating point is in an approximately linear region of operation. When compared with an EO modulator that consumes voltage on the order of 10^3 V, an AO modulator requires only a couple of volts. But since the acoustic wave propagation is slow, the AO devices are often limited by the frequency response of the acoustic source, figure of merit, and acoustic attenuation. Most of the AO materials are lossy. Materials with high figures of merit normally have a high attenuation. The most commonly used AO materials are quartz, tellurium dioxide, lithium niobate, and gallium phosphide. The materials with lower figures of merit are also used, but they operate with a higher drive power. A practical limit for small devices is a drive power density of 100–500 W/mm², provided there is a proper heat sink. Bandwidths of up to 800 MHz are common in most commercial AO modulators.

AO modulators are used widely in a large number of applications, such as laser ranging, signal-processing systems, optical computing, medium-resolution high-speed optical deflectors, acoustic traveling-wave lens devices, and mode-locking. Figure 11.44 shows a system where AO modulators are used to support beam scanning of a laser printer. In the laser printers, a rotating drum with an electrostatically charged photosensitive surface (a film of cadmium sulfide or selenium on an aluminum substrate) is used so that a modulated laser beam can repeatedly scan across the rotating surface to produce an image. The most commonly used beam-scanning system utilizes a He–Ne laser, a modulator, and a rotating polygonal prism. The He–Ne laser is preferred over other lasers because the photosensitive layer is sensitive to its output. But since it is difficult to modulate a He–Ne laser internally, the modulation is done externally. We can also use an EO cell for the modulator, but an AO modulator is preferred because of its ability to operate with unpolarized light, and also because it requires a low-voltage power supply.

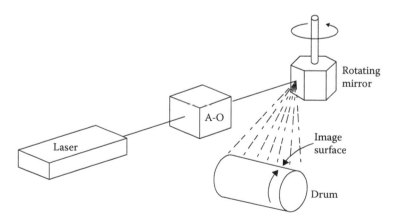

FIGURE 11.44 A beam-scanning laser printer.

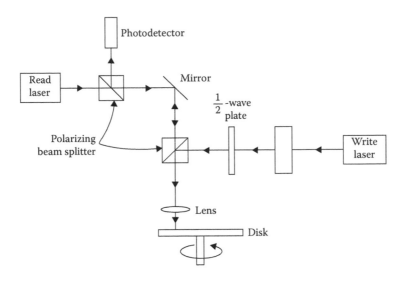

FIGURE 11.45   Direct-read-after-write optical disk system.

AO modulators are also used in systems involving optical disks. Quite like their audio counterparts, optical disks store information in optical tracks. But in the case of optical disks, there is neither a groove nor a continuous line present, but rather "pits," which are small areas providing an optical contrast with respect to their surroundings. The varied reflectance along the track represents the information stored. These disks are versatile in the sense that they can be used for both direct read-out and recording. Figure 11.45 shows one such direct-read-after-write optical disk system. The write laser usually has more power than the read laser. The more sophisticated systems are arranged to have angular and polarization separation of beam to ensure that the read beam does not interfere with the reflections of the write beam.

11.9 Summary

This chapter provides only a brief introduction and discussion of the basic role of photoconduction, photodetection, and electro-optic and acousto-optic modulation. Some of the more significant applications of these concepts have contributed to various detection, imaging, amplification, modulation, and signal-processing systems. which have also been discussed in this chapter. For details on many of the systems considered in this chapter, as well as for elaboration on other variations, one will need to use other reference sources and public-domain publications.

Suggested Reading

1. Banerjee PP, Poon TC. *Principles of Applied Optics*, Boston, MA: Irwin, 1991.
2. Berg NJ, Lee JN. *Acousto-Optic Signal Processing*, New York, NY: Marcel Dekker, 1983.
3. Karim MA. *Electro-Optical Devices and Systems*, Boston, MA: PWS-Kent Pub, 1990.
4. Karim MA, Awwal AAS. *Optical Computing: An Introduction*, New York, NY: John Wiley, 1992.
5. Lizuka K. *Engineering Optics*, New York, NY: Springer-Verlag, 1983.
6. Pollock CR. *Fundamentals of Optoelectronics*, Boston, MA: Irwin, 1995.
7. Tannas LE Jr. ed., *Flat-Panel Displays and CRTs*, New York, NY: Van Nostrand Reinhold, 1985.
8. Wilson J, Hawkes JFB. *Optoelectronics: An Introduction*, Englewood Cliffs, NJ: Prentice-Hall International, 1985.
9. Yu FTS. *Optical Information Processing*, New York, NY: John Wiley, 1983.
10. Yu FTS, Khoo IC. *Principles of Optical Engineering*, New York, NY: John Wiley, 1990.

12

Radiometry

12.1	Introductory Concepts	459
12.2	Standard Terminology for Radiometric Quantities	462
	Power • Exitance and Incidance • Radiance • Intensity	
12.3	Photon-Based Radiometric Quantities	466
12.4	Photometric Quantities and Units	468
12.5	Radiative Power Transfer in Three-Dimensional Space	474
	Geometrical Factors • Radiative Power • Sources and Collectors • Inverse Power Square Law • Power Transfer Equation for the Incremental Surface Areas • Power Transfer Equation in Integral Form	
12.6	Lambert's Law	489
	Directional Radiator • Lambertian Radiator • Relationship between the Radiance and Exitance for a Lambertian Radiator	
12.7	Power Transfer across a Boundary	496
	Specular and Diffuse Surface • BRDF • Surface or Fresnel Losses • Radiation Propagating in a Medium • Radiation Scattered in a Medium • Radiation Absorbed in a Medium • Emissivity • External and Internal Transmittance	
12.8	Optical Signal	507
	Imaging and Non-Imaging Systems • Signal Collection • Optical Noise and Stray Light • Signal-to-Noise Ratio • Detection of Radiation and Signal Conditioning • Radiometric Measurements	
12.9	Summary	514
	Acknowledgment	514
	References	514

12.1 Introductory Concepts

Radiometry refers to the measurement of radiation. The word radiation is closely associated with radio waves. These are the waves that Hertz generated in 1887, when electromagnetic waves were first generated in a laboratory using electronics. Based on the earlier theoretical work by Maxwell, we know today that the radiation of different wavelengths spans the whole electromagnetic spectrum; see Figure 12.1 and Table 12.1. Within the broad area of optical engineering, we are interested in the ultraviolet (UV), visible, infrared (IR), millimeter waves. Many of the concepts apply equally well to the X-ray region, which has recently been gaining in importance. Historically, humans have more narrowly focused their interest on the visible wavelength region, the portion of the electromagnetic spectrum where the human eyes are sensitive. This has resulted in the development of a special branch of radiometry, tailored to the human eyes as detectors, referred to as illumination engineering, together with the development of specific units related to human vision.

As we learned that radiation existed outside the visible region, we found it advantageous to apply a uniform terminology for the quantities related to the optical radiation and to the units applicable to

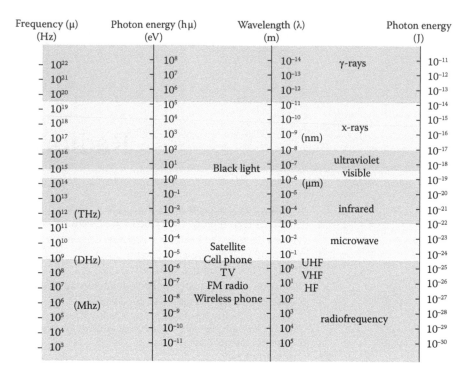

FIGURE 12.1 The electromagnetic spectrum.

the radiation of all wavelengths. Thus, we prefer to use the terminology for the optical radiation, that is based on power (the standard terminology in physics) for the shorter wavelengths, and that based on frequency (or wavelength), used traditionally in electrical engineering and infrared, for the longer wavelengths. We refer to those as radiometric quantities and units. The MKS units have been widely accepted in the international communities, based on the proposition that we can measure the radiative power at specific wavelengths or frequencies in watts [W]. When the visible radiation is considered just a portion of the electromagnetic spectrum, the radiometric terminology and units may also be applied. However, when referring to the effect of the visible radiation on the particular detector of the *human visual system*, the *photometric* terminology and units are preferred. They are discussed in Section 12.4.

The advantages of using the concept of power are obvious when we think of the radiation as the carrier of information from one point in space to a different point in space. The information may be coded in terms of the wavelength, or the spectral content, or, even better, in terms of the amount of the radiation at the specific wavelength that is being transferred. This quantity is the spectral radiative power.

While the amount of radiation may be characterized by its power, the wavelength region defines its spectral characteristics: we are familiar with the rainbow produced by water drops or dispersion generated with a glass prism. The concept of wavelength has been used advantageously in the infrared, or the long-wavelength region of the electromagnetic spectrum because this radiation has very low energy. The simplest way of visualizing the wavelength is by considering the generation of the radiation with an oscillating dipole, as in radio transmission. A wavelength is that distance in space for which the electromagnetic radiation has the same phase and the same algebraic sign of its derivative. For the radio waves, we actually prefer to use the concept of frequency, measuring the number of waves that pass a given position in space per unit time:

$$\nu = \frac{c}{\lambda}, \quad [\text{Hz}] \tag{12.1}$$

TABLE 12.1 The Electromagnetic Spectrum, Its Natural Sources, Detection, and Manmade Generators

Natural Sources	Detection	Artificial Sources	Wavelength (λ) [m]		Photon Energy [J]
Atomic nuclei	Scintillation and Geiger counters	Acelerators	10^{-14}	γ-rays	10^{-11}
			10^{-13}		10^{-12}
Inner electrons	Ionation Chamber Photographic film	X-ray tubes	10^{-12}	x-rays	10^{-13}
			10^{-11}		10^{-14}
			10^{-10}		10^{-15}
Outer electrons	Photomultiplier GaN photodetector Eye	Synchrotons Hg lamps LEDs	10^{-9}	ultraviolet visible	10^{-16}
			10^{-8}		10^{-17}
			10^{-7}		10^{-18}
			10^{-6}		10^{-19}
Molecular vibration	Bolometer HgCdTe photovoltaic Thermopile	Hot bodies	10^{-5}	infrared	10^{-20}
			10^{-4}		10^{-21}
Molecular rotation			10^{-3}		10^{-22}
			10^{-2}		10^{-23}
Electron spin	Resonant electronic circuits	Magnetron	10^{-1}	microwave	10^{-24}
Nuclear spin		Electronic resonator	10^{0}	radiofrequency	10^{-25}
			10^{1}		10^{-26}
		AC power lines	10^{2}		10^{-27}
			10^{3}		10^{-28}
			10^{4}		10^{-29}
			10^{5}		10^{-30}

(Vertical label spanning wavelength column: Independent lamp)

where c is speed of the radiation in [m/s], in vacuum = 2.9972×10^8 m/s, λ is the wavelength in [m], or usually in [μm] for convenience, and v is frequency in [s^{-1}] or [Hz]. We use a square bracket to denote the units. In radiometry, a consistent use of units helps avoid confusion due to varied, and oftentimes, inconsistent, terminology.

In the shorter wavelength regions, we are familiar with the effects of UV radiation, which is powerful enough to produce changes in material physical characteristics, ranging from burning the human skin to causing glass to darken. In the short-wavelength region, we characterize the radiation by energy carried by the smallest packet of such radiation, called a photon. Its energy is

$$Q_q = hv = \frac{hc}{\lambda_q}, \quad [J] \tag{12.2}$$

where Q is energy [J] and h is Planck's constant (= $6.6266176 \times 10^{-34}$ Js).

There are two ways of looking at the matter. In the first one, referred to as the microscopic point of view, we consider the atoms as individual objects, even if there are many of them. The second one, called the macroscopic view, deals with the collective behavior of very many atoms together, in an assembly.

Radiation may be viewed similarly. When we think of *the power of the radiation,* or the radiative power, we can define it macroscopically as the power of the radiation of a certain wavelength, or of a certain frequency. When we assume this view, we do not have to accept the concept of the minimum energy carried by a single photon. This point of view is most widely accepted in optical *engineering* applications. It deals with the type of power that is defined in mechanical and electrical engineering. It is the power that may be used to generate work; it is the power that generates heat when not used efficiently; it is the power that is the energy expanded per unit time. This concept of power is used in Section 12.2 to define the radiometric concepts,

$$P = \frac{dQ}{dt} \cdot [W],$$

(12.3)

where P is the power in [W], Q is energy in [J], t is the time in [s], and d(quantity A)/d(quantity B) = derivative of quantity A with respect to quantity B in units of [A]/[B].

Another concept of power may be introduced when we consider the quantum nature of light, with the number of quanta carrying energy per unit time to produce the "photon" power P_q. This power, and the quantities associated with it, are presented in Section 12.3. To distinguish them from the macroscopic, the more commonly used, quantities we denote them with the subscript q; that is,

$$P_q = \frac{dn}{dt}, \quad [\#/s]$$

(12.4)

where P_q is the photon power, the power transferred by N photons per unit time, and n = the number of photons in [#].

12.2 Standard Terminology for Radiometric Quantities

Unfortunately, standard terminology for radiation quantities does not exist. Every field and just about every application has found its particular nomenclature and symbols. Here, we adopt the terminology recommended by Nicodemus at the National Institute of Standards and Technology (NIST) [1–3]. Table 12.2 includes a list of the commonly used radiometric terms.

12.2.1 Power

We consider that the primary characteristic of the radiation is to propagate in free space: it cannot stand still. Due to its "motion" or "propagation," it is often referred to as the radiant flux, emphasizing the transfer of energy past some imaginary surface in space. When it encounters a material surface, it is either reflected or absorbed by it. We can say that the electromagnetic radiation carries the optical energy across space in time—*it transfers energy with the speed of light*. The amount of energy transferred per unit time is defined as power, given in Equation 12.3. An additional term that may be found in the literature, describing the fleeting nature of radiation, includes the radiative power.

The terms derived from the concept of power are related to the geometrical considerations and the spectral content of the radiative power. We refer to the spectral power as the power of wavelength λ found in the narrow wavelength interval $\delta\lambda$:

$$\delta P(\lambda) = P(\lambda)\delta\lambda, \quad [W]$$

(12.5)

where $\delta P(\lambda)$ is the infinitesimal element of power of the wavelength λ in the wavelength interval $\delta\lambda$ in [W], $\delta\lambda$ is the width of the infinitesimal wavelength interval in [µm], and $P_\lambda(\lambda)$ is the spectral power of wavelength λ in [W/µm].

We denote spectral quantities with the subscript λ, explicitly indicating a derivative with respect to the wavelength λ. The laser is one popular source of coherent, narrow wavelength-band radiation.

TABLE 12.2 List of Commonly Used Radiometric Terms

Spectral Term	Symbol	Defining Equation	Units
Radiant energy	U_λ		$J\ \mu m^{-1}$
Spectral radiant power	P_λ	$P_\lambda = \dfrac{dU_\lambda}{dt}$	$J(\mu m\ s)^{-1} = W\ \mu m^{-1}$
Spectral radiant exitance	M_λ	$M_\lambda = \dfrac{\partial P_\lambda}{\partial A_p}$	$W\ (m^2\ \mu m)^{-1}$
Spectral radiant incidance	E_λ	$E_\lambda = \dfrac{\partial P_\lambda}{\partial A_p}$	$W\ (m^2\ \mu m)^{-1}$
Spectral radiant intensity	I_λ	$I_\lambda = \dfrac{\partial P_\lambda}{\partial \omega}$	$W\ (sr\ \mu m)^{-1}$
Spectral radiance	L_λ	$L_\lambda = \dfrac{\partial^2 P_\lambda}{\partial A_p\ \partial \omega}$	$W\ (m^2\ sr\ \mu m)^{-1}$
Solid angle	ω	$d\omega = \dfrac{dA_p}{r^2}$	sr
Frequency	ν	$\nu = t^{-1}$	$s^{-1} = Hz$
Wavelength	λ	$\lambda = \dfrac{c}{\upsilon}$	μm (or m)
Wavenumber	m	$m = \lambda^{-1}$	m^{-1}

Equation 12.5 describes well the quasi-monochromatic power output of a laser source, due to its power output in the narrow wavelength range. The majority of sources, though, emit the radiation in a wide wavelength interval. For those, the spectral power is defined as follows:

$$P_\lambda(\lambda) = \frac{dP(\lambda)}{d\lambda}\ \left[W/\mu m\right] \tag{12.6}$$

Most of the natural sources, and, therefore, most of the naturally occurring radiation are broadband. Additionally, the electro-optical systems used to transfer the radiation do so effectively only when the radiative signal is present within two wavelengths λ_1 and λ_2, or within a wavelength interval $[\lambda_1, \lambda_2]$,

$$\Delta\lambda = \lambda_2 - \lambda_1.\ \left[\mu m\right] \tag{12.7}$$

The radiative power within a wavelength interval depends not only on the width of the wavelength interval $\Delta\lambda$, but also on specific choice of the wavelength limits λ_1 and λ_2. The radiative power within a wavelength interval is obtained by integrating Equation 12.6 over wavelength, from wavelength λ_1 to wavelength λ_2,

$$W_{[\lambda 1, \lambda 2]} = \int_{\lambda_1}^{\lambda_2} P(\lambda)\,d\lambda.\ [W] \tag{12.8}$$

We find it advantageous to display the wavelength subscripts explicitly for the wavelength interval to remind us that the integrated power (or any other radiometric quantity) is considered, evaluated, or integrated over a given wavelength interval. We refer to such power as the power in the wavelength interval $[\lambda_1, \lambda_2]$.

12.2.2 Exitance and Incidance

Matter, whose spatial extent is defined by its surfaces, generates the radiation. Matter may similarly absorb the radiation. The creation and destruction of radiation requires the presence of matter. The radiative power may be generated within the boundaries defined and limited by surfaces. Similarly, it may be incident on a surface.

The surface characteristics usually vary from point to point. Independently whether the surfaces are absorbing or emitting the radiation, the surface properties vary as a function of position. They generate or absorb a different amount of radiation depending on the surface spatial coordinates. We can thus define the power (area) density as the infinitesimal amount of power incident on the infinitesimal amount of area δA:

$$\delta P = p\delta A, \quad [W] \tag{12.9}$$

where δP in [W] is the infinitesimal radiative power incident on the infinitesimal area δA, p is the radiative power (area) density in [W/m²], and δA is the infinitesimal area in [m²].

Here we distinguish between radiative power (area) density and radiative power (volume) density. The volume power density is the relevant quantity when discussing the radiation in a cavity or the radiation generated by the volume, or bulk, radiators. From Equation 12.9, we define the radiative power (area) density as follows:

$$p = \frac{dP}{dA} \cdot \left[\frac{W}{m^2}\right] \tag{12.10}$$

This radiative quantity is often called irradiance, and denoted by E with units [W/m²]. In radiometry, we do not use this term, due to the possibility of ambiguities.

When we consider a specific spectral component of the radiative power, Equation 12.10 may be modified so that it refers to the spectral power area density

$$p_\lambda = \frac{dP_\lambda}{dA} \cdot \left[\frac{W}{\left(m^2 \mu m\right)}\right] \tag{12.11}$$

In optical engineering applications, the radiation leaving the surface (or, *exiting* the surface) is usually different from the radiation incident on the surface. For example, the spatial characteristics of radiation may be different when we use optical components to collect or redistribute the radiation. The radiation leaves the natural, unpolished surface dispersed in all directions. The radiation may be incident on the surface within only a narrow cone specified by the F-number of the optical system. This is a preferred design configuration when the surface is the detecting element in a radiometer.

Let us consider a beam of radiation incident on a specific surface. A portion of it is reflected from the surface, a portion is transmitted, and a portion is absorbed, depending on the material characteristics. The surface spectral reflectivity modifies the amount of reflected radiation, and, thus, its spectral distribution. The surface shape and its finish modify the directional characteristics of the reflected and transmitted radiation. For these reasons, the term irradiance is considered ambiguous within the radiometric community.

The nature of the surface shape, finish, and composition changes the spectral and directional characteristics of the reflected or transmitted radiation. As the surface modifies the radiation, we find it preferable to differentiate between the radiation before it is incident on the surface and after it is incident on and reflected off the surface. The power (area) density for the radiation incident on the surface is called the incidance, written with "a", and is denoted by E in [W/m²]:

$$E(x,y,z) = \frac{dP}{dA} \cdot \left[\frac{W}{m^2} \right] \tag{12.12}$$

The spectral incidance is similarly defined for each spectral component of radiation:

$$E_\lambda(x,y,z) = \frac{dP_\lambda}{dA} \cdot \left[\frac{W}{(m^2 \mu m)} \right] \tag{12.13}$$

The radiation leaving the surface, either generated by the matter or reflected from it, is described by the term exitance, written with "a", and is denoted by $M(x,y,z)$ with units [W/m²]:

$$M(x,y,z) = \frac{dP}{dA} \cdot \left[\frac{W}{m^2} \right] \tag{12.14}$$

The spectral exitance $M_\lambda(x, y, z)$ is defined for each spectral component of radiation,

$$M_\lambda(x,y,z) = \frac{dP_\lambda}{dA} \cdot \left[\frac{W}{(m^2 \mu m)} \right] \tag{12.15}$$

12.2.3 Radiance

The propagation of the radiation may be guided or manipulated using one or more optical components, resulting in a change in power density from one point to the next along the propagation path. The imaging components, such as lenses and mirrors, change the angular distribution of the radiation in order to modify the spatial extent and direction of propagation of radiation. Thus, the incidance is not a suitable parameter to characterize the radiation in an optical system.

The objects that generate radiation are called sources. They typically emit radiation with an angular dependence that is a consequence of the source physical characteristics, including shape, form, layout, and construction. Thus, the directional characteristics of emitted radiation limit the source capacity to produce useable radiative power. Additionally, the sources typically do not generate the power uniformly over their surface area, due to the surface non-uniformity.

Thus, a quantity that is suitable to characterize spatial and angular radiative properties of an extended (surface) area source is required. It is called the radiance, L in [W/(m² sr)]:

$$L(x,y,z,\theta,\varphi) = \frac{\partial^2 P}{(\partial A_p \, \partial \omega)}, \quad \left[\frac{W}{(m^2 sr)} \right] \tag{12.16}$$

where ω is the solid angle in steradians, [sr], and A_p is projected area in [m²]; $\partial A / \partial B$ denotes a partial derivative of quantity A with respect to quantity B. The use of partial derivatives is required when function depends on several variables.

The source radiance may also be given as a spectral quantity when the spectral power is being transmitted:

$$L_\lambda(x,y,z,\theta,\varphi) = \frac{\partial^2 P_\lambda}{(\partial A_p \, \partial \omega)} \cdot \left[\frac{W}{(m^2 sr \mu m)} \right] \tag{12.17}$$

Because of the directional nature of the emitted radiation from the extended area sources, the spectral radiance is the quantity used to describe the radiative characteristics of such a source.

In an optical system, the beam cross-section often decreases at the same time as the solid angle increases, resulting in simultaneous changes in area and solid angle in an optical beam. Thus, the radiance is also the most appropriate radiometric quantity to characterize the transfer of radiative power through an optical system.

12.2.4 Intensity

The radiation may propagate through the free space, but it cannot be generated or annihilated (absorbed) in free space. Matter is required for these two phenomena to take place. In particular, when talking about a radiative source, the establishment of energy levels in the matter is indispensable, so that its energy levels may be changed in order to generate the radiation. A material body is always of finite dimensions. As already mentioned in the previous section, the radiance is used to describe the spatial and directional characteristics of the radiation emitted from an extended-area source.

Often, though, we find it convenient to talk about point sources. While they do not exist in the physical world, we define the point source as the source whose spatial extent is not resolved by the resolution of the optical system used to image it. For this reason, the concept of the point source depends on the sensor resolution, rather than on the source spatial extent. We may consider stars as point sources, even though they are tremendously large in size. The human visual system resolves only the nearest star, the sun. Therefore, our sun is usually not considered a point source, while all the other stars are.

Even though the spatial characteristics of the area distribution of the radiation emitted by a point source may not be resolved with the sensor, the directional characteristics of its radiation are of uttermost importance. There may be another reason why we may want to talk about point sources – we are not interested in their spatial characteristics. This may be because we are far away from the source, because its spatial characteristics are within the uniformity tolerances of our application, or because the system analysis is significantly simplified when the source may be treated as a point. This is often the simplifying assumption when analyzing laser beam propagation: we define an imaginary point source such that the laser beam appears to originate from it.

We do not consider that the spatial distribution of the power emitted by the point source is important; however, its angular distribution is significant. Thus, we define the concept of intensity, using the symbol I [W/sr], the power per unit solid angle:

$$I(\theta,\varphi) = \frac{dP}{d\omega} \cdot \left[\frac{W}{sr}\right] \tag{12.18}$$

Often, the quantity of interest is the spectral intensity $I_\lambda(\theta, \varphi)$

$$I_\lambda(\theta,\varphi) = \frac{dP_\lambda}{d\omega} \cdot \left[\frac{W}{(sr\mu m)}\right] \tag{12.19}$$

The angular dependence of the power radiated by the point source into a solid angle is given by the spectral intensity. The intensity is the most appropriate quantity to characterize a point source. A radiator is said to be isotropic if its intensity is independent of direction.

12.3 Photon-Based Radiometric Quantities

There are two ways of looking at radiation. They were perceived mutually exclusive prior to the acceptance of the quantum theory at the beginning of the twentieth century; now, they are considered complementary. Radiation may be considered as a continuous flow of waves that carry energy (see Equation 12.1). The alternative, equally useful way of looking at radiation is that this same energy is carried in quanta, each containing a minimum packet of energy equal to the change between specific (electronic) energy states in an atom, molecule, or a solid. These changes between the energy levels, E_1 and E_2, are also referred to as energy transitions We use symbol "E" for energy levels in accordance to the standard terminology in physics.

$$\nabla E_q = E_1 - E_2 = h\nu = \frac{hc}{\lambda_q}, \quad [J] \tag{12.20}$$

where E_1 and E_2 are the two energy levels involved in the transition. Radiation may be considered to consist of packets and waves, both carrying the energy. This all-encompassing theory is referred to as the dual nature of the radiation. The smallest packet of radiation with energy ΔE_q is called a photon.

Instead of describing the radiation in terms of the energy transferred across an imaginary surface per unit time, or power, we can take advantage of the quantum representation to emphasize its other aspects. Sometimes, it is more appropriate or convenient to count the number of photons that pass an imaginary surface per unit time. The photon-based quantities are defined similarly to the power-based terms. We use a subscript q to distinguish the photon-based quantities from the power-based quantities. When it is necessary to emphasize the power-based quantities, the subscript e is used to indicate the energy or power-based units. Often, though, the subscript e is omitted for the power-based quantities because it is the usual one.

The photon flux P_q is the number of photons crossing an imaginary surface per unit time:

$$P_q = \frac{dn_q}{dt}. \quad \left[\frac{\#}{s}\right] \tag{12.21}$$

Here, dn_q/dt is the number of photons per unit time. It may be evaluated by considering that the power transferred is equal in both representations:

$$P = P_q(h\nu) = \left(\frac{dn_q}{dt}\right)(h\nu) = \left(\frac{dn_q}{dt}\right)\left(\frac{hc}{\lambda}\right). \quad [W] \tag{12.22}$$

To obtain the last two equalities, we used Equation 12.4. From here we obtain for the photon flux or photon power,

$$P_q = \frac{dn_q}{dt} = \frac{P}{(h\nu)} = P\left(\frac{\lambda}{hc}\right). \quad \left[\frac{\#}{s}\right] \tag{12.23}$$

The spectral photon flux is the power that includes only the photons of the same energy, or those with the same wavelength:

$$P_{q,\lambda} = \frac{dn_{q,\lambda}}{dt} = \frac{P_\lambda}{(h\nu)} = P_\lambda\left(\frac{\lambda}{hc}\right). \tag{12.24}$$

The photon incidence E_q is defined similarly:

$$E_q = \left(\frac{\partial}{\partial A}\right)\left(\frac{dn_q}{dt}\right) = \frac{E}{(h\nu)} = E\left(\frac{\lambda}{hc}\right). \quad \left[\frac{\#}{(s\,m^2)}\right] \tag{12.25}$$

The photon spectral incidance is, correspondingly,

$$E_{q,\lambda} = \left(\frac{\partial}{\partial A} \right) \left(\frac{dn_{q,\lambda}}{dt} \right) = \frac{E_\lambda}{(hv)} = E_\lambda \left(\frac{\lambda}{hc} \right) \cdot \left[\frac{\#}{\left(sm^2 \mu m \right)} \right] \tag{12.26}$$

Photon exitance M_q is defined as

$$M_q = \left(\frac{\partial}{\partial A} \right) \left(\frac{dn_q}{dt} \right) = \frac{M}{(hv)} = \frac{M}{(\lambda / hc)} \cdot \left[\frac{\#}{\left(sm^2 \right)} \right] \tag{12.27}$$

The photon spectral exitance $M_{q,\lambda}$ is given by

$$M_{q,\lambda} = \left(\frac{\partial}{\partial A} \right) \left(\frac{dn_{q,\lambda}}{dt} \right) = \frac{M_\lambda}{(hv)} = M_\lambda \left(\frac{\lambda}{hc} \right) \cdot \left[\frac{\#}{\left(sm^2 \mu m \right)} \right] \tag{12.28}$$

Additionally, the photon intensity $I_q(\theta,\varphi)$ may be defined as shown next.

$$I_q(\theta,\varphi) = \frac{dP_q}{d\omega} \cdot \left[\frac{\#}{sr} \right] \tag{12.29}$$

Likewise, photon spectral exitance $I_{q,\lambda}(\theta, \varphi)$ is correspondingly,

$$I_{q,\lambda}(\theta,\varphi) = \frac{dP_{q,\lambda}}{d\omega} \cdot \left[\frac{\#}{(s\ srmm)} \right] \tag{12.30}$$

In the photon-based formalism, the photon radiance is also the most appropriate quantity to characterize an extended-area source,

$$L_q(x, y, z; \theta, \varphi) = \frac{\partial^2 P_q}{(\partial A\ \partial \omega)} \cdot \left[\frac{\#}{(s\ m^2 sr)} \right] \tag{12.31}$$

Similarly, the photon spectral radiance $L_{q,\lambda}(x,y,z;\theta,\varphi)$ may be defined in the same way:

$$L_{q,\lambda}(x, y, z; \theta, \varphi) = \frac{\partial^2 P_{q,\lambda}}{(\partial A\ \partial \omega)} \cdot \left[\frac{\#}{(s\ m^2 sr\ \mu m)} \right] \tag{12.32}$$

This quantity is used often when analyzing sources that emit low amounts of radiation. This requires that individual photons be counted. It is important as the limiting noise in the background-limited photon detectors and quantum noise [4,5].

12.4 Photometric Quantities and Units

Photometric quantities, with their corresponding units, have been developed to deal with the use of a specific detector, that is, the human eye [6–8]. They incorporate the response of the human eye to the incident radiation. Thus, they are defined to deal with the radiation only within the wavelength interval where the human eye is sensitive. For a typical observer, this wavelength region is from 0.380 μm to 0.680 μm. Small responsibility has been reported up to 0.780 μm .

Photometric quantities and units have been developed to describe the sensation that a standard observer reports in response to the radiation of 1 W at a specific wavelength. Table 12.3 lists the basic photometric quantities, their defining equations, and units. In this system of units, therefore, there is no radiation if the average human observer cannot detect it.

Figure 12.2 displays the nominal eye response of a standard observer to a constant amount of radiation of 1 W, as a function of wavelength, for both photopic (the $K(\lambda)$ curve) and the scotopic (the $K'(\lambda)$ curve) vision. Photopic vision refers to vision under conditions of high illumination, such as daylight,

TABLE 12.3 Basic Photometric Quantities, Their Defining Equations, and Units

Term	Symbol	Defining Equation	Units
Radiative power (luminous flux)	P_v	$P_v = K_m \int_{0.380\mu m}^{0.760\mu m} P_\lambda V(\lambda)d\lambda \dfrac{dP}{dA}$	lm
Luminous intensity	I_v	$I_v = \dfrac{dP_v}{d\omega}$	lm/sr, cd
Luminance (luminous radiance)	L_v	$L_v = \dfrac{\partial^2 P_v}{\partial A_p \partial \omega}$	lm/(m² sr), cd/m²
Illuminance	E_v	$E_v = \dfrac{dP_v}{dA}$	lm/m²
Luminous exitance	M_v	$M_v = \dfrac{dP_v}{dA}$	lm/m²
Luminous efficacy function	$K(\lambda)$	$K(\lambda) = K_m V(\lambda)$	

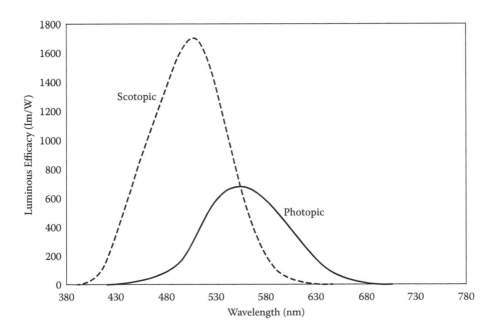

FIGURE 12.2 Spectral luminous efficacy functions $K(\lambda)$ and $K'(\lambda)$ for photopic and scotopic vision.

and includes color vision. Scotopic vision does not permit color recognition and is effective under conditions of low illumination.

The $K(\lambda)$ curve is also called the luminous efficacy, as it describes the ability, or the efficiency, of a human eye to sense the incident radiation of different wavelengths in the visible range. Specifically, the meaning of this curve is as follows: the response of a human observer to the constant monochromatic radiation is solicited as the wavelength of the incident radiation is varied.

The standard human observer reports the sensation of much more radiation when the wavelength of the incident light is 0.55 μm than when the wavelength is only 0.4 μm, for example, even though the incident (physical) radiative power is the same. Specifically, the observers were able to quantify their perceptions that the radiation of one given wavelength is sensed as twice as intense as radiation of another wavelength, even though the incident radiation had the same (physical) power in both cases. Thus, the response curve changes as a function of the wavelength. There is no response to the radiation outside the interval [0.380–0.760 μm]. Consequently, photometric units may not be used to characterize the radiation outside this interval; that is, any radiation or related quantity outside the visible interval expressed in photometric units has a value of zero.

The efficacy curves $K(\lambda)$ normalized to 1 at their peak responsivity are referred to as the relative spectral luminous efficiency for the CIE-standard photometric observer. They are designated as $V(\lambda)$ and $V'(\lambda)$ curves, respectively, and are shown in Figure 12.3. At the wavelength of 0.555 μm, were theh $V(\lambda)$ curve attains a peak, 1 W of (physical) radiative power equals 683 lumens. At other wavelengths, the curve provides the corrective factor by which the source spectral power in watts needs to be multiplied to obtain the source luminous power in lumens.

More precisely, the luminous power P_v in lumens [lm] is the radiative power detected by a standard human observer:

$$P_v = K_m \int_{\lambda_1=0.380\mu m}^{\lambda_2=0.760\mu m} P_\lambda V(\lambda) d\lambda, \quad [\text{lm}] \tag{12.33}$$

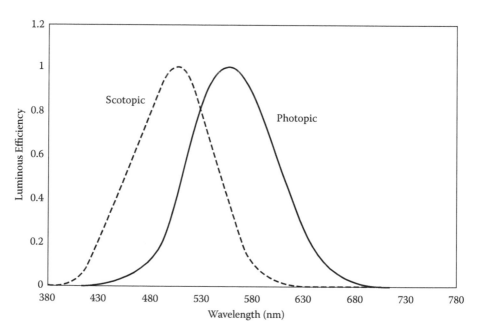

FIGURE 12.3 Normalized spectral luminous efficiency $V(\lambda)$ and $V(\lambda)$ for the standard photometric observer, for scotopic and photopic vision.

where P_v is the luminous power in lumens [lm] or talbots per second [talbot/s], K_m is the maximum luminous efficacy, equal to 683 lm/W, $V(\lambda)$ is the relative spectral luminous efficacy for the CIE-standard photometric observer. The tabulated values are given in Table 12.4; and $\lambda_1 = 0.380$ μm, $\lambda_2 = 0.760$ μm define the wavelength interval within which $V(\lambda)$ and $V'(\lambda)$ have non-zero values. The subscript v is used

TABLE 12.4 Luminous Efficiency Functions $V(\lambda)$ and $V'(\lambda)$ for Photopic and Scotopic Vision

Wavelength λ (nm)	Photopic $V(\lambda)$	Scotopic $V'(\lambda)$
380	0.00004	0.000589
390	0.00012	0.002209
400	0.0004	0.00929
410	0.0012	0.03484
420	0.0040	0.0966
430	0.0116	0.1998
440	0.023	0.3281
450	0.038	0.455
460	0.060	0.567
470	0.091	0.676
480	0.139	0.793
490	0.208	0.904
500	0.323	0.982
510	0.503	0.997
520	0.710	0.935
530	0.862	0.811
540	0.954	0.650
550	0.995	0.481
560	0.995	0.3288
570	0.952	0.2076
580	0.870	0.1212
590	0.757	0.0655
600	0.631	0.03315
610	0.503	0.01593
620	0.381	0.00737
630	0.265	0.003335
640	0.175	0.001497
650	0.107	0.000677
660	0.061	0.0003129
670	0.032	0.0001480
680	0.017	0.0000715
690	0.0082	0.00003533
700	0.0041	0.00001780
710	0.0021	0.00000914
720	0.00105	0.00000478
730	0.00052	0.000002546
740	0.00025	0.000001379
750	0.00012	0.000000760
760	0.00006	0.000000425
770	0.00003	0.0000002413
780	0.000015	0.0000001390

to designate the quantities related to the human visual system and for dealing with the illumination (the radiation as sensed by the human eye). The source luminous power depends on its spectral distribution, weighted with respect to that of the spectral luminous efficacy.

The luminous energy ε_v in lumen-second [lm s] (also called talbot) is the radiative energy detected by the standard observer:

$$\varepsilon_v = K_m \int_{\lambda_1=0.380\mu m}^{\lambda_2=0.760\mu m} Q_\lambda V(\lambda)d\lambda, \quad [\text{lms}]. \tag{12.34}$$

The luminous energy is related to the luminous power in the same way as energy is related to power. Luminous energy in lumen-second is luminous power integrated over time.

$$\varepsilon_v = \int P_v dt. \quad [\text{lms}] \tag{12.35}$$

Luminous power (volume) density $p_{v,V}$ in lumens per cubic meter [lm/m³] is luminous power per unit volume. It is also called luminous flux (volume) density.

$$p_{v,V} = \frac{dP_v}{dV} \left[\frac{\text{lm}}{\text{m}^3}\right] \tag{12.36}$$

Here V is the volume. The luminous power (area) density or luminous irradiance I_V is the luminous power per unit area:

$$I_V = \frac{dP_v}{dA} \cdot \left[\frac{\text{lm}}{\text{m}^2}\right], \quad [\text{lux}] \tag{12.37}$$

There are two units in use for the luminous power density; lumen per square meter [lm/m²] is also called lux. For smaller areas, we use lumen per square centimeter [lm/cm²], also called phot.

Luminous exitance M_v is luminous power (area) density for the luminous radiation per unit area leaving the surface:

$$M_v = \frac{dP_v}{dA} \cdot \left[\frac{\text{lm}}{\text{cm}^2}\right], \quad [\text{phot}] \tag{12.38}$$

Luminous exitance is measured in lumens per square meter [lm/m²]. A preferred unit is 1 phot [phot], equal to 1 lumen per square centimeter, [lm/cm²], [phot].

Illuminance E_v is the luminous power (area) density or luminous incidance for the luminous radiation incident per unit surface area:

$$E_v = \frac{dP_v}{dA} \cdot \left[\frac{\text{lm}}{\text{ft}^2}\right], \quad [ft-c]. \tag{12.39}$$

The established unit for the illuminance is lumen per square foot, also known as foot-candle [ft-c]. Taking into consideration that the definition assumes the human eye as a detecting element, the illuminance measures the same physical quantity as luminous irradiance, Equation 12.37.

Luminous intensity I_v is the luminous power per unit solid angle:

$$I_v = \frac{dP_v}{d\omega} \cdot \left[\frac{\text{lm}}{\text{sr}}\right] \tag{12.40}$$

Its units are lumens per steradian [lm/sr], commonly known as candela [cdla].

Luminance, luminous radiance, or photometric brightness L_v is the luminous radiation, either incident or exiting, per unit area per unit solid angle:

$$L_v = \frac{\partial^2 P_v}{(\partial A_p \, \partial \omega)} \cdot \left[\frac{\text{lm}}{(m^2 \text{sr})}\right] \tag{12.41}$$

The most common unit for the luminance is lumen per square meter per steradian [lm/(m² sr)], also called nit. This unit is also referred to as candela per square meter [cdla/m²], considering that lumen per steradian is a candela. A smaller unit is candela per square centimeter [cdla/cm²], also known as stilb [stlb].

An additional photometric unit for characterizing the luminance is a foot-lambert [ft-lb], generally applicable to describing diffuse materials (those that behave like Lambertian reflectors and radiators). One foot-lambert is equal to $(1/\pi)$ candela per square foot. A smaller unit is a lambert, which is equal to $(1/\pi)$ candela per square centimeter.

Here, we note once again that a great number of units have evolved over time in various branches of photometric studies. Table 12.5 lists the conversion factors between illuminance and luminance.

Now that we have established terminology used in diverse branches of radiometric endeavors, we can use it to present important laws that govern the propagation of radiation. Without loss of applicability to other systems of units and names, we employ physics-based radiometric quantities.

TABLE 12.5 Conversion Factors for Different Units Used for Illuminance and Luminance

1. Illuminance Conversion Factors

	Lux	Phot	Foot-Candle	Lumen
1 lux	1	10^{-4}	9.290×10^{-2}	$= 1\text{lm/m}^2$
1 phot	10^4	1	9.290×10^{-2}	$= 1\text{lm/cm}^2$
1 foot-candle	10.76	1.076×10^{-3}	1	$= 1\text{lm/ft}^2$

2. Luminance Conversion Factors

	Nit	Stilb	Apostilb	Lambert	Foot-Lambert	Candle/ft²	Candle/in²	Candle
1 nit	1	10^{-4}	3.142	3.142×10^{-4}	2.919×10^{-1}	9.290×10^{-2}	6.452×10^{-4}	cd/m²
1 stilb	10^4	1	3.142×10^4	4.142	2.919×10^3	9.290×10^2	6.452	cd/cm²
1 apostilb	0.3253	3.253×10^{-5}	1	10^{-4}	9.290×10^{-2}	2.957×10^{-2}	2.054×10^{-4}	$(1/\pi)$ cd/m²
1 lambert	3253	3.253×10^{-1}	10^4	1	9.290×10^2	2.957×10^2	2.054	$(1/\pi)$ cd/cm²
1 foot-lambert	3.426	3.426×10^{-4}	10.76	1.076×10^{-3}	1	0.3253	2.210×10^{-3}	$(1/\pi)$ cd/ft²
1 candle/ft²	1.076	1.076×10^{-3}	33.82	0.3382	3.142	1	6.944×10^{-3}	cd/ft²
1 candle/in²	1550	1.550×10^{-1}	4869	0.4868	4.524×10^2	1.44×10^2	1	cd/in²

12.5 Radiative Power Transfer in Three-Dimensional Space

12.5.1 Geometrical Factors

Proper functioning of the radiation source, as well as detectors, requires the presence of matter. A material object is limited by its surfaces: therefore, it is desirable to define the geometrical characteristics of the radiation-emitting and radiation-absorbing surfaces. Toward this goal, we first review some basic, but important geometrical concepts.

We start with the distinction between a surface and an area. A geometrical surface is an assembly of points in a three-dimensional (3D) space with points in contact with one another. The points separate two regions with different physical properties. In optics, materials are differentiated by their complex indices of refraction that, in general, are spectrally dependent. The real part of the complex index of refraction is the index of refraction as used in the optical design applications; its complex part is related to the absorption coefficient. Figure 12.4a shows a surface separating two regions of space.

Sometimes, we find it convenient to further extend this rather theoretical description of a surface by defining it as a continuous assembly of points separating two regions of space. Therefore, there exists no requirement that the surface separate two regions of space having different physical characteristics. An area, on the other hand, is a purely geometrical property of a surface, characterized by its orientation or position.

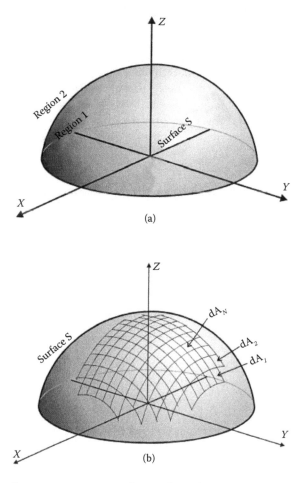

FIGURE 12.4 (a) Surface S separates two regions of space. (b) Surface S, separating two regions of space, is composed of small area elements.

Let us now consider a special case presented in Figure 12.4a: region 1 of space is filled with (made of) opaque material, raised to some temperature above that of its background. Consequently, the surface emits radiation into the region of space according to Planck's radiation law; thus, it is a radiative source. Region 2 is a vacuum, devoid of any material presence. The light emitted from the source propagates in a straight line through the vacuum, until it encounters an obstruction in the form of a physical object.

12.5.1.1 Area and Projected Area

12.5.1.1.1 Area

A surface of an arbitrary shape may be thought of as being covered by small, infinitesimal elements of area inclined with respect to one another (see Figure 12.4b). One of the intrinsic characteristics of a surface is its total area. It is obtained by adding up all the infinitesimal elements of the area that encompass the surface, i.e., by performing the integration over the surface. The total, actual area is obviously larger than the *apparent* surface area, seen from a particular direction.

In radiometry, we are in fact more interested in how an area appears from a certain direction, or from a particular view point, rather than in its exact, absolute size. When an element of area serves as a radiation collector, the intrinsic characteristics of its surface area are less significant than how large this surface "appears" to the incoming radiation. Similarly, the actual area of a radiative source is not as important as the "projected" area, seen from the direction of the collector.

We consider the geometrical configuration of Figure 12.5, showing a differential element of area ΔA in space and an arbitrary point **P**. The line connecting the center of the area and the point **P** is referred to as the line of sight between the area and the point **P**. We characterize the area by its unit normal **N**. (Vectors are denoted with a bold letter.)

The line directed from the center of the area to the point of observation **P** is referred to as the direction of observation, characterized by a unit vector **S**. It may be considered as defining the z-axis of a hypothetical spherical coordinate system, erected at the center of the area ΔA. Thus, θ is the angle between the direction of observation and the normal to the area ΔA, **N**. The cosine of the angle θ then is the dot or the scalar product of the unit normal to the area **N** and the direction of the line of sight **S**:

$$\cos \theta = \mathbf{N} \cdot \mathbf{S}, \quad \left[\text{unitless} \right] \tag{12.42}$$

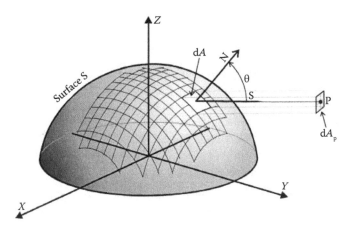

FIGURE 12.5   Area and projected area, as seen from the point of observation P.

12.5.1.1.2 Projected Area

We define the projected area ΔA_p as the projection of the area ΔA on the plane normal to vector **S**.

$$\Delta A_p = \Delta A \ \cos\theta. \ \left[\text{m}^2\right] \tag{12.43}$$

In the plane normal to vector **S**, an area of size ΔA_p will appear to the observer at point P as having the same size as the area ΔA. The plane that is normal to the unit vector S through the point P on the area ΔA is referred to as a projection plane. The area ΔA projected in this plane is called the projected area ΔA_p. The degree of its decrease in magnitude depends only on the orientation of the area ΔA with respect to the direction of observation. When we substitute Equation 12.42 into Equation 12.43, we obtain

$$\Delta A_p = \Delta A \mathbf{N} \cdot \mathbf{S}. \ \left[\text{m}^2\right] \tag{12.44}$$

This equation applies equally well to infinitesimal areas. The infinitesimal area is that assembly of points that all lie in the same plane, that is, the same unit normal may characterize them all:

$$dA_p = dA \ \cos\theta_A = dA\mathbf{N} \cdot \mathbf{S}. \ \left[\text{m}^2\right] \tag{12.45}$$

In this equation, the subscript A on θ has been included explicitly to emphasize that each differential area has its own normal (see Figure 12.4a). If a surface consists of many elements of area with different orientations, then the projected area is an integral over the individual projected surfaces. Each infinitesimal element of the area contributes an infinitesimal projected area to the total projected area. The total projected area is obtained by integrating over all infinitesimal projected surface areas:

$$A_p = \int_{\text{Surface}} dA_p. \left[\text{m}^2\right] \tag{12.46}$$

Substituting the expression for the infinitesimal projected area from Equation 12.44, we obtain

$$A_p = \int_{\text{Surface}} dA \cos\theta_A = \int_{\text{Surface}} \mathbf{N}_A \cdot \mathbf{S} dA. \ [\text{m}^2] \tag{12.47}$$

Here, we show explicitly that the angle θ depends on the orientation of each individual element of area dA and that, in general, this angle varies with position. Next, some familiar examples of projected areas are presented for a few well-known surfaces: the projected area of a cone is a triangle or a circle, depending on the direction of observation; that of a sphere is a circle; that of an ellipsoid of revolution is an ellipse or a circle; that of a cube is a square or a rectangle. There exist more complicated projections of these geometrical entities when the direction of observation does not coincide with one of the principal axes.

The difference between the area and the projected area is important for angles θ larger than 10°. In many radiometric applications, the angle θ is quite small. The approximation of replacing the projected area with the area is acceptable for a vast majority of cases. The error in using the area instead of the projected area is less than 5% for values of angle θ equal to or less than 10°.

12.5.1.2 Solid Angle and the Projected Solid Angle

The solid angle is an even more important concept in radiometry than the projected area, because it describes real objects in 3D space. It is an extension into three dimensions of a two-dimensional angle. For this reason, it will help us to appreciate the idea of a solid angle if we first review the formal definition of an angle.

12.5.1.2.1 *Angle*

Let us refer to Figure 12.6. We define the angle with the help of a point P located at the center of curvature of a unit circle and an arc $\underline{AB}$ on this circle. Angle α, measured in radians, is equal to the length of the arc on the unit circle with the center of the curvature at the point P; that is,

$$\alpha = AB =\sim s_u, \quad [\text{rad}] \tag{12.48}$$

where $\underline{AB}$ is the length of the arc. The subscript u refers to the unit radius of the circle: s_u is the length of a curve, along the arc on the unit circle. In a more general case, when the distance PA is not unitary, but rather it has a general value r, the angle α may be interpreted as the following ratio:

$$\alpha = \frac{AB}{r}. \quad [\text{rad}] \tag{12.49}$$

The infinitesimal angle $d\alpha$ is even more easily interpreted with the help of Figure 12.7. When the angle $d\alpha$ is infinitesimally small, the corresponding arc ds approaches a straight line dd, which is parallel to the tangent at the midpoint of the arc. Then, the tangent may replace the length of the curve at this point:

$$d\alpha = \frac{dd}{r}. \quad [\text{rad}] \tag{12.50}$$

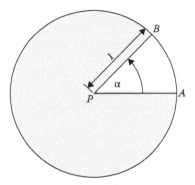

FIGURE 12.6 Definition of an angle in radians as a length on a unitary circle.

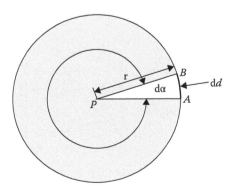

FIGURE 12.7 Definition of a differential angle in radians as a length on a unitary circle.

For the differential quantities, we have a similar relationship:

$$\Delta\alpha = \frac{\Delta d}{r}. \; [\text{rad}] \tag{12.51}$$

The unit for the angle is radian [rad]. In practice, it is defined as follows: a circle of unit radius subtends an angle of 2π radians. Similarly, we know that a circle of unit radius subtends an angle of 360 degrees [°] or [deg]. Thus, we obtain

$$1\,\text{rad} = \left(\frac{180}{\pi}\right)\text{deg}, \tag{12.52}$$

and

$$1\text{deg} = \left(\frac{\pi}{180}\right)\text{rad}. \tag{12.53}$$

Referring back to Figure 12.7, we reiterate that the arc $\underline{AB}$ subtends an angle α at the point P. The point of observation **P** may be thought of as coinciding with the center of curvature of the circle. Also, the arc or the element of path along the arc Δs is the distance that the angle subtends at the point of observation P. Figure 12.8 shows a number of arcs or (incomplete) circles with different radii of curvature, r, all centered at P. The radius r_1 is smaller than the unitary radius r_2, while the radii r_3 and r_4 are larger than it. We note that the arcs A_1B_1, A_2B_2, A_3B_3, A_4B_4 all subtend the same angle α. Thus, we conclude correctly that the two definitions presented above are completely equivalent.

We further see that the segment d_2, the straight-line segment connecting point A_2 with B_2, subtends the same angle as the arc $\underline{A_2B_2}$. Thus, for all practical purposes, we can replace the arc length with the arc segment in the functional definition of the angle. So, we obtain a new definition for an angle:

$$\alpha = A_2B_2 = A_uB_u \; \left[\text{rad}\right], \tag{12.54}$$

where $\underline{A_uB_u}$ is the distance along the arc on a circle with a unitary radius. For the arc $\underline{AB}$ on a circle with an arbitrary radius r, we obtain

$$\alpha = \frac{AB}{r}. \; \left[\text{rad}\right] \tag{12.55}$$

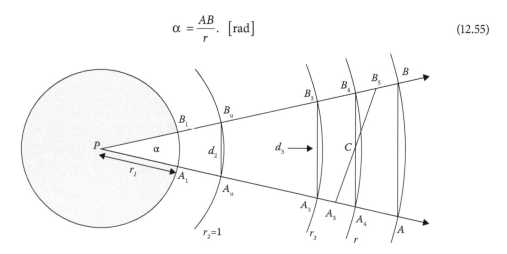

FIGURE 12.8   Generalized definition of a solid angle as an area on a sphere of arbitrary radius.

Line segments may be described according to their construction as tangents at the midpoint of the arc $\underline{AB}$. The tangent to an arc is normal to the radius at the contact point. Similarly, the line segment C is normal to the line connecting its midpoint with the point of observation **P**. The point of observation **P** defines the location of the vertex of the angle.

We now examine in more detail the general line segment $\underline{A_s B_s}$, skewed or inclined with respect to the segment $A_4 B_4$, but intersecting it at its midpoint C. We note that the line segment $\underline{A_s B_s}$ subtends the same angle α at the point **P** as the line segment $\underline{A_4 B_4}$, parallel to the tangent at the midpoint C of the arc $\underline{A_4 B_4}$.

We may generalize this observation as follows: any arbitrary curve segment whose endpoints lie on lines PA and PB, respectively, and whose midpoint C defines the distance to the origin CP equal to radius r, may be projected on the tangent at C with the same projected length AB. All such curve segments subtend the same angle α, given as

$$\alpha = AB/r. \quad [\text{rad}] \tag{12.56}$$

A segment whose endpoints lie on lines PA and PB, respectively, and whose midpoint C defines a distance to the origin CP equal to radius r, may be projected on the tangent at C, having the same length, according to

$$d = AB = A_s B_s \cos\theta_s. \quad [\text{rad}] \tag{12.57}$$

We now see that an angle may be defined using only two geometrical quantities – a point and a line segment oriented in an arbitrary, but specified direction, as illustrated in Figure 12.9. These two quantities in space determine a single plane in which the angle is defined. The angle α is given as follows:

$$\alpha = \frac{A_s B_s \cos\theta_s}{\text{PC}}. \tag{12.58}$$

This, indeed, is the definition of an angle that we will find most useful in the interpretation of a solid angle in three dimensions, a basic concept in radiometric analysis. By examining Equation 12.57 and Figure 12.9, we conclude that the general definition of an angle does not call for the circle or its radius.

Often, the geometrical definitions in the 3D space are simplified in the vector notation. A line segment d_s defines an angle at the point **P** at a distance r, equal to

$$\alpha = (\mathbf{N}_s \cdot \mathbf{S} d_s)/r. \quad [\text{rad}] \tag{12.59}$$

We employed two unitary vectors in this definition: $\mathbf{N}_d$ is a unitary vector, normal to the line segment d_s, that defines a line segment in a plane; $\mathbf{S}$ is a unitary vector directed from the central point on the line segment to the point of observation **P**.

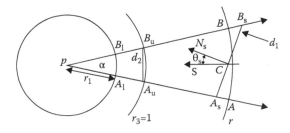

FIGURE 12.9 Definition of an angle using a line, its normal, and the point of observation.

12.5.1.2.2 Field of View

The field of view of an optical instrument is usually given as an angle. Its numerical value may be given as half of the apex angle of a cone whose base is the width (diagonal) of the focal plane array and height is the focal distance. The cone may be inscribed inside a square field of view subtended by a (square) focal plane array. It is sometimes given as a half-angle of the cone circumscribed outside the square field of view subtended by a (square) focal plane array. One half of the length of the array diagonal then defines the half field of view. These two options are illustrated in Figure 12.10. In both cases, the assumption is generally made that the optical system is rotationally symmetric, and that the field of view may be defined in a single (meridional) plane.

The casual use of the term "field of view" often leaves out "half" when referring to this concept.

12.5.1.2.3 Solid Angle

The solid angle ω is an angle in a 3D space. It is subtended by an area at a distance r from a point of observation P, in the same manner as an oriented line at distance r subtends an angle α in a plane.

The unit of a solid angle is a steradian [sr]. A (full) sphere subtends 4π steradians. An infinitesimal area dA at a distance r from the origin subtends an infinitesimal solid angle $d\omega$ equal to the size of the projection of this small area dA on a unit sphere. This is illustrated in Figure 12.11:

$$d\omega = \frac{dA}{r^2}. \quad [\text{sr}] \tag{12.60}$$

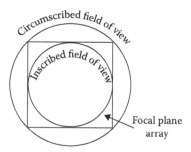

FIGURE 12.10   Inscribed and circumscribed field of view of an optical instrument with respect to a square focal plane array.

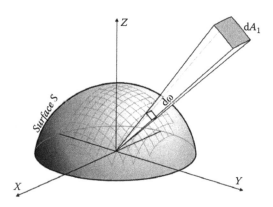

FIGURE 12.11   Differential solid angle subtended by an element of area dA at a distance r from the point of observation P.

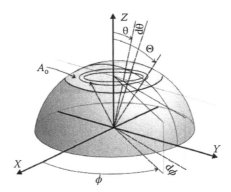

FIGURE 12.12 Optical surface of area A_o subtends a circular cone with a half-angle (apex angle) Θ at the image.

The concept of a solid angle in 3D space is similar to that of an angle in two dimensions. The definition of a solid angle is given correctly only in its differential form. The infinitesimal element of a solid angle may also be given in spherical coordinates:

$$d\omega = \sin\theta d\theta d\phi. \; [\text{sr}] \tag{12.61}$$

The factor r^2 in the numerator and the denominator has been canceled in the derivation leading to Equation 12.61.

We are most often interested in the solid angle subtended by an optical component, such as a lens, or a mirror, at a particular point along the optical axis as, for example, at the image (or object) location. The solid angle is completely characterized by the area that an optical element subtends at a particular point of observation. It may be found by integrating Equation 12.61 over the appropriate limits. In an optical system with cylindrical symmetry, an optical element with the area A subtends a circular cone with the apex half-angle Θ at the image. We orient the z-axis along the cone axis of symmetry, as illustrated in Figure 12.12:

$$\omega = \int_0^{2\pi} \left[\int_0^{\Theta} \sin\theta d\theta \right] d\phi = 2\pi(1-\cos\Theta). \; [\text{sr}] \tag{12.62}$$

In an optical system without an axis of symmetry, the solid angle must be found by numerical integration.

12.5.1.2.4 Projected Solid Angle

The projected solid angle Ω is neither a physical nor a geometrical concept. Rather, it is a convenient mathematical abstraction: it appears in many radiometric expressions, even though it has no physical significance. It may best be understood by extending the analogy between the area and the projected area to the concept of the solid angle: the projected solid angle is related to the solid angle in the same manner as the projected area is related to the area. A factor of $\cos\theta$, or obliquity factor, has to multiply an area to convert it to the "projected" area. The same obliquity factor, $\cos\theta$, has to multiply the solid angle to obtain the projected solid angle. It is visualized in the following manner, as illustrated in Figure 12.13.

First, we consider the solid angle as being equal to the area on a unit sphere, in units of steradians. Then, we recall that a projected area depends on the direction of observation from the point P, defining the angle θ. The projected solid angle is smaller in comparison with the solid angle by the factor of $\cos\theta$, the angle between the normal to the surface and the direction of observation.

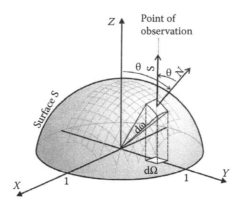

FIGURE 12.13 A projected solid angle as an area on a unit hemisphere projected on the base of the hemisphere.

The element of the projected solid angle $d\Omega$ is given in spherical coordinates:

$$d\Omega = \cos\theta d\omega = \sin\theta\cos\theta d\theta d\phi. \quad [sr] \tag{12.63}$$

The projected solid angle Ω for a circular disc subtending the apex semi-angle of Θ, shown in Figure 12.12, is obtained by integrating Equation 12.63 over angles:

$$\Omega = \int_0^{2\pi}\left[\int_0^{\Theta}\sin\theta\cos\theta\, d\theta\right]d\phi = \pi\sin^2\Theta. \quad [sr] \tag{12.64}$$

The numerical value of the projected solid angle is equal to the value of the solid angle with less than 10% error for the values of angle Θ equal to or less than $10°$. The unit of the projected solid angle is obviously the same as that of the solid angle, steradians, [sr].

A projected solid angle may be visualized as the projection of the area on a unit hemisphere onto the base of the hemisphere. With this interpretation, we can easily appreciate that the solid angle ω, subtended by the hemisphere, is equal to 2π, the area of a half-sphere. This solid angle corresponds to the apex semi-angle Θ of $\pi/2$. However, the projected solid angle, subtended by the hemisphere, is equal to only π, the area of the base of the hemisphere (the area of the circle of unitary radius) It may be obtained from Equation 12.64 by letting Θ be equal to $\pi/2$.

12.5.2 Radiative Power

Radiation carries energy from one point in space to another: being nothing other than a portion of the electromagnetic spectrum, it travels rapidly across space, transferring the energy with "the speed of light." The only time when the radiation does not move, it is either being generated or absorbed, which is done instantaneously for all practical considerations. The amount of radiation emitted by a specific surface element and the direction of the radiation propagation depends on the source's geometrical characteristics.

12.5.3 Sources and Collectors

Radiation is generated within a specific region of space. After having been created, it propagates somewhere else. A radiative source is a physical body of finite extent, occupying a specific region of space that generates the radiative energy. From each source point, the radiation propagates in all directions not obstructed by (another) physical body, in accordance with the source's radiance – the amount of radiative power emitted per unit source area per unit solid angle. In a homogeneous, transparent medium, the radiation propagates in a straight line. (In this context, we are considering the vacuum, the absence of material, a special case of homogeneous transparent medium.) The radiation may be absorbed, reflected (scattered) or, most commonly both, after it has encountered another physical body or particle. At this instant, the radiation no longer propagates in a straight line.

The physical parameters that determine the source spectral distribution and the amount of the emitted radiation are the source material, its composition, its finish, and its temperature. The source surface geometry additionally determines the directional characteristics of the emitted radiation.

The area that intercepts and collects the radiation is referred to as a collector. It is not to be confused with a detector. A detector is a transducer that changes the absorbed radiation into some other physical parameter, amenable to further processing. This includes, for example, the electrical current, voltage, polarization, displacement, and so on.

We consider two small area elements ΔA_s and ΔA_c, each with its arbitrary orientation in space, whose centers are separated by a distance d, shown in Figure 12.14. We may think of the element of area ΔA_s to be a part of the source with the area A_s, and the element of area ΔA_c to be a part of the collector with the area A_c, illustrated in Figure 12.15. We place the source and the collector into their respective coordinate systems. The element of area of the source is located at coordinates $S(x_s, y_s, z_s)$ and the element of area of the collector at the point $C(x_c, y_c, z_c)$.

The element of source area ΔA_s is considered so small that it is emitting the radiation whose radiance does not change as a function of position over the element of surface area ΔA_s or the direction of emission. This means that the radiance is constant over the surface element ΔA_s. Similarly, the elements of the area ΔA_s and ΔA_c are so small that all the points on the element of the area of the collector ΔA_c may be considered equidistant from all the points on the element of the area of the source ΔA_s. The orientation of the planar surface elements ΔA_s and ΔA_c, respectively, is defined by their surface normal, $\mathbf{N}_s$ and $\mathbf{N}_c$.

We connect the centers of the incremental area elements with a straight line of length d, which indeed depends on the specific choice of the area elements. The line connecting them is also called *the line of sight* from the source to the collector, or from the collector to the source, depending on the "point of view" that we wish to assume. We consider the direction of the line connecting these two points as being from S to P at the source, and as being from P to S at the collector: θ_s is the angle that the normal at the source subtends with the line of sight at the source; similarly, θ_c is defined as the angle that the normal at the collector subtends with the line of sight at the collector.

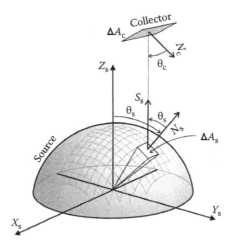

FIGURE 12.14 Geometry for the power transfer from the incremental source to the incremental image area.

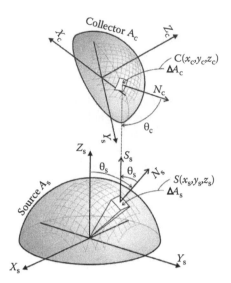

FIGURE 12.15 The (spectral) power transfer equation gives the amount of (spectral) power generated and emitted by a source of area A_s oriented in the direction N_s and intercepted by a collector of area A_c, oriented in direction N_c.

In many, but not all, power transfer applications, the distance d is sufficiently large with respect to the size of the element of the area ΔA_s that the radiation-emitting area ΔA_s may be considered a point source from the point on the collector $C(x_c, y_c, z_c)$.

12.5.4 Inverse Power Square Law

The inverse square law describes the power density (the power per unit area, also called the incidance, E) falloff as a function of radial distance from the radiating point source. This law may be easily understood if we consider the radiation emitted from a point source into a full sphere, as illustrated in Figure 12.16. By conservation of power, we see that the same amount of power passes through the sphere of radius r_1 as through the sphere of radius r_2:

$$P = \left(4\pi r_1^2\right) E_1 = \left(4\pi r_1^2\right) E_2.$$

(12.65)

Equation 12.65 is a conservation of energy, expressed in terms of power (the energy per unit time). We can solve it for the power density across the sphere of radius r_2, denoted as E_2 in Equation 12.65:

$$E_2 = E_1 \left(r_1^2 / r_2^2\right) = (E_1 r^2) / r_2^2 = 4\pi I_s / r_2^2 = P / r_2^2.$$

(12.66)

For a point source, the incidance E decreases with the distance from the source according to the inverse power square law; I_s is the intensity of the presumed point source at the center of the sphere. It was defined previously as the power per unit solid angle, the radiative quantity most appropriate for the characterization of point sources. Upon regrouping the terms in Equation 12.66, we obtain several different ways of expressing the inverse square law. The *power* in this case refers to the exponent of the sphere radius, as opposed to its normal use in radiometry.

While Equation 12.66 has been derived for simplicity for a full sphere, the same relationship may also be shown to be true for cones bounded by areas and subtending solid angles.

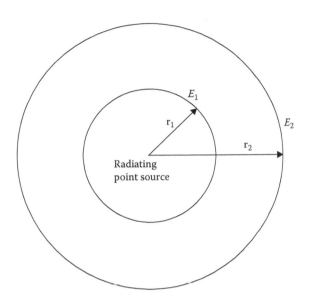

FIGURE 12.16 The radiative power passing through the sphere of radius r_1 is equal to that through the sphere of radius r_2. They both are equal to the power radiated by the point source, due to the absence of additional energy sources and sinks. This is the law of conservation of energy.

12.5.5 Power Transfer Equation for the Incremental Surface Areas

The radiance, that is, the incremental power that the source area ΔA_s emits into the incremental solid angle $\Delta\omega$, may be found by considering the definition of radiance. In this section only, we show explicitly the dependence of the radiometric quantities, radiance in particular, on the independent variables.

Referring to Figure 12.17, we define two coordinate systems to describe the source radiance. The source is shown as an extended body in a coordinate system (x_s, y_s, z_s) with an irregularly shaped boundary. The center of the incremental area ΔA_s is located at the point S with coordinates (x_s, y_s, z_s). This point forms one end of the line of sight in the direction toward the collector (x_c, y_c, z_c). The radiation is emitted from projected source area into an incremental projected solid angle $\Delta\Omega_s(\theta_s, \varphi_s)$, subtended by a collector. The subscript s on the (projected) solid angle indicates that the apex of the cone defining the solid angle is at the source. The direction of the line-of-sight forms two angles with respect to the normal to the incremental area ΔA_s, $\mathbf{N}_s$. These two angles are the angle coordinates in a spherical coordinate system that might be erected at each point $S(x_s, y_s, z_s)$ whose z-axis is along the normal to the incremental area ΔA_s.

We replace the differential quantities with the increments in the definition of radiance to obtain

$$L_\lambda(x, y; \theta, \phi) = \frac{\Delta^2 P_\lambda(x, y; \theta, \phi)}{\Delta A(x, y)\Delta\Omega(\theta, \phi)}. \tag{12.67}$$

We observe that the number of increments in the numerator is balanced with that in the denominator. We keep in mind that the radiance is defined for every point on the source $S(x_s, y_s, z_s)$ and for every direction of emission (θ_s, φ_s). Here, we note again that the subscript on the solid angle $\Omega_s(\theta_s, \varphi_s)$ refers to the point of observation, i.e., the apex of the cone that the solid angle may be seen to subtend. The spherical coordinate system (θ_s, φ_s) may be erected at any source point $S(x_s, y_s)$. From Equation 12.67, we solve for the double increment of spectral power $\Delta^2 P_\lambda$:

$$\Delta^2 P_\lambda(x, y; \theta, \phi) = L_\lambda(x, y; \theta, \phi)\Delta A(x, y)\Delta\Omega(\theta, \phi). \tag{12.68}$$

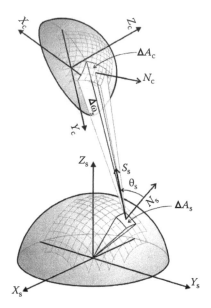

FIGURE 12.17 Incremental power transferred into the incremental solid angle $\Delta\omega_s(\theta_s, \varphi_s)$.

We refer to Figure 12.17 to obtain the full appreciation of the significance of Equation 12.68. It gives us the (doubly) incremental amount of power that an incremental area ΔA_s of the source with spectral radiance L_λ emits into the incremental, projected solid angle $\Delta\Omega_s$ in the direction of (θ_s, φ_s). The assumption is made again that the incremental projected solid angle $\Delta\Omega_s$ is so small that the radiance may be considered constant within it.

12.5.6 Power Transfer Equation in Integral Form

If the incremental source area ΔA_s and the incremental projected solid angle $\Delta\Omega_s$ are simultaneously allowed to decrease to infinitesimally small quantities, we obtain the expression for the infinitesimal amount of power emitted by the infinitesimal surface area dA_s into the infinitesimal element of a projected solid angle $d\Omega_s$. Formally, this means that all the incremental (physical) quantities are changed into the infinitesimal (mathematical) quantities:

$$d^2 P_\lambda(x, y; \theta, \phi) = L_\lambda(x, y; \theta, \phi) \, dA(x, y) \, d\Omega(\theta, \phi). \tag{12.69}$$

The infinitesimal amount of radiative power dP_λ, emitted by the total source area A_s into an infinitesimal projected solid angle, may then be obtained by integrating both sides of Equation 12.69 over the source area:

$$\left[\int_{A_s} dP_\lambda(x_s, y_s; \theta_s, \phi_s) \right] = \int_{A_s} L_\lambda(x_s, y_s; \theta_s, \phi_s) \, dA_s(x_s, y_s) \, d\Omega_s(\theta_s, \phi_s). \tag{12.70}$$

On the left side of Equation 12.70, one power of the infinitesimal symbol "d" may be integrated (out) over the source area:

$$dP_\lambda(\theta, \phi) = \int_{A_s} dP_\lambda(x, y; \theta, \phi) \quad \left[\frac{W}{\mu m} \right]. \tag{12.71}$$

The infinitesimal projected solid angle that the collector subtends at the source, $d\Omega_s\,(\theta_s, \varphi_s)$, is independent of the source Cartesian coordinates. Therefore, it is placed outside the integral. With this assumption, and using Equations 12.70–12.71 is simplified.

$$dP_\lambda(\theta,\phi) = d\Omega(\theta,\phi) \int_{A_s} L_\lambda(x,y;\theta,\phi)\,dA(x,y)\quad \left[\frac{W}{\mu m}\right]. \tag{12.72}$$

For this reason, the spectral power $P_\lambda(\theta_s, \varphi_s)$ no longer shows an explicit dependence on the source Cartesian coordinates x_s, y_s. We next integrate Equation 12.72 over a solid angle. Then, we obtain total spectral power on the left and radiance integrated over spatial coordinates of the source and angular coordinates of the collector.

$$P_\lambda = \int_{\Omega_s} \int_{A_s} L_\lambda(x,y;\theta,\phi)\,dA(x,y)\,d\Omega(\theta,\phi)\quad \left[\frac{W}{\mu m}\right]. \tag{12.73}$$

The infinitesimal amount of radiative power, emitted by the area A_s of the source with radiance L_λ into an infinitesimal solid angle $d\Omega_s$ is obtained by evaluating the integral in the parenthesis over the source coordinates. This quantity may only be found if the functional form of the source radiance is known.

Special Case 1: Radiance Independent of the Source Coordinates

We assume that the radiance is independent of the source position in order to simplify further the expression for the power transfer in Equation 12.72:

$$L_\lambda(x_s, y_s;\, \theta_s, \varphi_s) = L_\lambda(\theta_s, \varphi_s).\quad \left[W/(m^2 sr\mu m)\right]. \tag{12.74}$$

For the case that the source is characterized by the radiance that is independent of the source position, Equation 12.72 is appreciably simplified. The radiance may be placed outside the integral:

$$dP_\lambda(\theta,\phi) = L_\lambda(0,0;\theta,\phi)\,d\Omega(\theta,\phi)\int_{A_s} dA(x,y)\quad \left[\frac{W}{\mu m}\right] \tag{12.75}$$

The integral over the differential of the source area is just the source area,

$$A_s = \int_{A_s} dA_s(x_s,y_s).[m^2]. \tag{12.76}$$

The infinitesimal power dP_λ emitted into the infinitesimal projected solid angle $d\Omega_s$ by the source of area A_s, whose radiance is independent of the source coordinates, is obtained upon substitution of Equation 12.76 into Equation 12.75:

$$dP_\lambda(\theta,\phi) = L_\lambda(0,0;\theta,\phi)\,A_s\,d\Omega(\theta,\phi)\quad \left[\frac{W}{\mu m}\right]. \tag{12.77}$$

The power is still dependent on the emission direction. This brings the special case to an end.

We now return to the general expression for the radiation transfer presented in Equation 12.73. In order to find the power emitted by the source of area A_s into the projected solid angle Ω_s, we integrate both sides of Equation 12.72 over the solid angle Ω_s:

$$\int_{\Omega_s} dP_\lambda(\theta_s,\varphi_s) = \int_{\Omega_s}\left[\int_{A_s} L_\lambda(x_s,y_s;\theta_s,\varphi_s)\,dA_s(x_s,y_s)\right]d\Omega_s(\theta_s,\varphi_s).\quad \left[\frac{W}{\mu m}\right] \tag{12.78}$$

First, we comment on the left side of Equation 12.78. The integral over the infinitesimal power emitted into the projected solid angle Ω_s is just the total power emitted by the source:

$$P_\lambda = \int_{\Omega_s} d[P_\lambda(\theta_s,\varphi_s)]. \quad \left[\frac{W}{\mu m}\right]$$ (12.79)

The right side of Equation 12.78 is more difficult to evaluate because it depends on the source character- istics. The functional form of the dependence of the radiance on the set of four coordinates is, in general, not known. Therefore, we leave it in the form of the integral over the solid angle:

$$P_\lambda = \int_{\Omega_s}\left[\int_{A_s} L_\lambda(x_s,y_s;\theta_s,\varphi_s)dA_s(x_s,y_s)\right]d\Omega_s(\theta_s,\varphi_s). \quad \left[\frac{W}{\mu m}\right]$$ (12.80)

We have chosen a particular order of integration, indicated by the square bracket in Equation 12.80, although we could have chosen a different order by first integrating over the projected solid angle and then over the surface area. By examining Equation 12.80, we see that the order of the integration may be interchanged. Thus, the general equation for the power transfer is usually written without prescribing the order of integration:

$$P_\lambda = \int_{\Omega_s}\int_{A_s} L_\lambda(x_s,y_s;\theta_s,\varphi_s)dA_s(x_s,y_s)d\Omega_s(\theta_s,\varphi_s). \quad \left[\frac{W}{\mu m}\right].$$ (12.81)

This equation usually appears in a much-simplified form; it does not show explicitly that the source area is a function of the source coordinates and that the solid angle depends on the spherical coordinates erected at a source point, with the z-axis along the line of sight to the collector. Its familiar, but less explicit form is given next:

$$P_\lambda = \int_{\Omega_s}\int_{A_s} L_\lambda(x_s,y_s;\theta_s,\varphi_s)dA_s\,d\Omega_s. \quad \left[\frac{W}{\mu m}\right].$$ (12.82)

The radiance is usually given as a function of position and angle. It depends on the source Cartesian coordinates and the spherical coordinate system subtended at the specific source point. As this is implic- itly understood, the subscripts are not shown explicitly.

$$P_\lambda = \int_{\Omega_s}\int_{A_s} L_\lambda(x,y;\theta,\varphi)d\,A\,d\Omega. \quad \left[\frac{W}{\mu m}\right].$$ (12.83)

Equation 12.81 is an exact, informative, but still somewhat busy equation. As before, we try to simplify the power transfer equation, Equation 12.81, by incorporating some simplifying assumptions.

12.5.6.1 Special Case 1: Radiance Independent of the Source Coordinates

Once again, we make the assumption that the radiance is independent of the source position, given previously in Equation 12.74. We evaluate Equation 12.81 for the radiance that is independent of the source coordinates. The integral over the area of the source is just the source area, as in Equation 12.76. Then we can factor it out in Equation 12.83.

$$P_\lambda = A_s\int_{\Omega_s} L_\lambda(\theta_s,\varphi_s)d\Omega_s(\theta_s,\varphi_s). \quad \left[\frac{W}{\mu m}\right].$$ (12.84)

If we wish to further simplify the power transfer equation, we must make an additional assumption about the directional properties of radiance.

12.5.6.2 Special Case 2: Constant Angular Dependence for Radiance

We next assume that the radiance is also independent of the direction of observation, in addition to the source coordinates:

$$L_\lambda \left(x_s, y_s; \theta_s, \varphi_s \right) = L_\lambda . \ \left[\frac{W}{\left(m^2 sr \ \mu m \right)} \right] . \tag{12.85}$$

When the radiance is additionally independent of the angle coordinates, it may be placed outside the integral in Equation 12.84:

$$P_\lambda = L_\lambda A_s \int_{\Omega_s} d\Omega_s (\theta_s, \phi_s). \ \left[\frac{W}{\mu m} \right] . \tag{12.86}$$

The integral over the projected solid angle in Equation 12.86 is the projected solid angle:

$$\Omega_s = \int_{\Omega_s} d\Omega_s (\theta_s, \phi_s). \ [\text{sr}]. \tag{12.87}$$

Upon the substitution of Equation 12.87 into Equation 12.86, we obtain the simplest form of the power transfer equation, applicable to the case when the radiance is independent of the source coordinates and the direction of observation:

$$P_\lambda = L_\lambda \ A_s \ \Omega_s \quad [W \ / \ mm]. \tag{12.88}$$

For a surprisingly large number of radiometric problems, this equation produces adequate first-order results. For other applications, this equation represents an initial approximation, which may be used to estimate the order-of-magnitude results. This form is identical to the power transfer equation in the incremental form, given in Equation 12.68, but with the symbols for increments missing.

In an even greater number of radiometric problems, the distance between the source and the collector is so large that Equation 12.88 (or Equation 12.68) is given correctly for the incremental quantities. This implies that the transverse dimensions of the source and collector are appreciably smaller than their separation. Similarly, for such large separation distances, the radiance variation across the source and its angular dependence are justifiably assumed negligible in a large number of problems. Then, the power transfer equation (Equation 12.88) is applicable in its incremental form:

$$\Delta^2 P_\lambda = L_\lambda \Delta A_s \Delta \Omega_s . \ \left[\frac{W}{\mu m} \right] . \tag{12.89}$$

We may substitute alternate expressions for the solid angle to present Equation 12.89 in different forms.

12.6 Lambert's Law

12.6.1 Directional Radiator

In Section 12.2.3, we defined the radiance as the second derivative of the power with respect to both the projected solid angle and the area. The radiance depends on the Cartesian coordinates of the source and the spherical coordinates whose z-axis is along the line of sight to the collector area element.

These two coordinate systems are illustrated in Figure 12.17:

$$L\left(x_s, y_s; \theta, \phi\right) = \frac{\partial^2 P\left(x, y; \theta, \phi\right)}{\partial\Omega\, \partial A_s} \cdot \left[\frac{W}{m^2\, sr}\right] \tag{12.90}$$

We show the subscripts explicitly only for the Cartesian coordinates, indicating that we are in the plane of the source. The *spectral* radiance is indeed the most general radiometric quantity to characterize a radiative source:

$$L_\lambda\left(x_s, y_s; \theta, \phi\right) = \frac{\partial^2 P_\lambda\left(x, y; \theta, \phi\right)}{\partial\Omega\, \partial A_s} \cdot \left[\frac{W}{m^2 sr\, \mu m}\right] \tag{12.91}$$

While it is easy to understand that the radiance depends on the coordinates of the point on the source $S(x_s, y_s, z_s)$, its directional properties are a bit more difficult to visualize: the implication is that the source radiance has a different angular dependence for each source point. The direction of observation has been defined as the angle that the local normal to the surface makes with the line of sight (the line connecting the source with the collector). The angle θ, belonging to a spherical coordinate system erected at the source point $S(x_s, y_s, y_s)$, is then the angle of observation.

The angular dependence of the source radiance may be measured only with a great deal of difficulty even for just a few representative incremental source areas. In principle, we would like to know its value for all source *points*, but the point is an abstract, geometrical entity. The measurement of the radiation emission from an incremental area of the source into a specific direction toward the collector may be accomplished using two small apertures, as illustrated in Figure 12.18. The first aperture defines the size of the radiation-emitting surface area; the second aperture specifies the size of the incremental solid angle into which the radiation is being emitted.

The aperture that defines the incremental source area may actually interfere with the normal operation of the source, especially if it comes in contact with it. Even if this does not happen, the aperture may limit the solid angle into which the elemental area of the source is radiating. The important point to keep in mind here is that the very process of measurement is likely to introduce errors because of the finite size of the apertures and their unavoidable interference with the normal operation of the source.

Only a few sources fall into one of the two extreme cases of directional and nondirectional or isotropic radiators. The isotropic radiator emits the same amount of power in all directions. A point source in a homogeneous medium or a vacuum is an example of an isotropic radiator. A laser is an example of a directional source, emitting the radiation only within a very narrow angle.

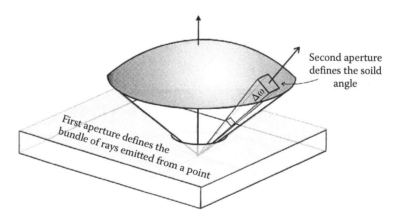

FIGURE 12.18   The experimental setup to measure the emission of radiation from an incremental area of the source.

Fortunately, most sources behave in a somewhat predictable manner that may be described sufficiently well with a few parameters for approximate radiometric analysis. The majority of natural sources tend to be adequately well described as Lambertian radiators.

12.6.2 Lambertian Radiator

Natural sources most often have directional characteristics; they emit strongly in the forward direction, that is, in the direction along the normal to the radiative surface. The amount of radiation usually decreases with increasing angle until its emission reduces to zero at the angle of 90 degrees with respect to the surface normal. This angular dependence is characteristic of the so-called Lambertian radiator. Most natural sources are similar to Lambertian radiators, even though there are very few perfect ones to be found.

A Lambertian radiator is a good example of why it is so difficult to measure radiative characteristics of a source. The finite apertures used in the measurement setup on the one hand limit the solid angle being measured, and, on the other hand, average the results due to their size. A Lambertian radiator is an extended area source characterized by a cosinusoidal dependence of the (spectral) radiance on the angle θ, according to the equation,

$$L_\lambda(x_s, y_s; \theta, \phi) = L_\lambda(\theta) = L_{\lambda 0} \cos\theta. \quad \left[\frac{W}{m^2 sr\, \mu m}\right] \tag{12.92}$$

This functional dependence is illustrated in Figure 12.19. Here $L_{\lambda 0}$ is the constant spectral radiance. So, the radiance of a Lambertian radiator is independent of the source coordinates. It has azimuthal symmetry. A Lambertian radiator may also be specified with the radiance, integrated over wavelength:

$$L(x_s, y_s; \theta, \phi) = L(\theta) = L_0 \cos\theta. \quad \left[\frac{W}{m^2 sr}\right] \tag{12.93}$$

Here, the integrated radiance L_0 is a constant. Thus, a Lambertian radiator is a source whose emission depends only on the cosine of the azimuthal angle. We may also say that a Lambertian source emits radiation according to the (Lambert's) cosine law. Such a source is also said to be a perfectly diffuse source. A diffuse source radiates in all directions within a hemisphere.

The amount of radiation emitted by incremental area ΔA_s in the direction θ is decreased by the obliquity factor $\cos\theta$, the very same factor that decreases the apparent size of the area viewed from this direction. These two factors cancel, making a spherical Lambertian lamp appears as a planar source with uniform exitance. It looks like a uniformly illuminated disc.

12.6.2.1 A Blackbody Radiator

The radiation that is established inside an evacuated cavity with completely absorbing walls at temperature T is referred to as a blackbody radiator (Figure 12.20a). It is completely isotropic, i.e., the radiation

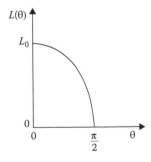

FIGURE 12.19 The functional dependence of the radiance of a Lambertian radiator.

Perfectly absorbing walls at temperature T

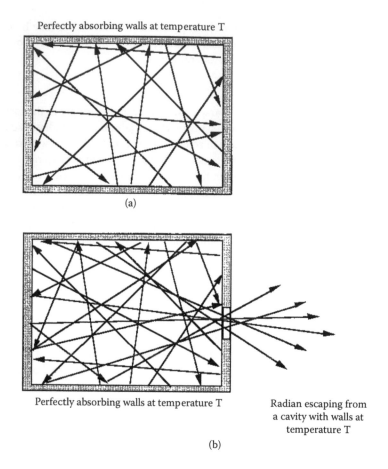

(a)

Perfectly absorbing walls at temperature T   Radian escaping from
a cavity with walls at
temperature T

(b)

FIGURE 12.20   (a) Blackbody radiation is established inside an evacuated cavity with completely absorbing walls at temperature T. (b) The radiation escaping from the cavity at equilibrium temperature T is referred to as a blackbody radiation.

has the same properties viewed from any direction. If an infinitesimally small hole of area δA_s is punched in a wall of the enclosure, as shown in Figure 12.20b, then the radiation escaping from the blackbody cavity is referred to as a blackbody radiation.

The radiation that escapes from an idealized blackbody cavity is Lambertian. The radiation that is incident on any area of the wall, including the area of the hole δA_s, is isotropic. However, the apparent size of the opening as seen by the radiation incident from the direction θ changes, according to the $\cos \theta$ obliquity factor. The reduced area has been called the projected area in the previous section. The small opening in the cavity where the isotropic blackbody radiation leaves the cavity with walls at temperature T is a Lambertian source – the projected area has an obliquity factor $\cos \theta$ when viewed from the direction θ from outside the enclosure.

12.6.2.2 Nonplanar Sources

In general, portions of a source may also lie outside the x–y plane. Nevertheless, if the source emission is given by Equation 12.93, it is considered a diffuse or a Lambertian radiator. Our sun is an example of such a Lambertian radiator: its emission surface is a sphere, with each surface element ΔA_s at the angle θ offering a projected area of $\Delta A_s \cos \theta$ (Figure 12.21). Thus, the sun appears as a disk of uniform radiance or brightness. (Brightness is the term used for the radiance when dealing with the visible radiation.)

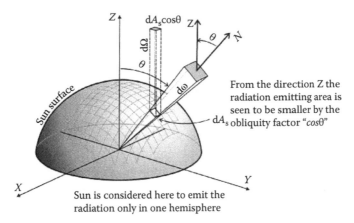

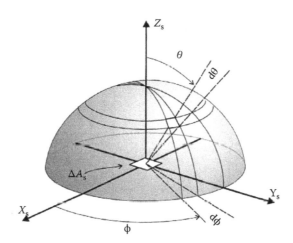

FIGURE 12.21 Our sun is an example of a Lambertian radiator: its emission surface is a sphere, with each surface element ΔA_s at the angle θ offering a projected area of $\Delta A_s \cos \theta$.

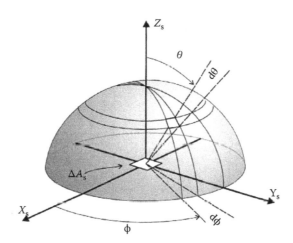

FIGURE 12.22 An increment of solid angle intercepts the radiation that a small area $\Delta A(x_s, y_s)$ on the $x_s–y_s$ plane emits as a Lambertian radiator into the hemisphere.

12.6.3 Relationship between the Radiance and Exitance for a Lambertian Radiator

The problem that we are tying to solve is the following: consider a small area $\Delta A(x_s, y_s)$ on the $x_s–y_s$ plane, radiating into a hemisphere as a Lambertian radiator. How much power does this area $\Delta A_s(x_s, y_s)$ emit into the space? Figure 12.22 illustrates this problem.

We first recall that to simplify analysis, an extended source emits the radiation only into a hemisphere. The increment of area ΔA_s is located in the $x–y$ plane; z is equal to zero, and, therefore, is omitted in the analytical development. We present the development for a general radiator. Finally, we evaluate the power emitted into the hemisphere for a Lambertian radiator, given in Equation 12.92:

$$L_\lambda\left(x_s, y_s, z_s; \theta, \phi\right) = L_\lambda\left(x_s, y_s, 0; \theta, \phi\right) = L_\lambda\left(x_s, y_s; \theta, \phi\right). \left[\frac{W}{m^2 sr\,\mu m}\right] \qquad (12.94)$$

The mathematical relationship between the radiance and exitance for an infinitesimal source area $\Delta A(x_s, y_s)$ has been found in Section 12.2.3. The (spectral) radiance of a source is its (spectral) exitance per unit solid angle:

$$L_\lambda\left(x_s, y_s; \theta, \phi\right) = \frac{dM_\lambda\left(x_s, y_s; \theta, \phi\right)}{d\omega}. \quad \left[\frac{W}{m^2 sr\ \mu m}\right] \tag{12.95}$$

Thus, the source (spectral) exitance is the (spectral) radiance integrated over the solid angle ω. From Equation 12.95, we solve for the spectral exitance dM_λ:

$$dM_\lambda\left(x_s, y_s; \theta, \phi\right) = L_\lambda\left(x_s, y_s; \theta, \phi\right) d\omega. \quad \left[\frac{W}{m^2\ \mu m}\right] \tag{12.96}$$

Another way of interpreting this problem is to emphasize its physical significance. How much power does an incremental surface area at $O(x_o, y_o)$, having Lambertian radiance, emit into an incremental solid angle subtended by the area dA on the unit sphere (equal to $d\omega$, by the definition of the solid angle)? We consider the specific point (x_o, y_o) on the source x_s–y_s plane:

$$M_\lambda\left(x_o, y_o\right) = M_\lambda\left(x_s, y_s\right). \quad \left[\frac{W}{m^2\ \mu m}\right] \tag{12.97}$$

We substitute Equations 12.97 and 12.92 into Equation 12.96:

$$dM_\lambda\left(x_s, y_s; \theta, \phi\right) = L_\lambda\left(\theta, \phi\right) d\omega. \quad \left[\frac{W}{m^2\ \mu m}\right]. \tag{12.98}$$

Then, we integrate both sides of this equation over the solid angle:

$$M_\lambda(x_o, y_o) = \int_{\text{Full hemisphere}} L_\lambda(\theta, \varphi)\, d\omega. \quad \left[\frac{W}{m^2 \mu m}\right] \tag{12.99}$$

Next, we substitute the differential angles for the differential solid angle and specify the limits of integration. First, we evaluate this integral for general angles:

$$M_\lambda(x_o, y_o) = \int_{\phi_{min}}^{\phi_{max}} \int_{\theta_{min}}^{\theta_{max}} L_\lambda(\theta, \phi) \sin\theta\, d\theta\, d\phi. \quad \left[\frac{W}{m^2\ \mu m}\right] \tag{12.100}$$

According to the illustration in Figure 12.22, we assume that the infinitesimally small radiation-emitting area δA is located in the plane of the source. A planar Lambertian radiator emits the radiation in all directions defined above the plane, into the whole hemisphere. (A source of infinitesimal area cannot emit in more than half of the hemisphere, because the infinitesimal area is assumed to be planar.)

Thus, the limits of integration in Equation 12.100 for the Lambertian radiative emitters are 0 to 2π for the coordinate ϕ and 0 to $\pi/2$ for the coordinate θ:

$$M_\lambda(x_o, y_o) = \int_0^{2\pi} \int_0^{\pi/2} L_\lambda(\theta, \phi) \sin\theta\, d\theta\, d\phi. \quad \left[\frac{W}{m^2\ \mu m}\right] \tag{12.101}$$

This integral is generally difficult to evaluate, unless the (spectral) radiance is a particularly simple function of θ and ϕ, such as the Lambertian radiator. Otherwise, we have to resort to the methods of numerical integration. There exist a number of approximations to the real radiation emitters of the form,

$$L_\lambda(\theta,\phi) = L_{\lambda o} cos^n\theta. \quad \left[\frac{W}{m^2 sr\,\mu m}\right] \tag{12.102}$$

Here n is a rational number and $L_{\lambda o}$ is a constant that depends only on wavelength. Fortunately, there are a large number of natural emitters that are well represented by this expression. The source characterized by the angular dependence given in Equation 12.97 reduces to a Lambertian radiator when $n = 1$. We substitute Equation 12.102 into Equation 12.101:

$$M_\lambda(x_o, y_o) = \int_0^{2\pi}\int_0^{\pi/2} L_{\lambda o} \cos^n\theta \sin\theta d\theta d\phi. \quad \left[\frac{W}{m^2\mu m}\right] \tag{12.103}$$

This integral is easy to integrate once we change the variable:

$$\sin\theta d\theta = -d[\cos\theta]. \tag{12.104}$$

The limits are changed correspondingly: when $\theta = 0$, $\cos\theta = 1$; when $\theta = \pi/2$, $\cos\theta = 0$. Using Equation 12.103, we get for (spectral) exitance, given in Equation 12.99,

$$M_\lambda(x_o, y_o) = L_{\lambda o}\int_0^{2\pi}\int_0^1 \cos^n\theta d[\cos\theta]d\phi. \quad \left[\frac{W}{m^2\,\mu m}\right] \tag{12.105}$$

We interchanged the limits of integration to eliminate the negative sign. With this change of variables, Equation 12.105 is easily evaluated:

$$M_\lambda\left(x_0, y_0\right) = \frac{2\pi I_{\lambda 0}}{(n+1)}. \quad \left[\frac{W}{m^2\,\mu m}\right] \tag{12.106}$$

We may assume that this relationship holds for any point x_o, y_o. For a Lambertian radiator, $n = 1$ (see Equation 12.92). Then, Equation 12.106 reduces to

$$M_\lambda\left(x_0, y_0\right) = \pi L_{\lambda 0}. \quad \left[\frac{W}{m^2\,\mu m}\right]. \tag{12.107}$$

The source coordinates are omitted in the familiar form of the relationship between the radiance and exitance for a Lambertian radiator. So, Equation 12.106 becomes

$$M_\lambda = \frac{2\pi L_{\lambda 0}}{(n+1)}. \quad \left[\frac{W}{m^2\,\mu m}\right] \tag{12.108}$$

Similarly, we obtain for a Lambertian radiator if we leave out the specific source coordinates.

$$M_\lambda = \pi L_{\lambda 0}. \quad \left[\frac{W}{m^2\,\mu m}\right] \tag{12.109}$$

This is a very significant relationship, stating that the spectral exitance of a Lambertian radiator into the full hemisphere is the spectral radiance multiplied by a factor π.

At first glance, this result may appear somewhat surprising, because the volume of the hemisphere equals 2π, rather than π. This apparent discrepancy can be understood quite easily when we remember functional dependence of a Lambertian radiator. We note that the amount of radiation is maximum for normal emission, decreasing with increasing angle θ to the point of being zero for the tangential emission of radiation ($\theta = 90°$). Thus, the exitance may be interpreted as the angle-average radiance over the whole hemisphere.

We have tried to indicate throughout this section that the relationships presented here are valid for the spectral and integrated quantities. If the relationship is valid for any wavelength, it is also valid for the sum of wavelengths, or their integrals. So, for the integrated exitance for a generalized Lambertian radiator, we obtain,

$$M = \frac{2\pi L_0}{(n+1)} \cdot \left[\frac{W}{m^2} \right] \tag{12.110}$$

Similarly, the integrated exitance is obtained for a Lambertian radiator:

$$M = \pi L_o \cdot \left[\frac{W}{m^2} \right] \tag{12.111}$$

Thus, for a Lambertian radiator the exitance is equal to π *times* the radiance. This may be contrasted with the amount of radiation emitted into a sphere by a point source of constant intensity I_o equal to $4\pi I_o$. The factor of 4 in the expression for the point source may be attributed to the different geometries of these two sources. A point source of constant intensity emits radiation uniformly in all directions into the solid angle 4π. An extended area source emits the radiation from its planar surface and experiences a decreased projected area. The projected area $\Delta A_s \cos \theta$ averaged over a hemisphere is $\Delta A_s/2$. The radiance of the point source differs from that of an extended source by a factor 2. The other factor of 2 may be explained away by the fact that 2 is the ratio of the solid angle of a sphere over that of a hemisphere. Another way of looking at this discrepancy is that a unitary hemisphere is projected on a plane as π.

12.7 Power Transfer across a Boundary

Radiometry deals with the transfer of information in the form of the electromagnetic radiation. The radiative power is transmitted from an incremental area on the source surface to the corresponding incremental area on the detector surface. This is accomplished with an optical system consisting of beam-shaping elements, such as mirrors, prisms, and gratings, whose function is to reshape and redistribute the radiation. These components are made of different materials with distinct complex indices of refraction. We refer to the boundary as that surface that separates two regions of space with different indices of refraction and absorption coefficients. Sometimes the boundary between gaseous regions is not well defined, as for example in the case of Earth's atmosphere with pressure increasing from top to the bottom of the atmosphere with thermal effects, Earth rotation, and atmospheric events forcing instantaneous and continuous redefinitions of boundary. Another noteworthy interface is the sea-atmosphere interface with *white* where the boundary is continuously changing in form and shape due to perpetual gas-water mixture re-distribution.

Oceans represent the first step in the food chain on the planet Earth; therefore, they are being monitored in the visible and near infrared part of the electromagnetic spectrum. To obtain precise bio-geo-chemical data for the whole ocean surface, in situ measurements are related to the in-orbit results for calibration [9]. Normalized nadir radiance just below the water surface is proportional to the volume scattering function, normalized to backscattering coefficient [10]. Hyper spectral imagers flying over low coastal regions detect low spatial frequency radiance noise due to sky radiance reflected off the ocean [11]. Water inherent optical properties may be used to connect angular effects in water leaving radiance [12]. Solar radiation is unpolarized when it reaches the top of the atmosphere. Scattering of the radiation from aerosol particles and water molecules results in partial polarization. Polarized radiance likewise provides data to deduce particle size and distribution in ocean waters [13].

In terms of its response to the incident radiation, a surface may be reflective, transmissive, or – most likely – both.

12.7.1 Specular and Diffuse Surface

A reflective surface may be specularly reflective, diffusive, or both. In fact, the majority of reflective surfaces are specularly reflective and diffusive at the same time. Surfaces are said to be specularly reflective when the specular component of the reflected light is the prominent one. For the diffusive surfaces the majority of the radiation is reflected uniformly into the hemisphere.

An optical surface is described as specularly reflective when it reflects the beam of incident radiation in accordance with the Snell's law of reflection. In this case, the angle that the reflected beam forms with the surface normal is equal to the angle that the incident beam makes with the surface normal. Also, the reflected beam lies in the plane defined by the incident beam and the surface normal, called the plane of incidence. An example of a reflective surface is a polished optical component, such as a plane or a curved mirror. A magnified image of a polished surface shows the actual surface lying within two bounding

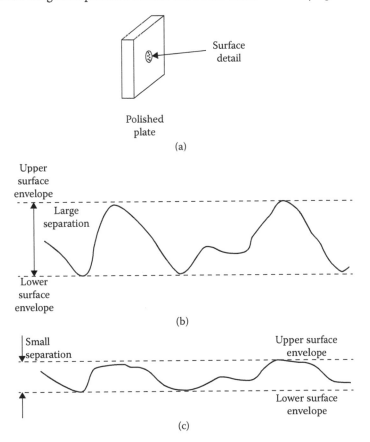

FIGURE 12.23 (a) A small surface detail on a polished plate. (b) A much magnified rough surface shows high peaks and low valleys. (c) A polished surface displays small separation between the upper surface envelope and the lower surface envelope.

surfaces that are separated by a very small distance and having approximately the same slope distrubu-tion. An unpolished surface has large deviations from the reference surface, with high peaks and deep valleys, and a wide distribution of slopes. The two types of surfaces are illustrated in Figure 12.23.

A polished surface has a prevailing amount of surface area at approximately the same height above the reference surface and has about the same slope; thus, it reflects the incident pencil of light as a collimated light beam, acting as a specular reflector. On the other hand, an unpolished or diffusing surface has a wide range of slope values and surface heights, creating a redistribution of the incoming collimated beam into all directions upon reflection. A non-specularly reflecting surface is often referred to as a scattering surface.

When the reflected radiation examined at a short distance from a surface shows no angular prefer-ence, then the surface is said to be a diffuse reflector. A perfectly diffuse reflector is also known as a Lambertian reflector, exhibiting only the cosine-θ dependence for a planar reflecting surface. This is due purely to the geometrical effects of the projected area. The radiation that appears to scatter in random directions from a rough surface nonetheless follows the laws of geometrical optics: due to the surface texture, Snell's law applies on a microscopic scale.

Light scattering is a physical phenomenon occurring when the electromagnetic radiation interacts with a boundary with rough surface features, or when it encounters a small particle or a localized charge inside the medium. In all these cases, the beam trajectory changes, no longer following the principles of geometrical optics. Thus, scattering is not just a surface phenomenon as it may take place in any medium where electromagnetic radiation may propagate: vacuum, gas, liquid, solid, including detector materials.

The angle that each scattered reflected ray makes with respect to its direction of propagation is referred to as a scattering angle. Thus, we use the term *scattering* to describe surface and volume effects, even though they are quite different phenomena. The surface scattering that is important in optical system fabrication and radiometric signal-to-noise ratio is a consequence of a reflection at micro-surfaces with random orientation. The scattering that is a volume effect occurs when a particle or charge interferes with radiation propagation and deviates it from its initial trajectory.

The optical surface is usually characterized as to its reflection characteristics because the finish of each surface impacts the imaging quality and performance of the entire optical system.

12.7.2 BRDF

The surface reflection properties may be formally described by a functional dependence that includes two angles of incidence (θ_i, ϕ_i), and two reflecting or scattering angles (θ_r, ϕ_r) that specify the direction

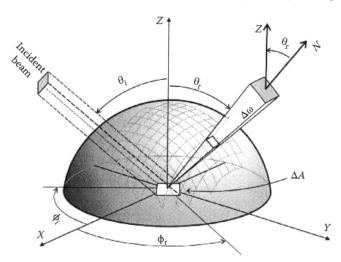

FIGURE 12.24 The reflectivity of a surface is determined by measuring the amount of the radiation incident as a collimated beam and reflected from an increment of area into an element of solid angle.

into which the light reflects or scatters (Figure 12.24). This function of four angles is called, the bidirectional reflectivity distribution function, BRDF(θ_i, ϕ_i; θ_r, ϕ_r). It is the surface reflection coefficient that relates the radiance reflected into an element of solid angle $\Delta\omega_r$ along a particular angular direction (θ_r, ϕ_r) with the beam incidence $M(\theta_i, \phi_i)$ on a small surface area ΔA being characterized:

$$BRDF\left(\theta_i, \phi_i; \theta_r, \phi_r\right) = \frac{L\left(\theta_r, \phi_r\right)\Delta\omega_r}{M\left(\theta_i, \phi_i\right)}. \quad \left[\text{unitless}\right] \tag{12.112}$$

Here, $\Delta\omega_r$ is the incremental solid angle subtended by the detecting surface. It is necessary to use radiance and incidance in the definition of the reflectance distribution function to quantify the amount of radiation exiting into a small solid angle at arbitrary direction. For a specularly reflecting surface, the BRDF is a delta function:

$$BRDF\left(\theta_i, \phi_i; \theta_r, \phi_r\right) = \delta\left(\theta_r - \theta_i, \phi_r - \phi_i\right) \qquad \text{when } \theta_r = \theta_i \text{ and } \phi_r = \phi_i$$

$$= 0 \qquad \text{otherwise} \qquad \left[\text{unitless}\right] \tag{12.113}$$

For a perfectly diffuse or a Lambertian surface, the BRDF has a cosine θ dependence,

$$BRDF\left(\theta_i, \phi_i; \theta_r, \phi_r\right) = \cos\theta_r \,\Delta\omega_r. \quad \left[\text{unitless}\right] \tag{12.114}$$

The subscript r is usually left out for diffuse surfaces. The component of the radiation that is not reflected according to the Snell's law in a reflective (polished) component is referred to as the scattered light. Fine polishing, careful handling, and storage in a dust-free environment may decrease the fraction of scattered light, though it can never be completely eliminated. It is one of the unavoidable major contributors to the stray light noise in an optical system.

The angular dependence of the radiation transmitted into the second medium may behave in one of two ways. It may obey the Snell's law for refraction if the surface is polished, as in the case of a lens or a prism; it may also be scattered into a broad cone of light if the surface is rough.

The material is said to be reflecting, when only a negligibly small amount of radiation is transmitted into it. A material is transparent if it is transmissive and the material is not absorptive. When material transmits the radiation, but it does not preserve the images, it is called translucent. With the exception of a few highly reflecting interfaces [14], a significant fraction of the radiation is usually transmitted into the second medium.

12.7.3 Surface or Fresnel Losses

When the incoming radiation passes from one transparent region with a given index of refraction n_1 to another transparent region with a different index of refraction n_2, only a small fraction of light is reflected, while the major part of the light is transmitted, as illustrated in Figure 12.25. In an optical system, the reflected light becomes stray light, degrading the signal and increasing the noise in the plane of detection [15]. An anti-reflection coating with index of refraction equal to about the geometrical mean of two neighboring materials decreases significantly the unwanted reflected radiation.

The portion of the radiative energy lost in crossing the boundary is referred to as the Fresnel loss. It depends only on the indices of refraction on each side of the interface and the beam angle of incidence. The fraction of energy reflected at a single surface separating two different regions of space, characterized by two indices of refraction n_1 and n_2, respectively, is simplified for the normally incident beam of light:

$$R_{12} = r_{12}^2 = \left(\frac{n_1 - n_2}{n_1 + n_2}\right)^2. \quad \left[\text{unitless}\right] \tag{12.115}$$

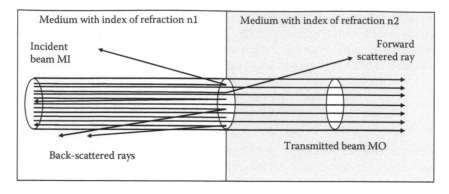

FIGURE 12.25 The surface or Fresnel loss is that fraction of incident power that is reflected back into the medium of incidence when the incident light passes from one transparent region with a given index of refraction n_1 to another one, also transparent, but with a different index of refraction n_2.

Here R is the energy (power) reflection coefficient, while r is the (electromagnetic) field reflection coefficient. The fraction of the radiative energy that remains in the main beam and is transmitted into the second medium, τ, is called the transmission coefficient. Using the conservation of energy and assuming that the surface does not absorb any energy, we get the single-surface transmission factor:

$$\tau_{12} = 1 - R_{12}. \quad [\text{unitless}] \tag{12.116}$$

Upon the substitution of Equation 12.115 into Equation 12.116, we obtain the result for the transmission coefficient in terms of indices of refraction:

$$\tau_{12} = 1 - \left(\frac{n_1 - n_2}{n_1 + n_2}\right)^2 = \frac{4 n_1 n_2}{\left(n_1 + n_2\right)^2}. \quad [\text{unitless}] \tag{12.117}$$

We note the symmetry in indices of two media, indicating that the transmission (losses) from medium 1 to medium 2 are equal to the transmission (losses) from medium 2 to medium 1:

$$\tau_{12} = \tau_{21} = \frac{4 n_1 n_2}{\left(n_1 + n_2\right)^2}. \quad [\text{unitless}] \tag{12.118}$$

The transmission losses of a complete optical system are a product of the losses at each individual surface.

$$\tau_{os} = \prod_{i=1}^{i=N-1} \tau_{i,i+1}. \quad [\text{unitless}] \tag{12.119}$$

π is a symbol for the product, while N is number of surfaces. Materials used in the IR spectral region have a high index of refraction, resulting in high transmission losses. Thus, designers in most optical systems prefer to employ reflective optical components. The reflection losses for an optical system consisting of a number of optical components made of the same material with the same index of refraction are equal to that of a single surface raised to the exponent of the number of surfaces, N.

$$\tau_{os} = \left(\tau_{i,i+1}\right)^N = \left[\frac{4 n_1 n_2}{\left(n_1 + n_2\right)^2}\right]^N. \quad [\text{unitless}] \tag{12.120}$$

When optical components with approximately the same index of refraction are employed, an average index of refraction is often assumed for the first-order analysis.

12.7.4 Radiation Propagating in a Medium

We consider again beam of light with incidance M_I incident on a medium. The amount of light transmitted into the medium M_0 is given by,

$$M_0 = \tau_{12}M_1 = (1 - R_{12})M_1 = (1 - r_{12}^2)M_I. \quad \left[\frac{W}{m^2}\right] \tag{12.121}$$

A collimated beam of light incident normally on a polished interface will continue as such even inside the medium, unless the medium is scattering, i.e., it includes particles that randomly deviate the beam direction.

Generally, the beam transmitted into the second medium M_0 consists of a component propagating in the direction of the incident beam M_T and a component scattered at the surface. The latter contributes to the stray light and is difficult to estimate beyond the Fresnel losses. At this time, computer packages are commercially available that allow models in detail of these processes. High quality optical surfaces decrease the effects of light scattering and stray light. Inside the medium, the radiation is additionally scattered and absorbed. Radiation may also scatter out of the beam in all directions, including backwards. The amount of absorbed radiation M_A represents additional losses. The scattered component M_S represents losses for the information-carrying main beam.

$$M_0 = M_T + M_S + M_A. \quad \left[\frac{W}{m^2}\right] \tag{12.122}$$

Any scattered radiation also increases the noise. These beams are illustrated in Figure 12.26. This relation expresses the conservation of radiative energy inside the medium. The radiation in the beam that has just crossed the material boundary is equal to the non-deviated, propagating beam, M_T, absorbed beam, M_A, and scattered beam, M_S. Basically, the scattered and the absorbed radiation draw their energy from the information-carrying transmitted beam, M_T.

12.7.5 Radiation Scattered in a Medium

The amount of radiation that is scattered within a medium may be described by an exponential function. We consider a beam of radiation incident on an imaginary surface at a distance z from the surface

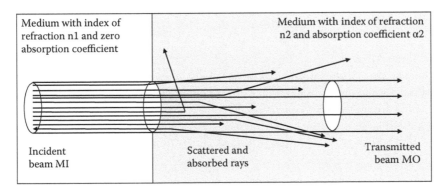

FIGURE 12.26 In a material, some rays are scattered and others are absorbed out of the main beam, resulting in a decrease in the exitance as a function of propagation distance.

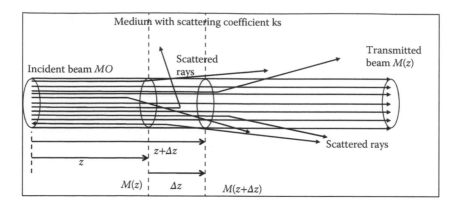

FIGURE 12.27   A beam of radiation is incident on an imaginary surface at a distance z from the surface. In the section of material of width Δz, a fraction of the radiation $\Delta M(z)$ is scattered outside the beam. This exitance is equal to the difference between the exitance at z, $M(z)$, and that at $z + \Delta z$, $M(z + \Delta z)$.

delineating the absorbing material, as illustrated in Figure 12.27. In the transverse section of material of width Δz, a fraction of the radiation $\Delta M(z)$ is scattered outside the beam. This exitance is equal to the difference between the exitance at z, $M(z)$, and that at $z + \Delta z, M(z + \Delta z)$; that is,

$$\Delta M(z) = M(z) - M(z + \Delta z). \quad \left[\frac{W}{m^2}\right] \tag{12.123}$$

The amount of radiation scattered out of the beam along the path from z to $z + \Delta z$ is proportional to the interval Δz and the exitance at z, $M(z)$. The proportionality constant is called the (linear) scattering coefficient, k_s, in $[m^{-1}]$. The negative sign indicates that the exitance remaining in the beam decreases with the propagation distance z:

$$\Delta M(z) = -k_s \Delta z M(z). \quad \left[\frac{W}{m^2}\right] \tag{12.124}$$

The volume scattering coefficient is important when considering the integrated amount of radiation scattered out of or into a 3D optical beam. In such a case, the radiation incident from all directions is evaluated. For this problem, the absorption differential equation has to be formulated in three dimensions.

Equation 12.124 is integrated after the differentials have been replaced with the infinitesimals:

$$\int_{M_0}^{M(z)} \frac{dM(z)}{M(z)} = -k_s \int_0^z dz \quad [\text{unitless}] \tag{12.125}$$

The lower limits on the definite integrals are set as follows: at $z = 0$, the exitance is intact, $M(0) = M_0$; at zero propagating distance, no light has been scattered out of the beam as yet. So, the definite integrals in Equation 12.125 are evaluated:

$$M(z) = M_0 e^{-k_s z}. \quad \left[\frac{W}{m^2}\right] \tag{12.126}$$

The scattered radiation is "lost" to the main beam as to its capacity to transfer information. It travels through the medium, until it is either absorbed by it or transmitted out of the medium. This part of the scattered radiation also contributes to the noise in the image plane. The scattered radiation is no longer associated with the image-carrying optical beam, so it reduces the signal. When the scattered radiation is finally incident on the image plane, most likely its location does not correspond to the conjugate object point. Thus, it increases the optical noise. Both of these effects reduce the signal-to-noise ratio.

Tissue is an example of a turbid medium where 3D scattering is prevalent, resulting in slow progression of beam in forward direction. This phenomenon has become known as photon migration, used in imaging, such as diffuse optical tomography, to examine a human or animal head in-vivo. Interestingly, absence of scattering medium in such case also provides significant insight about regions where a fluid might be accumulating [16]. Thermal waves likewise fail to propagate when void is encountered [17].

When there is no absorption, the scattering centers only scramble the beam, that is, redistribute the incidence spatially, so there is no geometric relation between the conjugate points of an image forming system. Such translucent material is often employed as a diffuser to increase illumination/irradiation uniformity.

The radiation absorbed within the matter similarly decreases the signal. It is considered next.

12.7.6 Radiation Absorbed in a Medium

We consider a beam of radiation incident on an imaginary surface at a distance z from the surface delineating the absorbing material, as illustrated in Figure 12.28. In material of width Δz, a fraction of the radiation incident on this imaginary surface $\Delta M(z)$ is absorbed from the beam. This exitance is equal to the difference between the exitance at z, $M(z)$, and that at $z + \Delta z$, $M(z + \Delta z)$:

$$\Delta M(z) = M(z) - M(z + \Delta z). \quad \left[\frac{W}{m^2}\right] \tag{12.127}$$

The amount of radiation absorbed from the beam along the path from z to $z + \Delta z$ is proportional to the width of the interval of propagation Δz and the exitance at z, $M(z)$. The proportionality constant is called the (linear) absorption coefficient, α, in [m^{-1}]. The negative sign indicates that the exitance remaining in the beam decreases with the propagation distance z:

$$\Delta M(z) = -\alpha \Delta z M(z). \quad \left[\frac{W}{m^2}\right] \tag{12.128}$$

Equation 12.128 is easily integrated, similarly to the case of scattering,

$$M(z) = M_0 e^{-\alpha z}. \quad \left[\frac{W}{m^2}\right] \tag{12.129}$$

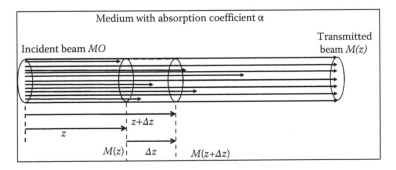

FIGURE 12.28 A beam of radiation is incident on an imaginary surface at a distance z from the surface. In the section of material of width Δz, a fraction of the radiation $\Delta M(z)$ is absorbed from the beam. This exitance is equal to the difference between the exitance at z, $M(z)$, and that at $z + \Delta z$, $M(z + \Delta z)$.

The absorbed radiation is likewise lost to the information-carrying beam. It increases the internal energy of the material, each absorbed photon ever so slightly raising its temperature. When much radiation is absorbed, temperature increase becomes measurable. This may be significant for those materials whose index of refraction, expansion or the absorption coefficient is temperature-dependent. Germanium is an example of such a material. With increased temperature, its absorption increases, resulting in further increase in absorption that results in elevated temperature, and a positive feedback results in a catastrophic event.

Some materials exhibit both absorption and scattering. It is easy to show that the exitance at the distance z in the medium $M(z)$ becomes

$$M(z) = M_0 e^{-(\alpha + k_s)z}. \quad \left[\frac{W}{m^2}\right] \tag{12.130}$$

When the absorption or scattering coefficients become appreciable, the material no longer functions as a transmissive medium.

When the thickness of material is much, much larger than $1/\alpha$, no radiation remains in the beam; that is, the radiation in a beam given in Equation 12.130 becomes very weak, approaching zero. All the radiation is scattered out of the beam or is absorbed. Due to random distribution of the absorption centers, some of the scattered radiation is absorbed and some exits the surface of the absorbing medium. This is called back-scattered radiation and provides an important signal in many applications, including non-contact skin monitoring [18]. Figure 12.29 illustrates some representative ray trajectories after a pencil of light is incident on a skin. We used a five-layer skin model to represent skin optical properties. A small portion of radiation returns outside the skin to provide spectral information about the state of blood oxygenation for remote oxymetry.

When no information-carrying radiation is traveling along the original direction of propagation, then Equation 12.122 becomes

$$M_0 = M_{BS} + M_A. \quad \left[\frac{W}{m^2}\right] \tag{12.131}$$

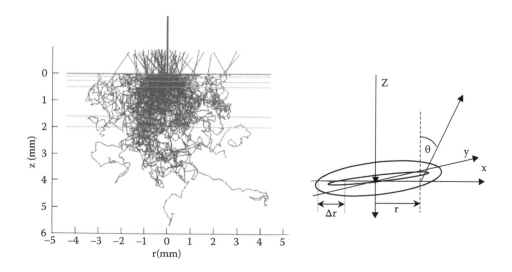

FIGURE 12.29 Monte Carlo simulation of backscattering signal reflected off a human skin (after Corral et al. 2014).

M_0 is the initial incidence in the beam; M_{BS} is the back-scattered incidence; M_A is the absorbed incidence. The back-scattered radiation returning in the direction opposite to the incident radiation may be added to Fresnel losses:

$$M_R = M_{BS} + M_{FL} \qquad \left[\frac{W}{m^2} \right]. \tag{12.132}$$

where M_{FL} is the incidence arising from the Fresnel losses. Diving with total incidence, this relationship is expressed in terms of energy reflection (R), backscattering (R_{BS}) and Fresnel loss coefficients (R_{FL}):

$$R = R_{BS} + R_{FL} \qquad [\text{unitless}]. \tag{12.133}$$

Then, all the remaining radiation is absorbed, according to the conservation of energy at the boundary:

$$A = 1 - R \qquad [\text{unitless}]. \tag{12.134}$$

A is absorptivity, in this case for a beam of light at normal incidence. It is a surface property of material. According to the Kirchhoff's law, it is also equal to emissivity. This location is used as reference in remote sensing.

12.7.7 Emissivity

Depending on the geometry of the interface, it is also possible to define hemispherical absorptivity. This is the fraction of the radiation absorbed when a unit area on the interface is irradiated from a solid angle equal to a half-sphere. Such condition corresponds to the whole sky irradiating a point on a flat surface on the Earth, for example in the middle of the Death Valley in California.

Hemispherical absorptivity is numerically equal to the hemispherical emissivity. Emissivity is a parameter that provides the degree of emission of radiation, relative to maximum possible. Emissivity, like all quantities discussed here, is a spectral quantity that characterizes a source or a radiator as to how close its radiative emission corresponds to the emission of an ideal blackbody radiator. Its maximum value is one when the source behaves as an ideal blackbody radiator. Its minimum value is zero when there is no radiation emitted. Neither of these extreme cases exists in nature.

We refer to a radiator with emissivity constant, i.e., independent of wavelength as a gray body. This is in reference to the visible palette of three-dimensional color space where the white represents all spectral components with highest saturation and gray levels are defined along the white-black axis.

Emissivity has been originally introduced to describe how closely an object behaves to a blackbody emitter at the same temperature. Recently, its use has been expanded to include descriptions of the Earth's surface as to its emission characterization [19]. Without calling it emissivity, the directional reflection and absorption of radiation within the forest canopy has been used to classify remotely forest constituents and their age [20].

12.7.8 External and Internal Transmittance

First, we consider the case of internal transmittance. We refer to Figure 12.29, which illustrates a collimated beam of light with incidence M_I incident on a plane slab of material of thickness d. M_0 is the incidence on the right side of the interface, inside the medium. We evaluate Equation 12.130 for the beam-propagation distance $z = d$ to find the internal transmittance in the case of the absorbing and scattering plate. Thus, we find from Equation 12.130 when z = 0,

$$M(d) = M_0 e^{-(\alpha + k_s)d}. \qquad \left[\frac{W}{m^2} \right] \tag{12.135}$$

The internal transmittance τ_i is defined as the ratio of the exitance at the end of the propagation distance d within the medium $M(d)$ to the exitance after the beam has entered the medium M_0. It is given as follows:

$$\tau_i = M(d) / M_0. \quad [\text{unitless}] \tag{12.136}$$

Thus, from Equation 12.135, the internal transmittance is equal to

$$\tau_i = e^{-(\alpha + k_s)d}. \quad [\text{unitless}] \tag{12.137}$$

The internal transmittance depends on the thickness d over which the beam is propagating and the material properties, the absorption and the scattering coefficients, α and k_s, respectively. If we know the plate thickness and the absorption and scattering coefficients, we can calculate the internal transmittance. The internal transmission of a plate is increased when the material has low absorption and scattering coefficients, for any plate thickness. When a plate of thickness d_1 is replaced by a plate of the same material but with a different thickness d_2, the internal transmittance is changed, multiplied by a factor $\exp[(\alpha+k_s)(d_1-d_2)]$.

The external transmittance is the ratio of the exitance leaving the plate M_D to the exitance incident on the plate M_I:

$$\tau_e = M_D / M_I. \quad [\text{unitless}] \tag{12.138}$$

The external transmittance includes, in addition to the internal transmittance, the Fresnel reflection losses at both boundaries. Using Equation 12.121 again, we obtain at each surface:

$$M_0 = \tau_{12}M_I \quad \left[\frac{W}{m^2}\right] \tag{12.139}$$

and

$$M_D = \tau_{21}M(d). \quad \left[\frac{W}{m^2}\right] \tag{12.140}$$

We now substitute Equations 12.134, 12.135, and 12.137 into Equation 12.138 to obtain an explicit result:

$$M_D = \tau_{21}\tau_{12}M_I e^{-(\alpha+k_s)d}. \quad \left[\frac{W}{m^2}\right] \tag{12.141}$$

This gives us the expression for the external transmittance, upon dividing with (incident) incidance,

$$\tau_e = \tau_{21}\tau_{12}e^{-(\alpha+k_s)d}. \quad [\text{unitless}] \tag{12.142}$$

Equation 12.142 is the most physically intuitive expression for the external transmittance, including dependence on material properties, such as index of refraction, absorption and scattering coefficients. External transmittance may also be given in terms of internal transmittance, when Equation 12.137 is substituted into Equation 12.142:

$$\tau_e = \tau_{21}\tau_{12}\tau_i. \quad [\text{unitless}] \tag{12.143}$$

This expression shows that the external transmittance is the product of the internal transmittance multiplied by the Fresnel losses at each surface. We keep in mind that we can only measure the external transmittance. We may calculate all parameters for a particular material if we measure external transmittance for a number of thicknesses. When a plate of thickness d_1 is replaced by plate of thickness d_2 the external transmittance is changed, multiplied by a factor $\exp[(d + k_s)(d_1-d_2)]$

For the normal angle of incidence, the Fresnel losses are given in terms of the indices of refraction, Equations 12.118 through 12.120. Then, Equations 12.142 and 12.143 become

$$\tau_e = \tau_{12}^2 e^{-(\alpha+k_s)d} = (1-R_{12})^2 e^{-(\alpha+k_s)d}, \quad [\text{unitless}] \tag{12.144}$$

and

$$\tau_e = \tau_{12}^2 \tau_i = (1-R_{12})^2 \tau_i. \quad [\text{unitless}] \tag{12.145}$$

Using Equation 12.117, we obtain another set of expressions:

$$\tau_e = \left[\frac{4n_1 n_2}{(n_1+n_2)^2}\right]^2 e^{-(\alpha+k_s)d}. \quad [\text{unitless}] \tag{12.146}$$

and

$$\tau_e = \left[\frac{4n_1 n_2}{(n_1+n_2)^2}\right]^2 \tau_i. \quad [\text{unitless}] \tag{12.147}$$

The internal transmittance is equal to one only in the case when the plate is made of material that is neither scattering nor absorbing. Then, the external transmittance of a plate is reduced to the inevitable Fresnel losses at the two surfaces When a plate of thickness d is sandwiched between two media with different indices of refraction (see Figure 12.30), the Fresnel losses at the second surface are modified.

12.8 Optical Signal

Radiometry is a fundamental field within the optical sciences: it establishes the physical basis as to why radiation may be measured at an incremental area at the detecting surface to give us information about

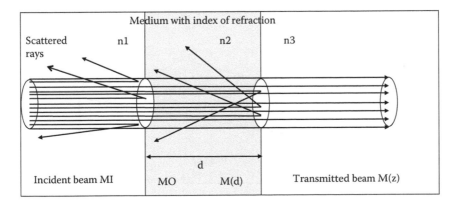

FIGURE 12.30 Transmission of radiation through a slab of thickness d and index of refraction n_2. Internal transmittance includes only the losses inside the medium. External transmittance includes, in addition to the internal transmittance losses, the Fresnel losses at both interfaces.

radiation at an incremental area of the source. Signal is the quantity of radiation that we may measure at a point of our choice in space and time to give us information about radiation at another location in space at a specific earlier time. When the measurement is actually performed over a finite area, the total amount of radiation incident on it is measured. In addition to the signal that originates at a finite area of the object of interest, the detector collects other radiation whose origin is not at the source area of interest. This quantity is referred to as noise. Most optical instruments, including the human eyeball and the associated visual nervous system, can only detect the signal when it is discernable above noise. Signal processing techniques are customarily implemented to increase the signal above the noise or to isolate the signal from the noise.

12.8.1 Imaging and Non-Imaging Systems

Every radiometric instrument includes an optical system to collect the radiation from an (incremental) area on the source or object and to bring it to the detector. A detector is a transducer that changes the radiative energy into some other form of energy. Nowadays, the electrical energy (current and voltage) is the preferred modality, because of its amenability to conditioning, processing, storage, and display. Human eyeball, excluding the photo-sensors – rods and cones – is a fine example of an optical instrument, with a rather small radiation-collecting area and high resolution in the central field area, allowing us to read the small print. We design optical instruments to increase resolution, such as optical and electron microscopes, and to subtend higher angle at the eyeball, including a magnifying glass and terrestrial telescope (see Volume 1).

One important requirement of an imaging instrument is that it preserves similarity between the object and the image: a small incremental area on the object is imaged on a small incremental area on the image. The relationships between distances on object and image are preserved through a common factor, magnification; distances are mapped by magnification and angles between straight lines are preserved. Corresponding incremental areas on the object and image space are referred to as conjugate areas. In geometrical optics and optical lens design, this discussion refers to *points* on the object and image rather than incremental areas. We keep in mind that in radiometry, radiation may be emitted only by the matter that occupies a finite spatial extend; therefore, we must associate an area on a surface with the emission of radiation from microscopic objects. For gasses, the emission is associated with incremental volume of space. Furthermore, the gas transitions and transitions arising from dopants in solids, such as rare earths [21], correspond to the changes in electronic state of the atom or molecule. For molecules, radiation is emitted as a consequence of the change in the vibrational and rotational states.

Radiometry provides the framework and tools to relate the radiation measured on the image incremental area with that originating at the conjugate object area. We may reformulate the definition of the signal for an image forming system: the signal is the radiation that originates at the object incremental area and is collected at the conjugate area on the image (See Figure 12.31). The radiation that originates

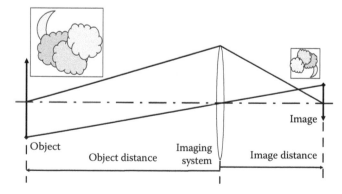

FIGURE 12.31 An imaging optical system preserves similarity between object and image. Moon behind three successively darker clouds is obtained in the image plane with similar levels of gray, same angles between lines, and proportionally reduced size, according to the magnification factor.

at an incremental area of the object that is not a conjugate area of the receiver is therefore noise. The signal may change from application to application, but it is always the quantity that we are interested in measuring and quantifying.

Non-imaging optical systems do not meet the requirements of an imaging system, listed above. They are often used to provide (uniform) illumination and to collect the radiation incident on a specific surface area, as for example, solar (energy) collectors. They are sometimes referred to as energy or power buckets. A simple radiometer usually incorporates an imaging system for collection of radiation, although it does not take advantage of its imaging features. An interferometer images the pupil planes, providing the information in the inverse space, i.e., the spatial frequency space.

12.8.2 Signal Collection

Humans started building optical instruments to extend capabilities of our vision. In addition to increasing the angular subtense of an object and its resolution, our optical instruments also collect more radiation by incorporating larger apertures, detect specific states of polarization, and respond to spectral regions outside the visual range. The larger collecting area is particularly important when weak signals are being detected, as in ever-larger telescopes that probe ever more distant and dim sources (See Telescope in Volume 1).

An optical system may be characterized by its aperture diameter and the effective focal length. These two parameters determine all the first-order optical parameters, including the image location. The former also defines the resolution, once the operating wavelength has been chosen, and the amount of collected radiative power. After the fundamental characteristics of an instrument have been specified, the primary objective of its design is to maximize the optical signal while at the same time minimizing the optical noise.

12.8.3 Optical Noise and Stray Light

The single most demanding consideration in the design of a radiometric system is noise minimization, after the preliminary system has been laid out. We limit the discussion to the transfer of radiation from the object incremental area to the conjugate area on the image. Then, optical noise includes radiation arising from the effects of its propagation, its interaction with the matter, radiation reflected at each surface, radiation scattered from the surface imperfections and volume inclusions, and the radiation undergoing total internal reflection. When imaging dim sources, out of field radiation also contributes to the background noise. For infrared systems, self-emission from surfaces at temperature T according to the Planck's law (see section 12.6.2) becomes a significant source of noise. These types of noises are depicted in Figure 12.32. All surfaces that the detector "sees," including itself in reflection, contribute

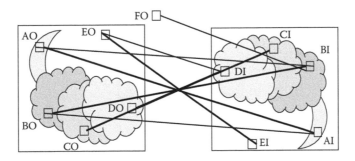

FIGURE 12.32 Imaging is accomplished when radiation from an object point is incident on its conjugate point: object points A through E are imaged on image points A through E. Optical noise is generated when ray originating at object points A, B, E is incident on image points B, A, D. Additional noise is introduced when a ray within the image (ray F) falls on the image plane (B).

to this so-called stray light. Each surface may be imaged – defocused on partially the detector plane contributing a set of superimposed ghost images that represent background noise. This pattern may interfere with the image features, especially for faint sources. For this reason, in infrared instruments and telescope facilities, the structures increasingly closer to the detector are often cooled down to progressively lower temperature to decrease their self-emission [4].

When the system analysis also includes the sensing surface, the read-out electronics, image processing, and display, each of these subsystems adds its own noise that must be evaluated and minimized (see Figure 12.33). However, maximizing the radiometric signal represents only the first step in signal-to-noise optimization. Noise minimization task starts with decreasing the radiative noise, considering that most noise types in imaging systems are signal-dependent. Usually they are additive and multiplicative.

12.8.4 Signal-to-Noise Ratio

According to good engineering practices, signal is to be maximized and the noise is to be minimized. However, these extremes may be achieved only for a degree under different circumstances. They are usually evaluated for a number of key parameters, often exhibiting different functional dependence. A ratio of these two quantities is therefore formed to characterize design with a single parameter, signal-to-noise ratio. Then, its value is maximized under the constraint conditions. According to the discrimination analysis, the signal-to-noise ratio is the key parameter that delineates regions of detection, recognition, and identification. For a simple analysis of natural scene, this quantity might be within the range of 40–100. For the detection of very faint objects, such as Neptune rings, signal-to-noise ratio was only 0.01, requiring prolonged exposure time and frame stacking, with extensive subsequent image analysis and enhancement.

In order to make use of the available resources, the signal-to-noise ratio is usually maximized under constraints of specific radiometric configuration, environmental conditions, and budget.

12.8.5 Detection of Radiation and Signal Conditioning

A complicated optical system may be described as an equivalent optical system with aperture and focal distance. The aperture diameter determines the radiation collecting area and resolution when the design wavelength of the instrument is also known. The sensitive surface of the detecting device is placed in the image plane. When the object is located "at infinity," the image plane coincides with the focal plane, giving rise to the popular terminology "focal plane array" for the two-dimensional detector array. Nowadays, these also incorporate the read-out and conditioning electronics. An end-to-end block diagram of an imaging system is presented in Figure 12.33.

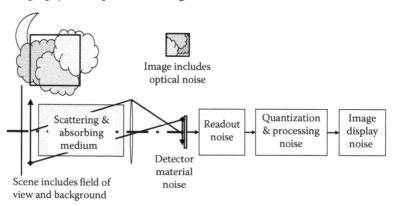

FIGURE 12.33 Complete system analyses includes noise contributions from the sensing surface and material, the read-out electronics, image processing, and display. Each of these subsystems adds its own noise that must be evaluated.

A detector is a device, a transducer of radiation, that changes the electromagnetic radiation to some other form of energy for further processing, storage, conditioning, and display. The old-fashioned photographic film, for example, employs chemical energy to produce changes. However, it may be used one-time only, while it permits the possibility of permanent storage after the processing. The modern charge-coupled devoices (CCD) may record and transfer energy into a storage device at a rate of several thousand frames per second. They may be reused innumerable number of times, and they are available in formats larger than mega-pixel arrays. The photo-sensors in the human eyeball change configuration between two isomers upon absorption of light energy. Within milliseconds, the lower energy configuration is re-established and we are ready to detect a new image. The human eyeball includes over one million light-sensing elements, although not all are connected for the purpose of image detection.

In addition to the temporal dependence and re-write capabilities, detector arrays may also be classified as measuring the radiative power or some derivative thereof. They are also differentiated as quantum and thermal detectors. The former have responsivity proportional to wavelength, while the responsivity of the latter is independent of wavelength. Most quantum detectors function within a specific spectral range. Another important consideration in the detector choice is the temperature range of its operability. Liquid-helium temperature is recommended for highly-sensitive, far-infrared operations, liquid nitrogen for established infrared detectors at about 10 microns, detector cooling incorporating Sterling system may be required for low-noise performance of CCD detectors, with spectral sensitivity within visible to near IR. For most everyday applications, CCDs function adequately at room temperature.

Within the last few decades, modern, room temperature, bolometer-based devices have been evolving rapidly. Each of these transducers has its own sources of noise, with one particular noise dominant for each detector and conditions of application.

12.8.6 Radiometric Measurements

Most instruments employed to measure the radiation do so at a specific location in space. Such measurement captures an instantaneous state of radiation. We should rather say that the radiative power is collected during a very short period, with the integration time of detector often being the determining factor in controlling the exposure duration. The radiation propagates until it encounters an interface with matter where it is at least partially absorbed. The surface sensitive area of the detector is by design highly absorptive.

Absorption of radiation produces changes in the state of the sensing material. The absorbed photon produces an ever so small temperature increase in the case of thermal detectors. It produces holes and electrons in the case of the quantum detectors, giving rise to current or voltage. When a new detector material is synthesized or when a finished focal plane device is being characterized as to the responsivity to the incident spectral radiation, this is performed for all pixels under specific experimental conditions. A blackbody simulator source with a few spectral filters may function as a reference illuminator. A blackbody simulator whose cavity opening fills the detector field-of-view provides a known spectral incidance at the focal plane. A recent example of detector calibration with a modulated blackbody source at 1200 K measured spectral response of a new long wave IR type II strained layer superlattice pBIBn photodetector [22]. IR focal plane arrays are often characterized for low photon levels, using spectral filters. Responsivity has been measured at 7.5 μm to be about 70 nV per photon for a type II strained layer superlattice detector array [23]. An even better quantum efficiency of about 60% has been measured at 2 μm incorporating InAs/GaSb superlattice photodetectors [24]. Interestingly, one disadvantage of infrared cameras incorporating microbolometers that function at ambient temperatures without cooling is that their performance is still temperature dependent to some degree. Thus, focal plane calibration must be performed over a range of expected temperatures [25] or an internal shutter is calibrated to function as a reference source [26].

Most detectors provide variable and slowly deteriorating output under different environmental conditions and upon aging. Accurate absolute measurements require that the detector output be frequently calibrated. In many experiments, less accurate relative measurements are implemented. However, measurement errors are decreased in ratio-ing techniques. These might include taking the ratio of signals recorded at different spectral intervals to decrease the measurement error inherent to non-calibrated measurements.

12.8.6.1 Absolute Radiometric Measurements

Absolute radiometric measurements are necessary for long-term evaluation of physical phenomena that may be performed under changing environmental conditions, separated in time and space, and often with different instruments. Absolute measurements are obligatory when results are to be compared and trends evaluated. Such measurements require calibration with a certified reference source, most often a blackbody simulator source. Oftentimes a transmissive cell is employed to provide a transmission reference for liquids and gases both in laboratory and in-space settings.

There are no blackbodies, as this concept was introduced as a theoretical construct to derive a quantization of the electromagnetic field inside a closed cavity with perfectly absorbing walls. The blackbody simulator is then a physical embodiment of this theoretical concept, with a relatively small cavity and a relatively large opening. Thus, its physical embodiment deviates from the physical dimensions of the theoretical blackbody in order to provide access to the emitted radiation. In practice, the cavity walls are not completely absorbing, and temperature is not completely uniform or kept at exactly specified value. The blackbody simulators vary in complexity as to their ability to approximate the ideal concept, in particular their ability to maintain uniform wall temperature and have walls with high absorptivity. Material with absorptivity, or emissivity, of one does not yet exist. Researchers have tried to increase its emissivity by shaping the cavity. More accurate blackbody performance is achieved by increasing the cavity volume against the diameter of the opening or decreasing the cavity opening while maintaining the cavity volume [2].

The best reference radiation source that a laboratory may incorporate is a blackbody simulator with traceability to the United States NIST. The NIST is a national facility responsible for developing, maintaining, and making available consulting services to perform comparison on the reference sources. Thus, they engineer the standards and facilitate the traceability to their high-quality equipment. They only certify that a commercially available blackbody simulator exemplar has been compared at their facility on a specific date, and they provide the performance results. This limited, yet best available, certification illustrates the challenge of radiation standards and expectation of their degradation with time away from the certification facility. The researcher then acquires calibration equipment that has been calibrated against such secondary standards. He incorporates it into his experimental setup and laboratory practices. Figure 12.34 illustrates how a reference blackbody simulator might be used in an experimental setup for continuous in-situ calibration.

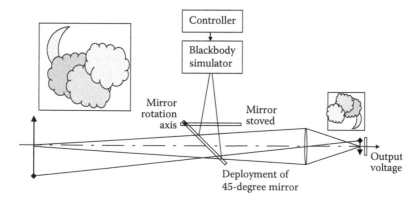

FIGURE 12.34 Calibration of surface temperature with an in-situ blackbody simulator, employing a beam splitter, moveable mirror, or reflective chopper.

Simpler blackbody simulators exist in a form of a heated plate or a filament to provide reference for extended periods without re-calibration. They feature reliable operation, such as in satellite-based radiometric measurements [27]. Their accuracy is lesser, because the emissive surface is exposed to a half-sphere solid angle, resulting in non-uniform temperature distribution over a plate surface. Look-up tables may be used to correct for errors in temperature and to implement corrections for emissivity. Figure 12.34 illustrates a potential experimental setup to measure surface temperature or to calibrate detector responsivity using a blackbody simulator. Despite imperfect emissivity, temperature may then be related to radiance using Planck's law or integrated radiance over a spectral band [3].

12.8.6.2 Relative Radiometric Measurements

It is universally accepted that all radiometric measurements include error. Its magnitude and its importance for the validity of the measurement depend on the specific application. The error may sometimes only be estimated, as there does not exist a way to determine it accurately even at a superior experimental facility. Other times, the magnitude of the error does not significantly impact the measurement outcome: the experimental techniques have been developed to decrease the impact of measurement error on the validity of results.

Ratioing techniques conveniently sidestep the radiometric uncertainty in the measurements when the signal-to-noise ratio in each spectral band is adequate. The signal of interest is constructed by dividing one measured radiometric quantity with respect to the other, requiring two simultaneous radiometric measurements. Visual and infrared thermometry is an example of an experiment where the radiation is measured in two wavelength bands to find the blackbody curve that corresponds to the object temperature [28]. The limitation of this technique is that the experimentalist must assume that emissivity is the same for both spectral bands, and the measurements must be performed at spectrally distant wavelength bands. Even better results are obtained with differential thermometry [29].

Finding ratio is a particularly elegant method to incorporate in temperature sensors. They traditionally employ materials that have temperature-dependent emission in several spectral bands. The ratio of band exitance at two distinct spectral lines is often also temperature-dependent. Designers find it convenient to use such ratio in spectral intervals where exitance ratio is most sensitive to changes in temperature. Thus, they are more appropriately described as line or band emitters than blackbody emitters, so the absolute calibration methods would not serve well [28,29]. With the availability of highly sensitive 2D arrays in mid- and far-IR, ratio-ing imaging techniques may be applied in selected applications [30,31].

In biomedical applications of radiometry, optical techniques have been promoted due to their capabilities for non-invasive measurements and in-vivo applications [32]. Tissue and blood, in particular, have different optical properties in vivo, then in absence of body activities. Furthermore, the patient status needs to be assessed in time to apply corrective action with minimal time delay. Finally, the study of human subject does not allow for measurement verification process, by *post factum* dissection for verification of measurement procedures.

Oximetry is an application of measurement of amounts of radiation in two spectral bands, based on different spectral absorption of oxygenated and de-oxygenated hemoglobin. In this application, the transmitted radiation is measured at two spectral bands to assess the degree of blood oxygen saturation. Then, a simple fraction of quantities involving transmission at two carefully selected spectral bands is calculated. Two-dimensional standard deviation distributions may be used to select the most appropriate pair of spectral bands [18]. Recently, with the development of ever more sensitive cameras in visible – near infrared spectral region, proposals have been made to use these measurements in reflection mode for remote monitoring of vital signs [33]. This development is desirable for non-cooperative patients, including infants and animals.

12.9 Summary

Many a radiometry student has observed that radiometry is a boring subject and questioned the relevance of definition of, let us say, radiance. We believe that we know radiometry intuitively, because we rely most strongly on the sense of vision to acquire information about the world around us. The facts are just the opposite. Humans are very poor observers: we only need to recall the "gorilla-in-the-basketball-court" experiment. No artificial vision system would ignore the gorilla, especially not one designed with the knowledge of basic radiometric concepts. But human observers failed completely at seeing the gorilla, while their attention was focused on something else.

Radiometry formalizes the concepts that all humans, all engineers, and all optical engineers use on a daily basis, employing different terminology and units that depend on the field of endeavor, and application. We graph results on axes that most often are not even denoted. However, they become normalized when the figure captions state "arbitrary units." In fact, scientists have no difficulty introducing novel results, presenting them in intuitive graphs and most of the time, the readers actually understand them. Only occasionally we do not understand them because "arbitrary" and "normalized" intensity and irradiance have a different intuitive meaning to a different user. This may lead to small and huge misunderstandings, depending on the project's (financial) magnitude and impact.

In the international year of light (2016), we could start using the same language to denote the same quantity or the same physical concept. This uniformity of language and expression would enhance information transfer from one user to another. Clarity of communication might avoid expensive instrument building and rebuilding when the designer appeared to break the radiance conservation law and everybody was in awe upon seeing promising results during the design stage. Some such errors could have been avoided if everybody on the engineering team had used the same terminology and understood the same concept when using the same words for it.

Acknowledgment

Special appreciation is expressed to Gonzalo Paez for preparing a number of illustrations in the first edition of this chapter.

References

1. Nicodemus F. Self Study *Manual on Optical Radiation Measurements, Part 1, Concepts*, Superintendent of Documents, Washington, DC: US Government Printing Office, 20402; 1976.
2. Scholl MS. Temperature calibration of an infrared radiation source, *Appl. Opt.*, **19** (21), 3622–3625; 1980.
3. Scholl MS. Errors in radiance simulation and scene discrimination, *Appl. Opt.*, **21** (10), 1839–1843; 1982.
4. Scholl MS. Stray light issues for background-limited far-infrared telescope operation, *Opt. Eng.*, **33** (3), 681–684; 1994.
5. Strojnik M, Scholl MK. Extrasolar planet observatory on the far side of the Moon, *J. Appl. Remote Sens.*, **8**(1), 084982; 2014.
6. Wyszecki G, Stiles WS. *Color Science*, New York, NY: John Wiley & Sons, 1982.
7. Strojnik M, Peez G, Scholl MK. Understanding human visual system and its impact on designs of intelligent instruments, *Honoring John Caulfield*, SPIE 8833, paper 3; 2013.
8. Scholl MS, Trimmier JR. Luminescence of YAG: Tm; Tb, *J. Electrochem. Soc.*, **133** (3), 643–648; 1986.
9. Hlaing S, Gilerson A, Foster R, Wang M, Arnone R, Ahmed S. Radiometric calibration of ocean color satellite sensors using AERONET-OC data, *Opt. Express*, **22** (19), 23385–23401; 2014.
10. Hirata T, Hardman-Mountford N, Aiken J, Fishwick J. Relationship between the distribution function of ocean nadir radiance and inherent optical properties for oceanic waters. *Appl. Opt.*, **48** (17), 3129–3138; 2009.

11. Kim M, Park PY, Kopilevich Y, Tuell G, Philpot W. Correction for reflected sky radiance in low-altitude coastal hyperspectral images, *Appl. Opt.*, **52** (32), 7732–7744; 2013.

12. Lee ZP, Du K, Voss KJ, et al. An inherent-optical-property-centered approach to correct the angular effects in water-leaving radiance, *Appl. Opt.*, **50** (19), 3155–3167; 2011.

13. Tonizzo A, Gilerson A, Harmel T, et al. Estimating particle composition and size distribution from polarized water-leaving radiance, *Appl. Opt.*, **50** (25), 50058; 2011.

14. Scholl MS, Peez Padilla G. Using the y, y-bar diagram to control stray light noise in IR systems, *Infrared Phys. Technol.*, **38**, 25–30; 1997.

15. Scholl MS. Figure error produced by the coating-thickness error, *Infrared Phys. Technol.*, **37**, 427–437; 1996.

16. Grabtchak S, Palmer TJ, Vitkin A, Whelan WM. Radiance detection of non-scattering inclusions in turbid media, *Biomed. Opt. Express*, **3** (11), 3001–3011; 2012.

17. Ramirez-Granados JC, Peez G, Strojnik M. Three-dimensional reconstruction of subsurface defects by using pulsed thermography videos, *Appl. Opt.*, **51** (16), 1153–1161; 2012.

18. Vasquez-Jaccaud C, Peez G, Strojnik M. Wavelength selection method with standard deviation: Application to pulse oximetry, *Ann. Biomed. Eng.*, **39** (7), 1994–2009; 2011.

19. Strojnik M, Peez G. High-resolution bispectral imager at 1000 frames per second, *Opt. Express*, **23** (19), A1259–A1269; 2015.

20. Kozoderov VV, Kondranin TV, Dmitriev EV, Sokolov AA. Retrieval of forest stand attributes using optical airborne remote sensing data, *Opt. Express*, **22** (13), 15410–15423; 2014.

21. Castrellon J, Paez G, Strojnik M. Radiometric analysis of a fiber optic temperature sensor, *Opt. Eng.*, **41** (6), 1255–1261; 2002.

22. Treider LA, Morath CP, Cowan VM, Tian ZB, Krishna S. Radiometric characterization of an LWIR, type-II strained layer superlattice pBiBn photodetector, *Infrared Phys. Technol.*, **70**, 70–75, May, 2015.

23. Hubbs JE, Nathan V, Tidrow M, Razeghi M. Radiometric characterization of long-wavelength infrared type II strained layer superlattice focal plane array under low-photon irradiance conditions, *Opt. Eng.*, **51** (6), 064002; 2012.

24. Giard E, Ribet-Mohamed I, Delmas M, Rodriguez JB, Christol P. Influence of the p-type doping on the radiometric performances of MWIR InAs/GaSb superlattice photodiodes, *Infrared Phys. Technol.*, **70**, 103–106; 2014.

25. Nugent PW, Shaw JA, Pust NJ. Correcting for focal plane array temperature dependence in microbolometer infrared cameras lacking thermal stabilization, *Opt. Eng.*, **52** (6) 061304, June, 2013.

26. Nugent PW, Shaw JA, Pust NJ. Radiometric calibration of infrared imagers using an internal shutter as an equivalent external blackbody, *Opt. Eng.*, **53** (12), 123106; December, 2014.

27. Huang F, Shen X, Liu J, Li G, Ying J. Research on radiometric calibration for super wide-angle staring infrared imaging system, *Infrared Phys. Technol.*, **61**, 9–13; 2013.

28. Peez G, Strojnik M. Cavity effects in coiled coil IR reference source, *Infrared Phys. Technol.*, **49**, 202–204; 2007.

29. Aranda A, Strojnik M, Paez G, Moreno G. Two-wavelength differential thermometry for microscopic extended source, *Infrared Phys. Technol.*, **49**, 205–209; 2006.

30. Castrellon J, Peez G, Strojnik M. Remote temperature sensor employing erbium-doped silica fiber, *Infrared Phys. Technol.*, **43**, 219–222; 2002.

31. Ren H, Liu R, Yan G, et al. Performance evaluation of four directional emissivity analytical models with thermal SAIL model and airborne images, *Opt. Express*, **23** (7), A346–A360; 2015.

32. Strojnik M, Peez G. Spectral dependence of absorption sensitivity on concentration of oxygenated hemoglobin: Pulse oximetry implications, *J. Biomed. Opt.*, **18** (10), 108001; 2013.

33. Corral F, Peez G, Strojnik M. A photoplethysmographic imaging system with supplementary capabilities, *Optica Applicata*, **44**, 191–204; 2014.

13

Color and Colorimetry

13.1 Color Vision 517
  Brief Review of Different Theories • Color Blindness • Other
  Chromatic Visual Effects
13.2 Trichromatic Theory..520
  Matching Color Functions • Tristimulus Values • Chromaticity
  Coordinates
13.3 Chromaticity CIE Diagram 525
13.4 Applications of the Chromaticity Diagram 529
13.5 Influence of the Illumination and Filters on the Color 530
13.6 Color Mixtures 531
  Color Addition • Color Subtraction
13.7 Other Color Representations..534
13.8 Light Sources and Illuminants.....................................536
13.9 Colorimeters 537
13.10 Light Sources 537
Suggested Reading...540

Daniel
Malacara-Doblado

13.1 Color Vision

The mechanisms through which we see color have not been fully understood despite the continuous investigations that have been carried out for several centuries. Nature has a large variety of colors that impress man. We could consider Newton in 1972 to be the first to formally study color when he carried out his experiments with a dispersive prism. It is interesting to know that like the man, some inferior animals, such as birds and reptiles, have color vision, whereas other more evolved animals, such as dogs and cats, do not have it.

We know that the color of a monochromatic light beam depends on its wavelength, so we would be tempted to say that color vision exists because we have detectors in the eye for each possible wavelength, each of them producing a different feeling. However, this hypothesis is very soon discarded, because after very simple considerations, we conclude that more than 200 different receivers would be necessary, which is absurd.

The hypothesis that there should be a limited number of detectors is confirmed by the fact that a mixture of two different colors produces the sensation of a third color distinct from either component. For example, a red light beam superimposed on a green beam gives the same sensation as a yellow monochromatic light beam. The blend is so perfect that it is completely impossible at first glance to differentiate pure monochromatic yellow light from the red and green blend.

It is interesting to note that while the eye fuses two different frequencies to produce the sensation of a third, the ear instead does not do the same with the sounds, since two different sound frequencies never merge to produce the sensation of a third. This is why the eye is said to be a synthesizer, while the ear is an analyzer.

13.1.1 Brief Review of Different Theories

In 1802, Thomas Young hypothesized that the human eye has only three different types of receptors: one sensitive to red, one to yellow, and one to blue. Young then assumed that any other color would produce combined sensations of two or three detectors. The white would therefore be the result of the simultaneous stimulation of the three detectors, with equal intensity.

Unfortunately, Young's theory was not accepted in his time but 50 years later Herman von Helmholtz and James Clerk Maxwell revived this theory, postulating that there could be a very large number of thirds of stimuli that, through combinations of them, it was possible to reproduce any color. Maxwell made a very convincing experiment to test Young-Helmholtz's theory. He took three black and white photographs of the same colored object. In each of the three shots, he placed a different color filter in front of the camera, one red, one green and one blue. With these three shots, he obtained three transparencies in positive, black and white. He then used three projectors, each with one of the slides and the corresponding color filter in front of each projector, as shown in Figure 13.1. The result was surprising, as the projection showed all the original colors of the object, including many different colors to those of each of the color filters.

This theory, known as the Young-Helmholtz theory, is complemented by the hypothesis of Arthur Koening, which in 1890 led to the existence of three primary detectors, one for red, one for green, and one for blue. Young-Helmholtz's theory explains color vision very well, with only a few small problems. For example, it is very difficult to choose the three fundamental colors. It is interesting that if we ask a person with normal vision what he believes are the fundamental colors, he will always choose four instead of three: blue, green, yellow and red. In addition, for some unknown reason, aesthetically the red seems to oppose the green and the yellow to the blue.

Noting these facts, Ewald Hering proposed, in 1870, a different theory, which also assumed three different types of receptors. A receiver responds simultaneously to yellow and blue, followed by a classifier in the brain to separate these two colors. When blue light arrives at this detector, only the blue portion of this detector is stimulated and the yellow portion is closed. In the same way, there is a second receiver for red and green, and a third for white and black. This theory has its appeal, so it enjoyed popularity for some time; finally, it was rejected when confirming that it does not cling to the reality because it does not explain many observations.

Young-Helmholtz theory received strong support when W. A. H. Rushton identified two red and green visual pigments in the eye retina, and finally, in 1964, Edward Mac Nichol isolated and measured the three visual pigments, which are mentioned in Table 13.1.

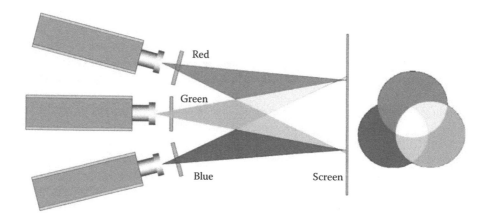

FIGURE 13.1 Maxwell's experiment to test Young-Helmholtz theory.

TABLE 13.1   Visual Pigment

Color	Pigment	Reflectance Peak
Red	Eritrolabe	577 nm (yellow)
Green	Clorolabe	540 nm (green)
Blue	–	477 nm (blue)

TABLE 13.2   Chromatic Defect

Classification	Defect	Explanation
Monochrome		It only has one pigment
Dichromate	Prontanope	Red pigment absent
	Deuteranope	Green pigment absent
	Tritanope	Blue pigment absent
Trichromate	Protanomaly	Defective red pigment
	Deuteranomaly	Defective green pigment
	Trinomaly	Defective blue pigment

TABLE 13.3   Abundance of Color Defect

Classification	Defect	Men	Women
Monochrome		0.003%	0.002%
Dichromate	Prontanope	1.01%	0.02%
	Deuteranope	1.27%	0.01%
	Tritanope	0.005%	0.005%
Trichromate	Protanomaly	1.08%	0.03%
	Deuteranomaly	4.63%	0.36%
	Trinomaly	0.0%	0.0%

Nevertheless, Young-Helmholtz's theory is not perfect, as it cannot explain some important observations. A good theory must be able to give a satisfactory explanation of numerous visual effects; some of them will be described in the following sections.

13.1.2 Color Blindness

The inability to differentiate the colors, either partially or totally, is called color blindness. John Dalton, who suffered from it, was the first to study it seriously in 1798. About 8% of men and only 0.5% of women are color blind to some degree. There are many types and degrees of color blindness. Most people with abnormal color vision simply have less sensitivity than most people to small color variations, but they can still differentiate very different colors. Color blindness can be classified and explained by the absence or defect of one or several visual pigments, as shown in Table 13.2.

The tritanope does not distinguish blue from yellow. Prothanopes and deuteranopes are hereditary, they do not see much difference between many colors, although they are very different. These defects can be graphically described by means of the chromaticity diagram, which will be studied later in this chapter and can be explained by Young-Helmholtz's theory. The abundance of chromatic defects is shown in Table 13.3. The reason for the difference between men and women is that visual chromatic defects are hereditary, linked to sex, where the X chromosome is the one that can be defective. The man has a chromosome X and one chromosome Y so, it is enough that the X is defective to be affected by the defect. In women, however, it is necessary that their two X chromosomes are defective to suffer the defect, although with one, it is already possible to be a carrier.

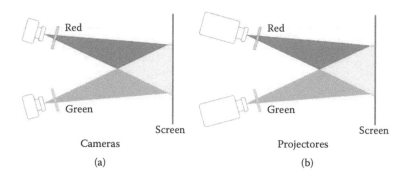

FIGURE 13.2   Edwin Land's experiment: (a) taking photographs and (b) projecting image.

13.1.3 Other Chromatic Visual Effects

If we see a colored figure for more than half a minute and then turn to see a sheet of white paper illuminated with white light, the same image that we just observed will be projected on its surface, but with the inverted contrast and complementary colors. It is said that two colors are complementary if when they overlap, they both produce white light. The image that is produced has been called post-image (afterimage). This effect can also be explained by Young-Helmholtz's theory.

Edwin H. Land in the 1970s found that to observe a fully colored scene under certain conditions only two colors are needed. Land proved this by taking two black and white slides of a scene with very varied colors. When taking the photographs, a colored filter of different color was placed in front of the camera for each slide, as shown in Figure 13.2a. The same scene was then projected using two projectors, placing in front of them the filters that were used in front of the cameras, as shown in Figure 13.2b.

Strictly speaking, the combined image of the two projectors contains only the two colors of the filters, but for an observer the image appears to have many of the colors of the original scene. This result is very surprising and is clearly against Young-Helmholtz's theory. Land's explanation with his Retinex (color retina) theory of color assumes that the visual information provided by the retina is subsequently processed in the cerebral cortex. Apparently, what happens is that in the cerebral cortex an integration of all the colors of the scene is made, compensating for any predominance of any color.

13.2 Trichromatic Theory

Although trichromatic theory has its problems, it is nevertheless sufficiently adequate to deal with practical problems of color. In this section, we will develop this theory in a quantitative way.

13.2.1 Matching Color Functions

Let us suppose we try to reproduce all the colors of the spectrum, that is to say, the monochromatic ones, by means of additive combinations of three colors of the previously selected spectrum, projecting them on a screen. These colors can be any, but we will choose a red (700.0 nm), a green (546.1 nm), and a blue (435.8 nm). The quantities of light from each of these light beams can be measured in powers or luminances, but instead other units will be chosen. In addition, these units will be different for each of these three components. These amounts of light will be represented by and will be called color equalization functions. Once the light quantities in watts or lumens of each of these components has been found for all the colors of the spectrum, the new units are defined by multiplying the values of these quantities by the appropriate constant such that the following equation is satisfied:

TABLE 13.4 Radiances and Illuminances of the Tristimulus Units (Proportional to These Values)

Color	Wavelength (nm)	Radiance Tristimulus Unit (μW)	Luminance Triestimulus Unit (Lumens)
Red	700.0	1.000	1.0000 (L_B)
Green	546.1	0.019	4.5907 (L_G)
Blue	435.8	0.014	0.0601 (L_B)

$$\sum_{i=1}^{n}\bar{r}(\lambda)=\sum_{i=1}^{n}\bar{g}(\lambda)=\sum_{i=1}^{n}\bar{b}(\lambda).\tag{13.1}$$

As will be seen later, the advantage of imposing this condition is that to obtain ideal white light with a spectral power $P(\lambda)=1$. Three equal quantities are needed $\bar{r}(\lambda)=1/3, \bar{g}(\lambda)=1/3, \bar{b}(\lambda)=1/3$ of each of these three colors. With this selection of units, the radiances and luminances of a unit of each of these three colors are as shown in Table 13.4.

When performing this experiment of equalizing the monochromatic colors the visual field, the sample where the monochromatic matched, preserves the same luminous power per unit area. There its irradiance is constant when changing the wavelength. The matched field is adjusted so that the brightness of the two fields is the same. It is easy to see that with these experimental conditions the relative light efficiency of the human eye can be written as the following linear combination of these three functions:

$$V(\lambda)=L_R\bar{r}(\lambda)+L_G\bar{g}(\lambda)+L_B\bar{b}(\lambda),\tag{13.2}$$

where L_R, L_G y L_B are the luminances per unit of each of the color matching functions, with the values shown in Table 13.4.

In order to ensure that the relation 13.1 is satisfied, two constants are sufficient, but three have been used in order to impose the condition that in the relation 13.2, the value of $V(\lambda)$ is normalized in such a way that its value maximum is equal to one.

When trying to do the matching of colors, it is found that for certain wavelengths this becomes impossible. It may happen, for example, that the color we have with the sum of the three colors is still very reddish compared to the pattern sample, and yet the red beam brightness has already been reduced to zero. We cannot remove more red, but we can add it to the pattern sample. Mathematically, this can be interpreted as if it had become negative in our sum of stimuli. The values of the color matching functions thus obtained are shown numerically tabulated in Table 13.5 and plotted in Figure 13.3.

13.2.2 Tristimulus Values

The spectral light distribution of a body clearly defines its color, but not the other way around. In other words, two very different spectral distributions can give the same color, but two equal colors do not necessarily have the same spectral distribution.

The laws of Grassmann tell us that the sum of two lights with the same color gives us another light with the same color as those that add up, although their spectral distributions are different, and therefore that of the resultant is different from the one of the two components.

A good way to specify the color of an object, not necessarily monochromatic independently of its spectral distribution, is by means of its tristimulus values, as will be seen now. These values tell us how many units of each of the three primary colors are needed to match the color of the object. Of course, if the object to be matched is monochromatic, it is easy to see that the values of the tristimuli

TABLE 13.5   Color Matching Functions

Wavelength	$\bar{r}(\lambda)$	$\bar{g}(\lambda)$	$\bar{b}(\lambda)$
380	0.00003	−0.00001	0.00117
390	0.0001	−0.00004	0.00359
400	0.0003	−0.00014	0.01214
410	0.00084	−0.00041	0.03707
420	0.00211	−0.00110	0.11541
430	0.00218	−0.00119	0.24769
440	−0.00261	0.00149	0.31228
450	−0.01213	0.00678	0.3167
460	−0.02608	0.01485	0.29821
470	−0.03933	0.02538	0.22991
480	−0.04939	0.03914	0.14494
490	−0.05814	0.05689	0.08257
500	−0.07173	0.08536	0.04776
510	−0.08901	0.1286	0.02698
520	−0.09264	0.17468	0.01221
530	−0.07101	0.20317	0.00549
540	−0.03152	0.21466	0.00146
550	0.02279	0.21178	−0.00058
560	0.0906	0.19702	−0.00130
570	0.16768	0.17087	−0.00135
580	0.24526	0.1361	−0.00108
590	0.30928	0.09754	−0.00079
600	0.34429	0.06246	−0.00049
610	0.33971	0.03557	−0.00030
620	0.29708	0.01828	−0.00015
630	0.22677	0.00833	−0.00008
640	0.15968	0.00334	−0.00003
650	0.10167	0.00116	−0.00001
660	0.05932	0.00037	0
670	0.03149	0.00011	0
680	0.01687	0.00003	0
690	0.00819	0	0
700	0.0041	0	0
710	0.0021	0	0
720	0.00105	0	0
730	0.00052	0	0
740	0.00025	0	0
750	0.00012	0	0
760	0.00006	0	0

are equal to the color matching functions described above. Note that what is important in these values of tristimuli is their relative value, not their absolute value.

A luminous body with emitted $P(\lambda)$ or a colored opaque with a white light illuminated $P(\lambda)$ would have a color determined by the spectral distribution of this function $P(\lambda)$ defined as the emitted or reflected light energy/time power by the object in a unit interval of wavelengths $\Delta\lambda$ centered at the wavelength λ.

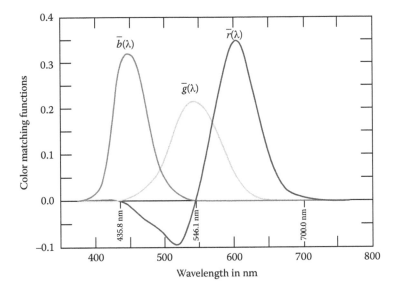

FIGURE 13.3 Color matching functions.

Consider the light emitted in this interval $\Delta\lambda$ with an emitted $P(\lambda)$. The color of this narrow interval will be defined by the three values of the tristimulus $\bar{r}(\lambda), \bar{g}(\lambda)$ and $\bar{b}(\lambda)$. If we add the stimuli of all the intervals $\Delta\lambda$, we obtain the tristimulus values of this non-monochromatic light source.

$$R = \int_0^\infty P(\lambda)\bar{r}(\lambda)d\lambda, \tag{13.3}$$

$$G = \int_0^\infty P(\lambda)\bar{g}(\lambda)d\lambda \tag{13.4}$$

and

$$B = \int_0^\infty P(\lambda)\bar{b}(\lambda)d\lambda. \tag{13.5}$$

In practice, these integrals are replaced by summations; for this, the values of the equalization functions are used in Table 13.5 as follows:

$$R = \sum_{i=1}^n P(\lambda)\bar{r}(\lambda), \tag{13.6}$$

$$G = \sum_{i=1}^n P(\lambda)\bar{g}(\lambda) \tag{13.7}$$

and

$$B = \sum_{i=1}^n P(\lambda)\bar{b}(\lambda). \tag{13.8}$$

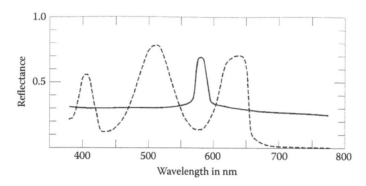

FIGURE 13.4 Spectral reflection of two hypothetical objects that have the same color under illumination with illuminant D_{65}.

It is easy to see that several different spectral distributions $P(\lambda)$ can give rise to the same color. That is, we can obtain the same values of R, G, and B with different $P(\lambda)$. This phenomenon is known as metamerism and is illustrated in Figure 13.4, where three completely different $P(\lambda)$ distributions produce the same color.

13.2.3 Chromaticity Coordinates

We can now, according to Maxwell, define the so-called chromaticity coordinates as the normalized values of the tristimuli, which are obtained by dividing each of them by the sum of the three tristimuli, as follows:

$$r = \frac{R}{R+G+B},\tag{13.9}$$

$$g = \frac{G}{R+G+B}\tag{13.10}$$

and

$$b = \frac{B}{R+G+B}.\tag{13.11}$$

This standardization has two advantages. One of these is that these dimensionless quantities represent the color of the colored object, being the same for two objects with the same color, regardless of their luminosity. The second advantage is that only two quantities are required to specify the color, since the third can be determined very quickly and easily with the following relation:

$$r + g + b = 1.\tag{13.12}$$

It is thus possible to completely determine the color with only the values of r and g. Then, applying this normalization to the color equalization functions gives the chromaticity coordinates. If the value of g is plotted against r, the graph of Figure 13.4 is obtained. Here, the colors of the spectrum are represented on a curve with an approximate horseshoe shape. The straight line joining the lower ends of the horseshoe represents the purple colors, which are combinations of red and violet in different proportions. The point P in the center represents the white color. Two complementary colors would then be placed symmetrically with respect to point P.

If points A and B represent the colors of two light sources, with the appropriate proportions of A and B, then all colors falling within the line AB can be obtained. This rule to obtain the color resulting from the additive combination of two colors A and B is a direct consequence, or rather another interpretation, of Grassmann's laws. For the same reasons, with three colors A, B, and C, we can get any color inside the triangle ABC.

13.3 Chromaticity CIE Diagram

In the chromaticity diagram just described, it is seen that a large range of colors requires negative r-stimuli, which, as we saw above, means that the color *r* has to be added to the sample against which the comparison is made, in place of the mixture to be obtained. This naturally has many practical disadvantages. In order not to subtract colors, and thus alter our pattern sample, for all mixtures to be additive, we have defined three new stimuli, indicated by the X, Y, Z positions in Figure 13.5, of such so that the whole horseshoe is within the triangle formed by these three stimuli, these in turn formed by stimuli *r* and *g*, as shown in Table 13.6.

As can be seen, these stimuli X, Y, Z are not physically real, since they contain negative values of some tristimulus value, but it has great computational advantages to define them as such. These stimuli have been determined with the following conditions:

1. All actual colors must be within the triangle XYZ.
2. The line XY must be tangent to the *r-g* curve in the region of the red colors, where it is almost a straight line. This means that these red colors will have no Z stimulus, but only combinations of X and Y.

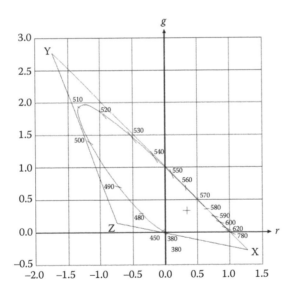

FIGURE 13.5 A chromaticity diagram for *r* versus *g*.

TABLE 13.6 Values of the Stimuli *r, g, b*, for the New Stimuli *X, Y, Z*

Stimulus	*r* value	*g* value	*b* value
X	1.275	−0.278	0.003
Y	−1.739	2.767	−0.028
Z	−0.743	0.141	1.602

3. The line ZX must be the locus of the points with zero luminance. This line of zero luminance in any chromaticity diagram is called alychne. The equation for this line can be written as

$$L_R r + L_G g + L_B b = 0. \tag{13.13}$$

L_R, L_G, and L_B are in Table 13.4. This means that the colors on the line XZ that do not have Y stimulus are the colors with zero luminance. The tristimulus $\bar{y}(\lambda)$ resulting from this selection of stimuli is identical to the relative light efficiency function of the eye, that is,

$$\bar{y}(\lambda) = V(\lambda). \tag{13.14}$$

4. The line ZY must be tangent to the horseshoe at the top, and form the triangle with minimum area.

$$\bar{x}(\lambda) = 2.7689 L_R \bar{r}(\lambda) + 0.38159 L_G \bar{g}(\lambda) + 18.801 L_B \bar{b}(\lambda), \tag{13.15}$$

$$\bar{y}(\lambda) = L_R \bar{r}(\lambda) + L_G \bar{g}(\lambda) + L_B \bar{b}(\lambda), \tag{13.16}$$

$$\bar{z}(\lambda) = 0.12307 L_G \bar{g}(\lambda) + 93.066 L_B \bar{b}(\lambda). \tag{13.17}$$

The resulting color matching functions $\bar{x}(\lambda), \bar{y}(\lambda) \, y \, \bar{z}(\lambda)$ are shown in Figure 13.6. Like the functions $\bar{r}(\lambda), \bar{g}(\lambda) \, y \, \bar{b}(\lambda)$, they represent the amount of each of these stimuli that is necessary to obtain the desired monochromatic spectral color.

After obtaining these color matching functions, the tristimulus values are found by means of the following relationships:

$$X = \int_0^\infty P(\lambda) \bar{x}(\lambda) d\lambda, \tag{13.18}$$

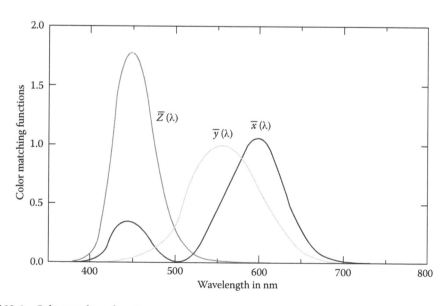

FIGURE 13.6   Color matching functions.

$$Y = \int_0^\infty P(\lambda)\bar{y}(\lambda)\mathrm{d}\lambda \qquad (13.19)$$

and

$$Z = \int_0^\infty P(\lambda)\bar{z}(\lambda)\mathrm{d}\lambda, \qquad (13.20)$$

where $P(\lambda)$ is the spectral emittance of the source. The luminance V of a luminous object can be found with the relation

$$V = \int_0^\infty P(\lambda)V(\lambda)\mathrm{d}\lambda, \qquad (13.21)$$

from where we can see that the tristimulus Y is equal to the luminance of the luminous object, that is,

$$Y = V(\lambda). \qquad (13.22)$$

As before, these tristimulus values can also be normalized to obtain the chromaticity coordinates, without units, and independent of the luminosity of the object, using the relationships

$$x = \frac{X}{X+Y+Z}, \qquad (13.23)$$

$$y = \frac{Y}{X+Y+Z} \qquad (13.24)$$

and

$$z = \frac{Z}{X+Y+Z}. \qquad (13.25)$$

Figure 13.7 shows these standard functions called chromaticity coordinates. If we graph now y against x, we obtain the horseshoe curve of Figure 13.8. This curve, called a chromaticity diagram, was proposed in 1931 by the CIE (Commission Internationale de l'Éclairage) and is similar to the curve described above, but with the great advantage that any real color can be represented with a sum of stimuli positive.

As in the diagram r, g, b, any color has a unique location within this diagram. The spectrally pure colors are on the curve, starting with the violet in the lower left, passing through the green at the top and ending with the red at the bottom right. In the lower line are the purples, which are combinations of red and violet.

At the center of the diagram, with the coordinates (⅓, ⅓), is the color white. However, a white color with constant energy per unit wavelength range does not exist in nature, so the CIE defined a more practical white color with the coordinates (0.3101, 0.3163).

Let us now consider a straight line from white to a spectrally pure color or to a purple, whatever it may be. Any color on this line can be considered as the sum of monochromatic or purple light, of the color defined by the end of the segment, with white light. It is customary to say that color is all the more saturated the closer it is to monochromatic or purple color.

Let us now consider two colors either A and B in the diagram. Any color on the line that joins them can be formed by an additive mixture of these two. Generally, any color within the triangle ABC can be formed with an additive mixture of the three colors, in the appropriate proportions, as in the diagram r, g.

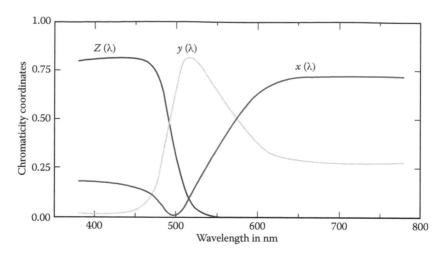

FIGURE 13.7   Chromaticity coordinates for spectrally pure colors.

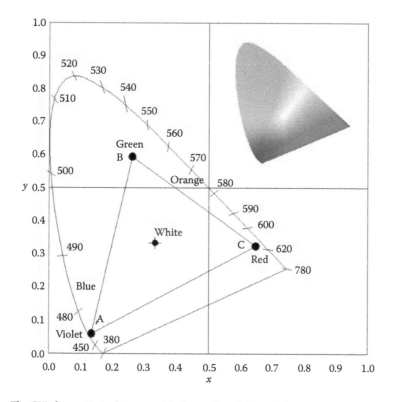

FIGURE 13.8   The CIE chromaticity diagram with three colors A, B, and C.

Two slightly different colors can be visually differentiated only if they have a certain minimum distance from each other in the chromaticity diagram. This distance is not constant, but depends on its position and orientation in the diagram. This tolerance or uncertainty is represented visually by means of the MacAdam ellipses, which are illustrated in Figure 13.9. The lengths of the axes of these ellipses indicate the minimum distance to differentiate two colors. For clarity, the ellipses have been drawn ten times larger.

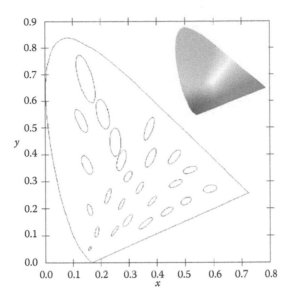

FIGURE 13.9 Color discrimination with McAdam ellipses. The ellipses are three times larger.

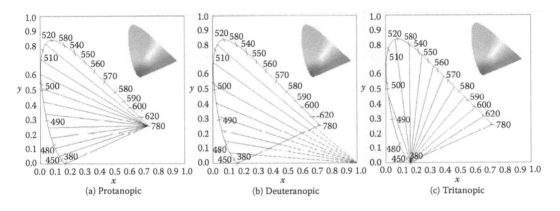

FIGURE 13.10 Lines of constant hue (a) protanopes, (b) deuteranopes, and (c) tritanopes.

13.4 Applications of the Chromaticity Diagram

Examining this diagram we see why it is impossible to choose three primary colors by means of which all others can be obtained. A better but not perfect selection would be to have the three colors at approximately the following wavelengths: red (700 nm), green (530 nm) and violet (410 nm). By means of this diagram, the reproducible fidelity of the color of any color image forming system can be judged quickly. For example, the blue, yellow, and red matches of color television have the positions indicated by points A, B, C, so a TV picture will only have colors within this triangle.

With this diagram, we can also more easily describe the colors that a colorblind confuses. Figure 13.10 shows on the chromaticity diagram lines that represent the geometric location of the colors that two types of color blinds confuse.

As a final example, Figure 13.11 presents the possible colors that a black body can take at different temperatures.

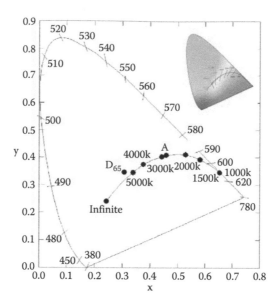

FIGURE 13.11   Colors of black body light sources for different temperatures.

13.5 Influence of the Illumination and Filters on the Color

In the color studies, several very important radiometric definitions are defined as follows:

a. The radiant flux Φ_R, is the total luminous power in watts carried by a light beam. The spectral radiant flux $\Phi_R(\lambda)$ is the total luminous power emitted in watts (joules/second), per unit wavelength, carried by a light beam, in such a way that

$$\Phi_R = \int_0^\infty \Phi_R(\lambda)d\lambda. \tag{13.26}$$

b. The irradiance E is the radiant flux per unit area, received by an illuminated body. The spectral irradiance $E(\lambda)$ is the radiant flux per unit area, per unit wavelength, received by an illuminated body, such that

$$E = \int_0^\infty E(\lambda)d\lambda. \tag{13.27}$$

c. The radiance L is the radiant flux per unit area, perceived in the direction of observation, by stereoradian, emitted by a luminous body. Thus, we can write

$$L = \frac{1}{\cos\theta} \frac{d^2 \Phi_R}{d A d\Omega}. \tag{13.28}$$

In a manner analogous to the above definitions, the spectral radiance $L(\lambda)$ is the radiant flux per unit area, per unit wavelength, perceived in the direction of observation, by steradian, emitted by a luminous body, in such a way what

$$L = \int_0^\infty L(\lambda)d\lambda. \tag{13.29}$$

d. The radiating exitance (or emittance) M of a luminous body is the radiant flux per unit area, emitted by a luminous body. The spectral radiance exit $M(\lambda)$ is the radiant flux per unit area, per unit wavelength, emitted by an illuminated body, in such a way that

$$M = \int_0^\infty M(\lambda)d\lambda. \tag{13.30}$$

e. The spectral reflectance $R(\lambda)$ is defined as the quotient of the spectral radiance exit $M(\lambda)$ emitted by an illuminated opaque body, divided by the spectral irradiance $E(\lambda)$ received by this body.

f. The spectral transmittance $T(\lambda)$ is defined as the quotient of the spectral radiance exit $M(\lambda)$ emitted by an illuminated colored transparent body divided by the spectral irradiance $E(\lambda)$ received by this illuminated colored transparent body.

For colored opaque bodies the spectral radiance exit $M(\lambda)$ can be written as

$$M(\lambda) = R(\lambda)E(\lambda). \tag{13.31}$$

It is easy to see that if two bodies have different power spectral distribution, they can have the same color when they are illuminated with some light, but different with another type of illumination. Two bodies have the same color under different lighting conditions only if they have the same reflectance curve.

Even if two light sources have different power spectral distributions, their colors can be matched by the appropriate filter, as long as their spectral distributions are large enough to overlap. This is widely used in photography to give a tungsten lamp the same color as day light or vice versa, depending on the type of film being used.

13.6 Color Mixtures

Two or more colors can be mixed to produce a third by two basically different methods. These methods are addition and color subtraction, which will be discussed in the following sections.

These addition and subtraction operations are more easily performed if the tristimulus values are known. These values can be easily obtained from the chromaticity coordinates, using Equations 13.22 through 13.25, as follows:

$$X = \frac{x}{y}V(\lambda), \tag{13.32}$$

$$Y = V(\lambda) \tag{13.33}$$

and

$$Z = \frac{z}{y}V(\lambda). \tag{13.34}$$

13.6.1 Color Addition

The addition of color is effected when the spectral power distributions of two or more light beams are added. One procedure to obtain the resulting color would be to sum these spectral distributions and then calculate their color. However, in general, only the colors are known, that is, their locations in the chromaticity diagram. The method to be followed is to first calculate the tristimulus values of each of the color sources with Equations 13.28 through 13.30. The next step, justified by Equations 13.18 through 13.20, is to add the tristimulus values as follows:

$$X_T = X_1 + X_2 + X_3 + ..., \tag{13.35}$$

$$Y_T = Y_1 + Y_2 + Y_3 + ... \tag{13.36}$$

and

$$Z_T = Z_1 + Z_2 + Z_3 + ... \tag{13.37}$$

The final step is to calculate the chromaticity coordinates with Equations 13.23 through 13.25, which for the case of two colors can only be written

$$x_T = \frac{X_1 + X_2}{X_1 + Y_1 + Z_1 + X_2 + Y_2 + Z_2} \tag{13.38}$$

and

$$y_T = \frac{Y_1 + Y_2}{X_1 + Y_1 + Z_1 + X_2 + Y_2 + Z_2}, \tag{13.39}$$

which, remembering that Y is equal to luminance $V(\lambda)$, it is possible to transform into

$$x_T = \left(\frac{V_1 / y_1}{V_1 / y_1 + V_2 / y_2} \right) x_1 + \left(\frac{V_2 / y_2}{V_1 / y_1 + V_2 / y_2} \right) x_2 \tag{13.40}$$

and

$$y_T = \left(\frac{V_1 / y_1}{V_1 / y_1 + V_2 / y_2} \right) y_1 + \left(\frac{V_2 / y_2}{V_1 / y_1 + V_2 / y_2} \right) y_2. \tag{13.41}$$

These results indicate that the resulting color of the sum of two light sources of color is another color on the straight line joining them, where their position is given by the center of gravity of the two colors, which have weights given by V_1/y_1 and V_2/y_2, respectively.

Figure 13.12 shows the colors obtained by superimposing three light beams: red, green, and blue, with similar luminances. Typical examples of color addition are as follows:

a. The formation of a color television image, where the image is formed by small points of three primary colors: red, green, and blue. As the eye cannot see each of these points individually, its three colors merge by joining the retina.
b. The projection of color images, using the method of Maxwell, with three projectors that each project an image of different color.
c. Mixtures of paints whose color particles are opaque, such as oil paints.

13.6.2 Color Subtraction

Color subtraction occurs when spectral power distributions multiply rather than add up. Unfortunately, from this definition, it is easy to see that it is not enough to know the chromaticity coordinates of both colors, as in the case of addition, but also to know their spectral power distributions.

To combine two colors by subtraction, their spectral distribution powers have to be wide, or in other words, they cannot be monochromatic. In the case of the addition of two monochromatic colors produce

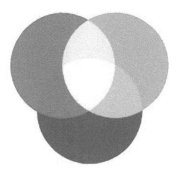

FIGURE 13.12 Three colors (red, green, and blue) combined by color addition.

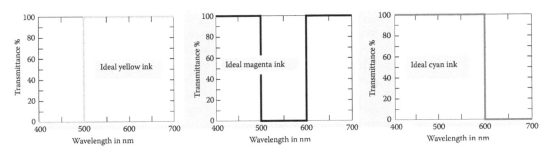

FIGURE 13.13 Spectral characteristics of three ideal filters that can be used to reproduce color by subtraction.

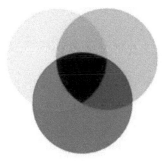

FIGURE 13.14 Three colored filters (cyan, magenta, and yellow) combined by color subtraction.

a third color, whereas in the case of subtraction the result would be black. When subtraction is done, the resulting color is not in the straight line that binds them, as in the case of addition.

To better understand this process, let us consider three ideal color filters, whose transmissions are shown in Figure 13.13. The transmission of these filters in their region of maximum transparency is 100%. Its transmissions in the regions of minimum transparency are variable, depending on its thickness. Overlapping these filters, as shown in Figure 13.14, it can be seen that:

a. The thickness of the magenta filter controls the transmission of the combination between 500 and 600 nm.
b. The thickness of the yellow filter controls the transmission of the combination in the region between 400 and 500 nm.
c. The thickness of the cyan filter controls the transmission of the combination in the region between 600 and 700 nm.

Therefore, from this, you can see that the magenta filter controls the transparency of the combination to green light, the yellow filter to blue light and the filter cyan to red light. From this, it can be concluded

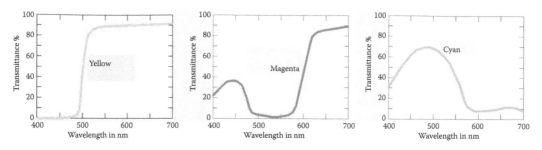

FIGURE 13.15 Three practical filter transmissions for color subtraction.

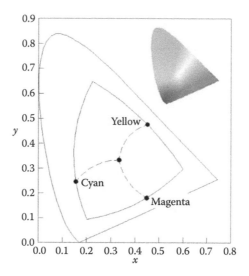

FIGURE 13.16 Colors obtained by subtraction using cyan, magenta, and yellow.

that the subtractive combination of these three filters is equivalent to the additive combination of red, blue, and green. Figure 13.15 shows the actual filters used in color photography which, as we can see, are very similar to the ideal filters in Figure 13.13. The color positions of these filters in the chromaticity diagram are shown in Figure 13.16, where the colors obtainable with them are illustrated with a dotted line.

This color subtraction process occurs when multi-colored transparent filters are superimposed, or when multi-colored transparent liquids are mixed, or when transparent liquids of different color are mixed. If liquids are opaque, as in the case of oil paints, the combination is additive rather than subtractive because light does not penetrate but is reflected in each particle of color. There are intermediate cases in which the mixture is made with an intermediate process between the additive and the subtractive. A typical example of color subtraction is color photography.

13.7 Other Color Representations

As we have seen in the CIE diagram of color, two colors seemingly equal in hue are not alike distant in the diagram. This is evident in Figure 13.9, where MacAdam ellipses are constant size. This represents a practical problem that some other representations of color have tried to solve.

The artist and master Albert H. Munsell in 1905 described color in a three-dimensional space, with polar coordinates in the x-y plane, as value (luminosity) shown in Figure 13.17. The height represents the luminosity so that the base of the cylinder is totally black and the intersection of the axis with the cylinder cover is white. The radial distance of the axis is the chroma or saturation of the color, so that there is no color in the

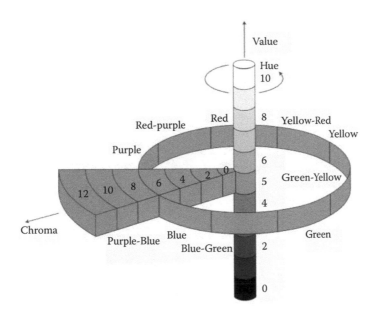

FIGURE 13.17 The Munsell color space.

axis and in the periphery, with a value of 10, we have very saturated colors like those of the spectrum. The angular coordinate would be the tone. The angular distribution of colors is formed in such a way that equal angle represents equal difference apparent in the tonality. The shades in this diagram were defined subjectively. In addition, the colors do not vary continuously, but by jumps, in samples of a certain color.

Another similar system is the CIELAB, widely used in the graphic arts. The origins of this system go back to Edwald HerIng, who, in 1870, postulated the existence of three receptors in the eye, one for luminosity and two for color. The two receptors for color according to this theory are: one to discriminate between two antagonistic colors, green and red, and another to discriminate between another pair of antagonistic colors, blue and yellow. The colors in this system cielab as in the Munsell representation, is in a three-dimensional space, where the luminosity L^* is the vertical axis. One of the two axes, called A^*, in the horizontal plane, is for the colors red and green. In the positive direction of this, it increases the content of red and decreases that of green and vice versa in the negative direction. On the other axis, the B^*, the relative contents of the blue and yellow colors are defined. A diagram with the approximate representation of the colors in this system is shown in Figure 13.18. In this system, the representation of colors if it is quantitative continuous, not discreet and by jumps like in the system of Munsell. Otherwise, the systems are so similar that the coordinates for each color are similar in both representations.

Another system to represent color is the CIELUV, widely used in color television. This is obtained with a transformation of chromaticity diagram x-y of the color, with the transformations equations:

$$u = \frac{4x}{-2x + 12y + 3} \tag{13.42}$$

and

$$v = \frac{6y}{-2x + 12y + 3}. \tag{13.43}$$

Thus, we obtain the diagram of Figure 13.18, where the colors are represented by the coordinates u^*, v^*. This transformation is not linear, as expected; the relative distances between colors have been modified. Grassmann's laws are no longer valid in this diagram, but in return, the representation is much more uniform, as desired.

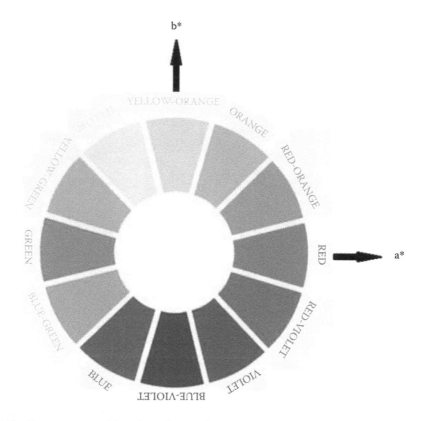

FIGURE 13.18   Representation of the colors of a given luminance in the color system L*, a*, b* in the plane a*, b*.

13.8 Light Sources and Illuminants

A light source is characterized by its spectral distribution of the light power. An illuminant is an ideal light source with certain characteristics that allow it to function as a reference. In daily life, we find very different light sources, both natural and artificial. The light of day, against what we could imagine, is not very constant, even on clear days, because it depends on the hour. At dawn and dusk, it is more reddish than at noon. On the other hand, as photographers know, indoors, especially at dusk, everything appears more bluish than outdoors. At night, with artificial light sources, the situation is even more variable. On the one hand, we have tungsten incandescent lamps, whose spectrum is similar to that produced by a hot black body. On the other hand, are fluorescent lamps. Because of metamerism, two objects may look exactly the same color as they are illuminated by a light source, but of very different color when illuminated by a different one. Therefore, to be reasonably certain that two objects have the same color, regardless of which light source illuminates them, it is necessary to use various types of illumination when observing them. The most commonly used light sources are those that approach the following illuminants:

a. Illuminant type A, which can be reproduced with an incandescent tungsten lamp with a color temperature of 2888 K.
b. Cold white fluorescent lamp, to simulate in the interior of commercial stores.
c. Illuminant D65, which emulates in daylight. It can be approximated in practice using an incandescent tungsten lamp as the type A light source, but with a suitable filter on the front, to give it a color temperature of 6500 K.

These illuminants have the spectra $P(\lambda)$, illustrated in Figure 13.19, and have the enormous advantage that they provide constant reference illumination conditions for industrial applications.

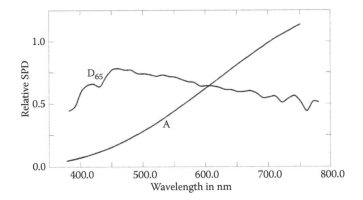

FIGURE 13.19 Spectra for illuminants D65 and A.

13.9 Colorimeters

A colorimeter is an instrument designed and built to measure color. These can be of very different types but basically there are three:

1. *Visual colorimeters.* They compare the color of two fields side by side. One of the fields has the color to be measured and the other is of adjustable color, illuminated by well-defined and well-known light sources. The luminous powers of these three colors are adjusted until the two fields appear the same color. In this way, the color is determined by the amounts of light energy provided by each of the sources.
2. *Spectrophotometers.* Another way to measure color is to determine the spectral distribution $P(\lambda)$ by means of a movable or multiple photometer, coupled to a spectroscope, to measure the entire spectrum of light. The tristimuli X, Y, and Z are then calculated by the integrals in Equations 13.18 through 13.20, using the known values of the color matching functions.
3. *Colorimeters of tristimuli.* In order to calculate the tristimuli X, Y, and Z, it is not necessary to know the spectral distribution $P(\lambda)$ of the sample to be measured and then to do the integration numerically. The luminous power of the object to be measured is measured using a photometer with a front filter. The three filter-photometer combinations have spectral responses very close to the color matching functions $\bar{x}(\lambda),= \bar{y}(\lambda)$ and $\bar{z}(\lambda)$. Thus, the luminous power measured will be the integral in Equation 13.18, that is, the value of tri-stimulus X. The other values of Y and Z tristimulus are similarly measured. The critical and most difficult part in practice is the elaboration of the three filters with response similar to the spectral curves of the color matching functions, but can be achieved by suitable combinations of filters.

13.10 Light Sources

Light sources can be of very different nature, with very different characteristics of color, luminous efficiency, and ways to produce light. The following are some of the most important sources.

a. *Incandescent lamp.* Incandescent lamps emit their light by heating a filament of tungsten by means of an electric current that runs secularly through it. Its emission is very similar to the one that emits a black body like the one studied in chapter VI, only multiplied by a constant factor. At common tungsten temperatures, the peak of its emission is in the infrared. Although the sun, which is the most important natural light source, also emits as a black body, the color of a tungsten lamp is much redder because its temperature is much lower. Figure 13.20 illustrates the spectral distribution of an incandescent lamp. Of course, a lamp with a higher temperature, that is, with more power, not only has more luminosity, but its color is whiter.

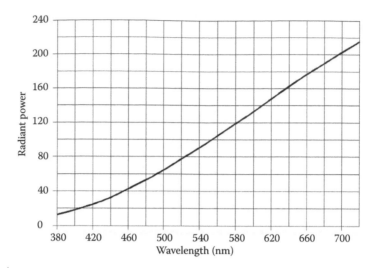

FIGURE 13.20   Spectral distribution of an incandescent lamp.

Tungsten lamps have an inert gas therein, typically nitrogen. In some lamps, in addition to nitrogen, they have a small amount of iodine or bromine for the purpose of increasing lamp life. This is achieved by a process known as the halogen cycle, which deposits the evaporated tungsten back into the filament. Thus, in addition to increasing the life of the lamp, prevents the glass globe from opaque with evaporated tungsten. However, this halogen cycle requires slightly higher tungsten temperatures. As glass does not easily withstand higher temperatures, molten silicon is used instead. A consequence of the highest temperature is that the color is a little more white.

As most of the emission of these lamps is in the infrared, much of the energy emitted manifests as heat and not as light, making these lamps very inefficient.

b. *Lamps of electric discharge in gas vapor of metal.* These lamps have in the interior of globe or tube of glass a gas, like hydrogen, neon, helium or any other. They may also contain a metal vapor, for example of mercury or of sodium. Within the cavity are two metal electrodes to which a voltage is applied, sometimes DC, but most often alternating (AC). If the voltage is high enough, the current discharge begins immediately. Otherwise, a higher voltage pulse is sometimes applied instead of the operating pulse to start the discharge. Some other times, the electrodes are heated. If DC is used the electrode to be heated is the negative, which is, the cathode, because it is the one that emits the electrons. If using AC, the two electrodes are heated.

Once the current begins, the electrons collide with the atoms of the gas or vapor and form a flow of positive ions in the opposite direction, thus transferring their energy to the atoms of the gas, causing them to emit light. If the pressure of the gas or vapor is low, on the order of a few millimeters of mercury (torr), the light emitted is the characteristic spectrum of the gas, with very narrow spectral lines. If the pressure is increased to a few hundred torrs, the electric current can be much greater, and thus the luminosity of the lamp, but that produces a great widening of the spectral lines, making the light emitted by each less monochromatic spectral line. Sodium or mercury lamps used on the streets are of this type.

The great problem with sodium or mercury vapor lamps is that despite the widening of the spectral lines, their light is far from being white and each has a characteristic color. The colors are very disturbed with any of these lamps.

Mercury lamps emit many spectral lines in violet and ultraviolet. However, ultraviolet does not come out because the glass that contains it is not transparent at these wavelengths. Ultraviolet is desired in the so-called germicidal or black light lamps, instead of molten silicon (quartz) glass, which is transparent to ultraviolet.

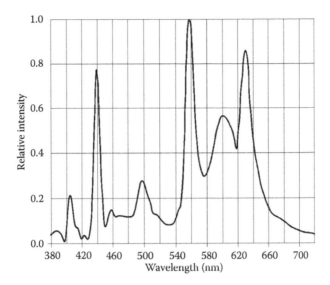

FIGURE 13.21   Spectral distribution of a fluorescent lamp.

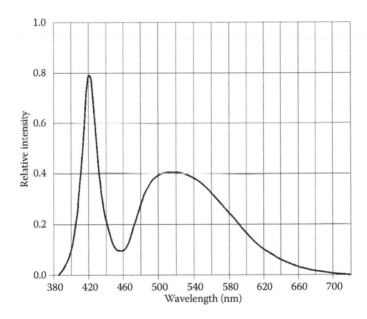

FIGURE 13.22   Spectral distributions of three LEDs with different colors.

   c. *Fluorescent lamps.* In order to improve the efficiency of low pressure mercury vapor lamps, the inner surface of the tube is coated with a fluorescent powder. This powder is a mixture of rare earth salts. As ultraviolet light from the lamp reaches the dust, this energy is transformed into visible light, transforming the lamp into a continuous visible spectrum, similar to daylight. Figure 13.21 shows the spectrum of a fluorescent lamp.

   d. *Light emitting diodes.* Light emitting diodes are semiconductor diodes that emit light when they are circulated by an electric current. Like solid state rectifier diodes, the electric current flows only in one direction. The LEDs emit a fairly broad, but relatively monochromatic, spectral line, as shown in Figure 13.22.

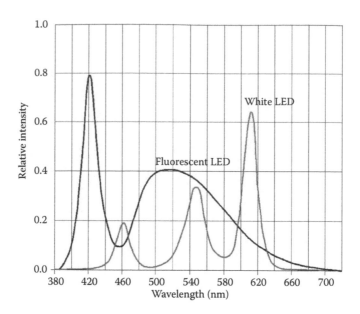

FIGURE 13.23 Compared spectral distributions of a white LED and a fluorescent LED.

But they have been able to manufacture white light by two methods. The most common method is to combine three LEDs of different colors, red, green, and blue. If their relative intensities are adjusted properly, the desired color can be obtained. The second method to produce white light is to use a phosphorescent material similarly to fluorescent lamps. Figure 13.23 shows the spectral distributions of a white LED consisting of three red, green, and blue diodes and one fluorescent LED.

The luminous efficacy of LEDs is one of the highest, exceeding 100 lumens per watt. In comparison, the efficiencies of fluorescent lamps of around 60–90 lumens per watt and for incandescent lamps are much lower. LEDs also have the additional advantage of having a very long service life of several thousand hours.

Suggested Reading

1. Thimann KV. Autumn colors, in *Scientific American*, October 1950.
2. Smith N. Color television, in *Scientific American*, December 1950.
3. Milne LJ, Milne MJ. How animals change color, in *Scientific American*, March 1952.
4. Timbergen N. Defense by Color, in *Scientific American*, October 1957.
5. Land EH. Experiments in Kenneth Color Vision, in *Scientific American*, May 1959.
6. Evans RM. Maxwell's color photograph, in *Scientific American*, November 1961
7. Rushton WAH. Visual pigments in man, in *Scientific American*, November 1962.
8. Brindley GS. Afterimages, in *Scientific American*, October 1963.
9. Clevenger S. Flower pigments, in *Scientific American*, June 1964.
10. Mac Nichol Jr. ER. Three pigment color vision, in *Scientific American*, December 1964.
11. Neisser U. The processes of vision, in *Scientific American*, September 1968 reprint in *Lasers and Light*, Arthur L. Schawlow (comp.), W. H. Freeman and Co., San Francisco, 1969.
12. Gregory RL. Visual Illusions, in *Scientific American*, November 1968.
13. Michael CR. Retinal processing of visual images, in *Scientific American*, May 1969.
14. Young RW. Visual cells, in *Scientific American*, October 1970.
15. Pettigrew JD. The neurophysiology of binocular vision, in *Scientific American*, August 1972.

16. Beck J. The perception of surface color, in *Scientific American*, August 1975.

17. Rushton WAH. Visual pigment and color blindness, in *Scientific American*, March 1975.

18. Favreau OE, Corballis MC. Negative after-effects in visual perception", in *Scientific American*, December 1976.

19. Land EH. The retinex theory of color vision, in *Scientific American*, December 1977.

20. Nassau K. The causes of color, in *Scientific American*, October 1980.

21. Nijhout HF. The color patterns of butterflies and moths, in *Scientific American*, November 1981.

22. Levine JS, Mac Nichol Jr. ER. Color vision in fishes, in *Scientific American*, February 1982.

23. Newman EA, Hartline PH. The infrared vision of snakes, in *Scientific American*, March 1982.

24. Brou P, Sciacia TR, Linden L, Lettvin JY. The color of things, in *Scientific American*, September 1986.

25. Vasyntin UV, Tishchenko AA. Space coloristics, in *Scientific American*, July 1989.

26. Romer GB, Delamoir J. The first color photographs, in *Scientific American*, December 1989.

27. Nathans J. The genes for color vision, in *Scientific American*, December 1989.

28. Jacobs GH, Nathans J. The evolution of primate color vision, in *Scientific American*, April 2004.

29. Billock VA, Tsou BH. Seeing forbidden colors, in *Scientific American*, February 2002.

30. Werner JS, Pinna B, Spillmann L. Illusory color and the brain, in *Scientific American*, March 2003.

31. Ball P. Understanding how animals create dazzling colors could lead to brilliant new nanotechnologies, in *Scientific American*, May 2012.

32. Howard WE. Better displays with organic films, in *Scientific American*, February 2004.

33. Luria SM. Color vision, in *Physics Today*, March 1966.

34. Mac Adam DL. Color Photography, in *Physics Today*, January 1967.

35. Mueller CG, Rudolph M. *Light and Vision*, Time-Life International, 1963.

36. Minnaert MJ. *The Nature of Light and Color in the Open Air*, New York, NY: Dover Publications, 1954.

37. Williamson SJ, Cummins HZ. *Light and Color in Nature and Art*, New York, NY: John Wiley & Sons, 1983.

38. Malacara D. *Color Vision and Colorimetry, Theory and Applications*, Second edition, Bellingham, WA: SPIE Pres, 2011.

14

The Human Eye and Its Aberrations

14.1	Introduction	543
14.2	The Human Eye	543
	Schematic Eye Models	
14.3	Chromatic Aberration	547
14.4	Defocus and Axial Astigmatism	549
14.5	Spherical Aberration and Irregular Astigmatism	550
14.6	Ocular Aberrometry	551
14.7	Higher-Order Aberrations	552
14.8	Correction of Higher-Order Aberrations	555
14.9	Peripheral Aberrations	556
14.10	Summary	556
	References	557

Jim Schwiegerling

14.1 Introduction

The human eye is a marvel of evolution. Its extraordinary imaging capabilities cover a field of view of nearly 180° and it can adjust its power (at least when young) to focus on objects ranging in distance from 70 mm to infinity. However, as with all optical systems, the human eye suffers from aberrations. These aberrations naturally change with field position and with object distance. They ultimately limit the quality of the image formed on the retina. In addition, aberrations are unique to the individual. As with all other traits, the dimensions of the ocular surfaces vary between individuals, so any analysis of ocular aberrations is going to reflect this variation. This chapter examines the inherent aberrations of the human eye and describes some of the mechanisms humans have developed both through evolution and through external appliances that can mitigate degradations in image quality induced by aberrations. A brief overview of the components of human eye as well as a schematic eye model are provided below. The schematic eye model is suitable for any ray tracing package and can be used to understand the inherent aberrations of the "average" eye. With this model, both chromatic and monochromatic aberrations are explored along with their nominal correction.

14.2 The Human Eye

A wide variety of books are available describing the optical properties of the human eye [1–3]. Below a brief summary of the various elements and their optical function is provided. Figure 14.1 shows a diagram of the human eye. This drawing shows a top view of the right eye. The primary refractive components of the eye are the cornea, which is the clear membrane on the front of the eye, and the crystalline lens, which lies immediately behind the iris. The two components are separated by a watery

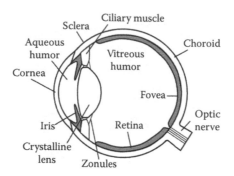

FIGURE 14.1   A top view of the right human eye.

fluid called the aqueous humor. The cornea is an aspheric meniscus lens that provides the bulk of the refraction of the light entering the eye. Its anterior surface has a refractive interface with air and consequently a large change in refractive index occurs at this boundary. Approximately two-thirds of the refractive power of the eye occurs at this surface. The posterior surface of the cornea has less impact on the incident rays since the index difference at the cornea/aqueous boundary is small. The crystalline lens is a gradient index component that is also flexible. The crystalline lens is suspended in the eye by the zonules, which connect the lens periphery to the ciliary muscle. When the ciliary muscle is relaxed, the muscle expands, the tension on the zonules increases and the lens equator is stretched. This stretching has the effect of flattening the radii of the crystalline lens surfaces, thereby reducing its overall power. When the ciliary muscle constricts, the tension on the zonules is reduced and the crystalline lens surfaces become more curved, increasing its power. Accommodation is the term describing the adjustment of crystalline lens power. In the young eye, the crystalline lens power can change by upwards of 14 diopters (D). This means that an eye focused at infinity when accommodation is relaxed, can focus on an object 70 mm away when fully accommodated. Unfortunately, the range of accommodation reduces with age, reducing to about 3 D at age 40 and to near zero after age 50. The iris lies immediately anterior to the crystalline lens. The iris acts as the aperture stop of the system and typically ranges in size from 2 mm under bright lighting conditions to 8 mm under extremely dim lighting. The retina lines the inside of the eyeball and serves as the image surface. The retina consists of an array of photosensitive cells and the corresponding neural wiring needed to record and transmit the image to the brain. There are two types of photosensitive cells: rods and cones. The cones, in turn, come in three varieties, which are sensitive to red, green and blue wavelength bands in the visible spectrum. Taken as a whole, the cones are sensitive to wavelengths between 380 nm and 780 nm. The peak sensitivity occurs at 555 nm, and the sensitivity falls off toward the red and blue ends of the spectrum. The rods are sensitive to only a single waveband ranging from 380 nm to 680 nm, and peaking at 505 nm. The cones are primarily concentrated in the fovea, which is the central portion of our visual field. The density of cones reduces quickly outside a field of 10° and they become more widely dispersed and larger in size in the remainder of the retina. The rods are non-existent in the fovea, but peak in density for field angles of 20° and then gradually reduce in density toward the periphery. The crystalline lens and the retina are separated by a gelatinous material called the vitreous humor. The image space of the eye therefore has a different value than the object space, so the principal planes of the eye do not coincide with the nodal points. The index of the vitreous humor also needs to be taken into account when calculating many of the paraxial quantities of the eye. The following section defines a schematic eye model that is useful for understanding these paraxial quantities and cardinal points. Finally, in general the surfaces of the cornea and the crystalline lens are not rotationally symmetric, nor are they necessarily coaxial. Eye models typically ignore this effect to simplify the calculations of paraxial quantities.

14.2.1 Schematic Eye Models

A wide array of schematic eye models have been proposed. They vary widely in sophistication and in their ability to predict optical effects in the eye. These range from *Exact* eye models, which typically have four to six spherical optical surfaces, representing the anterior and posterior cornea, the anterior and posterior crystalline lens, and possibly the anterior and posterior surfaces of a high index crystalline lens core. Examples of such "Exact" models include: the Gullstrand number 1 model [4] and the Le Grand full theoretical eye [2]. Simplified schematic eye models, where the cornea is considered as only a single refracting surface, have been proposed as well. Examples include the Gullstrand-Emsley eye [5] and the Bennett-Rabbetts eye [6]. The "Exact" and simplified eye models above avoid the complexity of the gradient index structure of the crystalline lens by using instead a homogeneous effective refractive index. The Gullstrand number 1 model uses a homogeneous high index core, surrounded by a lower homogeneous index shell. The other eye models simply use a single homogeneous index for the whole of the crystalline lens. Another simplification of schematic eye models is to consider a single refracting surface model, or a so-called reduced eye. Examples of reduced eyes include the Emsley eye [5] and the Indiana chromatic eye [7]. Here, the complexity of the gradient index crystalline lens is avoid completely by eliminating the crystalline lens altogether.

The "Exact," simplified and reduced eyes are useful for predicting paraxial properties of the human such as image size, cardinal point location and overall power. The Indiana chromatic eye goes a step further and predicts the longitudinal chromatic aberration of the eye. However, all of these eyes overpredict the amount of spherical aberration in the human eye. The high levels of spherical aberration predicted by these models are due to use of spherical surfaces for the refractive elements. Consequently, these models should *not* be used to model any aberration properties of the eye.

The human eye uses aspheric surfaces to reduce the overall spherical aberration of the eye. Finite schematic eye models have been proposed, which incorporate aspheric surfaces and better mimic the aberration aspects of the eye. Examples of finite schematic eye models include the Lotmar [8] and Kooijman [9] eye models, which are just variations of the Le Grand full theoretical eye with aspheric surfaces. Other examples include the Navarro eye [10,11], the Arizona eye [3], and the Liou-Brennan eye [12]. The finite schematic eye models provide a much better estimate of the aberration content of the eye and consequently should be used in situations where this information is relevant. The Navarro and Arizona eyes predict similar levels of spherical aberration and this level is consistent with modern measures of ocular spherical aberration. The Liou-Brennan model tends to predict much lower levels of spherical aberration and caution should be used with this model. Finally, both the Navarro and Arizona eye models provide variable surface parameters and material properties to model a continuous range of accommodation. These aspects provide utility when modeling the imaging properties for objects at different distances.

The Navarro and Arizona eye models maintain the previous simplification of using a uniform effective index for the crystalline lens. There are eye models that incorporate a full gradient index crystalline lens. The Liou-Brennan model uses two gradient structures to model the anterior portion and the posterior portion of the crystalline lens. Additional eye models with a single continuous gradient index structure have been proposed [13–15]. However, the data for the refractive index distribution, as well as how this distribution changes with accommodation and age is limited. For this reason, the homogeneous index models typically suffice for most modeling purposes. Here, the Arizona eye model will be updated to incorporate more recent data on the crystalline lens properties. This refined model, which will be called the AZEye15 model, will then be used to explore the aberrations of the human eye.

The previous version of the Arizona Eye model simply provided the *d*-light refractive index and the Abbe number. By switching to a full dispersion formula, the material properties are better predicted, especially for near-infrared wavelengths where many diagnostic instruments operate. For the AZEye15, the refractive indices of the ocular materials for the cornea, aqueous and vitreous are based on a study from Atchison and Smith [16]. The Atchison and Smith dispersion for the crystalline lens has been modified to raise its *d*-light index, but maintain its Abbe number. Since the crystalline lens index is an

TABLE 14.1 Cauchy Dispersion Coefficients for the Various Ocular Materials Assuming λ Is in nm

Coefficient	Cornea	Aqueous	Crystalline Lens[a]	Vitreous
a_0	1.363471	1.324294	1.420137	1.323638
a_1	6017.972	6083.028	6764.888	5565.628
a_2	-6.770081×10^8	-7.076535×10^8	-6.178756×10^8	-5.823028×10^8
a_3	5.916596×10^{13}	6.160249×10^{13}	5.974030×10^{13}	5.041691×10^{13}

[a] Note that the index of the crystalline lens is for the unaccommodated state.

effective index, there is flexibility in the choice of values for this element. This modification is performed so that best focus is achieved when average dimension values discussed further below are used. The refractive indices as a function of wavelength are given by a Cauchy-type dispersion formula as

$$n(\lambda) = a_0 + \frac{a_1}{\lambda^2} + \frac{a_2}{\lambda^4} + \frac{a_3}{\lambda^6}, \tag{14.1}$$

where $a_0 \ldots a_3$ are material dependent coefficients. Table 14.1 provides the coefficients for the various ocular materials for λ expressed in nm.

The shapes of the surfaces and spacing between them are based on work by Rozema et al. [17]. Specifically, the right eye averages for the anterior and posterior conic constants, the anterior chamber depth (i.e. the distance from the posterior cornea to the anterior crystalline lens surface), the crystalline lens radii and thickness, and the retinal radius have been incorporated from this study. Furthermore, the overall axial length of the eye has been used from this study as well. The anterior and posterior radii of the cornea and the corneal thickness have been kept from the earlier version of the Arizona eye model, as these numbers are historically consistent with multiple models and the values are nearly identical to the average values found in Rozema et al.

To incorporate the effects of accommodation into the AZEye15 model, the results from Ramasubramanian and Glasser have been used [18]. These investigators used corrected ultrasound bio-microscopy to examine the changes in anterior chamber depth, crystalline lens radii and thickness as a function of accommodation. The overall length of the eye is held constant with accommodation, so the distance from the posterior crystalline lens to the retina is adjusted with accommodation as well.

In terms of the asphericity of the crystalline lens surfaces, the values provided in the literature vary widely. For the AZEye15 model, the anterior conic constant value for Subject 1 in Ortiz et al. is used [19]. The posterior conic constant is then varied so that the ocular spherical aberration (discussed in detail below) matches clinical data. Since the radii of the crystalline lens change with accommodation, it is expected that the conic constant values will change as well. There is limited information regarding these changes. The rate suggested by Dubbelman et al. [20] has been used for the anterior crystalline lens conic constant. The change with accommodation of the posterior crystalline lens conic constant is again chosen to match clinical levels of spherical aberration.

Table 14.2 defines all of the parameters of the AZEye15 model. The variable A is the amount of accommodation in units of diopters. The value of A is the reciprocal of the distance (in meters) from the object to the corneal vertex. For example, to focus on an object at a distance of 333 mm, $A = 1/0.333$ m = 3 D. The refractive indices in the final column of Table 14.2 are calculated from Equation 14.1 and the corresponding coefficients summarized in Table 14.1.

Based on the preceding model eye parameters, the paraxial properties of the AZEye15 model can be calculated. The front focal length is $f_F = -16.222$ mm and the rear focal length $f_R' = 21.674$ mm, resulting in an overall ocular power of 61.64 D. The overall length of the eye is 23.656 mm, and the Petzval radius is -16.890 mm. The entrance pupil is located 2.866 mm behind the corneal vertex and the exit pupil is located 3.540 mm behind the corneal vertex. The magnifications of the entrance and exit pupils with respect to the iris diameter are 1.124× and 1.043×, respectively. Table 14.3 summarizes the locations

TABLE 14.2 The Parameters for the AZEye15 Model

Surface	Radius (mm)	Conic Constant	Thickness (mm)	Index
Anterior Cornea	7.8	−0.193	0.55	$n_{cor}(\lambda)$
Posterior Cornea	6.5	−0.100	2.87−0.0554A	$n_{aq}(\lambda)$
Iris	∞	0.000	0.00	$n_{aq}(\lambda)$
Anterior Crystalline Lens	10.43−0.8538A	−2.57−0.63A	4.07 + 0.0757A	$n_{lens}(\lambda)$ +0.000876A −0.000513A^2
Posterior Crystalline Lens	−6.86 + 0.2223A	−2.1364 −0.555A +0.1386A^2	16.16566−0.0203A	$n_{vit}(\lambda)$
Retina	−12.00	–	–	–

TABLE 14.3 Cardinal Points for the Unaccommodated AZEye15 Model

–	From Anterior Cornea Vertex (mm)	From Posterior Crystalline Lens Vertex (mm)	From Retina (mm)
Front Principal Plane	1.609	−5.881	−22.046
Rear Principal Plane	1.981	−5.509	−21.674
Front Focal Plane	−14.613	−22.103	−38.268
Rear Focal Plane	23.656	16.166	0.000
Front Nodal Point	7.061	−0.428	−16.594
Rear Nodal Point	7.434	−0.056	−16.222

of the cardinal points of the unaccommodated eye model with respect to the corneal vertex, the posterior crystalline lens vertex, and the retina.

14.3 Chromatic Aberration

The longitudinal chromatic aberration (LCA) of the eye has been measured by multiple groups and is remarkably consistent across these studies [7,21,22]. The ocular materials have refractive indices and an Abbe number similar to that of water. Consequently, the LCA of the eye should be expected to be similar to water. In fact, this is the case except for short wavelengths. Ocular LCA is typically provided in terms of diopters, with the assumption that LCA = 0 D at a reference wavelength. The average sodium doublet D-line, $\lambda_D = 589.29$ nm is a common choice for the reference wavelength. The LCA is defined as

$$LCA(\lambda) = \frac{n_{vit}(\lambda_D)}{\overline{P_D'F_D'}} - \frac{n_{vit}(\lambda_D)}{\overline{P_D'F_\lambda'}}, \tag{14.2}$$

where $\overline{P_D'F_D'}$ is the distance from the rear principal plane to the rear focal plane (i.e. the rear focal length) at the reference wavelength and $\overline{P_D'F_\lambda'}$ is the distance from the rear principal plane for the reference wavelength to the rear focal plane of the wavelength of interest. Note, that the last distance is not equivalent to the rear focal length at the wavelength of interest. Figure 14.2 shows the LCA of the AZEye15 model. The LCA, as well as the clinical data, is well-fit by the following expression

$$LCA(\lambda) = 1.68524 - \frac{633.46}{\lambda - 214.102}, \tag{14.3}$$

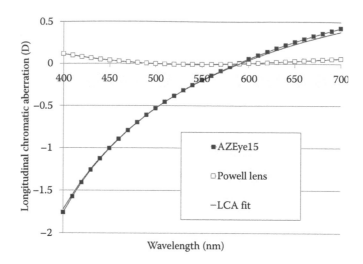

FIGURE 14.2 The longitudinal chromatic aberration (LCA) of the eye. The black squares show the values predicted from the AZEye15 model. The solid line is the fit to clinical LCA data given in Equation 14.3. The white squares show the residual LCA after placing a Powell-type lens in front of the AZEye15 model.

TABLE 14.4 Parameters of Powell-Type Achromatizing Lens

Radius (mm)	Thickness (mm)	Material	Semi-Diameter (mm)
∞	4.0	N-LAF21	7.5
−15.0	1.9	SF56A	7.5
∞	–	–	7.5

where λ is in nm. This fit is shown in Figure 14.2 as well. The LCA can be considered from an object space point of view as well. An eye that is corrected for the reference wavelength has the retina located at the rear focal point corresponding to that wavelength. For wavelengths shorter than the reference wavelength, the retina is conjugate to real object points. The LCA describes the vergence (or reciprocal of the object distance) for these points. For example, at a wavelength of 415 nm, the LCA is about –1.50 D. This corresponds to an object distance of $-1/1.50 = -2/3$ m or a point approximately 66 cm in front of the eye is conjugate to the retina at this wavelength. For wavelengths longer than the reference wavelength, the LCA is positive, corresponding to virtual object points behind the eye being conjugate to the retina at these wavelengths.

A variety of lens designs have been proposed to correct the LCA of the eye [23–25]. These lenses are typically zero-power cemented-doublet or cemented-triplet configurations. For example, Powell [25] provides a description of a zero-power cemented-doublet design for achromatizing the eye. To achieve the zero-power aspect of the lens, the two outer faces of the double are made plano and the d-line refractive indices of the two lenses are chosen to be nearly identical. Differences in dispersion of the two lens materials and the shape of the buried surface are then used to null the LCA of the eye. Table 14.4 summarizes the design of the Powell-type achromatizing lens, where the materials have been updated to currently available glasses. Figure 14.2 also shows the residual LCA when the Powell lens in Table 14.4 is placed in front of the AZEye15 model.

Correcting the LCA of the eye typically does not provide any visual benefit. Bradley et al. [26] provide a thorough summary of the limitations of correcting ocular LCA. The eye suffers from transverse chromatic aberration (TCA) as well, and the proposed lens designs do not correct for this effect leading to color fringing of objects viewed through the lenses. Also, being a non-rotationally symmetric system, the eye can suffer from x-tilt and y-tilt on axis, meaning that a polychromatic point source would be smeared into a spectrum even for on-axis viewing. Finally, the visual system has developed techniques to mitigate much of the effects of chromatic aberration. The blue-sensitive cones are much more widely dispersed so that the out-of-focus short wavelengths have limited sampling. In addition, the primary luminance signal for the visual system comes from the red- and green-sensitive cones, which have much lower levels of chromatic aberration between them. Consequently, correction of ocular chromatic aberration is typically not performed and visual instruments are typically corrected for color independently of the eye.

14.4 Defocus and Axial Astigmatism

Defocus of the human eye is termed *spherical refractive error*. Defocus accounts for the bulk of the loss in visual performance of the eye. If the wavefront error $W(r,\theta)$ in the presence of defocus is given by

$$W(r,\theta) = W_{020}r^2, \tag{14.4}$$

where (r,θ) are the polar coordinates in the exit pupil, then the spherical refractive error, φ_s, of the eye is given by

$$\varphi_s = -2000W_{020}, \tag{14.5}$$

where φ_s is in units of diopters for W_{020} in units of mm^{-1}. For $W_{020} > 0$, the spherical refractive error is negative and the eye is said to be *myopic*. In this case, a real object point in front of the eye is conjugate to the retina. For $W_{020} < 0$, the spherical refractive error is positive and the eye is said to be *hyperopic*. In this case, a virtual object point lying behind the eye is conjugate to the retina. Finally, an eye where the retina is conjugate to infinity is described as *emmetropic*.

The distribution of spherical refractive error is not Gaussian. The vast majority of people are emmetropic or slightly hyperopic, but the distribution is skewed with a large tail toward myopia [27]. Most people fall in the range $-5.00\,D < \varphi_s < +2.00\,D$. Astigmatism is also present in many cases. Since the eye is not rotationally symmetric, astigmatism can appear on axis. To distinguish this effect from the traditional off-axis astigmatism, the on-axis astigmatism is referred to as *axial astigmatism* and the off-axis or Seidel-type astigmatism is referred to as *oblique astigmatism*. Axial astigmatism arises from one or more of the ocular surfaces being toric in shape. Consequently, the power of the eye is different along different meridians. The *cylindrical refractive error* describes the power difference of the eye between its maximum and minimum meridians. Most people have less than 1 D of axial astigmatism, but the absolute value can reach upwards of 5 D of axial astigmatism.

Often spherical and cylindrical refractive errors are not deemed as ocular aberrations since they are readily corrected by external appliances, such as spectacle and contact lenses. Spectacle lenses are, in general, sphero-cylindrical lenses whose power is dictated by the spherical and cylindrical refractive error of the wearer. The spectacle lens is placed at the front focal point of the eye which minimizes changes in magnification of the retinal image. While this arrangement corrects for on-axis errors, the eye can rotate and look through different portions of the spectacle lens. Consequently, the spectacle lens bending is chosen to minimize aberrations as the eye looks through more peripheral regions of the lens. For the bulk of the refractive errors seen in the population, there are two bending solutions that minimize astigmatism when the eye rotates and views through the lens periphery [28]. Both solutions are meniscus-type lenses with the concavity toward the eye. Most commercial spectacles are of the form of the solution having the flatter curvature, as this is deemed more cosmetically appealing. Details and modern aberration analysis regarding the design of both spherical and aspheric spectacle lenses are found throughout the literature [29–35]. Contact lenses offer an alternative to spectacle lenses for correcting spherical and cylindrical refractive error. Again, the power of contact lenses is fixed by the errors of the eye. However, since contact lenses are placed on the cornea and move with the eye as it rotates, their bending is not constrained by aberration reduction. Instead, the bending of the contact lens is based on properly conforming to the shape of the cornea. Cylindrical refractive error correction with contact lenses is a bit more challenging than spherical refractive error, since the contact lens needs to have a toric component and the principal meridian of the contact needs to align with the principal meridian of the cylindrical refractive error of the eye.

14.5 Spherical Aberration and Irregular Astigmatism

While much of the image degradation caused by aberrations is attributed to defocus and axial astigmatism, there are additional aberrations in the eye that ultimately limit visual performance. Historically, these aberrations have been lumped together and called *irregular astigmatism*. The typical eye, either emmetropic or with its spherical and cylindrical refractive error corrected can achieve a visual acuity of 20/20. This metric corresponds to an angular resolution of 1 arcmin. However, based on diffraction-limited performance and the size of the photoreceptors sampling the retinal image, the ultimate performance of the eye is expected to be a visual acuity of about 20/10 or an angular resolution of 0.5 arcmin. In eyes with normal levels of irregular astigmatism, vision typically is limited to the 20/20 level. Furthermore, some eyes in the presence of disease, damage, or simply large variations in ocular surface shape have high levels of residual aberrations and consequently a poorer visual acuity. The components of irregular astigmatism have been further dissected to understand their implications toward visual performance and to examine further means of correcting these errors. Spherical aberration is the first aberration to be extracted from the residual aberrations.

Spherical aberration has long been known to cause visual effects in the eye [21,23,36–39]. Historically, one technique for measuring the ocular spherical aberration was to measure the spherical refractive error through a series of annular masks. The masks select specific zones within the pupil and spherical aberration can in general be considered a defocus that is dependent upon pupil zone. In making these measurements, the unaccommodated eye typically has under-corrected spherical aberration meaning that the marginal rays focus in front of the retina when the paraxial rays focus on the retina. Furthermore, the level of spherical aberration changes with accommodation, tending toward zero for 2–3 D of accommodation and becoming over-corrected for higher accommodation levels. Ocular spherical aberration is thought to be the cause of night myopia. Night myopia is the tendency for the eye to become more myopic under dim lighting conditions as the pupil dilates. For bright lighting conditions, the corrected eye should have its paraxial focus on the retina and spherical aberration will have little impact due to a constricted pupil. However, under dimmer conditions, the pupil size expands introducing larger levels of spherical aberration. The eye appears more myopic since best focus under these conditions is somewhere between the paraxial and the marginal focus where the blur or rms spot size on the retina is minimized.

In the vision literature, ocular spherical aberration is typically described differently than in more traditional aberration theory texts. If the wavefront error $W(r,\theta)$ in the presence of spherical is given by

$$W(r,\theta) = W_{040}r^4,\tag{14.6}$$

where (r,θ) are the polar coordinates in the exit pupil, then the ocular longitudinal spherical aberration (LSA) is given by

$$LSA = 4000W_{040}r^2,\tag{14.7}$$

where LSA is in units of diopters for W_{040} in units of mm^{-3}. For $W_{040} > 0$, the spherical aberration under-corrects, but the sign convention typically considers the LSA as positive in this case.

Beyond spherical aberration, the other aberrations of irregular astigmatism were historically difficult to assess. The advent of various forms of ocular wavefront sensing or aberrometry has led to the further understanding of the aberrations of the human eye.

14.6 Ocular Aberrometry

A variety of techniques have been developed to perform ocular aberrometry, or the measurement of the wavefront error of the eye. There are several thorough reviews of the various technologies, so only a brief summary of the techniques is provided here [40,41]. Specifically, the spatially-resolved refractometer, the Tscherning aberrometer and the Shack-Hartmann wavefront sensor are described below. Additional technologies such as laser ray tracing [42] and pyramid sensors [43] have been used to measure ocular wavefront error.

The spatially resolved refractometer is based on the 400-year-old work of a Jesuit priest named Christopher Scheiner [44]. The Scheiner disk is a mask with two small holes through which an observer views a distant point source. With no aberrations, the light is effectively collimated and passes through each hole to focus at the same point on the retina. However, in the presence of aberrations, the spots on the retina do not coincide. The separation of these spots is related to the transverse ray error. Smirnov [45] demonstrated that the Scheiner disk principle could be used to measure the wavefront aberrations of the eye. Figure 14.3 illustrates the Smirnov's system. Two parallel beams are directed into the eye of a subject at two different pupil locations. In the presence of aberrations, in general, two separate spots will be perceived by the subject. The subject then tilts one of the beams to cause the spots to merge. The angle of this beam is related to the slope of the wavefront at the entrance point in the pupil. The process is repeated for multiple points within the pupil to capture wavefront gradient information.

An alternative aberrometry technique was developed called the Tscherning technique and dates back to the late 1800s [46]. Tscherning placed a grid of equally spaced lines over a +5.00 D lens. Subjects viewing a distant point source through the lens perceived a distorted shadow of the grid on their retinas. By drawing the distorted grid, an analysis of individual wavefront aberrations could be performed. This technique was later improved and automated to create a system for measuring ocular wavefront error [47–50]. The concept of the Tscherning technique is illustrated in Figure 14.4. Incident collimated light passing through a grid pattern and a positive powered lens. The lens forces the incident light to focus in front of the retina and diverge again prior to striking the retina. A camera (not shown) is then used to photograph the distorted pattern formed on the retina. The distortion is related to the transverse ray error and proportional to the gradient of the wavefront error.

The most widespread technique for measuring ocular wavefront aberrations is based on the Shack-Hartmann sensor. There are several reviews describing the Shack-Hartmann system as an evolution from the Hartmann screen for optical metrology to modern systems used in astronomical optics and ocular wavefront sensing [51–53]. In the late 1960s, Shack and his colleagues developed the modern day version of the Shack-Hartmann sensor [54]. The original application of the technology was to measure the aberrations caused by atmospheric turbulence to improve the quality of observation from

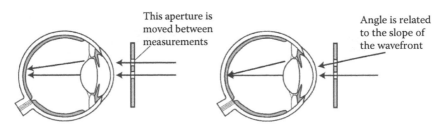

FIGURE 14.3 The spatially-resolved refractometer sends two parallel beams into the eye through two separate locations in the pupil. The observer then adjusts the slope of one of the beams to fuse the spots on the retina.

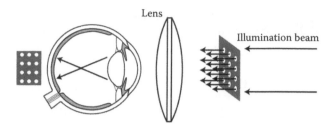

FIGURE 14.4 The Tscherning aberroscope forms a shadow of a grid pattern on the retina. Ocular aberrations distort the shape of the grid.

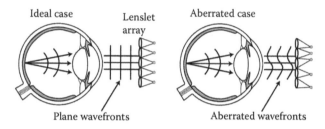

FIGURE 14.5 The Shack-Hartmann wavefront sensor projects the wavefront emerging from the pupil onto a lenslet array. The local wavefront slope is encoded in the deviations in the focal spots.

ground-based telescopes. In the early 1990s, Liang and colleagues [55] applied the Shack-Hartmann technique to the eye, and it is this configuration that makes up modern systems. Figure 14.5 shows the Shack-Hartmann system. A narrow beam of light is directed into the eye and forms a small point of light on the retina. This light scatters off the retina and forms an aberrated wavefront that emerges from the eye. This wavefront is projected onto a lenslet array and each lenslet focuses a small portion of the wavefront onto a sensor. In the aberration-free case, shown on the left in Figure 14.5, the emerging wavefront is planar. Each lenslet samples a portion of this flat wavefront and focuses it to a spot on the axis at the rear focal point of the lenslet. The resultant spot pattern from all of the lenslets is uniformly spaced. In the aberrated case, shown on the right in Figure 14.5, the emerging wavefront is no longer planar. Since the lenslet apertures are small, the sampled portion of the aberrated wavefront is approximately planar of the lenslet aperture, but the incident wavefront may be tilted. The lenslet still forms a spot in its rear focal plane, but the spot is no longer on the lenslet axis. The spot is instead shifted in an amount related to the local gradient of the incident wavefront.

The preceding techniques for measuring the ocular wavefront error have revolutionized the understanding of the eye's optical performance. Modern devices provide a rapid means to extract individual wavefront gradient information. Typically, a modal fit of the gradient data is used to represent the wavefront error as a linear expansion of Zernike polynomials. The expansion coefficients describe the amount of each aberration present [56]. Aberrations are no longer lumped into irregular astigmatism, but can instead be expanded into its various components and the impact of these components on visual performance assessed. The higher-order aberrations, those above simple defocus and axial astigmatism, are now routinely measured for both diagnostic purposes and correction.

14.7 Higher-Order Aberrations

Ocular wavefront sensors have been used to study the aberrations of normal and pathological eyes to understand the limitations on vision these aberrations produce. Large population studies have been performed to assess the mean and variation of the higher-order aberrations in terms of their Zernike

TABLE 14.5 The Ophthalmic Standard Zernike Polynomials

j	n	m	$Z_n^m(\rho,\theta)$
0	0	0	1
1	1	−1	$2\rho\sin\theta$
2	1	1	$2\rho\cos\theta$
3	2	−2	$\sqrt{6}\rho^2\sin2\theta$
4	2	0	$\sqrt{3}(2\rho^2-1)$
5	2	2	$\sqrt{6}\rho^2\cos2\theta$
6	3	−3	$\sqrt{8}\rho^3\sin3\theta$
7	3	−1	$\sqrt{8}(3\rho^3-2\rho)\sin\theta$
8	3	1	$\sqrt{8}(3\rho^3-2\rho)\cos\theta$
9	3	3	$\sqrt{8}\rho^3\cos3\theta$
10	4	−4	$\sqrt{10}\rho^4\sin4\theta$
11	4	−2	$\sqrt{10}(4\rho^4-3\rho^2)\sin2\theta$
12	4	0	$\sqrt{5}(6\rho^4-6\rho^2+1)$
13	4	2	$\sqrt{10}(4\rho^4-3\rho^2)\cos2\theta$
14	4	4	$\sqrt{10}\rho^4\cos4\theta$
15	5	−5	$\sqrt{12}\rho^5\sin5\theta$
16	5	−3	$\sqrt{12}(5\rho^5-4\rho^3)\sin3\theta$
17	5	−1	$\sqrt{12}(10\rho^5-12\rho^3+3\rho)\sin\theta$
18	5	1	$\sqrt{12}(10\rho^5-12\rho^3+3\rho)\cos\theta$
19	5	3	$\sqrt{12}(5\rho^5-4\rho^3)\cos3\theta$
20	5	5	$\sqrt{12}\rho^5\cos5\theta$

Note: J is the single index value and N and M are the double index values referring to the radial order and meridional frequency of the functions.

expansion coefficients [57–62]. There are several indexing, ordering, and normalization schemes for the Zernike polynomials. When applied to ocular wavefront error, the Zernike expansion scheme has been standardized, using double-index labels to avoid ambiguity [63–65]. This scheme is different than other Zernike representations found in the literature and lens design software packages, so care should be taken when comparing data from other sources. Table 14.5 summarizes the ophthalmic standard Zernike $Z_n^m(\rho,\theta)$ polynomials through 5th order. The coordinates (ρ,θ) are normalized polar coordinates in the pupil.

In population studies, the higher-order aberrations, with the exception of the Zernike spherical aberration $Z_4^0(\rho,\theta)$, tend to have a mean near zero and a standard deviation straddling zero. Individuals may have unique and elevated levels of some higher-order aberrations, but these aberrations when examined over the entire population are random and average to zero. The exception is spherical aberration. Figure 14.6 illustrates the mean and standard deviation of the higher-order aberrations as reported by Salmon and van de Pol [61]. The aberration coefficients are for an unaccommodated eye and a 5-mm pupil diameter.

Since the non-rotationally symmetric higher-order aberrations have a mean of approximately zero, modern schematic eye models typically maintain their rotational symmetry and only model the spherical aberration term found in the population. Figure 14.7 shows the LSA based on the measured spherical aberration of a series of clinically measured aberrometry [57–62]. The shape is parabolic with pupil

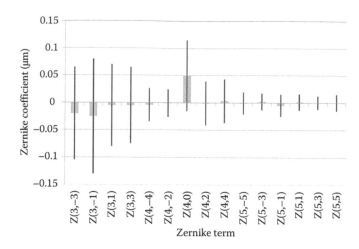

FIGURE 14.6 Histogram of mean (gray boxes) higher-order Zernike expansion coefficients with ± one standard deviation (black bars). Adapted from Salmon and van de Pol. (From Salmon TO and van de Pol C., *J. Cataract. Refract. Surg.,* 32, 2064–2074, 2006.)

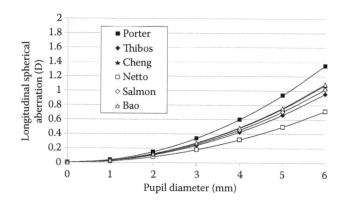

FIGURE 14.7 The LSA of the human eye based on several population studies. (From Porter J. et al., *J. Opt. Soc. Am. A,* 18, 1793–1803, 2001; Thibos LN. et al., *J. Opt. Soc. Am. A,* 19, 2329–2348, 2002; Cheng H. et al., *J. Vis.,* 2004; 4:272–280, 2004; Netto MV. et al., *J. Refract. Surg.,* 21, 332–338, 2005; Salmon TO and van de Pol C., *J. Cataract. Refract. Surg.,* 32, 2064–2074, 2006; Bao J. et al., *J. Optom.,* 2, 51–58, 2009.)

diameter as expected from Equation 14.7, but there is some variation between the studies. The AZEye15 model is designed to mimic the average spherical aberration of these population studies with LSA slightly greater than 1 D for a 6-mm diameter pupil. Furthermore, the LSA changes with accommodation. The study by Cheng et al. investigated the changes in higher-order aberrations with accommodation [59]. Again, the higher-order aberrations with accommodation, with the exception of spherical aberration, tend to be negligible on average. Spherical aberration, however, tends to reduce toward zero with mild accommodation and then becomes over-corrected with high accommodation. The AZEye15 model is designed to produce these spherical aberration effects with accommodation. Figure 14.8 shows the LSA of the AZEye15 model for various levels of accommodation. Of special note is that the ocular spherical aberration is about zero for an accommodation level between 2–3 D. This means that the visual system tolerates some image degradation for distant objects in exchange for optimal visual performance for object at 33–50 cm, exactly our natural reading position.

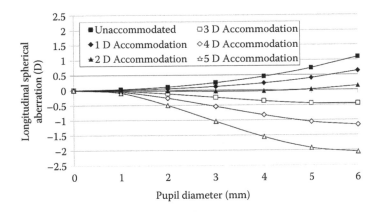

FIGURE 14.8 The LSA of the AZEye15 model with various levels of accommodation.

14.8 Correction of Higher-Order Aberrations

In recent years, there has been much interest in customizing the correction of higher-order aberrations. Ocular aberrometry enabled the rapid measurement of individual aberrations and its development coincided with the development of technologies that enable the correction of these aberrations. Smirnov [45] first suggested using customized contact lenses to fix aberrations, but the lens fabrication technology was not available at the time. Currently, contact lenses and/or their molds can be fabricated on a single-point diamond-turning lathe. These lathes can be equipped with an oscillating tool head that enables the cutting tool to move in and out as the part rotates. Non-rotationally symmetric surfaces can be cut with this technology, enabling a custom phase profile to be carved into the lens surface that compensates for the wearer's aberrations. Such custom contacts have been demonstrated to provide visual benefit over conventional corrections, especially in wearers with large aberrations [66–68]. The main limitations of customized contact lenses are the rotational and translational stability of the lens on the eye. Registering the lens compensating pattern to the eye's aberration is crucial [69,70].

Correction of ocular aberrations has also been performed using deformable mirror technology. This technique not only enhances an individual's vision, but also allows unprecedented imaging of the retina. Liang et al. provided the first images of human photoreceptors *in vivo* [71]. Aberration measurement and correction with adaptive optics has since been applied to not only conventional retina imaging, but also to scanning laser ophthalmoscopes and optical coherence tomography to provide high resolution axial and lateral scans on the retina and its underlying structure. These applications provide insight into the retinal architecture, color vision and disease progression.

Correction of higher-order aberrations is also commonly performed in refractive surgery. Laser in situ keratomileusis (LASIK) is a surgical technique originally developed for correcting myopia, hyperopia and axial astigmatism. The surgery involves forming a thin flap in the anterior surface of the cornea and using an excimer laser to ablate or sculpt the underlying tissue to a new shape. Once the flap is replaced, the shape of the cornea has been modified to correct the patient's refractive error [72]. This procedure has been extended to create custom ablation patterns to correct for an individual's aberrations. Again the patient's wavefront error is measured with aberrometry. The compensating phase profile is then ablated into the cornea providing enhanced vision over conventional procedures, which just correct spherical and cylindrical refractive error. Customized LASIK is now a routine option for patient's seeking vision correction.

14.9 Peripheral Aberrations

The ocular aberrations in the preceding discussion have been "on-axis," with the quotes being used because the eye is a non-rotationally symmetric system and a true optical axis is ill-defined. The line of sight is a better definition of an axis for the eye [73]. The line of sight connects the fixation point to the center of the entrance pupil, and the center of the exit pupil to the fovea. The line of sight collapses to the optical axis in the case of rotational symmetry. The ocular aberrations also have a field dependence for objects located away from the fixation point. These peripheral aberrations have been of limited interest until recent years because the resolution of the retina falls off dramatically when moving away from the fovea due to reduced sampling density of the photoreceptors. Consequently, peripheral aberrations were largely ignored because even if the image quality falling on the retina could be improved by modifying the optics, the retina would not be able to record the improved image. There is, however, a growing body of evidence that suggests that peripheral aberrations can stimulate eye growth in pediatric eyes leading to myopia [74]. For this reason, measurement and correction of peripheral aberration has become increasingly of interest in recent years.

Earlier work in measuring peripheral aberrations essentially measured the spherical and cylindrical refractive error for different field positions [75–77]. Oblique astigmatism is found to be the dominant artifact in peripheral vision. The oblique astigmatism in diopters on average is given by

$$\text{Oblique Astigmatism} = 0.00266\theta^2 - 2.09 \times 10^{-7}\theta^4, \tag{14.8}$$

where θ is the field angle in degrees [1]. The eye can also be myopic or hyperopic in the periphery dependent upon the relationship between the retinal curvature and the Petzval surface of the eye. More recently, the Shack-Hartmann wavefront sensor has been used to measure peripheral aberrations [78–81]. These studies confirm the earlier presence of the dominance of defocus and astigmatism in the periphery. Furthermore, these studies illustrate that coma tends to be the next major aberration of importance, with the other aberrations being well-balanced between the cornea and the crystalline lens. The induction of coma is attributed mainly to the ray bundles passing through the periphery of the anterior cornea, which is a prolate asphere. Finally, these studies demonstrate that corrective techniques, such as contact lenses and LASIK, can modify the peripheral aberrations. A variety of technologies have been proposed to reduce the development and progression of myopia which exploit modifying the peripheral aberrations of the young eye. These technologies include aspheric spectacle lenses [82], contact lenses [83], and orthokeratology [84,85]. Results with these technologies have been mixed and additional research is needed to control myopia progression.

14.10 Summary

The eye, like any optical system, suffers from aberrations. These aberrations ultimately limit the quality of the image formed on the retina. Since the eye in not rotationally symmetric, it suffers from the full spectrum of aberrations both on and off axis. Furthermore, the dispersive properties of the ocular media produce chromatic aberrations. Techniques have been offered to correct or reduce the effects of these aberrations. Correction of LCA offers little visual benefit since the eye still suffers from transverse chromatic aberration and chromatic tilt. Correction of defocus and astigmatism has been long provided by spectacles and contact lenses and offers dramatic improvements in visual performance. More recently, surgical procedures, such as LASIK, have provided similar improvements directly in the corneal surface. Correction of higher-order aberrations can offer even further improvements in visual acuity, especially in people with large inherent aberrations. This correction also allows unprecedented views of the retina to better understand the architecture of the retina. Correction of peripheral aberrations potentially offers benefit in delaying or reducing the development of myopia in the young eye. The continuous research into the measurement and correction of the aberrations of the human eye has led to dramatic improvements in vision and quality of life.

References

1. Atchison DA, Smith G. *Optics of the Human Eye.* (Oxford: Butterworth-Heinemann, 2000.)
2. Le Grand Y, El Hage SG. *Physiological Optics.* (Berlin: Springer-Verlag, 1980).
3. Schwiegerling J. *Field Guide to Visual and Ophthalmic Optics.* (Bellingham, WA: SPIE Press, 2004.)
4. Gullstrand A. Appendix II: Procedure of rays in the eye. Imagery – laws of the first order. In: *Handbuch der Physiologischen Optik,* Ed: H. von Helmholtz, Vol. 1, 3rd Ed. (English translation by Southall JP, Washington DC: Optical Society of America, 1924.)
5. Emsley HH. *Visual Optics,* Vol. 1, 5th Ed. (London: Butterworths, 1953.)
6. Bennett AG, Rabbetts RB. *Clinical Visual Optics,* 2nd Ed. (Oxford: Butterworth-Heinemann, 1989.)
7. Thibos LN, Ye M, Zhang X, Bradley A. The chromatic eye: A new reduced-eye model of ocular chromatic aberration in humans. *Appl. Opt.,* 1992; 31:3594–3600.
8. Lotmar W. Theoretical eye model with aspheric surfaces. *J. Opt. Soc. Am.,* 1971; 61:1522–1529.
9. Kooijman AC. Light distribution on the retina of a wide-angle theoretical eye. *J. Opt. Soc. Am.,* 1983; 73:1544–1550.
10. Navarro R, Santamaría J, Bescós J. Accommodation-dependent model of the human eye with aspherics. *J. Opt. Soc. Am. A,* 1985; 2:1273–1281.
11. Escudero-Sanz I, Navarro R. Off-axis aberrations of a wide-angle schematic eye model. *J. Opt. Soc. Am. A,* 1999; 16:1881–1891.
12. Liou H-L, Brennan NA. Anatomically accurate finite model eye for optical modeling. *J. Opt. Soc. Am. A,* 1997; 14:1684–1695.
13. Smith G, Atchison DA. The gradient index and spherical aberration of the lens of the human eye. *Ophthalmic. Physiol. Opt.,* 2001; 21:317–326.
14. Siedlecki D, Kasprzak H, Pierscionek BK. Schematic eye with a gradient-index lens and aspheric surfaces. *Opt. Lett.,* 2004; 29(11):1197–1199.
15. Huang Y, Moore DT. Human eye modeling using a single equation of gradient index crystalline lens for relaxed and accommodated states. *Proc. SPIE,* 2006; 6342:63420D.
16. Atchison DA, Smith G. Chromatic dispersions of the ocular media of human eyes. *J. Opt. Soc. Am. A,* 2005; 22:29–37.
17. Rozema JJ, Atchison DA, Tassignon M-J. Statistical eye model for normal eyes. *Invest. Ophthalmol. Vis. Sci.,* 2011; 52:4525–4533.
18. Ramasubramanian V, Glasser A. Prediction of accommodative optical response in presbyopic subjects using ultrasound biomicroscopy. *J. Cataract. Refract. Surg.,* 2015; 41:964–980.
19. Ortiz S, Pérez-Merino P, Gambra E, de Castro A, Marcos S. In vivo human crystalline lens topography. *Biomed. Opt. Exp.,* 2012; 3:2471–2488.
20. Dubbelman M, Van der Heijde GL, Weeber HA. Change in shape of the aging human crystalline lens with accommodation. *Vis. Res.,* 2005; 45:117–132.
21. Wald G, Griffin DR. The change in refractive power of the human eye in dim and bright light. *J. Opt. Soc. Am.,* 1947; 37:321–336.
22. Bedford RE, Wyszecki G. Axial chromatic aberration of the human eye. *J. Opt. Soc. Am.,* 1957; 47:564–565.
23. Van Heel ACS. Correcting the spherical and chromatic aberrations of the eye. *J. Opt. Soc. Am.,* 1946; 36:237–239.
24. Lewis AL, Katz M, Oehrlein C. A modified achromatizong lens. *Am. J. Optom. Physiol. Opt.,* 1982; 59:909–911.
25. Powell I. Lenses for correcting chromatic aberration of the eye. *Appl. Opt.,* 1981; 20:4152–4155.
26. Bradley A, Zhang X, Thibos LN. Achromatizing the human eye. *Optom. Vis. Sci.,* 1991; 68:608–616.
27. Zadnik K, Manny RE, Yu JA, et al. Ocular component data in schoolchildren as a function of age and gender. *Optom. Vis. Sci.,* 2003; 80:226–236.
28. Tscherning M. *Physiological Optics, Dioptrics of the Eye, Functions of the Retina, Ocular Movements and Binocular Vision.* (English translation by Weiland C, Philadelphia, PA: The Keystone, 1900.)

29. Atchison DA. Third-order theory and aspheric spectacle lens design. *Ophthalmic. Physiol. Opt.*, 1984; 4:179–184.

30. Malacara Z, Malacara D. Tscherning ellipses and ray tracing in ophthalmic lenses. *Am. J. Optom. Physiol. Opt.*, 1985; 62:447–455.

31. Malacara D, Malacara Z. Tscherning ellipses and ray tracing in aspheric ophthalmic lenses. *Am. J. Optom. Physiol. Opt.*, 1985; 62:456–462.

32. Atchison DA. Modern optical design assessment and spectacle lenses. *Opt. Acta*, 1985; 32:607–634.

33. Atchison DA, Smith G. Spectacle lenses and third-order distortion. *Ophthalmic. Physiol. Opt.* 1987; 7:303–308.

34. Malacara Z, Malacara D. Aberrations of sphero-cylindrical ophthalmic lenses. *Optom. Vis. Sci.*, 1990; 67:268–276.

35. Atchison DA. Spectacle lens design: A review. *Appl. Opt.*, 1992; 31:3579–3585.

36. Ivanoff A. On the influence of accommodation on spherical aberration in the human eye, an attempt to interpret night myopia. *J. Opt. Soc. Am.*, 1947, 37:730–731.

37. Koomen M, Tousey R, Scolnik R. The spherical aberration of the eye. *J. Opt. Soc. Am.*, 1949; 39:370–376.

38. Koomen M, Scolnik R, Tousey R. A study of night myopia. *J. Opt. Soc. Am.*, 1951; 41:80–90.

39. Ivanoff A. About the spherical aberration of the eye. *J. Opt. Soc. Am.*, 1956; 46:901–904.

40. Howland HC. Ophthalmic wavefront sensing: History and methods. In: *Wavefront Customized Visual Correction. The Quest for Super Vision II*, Eds: Krueger RR, Applegate RA, MacRae SM. (Thorofare, NJ: Slack, 2004.)

41. Schwiegerling J. The optics of wavefront technology. In: *Duane's Clinical Ophthalmology*, Eds: Tasman W, Jaeger EA. (Philadelphia, PA: Lippincott, 2004.)

42. Molebny VV, Pallikaris IG, Naoumidis LP, Chyzh IH, Molebny SV, Sokurenko VM. Retinal ray-tracing technique for eye refraction mapping. *Proc. SPIE*, 1997; 2971:175–183.

43. Ragazzoni R. Pupil plane wavefront sensing with an oscillating prism. *J. Mod. Opt.*, 1996; 43:289–293.

44. Scheiner C. *Oculus, sive fundamentum opticum*. Innspruk, 1619.

45. Smirnov MS. Measurement of the wave aberration of the human eye. *Biophys.*, 1961; 6:52–66.

46. Tscherning M. Die monochromatischen Aberrationen des menschlichen Auges. *Z. Psychol. Physiol. Sinne.*, 1894; 6:456–471.

47. Howland HC, Howland B. A subjective method for measurements of monochromatic aberrations of the eye. *J. Opt. Soc. Am.*, 1977; 67:1508–1518.

48. Walsh G, Charman WN, Howland HC. Objective technique for the determination of monochromatic aberrations of the human eye. *J. Opt. Soc. Am. A*, 1984; 1:987–992.

49. Mierdel P, Kaemmerer M, Mrochen M, Krinke H-E, Seiler T. Automated ocular wavefront analyzer for clinical use. *Proc. SPIE*, 2000; 3908:86–92.

50. Mrochen M, Kaemmerer M, Mierdel P, Krinke H-E, Seiler T. Principle of Tscherning aberrometry. *J. Refract. Surg.*, 2000; 16:S570–S571.

51. Platt BC, Shack RV. History and principles of Shack-Hartmann wavefront sensing. *J. Refract. Surg.*, 2001; 17:S573–S577.

52. Schwiegerling J, Neal DR. Historical development of the Shack-Hartmann wavefront sensor. In: *Robert Shannon and Roland Shack, Legends in Applied Optics*, Eds: Harvey JE, Hooker RB. (Bellingham, WA: SPIE Press, 2005.)

53. Schwiegerling J. History of the Shack Hartmann wavefront sensor and its impact in ophthalmic optics. *Proc. SPIE*, 2014; 91860:91860U.

54. Shack RV, Platt BC. Production and use of a lenticular Hartmann screen. *J. Opt. Soc. Am.*, 1971; 61:656.

55. Liang J, Grimm B, Goelz S, Bille JF. Objective measurement of the wave aberrations of the human eye with the use of a Hartmann-Shack wavefront sensor. *J. Opt. Soc. Am. A*, 1994; 11:1949–1957.

56. Cubalchini R. Modal wave-front estimation from phase derivative measurements. *J. Opt. Soc. Am.*, 1979; 69:972–977.

57. Porter J, Guirao A, Cox IG, Williams DR. Monochromatic aberrations of the human eye in a large population. *J. Opt. Soc. Am. A*, 2001; 18:1793–1803.

58. Thibos LN, Hong X, Bradley A, Cheng X. Statistical variation of aberration structure and image quality in a normal population of healthy eyes. *J. Opt. Soc. Am. A*, 2002; 19:2329–2348.

59. Cheng H, Barnett JK, Vilupuru AS, et al. A population study on changes in wave aberrations with accommodation. *J. Vis.*, 2004; 4:272–280.

60. Netto MV, Ambrósio R, Shen TT, Wilson SE. Wavefront analysis in normal refractive surgery candidates. *J. Refract. Surg.*, 2005; 21:332–338.

61. Salmon TO, van de Pol C. Normal-eye Zernike coefficients and root-mean-square wavefront errors. *J. Cataract. Refract. Surg.*, 2006; 32:2064–2074.

62. Bao J, Le R, Wu J, Shen Y, Lu F, He JC. Higher-order wavefront aberrations for populations of young emmetropes and myopes. *J. Optom.*, 2009; 2:51–58.

63. Thibos LN, Applegate RA, Schwiegerling JT, Webb R. Standards for reporting the optical aberrations of eyes. *J. Refract. Surg.*, 2002; 18:S652–S660.

64. ANSI Z80.28. Ophthalmics – Methods of reporting optical aberrations of eyes, American National Standards Institute, Washington DC, 2010.

65. ISO 24157. Ophthalmic optics and instruments – Reporting aberrations of the human eye. International Standard Organization, Geneva, Switzerland, 2008.

66. Sabesan R, Jeong TM, Carvalho L, Cox IG, Williams DR, Yoon G. Vision improvement by correcting higher-order aberrations with customized soft contact lenses in keratoconic eyes. *Opt. Lett.*, 2007; 32:1000–1002.

67. Marsack JD, Parker KE, Niu Y, Pesudovs K, Applegate RA. On-eye performance of custom wavefront-guided soft contact lenses in a habitual soft lens-wearing keratoconic patient. *J. Refract. Surg.*, 2007; 23:960–964.

68. Marsack JD, Parker KE, Applegate RA. Performance of wavefront-guided soft lenses in three keratoconus subjects. *Optom. Vis. Sci.*, 2008; 85:1172–1178.

69. Guirao A, Cox IG, Williams DR. Method for optimizing the correction of the eye's higher-order aberrations in the presence of decentration. *J. Opt. Soc. Am. A*, 2002; 19:126–128.

70. Shi Y, Queener HM, Marsack JD, Ravikumar A, Bedell HE, Applegate RA. Optimizing wavefront-guided corrections for highly aberrated eyes in the presence of registration uncertainty. *J. Vis.*, 2013; 13:1–15.

71. Liang J, Williams DR, Miller DT. Supernormal vision and high-resolution retinal imaging through adaptive optics. *J. Opt. Soc. Am. A*, 1997; 14:2884–2892.

72. Mrochen M, Kaemmerer M, Seiler T. Wavefront-guided laser in situ keratomileusis: Early results in three eyes. *J. Refract. Surg.*, 2000; 16:116–121.

73. Schwiegerling J. Eye axes and their relevance to alignment of corneal refractive procedures. *J. Refract. Surg.*, 2013; 29:515–516.

74. Mutti DO, Hayes JR, Mitchell GL, et al. Refractive error, axial length, and relative peripheral refractive error before and after the onset of myopia. *Invest. Ophthalmol. Vis. Sci.*, 2007; 48:2510–2519.

75. Ferree CE, Rand G, Hardy C. Refraction for the peripheral field of vision. *Arch. Ophthalmol.*, 1931; 5:717–731.

76. Rempt F, Hoogerheide J, Hoogenboom W. Peripheral retinoscopy and the skiagram. *Ophthalmologica*, 1971; 162:1–10.

77. Hoogerheide J, Rempt F, Hoogenboom WPH. Acquired myopia in young pilots. *Ophthalmologica*, 1971; 163:209–215.

78. Mathur A, Atchison DA, Scott DH. Ocular aberrations in the peripheral visual field. *Opt Lett.*, 2008; 33:863–865.

79. Mathur A, Atchison DA, Tabernero J. Effect of age on components of peripheral ocular aberrations. *Optom. Vis. Sci.*, 2012; 89:967–976.

80. Atchison DA. The Glenn A. Fry award lecture 2011: Peripheral optics of the human eye. *Optom. Vis. Sci.*, 2012; 89:954–966.

81. Mathur A, Atchison DA. Effect of orthokeratology on peripheral aberrations of the eye. *Optom. Vis. Sci.*, 2009; 86:476–484.
82. Sankaridurg P, Donovan L, Varnas S, et al. Spectacle lenses designed to reduce progression of myopia: 12-month results. *Optom. Vis. Sci.*, 2010; 87:631–641.
83. Shapiro EI, Kivaev AA, Kazakevich BG. Use of contact lenses in progressive myopia. *Vestn. Oftalmol.*, 1990; 106:30–33.
84. Cho P, Cheung SW, Edwards M. The longitudinal orthokeratology research in children (LORIC) in Hong Kong: A pilot study on refractive changes and myopic control. *Curr. Eye Res.*, 2005; 30:71–80.
85. Walline JJ, Jones LA, Sinnott LT. Corneal reshaping and myopia progression. *Br. J. Ophthalmol.*, 2009; 93:1181–1185.

15

Incoherent Light Sources

15.1 Introduction 561
        Basic Concepts
15.2 Tungsten Filament Sources 562
        Optical Applications of Tungsten Lamps
15.3 Arc Sources...566
15.4 Discharge Lamps 568
15.5 Fluorescent Lamps...570
15.6 Light-Emitting Diodes..572
        Optical Characteristics • Electrical Characteristics • Applications
References..576

Zacarías
Malacara-Hernández

15.1 Introduction

A light source is a necessary component in most optical systems. Except for those systems that use natural light, all others must include an artificial light source. In more than 130 years, a very large variety of light sources have been developed, and still new ones are currently being built. Five main types of artificial sources are available:

1. Light sources that emit from a thermally excited metal. Most of these sources are made from a tungsten filament. The spectrum of light corresponds to a quasi-blackbody emitter at the emitter temperature.
2. Light emitted by an electrically produced arc in a gap between two electrodes. The arc can be produced in open air, although most modern arc lamps are enclosed within a transparent bulb in a controlled atmosphere. The spectrum is composed of individual lines from the gas, superimposed to a continuous spectrum emitted by the hot electrode.
3. Light produced by the excitation of a material by ultraviolet energy in a long discharge tube, like those generically known as fluorescent lamps.
4. Light emitted due to a recombination of charge carriers in a semiconductor gap. A semiconductor pair is needed for the light to be emitted. These light sources receive the generic name of *light-emitting diodes* (LEDs).
5. Light emitted as a result of the stimulated radiation of an excited ensemble of atoms. This emission results in laser radiation with light having both spatial coherence (collimated light) and temporal coherence (monochromaticity). Due to its importance in optical instruments, these devices are described in chapter 16 in this book and will not be treated any more.

15.1.1 Basic Concepts

1. *Luminous efficacy.* All light sources emit only a small amount of visible power from their input power. *Luminous efficacy* ε is defined as the ratio of the total luminous flux ϕ to the total power input P, measured in lumen per watt (Lm/W); that is,

$$\varepsilon = \frac{\phi}{P} \qquad\qquad (15.1)$$

2. Assuming an ideal white source, namely one with constant output power over all the visible portions, the luminous efficacy will be about 220 Lm/W [1].

3. *Color temperature.* For a blackbody emitter, the color temperature corresponds to the spectral energy distribution for a blackbody at that temperature. The Kelvin temperature scale is used to describe color temperature for a source.

4. *Correlated color temperature.* When the emitter is not a perfect blackbody, but the color appearance resembles that of a blackbody, the correlated color temperature is the closest blackbody temperature found in the *CIExy* color diagram.

5. *Color rendering index (CRI).* This is a property of a light source to reproduce colors as compared with a reference source [2]. This figure reflects the capability of a light source to faithfully reproduce colors. The color rendering index is 100 for daylight.

15.2 Tungsten Filament Sources

The tungsten filament source, which is now more than 130 years old, was also the first reliable light source for optical devices. The basic lamp has evolved since its invention by Edison. A historical account is described by Anderson [3], Josephson [4] and Elenbaas [5]. The basic components for a tungsten lamp are (Figure 15.1): an incandescent electrically heated filament, supporting metal stems for the filament, two of them used to conduct electrical current to the filament, a glass envelope, filling gas, and a base to support and make the electrical contact. These basic components, with variations according to their applications, are now considered.

1. *Filament.* Electrical power heats the filament with a spectral distribution that follows a gray body. Since high temperatures are attained, filaments must support the highest possible temperatures. The higher the temperature, the higher the light efficacy. Carbon is capable of sustaining the highest temperature; unfortunately, it also evaporates too fast. After testing several materials, such as osmium and tantalum, tungsten is the most used metal for filaments. High temperature has the effect of vaporizing the filament material, until after some time, the thinning filament

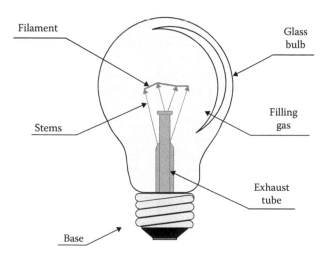

FIGURE 15.1 Basic components for a tungsten lamp.

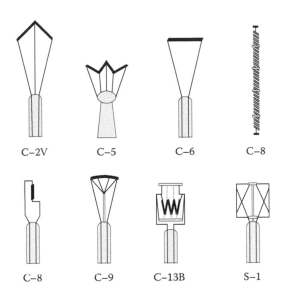

C–2V   C–5   C–6   C–8

C–8   C–9   C–13B   S–1

FIGURE 15.2 Common tungsten filaments and their designation.

breaks. The evaporation rate is not constant and depends on the impurity contents [6]. Most optical applications require a point-like source. Ideally, filaments should have low extension. To reduce the emitter extension, and at the same time increase the emissivity, filaments are coiled and, in some cases, double coiled. The reason is that by coiling the filament, the surface exposed to cool gas is reduced, decreasing the convection cooling [7]. Some typical filaments are shown in Figure 15.2. Additional filament shapes can be seen in the book by Rea [1]. Where light must be concentrated, filaments have low extension, as in projection, searchlights, or spotlights. If light is needed to cover a large area, large filaments are used instead. Some optical instruments require a line source. Straight filaments are used: for example, in hand-held ophthalmoscopes. Filaments operate at a temperature of about 3200 K. Tungsten emissivity ranges typically from 0.42 to 0.48. Spectral emissivity is reported in the literature [8,6].

2. *Wire stems.* Besides carrying electrical current to the filament, stems are used to hold the filament in position. Several different kinds of stems are used, depending on the filament and bulb shape. Mechanical vibration from the outside can create vibration modes in the filament and stems, reducing its life by metal fatigue. Some lamps have a design to reduce vibration but they must be mounted according to manufacturer specifications. Lead-in wires are chosen to have a similar thermal coefficient to match that from the glass at the electric seal in order to avoid breaking the glass.

3. *Filling gas.* The purpose of filling with gas is twofold: (i) to provide an inert atmosphere and avoid filament burning and (ii) to exert a pressure over the filament and delay the evaporation process. A negative effect of the gas is convective cooling, reducing the lamp efficacy. Low current lamps are made with a vacuum instead of a filling gas because the gas may give a negative effect for small filaments. Nitrogen and argon are the most commonly used gases, but some small lamps use the more expensive krypton gas. A common lamp is the quartz-halogen lamp [9]. A mixture of a halogen and a gas are enclosed in a low-volume quartz envelope. As the lamp burns, the tungsten filament is evaporated over the inner bulb surface, but due to the high temperatures attained by the quartz envelope, a reaction occurs between the halogen and the evaporated tungsten, removing it from the bulb. At the arrival of this mixture to the filament, the tungsten is captured again by the filament. Due to this effect, called the halogen cycle, the filament has a much longer life

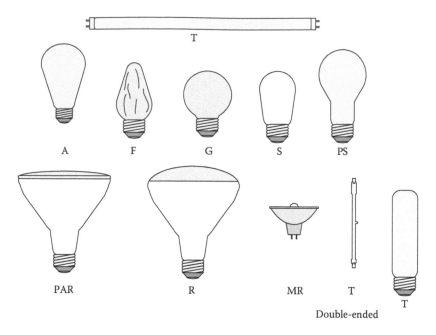

FIGURE 15.3   Common bulbs and their designation.

than any standard filament lamp for a filament at a higher temperature. Actually, the hot quartz envelope reacts easily with the stems, eroding the wire, and the breakage is usually produced in the stem. Quartz halogen lamps do not usually show bulb blackening. Due to the high chemical reactivity of the quartz bulb, care must be taken to avoid any grease deposition over the bulb surface; otherwise, a hot center is developed, resulting in the bulb breaking. It has been found that halogen vapor absorbs some radiation in the yellow-green part of the spectrum and emits in the blue and ultraviolet. (Studer, 1964)

4. *Glass envelope.* The glass bulb prevents the oxygen from burning the filament and allows a wide light spectrum to leave from the lamp. Most lamps are made of soft lime-soda glass. Outdoor lamps have impact or heat-resisting glass. Halogen and tubular heat lamps are made of quartz. As mentioned, glass bulbs are chosen to transmit most of the visible spectrum. The transmission spectrum for a glass bulb is reported by Kreidl and Rood [11] and Wolfe [12]. Optical transmittance is affected by temperature. For lamps for use at near-infrared wavelengths, a window is placed at some distance from the source, like Osram WI 17/g [13]. Tungsten lamps are manufactured with diverse bulb shapes, designated by their shape: *A* for arbitrary shape, *R* for reflector, *T* for tubular and *PAR* for parabolic reflector (Figure 15.3). New shapes are brought to the market constantly, and others are discontinued. The most recent catalog from the manufacturer is recommended for current bulb shapes. Some lamps are made transparent, while others are frosted by acid etching. Some are covered with white silica powder on the inner surface for a better diffusing light. Acid-etched lamps do not absorb significant amounts of light; silica-covered lamps absorb about 2% of the light [7].

5. *Supporting bases.* Besides the electrical contact, the base must support the lamp in place. The traditional all-purpose screw base is used for the most general lighting applications. For most optical instruments, where the filament position is critical, the so-called prefocus base is used. Other bases are also used, such as bayonet-type bases. Quartz-halogen lamps are subjected to high temperatures, and ceramic bases are used in those cases.

15.2.1 Optical Applications of Tungsten Lamps

The main optical applications of tungsten lamps are as follows:

1. *Spectral irradiance standard lamps.* Since quartz halogen lamps have a high stability for a relative long period of time (about 3%), it has been proposed for use as a secondary standard for spectral irradiance [13]. Later, an FEL lamp has been proposed as a secondary standard [14,15]. This type of lamp is placed at a fixed distance and at a specified orientation, where the spectral irradiance is known. This lamp provides a handy reference to check for the calibration of some light sensors. Some laboratories make a traceable calibration for every lamp they sell [16–18].

2. *Standard Type A source.* The International Committee for Illumination (**CIE**) has defined a standard source called a *Type A* source. This source is a reference for a color specimen to be observed. The original definition describes a tungsten lamp with a quartz window operating at a correlated color temperature of 2856 K. Originally, this description was for a 75 W lamp under a fixed supply voltage. Any tungsten lamp with a carefully controlled filament temperature could be used; also, it can be purchased with a certification to be used for this purpose, like the Osram WI 40 [13].

3. *Projection lamps.* Old projection lamps were made with a screen-shaped filament, to form a blurred image of the filament on the film gate, producing an almost uniform illumination at the gate. This resulted in a large filament with a complicated support and a reduced color temperature. Newer projection lamps rely on lamps with a single coiled filament and an elliptic reflector (Figure 15.4). The advantages for the new system results in the following: (a) the integrated elliptical reflector has a thin film cold mirror to reduce heat at film gate; (b) short single-coiled filaments increase color temperature; (c) quartz halogen lamps increase life and reduce bulb blackening; and (d) nonuse of condenser lenses. The condenser system brings about 55%–60% of total flux to the objective, while an elliptical mirror without a condenser lens can bring up to 80% of total flux [19].

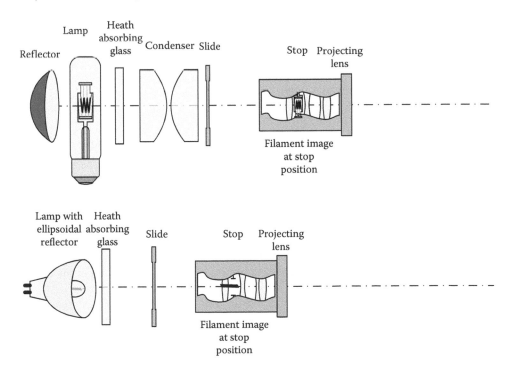

FIGURE 15.4 Two illumination systems for slide projectors.

Tungsten lamps for general illumination applications are set out of the market. Some countries are banning these lamps because of their low luminous efficacy. Some legacy optical instruments still use special tungsten lamps so they are manufactured for such purposes. Nevertheless, some researchers think that a redesigned tungsten lamp can offer new characteristics that were unavailable before. A tungsten lamp with a high directivity comparable to a carbon dioxide laser is described by Laroche et al. [20]. Also, by using nanotechnology, a tungsten lamp with a luminous efficacy comparable to that from an LED is discussed by Ilic et al. [21].

15.3 Arc Sources

Among other mechanisms that can produce light, there is the electron de-excitation in gases. An electron is excited by an electron or ion collision, which is achieved by the following means:

1. *Thermal electron emission by a hot cathode.* A heating incandescent filament emits some electrons, which statistically overcome the metal's work function. In the presence of an electrical potential, the electron is accelerated and, by collision, new ions are produced. This method is used to start a discharge.
2. *Secondary emission.* Several electrons are emitted from a metal that is struck by a colliding electron. Once a discharge begins, it is maintained by continuous ion bombardment. This mechanism is responsible for the arc maintaining in arc sources.
3. *Field emission.* A strong electrical field applied in a relatively cold cathode can be high enough for electrons to be emitted from the metal. This mechanism is used to start the discharge in an initially cold lamp.

The oldest form of an arc can be found in the now-obsolete carbon arc. Two carbon electrodes were brought into contact and in series to a limiting resistor. There are three types of carbon arcs: low intensity, high intensity, and flame. Low-intensity arcs operate by circulating a current high enough to reach the sublimation point for the carbon (3700 K). The emission characteristics of a carbon arc are those of a blackbody. Although simple in operation, carbon electrodes are consumed fast and new electrodes should be replaced. In old movie theater projectors, a device was made to maintain a constant current (constant luminance) on the electrodes. Because of its low reliability and short light cycle, arc lamps are not currently used; they have been replaced by short arc lamps. Emissivity for carbon arcs is about 0.98–0.99 [22].

Short arc lamps: modern arc sources are made from tungsten electrodes enclosed in a large spherical or ellipsoidal fused silica envelope. In this case, light is produced by electron de-excitation in gases. Gases are at about atmospheric pressure, but when hot, pressure may increase up to 50 atmospheres. Thoriated tungsten electrodes have a typical gap between 0.3 and 12 mm. Since short arc lamps have a negative resistance region, once the arc is started, very low impedance appears at the electrodes. Short arc lamps' lives are rated at more than 1000 hours, when the bulb blackens or the electrode gap increases and the lamp cannot start. Commercially, lamps are available from 75 W up to 30,000 W. Short arc or compact arc lamps are used in motion picture and television illumination, movie theater projection, solar simulators, and TV projection. A typical short arc lamp for projector is shown in Figure 15.5.

The starting voltage for a short arc lamp may rise up to 40 kV. To avoid a rapid destruction of the lamp due to a high current, ballast must be used. For ac operation, inductive ballasts are used, but for many applications where line current modulation is not allowed, an electronic current limiter must be provided after the start. Three short arc lamps are available: mercury and mercury—xenon, xenon lamps, and metal halide lamps.

1. *Mercury and mercury—xenon lamps.* Short arc mercury lamps have a low temperature and a pressure of about 20–60 Torr [19,1] of argon with traces of mercury. After the initial pulse starts

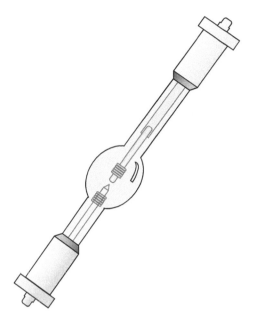

FIGURE 15.5   A typical short arc lamp.

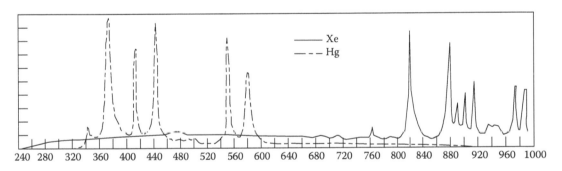

FIGURE 15.6   Relative spectral radiance for xenon (Xe) and mercury (Hg) short arc lamps.

the arc, mercury is vaporized, the pressure increases and the emission spectrum is of mercury, but with broad lines because of high gas pressure (Figure 15.6). It takes several minutes to reach full power, but if the lamp is turned off, it may take up to 15 minutes to cool down for restart. By adding xenon to the lamp, the warm-up time is reduced by about 50%. The spectral light distribution is the same as the mercury lamp. The luminous efficacy is about 50 Lm/W for a 1000 W lamp.

2. *Xenon lamps.* These lamps have a continuous spectral distribution in the visible portion of the spectrum, with some bright lines from the xenon at the infrared portion (Figure 15.6); maximum power is obtained a few seconds after the start. Correlated color temperature for a xenon lamp is around 5000 K. The luminous efficacy ranges from 30 to 50 Lm/W.

3. *Metal halide lamps.* The inclusion of rare earth iodides and bromides to a mercury short arc lamp results in a lamp with a full-spectrum emission, a high color-rendering index, and a high luminous efficacy. These lamps are used mainly in TV and movie lighting, and some fiber optics illuminating devices. Metal halide lamps are commonly used in projectors from computer output (Figure 15.7).

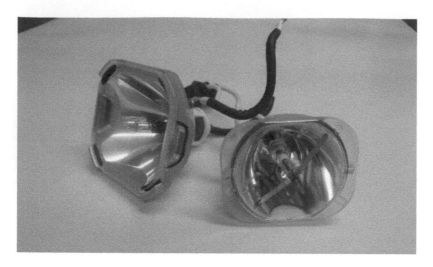

FIGURE 15.7   Lamps used for a projector from a computer source.

15.4 Discharge Lamps

High-intensity discharge lamps produce light by the electrical current flowing through a gas. It is necessary to reach the gas ionization for the gas to glow. As already mentioned, gas discharges have negative resistance characteristics; then, discharge lamps need a ballast to limit current once it is started. Most high-intensity discharge lamps operate from an ac supply. Three main types of lamps are produced for illumination purposes: mercury, metal halide, and high pressure.

1. *Mercury lamps.* Mercury lamps are made of two glass bulbs, one within the other. The inner bulb contains two tungsten electrodes for electrical contact. Argon gas is used to initiate the discharge, but small amounts of mercury are quickly vaporized to produce a broad line spectrum with the lines at 404.7, 432.8, 546.1, 577, and 579 nm. This results in a blue-green light. In contrast to a short arc lamp, the distance between the electrodes is several centimeters. The outer glass bulb serves as a filter for the ultraviolet light and contains some nitrogen to reduce pressure differences with the atmosphere. Sometimes the inner surface for this bulb is covered with a phosphor to convert ultraviolet radiation to visible light. The color is selected in the orange-red portion to improve the color rendering properties of the lamp. A version of this lamp includes a phosphor to convert the 253.7 nm UV radiation to near-UV light (black light). High-intensity mercury discharge lamps have a light efficacy of 30–65 Lm/W. Most mercury lamps operate with a 220 V supply voltage, but can also function with 127 V with an auto transformer-ballast.

2. *Metal halide.* The inclusion of some metal halide in a mercury lamp adds several spectral lines to an almost continuous spectrum [23]. The effect is a better color-rendering index and an improvement in the luminous efficacy (75–125 Lm/W). Scandium iodide and sodium iodide are two of the added materials.

3. *High-pressure sodium lamps.* Sodium vapor at high pressure is used for discharge lamps. Spectral width is strongly dependent on gas pressure, so that a high-pressure sodium lamp gives a broad spectrum dominant at the yellow line [24].

Most high-intensity discharge lamps are operated with inductive ballast; this drops the power factor up to 50%, but with a power factor correcting capacitor, it increases up to 90% [7]. The dimming of discharge lamps imposes several design restrictions for power factor correction and harmonics control. Electronic systems for this purpose are described in RCA [25].

TABLE 15.1 Characteristics for Spectral Lamps

Designation	Elements	Lamp Voltage (V)	Lamp Current (A)	Type of Current	Lamp Wattage (W)
Cd/10	Cadmium	15	1.0	ac	15
Cs/10	Cesium	10	1.0	ac	10
He/10	Helium	60	1.0	ac	55
Hg/100	Mercury	45	1.0	ac/dc	22
HgCd/10	Mercury + cadmium	30	1.0	ac	25
K/10	Potassium	10	1.0	ac	10
Na/10	Sodium	15	1.0	ac	15
Ne/10	Neon	30	1.0	ac	30
Rb/10	Rubidium	10	1.0	ac	10
Tl/10	Thallium	15	1.0	ac	15
Zn/10	Zinc	15	1.0	ac	15

Source: Osram, GmbH, Spectroscopic Lamps, Osram, Munchen, 2016.

Some discharge lamps are used in special optical applications not related to optical instruments. From these applications, we found:

1. *Low-pressure discharge lamps.* Small low-pressure discharge lamps are made for spectroscopic application. Due to its low pressure, spectral lines are sharp and can be used for spectral line calibrations. Table 15.1 shows some spectral lamps [26].
2. *Long-discharge lamps.* Low-pressure long-arc lamps are also made. They are used for solar simulators. Since they have low luminous efficacy (30 Lm/W), they are not used for general lighting. Long-arc xenon lamps produce a color similar to daylight, rich in UV content. These lamps are used for aging chambers, as recommended by ASTM G-181 and Gl55-98 and ISO 4582 standards.
3. *Flash lamps.* Flash lamps are discharge lamps with a long arc that emits a fast flash of light for a short time. Flash lamps are used in photography, stroboscopic lamps, warning lights in aviation and marine, and laser pumping. Flash lamps use low-pressure xenon as the active gas, although some have traces of hydrogen to change its spectral content. The tubes for photoflash are usually a long arc in a long straight tube; sometimes the tube is coiled or U-shaped. A ring electrode is wrapped to the tube to trigger the flash.

Lamp electrodes are at high impedance when cold, but the trigger electrode increases the gas conductivity to almost zero impedance. A typical circuit for a flash is shown in Figure 15.8. A capacitor is charged in a relatively long time; after charging the capacitor C_2, another capacitor, also previously charged, is discharged through a transformer to produce a high voltage at the trigger electrode. Capacitor C_1 is discharged through the lamp. The luminous efficacy is about 50 Lm/W for a typical flash lamp. Loading in joules for a flash tube is [1]:

$$\text{Loading} = \frac{CV^2}{2} \tag{15.2}$$

The flash tube impedance is

$$Z = \frac{\rho L}{A}, \tag{15.3}$$

where ρ is the plasma impedance in ohm-cm, L is the tube length, and A is the cross-sectional area in cm.

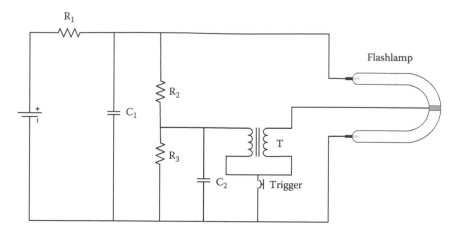

FIGURE 15.8   Flash lamp triggering circuit.

A version of a xenon arc lamp is used for photography. Xenon flash lamps have spectral distribution that closely resembles CIE D65 illuminant or daylight. Photographic flash lamps can stand more than 10,000 flashes. Flash tubes cannot be connected in parallel, since each lamp must have its own capacitor and trigger circuit. For multiple lamp operation, a slave flash lamp is designed to trigger with the light from another lamp. Flash lamps are synchronized to the camera in such a way that lamps are triggered when the shutter is fully opened. This is called the X-synchronization.

15.5 Fluorescent Lamps

Fluorescent lamps are a general kind of lamp that produces a strong mercury line at 253.7 nm from low-pressure mercury gas. This excites a fluorescent material to emit a continuum of visible radiation.

1. *Physical construction.* Fluorescent lamps are made mainly in tubular form; the diameter is speci-fied in eighths of an inch, starting with 0.5 inch designated to T-4 to 2.125 inches or T-17. The length, including lamp holders, ranges from 4 inches (100 mm) to 96 inches (2440 mm). Lamp tubes are usually made from soft lime soda glass, and at each end, a small tungsten filament is used as an electrode and as a preheater to start the lamp. Alternatively, fluorescent lamps are also made in circular form, U-shaped and quad or double-parallel lamps, helical coiled for compact fluorescent lamps. At each end of the tube, a base for electrical contact is provided. For circular lamps, at some point on the cycle, a connector for both ends is located. As a filling gas, low-pressure mercury with some argon or argon and krypton is added to initiate the discharge. At this low pressure (200 Pa), most of the mercury is vaporized, but this depends on the ambient temperature.

2. *Electrical characteristics.* When cold, the electrical impedance of the gas is very high, but as soon as the lamp ignites, the conductivity decreases suddenly to a very low value. A means must be pro-vided for the current limitation. The principle of the electrical operation is shown in Figure 15.9 Filaments, inductive ballast, and the starter are all in a single series circuit. Initially, the starter is closed, and both filaments in series are heated to vaporize the mercury gas and to ionize the gas. A rapid break in the starter circuit produces a high voltage from the ballast, enough to initiate the discharge. Multiple lamps can operate from a single ballast assembly designed for such a case. Electronic ballasts work on high line frequency (~20–60 kHz), since the efficacy increases about 10% for frequencies above 10 kHz. Electronic ballasts have better ballast efficiency, less weight, less noise, and some other advantages. There are two main modes of operation in fluorescent lamps, glow and arc. In the glow mode, electrodes are made of single-ended cylinders; the inside is cov-ered with an emissive material. In this mode of operation, the current in the lamp is less than 1 A,

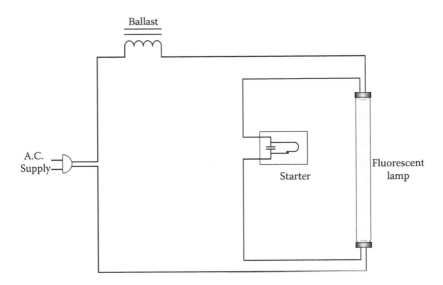

FIGURE 15.9 Basic circuit for a fluorescent lamp.

TABLE 15.2   Phosphor Properties for Fluorescent Lighting

Phosphor	Fluorescence Color
Zinc silicate	Green
Calcium Tungstate	Blue
Calcium borate	Pink
Calcium halo phosphate	Whites of various color temp.
Magnesium Tungstate	Bluish white

and the voltage through the lamp is about 50 V. In the arc mode, the electrodes are tungsten, at a temperature of 1000°C. Electrons are emitted thermally. The current increases up to more than 1.5 A and the lamp voltage across the lamp is around 12 V [1]. This mode of operation is more efficient than the glow mode, and most of the fluorescent lamps work on this principle.

3. *Optical characteristics.* The optical spectrum for a fluorescent lamp is composed mostly of the continuous emission from the fluorescent material and a small amount of the line spectra from mercury. Fluorescent materials for lamps determine the light color. Table 15.2 shows some common phosphors and its resulting color. Table 15.3 shows some of the most common lamp color designations and characteristics. CIE*xy* color coordinates for each phosphor are referenced in Philips [7] as well as the spectral distribution for several phosphors.

4. *Applications.* One of the most common applications for a fluorescent lamp is found in color evaluation boots. Some industries have standardized a light that is commonly found in stores. This lamp is the cool white lamp. For graphic arts industries, a correlated color temperature of 5000 K, is selected; then some lamps are specifically designed for this purpose, like the Ultralume 85® or equivalent. For a good color consistency, it is recommended to replace lamps well before they cease to emit light. Another use for fluorescent lamps is found in photography, but since the color is not matched for any commercial film, a color compensating filter must be used for most fluorescent lamps [27]. Since phosphors have a relatively low time constant, some flickering from line frequency can be observed. To avoid flickering in some fast optical detectors, like photodetectors and some video cameras, a high line frequency must be provided, such as the ones provided by electronic ballasts. Fluorescent lamps emit light in a cylindrically symmetric pattern.

TABLE 15.3 Fluorescent Lamp Designation Characteristics

Lamp Description	Designation	Light Output (%)	Color Rendering Index	Color Temperature (K)
Cool white	CW	100	67	4100
Cool white deluxe	CWX	70	89	4200
White	W	102	58	3500
Warm white	WW	102	53	3000
Warm white deluxe	WWX	68	79	3000
Daylight	D	83	79	6500
Colortone 50	C50	70	92	5000
Cool green	CG	83	70	6500
Sign white	SGN	75	82	5300
Natural	N	66	81	3400
Supermarket white	SMW	74	85	4100
Modern white deluxe	MWX	77	80	3450
Soft white	SW	68	79	3000
Lite white	LW	105	51	4100
Ultralume 83	83U	105	85	3000
Ultralume 84	84U	105	85	4100
Ultralume 85	85U	105	85	5000
Red	R	6	–	–
Pink	PK	35	–	–
Gold	Go	60	–	–
Green	G	140	–	–
Blue	B	35	–	–

For some purposes, this can be adequate, especially for diffuse illumination. In some other cases, lamps with an internal reflector have been designed to send most of the light in some preferred direction [28,1]. These lamps are used in desktop scanners and photocopiers.

Germicidal lamps are lamps with the same construction as any fluorescent lamp except that they have no fluorescent phosphor. This kind of lamp does not have high luminance since most (~ 95%) of the radiated energy is at the UV line of 253.7 nm. This radiation is harmful, since it produces burning to the eyes and skin. These lamps are used for air and liquid disinfection, lithography, and EPROM erasure. The so-called black light lamp has a phosphor that emits UV radiation at the UV-A band (350–400 nm) and peaks at 370 nm. Two versions are available for these lamps: unfiltered lamps with a white phosphor that emits a strong blue component and a filtered one, with a filter to block most of the visible part. Uses for these lamps are found in theatrical scenery lighting, counterfeit money detection, stamp examination, insect traps, fluorescence observing in color inspection boots, and mineralogy. These lamps are manufactured in tubular form, from T5 to T12, compact fluorescent lamps, and a high-intensity discharge lamp.

Because of ecological concerns, there is a movement against all kinds of mercury fluorescent lamps. Especially for the compact fluorescent lamps, since they have become very common. Many optical instruments, such as photocopiers, scanners, and computer displays have successfully replaced fluorescent lamps with LEDs. Many fluorescent or compact fluorescent lamp equipped optical instruments can be retrofitted with similar LED lamps.

15.6 Light-Emitting Diodes

Light-emitting diodes (LEDs) are light sources in which light is produced by the phenomenon of luminescence. An LED is made from a semiconductor device with two doped zones: one positive, or

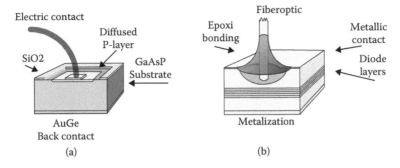

FIGURE 15.10 Cross section of (a) a typical LED and (b) a Burrus LED.

p-region, and the other negative, or n-region, as in any semiconductor diode. Electrons are injected in the n-region, while holes are injected in the p-region. At the junction, both holes and electrons are annihilated, producing light in the process. This particular case of luminescence is called electroluminescence and is explained in several books [29–34]. Recombined electrons and holes release some energy in either a radiative or non-radiative process. For the latter, a phonon is produced and no visible energy is produced, whereas for the first case, the energy is released in some form of electromagnetic radiation or photon. The emitted photon has an energy that is equal to the energy difference between the conduction and the valence band minus the binding energy for the isoelectronic centers for the crystal impurities in the semiconductor [29,34]. Hence, the photon has a wavelength

$$\lambda = \frac{1240}{\Delta E} \text{ nm,} \tag{15.4}$$

where ΔE is the energy transition in electron volts.

Historically, the first commercial LED was made in the late 1960s by combining three primary elements: gallium, arsenic, and phosphor (GaAsP), giving a red light at 655 nm. Galium phosphide LEDs were developed with a near-IR emission at 700 nm. Both found applications for indicators and displays, although the latter has poor luminance, since its spectral emission is in a region where the eye has a poor sensibility. Since light is produced in an LED at the junction, this device has to show this junction to the detecting area. Figure 15.10a shows a cross-section of an LED. These devices are also found in fiber optics. To make an efficient coupling to the fiber, most of the emitting area must be within the numerical aperture (acceptance cone for the fiber). The Burrus LED was developed for optical fibers and its cross-section is shown in Figure 15.10b. A super luminescent diode (SLD) is an edge-emitting LED that has high radiance like a laser, but has low temporal coherence. A SLD is an open cavity emitter, where the stimulated emission is responsible for part of the radiation [35]. Because of this, a SLD has a higher spatial coherence than an LED, but lower temporal coherence than a solid-state laser. These devices do not exhibit the common speckle pattern found in most lasers. Applications of SLD are include optical coherence tomography, white light interferometry, optical sensing and fiber optic gyroscopes.

15.6.1 Optical Characteristics

As mentioned, peak wavelength is a function of the band gap in the semiconductor. Several materials are used for making LEDs. Table 15.4 lists some materials used for making LEDs and their corresponding wavelengths.

Spectral bandwidth for most LEDs lies between 30 and 50 nm. Figure 15.11 shows the relative spectral radiance for some LEDs. Besides the listed monochromatic LEDs, a white LED can be achieved by any of four means: (1) a dichromatic LED, (b) a trichromatic LED, (3) a pentachromatic LED or (4) a violet LED

TABLE 15.4   Properties of Different Light Emitting Diodes

Material	Wavelength (µm)	Color
GaN	365–380	UV
InGaN	450–460	Blue
SiC	470	Deep blue
GaN	490	Blue
GaP	560	Green
GaP:N	565	Green
SiC	590	Yellow
GaAs$_{0.15}$P$_{0.85}$:N	590	Yellow
GaAsP$_{0.5}$	610	Amber
InGaAlP	618	Orange
GaAs$_{0.35}$P$_{0.65}$:N	630	Orange
GaAsP$_{0.6}$P$_{0.4}$	660	Red
GaAlAs	660	Red
GaP	690	Red
GaP:Zn,N	700	Deep red
GaAlAs	880	Infrared super bright
GaAs:Zn	900	Infrared
GaAs:Si	940	Infrared
InGaAsP	1300	Infrared

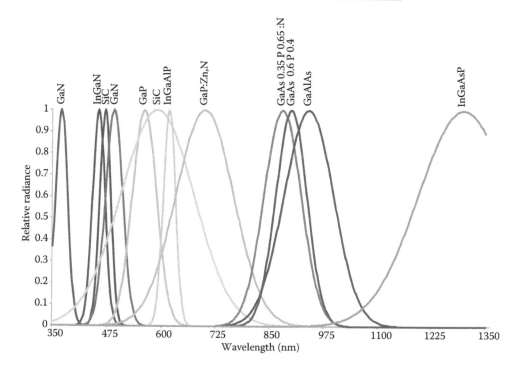

FIGURE 15.11   Relative spectral radiance for several LEDs.

with phosphorus [33]. The dichromatic LED is made monolithically in a single chip with two emitters at two complementary wavelengths. Such wavelengths should be chosen so that their power ratio gives a good luminous efficacy. Dichromatic white LEDs could have a good luminous efficacy but they have poor color rendering indices. They should be used for general illumination but not on those cases where color reproduction is critical. Trichromatic LEDs can either be made from monolithic or with discrete LEDs. The relative radiance for both the blue and red LEDs should be higher than the green one because of the eye's sensitivity to all three colors. A typical relative radiance for a white LED according to [36] is shown in Figure 15.12. Temperature, current variations and aging for the radiated power in an LED makes the color of a trichromatic variable [36].

White LEDs can also be made with four discrete color (usually red, green, cyan, and blue) LEDs. The addition of a cyan LED improves the color-rendering properties up to a value of 90. This feature also reduces the luminous efficacy and increases the cost. White LEDs for general illumination are now made from a violet emitting LED and a phosphor with broad spectrum in the pink spectral zones as shown in Figure 15.13.

At the semiconductor surface, light is emitted in all directions; hence at this point, an LED is a Lambertian emitter. Some LEDs have reflective electrical contacts, and the light is confined to emit along the junction. This device is called an edge-emitting LED, and is appropriate for fiber optics and

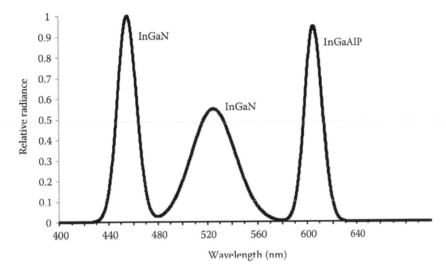

FIGURE 15.12 Relative spectral radiance for a trichromatic white LED. (After Chhajed S. et al., *J. Appl. Phys.*, 97, 2005.)

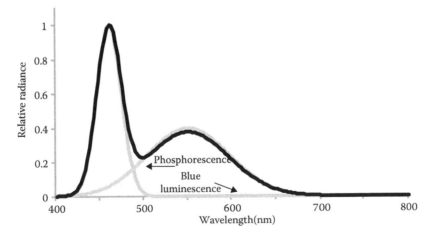

FIGURE 15.13 Relative spectral radiance for a blue LED with luminescent phosphor. (From Schubert EF. *Light-emitting Diodes*, Cambridge University Press, Cambridge, 2006.)

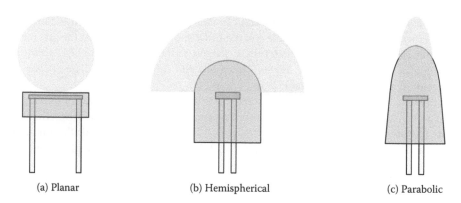

(a) Planar   (b) Hemispherical   (c) Parabolic

FIGURE 15.14   Goniometric radiation pattern for (a) planar, (b) spherical, and (c) parabolic LEDs.

integrated optics. Others LEDs emit at the thin surface. The goniometric radiation pattern for some LEDs is shown in Figure 15.14. A planar LED is essentially a Lambertian emitter, while others LED packages have a hemispherical package producing a hemispherical uniform radiation pattern. A parabolic profile for the package sends most of the radiation along the diode axis.

15.6.2 Electrical Characteristics

Electrically, an LED is a diode operated in direct polarization; by reversing polarization it does not emit light. To limit a current, a limiting resistor is placed in series with the source and the diode. Heat dissipated by the device may eventually end the diode's life. To avoid this, instead of operating the diode from direct current, a pulsed current operation dissipates less power for a given retinal perception. This is due to the eye retentivity to a rapid light pulse or light enhancement factor (General Instruments, 1983). Different LED materials have different threshold voltage: consult manufacturers' data for proper operating levels and maximum power dissipation.

15.6.3 Applications

The main applications of LEDs are found in economical fiber optic links. This is an economic alternative to lasers, where bit rate is lower than high-performance laser systems. LEDs for optical fiber communications can be purchased both connectorized or without the connector included. Wavelengths can be chosen to fit the optimum transmission for the fiber. Some LED combinations can give red, green, and blue light. Some colorimeters, as an alternative to tristimulus colorimetry, can use these LEDs instead of filters [38]. Since spectral distribution is not similar to tristimulus values, some errors exist in the color measurement. Also, LED triads are used in some color document scanners for image capture. Now illumination LEDs are used as a communication media in the so called LIFI instead of WIFI. This communication system has become more frequently used.

High efficiency—low-cost LEDs are now produced. They can be used for illumination as well as indicators. Currently, a green LED can have luminous efficiency of about 100 Lm/w, while a blue LED has only about 0.01 Lm/w.

References

1. Rea MS. (ed.). *Lighting Handbook. Reference and Application*, 8th Ed. New York, NY: Illuminating Engineering Society of America, 1993.
2. Nickerson D. Light sources and color rendering. *J. Opt. Soc. Am.*, 1960; 50(1):57–69.

3. Anderson JM, Saby JS. The electric lamp: 100 Years of applied physics. *Phys. Today*, 1979; 32(10):33–40.

4. Josephson M. The invention of the electric light. *Sci. Am.*, 1959; 201:98–118.

5. Elenbaas W. *Light Sources.* New York, NY: Crane Rusak, 1972.

6. Larrabee RD. Spectral emisivity of tungsten. *J. Opt. Soc. Am.*, 1959; 49(6):619–625.

7. Philips Lighting. *Lighting Handbook.* Eindhoven, the Netherlands: Philips Lighting, 1984.

8. Forsythe WE, Adams EQ. Radiating characteristics of tungsten and tungsten lamps. *J. Opt. Soc. Am.*, 1945; 35(2):108–113.

9. Zubler EG, Mosby FA. An iodine incandescent lamp with virtually 100 per cent lumen maintenance. *Illuminating Engineering*, 1959; 54(12):734–740.

10. Studer FJ, VanBeers. Modification of spectrum of tungsten filament quartz—Iodine lamps due to iodine vapor. *J. Opt. Soc. Am.*, 1964; 54(7):945–947.

11. Kreidl NJ, Rood JL. Optical materials. *Applied Optics and Optical Engineering*, Vol. 1, Kingslake R. (ed.). New York, NY: Academic Press, 1965.

12. Wolfe WL. Properties of optical materials. In *Handbook of Optics*, 1st Ed. Driscoll W. G. (ed.). New York, NY: McGraw-Hill, 1978.

13. Osram. Lamps for Scientific Purposes. Gas Filled Incandescent Lamps. Osram, 2016.

14. Stair R, Schneider WE, Jackson JK. A new standard of spectral irradiance. *Appl. Opt.*, 1963; 2(11):1151–1154.

15. Grum F, Becherer RJ. Radiometry optical radiation measurements. In *Radiometry*, Vol. 1, Grum Franc. (ed.). New York, NY: Academic Press, 1979.

16. Yoon WH, Gibson CE. NIST Measurement Services: Spectral Irradiance Calibrations National Institute of Standards and Technology; U. S. Department of Commerce. Gaithersburg, MD: 2011. NIST Pub 250–289.

17. Huang LK, Cebula RP, Hilsenrath E. New procedure for interpolating NIST FEL lamp Irradiances. *Metrologia*, 1998; 35:381–386.

18. Estrada-Hernández A, Oidor I, Rosas E. Correlated color temperature determination in FEL type incandescent lamps. Fifth Symposium Optics in Industry. E. Rosas R, Cardoso, J. C. Bermúdez, O. Barbosa-García. (ed.): *Proc SPIE*, 2006; 6046:60461Q1.

19. Anstee PT. Light sources for photography. In *Photography for the Scientist*, 2nd Ed. Morton R. A. (ed.). London, England: Academic Press, 1984.

20. Laroche M, Arnold C, Marquier F, et al. Highly directional radiation generated by a tungsten thermal source. *Opt. Lett.*, 2005; 30(19):2623–2625.

21. Ilic O, Bermel P, Chen G, Joannopoulos JD, Celanovic I, Soljačić M. Tailoring high-temperature radiation and the resurrection of the incandescent source. *Nat. Nanotechnol.*, 2016; 11:320–324.

22. Null MR, Lozier WW. Carbon arc as a radiation standard. *J. Opt. Soc. Am.*, 1962; 52(10):1156–1162.

23. Reiling GH. Characteristics of mercury vapor - Metallic iodide arc lamps. *J. Opt. Soc. Am.* 1964; 54:532–540.

24. van Vliet JAJM, de Groot JJ. High pressure sodium discharge lamps. *IEE Proc.* 1981; 128:415–441.

25. RCA. *Solid-State Power Circuits: Designer's Handbook*, Vol. 52, Sommerville, NJ: RCA, 1971.

26. Osram, GmbH. *Spectroscopic Lamps.* Munchen: Osram 2016.

27. Thomas W. *SPSE Handbook of Photographic Science and Engineering.* New York, NY: Wiley, 1973.

28. Eby JE, Levin RE. Incoherent light sources. In *Applied Optics an Optical Engineering*, Vol. 7, Kingslake R. (ed.). New York, NY: Academic Press, 1979.

29. Hewlett-Packard. *Optoelectronics/Fiber Optics Applications Manual*, 2nd Ed. New York, NY: McGraw-Hill, 1981.

30. Motorola Semiconductor Products. *Optoelectronics Device Data*, 2nd Ed. Phoenix, AZ: Motorola Semiconductor Products, 1981.

31. Seipel RG. *Optoelectronics.* –Reston, VA: Reston Pub., 1981.

32. Sze SM, Ng KK. *Physics of Semiconductor Devices,* 3rd Ed. New York, NY: Wiley Interscience, 2006:832.

33. Schubert EF. *Light-emitting Diodes*, 2nd Ed. Cambridge: Cambridge University Press, 2006:422.
34. Saleh BEA, Teich MC. *Fundamentals of Photonics*, 2nd Ed. Hoboken, NJ: John Wiley & Sons, 2007:1161.
35. Shidlovski V. *Superluminiscent Diodes. Short Overview of Device Operation Principles and Performance Parameters*. SuperlumDiodes, 2004. July 2016. https://www.superlumdiodes.com/pdf /sld_overview.pdf.
36. Chhajed S, Xi Y, Li Y-T, Gessmann Th, Schubert EF. Influence of junction temperature on chromaticity and color-rendering properties of trichromatic white-light sources based on light-emitting diodes. *J. Appl. Phys.*, 2005; 97:054506.
37. General Instruments. *Catalog of Optoelectronic Products 1983*. Palo Alto: Optoelectronic Division, General Instruments, 1983.
38. Yuan K, Yan H, Jin S. Integral colorimeter based on compound LED illumination. *Chin. Opt. Lett.*, 2014; 12(2):023302-1–023302-4.

16

Lasers

	16.1	Types of Lasers, Main Operation Regimes, and Examples 579
		Solid-State Lasers • Semiconductor Lasers • Gas Lasers • Liquid
		Lasers • Plasma Lasers • Free Electron Lasers • Fiber
		Lasers • Temporal Laser Operation
	16.2	Laser Medium.. 581
		Unsaturated Gain Coefficient • Line-Shape Function • Saturated
		Gain Coefficient
	16.3	Resonators and Laser Beams...586
		Stability, Resonator Types, and Diffraction Losses • Axial
		Modes • Transverse Modes • Resonator Quality Parameter
	16.4	Gaussian Beams 591
	16.5	Laser Applications and Instruments..594
		Laser Doppler Velocimetry • LIDAR • Laser Thermometry • Laser
		Applications to the Electronic Industry • Laser Applications to Art
Vicente Aboites	Appendix 16A..595
and Mario Wilson	References...596
	Further Reading... 600

16.1 Types of Lasers, Main Operation Regimes, and Examples

Based on the quantum idea used by Max Planck [1] to explain blackbody radiation emission, in 1917, Albert Einstein proposed the processes of stimulated emission and absorption of radiation [2]. Light amplification by stimulated emission of radiation (LASER) [3] was first demonstrated by Maiman [4] in 1960 using a ruby crystal pumped with a xenon flash lamp. Since then, laser coherent emission has been generated in thousands of substances using a wide variety of pumping mechanisms. Lasers are normally classified according to their active medium: solid, liquid, and gas. Table 16.1 shows examples of some of the most important used lasers according to the nature of the active media; their typical operation wavelengths and temporal operation regimes are also shown.

16.1.1 Solid-State Lasers

There are essentially two types of solid lasers media: impurity-doped crystals and glasses. They are almost exclusively optically pumped with flash lamps, continuous wave arc lamps or with other laser sources, such as semiconductor lasers. Well-known examples are the Nd^{3+}:YAG and the Nd^{3+}:glass lasers.

TABLE 16.1   Representative Examples of Laser Sources

Class	Laser Medium	Nominal Operating Wavelength (nm)	Typical Output Power or Energy	Typical Temporal Regime
Solid	Nd^{3+}:YAG	1064	10–100 W	CW
	Nd^{3+}:glass	1064	50 J	Pulsed
	Ti^{3+}:Al_2O_3	660–1180	10 W	CW
Semiconductor	InGaAsP	1300	10 mW	CW
Gas	He–Ne	633	5 mW	CW
	Ar^+	488	10 W	CW
	KrF	248	0.5 J	Pulsed
	CO_2	10600	100–1000 W	CW
	Kr^+	647	0.5 W	CW
	HCN	336.8×10^3	1 mW	CW
Liquid	Rhodamine-6G	560–640	100 mW	CW
Plasma	C^{6+}	18.2	2 mJ	Pulsed
FEL	Free electrons	300–4000	1 mJ	Pulsed

16.1.2 Semiconductor Lasers

Even though these are also "solid-state lasers", for classification purposes they are generally considered different due to the difference in the inversion mechanism. These lasers are characterized in terms of the way that the hole–electron pair population inversion is produced. They can be optically pumped by other laser sources, by electron beams, or more frequently by injection of electrons in a *p–n* junction. A common example is the gas laser.

16.1.3 Gas Lasers

There are essentially six different types of gas lasers, including

1. Electronic transitions in neutral atomic active media
2. Electronic transitions in ionized atomic active media
3. Electronic transitions in neutral molecular active media
4. Vibrational transitions in neutral molecular active media
5. Rotational transitions in neutral molecular active media
6. Electronic transitions in ionic molecular active media

These lasers are pumped by several methods including continuous wave (CW), pulsed, and RF electrical discharges, optical pumping, chemical reaction, and gas-dynamic expansion. Common examples for each of the above lasers are Ne–He, Ar^+, KrF, CO_2, CH_3F, and N_2^+.

16.1.4 Liquid Lasers

The active medium is a solution of a dye compound in alcohol or water. There are three main types: organic dyes, which are well known because of their tunability; rare-earth chelate lasers using organic molecules; and lasers using inorganic solvents and trivalent rare earth ion active centers. Typically, they are optically pumped by flash lamps or using other lasers. Common examples are Rh6G, TTF, and $POCl_4$.

16.1.5 Plasma Lasers

These lasers use a plasma typically generated by a high-power, high-intensity laser (or a nuclear detonation) as an active medium. They radiate in the UV or X-ray region of the spectrum. A typical example is the C^{6+} laser.

TABLE 16.2 Main Temporal Operation Regimes

Temporal Operation	Technique	Pulse Width (s)
Continuous wave (CW)	Continuous pumping; resonator Q is held constant	∞
Pulsed	Pulsed pumping; resonator Q is held constant	10^{-8}–10^{-3}
Q-switched	Pumping is continuous or pulsed; resonator Q varies between a low and a high value	10^{-8}–10^{-6}
Mode-locked	Excitation is continuous or pulsed; a modulation rate related to the transit time in the resonator is introduced	10^{-12}–10^{-9}

16.1.6 Free Electron Lasers

These lasers make use of a magnetic "wiggler" field produced by a periodic arrangement of magnets of alternating polarity. The active medium is a relativistic electron beam moving in the wiggler field. The most important difference in relation to other lasers is that the electrons are not bound to any active center, such as an atom or a molecule. The amplification of an electromagnetic field is due to the energy that the electromagnetic laser beam takes from the electron beam.

16.1.7 Fiber Lasers

These lasers use a waveguide as a gain medium, in this case a fiber optic. Depending on the case, such devices utilize the various special properties of fibers, such as the large amplification bandwidth, the high gain efficiency, the geometry and waveguiding, which allow for high output powers combined with good beam quality, and potentially low cost.

16.1.8 Temporal Laser Operation

Any laser can be induced to produce output radiation with specific temporal characteristics. This can be achieved by proper design of the excitation source and/or by controlling the Q of the laser resonator. Table 16.2 describes the most common temporal operation regimes. Tables 16.3 and 16.4 show examples and the performance of some important CW and pulsed lasers, respectively.

16.2 Laser Medium

The amplification of electromagnetic radiation takes place in a laser medium, which is pumped in order to obtain a population inversion. Next the basic terms used to characterize a laser medium are described.

16.2.1 Unsaturated Gain Coefficient

The unsaturated gain coefficient or unsaturated gain per unit length α is given by

$$\alpha = \frac{c^2}{8\pi n^2 f^2 \tau_R}[N_2 - (g_2/g_1)N_1]g(f),\qquad(16.1)$$

where $g(f)$ is the *line shape function*; N_2, N_1, g_2, and g_1 are the population inversion densities and the degeneracies of levels 22.2 and 22.1, respectively; and n, f, c, and τ_R are the refractive index, the frequency of the laser radiation, the speed of light, and the *radiative lifetime* of the upper laser level, respectively, which is given as

$$\tau_R = \left(\frac{\varepsilon_0}{2\pi}\right)\frac{m_e c^3}{f_{12}e^2 f_0^2},\qquad(16.2)$$

TABLE 16.3 Properties and Performance of Some Continuous Wave (CW) Lasers

Parameter	Unit	Neon-Helium	CO2	Liquid (Rhodamine-6G)	Solid (Ti:sapphire)
Excitation method		Dc discharge	RF Excited		Ar+ Laser pump
Gain medium composition		Neon–helium	$CO_2(1):N_2(1):HE(3):$ Xe(0.5)	R6G:sol–gel	Ti:Al$_2$O$_3$
Wavelength	nm	632.8	10,600	560	790
Laser cross-section	$\times 10^{-19}$ cm^2	3×10^6	1.5×10^{-16}	1.8×10^3	2.8
Radiative lifetime (upper level)	µs	≈ 0.1	4×10^3	6.7×10^{-3}	3.2
Decay lifetime (lower level)	µs	≈ 0.1	$\approx 4 \times 10^3$	6×10^{-3}	
Gain bandwidth	nm	2×10^{-3}	1.6×10^{-2}	3.4×10^{-3}	180
Type, gain saturation		Inhomogeneous	Homogeneous	Homogeneous	Homogeneous
Homogeneous saturation flux	W cm^{-2}	—	≈ 20	—	0.9
Inversion density	cm^{-3}	$\approx 1 \times 10^9$	2×10^{10}	$\approx 2 \times 10^{10}$	—
Small signal gain coefficient	cm^{-1}	$\approx 1 \times 10^{-3}$	$\approx 3 \times 10^{-2}$	1×10^{-2}	
Pump power	W	—	900	4×10^{-3}	5.5
Output power	W	0.03	46	2.0×10^{-3}	0.27
Efficiency	%	0.1	12	60	12
Reference		Xianshu et al. [5]	Chernikov et al. [6]	Lo et al. [7]	Shieh et al. [8]

where ε_0, e, m_e, f_0, and f_{12} are the permittivity in vacuum, the electronic charge, the electronic mass, the frequency at the line center, and the oscillator strength of the transition between levels 22.2 and 22.1.

The stimulated transition cross-section σ is

$$\sigma = \frac{c^2}{8\pi n^2 f^2 \tau_R} g(f).$$ (16.3)

Therefore, the unsaturated gain per unit length can also be written as

$$\alpha = \left[N_2 - \left(\frac{g_2}{g_1} \right) N_1 \right] \sigma.$$ (16.4)

The increase in the intensity I per unit length dI/dz is described by the equation

$$\frac{dI}{dz} = \alpha I.$$ (16.5)

For an initial intensity $I(0)$ at $z = 0$, the beam intensity varies along the propagation distance z according to

$$I(z) = I(0)e^{\alpha z}.$$ (16.6)

For a laser medium of length l the *total unsaturated gain* G_{db} (in decibels) is

$$G_{db} = 4.34\alpha l.$$ (16.7)

TABLE 16.4 Properties and Performance of Some Pulsed Lasers

Parameter	Unit	Gas XeCl	Gas CO$_2$	Liquid (Rh6G:methanol)	Solid Nd:YVO	Solid Nd:YLF
Excitation method		E-beam	Traverse dc with RF preionization	Frequency-doubled Nd:YAG	Diode pump laser	Diode pump laser
Gain medium composition		XeCl	CO$_2$(1):N$_2$(1):H$_3$(3)	Rh6G:methanol	Nd(3%):YVO$_4$	Nd(1.5%):YLF
Wavelength	nm	308	10,600	563–604	1064	1047
Laser cross-section	cm^{-2}	4.5×10^{-16}	2×10^{-18}	1.8×10^{-16}	25×10^{-19}	0.4×10^{-19}
Radiative lifetime (upper level)	μs	11×10^3	4×10^3	6.5×10^{-3}	50	480
Decay lifetime (lower level)	μs	—	5×10^{-2}	6×10^{-3}	92	—
Gain bandwidth	nm	2	1	80	1	1.3
Homogeneous saturation flux	W cm^{-2}	—	0.2	2×10^{-3}	0.037	—
Inversion density	cm^{-3}	—	3×10^{17}	2×10^{16}	—	—
Small signal gain coefficient	cm^{-1}	—	2×10^{-2}	4×10^{-2}		25×10^{-2}
Pump power	W	—	—	2	2	4
Excitation current/ voltage	A/V	210×10^3/ 800×10^3	3.6×10^4/15×10^3	—	—	—
Excitation current density	A cm^{-2}	5.25×10^2	180	—	—	—
Pump power density	W cm^{-3}	2×10^6	5.75×10^5	—	—	—
Output pulse energy	J	136	87.8×10^{-3}	100×10^{-3}	53×10^{-9}	30×10^{-9}
Output pulse length	Ns	100	36.5	3×10^{-3}	0.0037	8×10^{-3}
Output pulse power	W	1.36×10^9	0.9×10^9	3.3×10^{10}	0.46	3.75×10^3
Reference		Jingru et al. [9]	Jiang et al. [10]	Christophe et al. [11]	Spuhler et al. [12]	Hönninger et al. [13]

From Equation 16.6, it is clear that when the total gain per pass αl is sufficiently small the following approximation is valid:

$$\frac{I(l)}{I(0)} \approx 1 + \alpha l \qquad (16.8)$$

and the gain is equal to tthe fractional increase in intensity:

$$\frac{\left[I(l) - I(0) \right]}{I(0)}. \qquad (16.9)$$

Therefore, the *percentage gain* is

$$G(\text{in percent}) = 100\alpha l. \qquad (16.10)$$

The *threshold population inversion* ΔN_{th} needed to sustain laser oscillation in a resonator with output mirror reflectivity R is

$$\Delta N_{th} = [N_2 - (g_2/g_1)N_1]_{th} = \left[\gamma - \left(\frac{1}{2L} \right) \ln R \right] \Big/ \sigma, \tag{16.11}$$

where γ is the absorption coefficient of the host medium at frequency f and L is the distance between the mirror resonator (assumed equal to the laser medium length l). The *threshold gain coefficient* to start laser oscillation is given as

$$\alpha_{th} = \gamma - \frac{\ln R}{2L}. \tag{16.12}$$

16.2.2 Line-Shape Function

To describe the gain distribution as a function of the frequency of the *lineshape function* $g(f)$ is used. The function $g(f)$ is normalized:

$$\int_{-\infty}^{\infty} g(f)df = 1. \tag{16.13}$$

The normalized Lorentzian lineshape function $g_L(f)$ is

$$g_L(f) = \frac{\Delta f}{2\pi \left[(f - f_0)^2 + \left(\frac{\Delta f}{2} \right)^2 \right]}. \tag{16.14}$$

For natural broadened transitions, the Lorentzian lineshape has a linewidth $\Delta f = \Delta f_N$, where

$$\Delta f_N = \frac{1}{\pi \tau_F} = \frac{1}{\pi} \left(\frac{1}{\tau_R} + \frac{1}{\tau_{NR}} \right). \tag{16.15}$$

In this expression, τ_F is the fluorescent lifetime and τ_{NF} is the nonradiative decay time constant given by

$$\frac{1}{\tau_{NR}} = N_b Q_{ab} \sqrt{\frac{8kT}{\pi}} (M_a^{-1} + M_b^{-1}), \tag{16.16}$$

where M_a, M_b, and Q_{ab} are the masses of the atoms (or molecules) of species a and b, respectively, and its collision cross-section, N is the number density of atoms (or molecules), k is the Boltzmann constant, and T the temperature. For collision-broadened transitions, the Lorentzian lineshape has a linewidth $\Delta f = \Delta f_{coll}$, where

$$\Delta F_{coll} = \frac{NQ}{\pi} \sqrt{\frac{16kT}{\pi M}}. \tag{16.17}$$

This bandwidth arises from elastic collisions between like atoms (or molecules) of atomic mass M and collision cross-section Q.

The normalized Gaussian lineshape function $g_G(f)$ is given as

$$g_G(f) = \frac{2(\ln 2)^{1/2}}{\pi^{1/2}\Delta f} e^{-[4(\ln 2)(f-f_0)^2/(\Delta f)^2]}.$$ (16.18)

For a Doppler-broadened transition, the linewidth $\Delta f = \Delta f_D$ is

$$\Delta f_D = 2f_0 \sqrt{\frac{2kT \ln 2}{Mc^2}}.$$ (16.19)

The Gaussian and Lorentzian lineshape are shown in Figure 16.1. The linewidths at half-maximum are shown as Δf_G (for Gaussian profile) and as Δf_L (for the Lorentzian profile).

16.2.3 Saturated Gain Coefficient

The saturated gain coefficient for a homogeneously broadened line is

$$\alpha_s = \frac{\alpha}{1 + (I/I_s)},$$ (16.20)

where I_s is the saturation intensity:

$$I_s = \frac{4\pi n^2 hf}{\lambda^2 g(f)} \left(\frac{\tau_R}{\tau_F} \right),$$ (16.21)

where τ_F is the fluorescence lifetime of the upper laser level, h is the Planck constant, and λ is the wavelength of the laser radiation.

The *saturated gain coefficient for an inhomogeneously broadened line* is

$$\alpha_s = \frac{\alpha}{[1 + (I/I_s)]^{1/2}},$$ (16.22)

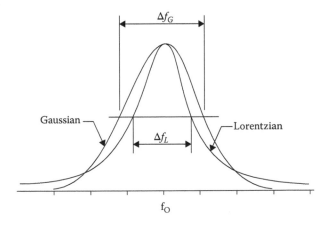

FIGURE 16.1 Gaussian and Lorentzian line shapes.

where the saturation intensity is

$$I_s = \frac{2\pi^2 n^2 h f \Delta f}{\lambda^2} \left(\frac{\tau_R}{\tau_F} \right). \tag{16.23}$$

The bandwidth Δf is the homogeneous linewidth of the inhomogeneously broadened transition.

16.3 Resonators and Laser Beams

Most lasers have an optical resonator consisting of a pair of mirrors facing each other. In this way it is possible to maintain laser oscillation due to the feedback provided to the active (amplifying) medium and it is also possible to sustain well-defined longitudinal and transversal oscillating modes. An *optical resonator* is shown in Figure 16.2. One of the mirrors has an ideal optical reflectivity of 100% and the other less than 100% (the laser beam is emitted through this second mirror). The mirrors are separated by a distance L and the radii of curvature of the mirrors are R_1 and R_2.

16.3.1 Stability, Resonator Types, and Diffraction Losses

A resonator is stable if the *stability condition*:

$$0 < g_1 g_2 < 1 \tag{16.24}$$

is satisfied, where g_1 and g_2 are the *resonator parameters*:

$$g_1 = 1 - \frac{L}{R_1} \quad \text{and} \quad g_2 = 1 - \frac{L}{R_2}. \tag{16.25}$$

The radius of curvature R is defined as positive if the mirror is concave with respect to the resonator interior and the radius of curvature is negative if the mirror is convex. In Figure 16.2, both R_1 and R_2 are positive.

Figure 16.3 draws the hyperbola defined by the stability condition (Equation 16.24) and is called the *stability diagram*. Since an optical resonator can be represented by its coordinates g_1 and g_2 in the stability diagram, the resonator is *stable* if the point (g_1, g_2) falls within the shaded region. Figure 16.3 also shows the location of several resonator types. Clearly, if

$$g_1 g_2 < 0 \quad \text{or} \quad g_1 g_2 > 1, \tag{16.26}$$

the resonator is *unstable* and upon multiple reflections a ray will diverge from the cavity axis.

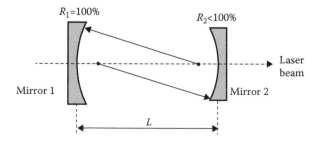

FIGURE 16.2   Optical resonator.

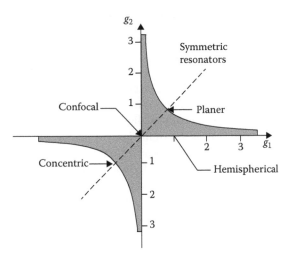

FIGURE 16.3 Stability diagram.

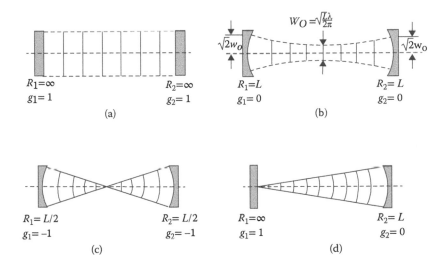

FIGURE 16.4 Resonator types: (a) plane parallel resonator, (b) symmetrical confocal resonator, (c) symmetrical concentric resonator, and (d) confocal-planar resonator.

The diffraction losses in a laser are characterized by the *resonator Fresnel number N*. This is a dimensionless parameter given as

$$N = \frac{a^2}{L\lambda}, \tag{16.27}$$

where α is the radius of the mirror resonator. A large Fresnel number implies small diffraction losses.

Some important *resonator types* (shown in Figure 16.4) are the following:

- Plane parallel resonator, as shown in Figure 16.4a. This resonator has the largest mode volume, but is difficult to keep in alignment and is used only with high gain medium. Its diffraction losses are larger than those of stable resonators with spherical mirrors.
- Symmetrical confocal resonator, as shown in Figure 16.4b. The spot sizes at the mirrors are the smallest of any stable symmetric resonator. With a Fresnel number larger than unity, the diffraction losses of this resonator are essentially negligible.
- Symmetrical concentric resonator, as shown in Figure 16.4c. As with the plane resonator, the exact concentric resonator is relatively difficult to keep in alignment; however, when L is slightly less than 2R, the alignment is no longer critical. The TEM$_{00}$ mode in this resonator has the smallest beam waist.
- Confocal-planar resonator, as shown in Figure 16.4d. This resonator is simple to keep in alignment, especially when L is slightly less than R. Also, small variations in the mirror spacing allow the adjustment of the spot size w$_2$, so that only the TEM$_{00}$ mode fills the laser medium or mirror. It is widely used in low-power gas lasers.

There are some useful empirical formulas to find the *diffraction losses* in a resonator. The one-way power loss per pass δ in a real finite-diameter resonator for two common cases is,

For a confocal resonator:

$$\delta \approx \pi^2 2^4 N exp\left(-4\pi N\right) \quad for\ N \geq 1$$

$$\delta \approx 1 - \left(N\pi^2\right) \quad for\ N \to 0$$

$$(16.28)$$

For a plane parallel resonator:

$$\delta \approx 0.33 N^{-3/2} \quad for\ N \geq 1. \tag{16.29}$$

16.3.2 Axial Modes

The *axial* or *longitudinal modes* of the resonator are the resonant frequencies of the cavity f_q, where q is the number of half-waves along the resonator axis:

$$f_q = q\left(c\,/\,2L\right). \tag{16.30}$$

The *frequency spacing* $f_{\Delta q}$ between two axial resonances of the laser cavity is

$$f_{\Delta q} = f_{(q+1)} - f_q = \frac{c}{2nL}. \tag{16.31}$$

For a TEM$_{mnq}$ mode, the *resonance frequency* of the qth axial mode with the *mn*th transverse mode is

$$f_{mnq} = \left[q + (m+n+1)\frac{\cos^{-1}\sqrt{g_1 g_2}}{\pi}\right]\frac{c}{2nL}. \tag{16.32}$$

The frequency spacing $f_{\Delta mnq}$ between transverse modes is

$$f_{\Delta mnq} = \left(\cos^{-1}\sqrt{g_1 g_2}\right)\frac{c}{2\pi nL}. \tag{16.33}$$

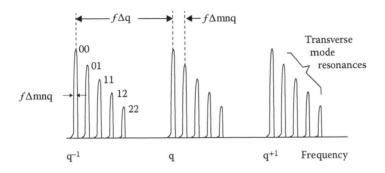

FIGURE 16.5 Frequency spacing between transverse modes.

The above expression is represented in Figure 16.5. The bandwidth of a resonant mode of frequency f_{mnq} is

$$\Delta f_{mnq} = \frac{1}{2\pi\tau_c} = \frac{c(\gamma L - \ln\sqrt{R})}{2\pi nL}. \tag{16.34}$$

Substituting in Equation 16.32, the parameters g_1 and g_2 for particular resonators, we obtain, for a plane parallel resonator,

$$f_{mnq} = \frac{qc}{2nL}; \tag{16.35}$$

for a symmetrical concentric resonator,

$$f_{mnq} = [q - (m+n+1)]\frac{c}{2nL}; \tag{16.36}$$

for a symmetrical confocal and a confocal-planar resonator,

$$f_{mnq} = \left[q + \frac{(m+n+1)}{2}\right]\frac{c}{2nL}. \tag{16.37}$$

16.3.3 Transverse Modes

For rectangular coordinates (x, y), the *transverse field distribution* $E(x, y)$ of a TEM_{mnq} mode is given as

$$E(x,y) = E_0 H_m\left(\frac{\sqrt{2}x}{w(z)}\right) H_n\left(\frac{\sqrt{2}y}{w(z)}\right)\exp\left(-\frac{x^2+y^2}{w^2(z)}\right), \tag{16.38}$$

where $H_n(x)$ are the Hermite polynomials defined by

$$H_n(x) = (-1)^2 e^{x^2}\frac{d^n}{dx^n}e^{-x^2}. \tag{16.39}$$

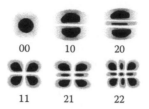

Rectangular

FIGURE 16.6   Transverse mode patterns TEM$_{mn}$.

The first four Hermite polynomials are

$$
\begin{aligned}
H_0\left(x\right) &= 1 \\
H_1\left(x\right) &= 2x \\
H_2\left(x\right) &= 4x^2 - 2 \\
H_3\left(x\right) &= 8x^3 - 12x;
\end{aligned}
$$

(16.40)

these polynomials obey the recursion relation

$$
H_{n+1}\left(x\right) = 2xH_n\left(x\right) - 2nH_{n-1}\left(x\right),
$$

(16.41)

which also provides a useful way of calculating the higher-order polynomials. Figure 16.6 shows some transverse mode patterns.

For polar coordinates (r,φ), the transverse field distribution $E(r,\varphi)$ of a TEM$_{mnq}$ mode is given as

$$
E(r,\phi) = E_0 \left(\frac{\sqrt{2}r}{w(z)} \right)^l L_p^l \left(\frac{2r^2}{w^2(z)} \right) \exp\left(-\frac{r^2}{w^2(z)} \right) \left(\left\{ \begin{matrix} \sin \\ \cos \end{matrix} \right\} l\phi \right),
$$

(16.42)

where $L_p^l(x)$ are the Laguerre polynomials defined by

$$
L_p^l(x) = e^x \frac{x^{-1}}{p!} \frac{d^p}{dx^p} (e^{-x} x^{p+1}).
$$

(16.43)

The first three Laguerre polynomials are

$$
\begin{aligned}
L_0^l(x) &= 1, \\
L_1^l(x) &= l+1-x, \\
L_2^l(x) &= \frac{1}{2}(l+1)(l+2) - (l+2)x + \tfrac{1}{2}(x^2);
\end{aligned}
$$

(16.44)

these polynomials obey the recursion relation:

$$
(p+1)L_{p+1}^l(x) = (2p+l+1-x)L_p^l(x) - (p+l)L_{p-1}^l(x),
$$

(16.45)

which also provides a useful way of calculating the higher-order polynomials. Figure 16.7 shows some transverse mode patterns. A mode may be described as a super-position of two like modes. For example, the TEM$_{01}^*$ is made up of rectangular TEM$_{01}$ and TEM$_{10}$ modes.

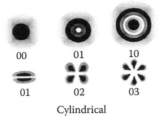

00   01   10

01   02   03

Cylindrical

FIGURE 16.7 Some transverse mode patterns TEM_{mn}.

16.3.4 Resonator Quality Parameter

The *quality factor Q* of a laser resonator is defined as

$$Q = 2\pi \frac{\text{energy stored in the resonator}}{\text{energy lost in one cycle}}, \tag{16.46}$$

$$Q = 2\pi f_{mnq} \frac{E}{[dE / dt]}. \tag{16.47}$$

The loss of energy dE/dt is related to the energy decay time or photon lifetime τ_c by

$$\frac{dE}{dt} = -\frac{E}{\tau_c}, \tag{16.48}$$

where

$$\tau_c = \frac{nL}{c(\gamma L - \ln \sqrt{R})}. \tag{16.49}$$

This expression can also be written as

$$Q = \frac{f_{mnq}}{\Delta f_{mnq}}. \tag{16.50}$$

16.4 Gaussian Beams

As can be seen from Equation 16.38 with $n = m = 0$, the transverse field distribution TEM_{00} has a bell-shaped Gaussian amplitude:

$$E(x, y) \propto E_o \exp\left(-\frac{x^2 + y^2}{w^2(z)}\right). \tag{16.51}$$

Taking as w_0 the $1/e$ transversal spot size at $z = 0$, the description of a *Gaussian beam* at any position z is given by the parameters $w(z)$ describing the spot size and $R(z)$ describing the wavefront radius of curvature. This is schematically shown in Figure 16.8.

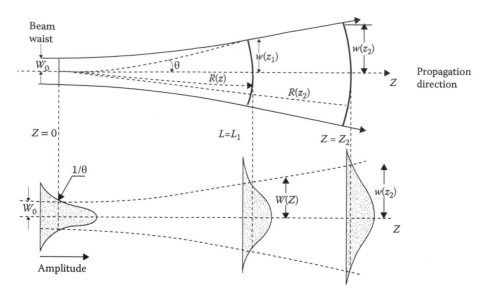

FIGURE 16.8   Gaussian beam showing the spot size $w(z)$ and the wavefront radius of curvature $R(2)$.

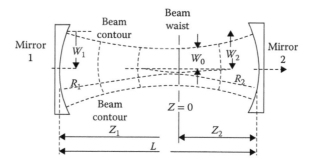

FIGURE 16.9   Optical resonator with parameters g_1 and g_2.

$$w(z) = w_0 \sqrt{1 + \left(\frac{\lambda z}{\pi w_0^2} \right)^2} = w_0 \sqrt{1 + \left(\frac{z}{z_R} \right)^2} \tag{16.52}$$

$$R(z) = z \left[1 + \left(\frac{\pi w_0^2}{\lambda z} \right)^2 \right] = z \left[1 + \left(\frac{z_R}{z} \right)^2 \right]. \tag{16.53}$$

In the above equations z_R is the *Rayleigh distance*, defined as

$$z_R = \frac{\pi w_0^2}{\lambda}. \tag{16.54}$$

For a laser resonator (Figure 16.9), with parameters g_1 and g_2, the spot size at the beam waist w_0 is given by

$$w_0 = \left(\frac{\lambda L}{\pi} \right)^{1/2} \frac{[g_1 g_2 (1 - g_1 g_2)]^{1/4}}{(g_1 + g_2 - 2 g_1 g_2)^{1/2}}. \tag{16.55}$$

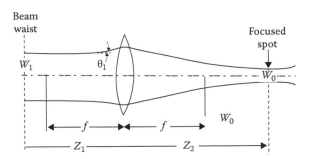

FIGURE 16.10 Focusing of a Gaussian beam by a thin lens.

The position of the beam waist w_0 relative to the resonator mirrors (Figure 16.9) is

$$z_1 = \frac{-g_2(1-g_1)L}{g_1+g_2-2g_1g_2}, \tag{16.56}$$

$$z_2 = \frac{g_1(1-g_2)L}{g_1+g_2-2g_1g_2} = z_1 + L. \tag{16.57}$$

The spot sizes w_1 and w_2 at each of the resonator mirrors are

$$w_1 = \left[\frac{L\lambda}{\pi}\right]^{1/2}\left[\frac{g_2}{g_1(1-g_1g_2)}\right]^{1/4}, \tag{16.58}$$

$$w_2 = \left[\frac{L\lambda}{\pi}\right]^{1/2}\left[\frac{g_1}{g_2(1-g_1g_2)}\right]^{1/4}. \tag{16.59}$$

The half-angle *beam divergence* in the far field (for $z \gg z_R$), shown in Figure 16.8, is given by

$$\theta = \frac{\lambda}{\pi w_0}. \tag{16.60}$$

The focusing of a Gaussian laser beam by a thin lens is shown in Figure 16.10. The position of the focused beam waist is given by

$$z_2 = f + \frac{(z_1-f)f^2}{(z_1-f)^2 + \left(\dfrac{\pi w_{01}^2}{\lambda}\right)^2}. \tag{16.61}$$

The spot size of the focused laser beam is obtained from

$$\frac{1}{w_2^2} = \frac{1}{w_1^2}\left(1-\frac{z_1}{f}\right)^2 + \frac{1}{f^2}\left(\frac{\pi w_1}{\lambda}\right)^2. \tag{16.62}$$

In most practical applications, this expression can be approximated to

$$w2 \cong f\theta_1 \qquad (16.63)$$

where (from Equation 16.60), $\theta_1 \equiv \lambda/\pi W_1$.

16.5 Laser Applications and Instruments

There are at least as many laser instruments as there are laser applications. This only mirrors the fact that all laser instruments (the laser itself) started as research tools. What follows is a general listing of some important laser applications and instruments along with some significant published results in the following areas: laser Doppler velocimetry and its medical applications; laser radar (LIDAR) and tunable laser radars for applications to pollution detection, laser thermometry, laser applications to the electronics and solid-state industry, and laser applications to art. For these and other areas not mentioned here, the reader is also referred to general references [14–30].

16.5.1 Laser Doppler Velocimetry

Laser Doppler velocimetry is a well-established technique widely applied in many engineering areas. Examples of recent research work applied to fluid dynamics can be found in references [9,31–38]. The use of this technique in medical applications is a wide and promising area of research where many new instruments are currently developed for specific problems [39–47].

16.5.2 LIDAR

The short and intense laser pulses produced by a Q-switched laser are ideal for optical ranging. These instruments are also called optical radar or LIDAR. Small solid-state lasers with the associated detection electronics are available in small and rugged units for military field applications. Laser ranging systems making use of tunable lasers whose wavelengths can be tuned to specific molecular or atomic transitions can be used for pollution detection, ranging of clouds, aerosol measurements, and so on. These instruments placed in orbiting satellites may also be used for weather forecasting applications [18,48–62].

16.5.3 Laser Thermometry

The fact that the laser can be used as a nonintrusive measurement instrument is widely used in many applications where a measurement instrument would be damaged or plainly destroyed by the studied system. Flames and air jets are good examples of systems that can be studied using laser-based techniques, such as spectroscopy and holography [63–72].

16.5.4 Laser Applications to the Electronic Industry

The electronics industry has widely benefited from the use of lasers. Many new laser instruments and applications are currently being designed. Nowadays, the laser is used in automatic microsoldering, in laser recrystallization of semiconductor substrates, and in laser ablation of thin-films deposition among other applications [73–90].

16.5.5 Laser Applications to Art

Lasers are being used for diagnostic, conservation, and restoration of great master-works. Until recently, the cleaning of painted surfaces required the use of solvents to remove old varnish or encrustations from the painted surfaces. This is a difficult process because the solvents may also damage the paint layer itself, causing soluble materials in the paint to diffuse out in a process known as leaching. This restoration process is now carried out using lasers without any effect on the original paint layers [91–96].

Appendix 16.A

List of Symbols

c	Speed of light $c = 2.998 \times 10^8$ m/s
e	Electronic charge $e = 1.602 \times 10^{-19}$ coulomb
f	Focal length of thin lens
f	Frequency of laser radiation
f_0	Frequency at line center
f_{12}	Oscillator strength of the transition between levels 2 and 1
Δf	Frequency bandwidth at half-maximum
f_{mnq}	Frequency of TEM_{mnq} mode
g_1, g_2	Degeneracies of lower and upper laser levels, respectively
g_1, g_2	Resonator g parameters for mirrors 1 and 2, respectively
$g(f)$	Lineshape function
I	Beam intensity
I_s	Saturation beam intensity
l	Transverse mode number (radial geometry)
l	Length of gain medium
L	Distance between resonator mirrors
m	Transverse mode number (rectangular coordinates)
m_e	Electronic rest mass $m_e = 9\ 109 \times 10^{-31}$ kg
M	Mass of atom (or molecule)
m_a, M_b	Mass of atom (or molecule) of species a and b, respectively
n	Transverse mode number in rectangular coordinates
n	Index of refraction
N	Number density of atoms (or molecules)
Ni	Number density of atoms (or molecules) in level i
ΔN_{th}	Threshold population inversion
N_0	Number density of laser atoms (or molecules)
p	Transverse mode number (radial geometry)
Q_{ab}	Collision cross-section
r	Radial coordinate
R_1, R_2	Radii of curvature of resonator mirrors 1 and 2, respectively
$R(z)$	Radius of curvature of wavefront
R	Reflectivity of output mirror
t	Time

T	Temperature
w_0	Spot size at beam waist
w_1, w_2	Spot sizes at mirrors 1 and 2, respectively
$w(z)$	Spot size at a distance z from beam waist
x, y, z	Rectangular coordinates
z_R	Rayleigh range

Greek Symbols

α	Unsaturated gain coefficient per unit length
α_s	Saturated gain coefficient per unit length
α_{th}	Threshold gain coefficient per unit length
γ	Absorption coefficient of laser medium
ε_0	Permittivity in vacuum
θ	Far-field beam divergence half-angle
λ	Wavelength of laser radiation
σ	Stimulated transition cross-section
τ_f	Fluorescence lifetime of the upper laser level
τ_{NR}	Nonradiative decay time constant
τ_R	Radiative lifetime of the upper laser level

References

1. Planck M. Uber das Gesetz der Enerpreverteilung in Normal Spectrum, *Ann. Physik*, **4**, 553–563; 1901.
2. Einstein A. Zur Quantentheorie des Strahlung, *Physikalische Zeitschrift*, **18**, 121–129; 1917.
3. Schawlow AL, Towes CH. Infrared and optical masers, *Phys. Rev.*, **112**, 1940–1949; 1958.
4. Maiman T, Hoskins HRH, D'haenens IJ, Asawa CK, Evtuhov V. Stimulated optical emission in fluorescent solids. II. spectroscopy and stimulated emission in ruby, *Phys. Rev.*, **123**, 1151–1159; 1961.
5. Xianshu L, Ychuan C, Taouyu L, et al. A development of red internal He-Ne lasers with near critical concave-convex stable resonator, *Proc. SPIE*, **2889**, 358–366; 1996.
6. Chernikov SB, Karapuzikov AL, Stojanov SA. RF Exited CO_2 Slab *Waveguide* Laser, *Proc. SPIE*, **2773**, 52–56; 1995.
7. Lo D, Lam SK, Ye C, Lam KS. Narrow Linewidth Operation of Solid-State Dye Laser Based on Sol-Gel Silica, *Optic Comm.*, **156**, 316–320; 1998.
8. Shieh J-M, Huang TC, Wang C-L, Pan C-L. Broadly tunable self-starting mode-locked Ti:sapphire laser with triple-strained quantum-well saturable Bragg reflector, *Optic Comm.*, **156**, 53–57; 1998.
9. Jingru L, Yuan X, Gan Y, et al. High power eximer laser and application, *SPIE-High-Power Lasers*, **2889**, pp. 98–103; 1996.
10. Jiang S, Gorai Y, Shun-Zhao X. Development of a new laser-doppler microvelocimetering and its application in patients with coronary artery stenosis, *Angiology*, **45**, 225–228; 1994.
11. Christophe JB, Robert B, David G, Nathalie M, Michel P. Stabilization of terahertz beats from a pair of picosecond dye lasers by coherent photon seeding, *Optic Comm.*, **161**, 31–36; 1999.
12. Spühler GJ, Paschotta R, Fluck R, et al. Experimentally confirmed design guidelines for passively Q-switched microchip lasers using semiconductor saturable absorbers, *J. Opt. Soc. Am.*, **12**, 376–388; 1999.

13. Hönninger C, Paschotta R, Morier-Genound F, Moser M, Keller U. Q-switching stability limits of continuous-wave passive mode locking, *J. Opt. Soc. Am.*, **16**, 46–56; 1999.

14. Dong Hou J, Tianjin Z, Chonghong L, et al. Study on mini high repetition frequency sealed-off TEA CO_2 laser, *Proc. SPIE*, **2889**, 392–397; 1996.

15. Greulich KO, Farkas DL. eds, *Micromanipulation by Light in Biology and Medicine*, Boston, MA: Birkhauser, 1998.

16. Grundwald E, Dever DF, Keehn PM. *Megawatt Infrared Laser Chemistry*, New York, NY: John Wiley & Sons, 1978.

17. Hinkley ED. ed., *Laser Monitoring of the Atmosphere*, Berlin: Springer-Verlag, 1976.

18. Jelalian AV. *Laser Radar Systems*, Cambridge: Artech House, 1992.

19. Kamerman XX. *Applied Laser Radar Technology*, Society for Photo-Optical Instrumentation Engineers, 1993.

20. Koebner HK. ed., *Lasers in Medicine*, New York, NY: John Wiley & Sons, 1980.

21. Kuhn KJ. *Laser Engineering*, New York, NY: Prentice Hall, 1998.

22. Metcalf HJ, Stanley HE. eds, *Laser Cooling*, Berlin: Springer Verlag, 1999.

23. Motz H. *The Physics of Laser Fusion*, San Diego, CA: Academic Press, 1979.

24. Papannareddy R. *Introduction to Lightwave Communication Systems*, Cambridge: Artech House, 1997.

25. Powell J. *CO_2 Laser Cutting*, Berlin: Springer Verlag, 1998.

26. Ready JF. *Industrial Applications of Lasers*, San Diego, CA: Academic Press, 1997.

27. Riviere CJB, Baribault R, Gay D, McCarthy N, Piché M. Stabilization of terahertz beats from a pair of picosecond dye lasers by coherent photon seeding, *Optic Comm.*, **161**, 31–36; 1999.

28. Stenholm S. *Laser in Applied and Fundamental Research*, London, UK: Hilger, 1985.

29. Ven Hecke GR, Karukstis KK. *A Guide to Lasers in Chemistry*, New York, NY: Jones and Bartlett, 1997.

30. Zuev VE. *Laser Beams in the Atmosphere*, New York, NY: Plenum, 1982.

31. Belorgey MB, Le J, Grandjean A. Application of laser doppler velocimetry to the study of turbulence generated by swell in the vicinity of walls or obstacles, *Coas. Eng.*, **13**, 183–187; 1989.

32. Karwe MV, Sernas V. Application of laser doppler anemometry to measure velocity distribution inside the screw channel of a twin-screw extruder, *J. Food Proc. Eng.*, **19**, 135–141; 1996.

33. Lemoine E, Wolff M, Lebouche M. Simultaneous concentration and velocity measurements using combined laser-induced fluorescence and laser doppler velocimetry: Application to turbulent transport, *Exp. Fluids*, **20**, 319–324; 1996.

34. Louranco LM, Krothapalli A. Application of on-line particle image velocimetry to high-speed flows, *Laser Anemometry*, **5**, 683–689; 1993.

35. Reisinger D, Heiser W, Olejak D. The application of laser doppler anemometry in a trisonic wind-tunnel, *Laser Anemometry*, **1**, 217–224; 1991.

36. Sanada M. New application of laser beam to failure analysis of LSI with multi-metal layers, *Microelectro. Reliab.*, **33**, 993–996; 1993.

37. Scharf R. A Two-component He-Ne laser-doppler anemometer for detection of turbulent reynold's stresses and its application to water and drag-reducing polymer solutions, *Means. Sci. Tech.*, **5**, 1546–1551; 1994.

38. Thiele B, Eckelmann H. Application of a partly submerged two-component laser-doppler anemometer in a turbulent flow, *Exp. Fluids*, **17**, 390–395; 1994.

39. Boutier A, Michel J. *Most—Laser Velocimetry in Fluid Mechanics*, Wiley Online Library, 2012.

40. Albrecht H-E. *Laser Doppler and Phase Doppler Measurement Techniques*, New York, NY: Springer, 2003.

41. Shepherd AP, Öberg PÅ. *Laser-Doppler Blood Flowmetry*, New York, NY: Springer, 2013.

42. Arbit E, DiResta RG. Application of laser doppler flowmetry in neurosurgery, *Neurosurgery Clin. N. Am.*, **7**, 741–744; 1996.

43. Tropea C. *Encyclopedia of Aerospace Engineering*, Wiley Online Library, 2010.

44. Porta PA, Curtin DP, McInerney JG. Laser doppler velocimetry by optical self-mixing in vertical-cavity surface-emitting lasers, *IEEE Photon. Technol. Lett.*, **14**(12), 1719–1721; 2002. doi:10.1109/LPT.2002.804666

45. Ruggero MA, Rich NC. Application of a commercially-manufactered doppler-shift laser velocimeter to the measurement of basilar-membrane vibration, *Hearing Res.*, **51**, 215–221; 1991.

46. Wagener JT, Demma N, Kubo T. 2 mm LIDAR for laser-based remote sensing: Flight demonstration and application survey, *IEEE Aerosp. Electron. Syst. Mag.*, **10**, 23–26; 1995.

47. Yokomise H, Wada H, Inui K. Application of laser doppler velocimetry to lung transplantation, *Transplantation*, **48**, 550–559; 1989.

48. Georgea AD, Thompson FR, Faaborga J. Using LIDAR and remote microclimate loggers to downscale near-surface air temperatures for site-level studies, *Remote Sens. Lett.*, **6**(12); 2015; 924–932. doi:10.1080/2150704X.2015.1088671

49. Arev NN, Gorbunov BF, Pugachev GV. Application of a laser ranging system to the metrologic certification of satellite radar measurement systems, *Meas. Techn.*, **36**, 524–527; 1993.

50. Bulatov V, Gridin VV, Schechter L. Application of pulsed laser methods to in situ probing of highway originated pollutants, *An. Chem. Acta*, **343**, 93–96; 1997.

51. Frejafon E, Kasparian J, Wolf JP. Laser application for atmospheric pollution monitoring, *Euro. Phys. J.*, **4**, 231–234; 1998.

52. Fried A, Drummond JR, Henry B. Versatile integrated tunable diode laser system for high precision: Application for ambient measurements of OCS, *Appl. Opt.*, **30**, 1916–1919; 1991.

53. Gaft M. Application of laser-induced luminescence in ecology, *Water Sci. Tech.*, **27**, 547–551; 1993.

54. Gavan J. Optimization for improvement of laser radar propagation range in detecting and tracking cooperative targets, *J. Elec. Waves Appl.*, **5**, 1055–1059; 1991.

55. Gordon JP, Zeiger HJ, Townes CH. Molecular microwave spectrum of HN_3, *Phys. Rev.*, **95**, 282–287; 1954.

56. Gherezghiher T. Choroidal and ciliary body blood flow analysis: Application of laser doppler flowmetry in experimental animals, *Exp. Eye Res.*, **53**, 151–154; 1991.

57. Hoffmann E, Ludke C, Stephanowitz H. Application of Laser-ICP-MS in environmental analysis, *Fresenius' J. An. Chem.*, **355**, 900–904; 1996.

58. Lefsky MA, Cohen WB, Parker GG. LIDAR remote sensing for ecosystem studies, *BioScience*, **52**(1), 19–27; 2002.

59. Dua S, Liub B, Liua Y, Liuc JG. Global–local articulation pattern-based pedestrian detection using 3D LIDAR data, *Remote Sens. Lett.*, **7**(1), 681–690; 2016. doi:10.1080/2150704X.2016.1177239

60. Wulfmeyer V, Bosenberg J. Single-mode operation of an injection-seeded alexandrite ring laser for application in water-vapor and temperature differential absorption LIDAR, *Opt. Lett.*, **21**, 1150–1153; 1996.

61. Li Z, Guo X. Remote sensing of terrestrial non-photosynthetic vegetation using hyperspectral, multispectral, SAR, and LIDAR data, *Prog. Phys. Geogr.*, **40**(2), 276–304; 2016.

62. Chang K-C, Huang J-M, Tieng S-M. Application of laser holographic interferometry to temperature measurements in buoyant Air Jets, *J. Heat Transfer*, **6**, 377–381; 1992.

63. Chen C-C, Chang K-C. Application of laser holographic interferometry to temperature measurements in premixed flames with different flame structures, *J. Chinese Inst. Eng.*, **14**, 633–638; 1991.

64. Robert GD, McNesby L, Miziolek WA. Application of tunable diode laser diagnostics for temperature and species concentration profiles of inhibited low-pressure flames, *Appl. Opt.*, **35**, 4018–4020; 1996.

65. Hencke H. The design and application of honeywell's laser-trimmed temperature sensors, *Meas. Control*, **22**, 233–236; 1989.

66. Kim IT, Kihm KD. Full-field and real-time surface plasmon resonance imaging thermometry, *Opt. Lett.*, **32** (23), 3456–3458; 2007. doi:10.1364/OL.32.003456

67. Kroll S. Influence of laser-mode statistics on noise in nonlinear-optical processes—Application to single-shot broadband coherent anti-stokes Raman scattering thermometry, *J. Opt. Soc. Am.*, **5**, 1910–1913; 1988.

68. Meier W, Plath I, Stricker W. The application of single-pulse CARS for temperature measurements in a turbulent stagnation flame, *Appl. Phys.*, **53**, 339–343; 1991.

69. Stanislav K, Roy S, Lakusta PJ, et al. Comparison of line-peak and line-scanning excitation in two-color laser-induced-fluorescence thermometry of OH, *Appl. Opt.*, **48**(32), 6332–6343; 2009. doi:10.1364/AO.48.006332

70. Su K-D, Chen C-Y, Lin K-C. Application of laser-enhanced ionization to flame temperature determination, *Appl. Spec.*, **45**, 1340–1345; 1991.

71. Wengab W, Luitenab AN. Ultra-sensitive thermometer based on a compact optical resonator, *Temperature*, **2**(1); 2015. doi:10.4161/23328940.2014.967598

72. Gillner A, Holtkamp J, Hartmann C, et al. Laser applications in microtechnology, *J. Mater. Process. Technol.*, **167**(2–3), 494–498; 2005. doi:10.1016/j.jmatprotec.2005.05.04

73. Choi D-H, Sadayuki E, Sugiura O. Lateral growth of Poly-Si film by excimer laser and its thin film transistor application, *Jap. J. Appl. Phys. Part 1*, **33**, 70–75; 1994.

74. Deneffe K, Van Mieghem P, Brijs B. As-deposited superconducting thin films by electron cyclotron resonance-assisted laser ablation for application in micro-electronics, *Jap. J. Appl. Phys. Part 1*, **30**, 1959–1963; 1991.

75. Engel C, Baumgartner P, Abstreiter G. Fabrication of lateral NPN- and PNP-structures on Si/SiGe by focused laser beam writing and their application as photodetectors, *J. Appl. Phys.*, **81**, 6455–6459; 1997.

76. Fujii E, Senda K, Emoto F. A laser recrystallization technique for silicon-TFT integrated circuits on quartz substrates and its application to small-size monolithic active-matrix LCD's, *IEEE Trans. Electron Devices*, **37**, 121–124; 1990.

77. Fujii E, Senda K, Emoto F. Planar integration of laser-recrystallized SOI-Tr's fabricated by lateral seeding process and Bulk-Tr's and its application to fabrication of a solid-state image sensor, *Elect. Comm. Jap. Part 2*, **72**, 77–81; 1989.

78. Garrett L, Argenal J, Rink A, Dan L. A fully automatic application for bonding surface assemblies: Laser microsoldering is an alternative method of integrated circuit manufacture, *Assembly Autom.*, **8**, 195–199; 1988.

79. Meijer J., Laser beam machining (LBM), state of the art and new opportunities, *J. Mater. Process. Technol.*, **149**(1–3), 2–17; 2004. doi:10.1016/j.jmatprotec.2004.02.003

80. Busch K, Lölkes S, Wehrspohn RB, Föll H. *Photonic Crystals: Advances in Design, Fabrication, and Characterization*, New York, NY: John Wiley & Sons, 2006.

81. Katz A, Pearton SJ. Single wafer integrated processes by RT-LPMOCVD Modules—Application in the manufacturing of InP-based laser devices, *Mat. Chem. Phys.*, **32**, 315–321; 1992.

82. Lamond C, Avrutin EA, Marsh JH. Two-section self-pulsating laser diodes and their application to microwave and millimetre-wave optoelectronics, *Int. J. Opt.*, **10**, 439–444; 1995.

83. Matsumura M. Application of excimer-laser annealing to amorphous, poly-crystal and single-crystal silicon thin-film transistors, *Phys. S. Sol. A: Appl. Res.*, **166**, 715–721; 1998.

84. Niino H, Kawabata Y, Yabe A. Application of excimer laser polymer ablation to alignment of liquid crystals: Periodic microstructure on polyethersulfone, *Jap. J. Appl. Phys., Part 2*, **28**, 2225–2229; 1989.

85. Sameshima T. Laser beam application to thin film transistors, *Appl. Surf. Sci.*, **96**, 352–355; 1996.

86. Seddon BJ, Shao Y, Fost J. The application of excimer laser micromachining for the fabrication of disc microelectrodes, *Electrochim. Acta*, **39**, 783–787; 1994.

87. Takai M, Natgatomo S, Kohda H. Laser chemical processing of magnetic materials for recording-head application, *Appl. Phys.*, **58**, 359–362; 1994.

88. Takasago H, Gofuku E, Takada M. An advanced laser application for controlling of electric resistivity of thick film resistors, *J. Elec. Mat.*, **18**, 651–655; 1989.

89. Yong J. Laser application in packaging of very large-scale integrated chips, *J. Vacuum Sci. Tech.*, **8**, 1789–1793; 1990.

90. Nevin A, Pouli P, Georgiou S, Fotakis C. Laser conservation of art, *Nat. Mater.*, **6**, 320–322; 2007. doi:10.1038/nmat1895

91. Anglos D, Solomidou M, Fotakis C. Laser-induced fluorescence in artwork diagnostics: An application in pigment analysis, *Appl. Spect.*, **50**, 1331–1335; 1996.

92. Castellini P, Paone N, Tomasini EP. The laser doppler vibrometer as an instrument for non-intrusive diagnostic of works of art: Application to Fresco paintings, *Opt. Las. Eng.*, **25**, 227–231; 1996.

93. Cruz A, Hauger SA, Wolbarsht ML. The role of lasers in fine arts conservation and restoration, *Opt. Photo. News*, **10**(7), 36–40; 1999.

94. Osticioli I, Wolf M, Anglos D. An optimization of parameters for application of a laser-induced breakdown spectroscopy microprobe for the analysis of works of art, *Appl. Spectrosc.*, **62**(11), 1242–1249; 2008.

95. Siano S, Agresti J, Cacciari I, et al. Laser cleaning in conservation of stone, metal, and painted artifacts: State of the art and new insights on the use of the Nd:YAG lasers, *Appl. Phy. A*, **106**,(2), 419–446; 2012.

96. Wey VD, Polder PL, Theo W, Gabreels J, Fons M. Peripheral nerve elongation by laser doppler flowmetry-monitored expansion: An experimental basis for future application in the management of peripheral nerve defects, *Plast. Reconstr. Surg.*, **97**, 568–571; 1996.

Further Reading

1. Kneubuhl FK, Sigrist MW. *Laser*, Stuttgart, Germany: B. G., Teubner, 1989.
2. Koechner W. *Solid-State Laser Engineering*, Berlin: Springer-Verlag, 1976.
3. Marschall TC. *Free Electron Lasers*, London: MacMillan, 1985.
4. Pressley RJ. ed., *CRC Handbook of Lasers*, Boca Raton, FL: The Chemical Rubber Co., 1971.
5. Prudhomme S, Seraudie A, Mignosi A. A recent three-dimensional laser doppler application at the T2 Transonic wind tunnel: Optimization, experimental results, measurement accuracy, *Laser Anemometry*, **3**, 57–63; 1991.
6. Saleh BEA, Teich MC. *Fundamentals of Photonics*, New York, NY: John Wiley, 1991.
7. Sargent M III, Scully MO, Lamb WE Jr. *Laser Physics*, New York, NY: Addison-Wesley, 1974.
8. Schafer FP. ed., *Dye Lasers*, Berlin: Springer-Verlag, 1977.
9. Schultz DA. ed., *Laser Handbook*, Amsterdam: North-Holland, 1972.
10. Siegman AE. *Lasers*, Mill Valley, CA: University Science Books,, 1986.
11. Silfvast WI. *Laser Fundamentals*, London: Cambridge University Press, 1996.
12. Smith WV, Sorokin PP. *The Laser*, New York, NY: McGraw-Hill, 1966.
13. Svelto O. *Principles of Lasers*, New York, NY: Plenum, 1976.
14. Tarasov L. *Physique des Processus dans les Generateurs de Rayonnement Optique Coherent*, Moscow: Mir, 1985.
15. Weber MJ. *Handbook of Laser Science and Technology*, Vol. 1, CRC Press, Boca Raton, FL: Boca Raton, 1982.

16. Weber MJ. *Handbook of Laser Science and Technology*, Supplement 1, Boca Raton, FL: CRC Press, 1991.

17. Wey VD, Polder PL, Theo W, Gabreels J, Fons M. Peripheral nerve elongation by laser doppler flowmetry-monitored expansion: An Experimental basis for future application in the management of peripheral nerve defects, *Plast Reconst. Surg.*, **97,** 568–571; 1996.

18. Witteman WJ. *The CO2 Laser*, Berlin: Springer-Verlag, 1987.

19. Yariv A. *Quantum Electronics*, 2nd ed., New York, NY: John Wiley, 1975.

17

Spatial and Spectral Filters

17.1   Introduction..603
17.2   Fundamental Ideas..603
17.3   Spatial and Spectral Filtering...607
17.4   Spatial Filters 609
17.5   Spectral Filters 615
        Absorption Filters • Refraction and/or Diffraction • Scattering
        • Polarization • Acousto-Optic Filters • Interference • Thin Films
        References..632

Angus Macleod

17.1 Introduction

A filter is a device that modifies a distribution of entities by varying its acceptance or rejection rate according to the value of a prescribed attribute. A true filter simply selects but does not change the accepted entities, although those rejected may be changed. Power may be supplied to aid in the filtering process. A simple form of filter is a screen that separates particles according to their size. The filters that are to be described here select according to the spectral characteristics or the spatial characteristics of a beam of light. The selected light is unmodified as far as wavelength or frequency is concerned although the direction may be changed. In most cases, the objective is to render the light more suitable for carrying out a prescribed task, with the penalty of a loss in available energy. The filters with which we are concerned in this chapter are all linear in their operation, meaning their response is independent of the actual magnitude of the input and for any value of the prescribed attribute the ratio of the magnitude of the output to that of the input is constant. We exclude tuned amplifiers, fluorescence filters, spatial modulators, and other active components as outside the scope of this chapter.

17.2 Fundamental Ideas

A propagating electromagnetic disturbance that is essentially in the optical region and that can be described as a single entity is usually loosely referred to as a beam of light. In the general case, a complete description of the attributes of an arbitrary light beam would be impossibly complicated. More often, however, the beams of light that concern us in optical systems have a regularity in character that allows us to make good use of them and, at the same time, allows us to describe them in reasonably uncomplicated terms. There are, of course, many properties that can be involved in the description of a regular beam of light. Here we are interested in only two of these, the spatial distribution and the spectral distribution of the energy carried by the beam. These two characteristics are not strictly completely separable in any given beam but in the simple cases that we use most often they can profitably be considered separately.

We imagine a beam of light that has the simplest spatial distribution possible. The beam can be considered to be propagating in a well-defined direction in an isotropic medium at levels of energy that are well below any that might produce nonlinear effects. We choose a set of coordinate axes such that the z-axis is along the direction of propagation of the beam of light, and then the x- and y-axes are normal to it. We can make measurements of the attributes of the beam in any plane parallel to the x- and y-axes that are normal to the direction of propagation. If we find that the attributes of the beam at any point are completely independent of the values of the x and y coordinates of that point and depend only on the values of the z-coordinate and the time, then we describe the beam of light as a *plane wave* propagating in the z-direction. See Heavens and Ditchburn [1] for a detailed description of fundamental optical ideas.

At any value of z, we can measure the temporal variation of the energy, or of the fields of the plane wave. This yields the temporal profile of the wave. When we compare the temporal profile of the wave at one value of z with that measured at another, we should expect to see the same general shape but with a separation in time equal to the separation in z divided by the velocity of the wave. However, we see this simple relationship only when the wave is propagating through free space. For all other media there is a distortion of the temporal profile that increases with the distance between the two measurement points. Furthermore, because of this effect it becomes impossible to assign a definite velocity to the wave. All media, except free space, exhibit an attribute known as *dispersion* and it is dispersion that is responsible for the profile shape change and the uncertainty in the velocity measurement.

Fortunately, we find that there is one particular temporal profile that propagates without change of shape even in dispersive media. This is a profile that can be described as a sine or cosine function. A wave that possesses such a profile is known as a *harmonic* or *monochromatic* wave. In the simplest form of the plane, monochromatic wave, there is a consistency in the directions of the electric and magnetic fields of the wave, which is known by the term *polarization*. Polarization is described elsewhere in this book. Here we adopt the simplest form known as *linear* or, sometimes, *plane*, where the directions of the electric and magnetic vectors are constant. Our simple wave is now a linearly polarized, plane, monochromatic wave. The electric field, magnetic field and direction of propagation are all mutually perpendicular and form a right-handed set.

Such a monochromatic wave travels at a well-defined velocity in a medium. The ratio of the velocity of the wave in free space to that in the medium is known as a refractive index, written n. The ratio of the magnetic field magnitude to the electric field magnitude of the wave is another constant of the medium known as the *characteristic admittance*, usually written as y. A great simplification is possible at optical frequencies. There, the interaction of a wave with a propagation medium is entirely through the electric field, which can exert a force on even a stationary electron. The magnetic field can interact only with moving electrons and, at optical frequencies, any direct magnetic interaction with the electrons is negligible. Then

$$y = nY \tag{17.1}$$

Y being the admittance of free space, 1/377 siemens. This relationship is usually expressed as

$$y = n \text{ free space units} \tag{17.2}$$

allowing the same number to be used for both quantities.

In the linear regime, the combination of two separate waves is quite simple. Whatever the characteristics and directions of the individual waves, the resultant electric and magnetic fields are simply the vector sum of the individual fields. This is the *Principle of Linear Superposition*. Irradiances, in general, do not combine in such a simple way.

Since the interaction with the medium is through the electric field, and since the magnetic field of a harmonic wave can be found readily from the direction of propagation and the characteristic admittance, *y*, it is normal to write the analytical expression for a harmonic wave in terms of the electric field only. Moreover, since the combinations of waves are linear, we can profitably use a complex form for the harmonic wave expression,

$$E = |E|\exp\{i(\omega t - \kappa z + \varphi)\} = \left[|E|\exp(i\varphi)\right]\cdot\left[\exp\{i(\omega t - \kappa z)\}\right] = E\exp\{i(\omega t - \kappa z)\}, \tag{17.3}$$

where the relative phase is usually incorporated into what is known as the complex amplitude, E, and the remaining exponential is known as the phase factor. ω and κ are the temporal and spatial angular frequencies, κ usually being known as the wavenumber. We can write

$$\omega = \frac{2\pi}{\tau} \text{ and } \kappa = \frac{2\pi n}{\lambda}, \tag{17.4}$$

where λ is the wavelength that the light has in free space.

The expression for the wave is usually written

$$E = E\exp\left\{i\left(\omega t - \frac{2\pi n z}{\lambda}\right)\right\}. \tag{17.5}$$

If the medium is absorbing, then there will be a fall in amplitude of the wave as it propagates. This can be accommodated in the expression by introducing a complex form for *n*.

$$n \rightarrow (n - ik) \text{ and } y \rightarrow (n - ik)Y \tag{17.6}$$

$$E = E\exp\left\{i\left(\omega t - \frac{2\pi(n - ik)z}{\lambda}\right)\right\} = E\exp\left(-\frac{2\pi k z}{\lambda}\right)\exp\left\{i\left(\omega t - \frac{2\pi n z}{\lambda}\right)\right\}. \tag{17.7}$$

The irradiance of the harmonic wave in its complex form is given by

$$\text{Irradiance} = I = \frac{1}{2}\text{Re}\{EH^*\} = \frac{1}{2}\text{Re}\{EH^*\} = \frac{1}{2}nY|E|^2. \tag{17.8}$$

Now let us return to the plane wave of arbitrary profile and let us insist that it be linearly polarized. In the first instance, we consider its propagation through free space. The electric field, the magnetic field and the direction of propagation will form a right-handed set. Let us freeze the electric field and magnetic field at a particular instant. By the Fourier decomposition process, we can construct a continuous set of harmonic profiles that, when added together at each point, will yield the instantaneous profile. In free space, there is no dispersion and we can show that the magnetic and electric field profiles are always in the ratio of the admittance of free space, Y. The magnetic field decomposition therefore mimics the electric and when we pair frequencies, we construct a complete set of monochromatic component waves that make up the primary wave. The set of attributes of the component waves as a function of wavelength or frequency is known as the spectrum of the primary wave and the process of finding the spectrum is sometimes called spectral decomposition. The spectrum is most often expressed in terms of the varying irradiance of the component waves per frequency or wavelength unit. Real optical sources exhibit fluctuations in the nature of their output. These translate into fluctuations, either of amplitude or phase or both, in the spectral components. Such fluctuations are best handled in a statistical context. The theory of coherence that deals with such ideas is discussed elsewhere in this book. Here we use only a very elementary treatment of waves.

Now let us remove our restriction of linear polarization. We can represent our wave now as made up of two major components linearly polarized in orthogonal directions. If there is no consistent relationship between these components, in other words they are unpolarized, then, although the irradiance spectrum of each component may be similar, the individual components will have no consistent phase relationship that will allow their combination into a consistent polarization state and neither the primary wave nor the component waves will exhibit polarization. If, however, there is a consistent polarization in the primary beam, then this will also be reflected in the components.

Now let the wave enter a dispersive material. Since the primary wave and the spectral component waves are entirely equivalent, we can look most closely at the latter and follow their progress through the medium individually. To find the net disturbance we combine them. Each experiences a refractive index and a characteristic admittance, which are functions of the properties of the medium. In a dispersive medium, these attributes show a dependence on frequency or wavelength. Thus, the components are not treated equally. As they propagate through the medium, therefore, their resultant changes, and this is the reason for the changes in the profile in dispersive media mentioned earlier.

A plane wave strictly has no center or axis and really does not correspond at all to our ideas of a light ray. It has a direction but there is nothing that corresponds to a lateral position or lateral displacement. Beams of light emitted from lasers have all the attributes of spectral purity and potential for interference that we would expect from a plane wave, yet they have a position as well as direction and act in many ways as if they were light rays rather than plane waves. These are better described as Gaussian beams. A Gaussian beam as in Figure 17.1 has a direction and phase factor not unlike the plane wave, but it is limited laterally and the irradiance expressed in terms of lateral distance from the center of the beam is a Gaussian expression. If we define the beam radius, ρ, as the diameter of the ring where the irradiance has fallen to $1/e^2$ of the irradiance in the center of the beam I_0 then we can write for the irradiance

$$I = I_0 e^{-2r^2/\rho^2}, \tag{17.9}$$

where r is the radius to the point in question. Strictly speaking, the Gaussian beams described by (Equation 17.9) are in the fundamental mode, the TEM_{00} mode. There can be higher-order modes that have more complicated variations of irradiance as a function of displacement from the axis.

Equation 17.9 is insufficient for a complete description of the beam. We need also the wavelength λ and the beam divergence (or convergence) ϑ, which may be defined as the angle between the axis and the $1/e^2$ irradiance ring. (The divergence is also sometimes defined as the total angle.) The Gaussian beam appears to emanate from or converge to a point. However, in the neighborhood of the point, the beam, instead of contracting completely, reaches a minimum size and then expands to take up the cone again. The minimum is called the beam waist and it has a radius that is a function both of the wavelength and the divergence of the beam.

$$\rho_0 = \frac{\lambda}{\pi\vartheta} \tag{17.10}$$

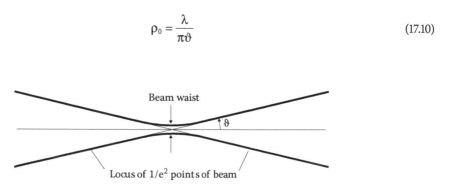

FIGURE 17.1 Schematic of the parameters of a Gaussian beam.

The diameter of the beam at distance d from the beam waist is given by

$$\rho^2 = \rho_0^2 + \vartheta^2 d^2. \tag{17.11}$$

This is a normal thin lens theory where the positions of the beam waists are taken as the image points apply to Gaussian beams. An excellent account of Gaussian beams and their manipulation is given by O'Shea [2].

17.3 Spatial and Spectral Filtering

Filtering is performed to render light more suitable for a given purpose. Spatial filters accept light on the basis of positional information in the filter plane, while spectral filters operate according to the wavelength or frequency. Although they are separate attributes of a light beam, they may be required to operate together to achieve the desired result. Frequently, for example, spatial filters are a necessary part of spectral filters. Rather less often, spectral filters may be involved in spatial filtering. We shall consider the two operations separately, but it will become clear in the section on spectral filtering that they are often connected.

A simple example of a common spatial filter is a field stop. This is usually an aperture that is placed in an image plane to limit the extent of the image that is selected. Normally the boundaries between that part of the image that is selected and that rejected should be sharp but this may not always be the case. Especially in the early part of the century, vignettes were a popular form of photographic portrait. There, a defocused field stop, usually elliptical in shape, graded the boundary of a head and shoulders image so that in the final print it was detached from the background and appeared to be surrounded by a gradually thickening white mist.

An improved aesthetic effect may demand a change in the color balance of illumination. An interferometric application may require a greater degree of spectral purity. A luminous light beam may also contain appreciable infrared energy, which will damage the object under illumination. An interaction with a material may demand a particular frequency. Operation of a spectrum analyzer may demand elimination of all light of frequency higher than a given limit. The signal-to-noise ratio of a measuring apparatus may need elimination of white background light and acceptance of a narrow emission line. A bright line creating problems in the measurement of a dim background may require removal. The list of possibilities is enormous. All these involve spectral filters that use wavelength or frequency as the operating criterion.

Spatial filters will usually either accept light, which is then used by the system, or reject it completely so that it takes no part in what follows. Spectral filters may have a similar role but they may also be required to separate the light into different spectral bands each of which is to be used separately. A bandpass filter is typical of the first role and a dichroic beamsplitter of the second.

We consider linear filters only, i.e. filters with properties independent of irradiance but dependent on wavelength, frequency or position. We exclude tuned receivers, wavelength shifting filters, tuned amplifiers and the like. The output of the filters is derived by the removal of light from the input beam and can be described by a response function that is the ratio of output to input as a function of wavelength or frequency or position and is a number between zero and one or between 0 and 100%. Usually the parameter characterizing input and output will be the irradiance per element of wavelength or frequency or area. The response will normally be a transmittance T or reflectance R. Typical ideal response functions for spectral filters are sketched in Figure 17.2. Spatial filters would be similar with wavelength replaced by position.

A beamsplitter will have two such responses. Usually one will be in reflection and the other in transmission.

There are many classifications of spectral filters but those that transmit narrow bands of wavelength are usually called narrow band pass filters, those that reflect narrow bands are called notch filters, those

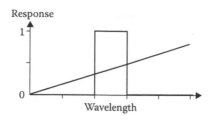

FIGURE 17.2 Typical ideal responses. One shows a response that varies gradually with wavelength, the other a response that accepts light with wavelength within a given band and rejects all others.

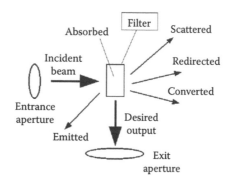

FIGURE 17.3 The character of the desired output light is unchanged, except perhaps in direction, from its character as part of the input beam. The rejected light may simply be redirected or may suffer a conversion, some possible mechanisms being illustrated here.

that transmit broad regions at wavelengths longer or shorter than a rapid transition between acceptance and rejection are called edge filters, and either long wave pass or short wave pass.

The light may be removed by redirection, by conversion or by both.

A process of absorption is really one of conversion. The light energy is converted into kinetic and potential energy of molecules, atoms and electrons. Some of this energy may then be reemitted in a changed form or dissipated as heat.

An important question that should always be asked about a filter is "Where does the unwanted energy go?".

Most filters exhibit the various modifications of the input shown in Figure 17.3. For example, an absorption filter to eliminate short wavelengths will reflect residual light (redirection), will convert some light (fluorescence), will scatter some light, and will emit thermal radiation - all unwanted. If this light is accepted back into the system, performance will suffer.

Figure 17.4 shows a sketch of a very simple instrument showing the stops. The image of the aperture stop in source space is the entrance pupil and in receiver space the exit pupil. Note that the performance of a filter inserted in the instrument cannot be separated from the details of stops and pupils. These are part of the overall system design.

In the design of an instrument that includes spatial or spectral filters, the behavior of the unwanted light, and especially any redirected light, is very important. For example, can rejected light be scattered back into the acceptance zone of the system?

Baffles, a special form of spatial filter, can help to prevent the return of unwanted energy and their correct design is as important as the design of the primary components such as lenses.

Since performance is system dependent, an optical filter is usually specified with regard to standard (and that usually means ideal) conditions. For spectral filters, for example, entrance and exit pupils are usually considered to be at infinity (light ideally collimated) and scattered, redirected, converted and

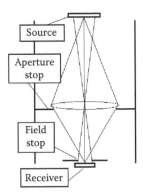

FIGURE 17.4   A very simple optical instrument showing an aperture and a field stop.

emitted light is assumed lost to the system. A real system will rarely have this ideal arrangement and therefore the performance may not correspond to the standard specified performance of the filter.

17.4 Spatial Filters

Any optical instrument that has a limited field of view is effectively a spatial filter. In the vast majority of instruments, a deliberate limitation of the field of view is desirable. If it is not well defined, then uncontrolled and unwanted input may be received that will interfere with the correct operation.

In a very simple case, a stop is placed in a field plane somewhere in the instrument. Microscopes are frequently fitted with a variable field stop in the form of an adjustable diaphragm that can be used to select a particular part of an image and eliminate the remainder, which is useful when precise measurements on a selected object have to be made.

Spatial filtering is often a vital instrumental feature. An instrument that is to be used to examine images near the sun, for example, may even be destroyed if light from the sun can reach the image plane. Such demanding spatial filtering tasks usually involve not just the elimination of the direct light by a suitably shaped aperture, but the elimination of scattered light as well, using complicated assemblies of spatial filters known as baffles. A solar telescope acquires an actual image of the sun, but when details of the limb must be investigated, the solar image must be removed; this is achieved by a special spatial filter called an occluding disk that fits the image of the disk of the sun exactly, but allows the surrounding regions to be examined. Elimination of stray light in this configuration is of critical importance.

Apart from the instrumental uses, probably the most common form of spatial filtering is in the collimation of light, which is the attempt to construct a beam of light that is as near as possible to a plane wave. Plane waves are necessary inputs to all kinds of optical instruments. To construct a plane wave we first of all create a point source. The point source emits a spherical wave and the spherical wave can be converted into a plane wave by passing it through a suitable lens. Unfortunately, a point source is an unattainable ideal. The best we can do is to make the source very small. We do this by first producing an image of a real source and, in the image plane, we introduce a diaphragm with a very small aperture. This aperture is then used as our point source. A lens is then placed such that the point source is at the focus. This is illustrated in Figure 17.5.

The radiance L of this source is the important quantity. This is the power in unit solid angle leaving the unit projected area of a source. It is measured in watt/steradian meter^2 (W sr^{-1} m^{-2}). The total power accepted by the collimating lens will be given by

$$\text{Power} = \frac{L \cdot A_S \cdot A_L}{R^2} \tag{17.12}$$

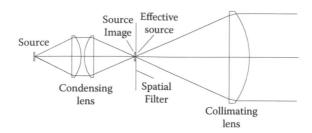

FIGURE 17.5   A simple collimator showing the light source, condenser, collimating lens, and spatial filter to produce the effective point source.

where A_S is the area of the effective source, A_L is the area of the collimating lens and R is the distance from effective source to collimating lens. If Ω_L and Ω_S are the solid angles subtended by the lens at the effective source and by the effective source at the lens, respectively, then we can write

$$\text{Power} = L \cdot A_S \cdot \Omega_L = L \cdot A_L \cdot \Omega_S. \tag{17.13}$$

$A_L\Omega_S$ and $A_S\Omega_L$ are equal and are known as the $A\Omega$ product. For a well-designed system, this should be a constant. Unfortunately, it is very easy to make a mistake in pairing the correct A with the correct Ω and so 17.12, which is completely equivalent, is a safer expression to start with.

The degree to which the light departs from perfect collimation may be defined as the semiangle of the vertex of the cone, which is the solid angle subtended by the effective source at the center of the collimating lens. It is easy to see that a factor of two improvement in this angle is accompanied by a factor of four reduction in power. Thermal sources, in particular, give disappointing performance when the collimated light must also be of narrow spectral bandwidth. A common use of collimators is in dispersive filters, such as grating or prism monochromators. There the dispersion can be arranged to be in one well-defined direction and the high degree of collimation is demanded only in this direction. This permits the use of a long slit as aperture in the spatial filter of the collimator. The degree of collimation can then be varied by changing the width of the slit but not its length. This makes the power proportional to the collimation angle rather than its square, and permits much more satisfactory use of thermal sources.

A Gaussian beam from a laser is already collimated with a degree of collimation defined by the divergence ϑ. This divergence may not be acceptable. Since ϑ and ρ_0 are interrelated, a reduction in ϑ implies an increase in ρ_0, and vice versa. A large degree of spatial magnification implies an object at or near the focal length of a positive lens and the image at near infinity. Since the Gaussian beam at the input to the collimator will almost certainly be some distance from its beam waist, it is necessary to insert a leading lens to form a waist at a suitable distance from the following collimating lens. Any beam that propagates through an optical system will tend to accumulate spurious stray light. It is usual, then, to insert a spatial filter with the purpose of cleaning up the beam. Any light representing a departure from regularity will tend to appear off-axis at the beam waist. A small diameter filter, usually called a pinhole, selects the ideal image and rejects the rest. Usually the pinhole is quite large compared with the Gaussian beam waist since the usual objective is the elimination of spurious light rather than the creation of the diffraction-limited spot itself. The assembly of input and collimator lenses is also sometimes called a beam expander, because this is effectively what it does. See O'Shea [2] for more information on designing systems using Gaussian beams.

A rather different type of spatial filter is sometimes used in wide-angle aircraft cameras. Figure 17.6 shows a typical arrangement. If we assume that the ground is flat and that its emission characteristic is independent of direction, which is a Lambertian surface, then the energy in an element of area on the photographic plate in the image plane will be proportional to the $A\Omega$ product for the areal element on

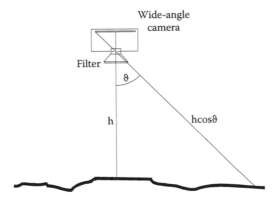

FIGURE 17.6 A wide-angle camera. The irradiance at the image plane is proportional to $1/\cos^4 \vartheta$, assuming constant radiance and a flat horizontal Lambertian surface for the terrain.

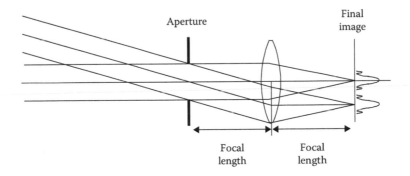

FIGURE 17.7 A simple telecentric imaging system with the object at infinity. The distribution of energy in the images is indicated to the right of the focal plane.

the ground and the pupil of the camera. If we assume the ground is completely flat, then this will be easily shown by

$$A\Omega = \frac{\left(A_{\text{pupil}}/\cos\vartheta\right)\left(A_{\text{element}}/\cos\vartheta\right)}{\left(h\cos\vartheta\right)^2} \propto \frac{1}{\cos^4\vartheta}. \tag{17.14}$$

However, since the surface is Lambertian and the focus of the camera is fixed, the energy on the unit element area on the photographic plate will follow the rule in Equation 17.14, even if the terrain height varies across the image. The fall off in energy with $\cos^4\vartheta$ applies to all cameras with a flat focal plane but it is only in wide-angle cameras that the problem can be severe. These may have acceptance angles higher than 45°. At 45°, the ratio of energies is 1:4 and such a large ratio cannot be accommodated in the optical design of the camera lens. A simple solution, again involving rejection of otherwise useful light, is a spatial filter fixed in front of the lens, usually on the hood, where there is an appreciable spatial separation of marginal from axial light. An evaporated film of Inconel or Nichrome with a radial distribution, so that it has a higher density in the center than at the periphery, is usually sufficient. This is often combined with the light yellow anti-haze filter, which is normal in high altitude photography. The correction is rarely exact.

A circular pupil with sharp boundaries is illuminated in collimated light that is imaged to a single spot as shown in Figure 17.7. If the rays obeyed the laws of geometry then the spot would be a single point, but diffraction causes a spreading of the point. The resulting distribution of energy in the focal plane is known as the point spread function. Note that the point spread function may be the result of aberrations in the system rather than diffraction. When an image is produced in the focal plane of any system, the result is a convolution of the image and the point spread function. The point spread function for a sharply defined circular pupil illuminated by a perfectly uniform collimated beam with no other aberrations is the Airy distribution. This has a pronounced outer ring that can be an undesirable feature in diffraction-limited systems. A special type of spatial filter, known as an apodizing filter, inserted in the plane of the pupil can modify this distribution. It operates by varying the distribution of irradiance across the aperture. As with all the other filters described, it is rejecting energy. A simple form of apodizing filter yields a linear fall in irradiance from a maximum in the center to zero at the outer edge. This gives a point spread function that is the square of the Airy distribution and the first diffraction ring is now much reduced. An apodizing filter that has a Gaussian distribution will produce a spot with a Gaussian distribution of energy. Although in all cases the central part of the spot is broader (we are rejecting some of the otherwise useful light), the resulting image can be much clearer.

The ideas of the previous paragraphs are carried much further in a class of spatial filters, which uses a property of a lens and its front and back focal planes. It can be shown that the light distribution in the rear focal plane is the two-dimensional spatial Fourier transform of the distribution in the front focal plane. If the front focal plane contains an image, then the rear focal plane will contain a distribution, which is the two-dimensional spectrum of the image in spatial frequency terms. Higher frequency terms in the image are translated into light in the back focal plane that appears at points further from the axis of the system. If, now, a second similar lens is added so that its front focal plane coincides with the current back plane, then the image will be recovered in the back focal plane of the second lens. Figure 17.8 shows the arrangement.

Since this system has at its final output the original image but has also at an intermediate plane the spatial spectrum of the image, it becomes possible to make modifications to the spectrum so that desirable changes may be made in the final image (Figure 17.9).

The spatial filters may be used in many different ways but a common application is in removing high spatial frequency structure from images. For instance the raster lines in a television frame could readily be removed using this technique. The apodizing filters mentioned previously are a special case of such spatial filters.

More advanced filters of this kind also operate to change the phase of selected frequencies but are completely beyond the scope of this chapter. Much more information is given by VanderLugt [3].

Phase contrast microscopy uses an interesting kind of spatial filter. Objects that are completely transparent and are immersed in a fluid of only slightly different refractive index are difficult to see. If the light that is transmitted by the phase objects is arranged to interfere with that transmitted by the medium, the result is a very slight difference in contrast because the phase difference is sufficient for just

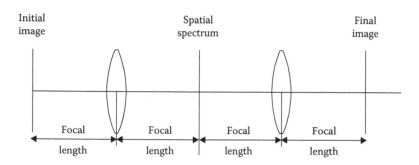

FIGURE 17.8 The two lens system showing the Fourier plane where the spatial spectrum is located and the final image plane.

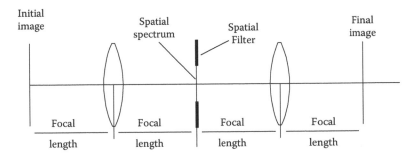

FIGURE 17.9 The two lens system of Figure 17.8 showing a spatial spectrum filter inserted in the Fourier plane. The filter is simply known as a spatial filter.

a slight change at the peak of a cosine fringe. Small changes in relative phase are much more visible if they are occurring on the side of a fringe peak where there is much larger variation with small phase differences. In the phase contrast microscope, the sample is illuminated by light derived from a ring source created by a spatial filter admitting a narrow circle only. Beyond the illuminated specimen, the optical system is arranged to produce an image of the circular light source. However, the light that passes through the objects that are the subject of examination is scattered and although it passes through the optical system, it does not form any image of the circular ring light source. Light that passes through the supporting medium of the specimen is scattered to a much smaller extent and forms a reasonable ring image. It is arranged that the ring is formed exactly over a narrow spatial filter, which alters the phase of the light by one quarter of a wavelength with respect to the scattered, unaffected light. All the light then goes on to form the final image where there is interference between the light that propagated through the medium and through the phase objects but now, because of the quarterwave phase difference, the variation is on the side of the cosine fringes and the variation in radiance large. This brilliant idea conceived by Fritz Zernike in the 1930s has inspired other optical instruments where a quarterwave phase difference reveals small phase excursions as much larger differences in the radiance of an image. See, for example, Hecht and Zajac [4].

A class of lasers uses what are known as unstable resonators. In such lasers, the gain is extremely high and so there is no need for the confining effect of the regular modes of stable resonators. It is possible with such constructions to make exceedingly efficient use of the gain medium. The penalty is that the output of the laser is rather far from ideal and certainly not at all like the regular Gaussian beam profile. In particular, much energy propagates rapidly away from the resonator axis making it difficult to use the beams efficiently. It has been found that a rather special type of spatial filter renders the output beams much nearer to the ideal Gaussian. The special type of spatial filter involved is the graded coating.

Even the unstable resonators must be in the form of a cavity. The light must be reflected back and forth between opposite mirrors, which need reflecting coatings for their correct operation. It has been shown that if the profile of the reflectance of the output mirror is Gaussian then the output beam will also have a Gaussian, or near Gaussian, profile. The Gaussian profile is achieved by a process of grading the thicknesses of the reflecting coating. Laser systems using such graded mirrors are already commercially available.

How do we make a reflector with a radial Gaussian profile? The reflectance of a quarterwave stack (see Section 17.5.7.6) as a function of the thicknesses of all the layers is shown in Figure 17.10. This means that a suitable radial grading of the thicknesses of the layers can result in a radial variation of reflectance at the reference wavelength, which is Gaussian. Similarly, Figure 17.11 shows another arrangement where just one of the layers is varied in thickness radially, the other layers remaining of constant thickness.

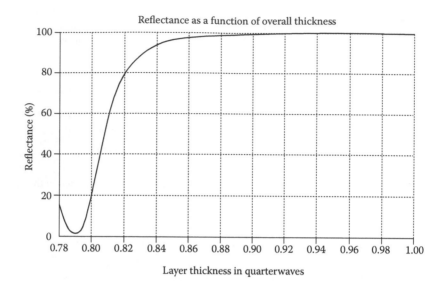

FIGURE 17.10   Variation of the reflectance at the reference wavelength of a quarterwave stack as a function of the thickness of the layers. The stack is made up of 13 alternating quarterwave layers of titania and silica.

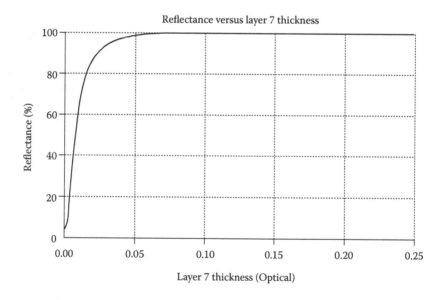

FIGURE 17.11   Variation of the reflectance at the reference wavelength of a quarterwave stack in which one of the layers is varied in thickness. The quarterwave stack is identical to that of Figure 17.10 but in this case only the thickness of the third layer from the incident medium is varied. The thickness scale is in units of the reference wavelength.

A Gaussian profile is ideal but the real mirrors are more or less successful in achieving it and provided the agreement is reasonable, exact correspondence is not necessary nor is it really practical. Usually masks are used during the appropriate part of the deposition process that assures the correct thickness distribution either of the chosen layer or even the entire multilayer.

Note that the variation represented by Figure 17.11 where only one layer is varied is easier than that of Figure 17.10 where the percentage variation to change from almost zero to almost 100% reflectance is quite small. On the other hand, the process that uses Figure 17.11 must mask only one of the layers, a difficult mechanical problem, whereas the other process masks all of them. Both types of graded coatings are actually used in practice. For a much more detailed description, see Piegari [5].

17.5 Spectral Filters

The operation of different filter types is often a mixture of several mechanisms, some of which include absorption, refraction, reflection, scattering, polarization, interference, and diffraction. We exclude neutral density, achromatic beamsplitters and similar coatings and filters. They cannot be described as spectral filters since their attempted purpose is to treat all spectral elements equally.

17.5.1 Absorption Filters

Semiconductors are intrinsically longwave pass absorption filters. Photons of energy greater than the gap between valence and conduction bands of the electrons are absorbed by transferring energy to electrons in the valence band to move them into the conduction band. High transmittance implies intrinsic semiconductors of high resistivity. Gallium arsenide, silicon, germanium, indium arsenide, and indium antimonide are all useful (Figure 17.12). Note that these semiconductors have high refractive indices so they must be antireflected in the pass region. They are rarely antireflected in the absorbing region and so a large amount of the rejected light is actually reflected (Figure 17.13).

Some colored filter glasses of the longwave pass type contain colloidal semiconductors.

Other colored glasses have metallic ions dispersed in them. In general, colored glass filters make excellent longwave pass filters but it is not possible to find shortwave pass filters with the same excellent edge steepness. In fact, shortwave pass filters present an almost universal problem. For more information, see Dobrowolski [6,7].

Strong absorbers are also strong reflectors. In the far infrared, the reststrahlen bands associated with very strong lattice resonances show strong reflectance and are sometimes used as filters. Beryllium oxide reflects strongly in the 8–12 μm atmospheric window. Because its thermal emittance is therefore very low in the atmospheric window, it is sometimes used as a glaze on electrical insulators to keep them warmer and therefore frost free during cold nights.

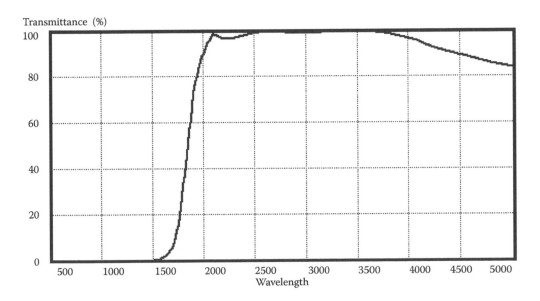

FIGURE 17.12 Transmittance of a germanium filter with antireflection coatings.

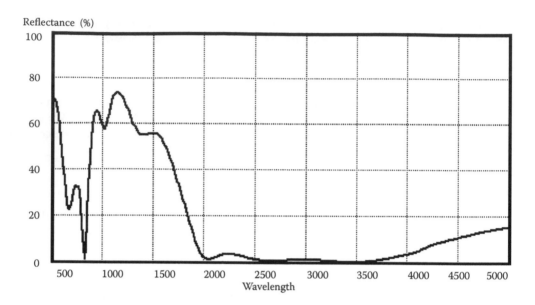

FIGURE 17.13   Reflectance of the same filter as in Figure 17.12.

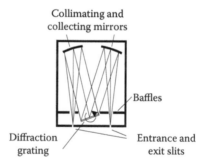

FIGURE 17.14   Sketch of a grating monochromator. Additional optics focus a light source on the entrance slit and collect the light from the exit slit. Note the use of baffles to reduce the scattered light. In high performance instruments, these are very important and much more complicated.

17.5.2 Refraction and/or Diffraction

A prism monochromator is a variable spectral filter of a very inexpensive nature. The prisms operate due to refraction that varies because of the dispersion of the index of the prism material. The optical arrangement usually assures that the spatial distribution of the component waves will vary according to wavelength or frequency. A spatial filter of a simple kind is then used to select the appropriate wavelength. Usually, although not always, the light source is in the form of a narrow slit that is then intentionally imaged to enhance chromatic effects in the image. A spatial filter, also in the shape of a slit is then used to select the light in the image having the correct wavelength.

Diffraction gratings are used in a similar way to refractive prisms. They are essentially multiple beam interference devices that use diffraction to broaden the beams to give reasonable efficiency over slightly more than an octave. Disadvantages are the low throughput because of the narrow entrance and exit slits, although they are superior in this respect to prisms, and the need for mechanical stability of a high degree. A further problem with diffraction gratings is that they admit multiple interference orders so that harmonics of the desired wavelength may also be selected. Order-sorting filters may be necessary to remove the unwanted orders. Figure 17.14 shows a sketch of a grating monochromator where the

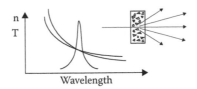

FIGURE 17.15 Schematic diagram of the principle of the Christiansen filter. Where the dispersion curves of the matrix and scattering particles cross, there is no scattering loss and the transmittance of the device is specular. Elsewhere the light is scattered. A spatial filter is then used to select the specular light and reject the scattered.

wavelength can be changed by rotating the diffraction grating, all other elements remaining unchanged. High performance monochromators invariably use mirrors rather than refracting elements for collimating and collecting since they have no chromatic aberrations.

17.5.3 Scattering

Christiansen filters consist of dispersed fragments or powder in a matrix. The dispersion curves of the two materials differ but cross at one wavelength at which the scattering disappears. The undesired scattered light is then removed from the desired specular light by a simple spatial filter (see Figure 17.15).

Christiansen filters have been constructed for virtually the entire optical spectrum. An excellent and compact account is given by Dobrowolski [6].

17.5.4 Polarization

Retardation produced by thickness d of a birefringent material is given by

$$\varphi = \frac{2\pi(n_1 - n_2)d}{\lambda} \tag{17.15}$$

so that a half wave retarder made of birefringent material is correct for only one wavelength. Let there be a polarizer, a retarder at 45° and an orthogonal polarizer in series. All the light transmitted by the first polarizer will be transmitted without loss through the remainder of the system provided the retardation is equivalent to an odd number of half wavelengths half wave plate. For any other value of retardation, the light will be stopped by the second polarizer, either totally, if the retardation is an integral number of wavelengths, or partially, if not. The irradiance is given by an expression of the style:

$$\text{Irradiance} \propto \sin^2 \frac{\varphi}{2}. \tag{17.16}$$

This effect can be used in filters either in a single element or in a series of such elements. The Lyot filter consists of a series with each member having an increased halfwave plate order and so a narrower, more rapidly changing response (see Figure 17.16). The responses of the various elements combine to give a very narrow bandwidth.

The Solc filter has only two polarizers and a set of identical retarder plates in between with the axes arranged in a fan. For a detailed theoretical analysis of such filters, see Yariv and Yeh [8].

17.5.5 Acousto-Optic Filters

In acousto-optic filters, the periodic strain caused by an acoustic wave alters the refractive index in step with the wave. This impresses a thick Bragg phase grating on the material. The light interacts with this

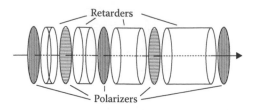

FIGURE 17.16 Schematic of a Lyot filter. Real filters have rather greater numbers of elements and must be tightly controlled in temperature.

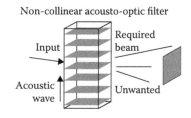

FIGURE 17.17 Schematic of a collinear acousto-optic filter.

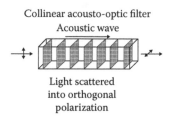

FIGURE 17.18 Schematic of a non-collinear acousto-optic filter.

grating, which effectively becomes a narrow band filter. In the collinear case, the light is polarized and a narrow band is scattered into the orthogonal plane of polarization where it is selected by a suitably oriented polarizer as in Figure 17.17. Variation of the frequency of the RF drive for the acoustic transducer varies the wavelength of the filter.

In the non-collinear type of filter, the Bragg grating deflects the light with high efficiency. The light need not be polarized. Unwanted light is obscured from the receiver by a spatial filter (not usually as crude as in the diagram, Figure 17.18).

Bandwidths of a few nm with an aperture of a few degrees with an area of almost 1 cm2 and with tuning over the visible region are possible. The collinear type has yielded bandwidths of 0.15 nm with tuning range over the visible and near ultraviolet. For more information, see Chang [9].

17.5.6 Interference

The bulk of filters that depend on interference for their operation are of the thin film class. Because they form a rather special and very important category, they are considered separately. Here we examine filters that operate by processes of interference, usually multiple beams, but are not made up of assemblies of thin films, nor can be included in the other classes we have considered.

Any interferometer is potentially a narrow band filter. Since the fringe positions normally vary with wavelength, a spatial filter can be used to select the light of a particular wavelength. The most common interference filter is probably based on a Fabry-Perot étalon. The fringes are localized at infinity. Thus, if a spatial filter is placed in the focal plane of an objective lens, then it can be used to select the appropriate wavelength. The étalon is usually arranged so that the central fringe is of the correct wavelength and then the spatial filter can be a simple circular aperture of the correct diameter. The filter can be tuned by varying the optical thickness of the spacer layer in the étalon and this can be achieved either by varying the refractive index, as for example by altering the gas pressure, or by physically moving the interferometer plates themselves. Over rather small tuning ranges, tilting the étalon is another arrangement that can be, but is less frequently, used.

Most filters that employ interference are of the thin film form. They are discussed separately. However, there is a variant of the thin film filter that is usually known as a holographic filter. The holographic filter is produced in a completely different way but in principle it can be thought of as a thin film reflecting filter that has been sliced across its width at an angle and then placed on a substrate. This is illustrated in Figure 17.19. We take a thin film reflecting filter that consists of alternate high and low index quarterwaves or, in the case of the hologram, a sinusoidal variation of index in the manner of a rugate filter (described in Section 17.5.7.6). The thin film reflecting filter is sliced by two parallel cuts at angle ϑ to the normal and the slice is laid over its substrate. The hologram now reflects strongly at the angle ϑ to the hologram surface. As in the rugate, the sinusoidal variation assures that there are no higher orders that are reflected at the same angle. A simple spatial filter that selects light propagating at the correct angle completes the filter. The variation in refractive index in the hologram is small and this means that the reflecting band is very narrow in terms of wavelength.

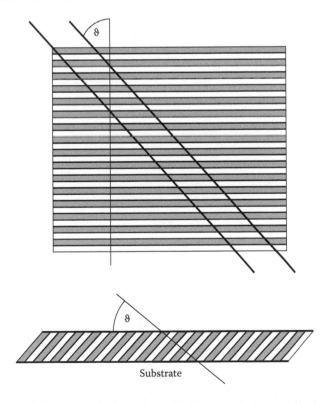

FIGURE 17.19 Diagram of the way in which a holographic filter may be imagined to be produced from a thin film filter. The thin film reflecting filter, usually a rugate filter, is sliced by two parallel cuts at angle ϑ to the normal. The slice is then placed on a substrate. Both filters reflect strongly, the thin film filter at normal incidence but the holographic filter at the angle ϑ to the hologram surface.

Another variant of the rugate filter is in the form of an optical fiber with a propagation constant that varies in a sinusoidal fashion along the core. This reflects a very narrow band of wavelengths and transmits all others. Such fiber filters are, therefore, very narrow notch filters.

17.5.7 Thin Films

A thin film is a sheet of material defined by surfaces that are sufficiently parallel for interference to exist between light reflected at the surfaces. Usually, this means that the films will be at most a few wavelengths in thickness. They are usually deposited from the vapor or liquid phase directly on a substrate. The optical thicknesses of the individual films and the wavelength determine the phase differences between the beams that interfere. Assemblies of many films exhibiting complicated but engineered interference effects are common. Frequently, the interference effects are supplemented by the intrinsic properties of the materials, especially their absorption behavior. Thin film optical coatings, therefore, operate by a mixture of interference and material properties.

We can classify the materials that are used in optical coatings into three principal classes: dielectrics, metals, and semiconductors. All have optical constants that can be expressed as $(n - ik)$. The extinction coefficient k is very small in dielectrics and as a first approximation we can neglect it so that it is n that principally characterizes a dielectric. In high performance metals like silver, it is n that is small and k that is large. Although it is a poorer approximation than neglecting k in a dielectric, as a first approximation we can assume that the metal is characterized by k. The main characteristics of these materials are summarized in Table 17.1. Note that these are somewhat idealized. For example, there is slight variation of n with λ in the case of the dielectrics but it is negligible compared with the enormous changes in k with λ in metals. Similarly there is a residual small finite n even in high performance metals.

Note that for interference coatings we require the presence of dielectric (or transparent semiconductor) layers. Metals by themselves are not enough, although they may act as broad band absorbers, reflectors, or beamsplitters.

TABLE 17.1   Principal Characteristics of Dielectrics, Metals, and Semiconductors

Dielectrics	Metals	Semiconductors
n real	$y = -ikY$	Usually classified as either metal or dielectric depending on the spectral region
$y = nY$	$k \propto \lambda$	
n independent of λ		
$\delta = \dfrac{2\pi nd}{\lambda} \propto \dfrac{1}{\lambda}$	$\beta = \dfrac{2\pi kd}{\lambda} = \text{constant}$	
$y = \text{constant}$	$y \propto \lambda$	
Progressive wave	Evanescent wave	
$Ee^{i(\omega t - \kappa z)}$	$Ee^{-\alpha z}e^{i\omega t}$	

$$\text{In all cases } R = \left| \frac{y_0 - y}{y_0 + y} \right|^2 \text{ and } T = \frac{4y_0 \,\mathrm{Re}\, y}{\left| y_0 + y \right|^2}$$

With Increasing Wavelength	
Dielectrics	Metals
Become weaker	Become stronger
T increases	R increases
Dielectrics have lower losses	Metals have higher losses

17.5.7.1 Quarterwaves and Halfwaves and Surface Admittance

The characteristic admittance of a medium is the ratio of the magnetic to electric fields of a plane harmonic wave as it propagates through it. When such a wave impinges on an optical surface, the reflected and transmitted amplitudes are a consequence of the boundary conditions, which are that the total tangential magnetic and electric fields are continuous across the boundary. A useful concept is the ratio of the total tangential magnetic to electric field. This is an admittance and characteristic of the surface and so is called the surface admittance. The reflectance in a medium of characteristic admittance y_0 of a surface presenting a surface admittance of Y is given by

$$R = \left| \frac{y_0 - Y}{y_0 + Y} \right|^2 \tag{17.17}$$

where Y can be complex but y_0 should be real.

The front surface of a semi-infinite slab of material is identical with the characteristic admittance of the material. Thin films exhibiting interference can be thought of as admittance transformers. A quarterwave thickness of dielectric material with characteristic admittance y_f transforms surface admittance Y following the quarterwave rule

$$Y_{\text{new}} = \frac{y_f^2}{Y} \tag{17.18}$$

A halfwave layer being two quarterwaves does not change Y and so is often referred to as absentee.

Since quarterwaves and halfwaves are used to a great extent in designs, a useful shorthand notation involves using a capital letter to refer to a quarterwave. If only two materials are involved in the design then the letter can conveniently be H for a quarterwave of high index and L for one of low. If there are more materials then other capital letters can be used. A multiplier in front of the symbol multiplies the thickness accordingly so that a high-index halfwave would be $2H$, or could also be HH. Substrates and incident media can be added, as in

Air | L HH $0.25L$ $0.25H$ | Glass

Note that sometimes the substrate will take the front position and the incident medium the rear depending largely on personal preferences.

17.5.7.2 Sensitivity to Tilting

In a device that operates by optical interference, the path differences vary with angle of incidence. Although at first sight it may seem anomalous, path differences actually are smaller when the angle of incidence moves away from normal. Thus, all filters of the thin film interference type exhibit a shift of their characteristics toward shorter wavelengths with increasing angle of incidence. Polarization effects also gradually become important. Tilting effects depend on the cosine of the propagation angle in the material and so for small tilts of a few degrees in air a good approximation is that the wavelength shift is proportional to the square of the angle of incidence:

$$\frac{\Delta\lambda}{\lambda} = \frac{1.5 \times 10^{-4}}{n_e^2} \vartheta^2 \tag{17.19}$$

where n_e is the effective index of the coating (a value normally in between the highest and lowest indices in the coating and yielding the same angular shift) and where ϑ is the angle of incidence.

Polarization is a little more complicated but provided we are dealing with isotropic materials in our structures, the linear nature of the calculations permits us to split our light into two orthogonally linearly polarized rays, *p*-polarized where the electric field is in the plane of incidence and *s*-polarized where it is normal to the plane of incidence. These two modes are known sometimes as the eigenmodes of polarization since there is no coupling between them. At higher angles of incidence, there is a growing sensitivity to polarization since the *p*- and *s*-performances diverge, the *s*-performance usually showing a strengthening and the *p*-performance a weakening of the characteristic leading to what is known as polarization splitting. This sensitivity can sometimes be reduced by clever design but it is virtually impossible to remove it entirely and sometimes advantage is taken of it in deliberately polarization-sensitive coatings.

17.5.7.3 Coatings of Metals and/or Dielectrics

Combinations of purely metal layers, unless very thin, show little interference effects and so multilayer filters are constructed either from purely dielectric layers, or from combinations of dielectrics and metals. Figures 17.20 and 17.21 show some idealized characteristics. Because of their nature, the natural type of filter for dielectric layers is one that reflects over a limited region at shorter wavelengths and transmits well to longer and longer wavelengths, in other words a long wave pass filter. Metal-dielectric systems,

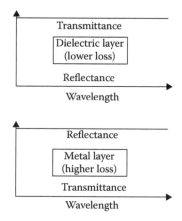

FIGURE 17.20 Ideal characteristics of a dielectric layer (top) and a metal layer (bottom). Dielectrics transmit well but reflect poorly. Metals reflect more and more strongly with increasing wavelength but have low transmittance.

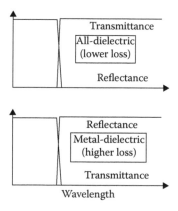

FIGURE 17.21 Idealized characteristics of an interference filter made up of dielectric layers (top) and of a filter made of metal and dielectric layers (bottom). The dielectric system is able to reflect by an interference process effective over a limited region only. Similarly, the metal is induced to transmit by an interference process that is likewise limited.

on the other hand, lend themselves to short wave transmittance and stronger reflection toward longer and longer wavelengths, which is a short wave pass filter. Any requirement that demands the opposite of these simple characteristics presents formidable difficulties.

Transparent conductors like ITO or ZnO:Al appear dielectric at shorter wavelengths and metallic at longer wavelengths (Figure 17.22). They are much used as heat-reflecting shortwave pass filters.

Decorative coatings represent particularly simple spectral filters. They normally operate in reflection and are arranged to reflect a portion of the visible region only.

17.5.7.4 Decorative Coatings

A decorative coating is a particularly simple spectral filter. Usually decorative coatings work in reflection and are arranged so that only a portion of the visible region is reflected and the remainder absorbed to produce a colorful effect. The sharper the transition between absorption and reflection, and the greater the reflectance, the better. Probably the simplest decorative coating is a single layer over a metal. Figure 17.23 sketches the arrangement. The optical constants of the metal and the overcoat will determine the relative amplitudes of the two rays, *A* and *B* that should be as close to each other as possible,

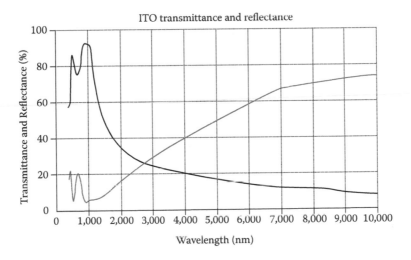

FIGURE 17.22 Calculated transmittance and reflectance of a layer of indium tin oxide showing high transmittance in the visible region and gradually increasing reflectance and reducing transmittance in the infrared. The refractive index in the visible region is around 2.0 and consequently interference fringes are pronounced.

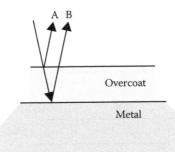

FIGURE 17.23 For low reflectance, the two rays, *A* and *B*, should be of equal amplitude and 180° out of phase.

while the phase thickness of the overcoat determines their relative phase. This implies that the overcoat should be of high index since the metal will usually reflect quite strongly. Absorbing materials can be used as the overcoat but normally the overcoat will be dielectric, since we want the reflected part of the spectrum to be as bright as possible.

High performance metals such as silver or aluminum reflect too strongly in this arrangement, but a titanium dioxide layer over a substrate of titanium works well. The oxide layer is arranged to be sufficiently thick to limit the width of the antireflecting effect.

One of the advantages of decorative coatings that depend on interference is their variation with angle of incidence, known as iridescence. Unfortunately the high-index layer in the above configuration inhibits the variation. A low index layer is much better in this application. To use a low-index layer it is necessary to enhance the outer reflectance, typically by adding a thin metal layer, usually chromium. Since the index of the layer is no longer limitation, a high performance metal like aluminum can now be used as the base giving still higher brightness for the resulting color. Such coatings are often deposited on the base of a decorative prism, which then shows large color variations with angle of incidence. A magenta color at normal incidence changing to green at oblique incidence is a popular arrangement.

An important application for such coatings is as a pigment in color-variable ink or paint. Now the thick metal is coated on either side with the two-layer antireflection coating and the structure is removed from the substrate and ground into a powder the flakes of which are some microns in size and large enough to preserve the interference effect. They are then incorporated as the pigment in an ink or paint that exhibits the same color and iridescence as the original coating. Anti-counterfeiting is an important application of such inks.

17.5.7.5 Metal-Dielectric Filters

The structures in the previous section were intended for reflection only. However, metal films can also transmit light if thin enough. A metal film, thin enough to transmit some light, can also be antireflected on either side by a system consisting of a phase matching layer and a reflector (Figure 17.24). The greater the thickness of the primary metal film, the greater the reflectance of the outer reflector must be. In this application, it is important to keep losses to a minimum and so the structures surrounding the metal are dielectric. We cannot use the thin chromium layers of the previous section. With high performance metals like silver, good transmittance within the antireflected region can result.

If the metal is quite thin, its reflectance is reduced and now the mismatch between high-admittance phase matching layers and the surrounding media can give a sufficiently high reflectance for the anti-reflection condition. This leads to a very simple design consisting of a thin silver layer surrounded by two thin high-index layers, sometimes titania, as shown in Figure 17.25. This is a structure that has been known for some time. Early filters of this type used gold and bismuth oxide but the performance of silver is superior. The metal assures the high reflectance at long wavelengths and the antireflection coating of the transmittance at shorter wavelengths, as in the idealized Figure 17.21. This is the basic heat reflecting filter that is much used in the thermal control of buildings. In practice, to inhibit oxidation of the silver

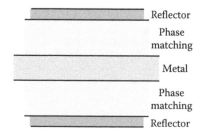

FIGURE 17.24 A central metal layer is antireflected by a system of phase matching layer and reflector on each side.

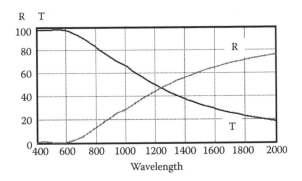

FIGURE 17.25 A three-layer heat reflecting filter consisting of a silver layer surrounded by two titania layers and a glass cover cemented over it. Glass|TiO2(20 nm)Ag(10 nm)TiO2(20 nm)|Glass.

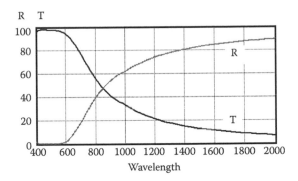

FIGURE 17.26 The design of Figure 17.25 with additional quarterwaves of magnesium fluoride added between the titania layers and the Glass media permitting the use of thicker silver. The additional silver makes the reflectance rise much more rapidly into the infrared and improves the performance of the coating.

layer, thin diffusion barriers of dense dielectric, a fraction of a nanometer thick, are inserted between the phase matching layers and the silver.

Such filters work well in the heat-reflecting application but for other filtering applications where steeper edges are required we need more silver in the central layer. This implies a more powerful outer matching reflector. Because we need transmission as well as reflection we cannot use the absorbing chromium layer and so we turn to quarterwave dielectric layers.

An additional quarterwave of magnesium fluoride on either side, making a five-layer coating, effectively reduces the admittance of the substrate and incident medium. This increases the contrast between the titania and the outer media and increases the reflectance permitting the use of a thicker silver layer. This thicker silver layer then reflects more strongly in the infrared and gives a steeper transition between transmitting and reflecting.

Increasing the outer reflectance still further, and therefore permitting still greater thickness of silver narrows, the transmittance zone steepens the edge between transmission and reflection. This increased reflectance is achieved by adding a further titania quarterwave layer to each side of the design of Figure 17.26. This gives the performance in Figure 17.27 where it is now much more that of a bandpass filter.

With a reflecting system consisting of four quarterwaves of titania and magnesium fluoride and the silver thickness increased to 60 nm, the coating is now a narrowband filter (Figure 17.28).

$$\text{Glass}|\ HLHLH'Ag\ H'LHLH\ |\text{Glass}$$

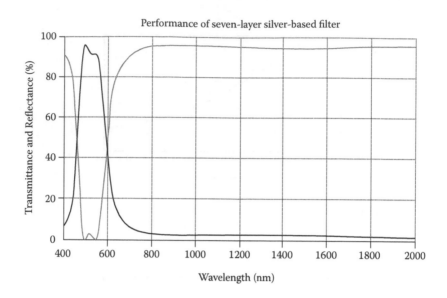

FIGURE 17.27 A bandpass filter consisting of seven layers. A reflector of titania and magnesium fluoride bounds the silver and phase matching layer system on either side. Glass|HLH'(34 nm)Ag(34 nm)H''(34 nm)LH|Glass. H and L represent quarterwaves at 510 nm.

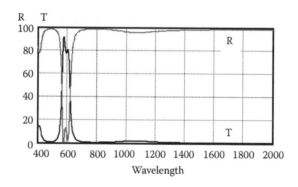

FIGURE 17.28 The performance of the eleven-layer silver-based narrowband filter.

Such filters are used by themselves as useful bandpass filters in their own right and also, because of their low infrared transmittance that is difficult to achieve in any other way, as blocking filters for use with other types of filters with less acceptable infrared performance, such as all-dielectric narrow band filters.

By a happy coincidence the symmetrical structure of the silver-based filters minimizes the electric field within the silver and contributes to the resulting high transmittance. The term induced transmission is sometimes used to describe this effect.

17.5.7.6 All-Dielectric Filters and Coatings

We have already noted that metal-dielectric structures lend themselves to coatings that reflect longer wavelengths and transmit shorter ones, while all-dielectric structures are more suited to the opposite. Thus, a coating to transmit the infrared and reflect a band in the visible would best be constructed from dielectric materials.

The quarterwave rule shows us that a coating consisting of a series of quarterwave layers alternating between high and low indices will yield high reflectance. Such a structure is known as a quarterwave stack and has a design on the lines of

$$\text{Air} \mid HLHLHLHLHLH \mid \text{Glass}$$

The reflectance of such structures shows a broad central fringe, forming a reflecting band with the expected high reflectance, surrounded by regions of lower reflectance exhibiting fringes with amplitude decaying with distance from the reflecting band. These regions form transmitting regions but with a superimposed ripple. Figure 17.29 shows the performance of a quarterwave stack set up to act as a long-wave pass filter.

Ripple in the long-wave pass band is clearly a severe problem. This can be reduced by changing the thicknesses of the outermost layers to eighth waves that act to match the quarterwave core to the outer media (Figure 17.30).

$$\text{Air} \left| \frac{H}{2} LHLHLHLHLIHL \frac{H}{2} \right| \text{Glass}$$

Losses in the central metal layers are the major limitations in the narrowband filters discussed in the previous section. We can use the quarterwave stack structure to replace the central metal layer and to reduce losses. This allows us to make still narrower filters, Figure 17.31.

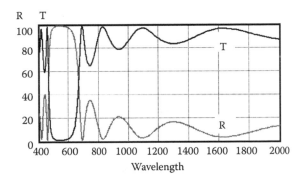

FIGURE 17.29 The performance of a quarterwave stack consisting of 11 alternate quarterwave layers of titania and silica with the titania outermost. The characteristic is a longwave pass filter but the pronounced oscillatory ripple is a problem.

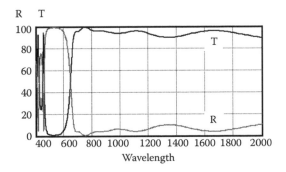

FIGURE 17.30 The greatly improved performance produced by the change in the thicknesses in the outermost layers of Figure 17.29 to eighth waves.

Figure 17.31 has the design,

$$\text{Air} \mid \textit{HLHLHHLHLHLHLHLHHLHLH} \mid \text{Glass}$$

Because the losses are so much lower, it is possible to achieve exceedingly narrow filters. The halfwave layers, those designated as *HH* and derived from the original phase matching layers, are called cavity layers because their function is really that of a tuned cavity. The filter of Figure 17.31 is a two-cavity filter. The greater the number of cavities, the steeper the passband sides become.

A narrowband filter design involving 59 layers and three cavities is shown in Figure 17.32.

$$\text{Glass} \mid \textit{HLHLHLHLH LL HLHLHLHLHLHLHLHLHLHLH LL}$$

$$\textit{HLHLHLHLHLHLHLHLHLH LL HLHLHLHLH} \mid \text{Glass}$$

Antireflection coatings are not strictly spectral filters in the normal sense of the word because their purpose is not to filter but to reduce losses due to residual reflection. But they are essential components of filters and so a few words about them are in order. An antireflection coating with low loss implies the use of dielectric layers.

The simplest antireflection coating is a single quarterwave layer. The quarterwave rule permits us to write an expression for the reflectance of a quarterwave of material on a substrate:

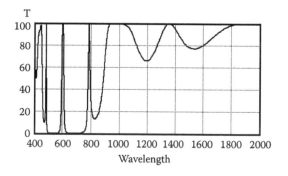

FIGURE 17.31 An all-dielectric narrowband filter based on the metal-dielectric arrangement of Figure 17.28. The central metal layer is now replaced by an all-dielectric quarterwave stack.

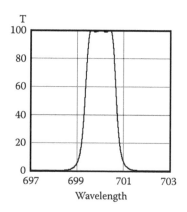

FIGURE 17.32 The performance of the 59-layer narrowband three-cavity filter.

$$R = \left\{ \frac{y_0 - y_f^2 / y_{sub}}{y_0 + y_f^2 / y_{sub}} \right\}^2 \tag{17.20}$$

and if $y_0 = y_f^2 / y_{sub}$, then the reflectance will be zero. Unfortunately, suitable materials with such a low refractive index as would perfectly antireflect glass, of admittance 1.52, in air, 1.00, are lacking. Magnesium fluoride, with admittance 1.38, is the best that is available. The performance, at 1.25% reflectance, is considerably better than the 4.25% reflectance of an uncoated glass surface (Figure 17.33).

This performance is good enough for many applications and the single-layer antireflection coating is much used. It has a characteristic magenta color in reflection because of the rising reflectance in the red and blue regions of the spectrum.

There are applications where improved performance is required. If zero reflectance at just one wavelength, or over a narrow range, then the two-layer V-coat is the preferred solution. This consists of a thin high index layer, say roughly one sixteenth of a wave in optical thickness of titania, followed by roughly five sixteenths of a wave of low index material, magnesium fluoride or, sometimes, silica. The thicknesses are adjusted to accommodate the particular materials that are to be used in the coating. A typical performance is illustrated in Figure 17.34.

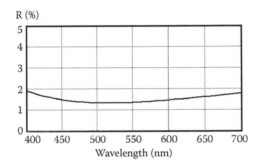

FIGURE 17.33 Computed performance of a single-layer antireflection coating consisting of a quarterwave of MgF$_2$ on glass with air as incident medium.

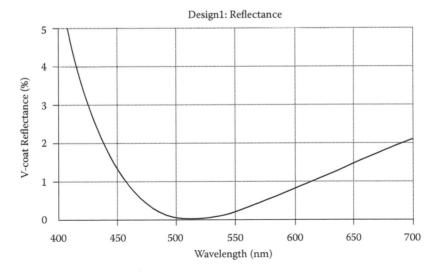

FIGURE 17.34 The performance of a typical V-coat, a two-layer antireflection coating to give (near) zero reflectance for just one wavelength.

Then there will be other applications demanding low reflectance over the visible region, 400 to 700 nm. These requirements are usually met with a four-layer design based on the V-coat with a halfwave flattening layer inserted just one quarterwave behind the outer surface. The design is of the form

$$\text{Air} \,|\, LHHL'H \,|\, \text{Glass}$$

A slight adjustment by refinement then yields the performance shown in Figure 17.35.

The ultimate antireflection coating is a layer that represents a gradual, smooth transition from the admittance of the substrate to the admittance of the incident medium. Provided this layer is thicker than a halfwave, the reflectance will be very low. This implies that for all wavelengths shorter than a maximum, the layer will be virtually a perfect antireflection coating. The profile of such a layer is shown in Figure 17.36. Figure 17.37 shows the performance as a function of g, which is λ_0/λ where λ_0 is the wavelength for which the layer is a halfwave. Best results are obtained if the derivative of admittance as a function of distance can be smoothed along with the admittance. This implies a law of variation that is a fifth-order polynomial.

For air incident medium, the inhomogeneous layer can be achieved only by microstructural variations, Figure 17.38. Processes involving etching, leaching, sol-gel deposition, and photolithography

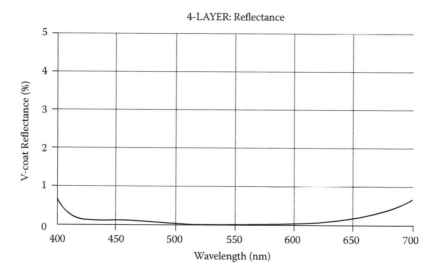

FIGURE 17.35 The performance of the four-layer two-material design over the visible region.

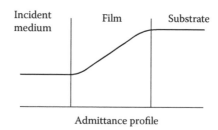

FIGURE 17.36 Sketch of the variation of optical admittance as a function of optical thickness through an inhomogeneous matching layer.

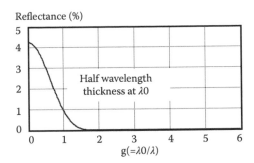

FIGURE 17.37 The reflectance of an inhomogeneous layer with fifth-order polynomial variation of admittance throughout. The substrate has admittance 5.1 and the incident medium, 1.00.

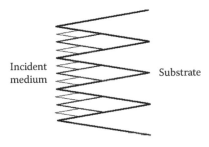

FIGURE 17.38 Microstructural approach to an inhomogeneous layer.

have all been used. This does give the expected reduction in reflectance, but as the wavelength gets shorter, the microstructural features become comparatively larger and so there is a short wave limit determined by scattering. For a low reflectance range of reasonable width, the features must be very long and thin.

Because of the inherent weakness of the film, this solution is limited to very special applications.

The quarterwave stack, already mentioned, is a useful rejection filter for limited spectral regions. Unfortunately, for some applications, the interference condition, which assures high reflectance, repeats itself for discrete bands of shorter wavelengths. Thus, the first-order reflection peak at 10 is accompanied by peaks at $\lambda_0/3$, $\lambda_0/5$ and so on. Potential peaks at $\lambda_0/2$, $\lambda_0/4$ ... are missing because, at these wavelengths, we have a series of halfwaves that are absentees (Figure 17.39). There are many applications where such behavior is undesirable. The inhomogeneous layer can be used to advantage in the suppression of these higher-order peaks because it can act as an antireflection coating that suppresses all reflection at wavelengths shorter than a long wavelength limit. A rugate filter, the word *filter* often being omitted, consists of essentially a quarterwave stack that has been antireflected by an inhomogeneous layer at each of the interfaces between the original quarterwave layers. The result is a cyclic variation of admittance that can be considered close to sinusoidal throughout the structure. All higher orders of reflectance are suppressed. The width of the zone of high reflectance, now only the fundamental, can be adjusted by varying the amplitude of the cycle of admittance. The smaller that this amplitude is, the narrower the reflectance peak is as a result. Of course when the amplitude is small, many cycles are needed to achieve reasonable reflectance. Rugate filters are used in applications where a narrow line must be removed from a background. They can simply be suppressing a bright laser line, as in Raman spectroscopic applications, or they can be actually reflecting the line and transmitting all others. Head-up displays can make use of narrowband beamsplitters of this type.

For more information on thin film filters, see Dobrowolski [10] or Macleod [11].

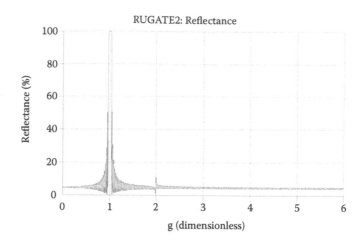

FIGURE 17.39 Performance of rugate filter comprising 40 cycles of sinusoidal admittance variation from 1.45 to 1.65. The horizontal axis is in terms of g, which is a dimensionless variable given by λ_0/λ where λ_0 is the reference wavelength. Note the absence of higher-order peaks. The slightly more pronounced oscillation at $g=2$ is a real feature of the design.

References

1. Heavens OS, Ditchburn RW. *Insight into Optics*. Chichester: John Wiley & Sons, 1991.
2. O'Shea DC. Elements of modern optical design. First ed. *Wiley Series in Pure and Applied Optics*, J W Goodman, Editor. New York, NY: John Wiley & Sons. 402; 1985.
3. VanderLugt A. Optical signal processing. First ed. *Wiley Series in Pure and Applied Optics*, J W Goodman, Editor. New York, NY: John Wiley & Sons. 604; 1992.
4. Hecht E, Zajac A. Optics. First ed. *Reading*. Boston, MA: Addison-Wesley Publishing Company. 565; 1974.
5. Piegari A. Graded coatings. *Thin Films for Optical Systems*, F Flory, Editor. New York, NY: Marcel Dekker. pp. 475–519; 1995.
6. Dobrowolski JA. Coatings and filters. *Handbook of Optics*, W G Driscoll and W Vaughan, Editors. New York, NY: McGraw-Hill Book Company. pp. 8.1–8. 124; 1978.
7. Dobrowolski JA, Marsh GE, Charbonneau DG, Eng J, Josephy PD. Colored filter glasses: An inter-comparison of glasses made by different manufacturers. *Applied Optics*. **16**(6): pp. 1491–1512; 1977.
8. Yariv A, Yeh P. Optical waves in crystals. First ed. *Pure and Applied Optics*, J W Goodman, Editor. New York, NY: John Wiley & Sons. 589; 1984.
9. Chang IC. Acousto-optic devices and applications. *Handbook of Optics*, M Bass, et al., Editors. New York, NY: McGraw-Hill. pp. 12.1–12.54; 1995.
10. Dobrowolski JA. Optical properties of films and coatings. *Handbook of Optics*, M Bass, et al., Editors. New York, NY: McGraw-Hill. pp. 42.1–42.130; 1995.
11. Macleod HA. *Thin-Film Optical Filters*. Second ed. Bristol: Adam Hilger. 519; 1986.

Optical Fibers and Accessories

18.1	Introduction	633
18.2	Optical Fibers	634

Single-Mode and Multimode Fibers • Fiber Modes • Fiber
Parameters • Optical Losses (Attenuation) • Dispersion, Fiber
Bandwidth • Typical Fibers and Fiber Parameters

18.3	Special Fibers	643

Erbium-Doped Fibers • Powerful Double-Clad Fibers • Infrared
Optical Fibers • Metallic and Dielectric Hollow
Waveguides • ZBLAN Fiber Based Up-Conversion Lasers • Plastic
Optical Fibers • Fiber Bundles

18.4	Fiber Optic Components	653

Optical Fiber Couplers • Wavelength-Division
Multiplexers • Switches • Attenuator • Polarization Fiber
Components • Polarizers • Polarization Splitter • Polarization
Controller • Fiber Bragg Gratings • Fiber Bragg Grating as a
Reflection Filter • Fiber Bragg Grating-Based Multiplexer • Chirped
Fiber Bragg Gratings • Fiber Connectors • Fiber Splicers

Andrei N.
Starodoumov

Suggested Reading 671

18.1 Introduction

Historically, light-guiding effects were demonstrated in the mid-nineteenth century when Swiss physicist D. Collodon and French physicist J. Babinet showed that light can be guided in jets of water for fountain displays. In 1854, the British physicist J. Tyndall demonstrated this effect in his popular lectures on science, guiding light in a jet of water flowing from a tank. In 1880, in Massachusetts, an engineer W. Wheeler patented a scheme for piping light through buildings. He designed a net of pipes with reflective linings and diffusing optics to carry light through a building. Wheeler planned to use light from a bright electric arc to illuminate distant rooms. However, this project was not successful. Nevertheless, the idea of light piping reappeared again and again until it finally converted into the optical fiber.

During the 1920s, J. L. Baird in England and C. W. Hansell in the United States patented the idea of image transmission through arrays of hollow pipes or transparent rods. In 1930, H. Lamm, a medical student in Munich, had demonstrated image transmission through a bundle of unclad optical fibers. However, the major disadvantage of this device was the poor quality of the transmitted images. Light "leaked" from fiber to fiber, resulting in degradation of the quality of the image.

Modern optical fibers technology began in 1954 when A. van Heel of the Technical University of Delft in Holland covered a bare fiber of glass with a transparent cladding of lower refractive index. This protected the reflection surface from contamination, and greatly reduced cross-talk between fibers.

The invention of the laser in 1957 provided a promising light source for an optical communication system and stimulated research in optical fibers. In 1966, Kao and Hockman pointed out that purifying glass could dramatically improve its transmission properties. In 1970, the scientists from Corning Glass reported fibers made from extremely pure fused silica with losses below 20 dB/km at 633 nm. Over the next few years, fiber losses dropped dramatically. In 1976 in a laboratory experiment, Bell Laboratories combined all the components needed for an optical communication system, including lasers, detectors, fibers, cables, splices, and connectors. Since 1980, the growth in the number of installed communications systems has been extremely rapid. The improvement of the quality of optical fibers and fast development of the fiber analogs of bulk optical elements stimulated non-telecommunication applications of fiber optics in such areas as aircraft and shipboard control, sensors, optical signal processors, displays, delivery of high-power radiation, and medicine.

Starting in the 1970s, a number of good books have been published that discuss the theoretical aspects of fiber optics. In the last ten years, practical guides of fiber optics for telecommunication applications have appeared. The objective of this chapter is to describe the principles of fiber optics, with an emphasis on basic fiber optical elements, such as fibers, spectrally selective and polarization-sensitive fiber elements, and couplers.

18.2 Optical Fibers

18.2.1 Single-Mode and Multimode Fibers

Optical fiber is the medium in which radiation is transmitted from one location to another in the form of guided waves through glass or plastic fibers. A fiber waveguide is usually cylindrical in form. It includes three layers: the center core that carries the light, the cladding layer covering the core, and the protection coating. The core and cladding are commonly made from glass, while the coating is plastic or acrylate.

Figure 18.1 shows a fiber structure. The core of radius a has a refractive index n_1. The core is surrounded by a dielectric cladding with a refractive index n_2 that is less than n_1. The silica core and cladding layers differ slightly in their composition due to small quantities of dopants, such as boron, germanium, or fluorine. Although light can propagate in the core without cladding, the latter serves several purposes. The cladding reduces scattering loss on a glass–air surface, adds mechanical strength to the fiber, and protects the core from surface contaminants. An elastic plastic material, which encapsulates the fiber (buffer coating), adds further strength to the fiber and mechanically isolates the fiber from adjacent surfaces, protecting it from physical damage and moisture.

Glass optical fibers can be *single-mode fibers* (SMFs) or *multimode fibers* (MMFs). A single-mode fiber sustains only one mode of propagation and has a small core of 3–10 μm. In an MMF, the core diameter is usually much larger than that of an SMF. In an MMF, light can travel many different paths (called modes) through the core of the fiber. The boundary between the core and the cladding may be sharp (*step-index profile*) or graduated (*graduated-index profile*). The core-index profile of an SMF is usually a step-index type (Figure 18.2a, while in an MMF it can be either step-index or a graded-index type

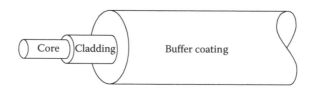

FIGURE 18.1	Fiber structure.

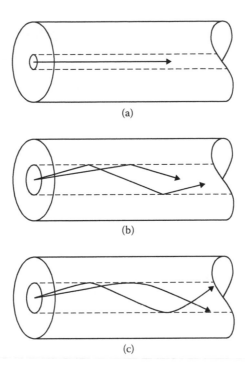

FIGURE 18.2 Single-mode and multimode fibers: (a) single-mode step-index fiber, (b) multimode step-index fiber, and (c) multimode graded-index fiber.

(Figure 18.2b and c). Step-index fiber has a core composed completely of one type of glass. Straight lines, reflecting off the core–cladding interface, can describe mode propagation in such a fiber. Since each mode travels at different angles to the axis of the fiber and has a different optical path, a pulse of light is dispersed while traveling along the fiber. This effect called *intermodal dispersion* limits the bandwidth of a step-index fiber.

In graded-index fibers, the core is composed of many different layers of glass. The index of refraction of each layer is chosen to produce an index profile approximating a parabola with maximum in the center of the core (Figure 18.2c). In such a fiber, low-order modes propagate close to the central part of the core with higher refractive index. Higher-order modes, although traversing a much longer distance than the central ray, does so in a region with less refractive indices and hence the velocities of these modes are greater. Thus, the effects of these two factors can be made to cancel each other out, resulting in very similar propagation velocities. A properly constructed index profile will compensate for different path lengths of each mode, increasing the bandwidth capacity of the fiber by as much as 100 times over that of step-index fiber.

Multimode fibers offer several advantages over single-mode fibers. The larger core radius of MMFs makes it easier to launch optical power into the fiber and facilitate the connecting together of similar fibers. Another advantage is that radiation from light-emitting diodes can be efficiently coupled to an MMF, whereas SMFs must generally be excited with laser diodes. However, SMFs offer higher bandwidths in communication applications.

18.2.2 Fiber Modes

When light travels through a medium with a high refractive index n_1 to a medium with a lower refractive index n_2, the optical ray is refracted at the boundary into the second medium. According to the Snell's

LP_{01} LP_{21}

FIGURE 18.3   Electric field distribution for LP_{01} mode (a) and LP_{21} mode (b).

law, $n_1 \sin \alpha_1 = n_2 \sin \alpha_2$; the angle of refraction α_2 is greater than the angle of incidence α_1 in this case. As the incident angle increases, a point is reached at which the optical ray is no longer refracted into the second medium ($\alpha_2 = \pi/2$). The optical radiation is completely reflected back into the first medium. This effect is called *total internal reflection*.

When total internal reflection occurs, a phase shift is introduced between the incident and reflected beams. The phase shift depends on polarization of the incident beam and on the difference of refractive indices. Two different cases should be considered: when electric field vector **E** is in the plane of incidence ($E_\parallel$), and when **E** is perpendicular to this plane ($E_\perp$). These two situations involve different phase shifts when total internal reflection takes place, and hence they give rise to two independent sets of modes. Because of the directions of **E** and **H** with respect to the direction of propagation down the fiber, the two sets of modes are called transverse magnetic (TM) and transverse electric (TE), which correspond to the $E_\parallel$ and $E_\perp$, respectively. Two integers, l and m, are required to completely specify the modes. This is because the waveguide is bounded in two dimensions. Thus, we refer to TM_{lm} and TE_{lm} modes.

In fiber waveguides, the core–cladding index difference is rather small and does not exceed a few percent. The phase shift acquired under total internal reflection is practically equal for both sets of modes. Thus, the full set of modes can be approximated by a single set called *linearly polarized* (LP_{lm}) modes. Such a mode in general has m field maxima along a radius vector and $2l$ field maxima around a circumference. Figure 18.3 shows an electric field distribution for LP_{01} and LP_{21} modes.

18.2.3 Fiber Parameters

The light can enter and leave a fiber at various angles. The maximum angle (acceptance angle θ_{max}) that supports total internal reflection inside the fiber defines the *numerical aperture* (NA). The numerical aperture can be expressed in terms of core–cladding refractive index difference as

$$NA = n_0 \sin \theta_{max} = \left(n_1^2 - n_2^2 \right)^{1/2} \approx n_1 \left(2\Delta \right)^{1/2}, \qquad (18.1)$$

where $\Delta = (n_1 - n_2)/n_1$ is the *relative* (or *normalized*) *index difference* and n_0 is the representation, which is valid when $\Delta < 1$. Since the numerical aperture is related to the maximum acceptance angle, it is commonly used to describe the fiber characteristics and to calculate source-to-fiber coupling efficiencies.

A second important fiber parameter is the normalized frequency V, which defines the number of modes supported by a fiber. This parameter depends on optical wavelength λ, a core radius a, and a core–cladding refractive index difference, and can be written as

$$V = \frac{2\pi a}{\lambda} NA. \qquad (18.2)$$

With the parameters V and Δ, index profile optical fibers can be classified more precisely. For $V < 2.405$, the fiber sustains only one mode and is a single mode. Multimode fibers have values $V > 2.405$ and can sustain many modes simultaneously. The number 2.405 corresponds to the first zero of the Bessel function, which appears in the solution of the wave equation for the fundamental mode in a cylindrical waveguide. A fiber can be multimode for short wavelengths and, simultaneously, can be single mode for longer wavelengths. The wavelength corresponding to $V = 2.405$ is known as the cutoff wavelength of the fiber and is given by

$$\lambda_c = \frac{2\pi a}{2.405} \text{NA.} \tag{18.3}$$

The fiber is single mode for all wavelengths longer than λ_c and is multimode for shorter wavelengths. By decreasing the core diameter or the relative index difference such that $V < 2.405$, the cutoff wavelength can be shifted to shorter wavelengths, and single-mode operation can be realized. Usually the fiber diameters are greater for longer wavelengths.

The small core diameter of an SMF makes it difficult to couple light into the core and to connect two fibers. To increase the effective core diameter, fibers with multiple cladding layers were designed. In such a fiber, two layers of cladding, an inner and an outer cladding surround its core with a barrier in between. The index profiles of single and multiple cladding fibers are shown in Figure 18.4a and b. The multiple cladding schemes permit the design of an SMF with a relatively larger core diameter, facilitating fiber-handling and splicing. The additional barrier provides an efficient control of dispersion properties of a fiber and permits us to vary the total dispersion of an SMF.

A more complicated index profile called a quadrupole clad (Figure 18.4c) gives even more flexibility in the handling of fiber dispersion. The quadrupole-clad fiber has low dispersion (< 1 ps/km-nm) over a wide wavelength range extending from 1.3 µm to 1.6 µm.

Figure 18.4d shows a triangle index profile. A fiber with triangle profile called a T-fiber provides a much higher second-order mode cutoff wavelength (for LP_{11} mode) and lower attenuation than a step-index fiber does.

If the parameter V increases much above 2.405, the step-index fiber can support a large number of modes. The maximum number of modes propagating in the fiber can be calculated as

$$N_{\text{mode}} \approx \frac{V^2}{2}. \tag{18.4}$$

When many modes propagate through a fiber, carrying the same signal but along different paths, the output signal is the result of interference of different modes. If a short pulse enters into the fiber, at the output this pulse has longer length because of the temporal delay between carrying modes. As has been mentioned above, this effect is called intermodal dispersion. The quality of the signal deteriorates as the fiber length increases, limiting the information capacity of the step-index fiber.

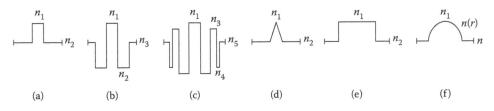

FIGURE 18.4 Typical refractive index profiles of fibers. (a) single cladding, (b) multiple cladding, (c) quadrople cladding, (d) triangular index profile, (e) rectangular index profile and (f) core profile.

To reduce the effect of intermodal dispersion, a graded-index fiber has been proposed. A typical core profile is shown in Figure 18.4f. The most often used index profile is the power law profile designed according to the following expression

$$n(r) = n_1 \left[1 - 2\Delta \left(\frac{r}{a} \right)^g \right]^{1/2}, \qquad (18.5)$$

where g is usually chosen to reduce the effect of intermodal dispersion. For optimum effect,

$$g = 2 - \left(\tfrac{12}{5} \right) \Delta. \qquad (18.6)$$

The refractive index of the cladding is maintained at a constant value.

18.2.4 Optical Losses (Attenuation)

Light traveling in a fiber loses power over distance. If P_0 is the power launched at the input of a fiber, the transmitted power P_t at a distance L is given by

$$P_t = P_0 exp(-\alpha L), \qquad (18.7)$$

where α is the attenuation constant, also known as the fiber loss. Attenuation is commonly measured in decibels (dB). In fibers, the loss is expressed as attenuation per 1 km length, or dB/km, using the following relationship:

$$\alpha_{dB} = -10 \frac{10}{L} \log \left[\frac{P_t}{P_0} \right] = 4.343\alpha. \qquad (18.8)$$

Fiber loss is caused by a number of factors that can be divided into two categories: *induced* and *inherent* losses. The induced losses may be introduced during manufacturing processes. These losses are caused by inclusions of contaminating atoms or ions, by geometrical irregularities, by bending and microbending, by splicing, by connectors, and by radiation. The fabrication process is aimed at reducing these losses as much as possible.

Bend losses occur due to the change of the angle of incidence at the core–cladding boundary and depend strongly on the radius of curvature. The loss will be greater (1) for bends with smaller radii of curvature, and (2) for those modes that extend most into the cladding. The bending loss can generally be represented by a loss coefficient α_B, which depends on fiber parameters and radius of curvature R, and is given as

$$\alpha_B = C \exp \left(-\frac{R}{R_c} \right), \qquad (18.9)$$

where C is a constant, R_c is given by $R_c = r/(NA)^2$, and r is the fiber radius. As is seen from Equation 18.9, the attenuation coefficient depends exponentially on the bend radius. Thus, decreasing the radius of curvature drastically increases the bending losses. The bending loss decreases as the core–cladding index difference increases. The optical radiation at wavelengths close to cutoff will be affected more than that at wavelengths far from cutoff.

Bending loss can also occur on a smaller scale due to fluctuations of core diameter and perturbations in the size of the fiber caused by buffer, jacket, fabrication, and installation practice. This loss is called microbending and can contribute significantly over a distance.

Splice and connector loss can also add to the total induced loss. The mechanism of these losses will be discussed below.

Light loss that cannot be eliminated during fabrication process is called inherent losses. The inherent losses have two main sources: (a) Rayleigh scattering and (b) ultraviolet and infrared absorption losses.

18.2.4.1 Scattering

Glass fibers have a disordered structure. Such a disorder results in variations in optical density, composition, and molecular structure. These types of disorder, in turn, can be described as fluctuations in refractive index of the material. If the scale of these fluctuations is of the order of $\lambda/10$ or less, then each irregularity can be considered as a point scattering center. The light is scattered in all directions by each point center. The intensity of scattered light and the power loss coefficient vary with wavelength as λ^{-4}. The term λ^{-4} is the characteristic wavelength-dependence factor of Rayleigh scattering. The addition of dopants into the silica glass increases the scattering loss because the microscopic inhomogeneities become more important. Rayleigh scattering is a fundamental process limiting the minimum loss that can be obtained in a fiber.

18.2.4.2 Absorption

The absorption of light in the visible and near-infrared regions at the molecular level arises mainly from the presence of impurities, such as transition metal ions (Fe^{3+}, Cu^{2+}) or hydroxyl (OH^{-}) ions. The OH radical of the H_2O molecule vibrates at a fundamental frequency corresponding to the infrared wavelength of 2.8 μm. Because the OH radical is an anharmonic oscillator, the overtones can occur, producing absorption peaks at 0.725 μm, 0.95 μm, 1.24 μm, and 1.39 μm. Special precautions must be taken during the fiber-fabrication process to diminish the impurity concentrations. The OH concentration should be kept at levels below 0.1 ppm if ultralow losses are desired in the 1.20–1.60 μm range.

Ultraviolet absorption produces an absorption tail in the wavelength region below 1 μm. This absorption decreases exponentially with increasing wavelength and is often negligible in comparison with Rayleigh scattering within the visible wavelength range.

At wavelengths greater than about 1.6 μm, the main contribution to the loss is due to transitions between vibrational states of the lattice. Although the fundamental absorption peaks occur at 9 μm, overtones and combinations of these fundamental vibrations lead to various absorption peaks at shorter wavelengths. The tails of these peaks result in typical values of 0.02 dB/km at 1.55 μm, and 1 dB/km at 1.77 μm. Figure 18.5 shows the total loss coefficient (solid curve) as a function of wavelength for a silica fiber. The absorption peak at 1.39 μm corresponds to OH radicals. A properly chosen fiber-fabrication process permits suppression of this peak. The inherent loss level is shown by a dashed curve. The minimum optical losses of 0.2 dB/km have been obtained with GeO_2-doped silica fiber at 1.55 μm.

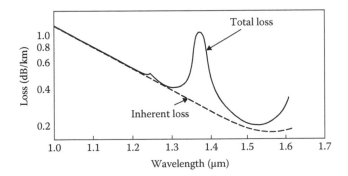

FIGURE 18.5   The total loss coefficient as a function of wavelength for a silica fiber.

18.2.5 Dispersion, Fiber Bandwidth

When an optical pulse propagates along a fiber, the shape of the pulse changes. Specifically, a pulse of light gets broader. There are three main sources of such changes: *intermodal* dispersion, *material* dispersion, and *waveguide* dispersion. Intermodal dispersion can be avoided by using single-mode fibers or can be diminished by using graded-index multimode fibers.

Material (or chromatic) dispersion is due to the wavelength dependence of the refractive index. On a fundamental level, the origin of material dispersion is related to the electronic absorption peaks. Far from the medium resonances, the refractive index is well approximated by the Sellmeier equation:

$$n^2(\omega) = 1 + \sum_{i=1}^{l} \frac{B_i \omega_i^2}{\omega_i^2 - \omega^2},$$

(18.10)

where ω_i is the resonance frequency and B_i is the strength of ith resonance. For bulk fused silica, these parameters are found to be $B_1 = 0.6961663$, $B_2 = 0.4079426$, $B_3 = 0.8974794$, $\lambda_1 = 0.0684043$ μm, $\lambda_2 = 0.1162414$ μm, $\lambda_3 = 9.896161$ mm, where $\lambda_i = 2\pi c/\omega_i$. Fiber dispersion plays an important role in propagation of short optical pulses since the different spectral components travel at different velocities, resulting in pulse broadening. Dispersion-induced pulse broadening can be detrimental for optical communication systems.

Mathematically, the effects of fiber dispersion are accounted for by expanding the mode-propagation constant $\beta(\omega)$ in a Taylor series near the central frequency ω_0:

$$\beta(\omega) = n(\omega)\frac{\omega}{c} = \beta_0 + \beta_1(\omega - \omega_0) + \frac{1}{2}\beta_2(\omega - \omega_0)^2 + \cdots,$$

(18.11)

where

$$\beta_m = \left(\frac{d^m \beta}{d\omega^m}\right)_{\omega_0} \quad (m = 0, 1, 2, \ldots).$$

(18.12)

The parameter β_1 characterizes a group velocity $v_g = d\omega/d\beta = (\beta_1)^{-1}$. Although individual plane waves travel with a phase velocity $v_p = \omega/\beta$, a signal envelope propagates with a group velocity. The parameter β_2 is responsible for pulse broadening.

The transit time required for a pulse to travel a distance L is $\tau = L/v_g$. Since the refractive index depends on wavelength, the group velocity is also a function of λ. The travel time per unit length τ/L may be written as

$$\frac{\tau}{L} = \frac{1}{c}\left(n - \lambda\frac{dn}{d\lambda}\right).$$

(18.13)

The difference in travel time $\Delta\tau$ for two pulses at wavelengths λ_1 and λ_2, respectively, is a measure of dispersion. In a dispersive medium, the optical pulse of a spectral width $\Delta\lambda$, after traveling a distance L, will spread out over a time interval

$$\Delta\tau = \frac{d\tau}{d\lambda}\Delta\lambda.$$

(18.14)

The derivative $d\tau/d\lambda$ describes the pulse broadening and can be expressed through parameter β_2 as

$$\frac{1}{L}\frac{d\tau}{d\lambda} = -\frac{\lambda}{c}\frac{d^2 n}{d\lambda^2} = -\frac{2\pi c}{\lambda^2}\beta_2 = D.$$

(18.15)

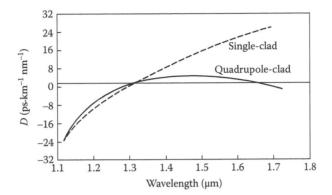

FIGURE 18.6 Dispersion parameter D as a function of wavelength for a single-clad fiber (dashed curve) and a quadrupole-clad fiber (solid curve).

The parameter D (in units ps/nm-km), also called dispersion parameter or dispersion rate, is commonly used in fiber-optics literature instead of β_2. The parameter β_2 is generally referred to as the group velocity dispersion (GVD) coefficient. For a bulk silica, β_2 vanishes with wavelength and is equal to zero at 1.27 µm. This wavelength is often called zero-dispersion wavelength λ_D. An interesting fact is that the sign of the dispersion term changes in passing through the zero-dispersion point.

Waveguide dispersion is due to the dependence of the propagation constant on fiber parameters when the index of refraction is assumed to be constant. The reason for this is that the fiber parameters, such as the core radius and the core–cladding index difference, cause the propagation constant of each mode to change for different wavelengths. The sign of waveguide dispersion is opposite to the sign of material dispersion at wavelengths above 1.3 µm. This feature is used to shift the zero-dispersion wavelength λ_D in the vicinity of 1.55 µm, where the fiber loss has a minimum value. Such *dispersion-shifted* fibers have found numerous applications in optical communication systems. It is possible to design *dispersion-flattened* optical fibers having low dispersion ($|D| \leq 1$ ps/nm-km) over a relatively large wavelength range. This is achieved by the use of multiple cladding layers. Figure 18.6 shows the dependence of the dispersion parameter D on the wavelength for a single-clad fiber (dashed curve) and a quadrupole-clad fiber (solid curve) with flattened dispersion in the wavelength range extending from 1.25 to 1.65 µm.

18.2.5.1 Fiber Bandwidith

Optical fiber bandwidth is a measure of the information-carrying capacity of an optical fiber. The fiber's total dispersion limits the bandwidth of the fiber. This occurs because pulses distort and broaden, overlapping one another and become indistinguishable at a receiver. To avoid an overlapping, pulses should be transmitted at smaller repetition rates (thereby reducing bit rate). The use of these terms (bandwidth and bit rate) is technically difficult because of two factors: link length and dispersion. To calculate a desired bandwidth or bit rate, the fiber provider must know the length of the link. In addition, the provider does not know the spectral bandwidth of the optical source to be used in the system. The spectral bandwidth of the light source determines the amount of chromatic dispersion in the link. Because of these two difficulties, instead of the terms "bandwidth" and "bit rate," two other terms are used: *bandwidth-distance product,* in MHz-km, for multimode fibers; *dispersion rate,* in ps/nm/km, for single-mode fibers.

The bandwidth-distance product is the product of the fiber length and the maximum bandwidth that the fiber can transmit. For example, a fiber with a bandwidth-distance product of 100 MHz-km, can transmit a signal of 50 MHz over a distance of 2 km or a signal of 100 MHz over a distance of 1 km. It should be noted that graded-index MMFs have an information-carrying capacity 30–50 times greater than step-index MMFs because of diminished intermodal dispersion. In single-mode fibers, the information-carrying capacity is approximately two orders of magnitude greater than that of graded-index MMF.

18.2.6 Typical Fibers and Fiber Parameters

Optical fibers used for telecommunications and other applications are manufactured with different core and cladding diameters. Fiber size is specified in the format "core/cladding." A 100/140 fiber means the fiber has a core diameter of 100 μm and a cladding diameter of 140 μm. A polymer coating covers the cladding and can be either 250 or 500 μm. For a tight-buffered cable construction, a 900 μm diameter plastic buffer covers the coating. Table 18.1 shows typical fiber core, cladding and coating diameters.

Most fibers have a glass core and glass or plastic cladding. These fibers can be classified in four types: all glass, plastic clad silica, hard clad silica, and plastic optical fibers.

TABLE 18.1 Fiber Parameters

Core/cladding/coating Diameter (μm)	Wavelength (nm)	Optical Loss (dB/km)	Bandwidth–Distance Product (MHz-km)	Numerical Aperture
2.4/65/190	400	60	—	0.13
3.3/80/200	500	22	—	0.13
4.0/125/250	630	10	—	0.12
5.5/125/250	820	3.5	—	0.12
6.6/125/250	1,060	2	—	0.13
6.6/80/200	1,300	1	—	0.16
7.8/125/250	1,550	1	—	0.16
9/125/250 or 500	780	—	<800	0.11
	850	—	2,000	
	1,300	0.5–0.8	20,000	
	1,550	0.2–0.3	4,000–20,000	
50/125/250 or 500	780	4.0–8.0	150–700	0.20
	850	3.0–7.0	200–800	
	1,300	1.0–3.0	400–1,500	
	1,550	1.0–3.0	300–1,500	
62.5/125/250 or 500	780	4.0–8.0	100–400	0.275
	850	3.0–7.0	100–400	
	1,300	1.0–4.0	200–1,000	
	1,550	1.0–4.0	150–500	
100/140/250 or 500	780	4.5–8.0	100–400	0.29
	850	3.5–7.0	100–400	
	1,300	1.5–5.0	100–400	
	1,550	1.5–5.0	10–300	
110/125/250 or 500	780	—	—	0.37
	850	15.0	17	
	1,300	—	—	
	1,550	—	—	
200/230/500	780	—	—	0.37
	850	12.0	17	
	1,300	—	—	
	1,550	—	—	

18.2.6.1 Glass Fibers

The most popular fibers are all-glass fibers, especially single-mode fibers. These fibers are widely used because of low attenuation rates and high information-carrying capacity. Single-mode fibers are less expensive than multimode fibers, but optoelectronic elements and connectors for single-mode systems are more expensive than those for multimode systems. The majority of single-mode fibers have a core diameter of 5–10 μm and a cladding diameter of 125 μm.

Most multimode telecommunications fibers are graded-index fibers with a cladding diameter of 125 μm. The typical core diameters with the 125 μm cladding are 50 μm, 62.5 μm, 85 μm, and 110 μm.

The 50/125-μm diameter fiber has a low numerical aperture of 0.2 but the highest bandwidth-distance product between MMF. The 62.5/125-μm fiber is the most popular for multimode transmission; its higher numerical aperture means that this fiber provides better light-coupling efficiency and is less sensitive to microbending losses. The large core diameter fibers, such as the 85/12, 110/125, and 100/140 μm fibers, have a good light-coupling ability, but have less bandwidth-distance product than fibers with small core diameters. Table 18.1 shows typical fiber optical losses, numerical apertures, and bandwidth-distance products. It should be noted that there are other fibers with larger core diameters, which find applications in fiber sensors and medicine.

18.2.6.2 Plastic-Clad Silica (PCS)

PCS consists of a step-index silica core surrounded by a soft plastic cladding of silicon rubber. This fiber combines the low attenuation of a glass core with a soft plastic cladding; however, this fiber needs a buffer coating to protect the soft cladding.

18.2.6.3 Hard-Clad Silica (HCS)

Hard-clad silica fiber includes a step-index silica core surrounded by a hard plastic cladding. This plastic cladding has important advantages compared to a glass cladding, providing a high strength of the fiber, small bending radius, and a high resistance to surface damage.

18.2.6.4 Plastic Optical Fibers (POF)

Plastic fibers have a step-index or graded-index core surrounded by a plastic clad. The core material of POF is normally made of acrylic resin, while the cladding is made of fluorinated polymer. The POF diameter is usually around 1 mm, which is many times larger than a glass fiber and the light-transmission core section accounts for 96% of the cross-sectional area. Compared with all glass fibers, POFs do not suffer from the problem of breakage. In Section 18.3.6, POFs are presented in more detail.

18.3 Special Fibers

18.3.1 Erbium-Doped Fibers

Incorporating rare-earth elements into glass gives the resulting material new optical properties that allow the material to perform amplification and generation of optical light. Doping can be done both for silica and for halide glasses.

Three commonly used rare-earth materials for silica fiber lasers are erbium, neodymium, and ytterbium. Erbium-doped fibers have been a key element in the transformation of modern optical communication systems. Erbium-doped fiber amplifiers and lasers operating at a wavelength of 1.55 μm have attracted the most attention because their amplification band coincides with the least-loss region of silica fibers used for telecommunication systems. In particular, erbium-doped fiber amplifiers (EDFAs) are used for the amplification of lightwave signals purely in the optical domain. They can be used as power amplifiers to boost transmitted power, as repeaters or in-line amplifiers to increase the transmission distance, or as preamplifiers to enhance receiver sensitivity. Figure 18.7 shows the basic configuration of an EDFA. The wavelength-division multiplexer (WDM) combines the light from the high-pump power

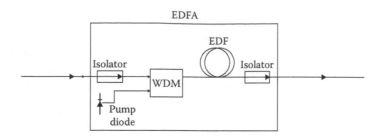

FIGURE 18.7 Basic configuration of an erbium-doped fiber amplifier: EDF = erbium-doped fiber, WDM = wavelength-division multiplexer.

TABLE 18.2 Parameters of an Erbium-Doped Fiber (3M)

Pumping Wavelength (nm)	Operating Wavelength (nm)	Core/cladding/jacket Diameter (μm)	Attenuation Maximum (dB/km)	Numerical Aperture
980–1480	1530–1560	5/125/245	15 (1200 nm)	0.28

laser diode (with wavelength of 980 nm or 1480 nm) and the signal to be amplified (in the wavelength region of 1530–1570 nm) into an Er-doped silica fiber. The optical isolators prevent any back reflections. To adjust the gain of EDFA, a part of the output signal is compared with the reference level. The produced control signal goes back to the pump diode to adjust the current.

The key element in an EDFA is a short-length (5–200 m) silica fiber doped with about 200 mole ppm or erbium, which corresponds to an erbium concentration of about 10^{19} ions/cm^3. The gain characteristics of EDFAs depend on the pumping scheme as well as on the various codopants (GeO_2, Al_2O_3, and P_2O_5) that are used to make the fiber core. Table 18.2 shows typical parameters of an erbium-doped fiber (available from 3 m). Efficient pumping may be obtained by using semiconductor lasers operating at 980 nm and 1480 nm, where the excited state absorption (ESA)—the excitation to higher levels than pump wavelength—is not present. A broadband gain of EDFAs permits amplification of multiple optical channels in the bandwidth, which ranges from 1 to 5 THz (~40 nm).

One of the advantages of EDFA amplifiers over electronic amplifiers is that EDFAs can amplify many signals at different wavelengths, which is used to expand the capacity of fiber-optic communication systems. Since the optical signals are directly amplified without conversion to electrical signals, the amplifier will work efficiently even at higher bit rates. This is in contrast to electronic repeaters, which work only at the fixed bit rate. A small signal gain of 40 dB can be achieved in EDFAs, while the noise, added by the amplifier, is close to the lowest level (3–4 dB). It should also be noted that the gain is polarization-insensitive, providing equal amplification for all polarization states of a signal. In long transmission systems, EDFAs are used to periodically restore the power level, after it has decreased due to attenuation in the fiber.

18.3.2 Powerful Double-Clad Fibers

The power levels generated by conventional fiber lasers, pumped with diode sources that couple light directly into the single-mode core, are relatively low, and currently limited to fractions of a watt. Double-clad fibers offer a solution to increasing the amount of pump power in a fiber laser. Double-clad fibers comprise a rare-earth-doped single-mode core within a multimode waveguide, which enables light pumping from a low-brightness multimode pump source, such as a diode array, to be efficiently absorbed by a single-mode core. The geometry of a high-power Yb^{3+} cladding-pumped fiber laser is shown in Figure 18.8. The inner rectangular silica cladding with refractive index $n = 146$ acts as a waveguide for

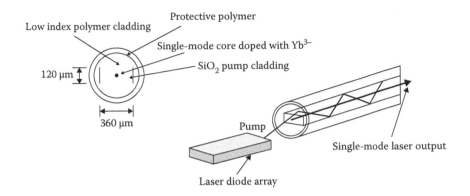

FIGURE 18.8 Double-clad fiber structure.

FIGURE 18.9 Cross-sectional view of various double-clad fibers.

the pump light. In Figure 18.8, the silica rectangular waveguiding region has dimensions 360×120 μm, and is referred to as the pump cladding. The non-circular shape of the pump cladding eliminates helical rays, which have poor overlap with the core. The pump-cladding region is typically surrounded by a low-index polymer ($n = 1.39$), which acts as a cladding for the inner cladding, providing a high numerical aperture (NA = 0.48) for the rectangular waveguide. This permits an efficient light coupling from a diode laser into the inner cladding. A second protective polymer surrounds the low-index polymer.

The pumping light is absorbed when optical rays cross the rare-earth-doped single-mode core. Then, the excited ions emit the light at lower frequency, which is amplified through stimulated emission in the single-mode core with much lower numerical aperture. Thus, pumping double-clad fiber lasers with low-brightness beams from a pump laser diode array may result in enhancing of the brightness by a factor in excess of 1000. This increased brightness is the significant advantage of double-clad fibers over both diodes and other types of fiber lasers. The necessary feedback elements for laser operation of the doped core may be formed by Bragg gratings directly written into the doped single-mode core, or by using mirrors deposited on or attached to the fiber ends. Slope efficiency approaches 70% and output power is limited only by the pump power.

Different geometry has been developed for double-clad fiber lasers, some of which are shown in Figure 18.9. The main purpose of the proposed design is to break a circular symmetry and to provide an efficient pump absorption in the core. Although the core can be doped with various rare-earth ions, such as Er^{3+}, Nd^{3+}, Tm^{3+}, Ho^{3+}, the highest output power has been obtained from Yb^{3+}-doped single-mode core. Using tens of meters of Yb-doped double-clad fiber, continuous wave (CW) power of 200 W has been achieved in a single-mode beam at 1064 nm (IRE-Polus Group).

Potential applications for high-power lasers are in medicine, laser cutting, pumping other lasers, and in satellite-to-satellite communications links. Polaroid developed a high-power double-clad Yb^{3+}-doped fiber laser for a printing system. High-power lasers are also of great interest in telecommunication networks, since they can provide the necessary pump power, for example, for the practical implementation of cascaded Raman lasers for optical amplification. Laboratory experiments have demonstrated that the use of Raman gain devices can quadruple a communication system capacity. Furthermore, in medical applications high-power 2 μm fiber sources may be useful in microsurgical applications. Also, medical spectroscopic applications in areas, such as dermatology and diagnostic imaging, should benefit from double-clad fiber lasers.

18.3.3 Infrared Optical Fibers

In 1979, optical fibers made from silica and silica-based glasses reached their limit of transparency. Transmission losses as low as 0.2 dB/km have been obtained. This value almost corresponds to the ultimate inherent (intrinsic) loss value for a silica-based fiber. However, the demand for further improvements in transmission capacity requires the realization of ultra-low loss optical fibers with losses far below those of the silica-based optical fibers. Moreover, fibers with low loss in mid-infrared are required in medicine and industrial applications. The solution is in non-silica infrared fiber materials, which offer the possibility of an ultra-low loss of less than 10^{-2} dB/km.

In principle, infrared optical fibers can be classified into two groups: dielectric optical fibers, based on the total internal reflection in solid cores; hollow waveguides, whose core regions are hollow. Optical materials studied to date for infrared fibers are halides, chalcogenides, and heavy-metal oxides.

18.3.3.1 Oxide Glass Fibers

Infrared oxide glass fibers are mainly based on heavy-metal oxides, such as GeO_2, GeO_2–Sb_2O_3, and TeO_2. The minimum losses typically occur at wavelengths of around 2–3 μm. In metal glasses, infrared absorption due to lattice vibration (Ge–O) can be shifted toward a longer wavelength, since their constituent metals (such as Ge) are heavier than Si in SiO_2 glass. As a result, an ultra-low loss in the infrared region is expected. Theoretical predictions give a value of 0.1 dB/km for minimum intrinsic loss of these fibers. However, the experimentally obtained values are more than one order of magnitude larger: in particular, losses of 4 dB/km have been reported in GeO_2–Sb_2O_3.

18.3.3.2 Fluoride Glass Fibers

Fluoride glasses are the most promising candidates for the ultra-low loss optical fibers in long-distance optical communication. The initial system discovered by Poulain and coworkers in France in 1974 were fluorozirconates, where ZrF_4 was the primary constituent (>50 mol%), BaF_2 was the principal modifier (~ 30 mol%) and various metal fluorides, such as ThF_4 and LaF_3, were tertiary constituents. Depending on the composition, fluoride glasses have various desirable optical characteristics, such as a broad transparency range extending from the mid-infrared (~ 7 μm) to near ultraviolet (0.3 μm), low refractive index and dispersion, low Rayleigh scattering, and the potential of ultra-low absorption and ultra-low thermal distortion. Recent progress in reducing transmission losses of fluoride fibers to less than 1 dB/km strongly encourages the realization of ultra-low loss fibers of 0.01 dB/km, or less. Intrinsic losses in fluoride glasses are estimated to be 0.01 dB/km at 2–4 μm. In particular, fibers based on ZrF_4–BaF_2 have intrinsic losses in the vicinity of 0.01 dB/km at around 2.5 μm.

The visible refractive index of most fluoride glasses lies in the range of 1.47–1.53, which is comparable to silicates but much lower than chalcogenides and can be tailored by varying composition. The zero of the material dispersion for fluorozirconates occurs in the region of 1.6–1.7 μm, as opposed to the loss minimum in the 3–4 μm regime. Nevertheless, the magnitude of the material dispersion is small through an extended range of wavelengths, including the minimum loss region, so that respectable values of pulse broadening, on the order of several picoseconds per angstrom-km are calculated for typical fluoride glasses near their loss minimum.

18.3.3.3 Chalcogenide Glass Fibers

Chalcogenides are compounds composed of chalcogen elements, that is, S, Se, and Te, and elements, such as Zn, Cd, and Pb. They are available in a stable vitreous state and have a wide optical transmission range. Chalcogenide glasses are advantageous because they exhibit no increase in scattering loss due to the plastic deformation that usually occurs in crystalline fibers. Chalcogenide glasses are divided into sulfides, selenides, and tellurides.

Sulfide glasses are divided into arsenic-sulfur and germanium-sulfur glasses. The optical transmission range of As-S and Ge-S glasses is almost the same, giving a broad transmission in the mid-infrared region. Transmission loss of 0.035 dB/m has been achieved for an As-S glass without a cladding, while in an As-S glass fiber with teflon FEP cladding, transmission loss of 0.15 dB/m has been obtained.

Various selenide glasses have been studied mainly in order to achieve lower loss at the wavelengths of 5.4 µm (CO laser) and 10.6 µm (CO_2 laser). Selenide glasses are divided into As-Se, As-Ge-Se (As rich) glasses, Ge-Se, Ge-As-Se (Ge rich) glasses, La-Ga-Ge-Se glasses and Ge-Sb-Se glasses. Selenide glasses have a wide transparency region compared with sulfide glasses. They exhibit a stable vitreous state, resulting in flexibility of the fiber and, thus, these fibers are the candidates for infrared laser power transmission and wide-bandwidth infrared light transmission, such as in radiometric thermometers.

Although the Se-based chalcogenide glasses have a wide transparency range, their losses at the wavelength of 10.6 µm for CO_2 laser power transmission are still higher than the 1 dB/m required for practical use. To lower transmission loss caused by lattice vibration, the atoms must be introduced to shift the infrared absorption edge toward a longer wavelength. Transmission losses of 1.5 dB/m at 10.6 µm in $Ge_{22}Se_{20}Te_{58}$ telluride glass have been reported.

18.3.3.4 Polycrystalline Fibers

The technology of fabrication of polycrystalline fibers by extrusion has allowed the preparation of TlBr-TlI (KRS-5) fibers with total optical loss of 120–350 dB/km. This crystalline material is known to transmit from 0.6 to 40 µm, and has a theoretical transmittance loss of ~10^{-3} dB/km at 10.6 µm due to low intrinsic scattering losses from Rayleigh and Brillouin mechanisms and a multiphonon edge that is shifted to longer wavelengths. Fiber-optic waveguides made from KRS-5, however, have optical losses at 10.6 µm that are orders of magnitude higher than those predicted by theory. In addition, the optical loss increases with time (~6 dB/m per year) via the mechanism of water incorporation. Transmission of a 138 W (70 kW/cm²) beam of a CO_2 laser through a 0.5 mm diameter and 1.5 m length fiber made from KRS-5 has been achieved with a transmitted power of 93%.

On the other hand, polycrystalline silver halide fibers, formed by extrusion of mixed crystals of $AgCl_xBr_{1-x}$ ($0 \leq x \leq 1$), are transparent to 10.6 µm CO_2 laser radiation. The total loss coefficient at this wavelength is in the range of 0.1–1 dB/m, and the fiber transmission is independent of power levels up to at least 50 W total power input, which is an input power density of about 10^4 W/cm². In addition, the fibers are nontoxic, flexible and insoluble in water. They have been successfully used in many applications such as directing high-power CO_2 laser radiation for surgical and industrial applications, IR spectroscopy, and radiometry.

18.3.3.5 Single-Crystalline Fibers

Single crystal fibers have the potential of eliminating all the deleterious effects of extruded fiber, and, therefore, of having a much lower loss than polycrystalline fibers. This material possesses a wide transparent wavelength region from visible to far-infrared, so that it is possible to transmit both visible and infrared light. Materials for single-crystalline fibers are almost the same as those for polycrystalline fibers: TlBr-TlI, AgBr, KCl, CsBr, and CsI have been mainly studied. The lowest losses at the CO_2 laser wavelength of 10.6 µm have been attained in thallium halide fibers (0.3–0.4 dB/m). Also, a transmission loss of 0.3 dB/m and maximum transmitted power of 47 W were obtained at a 10.6 µm wavelength by using a 1 mm diameter CsBr fiber.

18.3.4 Metallic and Dielectric Hollow Waveguides

In the mid-infrared region (10.6 µm and 5 µm), hollow-core waveguides have advantages over solid-core fibers in high power transmission of CO_2 and CO lasers. Lower loss and higher laser-induced-damage thresholds are attainable because of the highly transparent airy core. Various types of hollow infrared

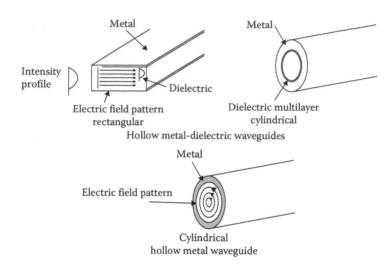

FIGURE 18.10   Hollow metal waveguides.

waveguides have been fabricated, including cylindrical and rectangular metallic, cylindrical dielectric, and rectangular metallic with inner dielectric coatings (Figure 18.10).

In cylindrical hollow waveguides, transverse electric modes TE_{0n} propagate with the least attenuation, while hollow rectangular waveguides propagate TE_{m0} and TM_{m0} modes most efficiently because these modes interact less with the waveguide inner walls. Metal material in hollow waveguides provides high infrared reflectivity. Transmission of more than 95%/m in straight rectangular hollow metal waveguides has been reported. Transmission losses of dielectric-coated metallic waveguides are expected to be low for the HE_{11} mode. Theory predicts that the power loss of the HE_{11} mode is 3 2 × 10^{-2} dB/m. The total transmission loss of fabricated waveguides is always below 0.5 dB/m, in contrast to 2.5 dB/m for a metal hollow waveguide with no dielectric layer. Metallic hollow fibers are not as flexible as conventional optical fibers, and so they are rather restricted.

In dielectric hollow-core fibers, the glass (e.g., oxide glasses, such as SiO_2 and GeO_2) acts as a cladding to the fiber core, which is air. To provide a condition of total internal reflection, and thus high transmission, the glass cladding must retain a refractive index of less than unity. This means selecting a dielectric with the anomalous dispersion of refractive index at an operating wavelength. Many inorganic materials, either in vitreous (i.e., amorphous) or crystalline form exhibit strong anomalous dispersion in the mid-infrared, providing their refractive indices are less than unity at certain frequencies. However, it should be noted that a refractive index of below unity at the middle region originates from the strong absorption due to lattice vibration and results in an increase of transmission losses and limits the length of the waveguide. A transmission loss of 0.1 dB/m at 10.6 µm is expected in a fiber with a cladding composition of 80 mol% GeO_2–10 mol% ZnO–10 mol% K_2O and a bore size of 1 mm. The experimental loss is about 20 times larger than the theoretical one. It is expected that lower transmission losses can be obtained by smoothing the inner surface.

For a given bore size, for example 1 mm, metal circular waveguides will have the highest attenuation, and this attenuation decreases significantly for those materials exhibiting anomalous dispersion. Hollow dielectric fibers with polycrystalline hexagonal GeO_2 cladding can have a loss below 0.5 dB/m, which is generally regarded as an acceptable loss for the CO_2 laser power delivery. High-power transmission hollow fibers are suitable for applications in laser surgery, cutting, welding, and heat treatment.

18.3.4.1 Liquid-Core Fibers

Liquid-filled hollow fibers contain liquid materials in the hollow cores. These fibers are advantageous because there are no stress effects leading to birefringence, and wall imperfections and scattering effects are negligible. The transmission losses in the transparent wavelength regions depend on the liquids used.

Liquid bromine has been inserted into a Teflon tube with attenuation of 0.5 dB/m at 9.6 μm. For silica glass tubes with a diameter of 125 μm filled with bromobenzene, the transmission loss of 0.14 dB/km at 0.63 μm has been reported. Although these fibers may be useful for the infrared region, fabrication and toxicity are still serious difficulties.

18.3.5 ZBLAN Fiber Based Up-Conversion Lasers

Frequency up-conversion is a term that is associated with a variety of processes whereby the gain medium, a trivalent rare-earth ion, in a crystal or glass host, absorbs two or more infrared pump photons to populate high-lying electronic states. Visible light is then produced by one-photon transitions to low-lying electronic levels. Up-conversion lasers appear to be attractive candidates for compact, efficient visible laser sources for applications in optical data storage, full-color displays, color printing, and biomedical instrumentation because of the relative simplicity (the gain and frequency conversion material are one and the same) of these devices.

In 1990, several CW, room temperature, up-conversion lasers in the ZBLAN single-mode fibers were demonstrated. Many up-conversion lasers based on doped optical fibers have produced output wavelengths ranging from the near infrared to the ultraviolet, but the greatest advantage offered by the optical fiber geometry is that room temperature operation is much easier to obtain than in bulk media. In addition to their simplicity and compactness, up-conversion fiber lasers are efficient and tunable. Slope efficiencies (pump power to output power conversion) of up to 50% have been obtained in the two-photon 550 nm Er^{3+} ZBLAN fiber laser and 32% in the three-photon pumped Tm^{3+} ZBLAN fluoride laser. Table 18.3 shows a summary of up-conversion fiber lasers that demonstrated generation at room temperature in rare-earth-doped fluorozirconate glass. The key to up-conversion laser operation in single-mode optical fibers has been the use of low phonon energy ($\hbar\omega < 660$ cm^{-1}) fluorozirconate glasses as hosts for the rare-earth ions.

Because the fluorozirconate host is a disordered medium, rare-earth-doped fibers fabricated from these glasses exhibit absorption and emission profiles that are broad compared with those characteristic of a crystalline host. The broad emission profile has a negative impact on the stimulated emission cross-section, but is more than compensated for by the advantages of the fiber laser geometry, the high pump intensities, and the maintenance of this intensity over the entire device length. The broad emission linewidths permit tuning continuously of the wavelength of generation over 10 nm.

18.3.6 Plastic Optical Fibers

Copper and glass have been the traditional solutions to data communication and they are well suited to specific applications. However, for high-speed data transmission, copper is unsuitable because of its susceptibility to interference. On the other hand, the small diameter and fragility of glass elevates a cost

TABLE 18.3   Rare-Earth-Doped ZBLAN Up-Conversion Fiber Lasers

Rare-Earth Ions	Pump Wavelength (nm)	Laser Wavelength (nm)	Slope Efficiency (%)
Er	801	544, 546	15
	970		>40
Tm	1064, 645	455	1.5
	1112, 1116, 1123	480, 650	32 (480 nm)
	1114–1137	480	13
Ho	643–652	547.6–594.5	36
Nd	582–596	381, 412	0.5 (412 nm)
Pr	1010, 835	491, 520, 605, 635	12 (491 nm)
Pr/Yb	780–885	491, 520, 605, 635	3 (491 nm)
			52 (635 nm)

of installation in fast-growing local-area networks. Plastic optical fibers (POFs) provide an alternative and fill some of the void between copper and glass.

POF shares many of the advantages and characteristics of glass fibers. The core material of POF is normally made of acrylic resin, while the cladding is made of fluorinated polymer. POF diameter is usually around 1 mm, which is many times larger than a glass fiber and the light-transmission core section accounts for 96% of the cross-sectional area. The first POF available had a step-index profile, in which high-speed transmission was difficult to achieve. In 1995, Mitsubishi Rayon announced the first graded-index plastic optical fiber (GIPOF) with transmission speeds in excess of 1 Gb/s. GIPOF in conjunction with high-speed 650 nm light-emitting diode (LED) provides an ultimate solution for high-speed, short- or moderate distance, low-cost, electromagnetic-interference-free data links demanded by desktop local-area network specifications, such as asynchronous transfer-mode-local area network (ATM-LAN) and fast ethernet.

Because of its bandwidth capability, POF is a much faster medium than copper, allowing for multitasking and multimedia applications that today's copper supports at slower, less productive speeds. The advantage of plastic optical fiber over glass fibers is in its low total-system cost. In multimode glass fiber, a precision technology is needed to couple the light effectively into a 62.5 μm core. The larger diameter of plastic fiber allows relaxation of connector tolerances without sacrificing optical coupling efficiency, which simplifies the connector design. In addition, the plastic fiber and large core diameter permit termination procedures other than polishing, which requires an expensive tool. For quick and easy termination of plastic fiber, a handheld hot-plate terminator is available, so that even a worker with no installation training can terminate and assemble links within a minute. Because POF transmission losses are higher than for silica fiber, it is not suitable for long distance, but is, however, suitable for home and office applications. Typical losses are of the order of 140–160 dB/km.

18.3.6.1 Applications of POF

The largest application of POF has been in digital audio interfaces for short-distance (5 m), low-speed communication between amplifiers and built-in digital-to-analog converter and digital audio appliances, such as CD/MD/DAT players and BS tuners. The noise-immune nature of POF contributes to creating sound of high quality and low jitter.

Because of its flexibility and immunity to factory floor noise interference, rugged and robust POF communication links have been successfully demonstrated in a tough industrial manufacturing environment.

Lightweight and durable POF networks could link the sophisticated systems and sensors used in automobiles, which would increase performance and overall efficiency. Also, POF could be used to incorporate video, minicomputers, navigational equipment, and fax machines into a vehicle. Another short-haul application for POF is home networking, where appliances, entertainment and security systems, and computers are linked to create a smart home.

18.3.7 Fiber Bundles

A fiber bundle is made up of many fibers that are assembled together. In a flexible bundle, the separate fibers have their ends fixed and the rest of their length unattached to each other. On the other hand, in a rigid or fused bundle, the fibers are melted together into a single rod. Fused bundles have a lower cost than flexible bundles, but because of their rigidity they are unsuitable for some applications.

The fiber cores of the bundle must occupy as much of the surface area as possible, in order to minimize the losses in their claddings. Thus, bundled fibers must have thin claddings to maximize the packing fraction, that is, the portion of the surface occupied by fiber cores. The optics of the fiber bundles is the same as that of a single fiber. In a good approximation, it may be assumed that a single light ray may represent the light entering the input end of the bundle. If a light ray enters the fiber at an angle θ within the acceptance angle of the fiber, it will emerge in a ring of angles centered on θ, as shown in Figure 18.11. It should be pointed out that in fiber bundles formed by step-index fibers with constant-diameter cores, the light entering the fiber emerges at roughly the same angle it entered.

FIGURE 18.11   The light rays emerge from the fiber in a diverging ring.

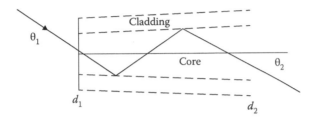

FIGURE 18.12   Light propagation through a tapered fiber.

If light focusing (and magnification and demagnification of objects) is needed, a tapered fiber may be used. Figure 18.12 shows a schematic representation of a tapered fiber. If a ray entering a fiber at an angle θ_1 meets criteria for total internal reflection, it is confined in the core. However, it meets the core–cladding boundary at different angles on each bounce so it emerges at a different angle θ_2,

$$d_1 \sin\theta_1 = d_2 \sin\theta_2, \tag{18.16}$$

where d_1 is the input core diameter and d_2 is the output core diameter. The same relationship holds for the fiber's outer diameter as long as the core and outer diameter change by the same factor d_2/d_1.

Bundles of step-index fibers can be used for imaging. In this case, each fiber core of the bundle will carry some segment of the image, so that the complete image is formed by different segments of the fibers of the bundle. As long as the fibers are aligned in the same way on both ends, the bundle will form an image of an object. Typical losses of fiber bundles are around 1 dB/m. Since fiber bundles are required in short-distance applications, such a large loss is not a limitation.

The majority of bundles are made from step-index multimode fibers, which are easy to make and have large numerical apertures. The higher NA of these fibers (* 0.4) gives large acceptance angles, which in turn decreases coupling losses.

The simplest application of optical fibers of any type is light piping, that is, transmission of light from one place to another. A flexible bundle of optical fibers, for example, can efficiently concentrate light in a small area or deliver light around corners to places it could not otherwise reach. Application of fiber bundles includes illumination in various medical instruments, including endoscopes, in which they illuminate areas inside the body.

Fiber bundles also may be used for beam-shaping by changing the cross-section of the light beam. It is possible to array the fibers in one of the ends of the bundle to form a circle or a line as shown in Figure 18.13. This may be important if the fibers are being used for illumination in medical instruments, where special arrangements of the output fibers help in the design and may result in more uniform illumination of the field of view.

The other application of optical fiber bundles is for beamsplitting or beam combining. For example, a Y-guide is formed by two fiber bundles combined into one bundle near a sample, while they are still separated at the other ends (Figure 18.14). If the NA of the individual fibers is large, the light-collection efficiency (the ratio between the input and output energy) of the Y-guide is fairly high.

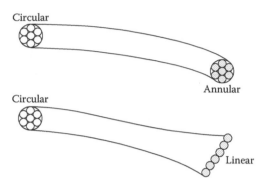

FIGURE 18.13   Beam-shaping by non-ordered bundles of fibers.

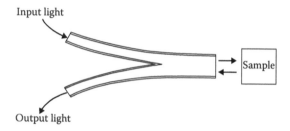

FIGURE 18.14   Y-fiber bundles.

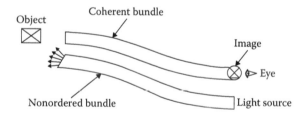

FIGURE 18.15   Illumination and image transmission through a non-ordered and coherent bundle of fibers.

In image transmission, the fibers must maintain identical relative positions on input and output faces. These are called ordered or coherent bundles. Images can be viewed through coherent fiber bundles by placing the bundle's input end close or directly on the object, or by projecting an image onto the input end. Light from the object or the image is transmitted along the bundle, and the input image is reproduced on the output face. Coherent fiber bundles are very valuable in probing otherwise inaccessible areas, such as inside machinery or inside the human body.

The easiest way to make coherent fiber bundles is to fuse fibers together throughout the length of the bundle. However, such fiber bundles are not usable in many situations because of the lack of flexibility. In flexible coherent bundles, the fibers bond together at the two ends, so they maintain their relative alignment, but they are free to move in the middle. Individual fibers, unlike fused bundles, are more flexible. The imaging transmission bundles are the basic building blocks of fiberscopes and endoscopes. The purpose of a coherent fiber bundle is to transmit the full range of an illuminated object. Normally, both the coherent and the non-ordered bundles are incorporated into an endoscope. A non-ordered bundle (Figure 18.15) illuminates the object inside the body. The imaging bundle must then transmit the color image of the object with adequate resolution.

18.4 Fiber Optic Components

18.4.1 Optical Fiber Couplers

The optical directional fiber coupler is a waveguide equivalent of a bulk beamsplitter and is one of the basic in-line fiber components. When two (or more) fiber cores are placed sufficiently close to each other, the evanescent tail of an optical field in one fiber extends to a neighboring core and induces an electric polarization in the second core. In its turn, the polarization generates an optical field in the second core, which also couples back to the core of the first fiber. Thus, the modes of different fibers become coupled through their evanescent fields, resulting in a periodical power transfer from one fiber to the other. If the propagation constants of the modes of the individual fibers are equal, then this power exchange is complete. If their propagation constants are different, then exchange of power between the fibers is still periodic, but incomplete.

The basic mechanism of a directional coupler can be understood on an example of a coupler formed by a pair of identical symmetric single-mode waveguides. The system of two coupled waveguides can be considered as a single waveguide with two cores. Such a system supports two modes, the fundamental being the symmetric mode and the first excited being the antisymmetric mode. These two modes have different propagation constants. Light launched in one waveguide excites a linear combination of the symmetric and antisymmetric modes (Figure 18.16) in both cores. The interference of two modes at the input is constructive in the first waveguide and is destructive in the second waveguide, resulting in the absence of the field in the last one. In the coupling region, the two modes propagate at different velocities, acquiring a phase difference. When the phase difference becomes π, then the superposition of these two modal fields will result in a destructive interference in the first waveguide and constructive in the second. Further propagation over an equal length will result in the phase difference of 2π, leading to a power transfer back to the first waveguide. Thus, the optical power exchanges periodically between the two waveguides. By an appropriate choice of the coupler length, one can fabricate couplers with an arbitrary splitting ratio.

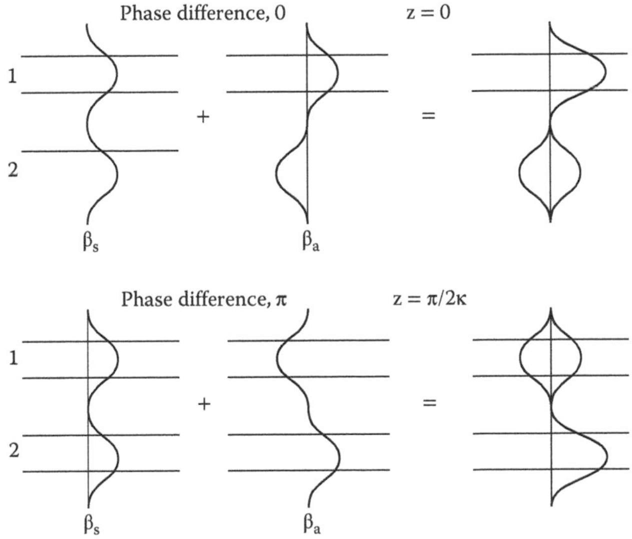

FIGURE 18.16 Symmetric and antisymmetric modes.

A power transfer ratio depends on the core spacing and interaction length. If $P_1(0)$ is the power launched into fiber 1 at $z = 0$, then the transmitted power $P_1(z)$ and the coupled power $P_2(z)$ for two non-identical single-mode fibers are given by

$$\frac{P_1(z)}{P_1(0)} = 1 - \frac{k^2}{\gamma^2}\sin^2\gamma z,$$

$$\frac{P_2(z)}{p_1(0)} = 1 - \frac{k^2}{\gamma^2}\sin^2\gamma z, \tag{18.17}$$

where

$$\gamma^2 = k^2 + \tfrac{1}{4}(\Delta\beta) \tag{18.18}$$

and $\Delta\beta = \beta_1 - \beta_2$ is the difference of propagation constants of the first and the second fiber, respectively, also called as a phase mismatch; k is the coupling coefficient, which depends on the fiber parameters, the core separation, and the wavelength of operation. If the two fibers are separated by a distance much greater than the mode size, then there would be no interaction between the two fibers.

In the case of two identical fibers, the phase mismatch is equal to zero and the power oscillates between two fibers. The coupling coefficient is given by

$$k(d) = \frac{\lambda_0}{2\pi n_1}\frac{U^2}{a^2V^2}\frac{K_0(Wd/a)}{K_1^2(W)}, \tag{18.19}$$

where λ_0 is the free space wavelength, n_1 and n_2 are the core and cladding refractive indices, respectively, a is the fiber core radius, d is the separation between the fiber axes, K_v is the modified Bessel function of order v, $k_0 = 2\pi/\lambda_0$, $n_e = \beta/k_0$, n_e is the mode effective index. Knowing a coupling coefficient, one can easily calculate the corresponding coupling length.

$$U = k_0 a\left(n_1^2 - n_e^2\right)^{1/2}$$

$$W = k_0 a\left(n_e^2 - n_2^2\right)^{1/2} \tag{18.20}$$

$$V = k_0 a\left(n_1^2 - n_2^2\right)^{1/2}$$

18.4.1.1 Parameters of a Coupler

The 2 × 2 coupler is shown schematically in Figure 18.17. For an input power P_i, transmitted power P_t, coupled power P_c, and back-coupled power P_r, we can determine the main characteristics of the coupler as follows:

$$\text{Power-splitting ratio } R(\%) = \frac{P_t}{P_c}\times 100;$$

$$\text{Excess loss } L_e\,(\text{dB}) = 10\log\left[\frac{P_i}{P_c + P_t}\right];$$

$$\text{Insertion loss } L_i\,(\text{dB}) = 10\log\left[\frac{P_i}{P_t}\right]; \tag{18.21}$$

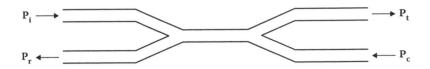

FIGURE 18.17 A 2 × 2 coupler.

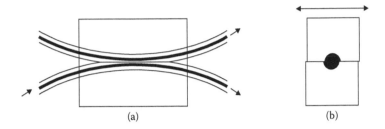

FIGURE 18.18 Polished fiber coupler. (a) side view and (b) transverse view.

$$\text{Directivity } D\,(\text{dB}) = 10 \log \left[\frac{P_r}{P_i} \right].$$

The popular power-splitting ratios between the output ports are 50% : 50%, 90% : 10%, 95 : 5%, and 99% : 1%; however, almost any value can be achieved on a custom basis. Excess loss in a fiber coupler is the intrinsic loss of the coupler when not all input power emerges from the operation ports of the device. Insertion loss is the loss of power that results from inserting a component into a previously continuous path. Couplers should have low excess loss and high directivity. Commercially available couplers exhibit excess loss ≤ 0.1 dB, and directivity of better than −55 dB.

A coupler is identified by the number of input and output ports. In the $N \times M$ coupler, N is the number of input fibers and M is the number of output fibers. The simplest couplers are fiber-optic splitters. These devices have at least three ports but may have more than 32 for more complex devices. In a three-port device (tee coupler), one fiber is called the common fiber, while the other two fibers are called input and output ports. A common application is to inject light into the common port and to split it into two independent output ports.

18.4.1.2 Fabrication of Fiber Couplers

Practical single-mode fibers have a thick cladding to isolate the light propagating in the core. Hence, to place two cores close to each other, it is necessary to remove a major portion of the cladding. Two methods have been developed for the fabrication process. The first one consists of polishing the cladding on one side of the core of both fibers, and then bringing the cores in close proximity. A technique for polishing the cladding away consists of fixing the fibers inside grooves in glass blocks, and polishing the whole block to remove the cladding on one side of the core (Figure 18.18a). The two blocks are then brought into contact (Figure 18.18b). Usually, the space between the blocks is filled with an index-matching liquid. By laterally moving one block with respect to the other, one can change the core separation, resulting in changes of the coupling constant k. This, in turn, changes the power-splitting ratio. Such couplers, called tunable, permit smooth tuning of the characteristics of a coupler.

Polished couplers exhibit excellent directivity, better than −60 dB. Their splitting ratio can be continuously tuned. The insertion losses of such couplers are very low (≈ 0.01 dB). One of the important characteristics of couplers is sensitivity to the polarization state of the input light. The polished couplers have the advantage of being low polarization sensitive. The variation in splitting ratio for arbitrary input polarization states can be less than 0.5%. The performance of such couplers can be affected by temperature variations because of the temperature dependence of the refractive index of the index-matching liquid.

Fabrication of polished fiber couplers is a time-consuming operation. Hence, more popular couplers today are fused directional couplers. Such couplers are easier to fabricate, and the fabrication process can be automated. In this type of coupler, two or more fibers with removed protecting coatings are twisted together and melted in a flame. After the pulling, the fiber cores approach each other, resulting in overlapping of the evanescent fields of the fiber modes. The coupling ratio can be monitored online as the fibers are drawn. Fused couplers exhibit low excess loss (typically less than 0.1 dB) and directivity better than –55 dB.

18.4.2 Wavelength-Division Multiplexers

A more complex coupler is the wavelength-division multiplexer. A WDM is a passive device that allows two or more different wavelengths of light to be split into multiple fibers or combined into one fiber. Let us consider a directional coupler made of two identical fibers with coupling coefficients k_1 and k_2 at wavelengths λ_1 and λ_2, respectively. As was shown above, the optical power at each wavelength exchanges periodically between the two waveguides. Since a coupling coefficient depends on wavelength, the period of oscillations at each wavelength is different. By an appropriate choice of the coupler's length L, the two conditions

$$K_1 L = m\pi, \qquad k_2 L = \left(m - \tfrac{1}{2}\right)\pi \tag{18.22}$$

can be satisfied simultaneously. In this case, if the total optical power at wavelength λ_1 and λ_2 is launched in port 1 (Figure 18.19), then the WDM will sort radiation at λ_1 and λ_2 between ports 3 and 4, respectively.

Two important characteristics of WDM device are cross-talk and channel separation. Cross-talk refers to how well the demultiplexed channels are separated. Optical radiation at each wavelength should appear only at its intended port and not at any other output port. Channel separation describes how well a coupler can distinguish wavelengths. Fused WDMs are more appropriate in applications where the wavelengths must be widely separated, for example in commercially available 980/1550 nm single-mode WDMs. Such a WDM exhibits an excess loss of 0.3 dB, an insertion loss of 0.5 dB, and a cross-talk better than 20 dB.

The communication WDM often needs a channel separation of about 1 nm. A sharp cutoff slope of the channel transmission characteristics is also required to achieve interchannel isolation (cross-talk) of 30 dB. The interference-filter-based WDMs are more suitable for this application. Figure 18.20 shows a multiplexer and demultiplexer using graded-index rod lenses and interference filters. The filter is designed so that it passes radiation at wavelength λ_2 and reflects at λ_1. The GRIN-rod lens collects the transmitted and reflected radiation into fibers.

18.4.3 Switches

Fiber-optical switches selectively reroute optical signals between different fiber paths. The performance of a switch is characterized by an insertion loss, cross-talk (back reflection), and speed. Figure 18.21

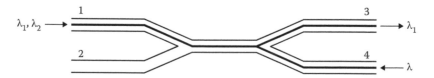

FIGURE 18.19   Wavelength-division multiplexer (WDM).

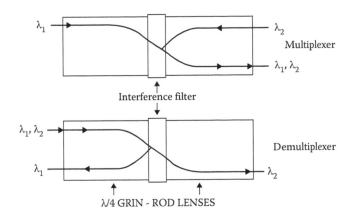

FIGURE 18.20 Interference-filter-based wavelength-division multiplexer (WDM).

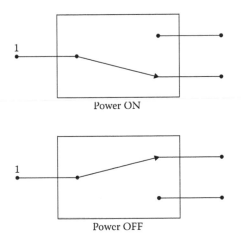

FIGURE 18.21 Typical 1 × 2 switch.

shows a typical 1 × 2 switch, which can have an output with two positions, "on" (port 2) and "off" (port 3). For such a switch, the insertion loss can be determined as

$$\text{Insertion loss} = -10\log\frac{P_2}{P_1};\tag{18.23}$$

the cross-talk determines the isolation between the input and the "off" port:

$$\text{Cross-talk} = -10\log\frac{P_3}{P_1},\tag{18.24}$$

where P_1, P_2, and P_3 are the input power, power at the output "on," and power at the output "off," respectively. Typical switches have low insertion loss (0.5 dB), and low cross-talk (55 dB). The speed of the switch depends on mechanisms involved in switching. Switches can be classified as optomechanical, electronic, and photonic (or optical) switches. Optomechanical switches include a moving optical element, such as a fiber, a prism, or a lens assembly in the path of a beam to deflect the light. The speed of

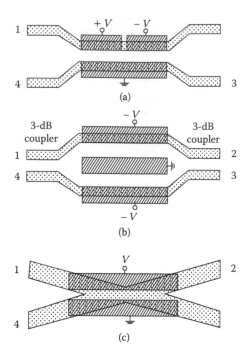

FIGURE 18.22   Photonic switching configurations: (a) directional couplers; (b) Mach–Zehnder interferometer; and (c) intersecting waveguides.

operation in this case is limited by millisecond speeds, so the optomechanical switches are not suitable for fast switching.

Electronic switches use an electronic-to-optical conversion technique, which can be fast enough for communication systems. However, the complexity of electronic-to-optical conversion limits the applications of electronic switching.

Photonic switching uses an integrated optic technology to operate in high-speed regimes. Electro-optic and acousto-optic phenomena are usually used to actuate the switching. The most advanced electro-optic waveguide technology utilizes $LiNbO_3$ crystals. This material provides necessary low insertion loss, high switching speed, and large bandwidth.

The waveguide configurations based on $LiNbO_3$ can be classified into directional couplers (Figure 18.22a), Mach–Zehnder interferometer (Figure 18.22b), and intersecting waveguides (Figure 18.22c). The directional coupler (Figure 18.22a) consists of a pair of optical waveguides placed in close proximity. Light propagating in one waveguide gets coupled to the second waveguide through an evanescent field. The coupling coefficient depends on refractive indices of the waveguides. By placing electrodes over the waveguides and varying applied voltage, the coupling coefficient can be efficiently tuned over a relatively large range. The length of the coupler is chosen in such a way that when no voltage is applied, the switch is in the cross state (the input and output ports are from different waveguides). With applied voltage, the switch changes to the bar state when the input and output ports are on the same waveguide. Splitting the electrode into two sections, one can tune both the cross and the bar states.

The Mach–Zehnder interferometer (Figure 18.22b) consists of a pair of 3 dB couplers connected by two waveguides. With no voltage applied to the electrodes, the optical paths of the two arms are equal, so there is no phase shift between light in the waveguides. In this case, the light enters in port 1 and goes out through port 3. By applying a voltage, a π phase difference is introduced between light in two arms, resulting in switching of optical power from port 3 to port 2. These switches are typically 15–20 mm long and require a phase shift voltage of 3–5 V. High cross-talk levels of 8–20 dB is a disadvantage of these switches.

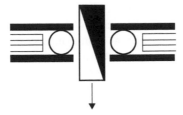

FIGURE 18.23 Attenuator.

The switch based on intersecting waveguides (Figure 18.22c) can be considered either as a modal interferometer or as a zero-gap directional coupler. Such a switch offers a topological flexibility, moderate voltage requirements, and simple electrode configurations.

In truly optical switching, a control optical pulse switches a signal pulse from one channel to the other. A control pulse changes the conditions of propagation for a signal pulse due to nonlinear optical effects. The response time of nonlinear optical effects (for example, the Kerr and Raman effects) in fibers and waveguides is in the femtosecond range, providing the highest switching speeds without the need for electronics. The insertion loss can be very low.

18.4.4 Attenuator

An optical attenuator is a passive device placed into a transmission path to control an optical loss. Both fixed and variable attenuators are available.

In fixed attenuators, an absorbing layer is inserted between two fibers. GRIN-rod or ball lenses are used to collimate the light between fibers. In variable attenuators, a wedge-shaped absorber element whose position can be adjusted accurately is placed between the fibers. Figure 18.23 shows a variable attenuator with a wedge-shape absorber (shadowed) and a wedge-shape transparent element. The latter is needed to avoid beam displacement.

18.4.5 Polarization Fiber Components

Polarization is a property that arises because of the vector nature of the electromagnetic waves. Electromagnetic light waves are represented by two vectors: the electric field strength **E** and the magnetic field strength **H**. Since the interaction of light with material media is mainly through the electric field strength **E,** the state of polarization of light is described by this field. Polarization refers to the behavior with time of the electric field vector **E,** observed at a fixed point in space. Time-harmonic transverse plane optical waves can be represented by

$$
\begin{aligned}
\mathbf{E}(\mathbf{r};t) &= \mathbf{x}E_x\left(x,y,z;t\right)+\mathbf{y}E_y\left(x,y,z,t\right) \\
E_x(x,y,z;t) &= E_x\left(x,y\right)\cos(\omega t - kz + \delta_x) \\
E_y\left(x,y,\ z;t\right) &= E_y\left(x,y\right)\cos(\omega t - kz + \delta_y)
\end{aligned}
\tag{18.25}
$$

where x and y are unit vectors along the transverse direction; E_x and E_y are the amplitudes of the waves along the transverse directions; $k\ (= 2\pi/\lambda)$ is the propagation constant of the wave; $\omega\ (= 2\pi\nu)$ is the angular frequency of the wave; and λ and ν are the wavelength and the frequency of the light, respectively. Phase shift $\delta_x - \delta_y$ between the orthogonal components defines the polarization state of light.

Whenever a light wave of arbitrary polarization propagates through optical media, the optical properties of such media induce changes in its polarization state. In isotropic media, the induced polarization in the medium is parallel to the electric field of the optical wave. In many media, however, the induced polarization depends on the magnitude and direction of the applied field. One of the most important consequences of this anisotropy is the phenomenon of birefringence in which the phase velocity of an optical wave propagating in the medium depends on the direction of polarization of its vector **E** (i.e., $k_x \neq k_y$). This anisotropy changes the state of polarization of the wave after propagating in the medium. For example, an input linear polarization state becomes elliptically polarized after propagating by some distance through the medium.

The ideal optical fiber is perfectly cylindrical and invariant by translation along the propagation axis, and hence isotropic. In particular, single-mode optical fibers have been designed for supporting only one mode. Because no real fiber has a perfect circular symmetry (due to fabrication defects or to environmental conditions), the fundamental mode splits into two submodes, orthogonally polarized and propagating with different velocities. This is of little consequence in applications where the fiber transmits signals in the form of optical power with pulse-code or intensity modulation, as the devices detecting the transmitted light are not sensitive to its polarization state. However, in modern applications, such as the fiber-optic interferometric sensor and coherent communication systems, polarization of the light is important. In these latter cases, it is necessary to be able to control polarization of light, and compensate the changes of the polarization along the fiber. There are various fiber-optics components that are used to control the polarization. Here, we focus on three types of fiber polarization components, namely, polarizers, polarization splitters, and polarization controllers.

18.4.6 Polarizers

A polarizer is an optical device that transmits (or reflects) a state of polarization and that suppresses any transmission (or reflection) of the orthogonal state of polarization. An ideal linear polarizer is a device that transforms any input state of polarization of light to a linear output state. A linear polarizer can also be defined as a device whose eigenpolarizations (for example, the two orthogonal polarization modes in a single-mode fiber) are linear with one eigenvalue (one of the orthogonal modes) equal to zero. In optical fibers, the main method used to eliminate one of the two orthogonal modes is a loss process, in which one of the modes is coupled toward the outer medium or providing larger radiation loss for one mode than the other.

There are two general classes of fiber polarizers: invasive and noninvasive. Invasive polarizers require polishing away of a small section of fiber to expose the core. Such polarizers can be prepared directly in the system fiber and require no splices. These polarizers utilize the differential attenuation of transversal electric (TE) and transversal magnetic (TM) modes. (In optical fibers, the fundamental mode is hybrid, i.e., both modes TE and TM exist and are orthogonally polarized.) Figure 18.24 shows a scheme of such a polarizer where a dielectric layer is deposited on the polished half-fiber, followed by the deposition of a metallic layer. The dielectric layer matches the propagation of one polarization to the TM plasmon wave (spatial oscillations of electric charge) at the metal interface, thus providing effective coupling to the

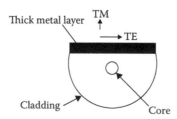

FIGURE 18.24 Transverse section of a metal film polarizer: transversal electric (TE) and transversal magnetic (TM) modes.

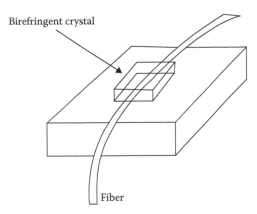

FIGURE 18.25 Polarizer with oriented crystal.

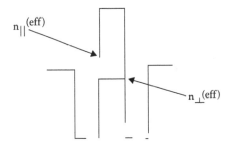

FIGURE 18.26 Fiber polarizer with W-structure index profile.

lossy mode. An alternative design is the so-called cut-off polarizer, in which the fiber is polished all the way down to the core and a very thin film metal is deposited on the polished fiber.

Another polarizer replaces the fiber cladding by a birefringent cladding with refractive indices such that polarization-selective coupling from the fiber core occurs. This method, therefore, uses the evanescent field of the guided waves that exist in the cladding. If the refractive index of the birefringent cladding is greater than the effective index of the guided wave for a given polarization, then this polarization is coupled out of the fiber core. Also, if the refractive index of the birefringent cladding as seen by the other guided polarization is lower than the effective index of the light wave in the core, then this polarization remains guided. When both these conditions are simultaneously achieved, one polarization radiates while the other remains guided in the fiber. Figure 18.25 shows a device of this type.

Noninvasive fiber polarizers are made directly in the fiber. Such a polarizer can be cut out of one fiber and spliced into devices or systems. Most noninvasive polarizers work by differential tunneling loss after high stress or geometrical birefringence splits the propagation constants. Figure 18.26 shows a polarizer that uses a W-structure index profile. In such a profile, the unfavored polarization is attenuated when its effective index falls below the refractive index of the second cladding. A noninvasive polarizer can also be made by bending a highly birefringent fiber. The principle of operation is similar to that of the W-structure polarizer because bending can be viewed as modifying the index profile. The two polarizations then suffer differential attenuation.

18.4.7 Polarization Splitter

Polarization-sensitive couplers or polarization state splitters are usually realized using two face-to-face half-couplers (Figure 18.27). The coupling coefficient is adjusted in such a way that light with

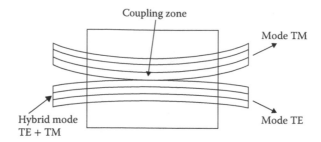

FIGURE 18.27   Linear polarization states splitter using two half-couplers.

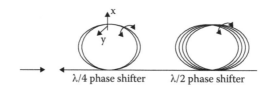

FIGURE 18.28   Half- and quarter-wave phase shifter realized with fiber loops.

one polarization state from the first half-coupler totally propagates in the fiber of the second coupler, whereas the coupling of the light of the other polarization state is not possible. In other words, there is a phase matching for one state of polarization between the guided modes of the two fibers and there is not for the other.

Polished directional couplers have been made with polarization-preserving fiber to achieve both polarization-preserving and polarization-selective coupling. In a polarization-preserving coupler, the propagation constants of the two polarizations are matched across the coupler; in a polarization-selective coupler, only one of the propagation constants is matched while the mismatch is as large as possible for the other polarization.

18.4.8 Polarization Controller

A common problem is to transform the state of polarization in a fiber from an arbitrary polarization state to linear polarization with a proper orientation for a polarization-sensitive optical component. Whenever an initially isotropic fiber is bent, with or without axial tension, it becomes linearly birefringent. This peculiarity permits the production of phase shifters of an angle $\pi/2$ or π, that is, the equivalent of a quarter- and half-wave plate of crystalline optics. Polarization controllers are made by bending ordinary nonbirefringent fiber in coils (Figure 18.28). Rotating the coil is equivalent to turning the bulk waveplate.

A section of birefringent fiber can also act as a high-order waveplate if both polarization axes are equally excited. This property is commonly used to convert linear polarization to elliptical polarization. The characteristics of this polarization element are strongly dependent on temperature, pressure, and applied stresses. The addition of a polarizer converts such a variable waveplate into a variable in-line attenuator. Variable waveplates separated by polarizers can also form a tunable Lyot filter.

In an all-fiber system, the fiber itself defines the optical path and is subject to perturbations that, in conjunction with the intrinsic birefringence, can affect the state of polarization in complicated and unpredictable ways. The birefringence perturbations can be originated by twisting (circular birefringence) and lateral stress (linear birefringence). In applications such as Mach–Zehnder interferometers and fiber-rotation sensors, it is important that this birefringence is compensated. Polarization

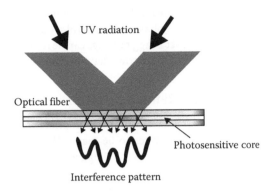

FIGURE 18.29 Holographic side exposure method.

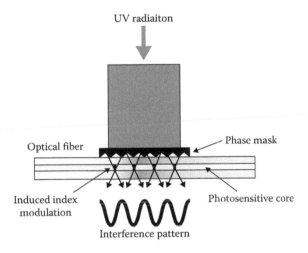

FIGURE 18.30 Fiber Bragg grating fabrication by the phase-mask method.

controllers involving electromagnetically induced stress twisting have been proposed for an accurate control of birefringence effects.

18.4.9 Fiber Bragg Gratings

Photosensitivity is a phenomenon specific to germanium-doped silica fibers in which the exposition of the fiber core to blue-green or ultraviolet (UV) radiation induces changes in the refractive index of the core. If the changes in the refractive index are periodic, a grating is formed. The stronger changes occur when the fiber is exposed to UV radiation close to the absorption peak at 240 nm of a germania-related defect.

Two widely used methods for grating fabrication are the holographic side exposure and the phase-mask imprinting. In the holographic side exposure, two beams from a UV laser interfere in the plane of the fiber (Figure 18.29). The periodicity of the interference pattern created in the plane of the fiber is determined by the angle between the two beams and the wavelength of the UV laser. The regions of constructive interference cause an increase in the local refractive index of the photosensitive core, while dark regions remain unaffected, resulting in the formation of a refractive Bragg grating.

In the phase-mask fabrication process, light from a UV source passes through a diffractive phase mask that is in contact with the fiber (Figure 18.30). The phase mask has the periodicity of the desired grating. Light diffracted in orders (+1, –1) of the mask interferes in the plane of the fiber, providing periodical modulation of the refractive index. The phase-mask technique allows fabrication of fiber gratings with variable spacing (chirped gratings). Fiber Bragg gratings can be routinely fabricated to operate over a wide range of wavelengths, extending from the ultraviolet to the infrared region.

18.4.10 Fiber Bragg Grating as a Reflection Filter

If a broadband radiation is coupled into the fiber, only an appropriate wavelength matching the Bragg condition is reflected, permitting a reflection filter to be made. The Bragg condition determines the wavelength of the reflected light λ_{Bragg}, referred to as the Bragg wavelength, as

$$\lambda_{\text{Bragg}} = 2\Lambda n_{\text{eff}}, \tag{18.26}$$

where n_{eff} is the refractive index of the mode and Λ is the period of the Bragg grating. A strong reflection can be understood from the fact that at each change in the refractive index, some light is reflected. If the reflections from points that are a spatial period apart are in phase, then the various multiple reflections add in phase, leading to a strong reflection. The peak reflectivity R of a grating may be calculated as

$$R = \tanh^2(\kappa L), \tag{18.27}$$

where L is the length of the fiber grating, and the coupling coefficient κ is defined by

$$\kappa = \frac{\pi n \delta n \eta}{\lambda_{\text{Bragg}} n_{\text{eff}}}, \tag{18.28}$$

where n is the cladding index, η is the overlap integral of the forward and backward propagating modes over the perturbed index within the core, and δn is the magnitude of the refractive index modulation.

The bandwidth of the reflection spectrum can be calculated as

$$\Delta\lambda = \frac{\lambda_{\text{Bragg}}^2}{\pi n_{\text{eff}} L}\left(\pi^2 + \kappa^2 L^2\right)^{1/2}. \tag{18.29}$$

The simplest application of fiber Bragg gratings is as reflection filters with bandwidths of approximately 0.05–20 nm. Multiple reflection gratings may be written into a piece of fiber to generate a number of reflections, each at a different wavelength. Fiber Bragg gratings may be used as narrow-band filters and reflectors to stabilize semiconductor lasers or DBR lasers, narrow-band reflectors for fiber lasers, simple broad- and narrow-band-stop reflection filters, radiation mode taps, band-pass filters, fiber grating Fabry–Perot etalons, in dispersion compensation schemes, narrow-linewidth dual-frequency laser sources, nonlinear pulsed sources, optical soliton sources, and applications in sensor networks.

18.4.11 Fiber Bragg Grating-Based Multiplexer

Fiber Bragg gratings may also be used as multiplexers, demultiplexers, or as add/drop elements. Figure 18.31 shows a simple scheme used for demultiplexing, which includes a fiber Bragg grating in conjunction with an optic circulator. The optical radiation at different wavelengths enters into port 1 of the circulator, comes out through port 2, and passes through the fiber grating with the reflection peak corresponding to the wavelength λ_1. As a result, the optical signal at λ_1 goes back to port 2 of the circulator and emerges at port 3.

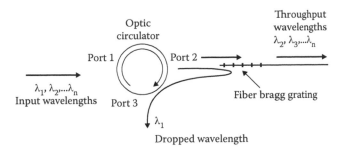

FIGURE 18.31 Demultiplexing of optical signals by using a fiber Bragg grating and an optic circulator.

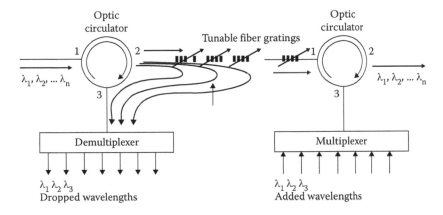

FIGURE 18.32 Tunable grating filters to add and drop different wavelengths.

Figure 18.32 shows a scheme used to add and drop several wavelengths. It includes three-port optic circulators with a series of electrically tunable fiber gratings for each wavelength. The demultiplexer separates the dropped wavelengths into individual channels and the multiplexer combines wavelengths into transmission fiber line. In the normal state, the gratings are transparent to all wavelengths. If a grating is tuned to a specific wavelength, this signal is reflected back to port 2 of the first circulator and exits from port 3. The signals at transmitted wavelengths enter into port 1 of the second optic circulator and exit from port 2. To add or reinsert wavelengths that were dropped, one injects these into port 3 of the second circulator. They first come out of port 1 and travel toward the tuned gratings. Each grating of the array reflects a specific wavelength back to port 1. The reflected signals exit from port 2, where all channels are recovered.

18.4.12 Chirped Fiber Bragg Gratings

In a chirped grating, the period of the modulation of the refractive index varies along the grating length. This type of grating can be used for compensation of the dispersion, which occurs when an optical signal propagates through a fiber. The dispersion associated with transmission through a Bragg grating may be used to compensate for dispersion of the opposite sign. Figure 18.33 shows a schematic of one of the methods used for dispersion compensation. In this method, the propagation delay through a grating is used to provide large dispersion compensation. The chirped grating reflects each wavelength from a different point along its length. Thus, the dispersion imparted by the grating depends on the grating length and is given by

$$\tau = \frac{2L}{v_g}, \tag{18.30}$$

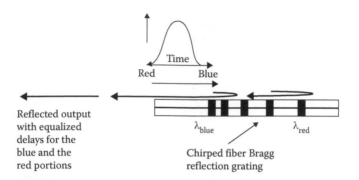

FIGURE 18.33 Chirped fiber Bragg grating for dispersion compensation.

where v_g is the group velocity of the pulse incident on the grating. Therefore, a grating with a linear wavelength chirp of $\Delta\lambda$ nm will have a dispersion of

$$D = \frac{\tau}{\Delta\lambda}\,\text{ps/nm}. \tag{18.31}$$

18.4.13 Fiber Connectors

A fiber connection is defined as the point where two optical fibers are joined together to allow a light signal to propagate from one fiber to the next continuing fiber. Fiber connection has three basic functions: precise alignment, fiber retention, and ends protection. To achieve low losses at the connection, a precise alignment is required because of the extremely small size of the fiber cores. Fiber retention prevents any misalignment or separation that may be caused by tension or manipulation of a cable. Ends protection ensures the optical quality of fiber ends against environmental factors, such as humidity and dust.

Fiber connections generally fall into two categories: the permanent, which uses a fiber splice; temporary (non-fixed), which uses a fiber-optic connector. Splices and connectors are used in different places. Typical uses of splices include pigtail vault splices, distribution breakouts, and reel ends. On the other hand, connectors are used as interfaces between terminal on LANs, patch panels, and terminations into transmitters and receivers. The quality of fiber connections whether splices or connectors is estimated from the point of view of signal loss or attenuation.

18.4.13.1 Interconnection Losses

Interconnection loss is the loss of signal or light intensity as it travels through the fiber joints. These losses are caused by several factors and can be classified into two categories: extrinsic and intrinsic losses. Intrinsic losses are related to the mismatches between fiber parameters and do not depend on applied technique as extrinsic losses do. The differences include variations in core and/or outer diameter, NA mismatch, and differences in the fiber's index profile, core ellipticity, and core eccentricity (Figure 18.34).

If the core diameter of the receiving fiber is smaller than that of the emitting fiber, some light is lost in the cladding of the receiving fiber. The same is true for NA mismatch. When connecting fibers have different NAs, testing will show a significant loss when the receiving fiber's NA is smaller than that of the emitting fiber, because a fiber with a lower NA cannot capture all light coming out from a fiber with a greater NA. If the core of either transmitting or receiving fiber is elliptic rather than circular, the collecting area of the receiving core is reduced, which results in less captured light in the receiving fiber. Any core eccentricity variations in both fibers also reduce the collecting area of the receiving fiber.

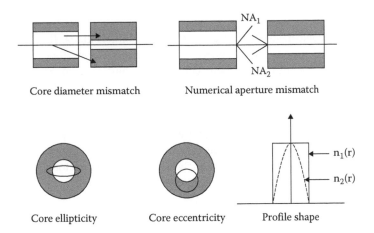

Core diameter mismatch   Numerical aperture mismatch

Core ellipticity   Core eccentricity   Profile shape

FIGURE 18.34   Intrinsic losses mechanism.

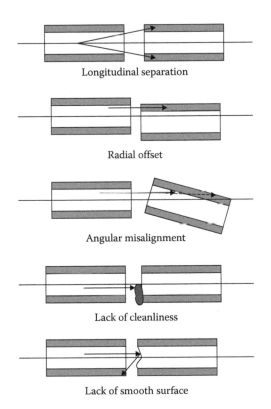

Longitudinal separation

Radial offset

Angular misalignment

Lack of cleanliness

Lack of smooth surface

FIGURE 18.35   Extrinsic losses mechanism.

Extrinsic losses are due to alignment errors and poor fiber-end quality, and depend on connection technique. Frequent causes of extrinsic losses include fiber-end misalignment, poor quality of the fiber-end face, longitudinal separation of fiber ends, contamination of fiber ends, and angular misalignment of bonded fibers. Figure 18.35 shows typical extrinsic loss mechanisms. In addition, back reflection caused by the abrupt change of refractive index at the end of each fiber also contributes to connection loss.

18.4.13.2 Connectors

A connector is a demountable device used to connect and disconnect fibers. The connector permits two fibers to be connected and disconnected hundreds of times easily. Connectors can be divided into two basic categories: expanded beam-coupled (lens-coupled) and butt-coupled. Expanded beam coupling requires a lens system in the connector to increase the size of the beam of light at one connector and to reduce it at the other. In butt coupling, the connectors are mechanically positioned close enough for the light to pass from one fiber to another.

Expanded beam connectors are frequently not used because of cost and performance disadvantages: they never achieve the low losses of butt connectors. In addition, such connectors use index-matching oil, which attracts dust and can affect the characteristics of the connector. Thus, fabrication of the majority of expanded beam connectors has been discontinued.

Although there are different types of butt connectors, the rigid ferrule designs are the most popular butt connectors used today. Regardless of type and manufacturer, a rigid ferrule connector comprises five basic structural elements: ferrule, retaining nut, backshell, boot, and cap (Figure 18.36).

A ferrule is a rigid tube within a connector with a central hole that contains and aligns a fiber. It can be made of different materials, such as ceramic, steel and plastic. Ceramic ferrules offer the lowest insertion loss and the best repeatability, while steel ferrules permit re-polishing and plastic ferrules have the advantages of low cost. In addition, ferrules can have different shapes—straight thin cylinder, conical, or stepped. The retaining nut provides the mechanism by which the connector is secured to an adapter. It can be made of either steel or plastic. Retaining nuts are threaded, bayonet, or push–pull to make a connection. The backshell is the portion of the connector in the back of the retaining nut for attaching the cable to the connector. The attachment is made of metal or plastic and provides the principal source of strength between the cable and connector. The boot is a plastic component that slides over the cable and the backshell of the connector. It limits the radius of curvature of the cable and relieves strains on the optical fiber. The cap is a plastic cover that protects the end of a connector ferrule from dust contamination and damage. In addition to these structural elements, some connectors have a crimp ring or other mechanism for attaching the cable to the backshell.

It should be noted that most fiber-optical connectors do not use the male–female configuration common to electronics connectors. Instead, a coupling device called a mating (or alignment) sleeve is used to mate the connectors (Figure 18.37). There are two kinds of mating sleeves: the flouting style adapter to connect two single-mode or multimode cables and the panel mount style adapter to mate fiber-optical transmitters and receivers to the optical cable via a connector. Other coupling devices are hybrid adapters that permit mating of different connector types.

One important criterion is connector performance. When selecting a connector, the following characteristics should be analyzed: insertion loss (usually 0.1 to 0.6 dB per connection); back reflection (−20 to −60 dB); repeatability of connection. In the short life of fiber-optic systems, fiber-optic connectors have gone through four generations. Fiber-optic connectors developed in the first generation were mostly screw-type (e.g., SMA 905 and SMA 906). Because of the lack of rotational alignment, these connectors have a large amount of variation in the insertion loss as the connector is unmated and remated.

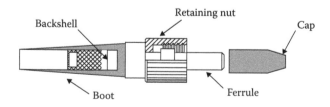

FIGURE 18.36   Connector.

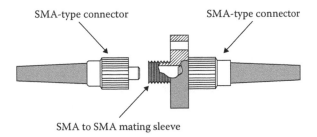

SMA-type connector   SMA-type connector

SMA to SMA mating sleeve

FIGURE 18.37   Mating (or alignment) sleeve.

TABLE 18.4   Connector Types

Connector	Fiber Type	Typical Loss (dB)	Applications	Popularity
SMA	MM	0.25–0.60	Military	Fading
Biconic	SM, MM	0.20–0.60	Telecommunications	Fading
FC	SM, MM	0 20–0.50	Telecommunications and Datacom	Widely used
D4	SM, MM	03–0.50	Telecommunications	Fading
ST	SM, MM	0.15–0.50	Inter/intra-building	Widely used
SC	SM, MM	0.20–0 40	Telecommunications	Growing
FDDI	SM, MM	0.20–0 50	Fiber optic networks	Widely used
ESCON	SM. MM	0.15–0.30	Fiber optic networks	Widely used
LC	SM, MM	0.10–0.15	Telecommunications and fiber optic Networks	Newer

Connectors of the second generation, such as bayonet type (e.g., straight tip [ST]), biconic, and FC types, solved many problems associated with earlier connectors, providing rotational alignment and greatly improving repeatability. ST is a keyed and contact connector with low average losses (0.5 dB). ST design includes a spring-loading twist-and-lock bayonet coupling that keeps the fiber and ferrule from rotating during multiple connections. However, this connector is not pull-proof.

The biconic connector is non-keyed, non-contact with rotational sensibility and susceptible to vibrations. Originally, it was developed as a multimode fiber connector; later, it was the first successful single-mode connector used by the telecommunications industry. Actually, biconic connectors are suitable for single-mode and multimode fibers.

In the FC connector, a new method called face contact (FC) has been introduced to reduce back reflections. FC provides a flat-end face contact between joining connectors with low average losses (0.4 dB). FC/PC is a new version of FC that includes a new polishing method called PC (physical contact). The PC uses a curved polishing that dramatically reduces back reflections. PC style offers very good performance for single-mode and multimode fiber connectors. They are commonly used in analog systems (CATV) and high bit-rate systems. The majority of the connectors of the second generation are threaded connectors. This is inconvenient and decreases packing density.

Connectors of the third and fourth generation tend to be push–pull types. This type of connector has shorter connection time and allows a significant increase in packing density. Since packing density is becoming more important, fourth-generation connectors are focused on ease of use and reduced size. One of the new designs is the LC connector. This connector offers an easy installation and high pack density because of a reduced connector's diameter.

Table 18.4 describes some of the widespread connector types.

18.4.14　Fiber Splicers

Splices are permanent connections between fibers. Splices are used in two situations: mid-span splices, which connect two lengths of cable; and pigtails, at the ends of a main cable, when rerouting of optical paths is not required or expected. Splices offer lower attenuation, easier installation, lower back reflection, and greater physical strength than connectors, and they are generally less expensive. In addition, splices can fit inside cable, offer a better hermetic seal, and allow either individual or mass splicing. There are two basic categories of splices: fusion splices and mechanical splices.

18.4.14.1　Fusion Splicing

The most common type of splice is a fusion splice, formed by aligning and welding the ends of two optical fibers together. Usually a fusion splicer includes an electric arc welder to fuse the fibers, alignment mechanisms, and a camera or binocular microscope to magnify the alignment by 50 times or more. The fusion parameters can usually be changed to suit particular types of fibers, especially if it is necessary to fuse two different fibers. After the splicing procedure, the previously removed plastic coating is replaced with a protective plastic sleeve.

Fusion splicing provides the lowest connection loss, keeping losses as low as 0.05 dB. Also, fusion splices have lower consumable costs per connection than mechanical splices. However, the capital investment in equipment to make fusion splices is significantly higher than that for mechanical splices. Fusion splices must be performed in a controlled environment, and should not be done in open spaces because of dust and other contamination. In addition, fusion splices cannot be made in an atmosphere that contains explosive gasses because of the electric arc generated during this process.

18.4.14.2　Mechanical Splices

A mechanical splice is a small fiber connector that precisely aligns two fibers together and then secures them by clamping them within a structure or by epoxying the fibers together. Because tolerances are looser than in fusion splicing, this approach is used more often with multimode fibers than single-mode fibers. Although losses tend to be slightly higher than those of fusion splices and back reflections can be a concern, mechanical splices are easier to perform and the requirements for the environment are looser for mechanical splicing than those for fusion splicing. Generally, the consumables for a mechanical splice result in a higher cost than consumables for fusion splices; however, the equipment needed to produce a mechanical splice is much less expensive than the equipment for fusion splices.

To prepare mechanical splices, the fibers are first stripped of all buffer material, cleaned, and cleaved. Cleaving a fiber provides a uniform surface, which is perpendicular to a fiber axis, needed for maximum light transmission to the other fiber. Then, the two ends of the fibers are inserted into a piece of hardware to obtain a good alignment and to permanently hold the fibers' ends. Many hardware devices have been developed to serve these goals. The most popular devices can be divided in two broad categories: capillary splices and V-groove splices.

Capillary splice is the simplest form of mechanical splicing. Two fiber ends are inserted into a thin capillary tube made of glass or ceramic, as illustrated in Figure 18.38. To decrease back reflections from

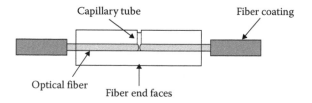

FIGURE 18.38　Capillary splice.

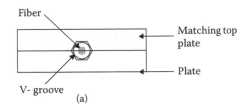

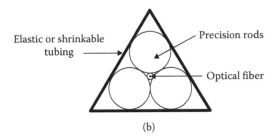

FIGURE 18.39 V-groove splice cross-sections: (a) V-groove using two plates and (b) V-groove using three rods.

the fiber ends, an index-matching gel is typically used in this splice. The fibers are held together by compression or friction, although epoxy may be used to permanently secure the fibers.

The V-groove splice is probably the oldest and still most popular method, especially for multifiber splicing of ribbon cable. This type of splice is either crimped or snapped to hold the fibers in place. Many types of V-groove splices have been developed using different techniques. The simplest technique confines the two fibers between two plates, each one containing a groove into which the fiber fits. This approach centers the fiber core, regardless of variation in the outer diameter of the fiber (Figure 18.39a).

The popular V-groove technique uses three precision rods tightened by means of an elastic band or shrinkable tube (Figure 18.39b). The splice loss in this method depends strongly on the fiber size (core and cladding diameter variations) and eccentricity (the position of the core relative to the center of the fiber). V-groove techniques properly carried out with multimode fibers result in splice losses of the order of 0.1 dB or less. Some of them can be applied to single-mode fibers as well.

The borderline connectors and splices are rather indefinite. One example is the rotary mechanical splice (RMS), which is a disconnectable splice made by attaching ferrules to the two fiber ends, joining the ferrules in a housing, and holding the assembly together with a spring clip. The assembly can be disconnected by removing the slip and is rated to survive 250 mating cycles. Rotary mechanical splices provide a simple and quick method of joining single-mode and multimode fibers with mean losses less than 0.2 dB without the need for optical or electronic monitoring equipment.

Splices, once completed, whether fusion or mechanical, are then placed into splicing trays designed to accommodate the particular type of splice in use. On the other hand, fiber-optic splices require protection from the environment, so they are stored in a splice enclosure. These special boxes are available for indoor as well as outdoor mounting. The outdoor type should be weatherproof, with a watertight seal. Additionally, splice enclosures protect stripped fiber-optic cable and splices from strain and help organize spliced fibers in multifiber cables.

Suggested Reading

1. Agrawal GP. *Nonlinear Fiber Optics*, 2nd edn, San Diego, CA: Academic Press, 1995.
2. Ghatak A, Thyagarajan K. *Introduction to Fiber Optics*, Cambridge: Cambridge University Press, 1998.

3. Goff DR. *Fiber Optic Reference Guide*, Boston, MA: Focal Press, 1996.

4. Hecht J. *Understanding Fiber Optics*, Indiana, IN: SAMS, 1987.

5. Keiser G. *Optical Fiber Communications*, Singapore: McGraw-Hill, 1991.

6. Midwinter J. *Optical Fibers for Transmission*, New York, NY: John Wiley & Sons, 1979.

7. Wilson J, Hawkes JFB. *Optoelectronics: An Introduction*, London, England: Prentice Hall, 1983.

8. Yeh C. *Handbook of Fiber Optics: Theory and Applications*, San Diego, CA: Academic Press, 1990.

9. Agrawal GP. *Nonlinear Fiber Optics*, 2nd edn, Academic Press, San Diego, 1995.

10. Armitage JR, Wyatt R, Ainslie BJ, Craig-Ryan SP. Highly efficient 980 nm operation of an Yb^{3+}-doped silica fiber laser, *Electron. Lett.*, **25**, 298–299; 1989.

11. Artjushenko VG, Butvina LN, Vojtsekhovsky VV, Dianov EM, Kolesnikov JG. Mechanisms of optical losses in polycrystalline fibers, *J. Lightwave Technol.*, **4**, 461–464; 1986.

12. Berman IE. Plastic optical fiber: A short-Haul solution, *Opt. Photonics News*, **9**, 29; 1998.

13. Bornstein A, Croitoru N. Experimental evaluation of a hollow glass fiber, *Appl. Opt.*, **25**(3), 355–358; 1986.

14. Desurviere E. *Erbium-Doped Fiber Amplifiers*, New York, NY: John Wiley & Sons, 1994.

15. DiGiovanni DJ, Muendel MH. High power fiber lasers, *Opt. Photonics News*, 26–30; 1999.

16. Drexhage MG, Moynihan CT. Infrared optical fibers, *Sci. Am.*, **259**, 76–81; 1988.

17. France PW, Carter SF, Moore MW, Day CR. Progress in fluoride fibers for optical communications, Reprinted from *British Telecom Technology Journal*, **5**(2), 28–44; 1987, in *Selected Papers on Infrared Fiber Optics*, SPIE, MS **9**, 59–75; (1990.

18. Garmire E, McMahon T, Bass M. Flexible Infrared Waveguides for High-Power Transmission, *IEEE J. Quantum Electron.*, **16**(1), 23–32; 1980.

19. Ghatak A, Thyagarajan K. *Introduction to Fiber Optics*, New York: Cambridge University Press, 1998.

20. Harrington JA, Standlee AG, Pastor AC, DeShazer LG. Single-crystal infrared fibers fabricated by traveling-zone melting, *Infrared Optical Materials and Fibers III, Proc. SPIE*, **484**, 124–127; 1984.

21. Hecht J. *Understanding Fiber Optics*, 1st edn, Carmel, Indiana: SAMS, 1992.

22. Hecht J. Perspectives: Fiber lasers prove versatile, *Laser Focus World*, 73–77; 1998.

23. Ikedo M, Watari M, Ttateishi F, Ishiwatari H. Preparation and characteristics of the TlBr-TlI fiber for a high power CO_2 laser beam, *J. Appl. Phys.*, **60**(9), 3035–3039; 1986.

24. Kaminov IP, Koch TL. *Optical Fiber Telecommunications IIIB*, San Diego, CA: Academic Press,1997.

25. Katsuyama T, Matsumura H. Low-loss te-based chalcogenide glass optical fibers, *Appl. Phys. Lett.*, **49**(1), 22–23; 1986.

26. Katsuyama T, Matsumura H. *Infrared Optical Fibers*, Bristol, UK: Adam Hilger, 1989.

27. Katzir A. *Lasers and Optical Fibers in Medicine*, San Diego, CA: Academic Press, 1993.

28. Keiser GE. A Review of WDM Technology and Applications, *Opt. Fiber Technol.*, **5**, 3–39; 1999.

29. Klocek P, Sigel GH Jr. *Infrared Fiber Optics*, Bellingham, Washington: SPIE Optical Engineering Press, 1989.

30. Lüthy W, Weber HP. High-power monomode fiber lasers, *Opt. Eng.*, **34**(8), 2361–2364; 1995.

31. Matsuura Y, Saito M, Miyagi M, Hongo A. Loss characteristics of circular hollow waveguides for incoherent infrared light, *JOSA A*, **6**(3), 423–427; 1989.

32. Minelly JD, Morkel PR. 320-mW Nd^{3+}-doped single-mode fiber superfluorescent source (Vol 6, pp. 624–626), in *Conference on Lasers & Electro-Optics (CLEO'93)*, United States, 7 May 1993.

33. Miyashita T, Manabe T. Infrared optical fibers, *IEEE J. Quantum Electron.*, **18**(10), 1432–1450; 1982.

34. Ono T. Plastic optical fiber: The missing link for factory, office equipment, *Photonics Spectra*, **29**(11), 88–91; 1995.

35. France PW. ed., *Optical Fiber Lasers and Amplifiers*, Glasgow: Blackie and Son, 1991.

36. Pask HM, Carman RJ, Hanna DC, Dawes JM., Ytterbium-doped silica fiber lasers: Versatile sources for the 1–1.2 μm region, *IEEE J. Selected Topics in Quantum Electron.*, **1**(1), 2–13; 1995.

37. Sa'ar A, Katzir A. Intrinsic losses in mixed silver halide fibers, **1048**, 24–32, in *Infrared Fiber Optics, Proc. SPIE*, 1989.

38. Sa'ar A, Barkay N, Moser F, Schnitzer I, Levite A, Katzir A. Optical and mechanical properties of silver halides fibers **843**, 98–104, in *Infrared Optical Materials and Fibers V, Proc. SPIE*, 1987.

39. Saito M, Takizawa M. Teflon-clad As-S glass infrared fiber with low-absorption loss, *J. Appl. Phys.*, **59**(5), 1450–1452; 1986.

40. Sakaguchi S, Takahashi S. Low-loss fluoride optical fibers for midinfrared optical communication, *J. Lightwave Technol.*, **5**(9), 1219–1228; 1987.

41. Tran DC, Sigel GH Jr, Bendow B. Heavy metal fluoride glasses and fibers: A review, *J. Lightwave Technol.*, **2**(5), 566–586; 1984.

42. Vasil'ev AV, Dianov EM, Dmitruk LN, Plotnichenko VG, Sysoev VK. Single-crystal waveguides for the middle infrared range, *Sov. J. Quantum Electron.*, **11**(6), 834–835; 1981.

43. Worrell CA. Infra-red optical properties of glasses and ceramics for hollow waveguides operating at 10.6 µm wavelength **843**, 80–87, in *Infrared Optical Materials and Fibers V, Proc. SPIE*, 1987.

44. Wysocki JA, Wilson RG, Standlee AG, et al. Aging effects in bulk and fiber TlBr-TlI, *J. Appl. Phys.*, **63**(9), 4365–4371; 1988.

45. Zyskind JL, Nagel JA, Kidorf HD. Erbium-doped fiber amplifiers for optical communications, in *Optical Fiber Telecommunications IIIB*, Kaminov, I. P. and T. L. Koch, eds, San Diego, CA: Academic Press, 1997.

46. Agrawal GP. *Nonlinear Fiber Optics*, 2nd edn, San Diego, CA: Academic Press, 1995.

47. Allard FC. *Fiber Optics Handbook: For Engineers and Scientists*, New York, NY: McGraw-Hill, 1990.

48. Azzam RMA, Bashara NM. *Ellipsometry and Polarized Light*, Amsterdam, the Netherlands: North-Holland, 1989.

49. Bergh RA, Digonnet MJF, Lefevre HC, Newton SA, Shaw HJ. Single mode fiber optic components **32**, 136–143, in *Fiber Optics-Technology, Proc. SPIE*, 1982.

50. Born M, Wolf E. *Principles of Optics*, 6th edn, Cambridge: Cambridge University Press, 1997.

51. Calvani R, Caponi R, Cisternino F. Polarization measurements on single-mode fibers, *J. Lightwave Technol.*, **7**(8), 1187–1196; 1989.

52. Chomycz B. *Fiber Optic Installations*, New York, NY: McGraw-Hill, 1996.

53. Culshaw B, Michie C, Gardiner P, McGown A. Smart structures and applications in civil engineering, *Proc. IEEE*, **84**(1), 78–86; 1996.

54. Ghatak A, Thyagarajan K. *Introduction to Fiber Optics*, Cambridge: Cambridge University Press, 1998.

55. Goff DR. *Fiber Optic Reference Guide*, Boston, MA: Focal Press, 1996.

56. Hayes J. *Fiber Optics Technician's Manual*, New York, NY: International Thomson Publishing Company, 1996.

57. Huard S. *Polarization of Light*, Masson, Belgium: John Wiley & Sons, 1997.

58. Kaminov IP. Polarization in optical fibers, *IEEE J. Quantum Electron.*, **17**(1), 15–22; 1981.

59. Kashyap R. Photosensitive Optical Fibers: Devices and Applications, *Opt. Fiber Technol.*, **1**, 17–34; 1994.

60. Keiser G. *Optical Fiber Communications*, Singapore: McGraw-Hill, 1991.

61. Keiser GE. A Review of WDM Technology and Applications, *Opt. Fiber Technol.*, **5**, 3–39; 1999.

62. Kersey AD. A review of recent developments in fiber optic sensor technology, *Opt. Fiber Technol.*, **2**, 291–317; 1996.

63. Miller C. *Optical Fiber Splicers and Connectors*, New York, NY: Marcel Dekker, 1986.

64. Murata H. *Handbook of Optical Fibers and Cables*, New York, NY: Marcel Dekker, 1996.

65. Noda J, Okamoto K, Yokohama I. Fiber devices using polarization-maintaining fibers, Reprinted from *Fiber and Integrated Optics*, **6**(4), 309–330; 1987, in *Selected Papers on Single-Mode Optical Fibers* (Brozeit, A., K. D. Hinsch, and R. S. Sirohi, eds), *SPIE*, MS **101**, 23–33; 1994.

66. Pearson ER. *The Complete Guide to Fiber Optic Cable System Installation*, Albany, NY: Delmar Publishers, 1997.

67. Rashleigh SC. Origins and control of polarization effects in single-mode fibers, *J. Lightwave Technol.*, 1(2), 312–331; 1983.
68. Sirkis JS. Unified approach to phase-strain-temperature models for smart structure interferometric optical fiber sensors: Part 1, development, *Opt. Eng.*, 32(4), 752–761; 1993.
69. Stolen RH, De Paula RP. Single-Mode Fiber Components, *Proc. IEEE*, 75(11), 1498–1511; 1987.
70. Todd DA, Robertson GRJ, Failes M. Polarization-splitting polished fiber optic couplers, *Opt. Eng.*, 32(9), 2077–2082; 1993.
71. Yariv A. *Optical Electronics in Modern Communications*, 5th edn, New York, NY: Oxford University Press, 1997.
72. Agrawal GP. *Nonlinear Fiber Optics*, 2nd edn, San Diego, CA: Academic Press, 1995.
73. Desurviere E. *Erbium-Doped Fiber Amplifiers*, New York, NY: John Wiley & Sons, 1994.
74. Ghatak A, Thyagarajan K. *Introduction to Fiber Optics*, New York: Cambridge University Press, 1998.
75. Keiser GE. A Review of WDM Technology and Applications, *Opt. Fiber Technol.*, 5, 3–39; 1999.
76. Zyskind JL, Nagel JA, Kidorf HD. Erbium-Doped Fiber Amplifiers for Optical Communications, in *Optical Fiber Telecommunications IIIB*, Kaminov, I. P. and T. L. Koch, eds, San Diego, CA: Academic Press, 1997.
77. Armitage JR, Wyatt R, Ainslie BJ, Craig-Ryan SP. Highly efficient 980 nm operation of an Yb^{3+}-doped silica fiber laser, *Electron. Lett.*, 25(5), 298–299; 1989.
78. Bell Labs Ultra-High-Power, Single-Mode Fiber Lasers, Technical Information.
79. DiGiovanni DJ, Muendel MH. High Power Fiber Lasers, *Opt. Photonics News*, 26–30; 1999.
80. Hecht J. Perspectives: Fiber lasers prove versatile, *Laser Focus world*, 73–77; 1998.
81. Kaminov IP, Koch TL. *Optical Fiber Telecommunications IIIB*, San Diego, CA: Academic Press, 1997.
82. Luthy W, Weber HP. High-Power Monomode Fiber Lasers, *Opt. Eng.*, 34(8), 2361–2364; 1995.
83. Minelly JD, Morkel PR. 320-mW Nd^{3+}-doped single-mode fiber superfluorescent source (Vol. 6, pp. 624–626), in Conference on Lasers & Electro-Optics (CLEO'93), United States, 7 May 1993.
84. Pask HM, Carman RJ, Hanna DC, Dawes JM. Ytterbium-doped silica fiber lasers: Versatile sources for the 1–1.2 μm region, *IEEE J. Selected Topics in Quantum Electron.*, 1(1), 2–13; 1995.
85. Artjushenko VG, Butvina LN, Vojtsekhovsky VV, Dianov EM, Kolesnikov JG. Mechanisms of optical losses in polycrystalline fibers, *J. Lightwave Technol.*, 4(4), 461–464; 1986.
86. Drexhage MG, Moynihan CT. Infrared Optical Fibers, *Sci. Am.*, 110–115; 1988.
87. France PW, Carter SF, Moore MW, Day CR. Progress in fluoride fibers for optical communications, Reprinted from *British Telecom Technology Journal*, 5(2), 28–44; 1987; in *Selected Papers on Infrared Fiber Optics*, SPIE, MS 9, 59–75; 1990.
88. Harrington JA, Standlee AG, Pastor AC, DeShazer LG. Single-crystal infrared fibers fabricated by traveling-zone melting, *Infrared Optical Materials and Fibers III, Proc. SPIE*, 484, 124–127; 1984.
89. Ikedo M, Watari M, Ttateishi F, Ishiwatari H. Preparation and characteristics of the TlBr-TlI fiber for a high power CO_2 laser beam, *J. Appl. Phys.*, 60(9), 30353039; 1986.
90. Kaminov IP, Koch TL. *Optical Fiber Telecommunications IIIB*, San Diego, CA: Academic Press, 1997.
91. Kanamori T, Terunuma Y, Takahashi S, Miyashita T. *J. Chalcogenide Glass Fibers for Mid-Infrared Transmission, Lightwave Technol.*, 2, 607–613; 1984.
92. Katsuyama T, Matsumura H. Low-loss te-based chalcogenide glass optical fibers, *Appl. Phys. Lett.*, 49(1), 22–23; 1986.
93. Katsuyama T, Matsumura H. *Infrared Optical Fibers*, Bristol, England: Adam Hilger, 1989.
94. Miyashita T, Manabe T. Infrared optical fibers, *IEEE J. Quantum Electron.*, 18(10), 1432–1450; 1982.
95. Sa'ar A, Katzir A. Intrinsic losses in mixed silver halide fibers, (Vol. 1048, pp. 24–32), in *Infrared Fiber Optics, Proc. SPIE*, Los Angeles, CA, United States, 2 June 1989.
96. Sa'ar A, Barkay N, Moser F, Schnitzer I, Levite A, Katzir A. Optical and mechanical properties of silver halides fibers (Vol 843, pp. 98–104), *Infrared Optical Materials and Fibers V, Proc. SPIE*, San Diego, CA, 1 January 1987.

97. Saito M, Takizawa M. Teflon-Clad As-S glass infrared fiber with low-absorption loss, *J. Appl. Phys.*, **59**(5), 1450–1452; 1986.
98. Sakaguchi S, Takahashi S. Low-loss fluoride optical fibers for midinfrared optical communication, *J. Lightwave Technol.*, **5**(9), 1219–1228; 1987.
99. Tran DC, Sigel GH Jr, Bendow B. Heavy metal fluoride glasses and fibers: A review, *J. Lightwave Technol.*, **2**(5), 566–586; 1984.
100. Vasil'ev AV, Dianov EM, Dmitruk LN, Plotnichenko VG, Sysoev VK. Single–crystal waveguides for the middle infrared range, *Sov. J. Quantum Electron.*, **11**(6), 834–835; 1981.
101. Wysocki JA, Wilson RG, Standlee AG, et al. Aging effects in bulk and fiber TlBr-TlI, *J. Appl. Phys.*, **63**(9), 4365–4371; 1988.
102. Bornstein A, Croitoru N. Experimental evaluation of a hollow glass fiber, *Appl. Opt.*, **25**(3), 355–358; 1986.
103. Garmire E, McMahon T, Bass M. Flexible Infrared Waveguides for High-Power Transmission, *IEEE J. Quantum Electron.*, **16**(1), 23–32; 1980.
104. Katsuyama T, Matsumura H. *Infrared Optical Fibers*, Bristol, England: Adam Hilger, 1989.
105. Klocek P, Sigel GH Jr. *Infrared Fiber Optics*, Bellingham, Washington: SPIE Optical Engineering Press, 1989.
106. Matsuura Y, Saito M, Miyagi M, Hongo A. Loss characteristics of circular hollow waveguides for incoherent infrared light, *JOSA A*, **6**(3), 423–427; 1989.
107. Worrell CA. Infra-red optical properties of glasses and ceramics for hollow waveguides operating at 10.6 µm wavelength (Vol. 843, pp. 80–87), in *Infrared Optical Materials and Fibers V, Proc. SPIE*, San Diego, CA, 1 January 1987.
108. Berman LE. Plastic optical fiber: A short-haul solution, *Opt. Photonics News*, **9**(2), 29; 1998.
109. Mitsubishi Rayon C.D., Ltd. Homepage.
110. Ono T. Plastic optical fiber: The missing link for factory, office equipment, *Photonics Spectra*, **29**(11), 88–91; 1995.
111. Hecht J. *Understanding Fiber Optics*, 1st edn, Carmel, Indiana: SAMS, 1992.
112. Katzir A. *Lasers and Optical Fibers in Medicine*, San Diego, CA: Academic Press, 1993.
113. Azzam RMA, Bashara NM. *Ellipsometry and Polarized Light*, Amsterdam, the Netherlands: North-Holland, 1989.
114. Bergh RA, Digonnet MJF, Lefevre HC, Newton SA, Shaw HJ. Single mode fiber optic components, (Vol. 32, pp. 136–143), in *Fiber Optics-Technology, Proc. SPIE*, Berlin, Heidelberg, 1982.
115. Born M, Wolf E. *Principles of Optics*, 6th edn, Cambridge: Cambridge University Press, 1997.
116. Calvani R, Caponi R, Cisternino F. Polarization measurements on single-mode fibers, *J. Lightwave Technol.*, **7**(8), 1187–1196; 1989.
117. Huard S. *Polarization of Light*, Masson, Belgium: John Wiley & Sons, 1997.
118. Kaminov IP. Polarization in optical fibers, *IEEE J. Quantum Electron.*, **17**(1), 15–22; 1981.
119. Noda J, Okamoto K, Yokohama I. Fiber devices using polarization-maintaining fibers, Reprinted from *Fiber and Integrated Optics*, 6(4), 309–330; 1987; in *Selected Papers on Single-Mode Optical Fibers*, (Brozeit A, Hinsch KD, Sirohi RS, eds), SPIE, MS **101**, 23–33; 1994.
120. Rashleigh SC. Origins and control of polarization effects in single-mode fibers, *J. Lightwave Technol.*, **1**(2), 312–331; 1983.
121. Stolen RH, De Paula RP. *Proc. IEEE*, **75**(11), 1498–1511; 1987.
122. Todd DA, Robertson GRJ, Failes M. Polarization-splitting polished fiber optic couplers, *Opt. Eng.*, **32**(9), 2077–2082; 1993.
123. Yariv A. *Optical Electronics in Modern Communications*, 5th edn, New York, NY: Oxford University Press, 1997.
124. Agrawal G. *Nonlinear Fiber Optics*, 2nd edn, San Diego, CA: Academic Press, 1995.
125. Culshaw B, Michie C, Gardiner P, McGown A. Smart structures and applications in civil Engineering, *Proc. IEEE*, **84**(1), 78–86; 1996.

126. Ghatak A, Thyagarajan K. *Introduction to Fiber Optics*, New York: Cambridge University Press, 1998.
127. Kashyap R. Photosensitive optical fibers: Devices and applications, *Opt. Fiber Technol.*, **1**, 17–34; 1994.
128. Keiser GE. A Review of WDM Technology and Applications, *Opt. Fiber Technol.*, **5**, 3–39; 1999.
129. Kersey AD. A review of recent developments in fiber optic sensor technology, *Opt. Fiber Technol.*, **2**, 291–317; 1996.
130. Sirkis JS. Unified approach to phase-strain-temperature models for smart structure interferometric optical fiber sensors: Part 1, development, *Opt. Eng.*, **32**(4), 752–761; 1993.

19

Isotropic Amorphous Optical Materials

19.1   Introduction 677
        Refractive Index and Chromatic Dispersion • Other Optical
        Characteristics • Physical and Chemical Characteristics • Cost
19.2   Optical Glasses...683
19.3   Vitreous Silica 687
19.4   Mirror Materials 688
        Low-Expansion Glasses • Very Low-Expansion Glasses • Glass
        Ceramics • Beryllium • Aluminum
19.5   Optical Plastics...690
19.6   Infrared and Ultraviolet Materials 691
19.7   Optical Coatings and Other Amorphous Materials 692
19.8   Conclusions...693
References...693

Luis Efrain
Regalado and Daniel
Malacara-Hernández

19.1 Introduction

The materials used in optical instruments and research can be characterized by many physical and chemical properties [1–7]. The ideal material is determined by the specific intended application. Optical materials can be crystalline or amorphous. Crystalline materials can be isotropic or anisotropic, but amorphous materials can only be isotropic, except at its boundaries because of the surrounding media. In this chapter, some of the main isotropic amorphous materials used to manufacture optical elements are described.

The optical materials used to make optical elements, such as lenses or prisms, have several important properties to be considered, most important of which are described below.

19.1.1 Refractive Index and Chromatic Dispersion

The refractive index of a transparent isotropic material is defined as the ratio of the speed of light in vacuum to the speed of light in the material. With this definition, Snell's law of refraction gives

$$n_1 \sin\theta_1 = n_2 \sin\theta_2, \qquad (19.1)$$

where n_1 and n_2 are the refractive indices in two transparent isotropic media separated by an interface. The angles θ_1 and θ_2 are the angles between the light rays and the normal to the interface at the point where the ray passes from one medium to the other.

The refractive index for most materials in the visible region of the electromagnetic spectrum can vary from values close to one to values greater than two, as shown in Table 19.1. Sapphire is sometimes used when high resistance to scratching or shattering is needed. It is highly transparent but slightly birefringent.

TABLE 19.1   Refractive Indices of Some Materials

Material	Refractive Index
Air	1.0003
Water	1.33
Acrylic	1.49
Crown glass	1.48–1.70
Flint glass	1.53–1.95
Diamond	2.42
Sapphire	1.76

TABLE 19.2   Spectral Lines Used to Measure Refractive Indices

Wavelength (nm)	Spectral Line	Color	Element
1013.98	t	Infrared	Hg
852.11	s	Infrared	Cs
706.52	r	Red	He
656.27	C	Red	H
643.85	C′	Red	Cd
589.29	D	Yellow	Na
587.56	d	Yellow	He
546.07	e	Green	Hg
486.13	F	Blue	H
479.99	F′	Blue	Cd
435.83	g	Blue	Hg
404.66	H	Violet	Hg
365.01	i	Ultraviolet	Hg

The refractive index n of a given optical material is not the same for all colors: the value is greater for smaller wavelengths. The refractive indices of optical materials are measured at some specific wavelengths, as shown in Table 19.2.

The chromatic dispersion is determined by the principal dispersion ($n_F - n_C$). Another quantity that determines the chromatic dispersion is the Abbe value for the line d, given by

$$V_d = \frac{n_d - 1}{n_F - n_C}. \tag{19.2}$$

The Abbe value expresses the way in which the refractive index changes with wavelength. Optical materials are mainly determined by the value of these two constants.

Two materials with different Abbe numbers can be combined to form an achromatic lens with the same focal length for red (C) and for blue (F) light. However, the focal length for yellow (d) can be different. This is called the secondary spectrum. The secondary spectrum produced by an optical material or glass is determined by its partial dispersion. The partial dispersion $P_{x,y}$ for the lines x and y is defined as

$$P_{x,y} = \frac{n_x - n_y}{n_F - n_C}. \tag{19.3}$$

An achromatic lens for the lines C and F without secondary spectrum for yellow light d can only be made if the two transparent materials being used have different Abbe numbers V_d, but the same partial dispersion number $P_{d,F}$.

19.1.2 Other Optical Characteristics

19.1.2.1 Spectral Transmission and Reflection

The light transmittance through an optical material is affected by two factors, that is, the Fresnel reflections at the interfaces and the transparency of the material. The Fresnel reflections in a dielectric material like glass are a function of the angle of incidence, the polarization state of the incident light beam, and the refractive index. At normal incidence in air, the irradiance reflectance ρ_λ is a function of the wavelength; thus, the irradiance transmittance T_R due to the reflections at the two surfaces of the glass plate is

$$T_R = [1 - \rho_\lambda^2]^2. \tag{19.4}$$

The spectral transparency has large fluctuations among different materials and is also a function of the wavelength. Any small impurities in a piece of glass with concentrations as small as ten parts per billion can introduce noticeable absorptions at some wavelengths. For example, the well-known green color of window glass is due to iron oxides. The effect of impurities can be so high that the critical angle for prism-shaped materials can vanish for non-optical grade materials. High-index glasses have a yellowish color due to absorptions in the violet and ultraviolet regions, from the materials used to obtain the high index of refraction. At the spectral regions where a material has absorption, the refractive index is not a real number but a complex number n^* that can be written as

$$n^* = n(1 - i\kappa_\lambda), \tag{19.5}$$

where κ_λ is the absorption index, which is related to an extinction coefficient α_λ by

$$\alpha_\lambda = \frac{4\pi n}{\lambda}\kappa_\lambda. \tag{19.6}$$

If α_λ is the extinction coefficient for the material, the irradiance transmittance T_A due to absorption is

$$T_A = e^{-\alpha_\lambda t}, \tag{19.7}$$

where t is the thickness of the sample. Figure 19.1 shows the typical variations of the absorption index κ with the wavelength for metals, semiconductors, and dielectrics [8]. Dielectrics have two characteristic

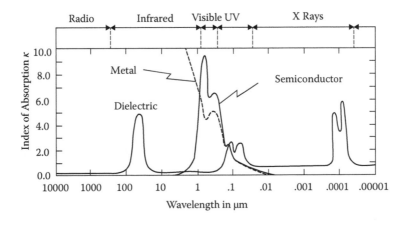

FIGURE 19.1 Absorption index as a function of the wavelength for optical materials. (Adapted from Kingery WD, et al., *Introduction to Ceramics,* John Wiley & Sons, New York, 1976.)

absorption bands. The absorption in the ultraviolet band is due to the lattice structure vibrations. Metals are highly absorptive at long wavelengths, due to their electrical conductivity, but they become transparent at short wavelengths. The absorption band in semiconductors is around the visible region.

The total transmittance T_λ of a sample with thickness t taking into consideration surface reflections as well as internal absorption is

$$T_\lambda = T_R T_A = [1 - \rho_\lambda^2]^2 e^{-\alpha_\lambda t}. \tag{19.8}$$

For dielectric materials far from an absorption region, the refractive index n_λ is real and the irradiance reflectance ρ is given by

$$\rho = \left[\frac{n_\lambda - 1}{n_\lambda + 1} \right]^2. \tag{19.9}$$

19.1.2.2 Optical Homogeneity

The degree to which the refractive index varies from point to point within a piece of glass or a melt is a measure of its homogeneity. A typical maximum variation of the refractive index in a melt is $\pm 1 \times 10^{-4}$, but more homogeneous pieces can be obtained. For the case of optical glasses, the homogeneity is specified in four different groups, as shown in Table 19.3.

19.1.2.3 Stresses and Birefringence

The magnitude of the residual stresses within a piece of glass depends mainly on the annealing of the glass. Internal stresses produce birefringence. The quality of the optical instrument may depend very much on the residual birefringence.

19.1.2.4 Bubbles and Inclusions

Bubbles are not frequent in good-quality optical glass, but they are always present in small quantities. When specifying bubbles and inclusions, the total percentage covered by them is estimated, counting only those ≥ 0.05 mm. Bubble classes are defined as shown in Table 19.4.

19.1.3 Physical and Chemical Characteristics

19.1.3.1 Thermal Expansion

The dimensions of most optical materials increase with temperature. The thermal expansion coefficient is also a function of temperature. If we plot the natural algorithm of L/L_0 versus the temperature T (°C), we obtain a graph (Figure 19.2) for glass BK7. Then, we can easily show that the slope of this curve at the temperature T is equal to the thermal expansion coefficient $\alpha(T)$ defined by

$$\alpha(T) = \frac{1}{L}\frac{dL}{dL}, \tag{19.10}$$

TABLE 19.3   Homogeneity Groups for Optical Glasses

Homogeneity Group	Maximum n_d Variation
H1	$\pm 2 \times 10^{-5}$
H2	$\pm 5 \times 10^{-6}$
H2	$\pm 2 \times 10^{-6}$
H3	$\pm 1 \times 10^{-6}$

TABLE 19.4   Bubble Classes for Optical Glasses

Bubble Class	Total Area of Bubbles (mm²)
B0	0–0.029
B1	0.03–0.10
B2	0.11–0.25
B3	0.26–0.50

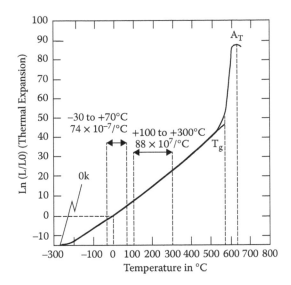

FIGURE 19.2 Thermal expansion of glass BK7 as a function of the temperature.

TABLE 19.5 Moh Hardness Scale

Hardness	Material Name	Chemical Compound
1	Talc	$Mg_3Si_4O_{10}(OH)_2$
2	Gypsum	$CaSO_4 \cdot 2H_2O$
3	Calcite	$CaCO_3$
4	Fluorite	CaF_2
5	Apatite	$Ca_5(PO_4)_3(F, Cl, OH)$
6	Felspar	$KAlSi_3O_8$
7	Quartz	SiO_2
8	Topaz	$Al_2SiO_4(OH, F)_2$
9	Ruby	Al_2O_3
10	Diamond	C

where L is the measured length at the temperature T and L_0 is the length at 0°C. For the case of glass, typically the expansion coefficient is zero at 0 K, increasing steadily until near room temperature; then, the coefficient continues to increase almost linearly until about 700 K, after which it increases very sharply.

It is customary to specify two values of the thermal expansion coefficient $\alpha(T)$ for two linear ranges of the plot of $\ln(L/L_0)$ vs. T, where this coefficient remains constant. One interval is from −30 to +70°C and the other from +100 to +300°C. The transformation temperature T_g is that where the glass begins to transform from a solid to a plastic state. The yield point A_T is the temperature at which the thermal expansion coefficient becomes zero.

Thermal expansion is nearly always very important, but for mirrors, especially astronomical mirrors, it becomes very critical, because any thermal expansion deforms the optical surface, and thus the formed image.

19.1.3.2 Hardness

The hardness of materials is important: hard materials are more difficult to scratch but also more difficult to polish. The Mohs scale, introduced more than 100 years ago by Friedrich Mohs (1773–1839), is based on which materials scratch others and ranges from 1 to 10, in unequal steps. Table 19.5 shows the materials used to define this scale.

There are many other ways to define the hardness of a material, but another common method used mainly for optical glasses is the Knoop scale, which is defined by the dimensions of a small indentation produced with a diamond under a known pressure. A rhomboidal diamond with vertex angles 172°30' and 130°00' with respect to the vertical direction is used to produce a mark on the polished surface. Then, the Knoop hardness HK is defined as

$$\text{HK} = 1.45 \frac{F_N}{d^2} = 14.23 \frac{F_K}{d^2},$$ (19.11)

where F_N is the force in newtons, F_K is the weight (force) of 1 kg, and d is the length of the longest diagonal of the indentation in millimeters. In glasses, a pressure of 1.9613 newtons (the weight of 200 g) is applied to the surface for 10 s. This number is important because it indicates the sensitivity to scratches and is also directly related to the grinding and lapping speeds.

The Rockwell hardness is defined by the depth of penetration under the load of a steel ball or a diamond cone. This method is used for materials softer than glass, such as metals or plastics. There are separate scales for ferrous metals, nonferrous metals, and plastics. The most common Rockwell hardness scales are B and C for metals: the B scale uses a ball, while the C scale uses a cone. The M and R scales are employed for polymers: the M scale is for harder materials and the R scale is for the softer ones.

19.1.3.3 Elasticity

Elasticity and rigidity are specified by Young's modulus E, which is related to the torsional rigidity modulus G and to Poisson's ratio μ. Young's modulus E in glasses is measured from induced transverse vibrations of a glass rod at audio frequencies. The torsional rigidity is calculated from torsional vibrations of the rods. The relation between the rigidity modulus, Poisson's ratio, and Young's modulus is

$$\mu = \frac{E}{2G} - 1.$$ (19.12)

Young's modulus is directly related to the hardness of the material, as can be seen in most glass manufacturer's specifications. Young's modulus is also important when considering thermal and mechanical stresses.

19.1.3.4 Density

The density is the mass per unit volume. This quantity is quite important in many applications: for example, in space instruments and in ophthalmic lenses. Crown glasses are in general less dense than Flint glasses.

19.1.3.5 Chemical Properties

Resistance to stain and corrosion by acids or humidity is also important, mainly if the optical glass is going to be used in adverse atmospheric conditions. The water vapor present in the air can produce stains on the glass surface, which cannot be wiped out. An accelerated procedure is used to test these properties of optical materials by exposing them to a water vapor saturated atmosphere. The temperature is alternated between 45°C and 55°C in a cycle of about one hour. Then, a condensate forms on the material during the heating phase and dries during the cooling phase. With this test, the climatic resistance of optical materials is classified in four different classes:

- Class CR1: after 180 hours (one week) there is no sign of deterioration of the surface of the material.
- Class CR4: after less than 30 hours there are signs of scattering produced by the atmospheric conditions.
- Classes CR2 and CR3 are intermediate to Classes CR1 and CR4.

In a similar manner, the resistance to acids and alkalis is tested and classified.

19.1.4 Cost

The cost of the optical material or glass to be used should also be considered when selecting a material for an optical design. The range of prices is extremely large. Glasses and plastics produced in large quantities are cheaper than specially produced materials.

19.2 Optical Glasses

Glass is a material in a so-called glassy state, structurally similar to a liquid. At ambient temperatures, it reacts to the impact of force with elastic deformations. Thus, it can be considered as an extremely viscous liquid. It is always an inorganic compound made from sand and sodium and calcium compounds. Plastics, on the other hand, are organic. Glass can also be in natural forms, such as obsidian, found in volcanic places. Obsidian was fashioned into knives, arrowheads, spearheads, and other weapons in ancient times.

Glass is made by mixing silica sand (SiO_2) with small quantities of some inorganic materials, such as soda and lime and some pieces of previously fabricated glass, called glass cullet. This glass cullet acts as a fluxing agent, which accelerates the melting of the sand. With the silica and the soda, sodium silicate is formed, which is soluble in water. The presence of the lime reduces the solubility of the sodium silicate. This mixture is heated to about 1500°C and then cooled at a well-controlled rate to prevent crystallization. To release any internal stresses, the glass is cooled very slowly at the proper speed in a process called annealing [9]. The quartz sands used to make glass have to be quite free from iron (less than 0.001%), otherwise a greenish-colored glass is obtained. The optical glass used for refracting optical elements, such as lenses or prisms has several important properties to be considered, the most important of which is described here. The refractive indices versus the wavelength for some optical glasses are shown in Figure 19.3. A diagram of the Abbe number V_d versus the refractive index n_d, for Schott glasses is shown in Figure 19.4. The glasses with a letter K at the end of the glass type name are crown glasses and those with a letter F are flint glasses.

The chromatic variation in the refractive index of glasses for the visual spectral range can be represented by several approximate expressions. The simplest and probably oldest formula was proposed by Cauchy [10]:

$$n = A_0 + \frac{A_1}{\lambda^2} + \frac{A_2}{\lambda^4}.$$

(19.13)

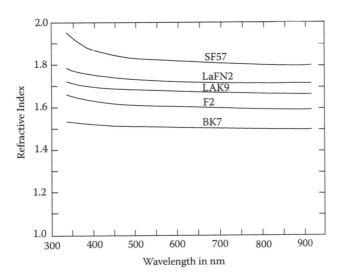

FIGURE 19.3 Refractive indices as a function of the wavelength for some optical glasses.

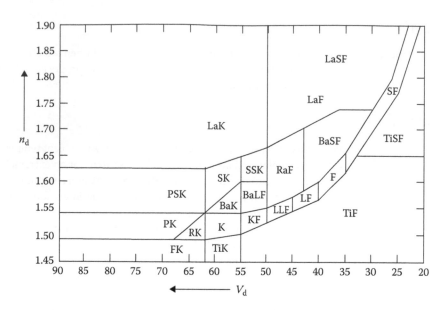

FIGURE 19.4   Abbe number versus refractive index chart for optical glasses.

This formula is accurate to the third or fourth decimal place in some cases. An empirically improved formula was proposed by Conrady [11] as

$$n = A_0 + \frac{A_1}{\lambda^2} + \frac{A_2}{\lambda^{3.5}} \tag{19.14}$$

with an accuracy of one unit in the fifth decimal place. Better formulas have been proposed by several authors: for example, by Herzberger [12].

From a series expansion of a theoretical dispersion formula, a more accurate expression was used by Schott for many years. Recently, Schott has adopted a more accurate expression for glasses called the Sellmeier formula [13], which is derived from the classical dispersion theory. This formula permits the interpolation of refractive indices in the entire range, from infrared to ultraviolet, with a precision better than 1×10^{-5}, and it is written as

$$n^2 = \frac{B_1\lambda^2}{\lambda^2 - C_1} + \frac{B_2\lambda^2}{\lambda^2 - C_2} + \frac{B_3\lambda^2}{\lambda^2 - C_3}. \tag{19.15}$$

The coefficients are provided by the glass manufacturers, using the refractive indices' values from several melt samples. The values of these coefficients for each type of glass are supplied by the glass manufacturers.

The refractive index and the chromatic dispersion of optical materials are not the only important factors to be considered when designing an optical instrument.

There is such a large variety of optical glasses that to have a complete stock of all types in any optical shop is impossible. Many lens designers have attempted to reduce the list to the most important glasses, taking into consideration important factors, such as optical characteristics, availability, and price. Table 19.6 lists some of the most commonly used optical glasses.

The location of these glasses in a diagram of the Abbe number V_d versus the refractive index n_d is given in Figure 19.5. Figure 19.6 shows a plot of the partial dispersion $P_{g,F}$ versus the Abbe number V_d for these

TABLE 19.6 Refractive Indices for Some Optical Glasses

Schott Name	V_d	n_C	n_d	n_F	n_g
BaF4	43.93	1.60153	1.60562	1.61532	1.62318
BaFN10	47.11	1.66579	1.67003	1.68001	1.68804
BaK4	56.13	1.56576	1.56883	1.57590	1.58146
BaLF5	53.63	1.54432	1.54739	1.55452	1.56017
BK7	64.17	1.51432	1.51680	1.52238	1.52668
F2	36.37	1.61503	1.62004	1.63208	1.64202
K4	57.40	1.51620	1.51895	1.52524	1.53017
K5	59.48	1.51982	1.52249	1.52860	1.53338
KzFSN4	44.29	1.60924	1.61340	1.62309	1.63085
LaF2	44.72	1.73905	1.74400	1.75568	1.76510
LF5	40.85	1.57723	1.58144	1.59146	1.59964
LAK9	54.71	1.68716	1.69100	1.69979	1.70667
LLF1	45.75	1.54457	1.54814	1.55655	1.56333
PK51A	76.98	1.52646	1.52855	1.53333	1.53704
SF2	33.85	1.64210	1.64769	1.66123	1.67249
SF8	31.18	1.68250	1.68893	1.70460	1.71773
SF10	28.41	1.72085	1.72825	1.74648	1.76198
SF56A	26.08	1.77605	1.78470	1.80615	1.82449
SK6	56.40	1.61046	1.61375	1.62134	1.62731
SK16	60.32	1.61727	1.62041	1.62756	1.63312
SK18A	55.42	1.63505	1.63854	1.64657	1.65290
SSKN5	50.88	1.65455	1.65844	1.66749	1.67471

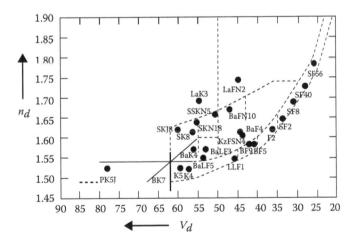

FIGURE 19.5 Abbe number versus refractive index for some common optical glasses as shown in Table 19.6.

glasses. Table 19.7 shows some physical properties for the glasses in Table 19.6. The visible transmittance for some optical glasses is shown in Figure 19.7, which can be considered as an expansion in the visible region of Figure 19.1.

Ophthalmic glasses are cheaper than optical glass, since their quality requirements are in general lower. They are produced to make spectacles. Table 19.8 lists some of these glasses.

Finally, Table 19.9 lists some other amorphous isotropic materials used in optical elements.

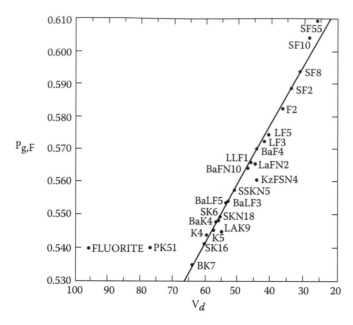

FIGURE 19.6   Abbe number versus relative partial dispersion of some common optical glasses as shown in Table 19.6.

TABLE 19.7   Physical Properties for Glasses in Table 19.6

Schott Name	Thermal Expansion Coefficient α (1/°C)	Density ρ (g/cm³)	Young's Modulus E (N/mm²)	Knoop Hardness HK
BaF4	7.9×10^{-6}	3.50	66×10^3	400
BaFN10	6.8×10^{-6}	3.76	89×10^3	480
BaK4	7.0×10^{-6}	3.10	77×10^3	470
BaLF5	8.1×10^{-6}	2.95	65×10^3	440
BK7	7.1×10^{-6}	2.51	81×10^3	520
F2	8.2×10^{-6}	3.61	58×10^3	370
K4	7.3×10^{-6}	2.63	71×10^3	460
K5	8.2×10^{-6}	2.59	71×10^3	450
KzFSN4	4.5×10^{-6}	3.20	60×10^3	380
LaF2	8.1×10^{-6}	4.34	87×10^3	480
LF5	9.1×10^{-6}	3.22	59×10^3	410
LaK9	6.3×10^{-6}	3.51	110×10^3	580
LLF1	8.1×10^{-6}	2.94	60×10^3	390
PK51A	12.7×10^{-6}	3.96	73×10^3	340
SF2	8.4×10^{-6}	3.86	55×10^3	350
SF8	8.2×10^{-6}	4.22	56×10^3	340
SF10	7.5×10^{-6}	4.28	64×10^3	370
SF56A	7.9×10^{-6}	4.92	58×10^3	330
SK6	6.2×10^{-6}	3.60	79×10^3	450
SK16	6.3×10^{-6}	3.58	89×10^3	490
SK18A	6.4×10^{-6}	3.64	88×10^3	470
SSKN5	6.8×10^{-6}	3.71	88×10^3	470

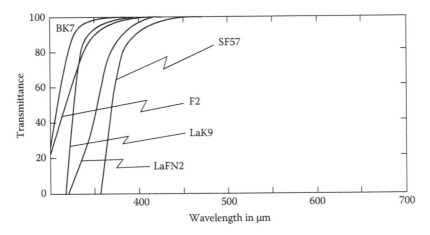

FIGURE 19.7 Internal transmittance as a function of the wavelength for some optical glasses (without the surface reflections), with a thickness of 5 mm.

TABLE 19.8 Refractive Indices for Some Ophthalmic Glasses

Glass Type	V_d	n_C	n_d	n_F	Density (g/cm³)
Crown	58.6	1.5204	1.5231	1.5293	2.54
Compatible flint	42.3	1.5634	1.5674	1.5768	3.12
Compatible flint	38.4	1.5926	1.5972	1.6082	3.40
Compatible flint	36.2	1.6218	1.6269	1.6391	3.63
Compatible flint	32.4	1.6656	1.6716	1.6863	3.86
Soft barium flint	33.2	1.6476	1.6533	1.6672	3.91
Soft barium flint	29.6	1.6944	1.7013	1.7182	4.07
Low-density flint	31.0	1.6944	1.7010	1.7171	2.99

TABLE 19.9 Refractive Indices for Some Optical Isotropic Materials

Material	V_d	n_C	n_d	n_F	n_g
Fused rock crystal	67.6	1.45646	1.45857	1.46324	1.46679
Synthetic fused silica	67.7	1.45637	1.45847	1.46314	1.46669
Fluorite	95.3	1.43249	1.43384	1.43704	1.43950

19.3 Vitreous Silica

Vitreous silica is a natural material formed by silicon dioxide (SiO_2): it is noncrystalline and isotropic. It is also known as fused quartz or fused silica. This material can be fabricated by the fusion of natural crystalline quartz or synthesized by thermal decomposition and oxidation of $SiCl_4$. The optical properties of these types of vitreous silica are similar but not identical.

The fusion of crystalline quartz can be made by direct fusion of large pieces. This fused quartz has a low ultraviolet transmittance due to metallic impurities. On the other hand, it has a high infrared transmittance due to its low content of hydroxyl. For this reason, this material is frequently used for infrared windows.

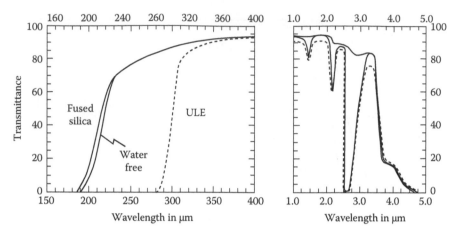

FIGURE 19.8 Ultraviolet and infrared transmittances for three types of vitreous silica, including surface reflections, with a thickness of 10 mm.

TABLE 19.10 Refractive Indices for Fused Quartz

Material	V_d	n_C	n_d	n_F	n_g
Fused silica Corning 7940	67.8	1.4564	1.4584	1.4631	1.4667

If the fusion is made with powdered quartz in the presence of chlorine gases, the transmission in the ultraviolet region is improved, but it is worsened in the infrared. The reason is that the water content increases.

If the hydrolysis of an organosilicon compound in the vapor phase is performed, a synthetic high-purity fused silica free of metals is obtained. It has a high transmission in the ultraviolet but the infrared transmission is low due to its high water content. Some manufacturers have improved their processes to reduce the water, and obtained good transparency in the ultraviolet as well as in the infrared. These good transmission properties make this fused quartz ideal for infrared and ultraviolet windows and for the manufacturing of optical fibers.

The transmittance of three different types of fused quartz are shown in Figure 19.8. Table 19.10 shows the refractive indices for this material. The thermal expansion of fused silica is 0.55×10^{-6}/°C in the range from 0 to 300°C. Below 0°C, this coefficient decreases, until it reaches a minimum near –120°C.

19.4 Mirror Materials

Glasses or other materials used to make metal [14] or dielectric coated mirrors do not need to be transparent. Instead, the prime useful characteristic is thermal stability. The thermal expansion coefficient for mirrors has to be lower than for lenses for two reasons: first, a change in the figure of a surface affects the refracted or reflected wavefront four times more in reflecting surfaces than in refracting surfaces; secondly, mirrors are frequently larger than lenses, especially in astronomical instrumentation. Some of the materials used for mirrors are shown in Table 19.11 with the thermal expansion coefficients plotted in Figure 19.9. They are described below.

19.4.1 Low-Expansion Glasses

The most well-known of these glasses are Pyrex (produced by Corning Glass Works), Duran 50 (Schott), and E-6 (Ohara). These are borosilicate glasses in which B_2O_3 replaces the CaO and MgO of soda-lime window glass. The softening temperature is higher in these glasses than in normal window glass.

TABLE 19.11 Physical Properties for Some Mirror Materials

Material	Thermal Expansion Coefficient α (1/°C)	Density ρ (g/cm³)	Young's Modulus E (N/cm²)
Pyrex	2.5×10^{-6}	2.16	5.4×10^6
ULE™	$\pm0.03 \times 10^{-6}$	2.21	6.7×10^6
Fused quartz	0.55×10^{-6}	2.19	7.2×10^6
Zerodur	$\pm0.05 \times 10^{-6}$	2.53	9.1×10^6
Beryllium	12.0×10^{-6}	1.85	28.7×10^6
Aluminum	24.0×10^{-6}	2.70	6.8×10^6

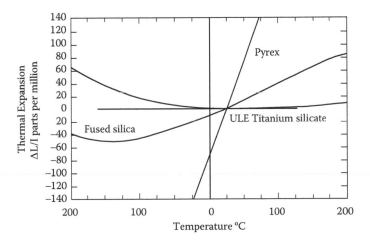

FIGURE 19.9 Thermal expansion for fused silica, ULE™ titanium silicate, and Pyrex.

19.4.2 Very Low-Expansion Glasses

The thermal expansion of fused silica can be lowered even more by shifting the zero expansion point to 300 K by adding 7% of titania-silica glass. This glass is a synthetic amorphous silica known as Corning ULE™ titanium silicate.

19.4.3 Glass Ceramics

Glass ceramic materials contain both crystalline and glass phases. They are produced as lithia-alumina glasses where a microcrystalline structure is produced with a special thermal procedure. As opposed to ceramic materials, glass ceramics are nonporous. In general, they are not transparent, although sometimes they could be, depending on the size of the crystals, which is about 0.05 µm. Frequently, they are turbid with an amber color. The Pyroceram technology developed by Corning led to the production of a glass ceramic with almost zero thermal expansion coefficient, which is manufactured by several companies, such as CER-VIT (produced by Owens-Illinois), Zerodur (Schott), and Cryston-Zero (Hoya).

19.4.4 Beryllium

Beryllium is a very light and elastic metal. Aluminum, copper, or most other metal mirrors are fabricated by casting the metal. However, for beryllium, this is not possible because cast grains are very large and brittle with little intergranular strength. The method of producing beryllium mirrors is by direct compaction of powder.

Hot isostatic pressing has been used in one single step for many years since the late 1960s [15], using pressures up to 50,000 psi to obtain a green compact blank. A multistep cold isostatic pressing in a rubber container followed by vacuum sintering or hot isostatic pressing has also been recently used [16].

Beryllium is highly toxic in powder form. It can be machined only in specially equipped shops using extreme precautions when grinding and polishing. In many respects–mainly its specific weight, elasticity, and low thermal distortion—this is the ideal material for space mirrors.

19.4.5 Aluminum

Aluminum is a very light metal than can also be easily polished and machined. Many mirrors are being made with aluminum for many applications. Even large astronomical mirrors for telescopes that do not need a high image resolution have been made with this material.

19.5 Optical Plastics

There are a large variety of plastics, with many different properties, used to make optical components [17–19]. Plastics have been greatly improved recently, mainly because of the high-volume production of plastics for CD-roms [20] and spectacle lenses [21,22]. The most common plastics used in optics, whose properties are given in Table 19.12, are as follows:

1. Methyl methacrylate, also called acrylic, is the most desirable of all plastics. It is the common equivalent to crown glass. It is relatively inexpensive and can be easily molded and polished. It also has many other advantages, such as transmission, scratch resistance, and moisture absorption.
2. Polystyrene or styrene is the most common equivalent to flint glass. It is the cheapest plastic and the easiest to mold, but it is difficult to polish. It is frequently used for color-correction lenses.
3. Polycarbonate has a very high impact strength as well as high temperature stability and very low moisture absorption. However, it is expensive, difficult to polish, and scratches quite easily.
4. Methyl methacrylate or allyldiglycol carbonate, commonly known as CR-39, is a thermosetting material that should be casted and cured with highly controlled conditions. It can be easily machined, molded, and polished. It is as transparent as acrylic. Its most frequent use is in ophthalmic lenses.

Some of the most important physical characteristics of these plastics are given in Table 19.13 and their spectral transmission is shown in Figure 19.10. (See also Figures 19.7 and 19.8.)

TABLE 19.12   Refractive Indices for Some Optical Plastics

Material	V_d	n_C	n_d	n_F
Acrylic	57.2	1.488	1.491	1.497
Polystyrene	30.8	1.584	1.590	1.604
Polycarbonate	30.1	1.577	1.583	1.597
CR-39	60.0	1.495	1.498	1.504

TABLE 19.13   Physical Characteristics for Some Optical Plastics

Material	Thermal Expansion Coefficient α (1/°C)	Density ρ (g/cm³)	Rockwell Hardness
Acrylic	65×10^{-6}	1.19	M97
Polystyrene	80×10^{-6}	1.06	M90
Polycarbonate	70×10^{-6}	1.20	M70
CR-39	100×10^{-6}	1.09	M75

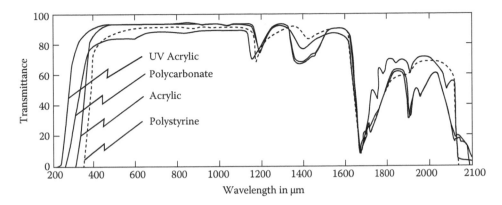

FIGURE 19.10 Ultraviolet and infrared transmittances of some plastics, including surface reflections, with a thickness of 3.22 mm.

TABLE 19.14 Characteristics of Some Infrared Materials

Material	n_d	Wavelength Cutoff (µm)
Vycor*	1.457	3.5
Germanate	1.660	5.7
$CaAl_2O_4$	1.669	5.8
As_2S_3	2.650	12.5
Irtran 1 MgF_2	1.389	7.5
Irtran 2 ZnS	2.370	14.0
Irtran 3 CaF_2	1.434	10.0
Irtran 5 MgO	1.737	8.0

Plastic lenses [23] are very cheap compared with glass lenses if made in large quantities. However, they cannot be used in hostile environments, where the lens is exposed to chemicals, high temperatures, or abrasion. Their use in high-precision optics is limited because of their physical characteristics. There is a lack of plastics with high refractive indices; their homogeneity is not as high as in optical glasses and this is a field needing intense investigation. Plastics have a low specific gravity and their thermal expansion is high.

19.6 Infrared and Ultraviolet Materials

Most glasses are opaque to infrared and ultraviolet radiation. If a lens or optical element has to be transparent at these wavelengths, special materials have to be selected.

The infrared spectral regions from red light to about 4 µm are quite important. The research on materials that are transparent at these regions is making them more available and better every day [24–31]. Infrared materials are in the form of glasses, semiconductors, crystals, metals, and many others. In the case of optical glasses, the absorption bands due to absorbed water vapor need to be avoided. Common optical glasses transmit until about 2.0 µm and their absorption becomes very high in the water band region 2.7–3.0 µm. Special manufacturing techniques are used to reduce the amount of water in glasses. Unfortunately, these processes also introduce some scattering. An example is the fused quartz manufactured with the name of Vycor*. Table 19.14 shows the properties of some of the many available infrared materials in the form of glasses or hot-pressed polycrystalline compacts (crystals and semiconductors are excluded).

TABLE 19.15   Characteristics of Some Ultraviolet Materials

Material	n_d	V_d	Wavelength Cutoff (nm)
Sodium chloride	1.544	42.8	250
Potassium bromide	1.560	33.4	210
Potassium iodide	1.667	32.2	250
Lithium fluoride	1.392	99.3	110
Calcium fluoride	1.434	95.1	120
Fused quartz	1.458	67.8	220
Barium fluoride	1.474	81.8	150

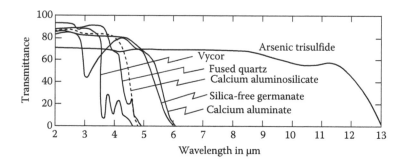

FIGURE 19.11   Infrared transmittance as a function of wavelength for several materials.

The ultraviolet region also has many interesting applications. Pellicori [32] has described the transmittances of some optical materials for use in the ultraviolet region, between 190 and 340 nm. In general, materials for the ultraviolet region are more rare and expensive than those for the infrared. Table 19.15 shows some of these materials: the first three are highly soluble in water.

Figure 19.11 shows the infrared transmittances for some other infrared materials.

19.7 Optical Coatings and Other Amorphous Materials

In thin-film work, the properties of materials are also very important. Most of the materials used in the production of optical coatings, single layer or multilayers, that is, mirrors, antireflective, broadband and narrow filters, edge filters, polarizers, and so on, must be performed with a knowledge of the refractive index and the region of transparency, hardness or resistance to abrasion, magnitude of any built-in stresses, solubility, resistance to attack by the atmosphere, compatibility with other materials, toxicity, price, and availability. Also, sometimes, the electrical conductivity or dielectric constant or emissivity for selective materials used in solar absorption (only from elements of groups IV, V, and VI of the periodic system) has to be known.

Most of the techniques used to prepare thin films yield amorphous or poly-crystalline materials in a two-dimensional region, because the thickness is always of the order of a wavelength. The most common materials used in optical coatings are metals, oxides, fluorides, and sulfides. Their optical properties may vary because the method of evaporation produces different density packing in thin films than in bulk materials. The dielectric function may be studied *in situ* or *ex situ* by analyzing the spectral transmittance and reflectance and by weighing the deposited materials.

A stack of thin films or multilayers with alternated high and low refractive index (sulphides and fluorides) may have the properties of new materials to produce high reflection mirrors, antireflection treatments, edge and interference filters among other optical devices. Because of their thickness dimensions, they are called actually nanomaterials [33].

The use of metal films coated over prisms or diffraction rulings introduce a momentum because of the phase shift introduced by the metallic absorption index and the frequency is modified by the refraction index.

Many applications have been published using plasma surface waves for studies of radiative interaction of interfaces in the region of non-radiative or evanescent waves [34].

This topic called Plasmonics [35] represents an important discovery, which is that of metamaterials with negative refractive indices in the optical frequencies and their application to a cloaking approach [36] or imaging with a perfect lens [37].

Sol–gel glasses are of high purity, but at the present stage of development there are only applications in large thin-film coatings as heat or IR reflecting coatings on windows or coatings on rear-view to reduce glare and reflections.

Inhomogeneous films or rugates [38] are a kind of multilayer film with sinusoidal or other functional refractive index profiles avoiding the presence of very abrupt interfaces. They act as a single layer with the same properties of some multilayers and show no harmonic peaks in the case of optical filters. These materials are mainly isotropic and amorphous.

Only deposits made on crystal substrates, at high substrate temperatures, with special techniques, such as Knudsen cells or molecular or liquid beam epitaxy (MBE, LBE), produce crystalline films epitaxially, but these are extreme conditions searched for very specific applications, such as semiconductor devices, and even the lattice parameters of the materials available are very restrictive.

19.8 Conclusions

The field of optical materials is a very active one and every day we have new materials that permit us to make much better optical instruments.

References

1. Barnes WP Jr. Optical materials—Reflective, in *Applied Optics and Optical Engineering*, Vol. VII, Shannon, R. R. and J. C. Wyant, eds, San Diego, CA: Academic Press, 1979.
2. Kavanagh AJ. Optical material, in *Military Standardization Handbook: Optical Design, MIL-HDBK 141*, Washington, DC: US Defense Supply Agency, 1962.
3. Kreidl NJ, Rood JL. *Optical materials, in Applied Optics and Optical Engineering*, Vol. I, Kingslake, R., ed., San Diego, CA: Academic Press, 1965.
4. Musikant S. *Optical Materials. An Introduction to Selection and Application*, New York, NY: Marcel Dekker, 1985.
5. Tropf WJ, Thomas ME, Harris TJ. Properties of crystals and glasses, in *Handbook of Optics*, 2nd ed, Part II, New York, NY: McGraw-Hill, 1995.
6. Villa F, Martinez A, Regalado LE. Correction masks for thickness uniformity in large area thin films, *Appl. Opt.*, **39**, 1602–1610; 2000.
7. Wolfe WL. Optical materials in *The Infrared and Electro Optics Systems* Handbook, Shumaker, D.L., Accetta, S.J., Eds., Vol. 3, Chap 1, Rogatto, W.D., Ed., Bellingham, WA: *SPIE Press*, 1993.
8. Kingery WD, Bowen HK, Uhlmann DR. *Introduction to Ceramics*, New York, NY: John Wiley & Sons, 1976.
9. Horne DF. *Optical Production Technology*, Bristol, England: Adam Hilger, 1972.
10. Calve JG. *Mémoire sur la Dispersion de la Lumiere*, the Czech Republic: Prague, 1936.
11. Conrady AE. *Applied Optics and Optical Design Part II*, Dover, New York, NY: Dover 1960.
12. Weber MJ. *CRC Handbook of Laser Science and Technology Supplement 2: Optical Materials*, Boca Raton, FL: CRC Press, LLC, 1995.
13. Pellicori SF. Transmittances of some optical materials for use between 1900 and 3400 Å, *Appl. Opt.*, **3**, 361–366; 1964.

14. Taylor HD. Metal mirrors in the large, *Opt. Eng.*, **14**, 559–561; 1975.
15. Paquin RA. Hot isostatic pressed beryllium for large optics, *Opt. Eng.*, **25**, 1003–1005; 1986.
16. Paquin RA. Hot isostatic pressed beryllium for large optics, *Proc. SPIE*, **1168**, 83–87; 1989.
17. Tatian B. Fitting refractive-index data with the Sellmeier dispersion formula, *Appl. Opt.*, **23**, 4477–4485; 1984.
18. Greis U, Kirchhof G. Injection molding of plastic optics, *Proc. SPIE*, **381**, 69–71; 1983.
19. Leonhardt L, Smith DR. Focus issue on Cloaking and transformation optics, *New Journal of Physics*, **10**, Nov 2008.
20. McCarthy DE. The reflection and transmission of infrared materials, part 3, spectra from 2 μm to 50 μm, *Appl. Opt.*, **4**, 317–320; 1965a.
21. Koppen W. Optical surfaces out of plastic for spectacle lenses, *Proc. SPIE*, **381**, 78–80; 1983.
22. Ventures GM. Improved plastic molding technology for ophthalmic lens & contact lens, *Proc. SPIE*, **1529**, 13–21; 1991.
23. Conrady AE. *Applied Optics and Optical Design Part II*, New York, NY: Dover, 1960.
24. Malitson IH. A redetermination of some optical properties of Calcium Fluoride, *Appl. Opt.*, **2**, 1103–1107; 1963.
25. McCarthy DE. The reflection and transmission of infrared materials, part 1, Spectra From 2 μm to 50 μm, *Appl. Opt.*, **2**, 591–595; 1963a.
26. McCarthy DE. The reflection and transmission of infrared materials, part 2, bibliography, *Appl. Opt.*, **2**, 596–603; 1963b.
27. Lytle JD. Aspheric surfaces in polymer optics, *Proc. SPIE*, **381**, 63–65; 1983.
28. Palmer AL. Practical design considerations for polymer optical systems, *Proc. SPIE*, **306**, 18–20; 1981.
29. McCarthy DE. The reflection and transmission of infrared materials, part 5, spectra from 2 μm to 50 μm, *Appl. Opt.*, **7**, 1997–2000; 1965c.
30. Peck WG, Tribastone C. Issues in large scale production of plastic lenses, *Proc. SPIE*, **2622**, 144–146; 1995.
31. Parker CJ. *Optical materials—Refractive, in Applied Optics and Optical Engineering*, Vol. VII, Shannon RR, Wyant JC, eds, San Diego, CA: Academic Press, 1979.
32. McCarthy DE. The reflection and transmission of infrared materials, part 4, bibliography, *Appl. Opt.*, **4**, 507–511; 1965b.
33. Novotny L, Hecht B. *Nano-Optics*, Cambridge, London: Cambridge University Press, 2006 Also Busch K, von Freymann G, Linden S, Mingaleev S, Tkeshelashvili L, and Wegener M, 2007.
34. Piazza L, Lummen TTA Carbone F. Simultaneous observation of the quantization and the interference pattern of a plasmonic near-field, *Nature Communications*, **6**, 6407; 2015.
35. Stefan A. Maier, *Plasmonics: Fundamentals and Applications*, New York, NY: Springer Science Business Media LLC, 2007.
36. McCarthy DE. The reflection and transmission of infrared materials, part 6, bibliography, *Appl. Opt.*, **7**, 2221–2225; 1965d.
37. Hersberger M, Salzberg CD. Refractive indices of infrared optical materials and color correction of infrared lenses, *J. Opt. Soc. Am.*, **52**, 420–427; 1962.
38. Rainwater D, et al., *New J. Phys.*, doi:10.1088/1367-2630/14/1/013054; 2012.

20

Anisotropic Materials

20.1	Introduction	695
20.2	Review of Concepts from Electromagnetism	696
20.3	Crystalline Materials and Their Properties	698
20.4	Crystals	699
	The Index Ellipsoid • Natural Birefringence • Wave Surface • Wavevector Surface	
20.5	Application of Electric Fields: Induced Birefringence and Polarization Modulation	711
20.6	Magneto-Optics	718
20.7	Liquid Crystals	719
20.8	Modulation of Light	722
20.9	Concluding Remarks	722
	References	724

Dennis H. Goldstein

20.1 Introduction

In this chapter, we discuss the interaction of light with anisotropic materials. An anisotropic material has properties (thermal, mechanical, electrical, optical, etc.) that are different in different directions. Most materials are anisotropic. This anisotropy results from the structure of the material, and our knowledge of the nature of that structure can help us to understand the optical properties.

The interaction of light with matter is a process that is dependent upon the geometrical relationships of the light and matter. By its very nature, light is asymmetrical. Considering light as a wave, it is a transverse oscillation in which the oscillating quantity, the electric field vector, is oriented in a particular direction in space perpendicular to the propagation direction. Light that crosses the boundary between two materials, isotropic or not, at any angle other than normal to the boundary, will produce an anisotropic result. The Fresnel equations illustrate this. Once light has crossed a boundary separating materials, it experiences the bulk properties of the material through which it is currently traversing, and we are concerned with the effects of those bulk properties on the light.

The study of anisotropy in materials is important to understanding the results of the interaction of light with matter. For example, the principle of operation of many solid-state and liquid crystal spatial light modulators is based on polarization modulation [1]. Modulation is accomplished by altering the refractive index of the modulator material, usually with an electric or magnetic field. Crystalline materials are an especially important class of modulator materials because of their use in electro-optics and in ruggedized or space-worthy systems, and also because of the potential for putting optical systems on integrated circuit chips.

We briefly review the electromagnetics necessary to the understanding of anisotropic materials, and show the source and form of the electro-optic tensor. We discuss crystalline materials and their properties, and introduce the concept of the index ellipsoid. We show how the application of electric and magnetic fields alters the properties of materials, and give examples. Liquid crystals are also discussed.

A brief summary of electro-optic modulation modes using anisotropic materials concludes the chapter.

20.2 Review of Concepts from Electromagnetism

Recall from electromagnetics (see, for example, [2–4]) that the electric displacement vector $\vec{D}$ is given by (MKS units):

$$\vec{D} = \varepsilon \vec{E}, \tag{20.1}$$

where ε is the permittivity and $\varepsilon = \varepsilon_0(1 + \chi)$, where ε_0 is the permittivity of free space, χ is the electric susceptibility, $(1 + \chi)$ is the dielectric constant, and $n = (1 + \chi)^{1/2}$ is the index of refraction. The electric displacement is also given by

$$\vec{D} = \varepsilon_0 \vec{E} + \vec{P}, \tag{20.2}$$

but

$$\vec{D} = \varepsilon_0(1 + \chi)\vec{E} = \varepsilon_0\vec{E} + \varepsilon_0\chi\vec{E}, \tag{20.3}$$

so $\vec{P}$, the polarization (also called the electric polarization or polarization density) is $\vec{P} = \varepsilon_0\chi\vec{E}$.

The polarization arises because of the interaction of the electric field with bound charges. The electric field can produce a polarization by inducing a dipole moment, that is, separating charges in a material, or by orienting molecules that possess a permanent dipole moment.

For an isotropic, linear medium,

$$\vec{P} = \varepsilon_0\chi\vec{E} \tag{20.4}$$

and χ is a scalar; but note that in

$$\vec{D} = \varepsilon_0\vec{E} + \vec{P} \tag{20.5}$$

the vectors do not have to be in the same direction and, in fact, in anisotropic media, $\vec{E}$ and $\vec{P}$ are not in the same direction (and so $\vec{D}$ and $\vec{E}$ are not in the same direction). Note that χ does not have to be a scalar nor is $\vec{P}$ necessarily related linearly to $\vec{E}$. If the medium is linear but anisotropic,

$$P_i = \sum_j \varepsilon_0\chi_{ij}Ej, \tag{20.6}$$

where χ_{ij} is the susceptibility tensor, that is,

$$\begin{pmatrix} P_1 \\ P_2 \\ P_3 \end{pmatrix} = \varepsilon_0 \begin{pmatrix} \chi_{11} & \chi_{12} & \chi_{13} \\ \chi_{21} & \chi_{22} & \chi_{23} \\ \chi_{31} & \chi_{32} & \chi_{33} \end{pmatrix} \begin{pmatrix} E_1 \\ E_2 \\ E_3 \end{pmatrix} \tag{20.7}$$

and

$$
\begin{pmatrix} \mathbf{D}_1 \\ \mathbf{D}_2 \\ \mathbf{D}_3 \end{pmatrix} = \varepsilon_0 \begin{pmatrix} 1 & 0 & 0 \\ 0 & 1 & 0 \\ 0 & 0 & 1 \end{pmatrix} \begin{pmatrix} \mathbf{E}_1 \\ \mathbf{E}_2 \\ \mathbf{E3} \end{pmatrix} + \varepsilon_0 \begin{pmatrix} \chi_{11} & \chi_{12} & \chi_{13} \\ \chi_{21} & \chi_{22} & \chi_{23} \\ \chi_{31} & \chi_{32} & \chi_{33} \end{pmatrix} \begin{pmatrix} \mathbf{E}_1 \\ \mathbf{E}_2 \\ \mathbf{E}_3 \end{pmatrix}
$$

$$
= \varepsilon_0 \begin{pmatrix} 1+\chi_{11} & \chi_{12} & \chi_{13} \\ \chi_{21} & 1+\chi_{22} & \chi_{23} \\ \chi_{31} & \chi_{32} & 1+\chi_{33} \end{pmatrix} \begin{pmatrix} \mathbf{E}_1 \\ \mathbf{E}_2 \\ \mathbf{E}_3 \end{pmatrix},
$$

(20.8)

where the vector indices 1, 2, 3 represent the three Cartesian directions.

This can be written

$$
\mathbf{D}_i = \varepsilon_{ij} \mathbf{E}_j,
$$

(20.9)

where

$$
\varepsilon_{ij} = \varepsilon_0 (1 + \chi_{ij}),
$$

(20.10)

is variously called the dielectric tensor, or permittivity tensor, or dielectric permittivity tensor. Equations 20.9 and 20.10 use the Einstein summation convention, that is, whenever repeated indices occur, it is understood that the expression is to be summed over the repeated indices. This notation is used throughout this chapter.

The dielectric tensor is symmetric and real (assuming the medium is homogeneous and nonabsorbing), so that

$$
\varepsilon_{ij} = \varepsilon_{ji},
$$

(20.11)

and there are at most six independent elements.

Note that for an isotropic medium with nonlinearity (which occurs with higher field strengths)

$$
\mathbf{P} = \varepsilon_0 (\chi \mathbf{E} + \chi_2 \mathbf{E}^2 + \chi_3 \mathbf{E}^3 + \ldots),
$$

(20.12)

where χ_2, χ_3, and so on, are the nonlinear terms.

Returning to the discussion of a linear, homogeneous, anisotropic medium, the susceptibility tensor

$$
\begin{pmatrix} \chi_{11} & \chi_{12} & \chi_{13} \\ \chi_{21} & \chi_{22} & \chi_{23} \\ \chi_{31} & \chi_{32} & \chi_{33} \end{pmatrix} = \begin{pmatrix} \chi_{11} & \chi_{12} & \chi_{13} \\ \chi_{12} & \chi_{22} & \chi_{23} \\ \chi_{13} & \chi_{23} & \chi_{33} \end{pmatrix}
$$

(20.13)

is symmetric so that we can always find a set of coordinate axes (i.e., we can always rotate to an orientation) such that the off-diagonal terms are zero and the tensor is diagonalized, thus,

$$
\begin{pmatrix} \chi_{11}' & 0 & 0 \\ 0 & \chi_{22}' & 0 \\ 0 & 0 & \chi_{33}' \end{pmatrix}.
$$

(20.14)

The coordinate axes for which this is true are called the principal axes, and these χ' are the principal susceptibilities. The principal dielectric constants are given by

$$
\begin{pmatrix} 1 & 0 & 0 \\ 0 & 1 & 0 \\ 0 & 0 & 1 \end{pmatrix} + \begin{pmatrix} \chi_{11} & 0 & 0 \\ 0 & \chi_{22} & 0 \\ 0 & 0 & \chi_{33} \end{pmatrix} = \begin{pmatrix} 1+\chi_{11} & 0 & 0 \\ 0 & 1+\chi_{22} & 0 \\ 0 & 0 & 1+\chi_{33} \end{pmatrix}
$$

$$
= \begin{pmatrix} \mathbf{n}_1^2 & 0 & 0 \\ 0 & \mathbf{n}_2^2 & 0 \\ 0 & 0 & \mathbf{n}_3^2 \end{pmatrix} \tag{20.15}
$$

where $\mathbf{n}_1$, $\mathbf{n}_2$, and $\mathbf{n}_3$ are the principal indices of refraction.

20.3 Crystalline Materials and Their Properties

As we have seen above, the relationship between the displacement and the field is

$$
\mathbf{D}_i = \varepsilon_{ij}\mathbf{E}_j, \tag{20.16}
$$

where ε_{ij} is the dielectric tensor. The impermeability tensor η_{ij} is defined as

$$
\eta_{ij} = \varepsilon_0 \left(\varepsilon^{-1}\right)_{ij}, \tag{20.17}
$$

where ε^{-1} is the inverse of the dielectric tensor. The principal indices of refraction, $\mathbf{n}_1$, $\mathbf{n}_2$, and $\mathbf{n}_3$ are related to the principal values of the impermeability tensor and the principal values of the permittivity tensor by

$$
\frac{1}{\mathbf{n}_1^2} = \eta_{ii} = \frac{\varepsilon_0}{\varepsilon_{ii}}; \quad \frac{1}{\mathbf{n}_2^2} = \eta_{jj} = \frac{\varepsilon_0}{\varepsilon_{jj}}; \quad \frac{1}{\mathbf{n}_3^2} = \eta_{kk} = \frac{\varepsilon_0}{\varepsilon_{kk}}. \tag{20.18}
$$

The properties of the crystal change in response to the force from an externally applied electric field. In particular, the impermeability tensor is a function of the field. The electro-optic coefficients are defined by the expression for the expansion, in terms of the field, of the change of the impermeability tensor from zero field value, that is,

$$
\eta_{ij}(\mathbf{E}) - \eta ij(0) = \Delta\eta_{ij} = r_{ijk}\mathbf{E}_k + s_{ijkl}\mathbf{E}_k\mathbf{E}_l + \mathbf{O}(\mathbf{E}^n), n = 3,\ 4,\ldots, \tag{20.19}
$$

where η_{ij} is a function of the applied field E, r_{ijk} are the linear, or Pockels, electro-optic tensor coefficients, and s_{ijkl} are the quadratic, or Kerr, electro-optic tensor coefficients. Terms higher than quadratic are typically small and are neglected.

Note that the values of the indices and the electro-optic tensor coefficients are dependent on the frequency of light passing through the material. Any given indices are specified at a particular frequency (or wavelength). Also note that the external applied fields may be static or alternating fields, and the values of the tensor coefficients are weakly dependent on the frequency of the applied fields. Generally, low- and/or high-frequency values of the tensor coefficients are given in tables. Low frequencies are those below the fundamental frequencies of the acoustic resonances of the sample, and high frequencies are those above. Operation of an electro-optic modulator subject to low (high) frequencies is sometimes described as being unclamped (clamped).

The linear electro-optic tensor is of third rank with 3^3 elements and the quadratic electro-optic tensor is of fourth rank with 3^4 elements; however, symmetry reduces the number of independent elements. If the medium is lossless and optically inactive,

ε_{ij} is a symmetric tensor, i.e., $\varepsilon_{ij} = \varepsilon_{ji}$,

η_{ij} is a symmetric tensor, i.e., $\eta_{ij} = \eta_{ji}$,

$\mathbf{r}_{ijk}$ has symmetry where coefficients with permuted first and second indices are equal, i.e., $\mathbf{r}_{ijk} = \mathbf{r}_{jik}$, and

S_{ijkl} has symmetry where coefficients with permuted first and second coefficients are equal and coefficients with permuted third and fourth coefficients are equal, i.e., $S_{ijkl} = S_{jikl}$ and $S_{ijkl} = S_{ijlk}$.

Symmetry reduces the number of linear coefficients from 27 to 18, and reduces the number of quadratic coefficients from 81 to 36. The linear electro-optic coefficients are assigned two indices so that they are $\mathbf{r}_{lk}$ where l runs from 1 to 6 and k runs from 1 to 3. The quadratic coefficients are assigned two indices so that they become s_{ij} where i runs from 1 to 6 and j runs from 1 to 6. For a given crystal symmetry class, the form of the electro-optic tensor is known.

20.4 Crystals

Crystals are characterized by their lattice type and symmetry. There are 14 lattice types. As an example of three of these, a crystal that has a cubic structure can be simple cubic, face-centered cubic, or body-centered cubic.

There are 32 point groups corresponding to 32 different symmetries. For example, a cubic lattice has five types of symmetry. The symmetry is labeled with point group notation, and crystals are classified in this way. A complete discussion of crystals, lattice types, and point groups is outside the scope of the present work, and will not be given here; there are many excellent references [5–11]. Table 20.1 gives a summary of the lattice types and point groups, and shows how these relate to optical symmetry and the form of the dielectric tensor.

In order to understand the notation and terminology of Table 20.1, some additional information is required, which we now introduce. As we have seen in the previous sections, there are three principal indices of refraction. There are three types of materials: those where the three principal indices are equal; those where two principal indices are equal; and those where all three principal indices are different. We will discuss these three cases in more detail in the next section. The indices for the case where there are only two distinct values are named the ordinary index (n_o) and the extraordinary index (n_e). These labels are applied for historical reasons [12]. Erasmus Bartholinus, a Danish mathematician, in 1669 discovered double refraction in calcite. If the calcite crystal, split along its natural cleavage planes, is placed on a type-written sheet of paper, two images of the letters will be observed. If the crystal is then rotated about an axis perpendicular to the page, one of the two images of the letters will rotate about the other. Bartholinus named the light rays from the letters that do not rotate the ordinary rays, and the rays from the rotating letters he named the extraordinary rays; hence, the indices that produce these rays are named likewise. This explains the notation in the dielectric tensor for tetragonal, hexagonal, and trigonal crystals.

Let us consider such crystals in more detail. There is a plane in the material in which a single index would be measured in any direction. Light that is propagating in the direction normal to this plane with equal indices experiences the same refractive index for any polarization (orientation of the **E** vector). The direction for which this occurs is called the optic axis. Crystals that have one optic axis are called uniaxial crystals. Materials with three principal indices have two directions in which the **E** vector experiences a single refractive index. These materials have two optic axes and are called biaxial crystals. This is more fully explained in Section 20.4.1. Materials that have more than one principal index of refraction are called birefringent materials and are said to exhibit double refraction.

Crystals are composed of periodic arrays of atoms. The lattice of a crystal is a set of points in space. Sets of atoms that are identical in composition, arrangement, and orientation are attached to each lattice point. By translating the basic structure attached to the lattice point, we can fill space with the crystal. Define

TABLE 20.1   Crystal Types, Point Groups, and the Dielectric Tensors

Symmetry	Crystal System	Point Group	Dielectric Tensor
Isotropic	Cubic	$\overline{4}3m$ 432 m3 23 m3m	$\varepsilon = \varepsilon_0 \begin{pmatrix} n^2 & 0 & 0 \\ 0 & n^2 & 0 \\ 0 & 0 & n^2 \end{pmatrix}$
Uniaxial	Tetragonal	$\overline{\dfrac{4}{4}}$ 4/m 422 4mm $\overline{4}2m$ 4/mmm	
	Hexagonal	$\overline{\dfrac{6}{6}}$ 6/m 622 6mm $\overline{6}m2$ 6/mmm	$\varepsilon = \varepsilon_0 \begin{pmatrix} n_0^2 & 0 & 0 \\ 0 & n_0^2 & 0 \\ 0 & 0 & n_e^2 \end{pmatrix}$
	Trigonal	$\overline{\dfrac{3}{3}}$ 32 3m $\overline{3}m$	
Biaxial	Triclinic	$\overline{\dfrac{1}{1}}$	
	Monoclinic	2 m 2/m	$\varepsilon = \varepsilon_0 \begin{pmatrix} n_0^2 & 0 & 0 \\ 0 & n_0^2 & 0 \\ 0 & 0 & n_e^2 \end{pmatrix}$
	Orthorhombic	222 2 mm **mmm**	

Source: Yariv A, Yeh P. *Optical Waves in Crystals*, Wiley, New York, 1984.

vectors **a**, **b**, and **c**, which form three adjacent edges of a parallelepiped, which spans the basic atomic structure. This parallelepiped is called a unit cell. We call the axes that lie along these vectors the crystal axes.

We would like to be able to describe a particular plane in a crystal, since crystals may be cut at any angle. The Miller indices are quantities that describe the orientation of planes in a crystal. The Miller indices are defined as follows: (1) Locate the intercepts of the plane on the crystal axes: these will be multiples of lattice point spacing. (2) Take the reciprocals of the intercepts and form the three smallest integers having the same ratio. For example, suppose we have a cubic crystal so that the crystal axes are the orthogonal Cartesian axes. Suppose further that the plane we want to describe intercepts the axes at the points 4, 3, and 2. The reciprocals of these intercepts are $\frac{1}{4}$, $\frac{1}{3}$, and $\frac{1}{2}$. The Miller indices are then (3, 4, 6). This example serves to illustrate how the Miller indices are found, but it is more usual to encounter simpler crystal cuts. The same cubic crystal, if cut so that the intercepts are 1, ∞, ∞ (defining a plane parallel to the *yz*-plane in the usual Cartesian coordinates) has Miller indices (1, 0, 0). Likewise, if the intercepts are 1, 1, ∞ (diagonal to two of the axes), the Miller indices are (1, 1, 0), and if the intercepts are 1, 1, 1 (diagonal to all three axes), the Miller indices are (1, 1, 1).

Two important electro-optic crystal types have the point group symbols $\overline{4}3\mathbf{m}$ (this is a cubic crystal, e.g., CdTe and GaAs) and $\overline{4}2\mathbf{m}$ (this is a tetragonal crystal, e.g., $AgGaS_2$). The linear and quadratic electro-optic tensors for these two crystal types, as well as all the other linear and quadratic electro-optic coefficient tensors for all crystal symmetry classes, are given in Tables 20.2 and 20.3. Note from these tables that the linear electro-optic effect vanishes for crystals that retain symmetry under inversion, that is, centrosymmetric crystals, whereas the quadratic electro-optic effect never vanishes. For further discussion of this point, see Yariv and Yeh [13].

TABLE 20.2   Linear Electro-Optic Tensors

Crystal Symmetry Class	Symmetry Group	Tensor
Centrosymmetric	$\overline{1}$ 2/m mmm 4/m 4/mmm $\overline{3}$ 3m 6/m 6/mmm m3 m3m	$\begin{pmatrix} 0 & 0 & 0 \\ 0 & 0 & 0 \\ 0 & 0 & 0 \\ 0 & 0 & 0 \\ 0 & 0 & 0 \\ 0 & 0 & 0 \end{pmatrix}$
Triclinic	1	$\begin{pmatrix} r_{11} & r_{12} & r_{13} \\ r_{21} & r_{22} & r_{23} \\ r_{31} & r_{32} & r_{33} \\ r_{41} & r_{42} & r_{43} \\ r_{51} & r_{52} & r_{53} \\ r_{61} & r_{62} & r_{63} \end{pmatrix}$
Monoclinic	$2\,(2\|\mathbf{x}_2)$	$\begin{pmatrix} 0 & r_{12} & 0 \\ 0 & r_{22} & 0 \\ 0 & r_{32} & 0 \\ r_{41} & 0 & r_{43} \\ 0 & r_{52} & 0 \\ r_{61} & 0 & r_{63} \end{pmatrix}$
	$2\,(2\|\mathbf{x}_3)$	$\begin{pmatrix} 0 & 0 & r_{13} \\ 0 & 0 & r_{23} \\ 0 & 0 & r_{33} \\ r_{41} & r_{42} & 0 \\ r_{51} & r_{52} & 0 \\ 0 & 0 & r_{63} \end{pmatrix}$
	$\mathbf{m}\,(\mathbf{m} \perp x_2)$	$\begin{pmatrix} r_{11} & 0 & r_{13} \\ r_{21} & 0 & r_{23} \\ r_{31} & 0 & r_{33} \\ 0 & r_{42} & 0 \\ r_{51} & 0 & r_{53} \\ 0 & r_{62} & 0 \end{pmatrix}$

(*Continued*)

TABLE 20.2 (*Continued*)   Linear Electro-Optic Tensors

Crystal Symmetry Class	Symmetry Group	Tensor
	$\mathbf{m}(\mathbf{m}\perp\mathbf{x}_3)$	$\begin{pmatrix} \mathbf{r}_{11} & \mathbf{r}_{12} & 0 \\ \mathbf{r}_{21} & \mathbf{r}_{22} & 0 \\ \mathbf{r}_{31} & \mathbf{r}_{32} & 0 \\ 0 & 0 & \mathbf{r}_{43} \\ 0 & 0 & \mathbf{r}_{53} \\ \mathbf{r}_{61} & \mathbf{r}_{62} & 0 \end{pmatrix}$
Orthorhombic	222	$\begin{pmatrix} 0 & 0 & 0 \\ 0 & 0 & 0 \\ 0 & 0 & 0 \\ \mathbf{r}_{41} & 0 & 0 \\ 0 & \mathbf{r}_{52} & 0 \\ 0 & 0 & \mathbf{r}_{63} \end{pmatrix}$
	2mm	$\begin{pmatrix} 0 & 0 & \mathbf{r}_{13} \\ 0 & 0 & \mathbf{r}_{23} \\ 0 & 0 & \mathbf{r}_{33} \\ 0 & \mathbf{r}_{42} & 0 \\ \mathbf{r}_{51} & 0 & 0 \\ 0 & 0 & 0 \end{pmatrix}$
Tetragonal	4	$\begin{pmatrix} 0 & 0 & \mathbf{r}_{13} \\ 0 & 0 & \mathbf{r}_{13} \\ 0 & 0 & \mathbf{r}_{33} \\ \mathbf{r}_{41} & \mathbf{r}_{51} & 0 \\ \mathbf{r}_{51} & \mathbf{r}_{41} & 0 \\ 0 & 0 & 0 \end{pmatrix}$
	$\bar{4}$	$\begin{pmatrix} 0 & 0 & \mathbf{r}_{13} \\ 0 & 0 & -\mathbf{r}_{13} \\ 0 & 0 & 0 \\ \mathbf{r}_{41} & -\mathbf{r}_{51} & 0 \\ \mathbf{r}_{51} & \mathbf{r}_{41} & 0 \\ 0 & 0 & \mathbf{r}_{63} \end{pmatrix}$
	422	$\begin{pmatrix} 0 & 0 & 0 \\ 0 & 0 & 0 \\ 0 & 0 & 0 \\ \mathbf{r}_{41} & 0 & 0 \\ 0 & -\mathbf{r}_{41} & 0 \\ 0 & 0 & 0 \end{pmatrix}$
	4mm	$\begin{pmatrix} 0 & 0 & \mathbf{r}_{13} \\ 0 & 0 & \mathbf{r}_{13} \\ 0 & 0 & \mathbf{r}_{33} \\ 0 & \mathbf{r}_{51} & 0 \\ \mathbf{r}_{51} & 0 & 0 \\ 0 & 0 & 0 \end{pmatrix}$

(*Continued*)

TABLE 20.2 (*Continued*)   Linear Electro-Optic Tensors

Crystal Symmetry Class	Symmetry Group	Tensor
	$\overline{4}2\mathbf{m}\,(2\,\|\,\mathbf{x}_1)$	$\begin{pmatrix} 0 & 0 & 0 \\ 0 & 0 & 0 \\ 0 & 0 & 0 \\ \mathbf{r}_{41} & 0 & 0 \\ 0 & \mathbf{r}_{41} & 0 \\ 0 & 0 & \mathbf{r}_{63} \end{pmatrix}$
Trigonal	3	$\begin{pmatrix} \mathbf{r}_{11} & -\mathbf{r}_{22} & \mathbf{r}_{13} \\ -\mathbf{r}_{11} & \mathbf{r}_{22} & \mathbf{r}_{13} \\ 0 & 0 & \mathbf{r}_{33} \\ \mathbf{r}_{41} & \mathbf{r}_{51} & 0 \\ \mathbf{r}_{51} & -\mathbf{r}_{41} & 0 \\ -\mathbf{r}_{22} & -\mathbf{r}_{11} & 0 \end{pmatrix}$
	32	$\begin{pmatrix} \mathbf{r}_{11} & 0 & 0 \\ -\mathbf{r}_{11} & 0 & 0 \\ 0 & 0 & 0 \\ \mathbf{r}_{41} & 0 & 0 \\ 0 & -\mathbf{r}_{41} & 0 \\ 0 & -\mathbf{r}_{11} & 0 \end{pmatrix}$
	$3\mathbf{m}\,(\mathbf{m}\perp\mathbf{x}_1)$	$\begin{pmatrix} 0 & -\mathbf{r}_{22} & \mathbf{r}_{13} \\ 0 & \mathbf{r}_{22} & \mathbf{r}_{13} \\ 0 & 0 & \mathbf{r}_{33} \\ 0 & \mathbf{r}_{51} & 0 \\ \mathbf{r}_{51} & 0 & 0 \\ -\mathbf{r}_{22} & 0 & 0 \end{pmatrix}$
	$3\mathbf{m}\,(\mathbf{m}\perp\mathbf{x}_2)$	$\begin{pmatrix} \mathbf{r}_{11} & 0 & \mathbf{r}_{13} \\ -\mathbf{r}_{13} & 0 & \mathbf{r}_{13} \\ 0 & 0 & \mathbf{r}_{33} \\ 0 & \mathbf{r}_{51} & 0 \\ \mathbf{r}_{51} & 0 & 0 \\ 0 & -\mathbf{r}_{11} & 0 \end{pmatrix}$
Hexagonal	6	$\begin{pmatrix} 0 & 0 & \mathbf{r}_{13} \\ 0 & 0 & \mathbf{r}_{13} \\ 0 & 0 & \mathbf{r}_{33} \\ \mathbf{r}_{41} & \mathbf{r}_{51} & 0 \\ \mathbf{r}_{51} & -\mathbf{r}_{41} & 0 \\ 0 & 0 & 0 \end{pmatrix}$
	6mm	$\begin{pmatrix} 0 & 0 & \mathbf{r}_{13} \\ 0 & 0 & \mathbf{r}_{13} \\ 0 & 0 & \mathbf{r}_{33} \\ 0 & \mathbf{r}_{51} & 0 \\ \mathbf{r}_{51} & 0 & 0 \\ 0 & 0 & 0 \end{pmatrix}$

(*Continued*)

TABLE 20.2 (*Continued*)   Linear Electro-Optic Tensors

Crystal Symmetry Class	Symmetry Group	Tensor
	622	$\begin{pmatrix} 0 & 0 & 0 \\ 0 & 0 & 0 \\ 0 & 0 & 0 \\ r_{41} & 0 & 0 \\ 0 & -r_{41} & 0 \\ 0 & 0 & 0 \end{pmatrix}$
Hexagonal	$\bar{6}$	$\begin{pmatrix} r_{11} & -r_{22} & 0 \\ -r_{11} & r_{22} & 0 \\ 0 & 0 & 0 \\ 0 & 0 & 0 \\ 0 & 0 & 0 \\ -r_{22} & -r_{11} & 0 \end{pmatrix}$
	$\bar{6}m2(m \perp x_1)$	$\begin{pmatrix} 0 & -r_{22} & 0 \\ 0 & r_{22} & 0 \\ 0 & 0 & 0 \\ 0 & 0 & 0 \\ 0 & 0 & 0 \\ -r_{22} & 0 & 0 \end{pmatrix}$
	$\bar{6}m2(m \perp x_2)$	$\begin{pmatrix} r_{11} & 0 & 0 \\ -r_{11} & 0 & 0 \\ 0 & 0 & 0 \\ 0 & 0 & 0 \\ 0 & 0 & 0 \\ 0 & -r_{11} & 0 \end{pmatrix}$
Cubic	$\bar{4}3m$ 23	$\begin{pmatrix} 0 & 0 & 0 \\ 0 & 0 & 0 \\ 0 & 0 & 0 \\ r_{41} & 0 & 0 \\ 0 & r_{41} & 0 \\ 0 & 0 & r_{41} \end{pmatrix}$
	432	$\begin{pmatrix} 0 & 0 & 0 \\ 0 & 0 & 0 \\ 0 & 0 & 0 \\ 0 & 0 & 0 \\ 0 & 0 & 0 \\ 0 & 0 & 0 \end{pmatrix}$

Source: Yariv A, Yeh P. *Optical Waves in Crystals*, Wiley, New York, 1984.

TABLE 20.3 Quadratic Electro-Optic Tensors

Crystal Symmetry Class	Symmetry Group	Tensor
Triclinic	$\dfrac{1}{\bar{1}}$	$\begin{pmatrix} s_{11} & s_{12} & s_{13} & s_{14} & s_{15} & s_{16} \\ s_{21} & s_{22} & s_{23} & s_{24} & s_{25} & s_{26} \\ s_{31} & s_{32} & s_{33} & s_{34} & s_{35} & s_{36} \\ s_{41} & s_{42} & s_{43} & s_{44} & s_{45} & s_{46} \\ s_{51} & s_{52} & s_{53} & s_{54} & s_{55} & s_{56} \\ s_{61} & s_{62} & s_{63} & s_{64} & s_{65} & s_{66} \end{pmatrix}$
Monoclinic	$\begin{matrix} 2 \\ \mathbf{m} \\ 2/\mathbf{m} \end{matrix}$	$\begin{pmatrix} s_{11} & s_{12} & s_{13} & 0 & s_{15} & 0 \\ s_{21} & s_{22} & s_{23} & 0 & s_{25} & 0 \\ s_{31} & s_{32} & s_{33} & 0 & s_{35} & 0 \\ 0 & 0 & 0 & s_{44} & 0 & s_{46} \\ s_{51} & s_{52} & s_{53} & 0 & s_{55} & 0 \\ 0 & 0 & 0 & s_{64} & 0 & s_{66} \end{pmatrix}$
Orthorhombic	$\begin{matrix} \mathbf{2mm} \\ 222 \\ \mathbf{mmm} \end{matrix}$	$\begin{pmatrix} s_{11} & s_{12} & s_{13} & 0 & 0 & 0 \\ s_{21} & s_{22} & s_{23} & 0 & 0 & 0 \\ s_{31} & s_{32} & s_{33} & 0 & 0 & 0 \\ 0 & 0 & 0 & s_{44} & 0 & 0 \\ 0 & 0 & 0 & 0 & s_{55} & 0 \\ 0 & 0 & 0 & 0 & 0 & s_{66} \end{pmatrix}$
Tetragonal	$\begin{matrix} 4 \\ \bar{4} \\ 4/\mathbf{m} \end{matrix}$	$\begin{pmatrix} s_{11} & s_{12} & s_{13} & 0 & 0 & s_{16} \\ s_{12} & s_{11} & s_{13} & 0 & 0 & -s_{26} \\ s_{31} & s_{31} & s_{33} & 0 & 0 & 0 \\ 0 & 0 & 0 & s_{44} & s_{45} & 0 \\ 0 & 0 & 0 & -s_{45} & s_{44} & 0 \\ s_{61} & -s_{61} & 0 & 0 & 0 & s_{66} \end{pmatrix}$
	$\begin{matrix} 422 \\ \mathbf{4mm} \\ \overline{\mathbf{4}}\mathbf{2m} \\ 4/\mathbf{mm} \end{matrix}$	$\begin{pmatrix} s_{11} & s_{12} & s_{13} & 0 & 0 & 0 \\ s_{12} & s_{11} & s_{13} & 0 & 0 & 0 \\ s_{31} & s_{31} & s_{33} & 0 & 0 & 0 \\ 0 & 0 & 0 & s_{44} & 0 & 0 \\ 0 & 0 & 0 & 0 & s_{44} & 0 \\ 0 & 0 & 0 & 0 & 0 & s_{66} \end{pmatrix}$
Trigonal	$\dfrac{3}{\bar{3}}$	$\begin{pmatrix} s_{11} & s_{12} & s_{13} & s_{14} & s_{15} & -s_{61} \\ s_{12} & s_{11} & s_{13} & -s_{14} & -s_{15} & s_{61} \\ s_{31} & s_{31} & s_{33} & 0 & 0 & 0 \\ s_{41} & -s_{41} & 0 & s_{44} & s_{45} & -s_{51} \\ s_{51} & -s_{51} & 0 & -s_{45} & s_{44} & s_{41} \\ s_{61} & -s_{61} & 0 & -s_{15} & s_{14} & \frac{1}{2}(s_{11}-s_{12}) \end{pmatrix}$

(Continued)

TABLE 20.3 (*Continued*)   Quadratic Electro-Optic Tensors

Crystal Symmetry Class	Symmetry Group	Tensor
	32 3m $\bar{3}$m	$\begin{pmatrix} s_{11} & s_{12} & s_{13} & s_{14} & 0 & 0 \\ s_{12} & s_{11} & s_{13} & -s_{14} & 0 & 0 \\ s_{13} & s_{13} & s_{33} & 0 & 0 & 0 \\ s_{41} & -s_{41} & 0 & s_{44} & 0 & 0 \\ 0 & 0 & 0 & 0 & s_{44} & s_{41} \\ 0 & 0 & 0 & 0 & s_{14} & \frac{1}{2}(s_{11}-s_{12}) \end{pmatrix}$
Hexagonal	6 $\bar{6}$ 6/m	$\begin{pmatrix} s_{11} & s_{12} & s_{13} & 0 & 0 & -s_{61} \\ s_{12} & s_{11} & s_{13} & 0 & 0 & s_{61} \\ s_{31} & s_{31} & s_{33} & 0 & 0 & 0 \\ 0 & 0 & 0 & s_{44} & s_{45} & 0 \\ 0 & 0 & 0 & -s_{45} & s_{44} & 0 \\ s_{61} & -s_{61} & 0 & 0 & 0 & \frac{1}{2}(s_{11}-s_{12}) \end{pmatrix}$
	622 6mm $\bar{6}$m2 6/mmm	$\begin{pmatrix} s_{11} & s_{12} & s_{13} & 0 & 0 & 0 \\ s_{12} & s_{11} & s_{13} & 0 & 0 & 0 \\ s_{31} & s_{31} & s_{33} & 0 & 0 & 0 \\ 0 & 0 & 0 & s_{44} & 0 & 0 \\ 0 & 0 & 0 & 0 & s_{44} & 0 \\ 0 & 0 & 0 & 0 & 0 & \frac{1}{2}(s_{11}-s_{12}) \end{pmatrix}$
Cubic	23 m3	$\begin{pmatrix} s_{11} & s_{12} & s_{13} & 0 & 0 & 0 \\ s_{13} & s_{11} & s_{12} & 0 & 0 & 0 \\ s_{12} & s_{13} & s_{11} & 0 & 0 & 0 \\ 0 & 0 & 0 & s_{44} & 0 & 0 \\ 0 & 0 & 0 & 0 & s_{44} & 0 \\ 0 & 0 & 0 & 0 & 0 & s_{44} \end{pmatrix}$
	432 m3m $\bar{4}$3m	$\begin{pmatrix} s_{11} & s_{12} & s_{12} & 0 & 0 & 0 \\ s_{12} & s_{11} & s_{12} & 0 & 0 & 0 \\ s_{12} & s_{12} & s_{11} & 0 & 0 & 0 \\ 0 & 0 & 0 & s_{44} & 0 & 0 \\ 0 & 0 & 0 & 0 & s_{44} & 0 \\ 0 & 0 & 0 & 0 & 0 & s_{44} \end{pmatrix}$
Isotropic		$\begin{pmatrix} s_{11} & s_{12} & s_{12} & 0 & 0 & 0 \\ s_{12} & s_{11} & s_{12} & 0 & 0 & 0 \\ s_{12} & s_{12} & s_{11} & 0 & 0 & 0 \\ 0 & 0 & 0 & \frac{1}{2}(s_{11}-s_{12}) & 0 & 0 \\ 0 & 0 & 0 & 0 & \frac{1}{2}(s_{11}-s_{12}) & 0 \\ 0 & 0 & 0 & 0 & 0 & \frac{1}{2}(s_{11}-s_{12}) \end{pmatrix}$

Source: Yariv A, Yeh P. *Optical Waves in Crystals*, Wiley, New York, 1984.

20.4.1 The Index Ellipsoid

Light propagating in anisotropic materials experiences a refractive index and a phase velocity that depends on the propagation direction, polarization state, and wavelength. The refractive index for propagation (for monochromatic light of some specified frequency) in an arbitrary direction (in Cartesian coordinates)

$$\vec{a} = x\hat{i} + y\hat{j} + z\hat{k} \tag{20.20}$$

can be obtained from the index ellipsoid, a useful and lucid construct for visualization and determination of the index. (Note that we now shift from indexing the Cartesian directions with numbers to using x, y, and z.) In the principal coordinate system, the index ellipsoid is given by

$$\frac{x^2}{n_x^2} + \frac{y^2}{n_y^2} + \frac{z^2}{n_z^2} = 1 \tag{20.21}$$

in the absence of an applied electric field. The lengths of the semi-major and semi-minor axes of the ellipse formed by the intersection of this index ellipsoid and a plane normal to the propagation direction and passing through the center of the ellipsoid are the two principal indices of refraction for that propagation direction. Where there are three distinct principal indices, the crystal is defined as biaxial, and the above equation holds. If two of the three indices of the index ellipsoid are equal, the crystal is defined to be uniaxial and the equation for the index ellipsoid is

$$\frac{x^2}{n_0^2} + \frac{y^2}{n_0^2} + \frac{z^2}{n_e^2} = 1. \tag{20.22}$$

Uniaxial materials are said to be uniaxial positive when $n_o < n_e$ and uniaxial negative when $n_o > n_e$. When there is a single index for any direction in space, the crystal is isotropic and the equation for the ellipsoid becomes that for a sphere,

$$\frac{x^2}{n^2} + \frac{y^2}{n^2} + \frac{z^2}{n^2} = 1. \tag{20.23}$$

The index ellipsoids for isotropic, uniaxial, and biaxial crystals are illustrated in Figure 20.1.

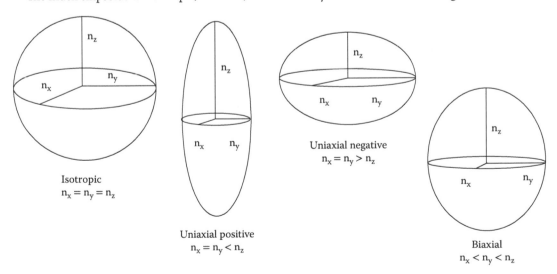

FIGURE 20.1 Index ellipsoids.

Examples of isotropic materials are CdTe, NaCl, diamond, and GaAs. Examples of uniaxial positive materials are quartz and ZnS. Materials that are uniaxial negative include calcite, $LiNbO_3$, $BaTiO_3$, and KDP (KH_2PO_4). Examples of biaxial materials are gypsum and mica.

20.4.2 Natural Birefringence

Many materials have natural birefringence, that is, they are uniaxial or biaxial in their natural (absence of applied fields) state. These materials are often used in passive devices, such as polarizers and retarders. Calcite is one of the most important naturally birefringent materials for optics, and is used in a variety of well-known polarizers, for example, the Nichol, Wollaston, or Glan-Thompson prisms. As we shall see later, naturally isotropic materials can be made birefringent, and materials that have natural birefringence can be made to change that birefringence with the application of electromagnetic fields.

20.4.3 Wave Surface

There are two additional methods of depicting the effect of crystal anisotropy on light. Neither is as satisfying or useful to this author as the index ellipsoid; however, both are mentioned for the sake of completeness and in order to facilitate understanding of those references that use these models. They are most often used to explain birefringence, for example, in the operation of calcite-based devices [14–16].

The first of these is called the wave surface. As a light wave from a point source expands through space, it forms a surface that represents the wavefront. This surface is composed of points having equal phase. At a particular instant in time, the wave surface is a representation of the velocity surface of a wave expanding in the medium; it is a measure of the distance through which the wave has expanded from some point over some time period. Because the wave will have expanded further (faster) when experiencing a low refractive index and expanded less (slower) when experiencing a high index, the size of the wave surface is inversely proportional to the index.

To illustrate the use of the wave surface, consider a uniaxial crystal. Recall that we have defined the optic axis of a uniaxial crystal as the direction in which the speed of propagation is independent of polarization. The optic axes for positive and negative uniaxial crystals are shown on the index ellipsoids in Figure 20.2, and the optic axes for a biaxial crystal are shown on the index ellipsoid in Figure 20.3.

The wave surfaces are now shown in Figure 20.4 for both positive and negative uniaxial materials. The upper diagram for each pair shows the wave surface for polarization perpendicular to the optic axes (also perpendicular to the principal section through the ellipsoid), and the lower diagram shows

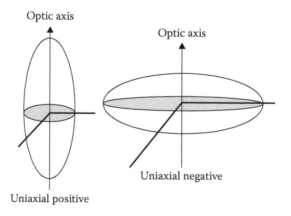

FIGURE 20.2 Optic axis on index ellipsoid for uniaxial positive and uniaxial negative crystals.

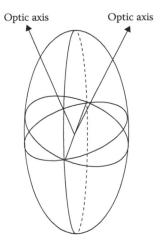

FIGURE 20.3 Optic axes on index ellipsoid for biaxial crystals.

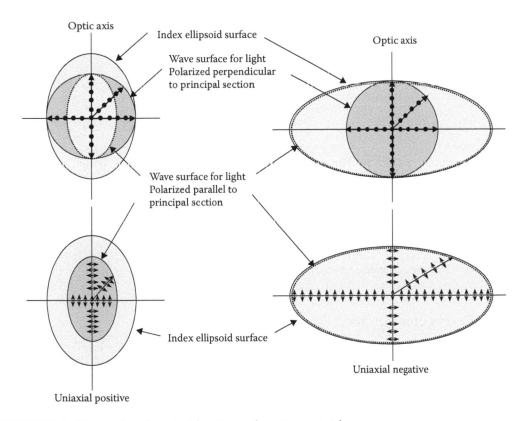

FIGURE 20.4 Wave surfaces for uniaxial positive and negative materials.

the wave surface for polarization in the plane of the principal section. The index ellipsoid surfaces are shown for reference. Similarly, cross-sections of the wave surfaces for biaxial materials are shown in Figure 20.5. In all cases, polarization perpendicular to the plane of the page is indicated with solid circles along the rays, whereas polarization parallel to the plane of the page is shown with short double-headed arrows along the rays.

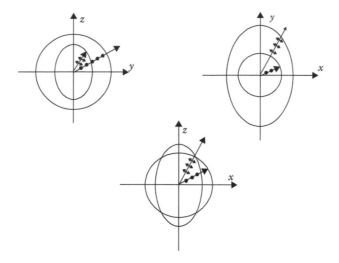

FIGURE 20.5 Wave surfaces for biaxial materials in principal planes.

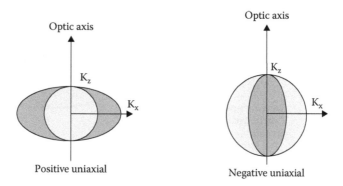

FIGURE 20.6 Wavevector surfaces for positive and negative uniaxial crystals.

20.4.4 Wavevector Surface

A second method of depicting the effect of crystal anisotropy on light is the wavevector surface. The wavevector surface is a measure of the variation of the value of **k**, the wavevector, for different propagation directions and different polarizations. Recall that

$$\mathbf{k} = \frac{2\pi}{\lambda} = \frac{\omega \mathbf{n}}{c},$$

(20.24)

so $\mathbf{k} \propto \mathbf{n}$. Wavevector surfaces for uniaxial crystals will then appear as shown in Figure 20.6. Compare these to the wave surfaces in Figure 20.4.

Wavevector surfaces for biaxial crystals are more complicated. Cross-sections of the wavevector surface for a biaxial crystal where $\mathbf{n_x} < \mathbf{n_y} < \mathbf{n_z}$ are shown in Figure 20.7. Compare these to the wave surfaces in Figure 20.5.

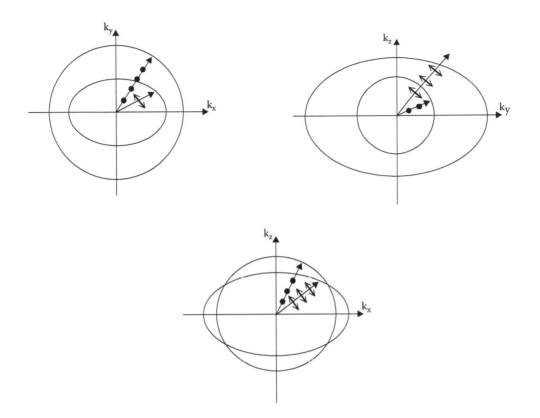

FIGURE 20.7 Wavevector surface cross-sections for biaxial crystals.

20.5 Application of Electric Fields: Induced Birefringence and Polarization Modulation

When fields are applied to materials, whether isotropic or anisotropic, birefringence can be induced or modified. This is the principle of a modulator; it is one of the most important optical devices, since it gives control over the phase and/or amplitude of light.

The alteration of the index ellipsoid of a crystal on application of an electric and/or magnetic field can be used to modulate the polarization state. The equation for the index ellipsoid of a crystal in an electric field is

$$\eta_{ij}(\mathbf{E})\mathbf{x}_i\mathbf{x}_j = 1 \tag{20.25}$$

or

$$(\eta_{ij}(0) + \Delta\eta_{ij})\mathbf{x}_i\mathbf{x}_j = 1. \tag{20.26}$$

This equation can be written as

$$\mathbf{x}^2\left(\frac{1}{\mathbf{n}_x^2} + \Delta\left(\frac{1}{\mathbf{n}}\right)_1^2\right) + \mathbf{y}^2\left(\frac{1}{\mathbf{n}_y^2} + \Delta\left(\frac{1}{\mathbf{n}}\right)_2^2\right) + \mathbf{z}^2\left(\frac{1}{\mathbf{n}_z^2} + \Delta\left(\frac{1}{\mathbf{n}}\right)_3^2\right)$$

$$+\, 2\,\mathbf{yz}\left(\Delta\left(\frac{1}{\mathbf{n}}\right)_4^2\right) + 2\mathbf{xz}\left(\Delta\left(\frac{1}{\mathbf{n}}\right)_5^2\right) + 2\mathbf{xy}\left(\Delta\left(\frac{1}{\mathbf{n}}\right)_6^2\right) = 1 \tag{20.27}$$

or

$$\mathbf{x}^2\left(\frac{1}{\mathbf{n}_x^2}+\mathbf{r}_{1k}\mathbf{E}_k+\mathbf{s}_{1k}\mathbf{E}_k^2+2\mathbf{s}_{14}\mathbf{E}_2\mathbf{E}_3+2\mathbf{s}_{15}\mathbf{E}_3\mathbf{E}_1+2\mathbf{s}_{16}\mathbf{E}_1\mathbf{E}_2\right)$$

$$+\mathbf{y}^2\left(\frac{1}{\mathbf{n}_y^2}+\mathbf{r}_{2k}\mathbf{E}_k+\mathbf{s}_{2k}\mathbf{E}_k^2+2\mathbf{s}_{24}\mathbf{E}_2\mathbf{E}_3+2\mathbf{s}_{25}\mathbf{E}_3\mathbf{E}_1+2\mathbf{s}_{26}\mathbf{E}_1\mathbf{E}_2\right)$$

$$+\mathbf{z}^2\left(\frac{1}{\mathbf{n}_x^2}+\mathbf{r}_{3k}\mathbf{E}_k+\mathbf{s}_{3k}\mathbf{E}_k^2+2\mathbf{s}_{34}\mathbf{E}_2\mathbf{E}_3+2\mathbf{s}_{35}\mathbf{E}_3\mathbf{E}_1+2\mathbf{s}_{36}\mathbf{E}_1\mathbf{E}_2\right)$$

$$+2\mathbf{yz}\left(\mathbf{r}_{4k}\mathbf{E}_k+\mathbf{s}_{4k}\mathbf{E}_k^2+2\mathbf{s}_{44}\mathbf{E}_2\mathbf{E}_3+2\mathbf{s}_{45}\mathbf{E}_3\mathbf{E}_1+2\mathbf{s}_{46}\mathbf{E}_1\mathbf{E}_2\right) \qquad (20.28)$$

$$+2\mathbf{zx}\left(\mathbf{r}_{5k}\mathbf{E}_k+\mathbf{s}_{5k}\mathbf{E}_k^2+2\mathbf{s}_{54}\mathbf{E}_2\mathbf{E}_3+2\mathbf{s}_{55}\mathbf{E}_3\mathbf{E}_1+2\mathbf{s}_{56}\mathbf{E}_1\mathbf{E}_2\right)$$

$$+2\mathbf{xy}\left(\mathbf{r}_{6k}\mathbf{E}_k+\mathbf{s}_{6k}\mathbf{E}_k^2+2\mathbf{s}_{64}\mathbf{E}_2\mathbf{E}_3+2\mathbf{s}_{65}\mathbf{E}_3\mathbf{E}_1+2\mathbf{s}_{66}\mathbf{E}_1\mathbf{E}_2\right)=1$$

where the $\mathbf{E}_k$ are components of the electric field along the principal axes and repeated indices are summed.

If the quadratic coefficients are assumed to be small and only the linear coefficients are retained, then

$$\Delta\left(\frac{1}{\mathbf{n}}\right)_l^2=\sum_{k=1}^{3}\mathbf{r}_{lk}\mathbf{E}_k \qquad (20.29)$$

and $k = 1, 2, 3$ corresponds to the principal axes x, y, and z. The equation for the index ellipsoid becomes

$$\mathbf{x}^2\left(\frac{1}{\mathbf{n}_x^2}+\mathbf{r}_{1k}\mathbf{E}_k\right)+\mathbf{y}^2\left(\frac{1}{\mathbf{n}_y^2}+\mathbf{r}_{2k}\mathbf{E}_k\right)+\mathbf{z}^2\left(\frac{1}{\mathbf{n}_z^2}++\mathbf{r}_{3k}\mathbf{E}_k\right)+2\mathbf{yz}\left(\mathbf{r}_{4k}\mathbf{E}_k\right)$$

$$+2\,\mathbf{zx}\left(\mathbf{r}_{5k}\mathbf{E}_k\right)+2\mathbf{xy}\left(\mathbf{r}_{6k}\mathbf{E}_k\right)=1 \qquad (20.30)$$

Suppose we have a cubic crystal of point group $\overline{4}3\mathbf{m}$, the symmetry group of common materials, such as GaAs. Suppose further that the field is in the z-direction. Then the index ellipsoid is

$$\frac{\mathbf{x}^2}{\mathbf{n}^2}+\frac{\mathbf{y}^2}{\mathbf{n}^2}+\frac{\mathbf{z}^2}{\mathbf{n}^2}+2\mathbf{r}_{41}\mathbf{E}_z\mathbf{xy}=1. \qquad (20.31)$$

The applied electric field couples the x-polarized and y-polarized waves. If we make the coordinate transformation

$$\mathbf{x}=\mathbf{x}'\cos 45°-\mathbf{y}'\sin 45°,$$

$$\mathbf{y}=\mathbf{x}'\sin 45°-\mathbf{y}'\cos 45° \qquad (20.32)$$

and substitute these equations into the equation for the ellipsoid, the new equation for the ellipsoid becomes

$$\mathbf{x}'^2\left(\frac{1}{\mathbf{n}^2}+\mathbf{r}_{41}\mathbf{E}_z\right)+\mathbf{y}'^2\left(\frac{1}{\mathbf{n}^2}+\mathbf{r}_{41}\mathbf{E}_z\right)+\frac{\mathbf{z}^2}{\mathbf{n}^2}=1 \qquad (20.33)$$

and we have eliminated the cross term. We want to obtain the new principal indices. The principal index will appear in Equation 20.33 as $1/n_x^2$, and must be equal to the quantity in the first parenthesis of the equation for the ellipsoid, that is,

$$\frac{1}{n_{x'}^2} = \frac{1}{n^2} + r_{41}E_z. \tag{20.34}$$

Consider the derivative of $1/n^2$ with respect to n: $d(n^{-2})/dn = -2n^{-3}$, or, rearranging, $dn = -\frac{1}{2}n^3 d(1/n^2)$. Assume $r_{41}E_z \ll n^{-2}$. Now $dn = n_{x'} - n$ and $d(1/n^2) = ((1/n_{x'}^2) - (1/n^2)) = r_{41}E_z$ and we can write $n_{x'} - n = -\frac{1}{2}n^3 r_{41}E_z$. The equations for the new principal indices are

$$n_{x'} = n - \tfrac{1}{2}n^3 r_{41}E_z$$

$$n_{y'} = n + \tfrac{1}{2}n^3 r_{41}E_z \tag{20.35}$$

$$n_{z'} = n.$$

As a similar example for another important material type, suppose we have a tetragonal (point group $\overline{4}2\mathbf{m}$) uniaxial crystal in a field along z. The index ellipsoid becomes

$$\frac{x^2}{n_o^2} + \frac{y^2}{n_o^2} + \frac{z^2}{n_e^2} + 2r_{63}E_z xy = 1. \tag{20.36}$$

A coordinate rotation can be done to obtain the major axes of the new ellipsoid. In the present example, this yields the new ellipsoid

$$\left(\frac{1}{n_o^2} + r_{63}E_z\right)x'^2 + \left(\frac{1}{n_o^2} - r_{63}E_z\right)y'^2 + \left(\frac{z^2}{n_e^2}\right) = 1. \tag{20.37}$$

As in the first example, the new and old z-axes are the same, but the new x' and y' axes are 45° from the original x- and y-axes (see Figure 20.8).

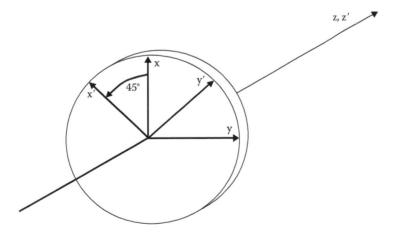

FIGURE 20.8 Rotated principal axes.

The refractive indices along the new x- and y-axes are

$$n'_x = n_o - \tfrac{1}{2}n_o^3 r_{63}E_z$$
$$n'_y = n_o + \tfrac{1}{2}n_o^3 r_{63}E_z \qquad (20.38)$$

Note that the quantity $n^3 rE$ in these examples determines the change in refractive index. Part of this product, $n^3 r$, depends solely on inherent material properties, and is a figure of merit for electro-optical materials. Values for the linear and quadratic electro-optic coefficients for selected materials are given in Tables 20.4 and 20.5, along with values for n and, for linear materials, $n^3 r$. While much of the information from these tables is from Yariv and Yeh [13], materials tables are also to be found in Kaminow [5,6]. Original sources listed in these references should be consulted on materials of particular interest. Additional information on many of the materials listed here, including tables of refractive index versus wavelength and dispersion formulae, can be found in Tropf et al. [20].

TABLE 20.4   Linear Electro-Optic Coefficients

Substance	Symmetry	Wavelength (μm)	Electro-Optic Coefficients r_{lk} (10^{-12}m/V)	Indices of Refraction	$n^3 r$ (10^{-12}m/V)
CdTe	$\bar{4}3$m	1.0	$r_{41} = 4.5$	$n = 2.84$	103
		3.39	$r_{41} = 6.8$		
		10.6	$r_{41} = 6.8$	$n = 2.60$	120
		23.35	$r_{41} = 5.47$	$n = 2.58$	94
		27.95	$r_{41} = 5.04$	$n = 2.53$	82
GaAs	$\bar{4}3$m	0.9	$r_{41} = 1.1$	$n = 3.60$	51
		1.15	$r_{41} = 1.43$	$n = 3.43$	58
		3.39	$r_{41} = 1.24$	$n = 3.3$	45
		10.6	$r_{41} = 1.51$	$n = 3.3$	54
ZnSe	$\bar{4}3$m	0.548	$r_{41} = 2.0$	$n = 2.66$	
		0.633	$r_{41}^a = 2.0$	$n = 2.60$	35
		10.6	$r_{41} = 2.2$	$n = 2.39$	
ZnTe	$\bar{4}3$m	0.589	$r_{41} = 4.51$	$n = 3.06$	
		0.616	$r_{41} = 4.27$	$n = 3.01$	
		0.633	$r_{41} = 4.04$	$n = 2.99$	108
			$r_{41}^a = 4.3$		
		0.690	$r_{41} = 3.97$	$n = 2.93$	
		3.41	$r_{41} = 4.2$	$n = 2.70$	83
		10.6	$r_{41} = 3.9$	$n = 2.70$	77
$Bi_{12}SiO_{20}$	23	0.633	$r_{41} = 5.0$	$n = 2.54$	82
CdS	6mm	0.589	$r_{51} = 3.7$	$n_o = 2.501$	
				ne = 2.519	
		0.633	$r_{51} = 1.6$	$n_o = 2.460$	
				ne = 2.477	
		1.15	$r_{31} = 3.1$	$n_o = 2.320$	
			$r_{33} = 3.2$	$n_e = 2.336$	
			$r_{51} = 2.0$		
		3.39	$r_{13} = 3.5$	$n_o = 2.276$	
			$r_{33} = 2.9$	ne = 2.292	

(Continued)

TABLE 20.4 (*Continued*)   Linear Electro-Optic Coefficients

Substance	Symmetry	Wavelength (μm)	Electro-Optic Coefficients r_{lk} (10^{-12}m/V)	Indices of Refraction	n^3r (10^{-12}m/V)
			$r_{51} = 2.0$		
		10.6	$r_{13} = 2.45$	$n_o = 2.226$	
			$r_{33} = 2.75$	$n_e = 2.239$	
			$r_{51} = 1.7$		
CdSe	6mm	3.39	$r_{13}^a = 1.8$	$n_o = 2.452$	
			$r_{33} = 4.3$	$n_e = 2.471$	
PLZT[b]	∞m	0.546	$n_e^3r_{33} - n_e^3r_{13} = 2.320$	$n_o = 2.55$	
$(Pb_{0.814}La_{0.124}Zr_{0.4}Ti_{0.6}O_3)$					
LiNbO$_3$	3m	0.633	$r_{13} = 9.6$	$n_o = 2.286$	
			$r_{22} = 6.8$	$n_e = 2.200$	
			$r_{33} = 30.9$		
			$r_{51} = 32.6$		
		1.15	$r_{22} = 5.4$	$n_o = 2.229$	
				$n_e = 2.150$	
		3.39	$r_{22} = 3.1$	$n_o = 2.136$	
				$n_e = 2.073$	
LiTaO$_3$	3m	0.633	$r_{13} = 8.4$	$n_o = 2.176$	
			$r_{33} = 30.5$	$n_e = 2.180$	
			$r_{22} = -0.2$		
		3.39	$r_{33} = 27$	$n_o = 2.060$	
			$r_{13} = 4.5$	$n_e = 2.065$	
			$r_{51} = 15$		
			$r_{22} = 0.3$		
KDP	$\bar{4}$2m	0.546	$r_{41} = 8.77$	$n_o = 1.5115$	
(KH$_2$PO$_4$)			$r_{63} = 10.3$	$n_e = 1.4698$	
		0.633	$r_{41} = 8$	$n_o = 1.5074$	
			$r_{63} = 11$	$n_e = 1.4669$	
		3.39	$r_{63} = 9.7$		
			$n_o^3r_{63} = 33$		
ADP	$\bar{4}$2m	0.546	$r_{41} = 23.76$	$n_o = 1.5079$	
(NH$_4$H$_2$PO$_4$)			$r_{63} = 8.56$	$n_e = 1.4683$	
		0.633	$r_{63} = 24.1$		
RbHSeO$_4$[c]		0.633			13,540
BaTiO$_3$	4mm	0.546	$r_{51} = 1,640$	$n_o = 2.437$	
				$n_e = 2.365$	
KTN	4mm	0.633	$r_{51} = 8,000$	$n_0 = 2.318$	
(KTa$_x$Nb$_{1-x}$O$_3$)				$n_e = 2.277$	
AgGaS$_2$	$\bar{4}$2m	0.633	$r_{41} = 4.0$	$n_o = 2.553$	
			$r_{63} = 3.0$	$n_e = 2.507$	

[a] These values are for clamped (high-frequency field) operation.

[b] PLZT is a compound of Pb, La, Zr, Ti, and O [17,18]. The concentration ratio of Zr to Ti is most important to its electro-optic properties. In this case, the ratio is 40:60.

[c] Salvestrini et al. [19].

TABLE 20.5   Quadratic Electro-Optic Coefficients

Substance	Symmetry	Wavelength (μm)	Electro-Optic Coefficients S_{ij} (10^{-18} m²/V²)	Index of Refraction	Temperature (°C)
BaTiO$_3$	**m3m**	0.633	$s_{11} - s_{12} = 2{,}290$	$n = 2.42$	$T > Tc (Tc = 120°C)$
PLZT[a]	∞**m**	0.550	$s_{33} - s_{13} = 26{,}000/n^3$	$n = 2.450$	Room temperature
KH$_2$PO$_4$ (KDP)	$\overline{4}$**2m**	0.540	$n_e^3 (s_{33} - s_{13}) = 31$	$n_o = 1.5115^b$	Room temperature
			$n_0^3 (s_{31} - s_{11}) = 13.5$	$n_e = 1.4698^b$	
			$n_0^3 (s_{12} - s_{11}) = 8.9$		
			$n_0^3 s_{66} = 3.0$		
NH$_4$H$_2$PO$_4$ (ADP)	$\overline{4}$**2m**	0.540	$n_e^3 (s_{33} - s_{13}) = 24$	$n_o = 1.5266^b$	Room temperature
			$n_0^3 (s_{31} - s_{11}) = 16.5$	$n_e = 1.4808^b$	
			$n_0^3 (s_{12} - s_{11}) = 5.8$		
			$n_0^3 s_{66} = 2$		

Source: Yariv A, Yeh P. *Optical Waves in Crystals,* Wiley, New York, 1984.
   [a] PLZT is a compound of Pb, La, Zr, Ti, and O [17,18]. The concentration ratio of Zr to Ti is most important to its electro-optic properties. In this case, the ratio is 65:35.
   [b] At 0.546 mm.

For light linearly polarized at 45°, the x and y components experience different refractive indices $\mathbf{n'_x}$ and $\mathbf{n'_y}$. The birefringence is defined as the index difference $\mathbf{n'_y} - \mathbf{n'_x}$. Since the phase velocities of the x and y components are different, there is a phase retardation Γ (in radians) between the x and y components of E, given by

$$\Gamma = \frac{\omega}{c}\left(\mathbf{n'_y} - \mathbf{n'_x}\right) d = \frac{2\pi}{\lambda} n_o^3 \mathbf{r}_{63} \mathbf{E}_z d, \tag{20.39}$$

where d is the path length of light in the crystal. The electric field of the incident light beam is

$$\vec{\mathbf{E}} = \frac{1}{\sqrt{2}} \mathbf{E}\left(\hat{\mathbf{x}} + \hat{\mathbf{y}}\right). \tag{20.40}$$

After transmission through the crystal, the electric field is

$$\frac{1}{\sqrt{2}} \mathbf{E}\left(e^{i\Gamma/2}\hat{\mathbf{x}}' + e^{-i\Gamma/2}\hat{\mathbf{y}}'\right). \tag{20.41}$$

If the path length and birefringence are selected such that $\Gamma = \pi$, the modulated crystal acts as a half-wave linear retarder and the transmitted light has field components

$$\frac{1}{\sqrt{2}} \mathbf{E}\left(e^{i\pi/2}\hat{\mathbf{x}}' + e^{-i\pi/2}\hat{\mathbf{y}}'\right) = \frac{1}{\sqrt{2}} \mathbf{E}\left(e^{i\pi/2}\hat{\mathbf{x}}' - e^{-i\pi/2}\hat{\mathbf{y}}'\right)$$

$$= \mathbf{E}\frac{e^{i\pi/2}}{\sqrt{2}}\left(\hat{\mathbf{x}}' - \hat{\mathbf{y}}'\right). \tag{20.42}$$

The axis of linear polarization of the incident beam has been rotated by 90° by the phase retardation of π radians or one-half wavelength. The incident linear polarization state has been rotated into the orthogonal polarization state. An analyzer at the output end of the crystal aligned with the incident (or unmodulated) plane of polarization will block the modulated beam. For an arbitrary applied voltage producing a phase retardation of Γ, the analyzer transmits a fractional intensity $\cos^2 \Gamma$. This is the principle of the Pockels cell.

Note that the form of the equations for the indices resulting from the application of a field is highly dependent upon the direction of the field in the crystal. For example, Table 20.6 gives the electro-optical properties of cubic $\overline{4}3m$ crystals when the field is perpendicular to three of the crystal planes. The new principal indices are obtained in general by solving an eigenvalue problem. For example, for a hexagonal material with a field perpendicular to the (111) plane, the index ellipsoid is

$$\left(\frac{1}{n_o^2}+\frac{r_{13}E}{\sqrt{3}}\right)x^2+\left(\frac{1}{n_o^2}+\frac{r_{13}E}{\sqrt{3}}\right)y^2+\left(\frac{1}{n_e^2}+\frac{r_{33}E}{\sqrt{3}}\right)z^2+2yzr_{51}\frac{E}{\sqrt{3}}+2zxr_{51}\frac{E}{\sqrt{3}}=1, \qquad (20.43)$$

and the eigenvalue problem is

$$\begin{pmatrix} \dfrac{1}{n_o^2}+\dfrac{r_{13}E}{\sqrt{3}} & 0 & \dfrac{2r_{51}E}{\sqrt{3}} \\[2ex] 0 & \dfrac{1}{n_o^2}+\dfrac{r_{13}E}{\sqrt{3}} & \dfrac{2r_{51}E}{\sqrt{3}} \\[2ex] \dfrac{2r_{51}E}{\sqrt{3}} & \dfrac{2r_{51}E}{\sqrt{3}} & \dfrac{1}{n_e^2}+\dfrac{r_{33}E}{\sqrt{3}} \end{pmatrix} V = \frac{1}{n'^2}V \qquad (20.44)$$

The secular equation is then

$$\begin{vmatrix} \left(\dfrac{1}{n_o^2}+\dfrac{r_{13}E}{\sqrt{3}}\right)-\dfrac{1}{n'^2} & 0 & \dfrac{2r_{51}E}{\sqrt{3}} \\[2ex] 0 & \left(\dfrac{1}{n_o^2}+\dfrac{r_{13}E}{\sqrt{3}}\right)-\dfrac{1}{n'^2} & \dfrac{2r_{51}E}{\sqrt{3}} \\[2ex] \dfrac{2r_{51}E}{\sqrt{3}} & \dfrac{2r_{51}E}{\sqrt{3}} & \left(\dfrac{1}{n_o^2}+\dfrac{r_{33}E}{\sqrt{3}}\right)-\dfrac{1}{n'^2} \end{vmatrix} = 0 \qquad (20.45)$$

and the roots of this equation are the new principal indices.

TABLE 20.6 Electro-Optic Properties of Cubic $\overline{4}3m$ Crystals

E Field Direction	Index Ellipsoid	Principal Indices
E perpendicular to (001) plane: $E_x = E_y = 0$ $Ez = E$	$\dfrac{x^2+y^2+z^2}{n_o^2}+2r_{41}Exy=1$	$n'_x = n_o + \frac{1}{2}n_o^3 r_{41}E$ $n'_y = n_o - \frac{1}{2}n_o^3 r_{41}E$ $n'_z = n_o$
E perpendicular to (110) plane: $E_x = E_y = E/\sqrt{2}$ $Ez = 0$	$\dfrac{x^2+y^2+z^2}{n_o^2}+\sqrt{2}r_{41}E(yz+zx)=1$	$n'_x = n_o + \frac{1}{2}n_o^3 r_{41}E$ $n'_y = n_o - \frac{1}{2}n_o^3 r_{41}E$ $n'_z = n_o$
E perpendicular to (111) plane: $E_x = E_y = E_z = E/\sqrt{3}$	$\dfrac{x^2+y^2+z^2}{n_o^2}+\dfrac{2}{\sqrt{3}}r_{41}E(yz+zx+xy)=1$	$n'_x = n_o + \dfrac{1}{2\sqrt{3}}n_o^3 r_{41}E$ $n'_y = n_o - \dfrac{1}{2\sqrt{3}}n_o^3 r_{41}E$ $n'_z = n_o - \dfrac{1}{\sqrt{3}}n_o^3 r_{41}E$

Source: Goldstein D. Polarization Modulation in Infrared Electrooptic Materials, PhD dissertation, 1990.

20.6 Magneto-Optics

When a magnetic field is applied to certain materials, the plane of incident linearly polarized light may be rotated in passage through the material. The magneto-optic effect linear with field strength is called the Faraday effect, and was discovered by Michael Faraday in 1845. A magneto-optic cell is illustrated in Figure 20.9. The field is set up so that the field lines are along the direction of the optical beam propagation. A linear polarizer allows light of one polarization into the cell. A second linear polarizer is used to analyze the result.

The Faraday effect is governed by the equation

$$\theta = \mathbf{V}\mathbf{B}d, \tag{20.46}$$

where $\mathbf{V}$ is the Verdet constant, θ is the rotation angle of the electric field vector of the linearly polarized light, $\mathbf{B}$ is the applied field, and d is the path length in the medium. The rotary power ρ, defined in degrees per unit path length, is given by

$$\rho = \mathbf{V}\mathbf{B}. \tag{20.47}$$

A table of Verdet constants for some common materials is given in Table 20.7. The material that is often used in commercial magneto-optic-based devices is some formulation of iron garnet. Data tabulations

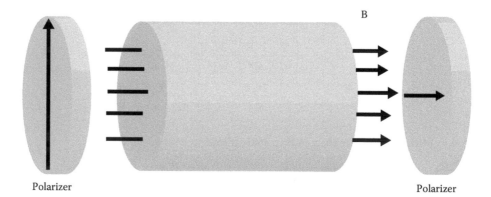

FIGURE 20.9 Illustration of a setup to observe the Faraday effect.

TABLE 20.7 Values of the Verdet Constant at $\lambda = 5893$ Å

Material	$T(°C)$	Verdet Constant (deg/G-mm)
Water[a]	20	2.18×10^{-5}
Air ($\lambda = 5780$ Å and 760mmHg)[b]	0	1.0×10^{-8}
NaCl[b]	16	6.0×10^{-5}
Quartz[b]	20	2.8×10^{-5}
CS_2[a]	20	7.05×10^{-5}
P[a]	33	2.21×10^{-4}
Glass, flint[a]	18	5.28×10^{-5}
Glass, crown[a]	18	2.68×10^{-5}
Diamond[a]	20	2.0×10^{-5}

[a] Yariv and Yeh [13].
[b] Hecht [12].

for metals, glasses, and crystals, including many iron garnet compositions, can be found in Chen [22]. The magneto-optic effect is the basis for magneto-optic memory devices, optical isolators, and spatial light modulators [23,24].

Other magneto-optic effects, in addition to the Faraday effect, include the Cotton–Mouton effect, the Voigt effect, and the Kerr magneto-optic effect. The Cotton–Mouton effect is a quadratic magneto-optic effect observed in liquids. The Voigt effect is similar to the Cotton–Mouton effect but is observed in vapors. The Kerr magneto-optic effect is observed when linearly polarized light is reflected from the face of either pole of a magnet. The reflected light becomes elliptically polarized.

20.7 Liquid Crystals

Liquid crystals are a class of substances that demonstrate that the premise that matter exists only in solid, liquid, and vapor (and plasma) phases is a simplification. Fluids, or liquids, are generally defined as the phase of matter that cannot maintain any degree of order in response to a mechanical stress. The molecules of a liquid have random orientations and the liquid is isotropic. In the period 1888–1890, Reinitzer, and separately Lehmann, observed that certain crystals of organic compounds exhibit behavior between the crystalline and liquid states [25]. As the temperature is raised, these crystals change to a fluid substance, which retains the anisotropic behavior of a crystal. This type of liquid crystal is now classified as thermotropic because the transition is effected by a temperature change, and the intermediate state is referred to as a mesophase [26]. There are three types of mesophases: smectic, nematic, and cholesteric. Smectic and nematic mesophases are often associated and occur in sequence as the temperature is raised. The term smectic derives from the Greek word for soap and is characterized by a more viscous material than the other mesophases. Nematic is from the Greek for thread and was named because the material exhibits a striated appearance (between crossed polaroids). The cholesteric mesophase is a property of the cholesterol esters, hence the name.

Figure 20.10a illustrates the arrangement of molecules in the nematic mesophase. Although the centers of gravity of the molecules have no long-range order as crystals do, there is order

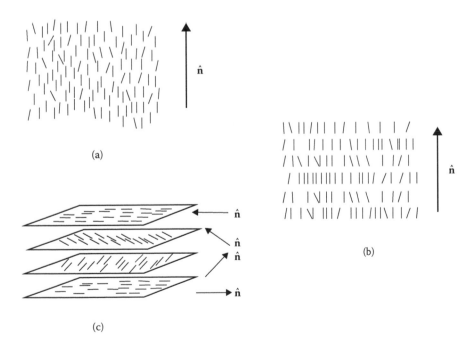

FIGURE 20.10 Schematic representation of liquid crystal order: (a) nematic, (b) smectic, and (c) cholesteric.

in the orientations of the molecules [27]. They tend to be oriented parallel to a common axis indicated by the unit vector $\hat{n}$.

The direction of $\hat{n}$ is arbitrary and is determined by some minor force, such as the guiding effect of the walls of the container. There is no distinction between a positive and negative sign of $\hat{n}$. If the molecules carry a dipole, there are equal numbers of dipoles pointing up as there are down. These molecules are not ferroelectric. The molecules are achiral, that is, they have no handedness, and there is no positional order of the molecules within the fluid. Nematic liquid crystals are optically uniaxial.

The temperature range over which the nematic mesophase exists varies with the chemical composition and mixture of the organic compounds. The range is quite wide: for example, in one study of ultraviolet imaging with a liquid crystal light valve, four different nematic liquid crystals were used [28]. Two of these were MBBA (*N*-(*p*-methoxybenzylidene)-*p*-*n* butylaniline) with a nematic range of 17 to 43°C, and a proprietary material with a range of –20 to 51°C.

There are many known electro-optical effects involving nematic liquid crystals [29,26,30]. Two of the more important are field-induced birefringence, also called deformation of aligned phases, and the twisted nematic effect, also called the Schadt–Helfrich effect. An example of a twisted nematic cell is shown in Figure 20.11.

Figure 20.11a shows the molecule orientation in a liquid crystal cell, without and with an applied field. The liquid crystal material is placed between two electrodes. The liquid crystals at the cell wall align themselves in some direction parallel to the wall as a result of very minor influences. A cotton swab lightly stroked in one direction over the interior surface of the wall prior to cell assembly is enough to produce alignment of the liquid crystal [31]. The molecules align themselves with the direction of the rubbing. The electrodes are placed at 90° to each other with respect to the direction of rubbing. The liquid crystal molecules twist from one cell wall to the other to match the alignments at the boundaries as

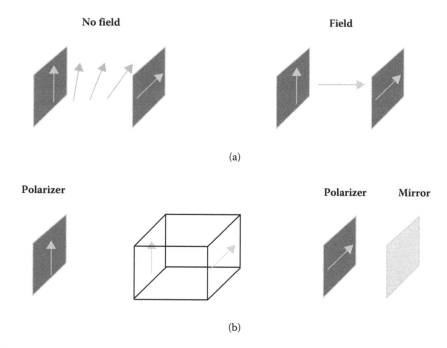

FIGURE 20.11   Liquid crystal cell operation: (a) molecule orientation in a liquid crystal cell, with no field and with field; (b) a typical nematic liquid crystal cell.

illustrated, and light entering at one cell wall with its polarization vector aligned to the crystal axis will follow the twist and be rotated 90° by the time it exits the opposite cell wall. If the polarization vector is restricted with a polarizer on entry and an analyzer on exit, only the light with the 90° polarization twist will be passed through the cell. With a field applied between the cell walls, the molecules tend to orient themselves perpendicular to the cell walls, that is, along the field lines. Some molecules next to the cell walls remain parallel to their original orientation, but most of the molecules in the center of the cell align themselves parallel to the electric field, destroying the twist. At the proper strength, the electric field will cause all the light to be blocked by the analyzer.

Figure 20.11b shows a twisted nematic cell as might be found in a digital watch display, gas pump, or calculator. Light enters from the left. A linear polarizer is the first element of this device and is aligned so that its axis is along the left-hand liquid crystal cell wall alignment direction. With no field, the polarization of the light twists with the liquid crystal twist, 90° to the original orientation, passes through a second polarizer with its axis aligned to the right-hand liquid crystal cell wall alignment direction, and is reflected from a mirror. The light polarization twists back the way it came and leaves the cell. Regions of this liquid crystal device that are not activated by the applied field are bright. If the field is now applied, the light does not change polarization as it passes through the liquid crystal and will be absorbed by the second polarizer. No light returns from the mirror, and the areas of the cell that have been activated by the applied field are dark.

A twisted nematic cell has a voltage threshold below which the polarization vector is not affected due to the internal elastic forces. A device 10 μm thick might have a threshold voltage of 3V [26].

Another important nematic electro-optic effect is field-induced birefringence or deformation of aligned phases. As with the twisted nematic cell configuration, the liquid crystal cell is placed between crossed polarizers. However, now the molecular axes are made to align perpendicular to the cell walls and thus parallel to the direction of light propagation. By using annealed SnO_2 electrodes and materials of high purity, Schiekel and Fahrenschon [30] found that the molecules spontaneously align in this manner. Their cell worked well with 20 μm thick MBBA. The working material must be one having a negative dielectric anisotropy so that when an electric field is applied (normal to the cell electrodes), the molecules will tend to align themselves perpendicular to the electric field. The molecules at the cell walls tend to remain in their original orientation and the molecules within the central region will turn up to 90°; this is illustrated in Figure 20.12.

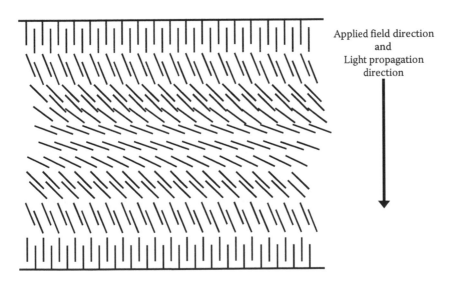

Applied field direction
and
Light propagation
direction

FIGURE 20.12 Deformation of liquid crystal due to applied voltage [29].

There is a threshold voltage typically in the 4–6V range [26]. Above the threshold, the molecules begin to distort and become birefringent due to the anisotropy of the medium. Thus with no field, no light exits the cell; at threshold voltage, light begins to be divided into ordinary and extraordinary beams, and some light will exit the analyzer. The birefringence can also be observed with positive dielectric anisotropy when the molecules are aligned parallel to the electrodes at no field and both electrodes have the same orientation for nematic alignment. As the applied voltage is increased, the light transmission increases for crossed polarizers [26]. The hybrid field-effect liquid crystal light valve relies on a combination of the twisted nematic effect (for the off state) and induced birefringence (for the on state) [32].

Smectic liquid crystals are more ordered than the nematics. The molecules are not only aligned, but they are also organized into layers, making a two-dimensional fluid. This is illustrated in Figure 20.10b. There are three types of smectics: A, B, and C. Smectic A is optically uniaxial; smectic C is optically biaxial; smectic B is most ordered, since there is order within layers. Smectic C, when chiral, is ferroelectric. Ferroelectric liquid crystals are known for their fast switching speed and bistability.

Cholesteric liquid crystal molecules are helical, and the fluid is chiral. There is no long-range order, as in nematics, but the preferred orientation axis changes in direction through the extent of the liquid. Cholesteric order is illustrated in Figure 20.10c.

For more information on liquid crystals and an extensive bibliography, see Wu [33,34].

20.8 Modulation of Light

We have seen that light modulators are composed of an electro- or magneto-optical material on which an electromagnetic field is imposed. Electro-optical modulators may be operated in a longitudinal mode or in a transverse mode. In a longitudinal mode modulator, the electric field is imposed parallel to the light propagating through the material, and in a transverse mode modulator, the electric field is imposed perpendicular to the direction of light propagation. Either mode may be used if the entire wavefront of the light is to be modulated equally. The longitudinal mode is more likely to be used if a spatial pattern is to be imposed on the modulation. The mode used will depend upon the material chosen for the modulator and the application.

Figure 20.13 shows the geometry of a longitudinal electro-optic modulator. The beam is normal to the face of the modulating material and parallel to the field imposed on the material. Electrodes of a material that is conductive yet transparent to the wavelength to be modulated are deposited on the faces through which the beam travels. This is the mode used for liquid crystal modulators.

Figure 20.14 shows the geometry of the transverse electro-optic modulator. The imposed field is perpendicular to the direction of light passage through the material. The electrodes do not need to be transparent to the beam. This is the mode used for modulators in laser beam cavities, that is, a CdTe modulator in a CO_2 laser cavity.

20.9 Concluding Remarks

The origin of the electro-optic tensor, the form of that tensor for various crystal types, and the values of the tensor coefficients for specific materials have been discussed. The concepts of index ellipsoid, the wave surface, and the wavevector surface were introduced. These are quantitative and qualitative models that aid in the understanding of the interaction of light with crystals. We have shown how the equation for the index ellipsoid is found when an external field is applied, and how expressions for the new principal indices of refraction are derived. Magneto-optics and liquid crystals were described. The introductory concepts of constructing an electro-optic modulator were discussed.

While the basics of electro- and magneto-optics in bulk materials have been covered, there is a large body of knowledge dealing with related topics which cannot be included here. A more detailed

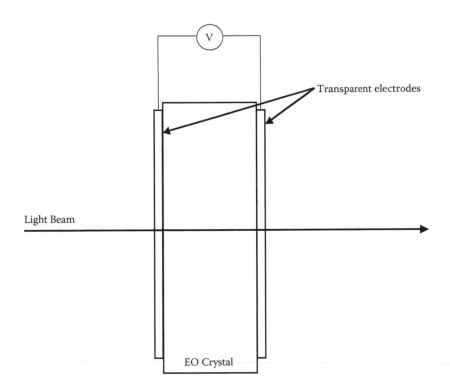

FIGURE 20.13 Longitudinal mode modulator.

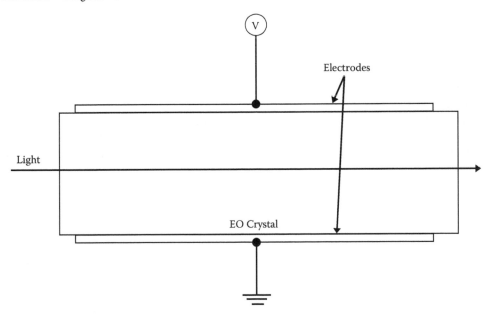

FIGURE 20.14 Transverse mode modulator.

description of electro-optic modulators is covered in Yariv and Yeh [37]. Information on spatial light modulators may be found in Efron [1]. Shen [35] describes the many aspects and applications of non-linear optics, and current work in such areas as organic nonlinear materials can be found in SPIE Proceedings [36,37].

References

1. Efron U. ed., *Spatial Light Modulator Technology, Materials, Devices, and Applications*, New York, NY: Marcel Dekker, 1995.
2. Jackson JD. *Classical Electrodynamics*, 2nd edn, New York, NY: Wiley, 1975.
3. Lorrain P, Corson D. *Electromagnetic Fields and Waves*, 2nd edn, New York, NY: Freeman, 1970.
4. Reitz JR, Milford FJ. *Foundations of Electromagnetic Theory*, 2nd edn, New York, NY: Addison-Wesley, 1967.
5. Kaminow IP. *An Introduction to Electrooptic Devices*, New York, NY: Academic Press, 1974.
6. Kaminow IP. Linear electrooptic materials. In *CRC Handbook of Laser Science and Technology, Volume IV, Optical Materials, Part 2: Properties*. Weber, M. J., ed., Boca Raton, FL: CRC Press, 1986.
7. Kittel C. *Introduction to Solid State Physics*, New York, NY: Wiley, 1971.
8. Lovett DR. *Tensor Properties of Crystals*, Bristol: Hilger, 1989.
9. Nye JF. *Physical Properties of Crystals: Their Representation by Tensors and Matrices*, Oxford: Oxford University Press, 1985.
10. Senechal M. *Crystalline Symmetries: An Informal Mathematical Introduction*, Bristol: Hilger, 1990.
11. Wood A. *Crystals and Light: An Introduction to Optical Crystallography*, London: Dover, 1977.
12. Hecht E. *Optics*, New York, NY: Addison-Wesley, 1987.
13. Yariv A, Yeh P. *Optical Waves in Crystals*, New York, NY, Wiley, 1984.
14. Born M, Wolf E. *Principles of Optics*, Oxford: Pergamon Press, 1975.
15. Jenkins FA, White HE. *Fundamentals of Optics*, New York, NY: McGraw-Hill, 1957.
16. Klein MV. *Optics*, New York, NY: Wiley, 1970.
17. Haertling GH, Land CE. Hot-pressed $(Pb,La)(Zr,Ti)O_3$ ferroelectric ceramics for electrooptic applications, *J. Am. Ceram. Soc.*, **54**, 1; 1971.
18. Land CE. Optical information storage and spatial light modulation in PLZT ceramics, *Opt. Eng.*, **17**, 317; 1978.
19. Salvestrini JP, Fontana MD, Aillerie M, Czapla Z. New material with strong electro-optic effect: Rubidium hydrogen selenate $(RbHSeO_4)$, *Appl. Phys. Lett.*, **64**, 1920; 1994.
20. Tropf WJ, Thomas ME, Harris TJ. Properties of crystals and glasses. In *Handbook of Optics, Volume II, Devices, Measurements, and Properties*, 2nd edn, New York, NY: McGraw-Hill, 1995.
21. Goldstein D. *Polarization Modulation in Infrared Electrooptic Materials*, PhD dissertation, 1990.
22. Chen Di. Data tabulations. In *CRC Handbook of Laser Science and Technology, Volume IV, Optical Materials, Part 2: Properties*, Weber, M. J., ed. Boca Raton, FL: CRC Press, 1986.
23. Ross WE, Psaltis D, Anderson RH. Two-dimensional magneto-optic spatial light modulator for signal processing, *Opt. Eng.*, **22**, 485; 1983.
24. Ross WE, Snapp KM, Anderson RH. Fundamental characteristics of the litton iron garnet magneto-optic spatial light modulator, *Proc. SPIE*, **388**, *Advances in Optical Information Processing*, **55–64**, Volume 388, 1983.
25. Gray GW. *Molecular Structure and the Properties of Liquid Crystals*, New York, NY: Academic Press, 1962.
26. Priestley EB, Wojtowicz PJ, Sheng P. eds, *Introduction to Liquid Crystals*, New York, NY: Plenum Press, 1974.
27. De Gennes PG. *The Physics of Liquid Crystals*, Oxford: Oxford University Press, 1974.
28. Margerum JD, Nimoy J, Wong SY. Reversible ultraviolet imaging with liquid crystals, *Appl. Phys. Lett.*, **17**, 51; 1970.
29. Meier G, Sackman H, Grabmaier F. *Applications of Liquid Crystals*, Berlin: Springer-Verlag, 1975.
30. Schiekel MF, Fahrenschon K. Deformation of nematic liquid crystals with vertical orientation in electrical fields, *Appl. Phys. Lett.*, **19**, 391; 1971.
31. Kahn FJ. Electric-field-induced orientational deformation of nematic liquid crystals: Tunable birefringence, *Appl. Phys. Lett.*, **20**, 199; 1972.

32. Bleha WP, et al. Application of the liquid crystal light valve to real-time optical data processing, *Opt. Eng.*, **17**, 371; 1978.

33. Wu S-T. Nematic liquid crystals for active optics. In *Optical Materials, A Series of Advances*, Vol. 1, S. Musikant, ed. New York, NY: Marcel Dekker, 1990.

34. Wu S-T. Liquid crystals. In *Handbook of Optics, Volume II, Devices, Measurements, and Properties*, 2nd edn, New York, NY: McGraw-Hill, 1995.

35. Shen YR. *The Principles of Nonlinear Optics*, New York, NY: Wiley, 1984.

36. Kuzyk MG. ed., Nonlinear optical properties of organic materials X. *Proc. SPIE*, 3147; 1997.

37. Moehlmann GR. eds., Nonlinear optical properties of organic materials IX. *Proc. SPIE*, 2852; 1996.

Optical Fabrication

21.1 Introduction..727

21.2 Shapes and Properties of Optical Surfaces....................................728
Optical Surface Shapes • Specifications of Optical Components

21.3 Traditional Methods of Optical Fabrication 739
Introduction • Support Methods for Fabrication • Diamond Machining and Generating • Fining • Polishing • Centering and Edging • Cleaning • Cementing • Current Trends in Optical Fabrication

21.4 Fabrication of Aspheres and Non-Traditional Manufacturing Methods..755
Conventional Methods for Fabricating Aspherics • Modern Methods of Asphere Fabrication

21.5 Conclusion...760

References...760

Web References..763

David Anderson
and Jim Burge

21.1 Introduction

The goal of most optical engineering is to develop hardware that uses optical components, such as lenses, prisms, mirrors, and windows. The purpose of this chapter is to summarize the principles and technologies used to manufacture these components, with the goal of helping the optical engineer understand the relationships between fabrication issues and specifications. While we describe here an overview of many of the basic methods of fabrication, we also provide references to other books and articles that provide a more complete treatment.

The field of optical fabrication covers the manufacture of optical elements, typically made from glass, but also from a wide range of other materials that the designer can choose from depending on performance and environmental requirements. These requirements often conflict with each other and can significantly affect the cost and manufacturability of system components when tradeoffs are made in the design. The most difficult aspect of the fabrication of many optical components comes from the tight tolerances required for high optical performance and precise mechanical mounting hardware. The optical system engineer must assign specifications that balance performance with fabrication costs. The tolerances must be tight enough to ensure acceptable system performance, yet not so tight that the parts cannot be made economically. For a particular project, the fabrication process is usually selected to achieve the specified tolerances. Parts with tighter requirements are nearly always more expensive and take longer. Understanding the capabilities and limitations of modern fabrication methods will help the designer develop realistic specifications and tolerances.

The optical engineer who is specifying the optical elements also needs to understand how the size and quantity affect the manufacturing process, quality, and cost. For example, special tooling is usually required for large and difficult parts, which drive the cost up. "Tooling" refers to any special equipment used for manufacturing an item. Tooling is not used up in the process, so it can be used repeatedly and is generally categorized as a "non-recurring cost." However, special tooling can also lead to an efficient process, reducing the per-item cost for parts made in large quantities. Like any industrial process, optical fabrication has significant economies of scale, meaning that items can be mass-produced more efficiently than they can be made one at a time. However, there is always a tradeoff between improved efficiency and tooling costs. If only a few elements are needed, it does not make sense to spend more on tooling than it would cost to make the parts by a less efficient method.

The capability of fabricating an optical component to a particular specification and tolerance is intimately related to how that specification is measured. The old adage "you can only make it as good as you can test it" surely holds here. Modern fabrication capability is sometimes limited by the available manufacturing method and sometimes by the test method. The specification that presses the limits of either will result in increased difficulty, time, and cost. But that pressure has other effects as well. Advances in the methods of optical fabrication have largely resulted from the ever-increasing difficulty of fabricating the required surfaces, which has had the concomitant effect of advancing test methods to verify those surfaces.

Through the use of various types of motors, sensors, switching devices, and computers, automation has begun to have a major impact on the capability and productivity of fabrication equipment. Numerically controlled (NC) machining has made the production of tooling and shaping of parts much more rapid and less costly. Generating has become highly automated with the application of position encoders and radius measuring hardware and software. Grinding/polishing machines are slowly having most of their subsystems automated, and new machine designs allow more efficient production of previously very difficult or impossible surfaces. However, for most precision optics made today, the optician's skill in the operation of the polishing machine still has a large impact on the results. Still, automation is making the fabrication process less skill-dependent and more "deterministic," a buzzword of modern optical fabricators. We are currently in the middle of a sea change in the methods and machinery used to fabricate optical surfaces largely driven by the rapid advance of digital technology.

New machines that require a somewhat different approach to fabricating custom optics are becoming more efficient and eventually will outperform current production methods. A single, high-precision machine can now rapidly generate, grind, polish and shape (figure) a single lens at a time. Metrology for each stage is integrated into the machine and corrections are applied automatically. Stiff, high-precision spindles with diamond wheels use shallow cuts to produce accurate surfaces with minimal subsurface damage. Various polishing configurations are used to bring the surfaces to final figure and finish. Although these machines are currently expensive compared with conventional labor-intensive methods, the future of production optics clearly lies in this direction. The development of these machines and the improvements in measurement systems has driven a wide range of deeper investigations into the grinding and polishing process. These investigations will inevitably lead to further developments in the automation of optics production in the years ahead.

21.2 Shapes and Properties of Optical Surfaces

21.2.1 Optical Surface Shapes

Most fabrication methods deal with the production of spherical surfaces. Figure 21.1 illustrates the basic geometry of a spherical surface. A description of a spherical surface for fabrication purposes requires only an aperture radius and a radius of curvature. Since a sphere has no optical axis but only a radius of curvature (or, equivalently, a center of curvature) that defines its shape, any section of that sphere looks like any other section. This fact has important consequences on how these surfaces are produced. Note that a flat surface is simply a spherical surface with an infinitely long radius of curvature. One fundamental

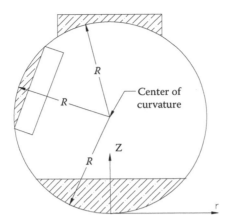

FIGURE 21.1 The basic geometry of a spherical surface is defined by only two parameters, its size or aperture radius r, and its radius of curvature R. The surface profile $z(r)$ is defined as $z(r) = R^2 - \sqrt{R^2 - r^2}$. A spherical surface does not have an optical axis, only a single point of symmetry at the center of curvature.

idea in optical fabrication that is described below is that randomly rubbing two surfaces together with an abrasive between generally results in two mating spherical surfaces with opposite curvatures.

Aspheric surfaces, on the other hand, lack this symmetry as shown in Figure 21.2. They have an optical axis that is defined as the line about which the surface is symmetrical. Because aspherical surfaces have, by definition, a varying radius of curvature, more information is needed to adequately describe them. In addition to the aperture radius r and the radius of curvature at the vertex R_v, in the case of conic sections, a third parameter is needed, the conic constant K. For more complicated aspherics, such as an off-axis section, the distance off-axis must be specified. Additionally, higher-order terms may also be included but, as described in Section 3, these surfaces are much more difficult to fabricate and test. Freeform optical surfaces can take any shape, with no symmetry whatsoever, making them especially difficult to fabricate. Also, note that because a sphere has no unique optical axis, there is no such thing as an "off-axis sphere"—just one that has a wedge between the front and rear surfaces.

The surface profile for a purely conic aspherical surface is given by Equation 21.1:

$$z(r) = \frac{r^2}{R + \sqrt{R^2 - (K+1)r^2}}, \tag{21.1}$$

where:

$z(r)$ = surface height
r = radial position ($r^2 = x^2 + y^2$)
R = radius of curvature
K = conic constant ($K = -e^2$ where e is eccentricity).

The types of conic surfaces, determined by the conic constant, are as follows, and are shown in Figure 21.2:

$K < -1$	Hyperboloid
$K = -1$	Paraboloid
$-1 < K < 0$	Prolate ellipsoid (rotated about its major axis)
$K = 0$	Sphere
$K > 0$	Oblate ellipsoid (rotated about its minor axis)

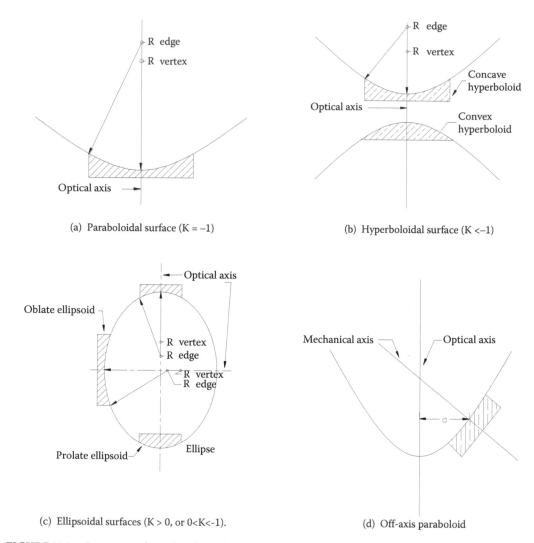

(a) Paraboloidal surface (K = –1)

(b) Hyperboloidal surface (K <–1)

(c) Ellipsoidal surfaces (K > 0, or 0<K<-1).

(d) Off-axis paraboloid

FIGURE 21.2 Common aspherical surfaces, defined as conic sections of revolution. These surfaces have a varying radius of curvature and require at least three parameters to adequately define them, the aperture radius r, the radius of curvature at the vertex R_v, and, in the case of conic sections, the conic constant K.

Sometimes additional, higher-order aspherical terms are added to the conic to adequately describe the surface as given in Equation 21.2.

$$z(r) = \frac{r^2}{R + \sqrt{R^2 - (K+1)r^2}} + a_4 r^4 + a_6 r^6 + \dots \tag{21.2}$$

Equation 21.1 gives the sag, which is the distance of the optical surface from a plane. However, in optical fabrication, we are often concerned with the deviation of this surface from a sphere. Using a Taylor expansion, the aspheric departure $S(r)$ is given in Equation 21.3.

$$S(r) = \frac{Kr^4}{8R^3} + \frac{1 \cdot 3 \left[(K+1)^2 - 1 \right] r^6}{2^3 3! R^5} + \frac{1 \cdot 3 \cdot 5 \left[(K+1)^3 - 1 \right] r^8}{2^4 4! R^7} + \dots \tag{21.3}$$

Although most aspherics are specified as conic surfaces and polynomial aspherics, there are some other common aspheric surfaces:

- Toroids—These surfaces are part of a torus, having a different radius of curvature for two orthogonal directions on the optical surface. These are used for astigmatism correction in eyeglasses, and are used at grazing incidence for focusing high-energy radiation. Toroids are made with special generators, and polished with a variation of the process for making spherical optics.
- Axicons—These surfaces are basically cones, generated by a tilted line rotated about an optical axis. Axicons are used in unstable resonator laser cavities and for special alignment tooling. These are nearly always made by molding or single-point diamond tooling.
- Freeform surfaces—These are general surfaces that do not fit any of the above parametric descriptions. They can be specified as departures from conventional surfaces using functions, such as Zernike polynomials, or they can be specified as an array of points coupled with a convention, such as a bicubic spline or NURBS, to provide a continuous smooth surface.

21.2.2 Specifications of Optical Components

When the optical designer is first developing the system concept, he should ask the question "How is this going to be made?" It makes no sense to design a system with components that cannot be manufactured accurately enough to meet the technical specifications or economically enough to meet the cost goals. This section discusses some of the fabrication issues that face optical designers. Different manufacturing methods and optical shops can be used depending on whether the order is for thousands of optics or only a few. Tolerances on the components can drive the fabrication method, so these must be carefully thought out. Size also plays an important role in deciding the fabrication method and reasonable specifications. The choice of material for the optics can also limit the choices of fabrication methods.

Optical designers tend to design systems that perform well according to simulations, then to expect this performance from the real optical components. The path from the design world to the real world takes place through the tolerancing of the relevant optical parameters by the engineer who then must translate these tolerances into a set of specifications of the various material, mechanical, and optical properties that the fabricator requires to make the component. Specifications for the system are often assigned as an afterthought to the design and because of this they tend to be too tight, resulting in unnecessary expense and longer delivery time since tighter tolerances makes the fabrication increasingly difficult. A better way to design is to understand and anticipate the fabrication limitations and incorporate them in the design of the component. This way the designer balances sensitivity to expected errors based on fabrication limitations as part of the optical design.

Defining the optimal value for the tolerances will be greatly aided by some discussion between the fabricator and the designer. The designer always wants tighter tolerances because they will give improved performance. These come at a cost, because the fabricator must work harder to meet these tolerances. So how good is "good enough"? The designer cannot decide this on his own because it depends on the incremental fabrication costs. The fabricator cannot define this because he does not have sufficient information to know how the manufacturing errors affect the system performance.

When the system tolerance analysis is performed, the engineer will assume some tolerances and perturb the simulation of the optical system to determine the effect of each parameter (such as radius of curvature or lens thickness) being at the edge of tolerance. The overall performance is estimated by combining the effect of all of the terms as a root sum square. This is where the fun begins. Usually, the optical designer finds one of two things from this exercise: that the system has excellent performance, in which case the assumed tolerances are too tight, or that the performance is not acceptable, in which case the assumed tolerances must be tightened. At this point, the designer should go to a fabricator and discuss which tolerances to adjust to give acceptable performance without driving up manufacturing costs. Fabricators are generally a very congenial lot and are usually very willing to discuss fabrication issues with designers.

Because the effects of the separate tolerances are uncorrelated and added as RSS, only the few largest terms contribute to the total. If the designer looks carefully at the individual terms, tolerances that do not affect the performance can be made looser than would be otherwise. Also, only a few critical parameters will need to be controlled to high accuracy.

The key to good tolerancing is to know the relationship between tolerances and cost. Unfortunately, this information can be hard to get, and can vary significantly from one shop to another and over time. This relationship depends on two things: how much extra work is required to achieve the tolerance and whether special equipment is required.

A simple example is the angle for a prism. Using standard shop practices, and paying no particular attention, the angle will be good to about 5 arc minutes. If the optician takes special care using common tools, the angle can be controlled to 1 arc minute. The added expense here is only the additional time required by the optician to measure the angle and adjust the process. For accuracy of 10 arc seconds, the optician will need more sophisticated measuring equipment and it will take more iterations of the measure/adjust cycle. Now, if the angle must be made to 1 arc second, only an experienced optician with good metrology can get there, and it will take him considerable effort and time. Optics having a requirement of 0.1 arc second will require a research effort to come up with a way of both making this part and validating it, so the cost may be extremely high and the delivery time quite long.

In some cases, the cost curves do not change with tolerance until the capacity of a machine is exceeded. A good example here is machining with numerically controlled machines. A good machine will give 10 µm accuracy over small distances, independent of the tolerance assigned by the designer. There would be no cost savings for assigning a looser tolerance. There would, however, be a sharp cost increase if a 9 µm tolerance were assigned and the machine were certified to 10 µm. This would drive the fabricator to another method, which could cost several times more. It is important to discuss the tolerances with your fabricator.

21.2.2.1 Material Specifications

The choice of the right optical material is the first and a very important choice that the engineer must face in specifying an optical component. Components can be made from an astounding number of substrate materials based on the various optical, mechanical, and environmental requirements of their use. However, the majority of components are for use in the visible and are made of glass since glass is very stable, relatively strong (though brittle), and, perhaps most importantly, can be polished to incredible smoothness. Although it is generally made from silicon dioxide plus other components, the incredibly complicated and high temperature processes involved in creating high purity glass with varying optical and mechanical properties has limited its production to just a few companies in the world. It is a great educational experience to study the catalogs from these companies as they discuss in detail the various properties of glass and how they are measured and specified. Being the dominant material in optical fabrication, we will focus on glass fabrication for the rest of this chapter.

Two broad categories of optical glass are those intended for use as mirrors where only one surface is optically important and the transmission properties less important, and those intended for use in transmission, such as lenses, windows, prisms, and so on, where the transmission properties are crucial. Shown in Table 21.1 are some of the principal material specifications of glass used in transmission and the range of tolerances of each that the designer can choose from with the expectation that as you move from base to high precision, the cost can escalate dramatically.

Properties important to the optical designer, such as index of refraction, dispersion, and transmissibility at various wavelengths, may not be of much importance to the fabricator but other properties, such as the coefficient of linear expansion, hardness, and resistance to chemical attack, while sometimes not important to an optical designer can be very important to the fabricator and opto-mechanical engineer. Some optical designs look just wonderful in the computer but are woefully impractical or

TABLE 21.1 Optical Material Tolerances (Using Schott Specifications, Others Are Equivalent)

Parameter	Base	Precision	High Precision
Refractive index departure from nominal	±0.0005 (Step 3)	±0.0003 (Step 2)	±0.0002 (Step 1)
Refractive index measurement	$\pm 3 \times 10^{-5}$	$\pm 2 \times 10^{-5}$	$\pm 1 \times 10^{-5}$
Dispersion departure from nominal	±0.5% (Step 3)	±0.3% (Step 3)	±0.2% (Step 3)
Refractive index homogeneity	$\pm 20 \times 10^{-6}$ (H1)	$\pm 5 \times 10^{-6}$ (H2)	$\pm 1 \times 10^{-6}$ (H4)
Stress birefringence (depends strongly on glass)	10 nm/cm	6 nm/cm	4 nm/cm
Bubbles/inclusions (>30 μm) (Area of bubbles per 100 cm³)	0.5 mm² (class B3)	0.1 mm² (class B1)	0.029 mm² (class B0)
Striae based on shadow graph test	Normal quality (has fine striae)	Precision quality (no detectable striae)	Precision quality (no detectable striae)

expensive to actually fabricate when the materials are difficult to work with or difficult to procure. Many optical glasses, while available in small batches or in small sizes, are simply not practical for the glass manufacturers to produce in larger quantities or size. It is always a good idea to check with the glass manufacturers or fabricators when using uncommon glasses to ascertain their availability. It is almost always a good idea to use commonly available glasses, such as NBK-7 from Schott or fused silica from Corning Glass for lenses or Zerodur from Schott or ULE from Corning for mirrors, since fabricators are familiar with their properties and have developed specific methods for fabrication using them.

Some material properties that can affect the fabrication method and cost include the following:

1. **Coefficient of thermal expansion (CTE).** All materials change dimension with temperature and optical materials exhibit a wide range of variation in this property, which can cause large problems during fabrication as well as in its ultimate use. Whenever an optical surface is worked on, a temperature gradient is set up in the bulk of the glass that causes its shape to deform and the higher the CTE the greater the deformation. Depending on its size and shape a fabricator may have to wait a long time for a part to thermally equilibrate so an accurate measurement of its surface figure can be made. There are three broad categories of thermal expansion: high, low, and "zero" with high expansion materials having a CTE above 5 ppm/°C (parts-per-million per °C) like NBK7 at 7.1 ppm/°C, low expansion materials with CTE's between 5 ppm/°C and 0.6 ppm/°C, such as borosilicate, at 3.2 ppm/°C down to fused silica at 0.6 ppm/°C, and "zero" expansion materials, such as Zerodur or ULE, with CTE's < 0.1 ppm/°C. There are also some materials with unique optical properties that look really good in designs, such as CaF or crystal quartz, that are extremely sensitive thermally and shatter seemingly just by looking at them, but with very careful handling can be successfully fabricated.

2. **Homogeneity.** Homogeneity is a measure of the variation in refractive index throughout a bulk volume of glass. It is generally expressed in terms of parts per million (ppm) and generally varies from about 20 ppm to less than 1 ppm for very high quality glass. Typically, it is around 2–5 ppm for standard glass types. For reflective optics, this value is irrelevant but for transmissive optics, it can be crucial and can be a big problem for the fabricator and user when tight transmitted wavefront specifications are required. Transmitted wavefront errors are the sum of the errors due to the surface figure errors and errors introduced by glass inhomogeneity. The fabricator can "fix up" a deformed wavefront caused by index variations by figuring one or more surfaces to compensate for it but that requires that the part be tested in transmission to measure the error. This can allow the use of lesser quality glass that is either cheaper and/or the only quality available where the surfaces can be figured to compensate and produce a high quality transmitted wavefront.

3. **Hardness.** The hardness of glass varies widely and affects how the fabricator will process the material. The rate of grinding and polishing will vary with softer materials generally having higher removal rates. Softer materials are also more prone to sleeks and scratches and producing very

low surface roughness values can be difficult. The hardness of the various glasses are published in the glass catalogs and should be checked to see whether it may be of concern to the fabricator.

4. **Weather, chemical, and stain resistance.** These are other properties of glass that vary widely and can be of concern with more exotic glasses. The glass companies publish measured values of weather, chemical, and stain resistances in their catalogs and like hardness should be checked to see whether it is a problem. A glass that stains easily or is attacked chemically will require the fabricator to take special precautions and vary some processes that can seriously impact the difficulty in fabrication and the user should be aware of its weather resistance.

Big differences in difficulty and cost come from more exotic materials, such as crystals and special metals. Some of these materials are extremely useful in optical systems, but their material properties make them difficult and expensive to fabricate. [1,2]. The best advice for difficult materials is to find a shop that specializes in processing that type of optic. Again, it is important to talk to the potential fabricator early in the design phase because some materials will impose hard size or quality constraints that need to be incorporated from the start. Also, you may be pleasantly surprised to find that there are better alternatives to the original material or process.

There are steep cost curves for fabricating difficult materials that depend largely on equipment and the state of the market. Like large optics, these markets are not large enough to have a wide selection of vendors competing for your business. The expertise for fabricating optics from less common materials tends to be with small companies that have developed particular specialties.

A different issue is the choice of substrate material for reflective optics. The light does not care what substrate the mirror is made of because it reflects off a coating on the surface and never goes through the mirror. The mirror substrate can be chosen according to the operating environment. Frequently, mirrors are made from low expansion glass because this takes an excellent polish, and it minimizes the sensitivity to thermal effects. Mirror substrates can be procured as light weighted structures to reduce the self-weight deformation.

Optical glass is purchased in several forms—rolled plate, blocks, strips, pressings, gobs, slabs, and rods. The choice of the bulk glass is made according to the fabrication plan and the material specifications. In general, glass for mass-produced optics is supplied in nearly the final shape to minimize the cost of additional processing. Glass blanks for production lenses and prisms are produced in large quantities as pressings oversized and irregular by about 1 mm. Precision pressings are available at higher cost, requiring as little as 0.1 mm of glass to be removed to shape the part. These are shown in Figure 21.3a.

Glass for high performance systems must be carefully selected to get the highest quality. Glass with tight requirements on internal quality is provided in blocks, shown in Figure 21.3b. These blocks are then polished on two or more surfaces and are inspected and graded for inclusions, striae, birefringence

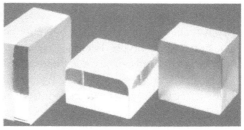

(a) Pressings, hot-molded and annealed.   (b) Block glass, with two opposite
These may have rough or fire-polished surfaces.   faces polished for test purposes.

FIGURE 21.3 Optical glass is commonly procured in (a) pressings and (b) blocks. Other common forms are slabs (six worked surfaces), rods, strips, and rolled sheet (unworked surfaces, cut to length), and gobs (roughly cylindrical). (Courtesy of Schott Glassworks.)

and refractive index variation. The blanks for the optics are then shaped from the glass blocks by a combination of sawing, cutting, and generating.

21.2.2.2 Mechanical Specifications

One of the most obvious and interesting observations about optical elements is that they are almost always circular, with a cylindrical outer edge. Optical design and theoretical reasons can be used to explain this but the development of fabrication methods clearly has been driven largely by this fact. Departures from this norm are becoming more frequent as fabrication technology has progressed. Large segmented telescope mirrors comprised of many hexagonal segments are a good example. Still, most fabrication is dominated by the production of circular optics having curved, usually spherical, surfaces. Figure 21.4 depicts a sketch of a typical lens element showing the most important features that should be specified for fabrication.

In Tables 21.2 and 21.3, we provide some rules of thumb for tolerancing the mechanical parameters of this type of small optic. Like any rules of thumb, these serve as useful guidelines, but the particular circumstances may be well outside these assumptions. Many of the numbers come from some excellent

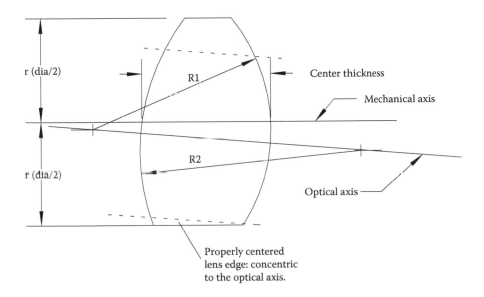

FIGURE 21.4 The anatomy of a lens.

TABLE 21.2 Rules of Thumb for Optical Element Tolerances

Parameter	Base	Precision	High Precision
Lens diameter	100 μm	12 μm	6 μm
Lens thickness	200 μm	50 μm	10 μm
Radius of curvature (tolerance on sag)	20 μm	1.3 μm	0.5 μm
Wedge (light deviation)	6 arc min	1 arc min	15 arc sec
Surface irregularity	5 fringes	1 fringe	0.25 fringe
Surface finish	50 Å rms	20 Å rms	5 Å rms
Scratch/dig	160/100	60/40	20/10
Dimension tolerances for complex elements	200 μm	50 μm	10 μm
Angular tolerances for complex elements	6 arc min	1 arc min	15 arc sec
Bevels (0.2–0.5 mm typical)	0.2 mm	0.1 mm	0.02 mm

TABLE 21.3   Rules of Thumb for Optical Element Mounting Tolerances

Parameter	Base	Precision	High Precision
Spacing (manual machined bores or spacers)	200 μm	25 μm	6 μm
Spacing (NC machined bores or spacers)	50 μm	12 μm	2.5 μm
Concentricity (if part must be removed from chuck between cuts)	200 μm	100 μm	25 μm
Concentricity (cuts made without de-chucking part)	200 μm	25 μm	5 μm

articles on the subject of tolerancing: Willey and Parks 1997; Willey 1984, 1983; Parks 1983, 1980; Smith 1985; Plummer 1979, and DeGroote-Nelson et al 2009 [3-10]. A complete set of such tolerances is maintained by Optimax: http://www.optimaxsi.com/innovation/optical-manufacturing-tolerance-chart/.

We define several classes of tolerances:

> **Base**—This is what the manufacturing process gives, without any special effort.
> **Precision**—Most shops can do this, at a cost increase of roughly 25% for that operation.
> **High precision**—At the limit for most shops, cost could increase 100% for that operation.

1. Diameter. This is usually specified as a value with a ± value from the nominal. Most of the time, these values are derived from mounting considerations. An important related specification is the clear aperture which is the diameter over which all other specifications must be held. Generally, outside of the stated clear aperture there are no specifications unless explicitly stated. This allows some relief at the edges, which are the most difficult to get right.
2. Center thickness. This value is the thickness of the glass at the mechanical center of the optic. It is particularly important for lenses since it is the value used in lens design programs to optically define the optic along with the radii of the surfaces. For larger optics, an accurate measure of this can be difficult depending on the accuracy required.
3. Radius of curvature. This value is the distance between the mechanical center or vertex of a spherical surface and its center of curvature. During fabrication, various instruments may be used depending on the operation and type of surface. When the glass is in the ground state, a spherometer is usually used to get to within around 0.5% of the radius. Test plates having a known radius can also be used on a finely ground surface if the surface is first "waxed" to make it weakly specular. Once the surface is polished, an interferometer having a real focus can be used to measure the radius of concave surfaces to very high accuracy, better than .05%. Convex surfaces can also be measured interferometrically if the interferometer's reference has a focal length longer than the radius to be measured. Otherwise, a concave test plate can be fabricated first, and then used to measure the convex surface radius.
4. Wedge or decentration. The wedge or decenter of an optic is the distance that the optical axis, defined by a line connecting the centers of curvature, is from the mechanical axis defined by the cylindrical outer edge. This is often described as ETD, edge thickness difference. In some cases, the wedge or decenter is specified to be measured with respect to some other surface, usually a flat mounting surface.
5. Two of the parameters listed above are concerned with what are called **cosmetic** properties of an optical surface, namely the scratch/dig and the surface microroughness. The term "cosmetic" might suggest something that is not very important but these two properties can cause significant difficulties for a fabricator and can affect the cost significantly if they are specified improperly. There is a great deal of ignorance and confusion over the specification and measurement of scratch/dig. A good description of the difficulties and methods of specification can be found in Web References 3 and 4. An even bigger problem for the designer concerns exactly how scratches and digs will affect the performance of a surface in any particular implementation. Many shops, especially those dealing with laser optics where defects can have significant impact

on performance, have developed their own scratch/dig specification where high intensity light is used for inspection and careful measurements of the defects are made. Very low defect surfaces can be made but at a price. This is one of the parameters where the choice of material may be very important as producing low defect surfaces varies considerably depending on the material. Also, a specification that is appropriate for a small laser optic might be completely inappropriate for a large telescope mirror.

Microroughness is the fine surface texture with scales << 1 mm that causes wide-angle scattering in optical systems. The skill of the fabricator to produce a highly polished surface is put to the test in the application of the surface microroughness specification. As with scratch/dig, an understanding of the complexities of measuring the microroughness is helpful in deciding the propriety of a particular roughness level and is something, like scratch/dig, that should be discussed with a fabricator. Various methods have been developed to measure the microroughness and a good discussion can be found in Bennett and Mattson [11]. The difficulty in measuring any of these parameters is a function of both its magnitude and its tolerance. Size really matters here and is what separates "large optic" shops from all the others. The effect of size on optical fabrication is quite interesting. There are numerous methods and plenty of shops that make production lenses to 50 mm. Optics in the range of 50–500 mm are not uncommon, but they require special tooling and they are usually made as single parts (with the exception of flats processed on a continuous polisher as described below). Optics greater than 500 mm, nearly always mirrors, are in a class by themselves and there are only a few places with equipment and expertise to handle these.

The advantage of using optics smaller than 50 mm is that there are so many of them! There are large numbers of companies set up to make these optics with high quality at good prices. The parts are small enough that many optics can be processed economically on a common block. The infrastructure is in place for grinding, polishing, edging, cleaning, and coating optics of this size. In fact, much of the processing can be totally automated.

Things get more difficult for larger optics. The market has not supported the development of efficient tools and processes for mass-producing optics in the 50–500 mm range. In fact, each new part in this range will need a special polishing support and set of polishing tools. These parts need to be processed one at a time, so they require significantly more labor than the small parts. The size of these parts is such that they can use a simple support, with either a few defining points or using a compliant pad.

In addition, the metrology for these larger optics can drive the cost up. Small optics are easily measured with test plates. The larger optics may need to use auxiliary optics for testing. The testing is not just for qualification, but it is an integral part of the fabrication sequence. The optician works these optics according to the results from the optical test.

Large optics (> 50 cm) are almost always mirrors, and have other unique difficulties due to their size and surface requirements. (For the same optical performance, a mirror surface must be four times better than a refractive surface. A reflected wavefront picks up errors two times those on the surface. The errors in a refracted wavefront are n-1 times the surface error, or about half.) For large optics, each processing and handling operation requires custom tooling. Ray Wilson [12] gives a good overview of manufacturing methods for very large modern telescopes. The support for large optics becomes difficult and extremely sensitive. Often, separate supports must be used for holding the optics during polishing than can be used for testing. The polishing forces from large laps can be substantial and must be resisted by the support. The self-weight deflection of large mirrors alone will quickly dominate the shape if it is not accommodated in the support.

The sheer size of large mirrors presents a challenge. The opticians may need to climb out onto the optical surface to clean and inspect a large mirror. Every handling operation must be carefully thought out and all of the tooling must be tested before it can be used safely. Unlike picking up small optics, large optics are extremely heavy. The forces are large, and the parts are extremely valuable, so all efforts to make sure every operation is completely safe are justified.

It is much more difficult to estimate the costs for large optics than for small ones because of the difficulties with large optics and the fact that each one is special. Large optics are only processed in a limited number of shops, so the costs will often depend on the current workload in the shop as much as it will on the technical difficulties. The best advice here is to plan ahead, and to design for optics that are identical to others already in production. Much of the cost for large optics is in the equipment, so considerable savings can be made by using existing tooling. A good example is the lightweight mirrors made at the University of Arizona. Figure 21.5 shows a primary mirror blank that is 8.4 meters across, which will be used as one of the twin telescopes in the Large Binocular Telescope. A large fraction of the cost of this mirror is due

FIGURE 21.5   8.4 m diameter, f/1.1 primary mirror blank for the Large Binocular Telescope. This optic, the largest in the world, requires considerable engineering and tooling to support each operation in the shop. This image shows the backside of the honeycomb mirror as it is supported vertically in the shop. *Photo by: Lori Stiles.*

to the engineering and fabrication of all of the equipment to process and handle this glass. Much of this equipment is specifically designed for this mirror and could not be used for an optic with a different shape.

21.2.2.3 Optical Specifications

Optical specifications generally refer to how accurately the surface of a lens or mirror needs to make with respect to some ideal reference surface, thus it is usually stated as the amount of departure from that ideal surface. This is one of the biggest cost drivers in fabrication due to the precision and accuracy to which modern optical surfaces must be made and the effort, equipment, and expertise required to perform the measurements. The range of accuracy required can vary by several orders of magnitude depending on the application and the wavelength of light for which the optic is used. For some applications, the departures are measured in microns, which is large in the context of optical surfaces. For visible light applications, a departure of a small fraction of a wavelength of light is common and is specified that way. Interferometry is the most common and precise method of measuring optical surfaces, usually at the He-Ne laser wavelength of 633 nm. Details of how these measurements are made can be found in this book as well as on the web and in many optical texts and journals.

Two of the most common ways of specifying the departure is by using either a peak-to-valley value, which is the value of the highest point of deviation to the lowest point or using the rms value of the departure. Modern digital interferometry where the departure is measured over many thousands or even millions of points makes the rms number much more relevant since any noise or extraneous error in the measurement can make the peak-valley number be out of spec where the rms value would be in spec. This was a very common way of specifying surface figure quality before the advent of digital interferometry when the estimation of peak-valley departure was made purely visually. Designers many times fail to appreciate the difficulty of making optics to a peak-valley specification and will usually pay a premium for it. Also, when specifying departures in terms of wavelengths of light, it is essential that the wavelength be specified along with whether the departure is from the ideal surface or the ideal wavefront, this distinction means a factor of 2 in reflection and nearly a factor of 2 in transmitted wavefront. It is probably best for clarity to specify the value in spatial units, such as nanometers rms surface, or nanometers rms wavefront.

21.3 Traditional Methods of Optical Fabrication

21.3.1 Introduction

Current optical fabrication methods are a curious blend of old and new. Pitch polishing with metal oxide polishing compound was developed centuries ago, and was used by Sir Isaac Newton. The basics have changed little, but the modern practices are more efficient with computer controlled machines, more accurate with laser interferometry, and more varied with advanced materials. There is still a considerable "art" component to these methods in most optical shops, especially for custom optics. Optical technicians require a high level of expertise that takes years to develop. However, with the recent application of computer controls to fabrication machines and the development of more deterministic shaping methods and processes, a revolution is underway. Both custom and production optics are being manufactured more efficiently due to these advances.

The optician's expertise must now include computer literacy, a requirement shared by many industries. Research has led to a greater understanding of the ground and polished surfaces, and ways of producing them. Diamond-turning and grinding technology, in combination with computer controlled machines, has had a large impact on both glass and metal fabrication methods. Pitch polishing is no longer the only way to finish a high quality optical surface. A great deal of work and progress continues to be made in the production of aspheric optics utilizing advances in all areas of fabrication. The direct milling of glass and other brittle materials is now accomplished not only with diamonds, but also with streams of ions.

We must note that much of the progress has resulted from various new or improved testing methods, particularly computer controlled interferometry and profilometry. The advanced measurement techniques, along with developments in fabrication described here, have made possible the production of optics that cover virtually the entire electromagnetic spectrum.

The explosion of new materials and processes available to the engineer and the fabricator has fragmented the industry to a large extent. Expertise can no longer be found in a single "optics house" for all optics needs. Nor can a single chapter begin to review all the existing methods and materials.

There are a few common steps for making optical elements, although each step will be done differently depending on the optic and the quantity:

- *Rough shaping*: The initial blank is manufactured, typically to within a few millimeters of final dimensions.
- *Support*: The optics must be held for the subsequent operations. Much of the difficulty in fabrication comes from the requirements of the support.
- *Generating*: The blank is machined, typically with diamond tools, to within 1–0.1 mm of finished dimensions.
- *Fining*: The optical surfaces are ground to eliminate the layer of damaged glass from generating and to bring the surface within a 1–5 μm from the finished shape.
- *Polishing*: The optical surfaces are polished, providing a specular surface, accurate to within 0.1 μm. Through repeated cycles of polishing, guided by accurate measurements, surfaces can be attained with 0.005 μm accuracy.
- *Centering and Edging*: The optic is aligned on a spindle and the outer edge is cut.
- *Cleaning*: The finished elements are cleaned and prepared for coating.
- *Bonding*: Frequently lenses and prisms are cemented to form doublets (2 lenses) or triplets (3 lenses).

Subsequent coating and mounting are usually handled by a different group of people and are not generally considered part of optical fabrication.

The first step in fabrication is to order the glass. For most cases, it is better to specify the optics and the glass requirements to the fabricator, and have them order the glass, rather than to purchase and supply the glass yourself. The fabricators are used to dealing with the glass companies and they will know best what form the material should come in. The fabricator will know how much glass to buy to cover samples for setup, tooling, process development, and so on, as well as the inevitable losses due to parts outside of tolerances. By letting the fabricator purchase the glass, you also reduce the number of interfaces for the project. The fabricators can then take responsibility for the overall performance of the optic, including the glass. If you supply the glass yourself, the tendency for the fabricator is to treat the optic as a set of surfaces being made on a substrate, which is out of their control. For example, if you need lenses with a particular focal length, the shop cannot take responsibility for this specification if the refractive index of the glass varies.

21.3.2 Support Methods for Fabrication

Most optical fabrication processes begin with the extremely important consideration of holding onto the part during subsequent fabrication steps. Numerous factors must be considered when choosing the support method: part size, thickness, shape, expansion coefficient, and the direction and magnitude of applied forces. The support should not stress the optic, otherwise when the part is finished and unmounted (or "deblocked"), it will distort by "springing" into its stress-free condition. However, the part must be held rigidly enough to resist the forces of the various surfacing methods. Often, the support is changed as the part progresses, due to different forces and the precision required for each step.

Most modern fabrication begins with fixed diamond abrasive on high-speed spindles (as discussed in Karow 1993, Piscotty 1995 [13,14]). The lateral forces can be large, so the part must be held quite firmly to a rigid plate or fixture. This plate, called the blocking body, or "block," can be made of various materials depending on the process. It is usually made of aluminum, steel, cast iron, or glass, with rigidity being the most important factor. The two principal methods for holding the part to the block are to use adhesives or mechanical attachments at the edge.

The ideal adhesive would provide a rigid bond with little stress, and it should allow the part to be easily removable. Most adhesives cannot achieve all three requirements well, so the optician must choose, depending on which consideration is most critical. For the generation processes using high-speed diamond tools, rigidity and ease of removal are usually the dominant criteria with higher stress being allowed. The effects of this stress are then removed in the subsequent processes of grinding and polishing, where a less stressful blocking method is employed.

Blocking of plano and spherical parts up to around 100 mm in diameter is done with a variety of waxes, both natural and synthetic. These are heated to a liquid before applying to the block, or heated by the block itself. The glass parts are then warmed and placed on the waxed block. For heat sensitive materials, the wax can be dissolved in solvent before applying to the block. The great advantage of waxes is that they hold the glass quite firmly and are also easily removable by dissolving them in common solvents. Some more recently developed waxes are forms of thermoplastic that are coming closer to the ideal bonding material. They melt at a relatively low temperature (55–90°C), do not require that the part be heated, and are soluble in hot water as well as common solvents. Most waxes, however, impart large stresses due to their shrinkage. This requires parts to be de-blocked after generating, and subsequently re-blocked with a less stressful substance for grinding and polishing.

Pitch remains the blocking material of choice when the parts cannot be highly stressed. Pitch is an outstanding material, and is used in the optics shop both for blocking and for facing polishing tools. Brown [15] gives an excellent reference on the properties of pitch. Pitch is a visco-elastic material that flows when stress is applied, even at room temperatures. Parts blocked with pitch will be stress-relieved if left long enough.

Cements, such as epoxies and RTV's, bond very well, but are extremely difficult to de-block and remove. There are also some UV curable cements that can provide low stress blocking and can be removed with hot water. For more information about these cements, contact the manufacturers of optical adhesives.

The optical contact method is used when the surface needs to be held precisely to the block. Windows with precise wedge angles and prisms use this method. The block is usually made from the same material as the part, and the mating surfaces must both be polished and clean. When the two surfaces are brought together, with a little finger pressure to force out the air, they will pull together in a tight bond due to the molecular forces. This blocking method can be used with parts of any thickness, but is difficult to apply to large surfaces due to the required cleanliness.

In production optics, where many parts with the same radius of curvature are produced, a number of the parts are blocked together as shown in Figure 21.6a. Often, the block is carefully machined so each part can be loaded into a recess, giving precise position relative to the block's center. This type of block is called a spot block, and is used widely in production shops. These spots can be machined directly into the block, as shown in Figure 21.6b, or separate lens seats can be machined that are screwed onto the block. The spot blocks are costly to make, but they can be used efficiently for making numerous runs of the same lens.

Limitations on block size are based on machine size limitations and on the radius of curvature. Most generators and grinding/polishing machines cannot handle anything beyond a hemisphere, limiting the number of parts to a block. Plano parts are limited only by the capacity of the machines in the shop. Hundreds of small plano parts can be fabricated on a single block.

Aspherics cannot be fabricated on blocks because the aspheric surface has an optical axis that coincides with its mechanical axis. Only a part that is centered to the machine spindle can be turned into an asphere. This is one reason aspherics are more expensive than spherical surfaces. Note, however, that off-axis aspherics can be made as a block; this is how most off-axis aspheres are made, by making a parent block large enough to encompass the off-axis section pieces. The parent is then aspherized in a symmetric way (as discussed below), after which the required off-axis aspheres are removed from the

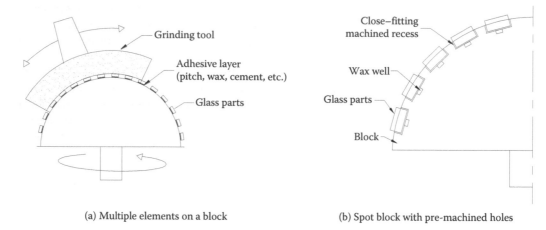

(a) Multiple elements on a block   (b) Spot block with pre-machined holes

FIGURE 21.6 Multiple parts may be made on the same block by adhering them to a common spherical block. A more accurate and repeatable method uses a spot block where pre-machined holes are provided for the lens blanks. The usual method for grinding and polishing is to have the block rotating while a matching spherical tool is stroked across it. This can also be inverted.

correct position on the block. Usually, the parent is manufactured into a single piece of glass, and the off-axis sections are cut from the parent after aspherizing.

These blocking techniques are used for production of a large number of parts. Even if only one part is required, it is usually wise to block many together so that spares are available. It generally does not pay to make just one spherical part if it is small (less than 100 mm). Designers should always try to use off-the-shelf elements for optics in this size range.

Optics larger than this are supported mechanically without the use of adhesives of any sort. Mechanical supports for larger optics have the same requirements as their adhesive counterparts in that they must hold the part firmly while introducing little stress. Like the smaller optics, large optics can be supported differently for different fabrication processes where the conflicting requirements of high rigidity and low stress must be balanced.

Mechanical supports during diamond generating must be quite rigid, since the forces placed on the part by the high-speed diamond tools are large. The generating support can allow larger distortions, which will be corrected later in grinding and polishing. Most generating machines have turntables with either magnetic or vacuum systems to hold moderately sized parts (up to about 500 mm). A magnetic system, commonly found on Blanchard type machines, uses steel plates that are placed around the periphery of the part. The electromagnetic turntable is switched on, firmly holding the plates and the part in position. In vacuum systems, the part is held on a shallow cup with an "O" ring seal. A vacuum is pulled on the cup, and the part is held in place by friction against the turntable.

For larger optics, the part may rest on a multi-point support system that is adjustable in tilt, and held laterally by three adjustable points at the edge of the part. These support systems can introduce large figure errors that need to be eliminated in subsequent grinding and polishing. Some machine turntables are machined to be extremely flat, even diamond turned in some cases, to reduce the amount of induced deformation.

During grinding and polishing, large parts are supported axially using pitch or other visco-elastic materials (such as Silly Putty), depending on the stiffness of the part. This type of support can flow to eliminate any induced stresses in the part. There are also several methods of achieving a well-defined set of axial forces for the case where the part is supported at a number of discrete points. Hindle type "whiffle-tree" supports or hydrostatic supports use mechanics or hydraulics to provide a unique, well-defined set of support forces. [16]. Figure 21.7 shows an application of a hydrostatic support for fabricating a fast lens. The required number and arrangement of the support points can be predicted using finite-element analysis. Lateral forces can be taken with metal brackets or tape applied tangent to the edge.

FIGURE 21.7 Hydrostatic support of a lens.

21.3.3 Diamond Machining and Generating

Following the blocking, the part is generated, which is a common term for machining by grinding with diamond impregnated tools. The generating can rapidly bring the part to its near-final shape, thickness, and curvature, with the surface smooth enough for fine grinding or direct polishing. The generating tool uses exposed diamond particles to chip away at the glass on the scale of tens of microns. Additional information on specific aspects of generating are Piscotty et al. 1995 [14], Ohmori 1995 [17], Stowers et al. *1988* [18], and Horne 1977 [19].

Most generating tools have a steel body, onto which a layer of material impregnated with diamond particles of a particular size distribution is bonded. The size is usually specified as a mesh number, which is approximately equal to 12 mm divided by the average diamond size (see Figure 21.8). A 600-mesh wheel has 20 μm diamonds. The specifications for the absolute sizes of the diamonds and their distribution are not standard and should be obtained from the vendor.

There are two basic configurations for diamond tooling as shown in Figure 21.9; a peripheral tool with the diamond bonded to the outer circumference of the tool, and a cup tool with the diamond bonded to the bottom of the tool in a ring. Peripheral wheels are used for shaping operations on the edge of the part, such as edging, sawing, and beveling. Cup wheels are used for working on the surface of the part, like cutting holes and generating curvature.

Small optics (<20 cm) and blocks of lenses are generated spherical using a cup wheel where the axis of rotation of the wheel is tilted with respect to the part so that it passes through the desired center of curvature. If the axis of rotation does pass through the center of curvature, it will cut a perfectly spherical shape into the part or block. Since all the parts on a block of lenses share the same center of curvature, they will all be cut to the same radius and thickness if properly blocked (Figure 21.10). This fact is key to the production of large quantities of smaller optics blocked together.

Plano optics are generated using a Blanchard type geometry. Here a cup wheel is used with the axis of rotation aligned to be perpendicular to the linear axis of a tool bed. The parts are translated under the spinning diamond wheel, and are ground flat to high precision. Multiple operations of this type must be performed for the different faces of prisms, and the relative orientation of the different cuts determines the accuracy of the prism (Figure 21.11).

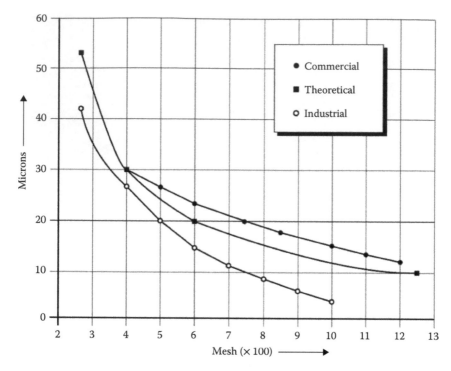

FIGURE 21.8 Correlation between mesh sizes and micron sizes. (Courtesy of Karow 1993.)

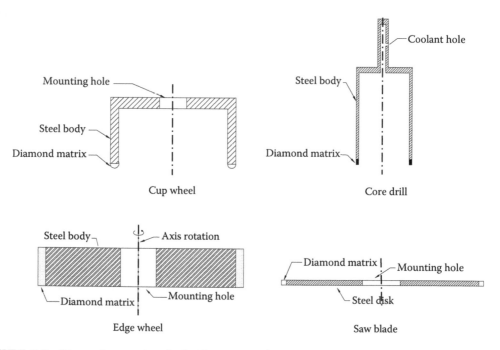

FIGURE 21.9 Diamond generating wheels. There are two basic types of diamond tooling used for cutting and generating, depending on whether the diamond is on the face or on the edge. The cup wheel and core drill are the most common face wheels used in cutting radii and drilling holes. The peripheral wheels, with the diamond on their edges, are used for edging and sawing.

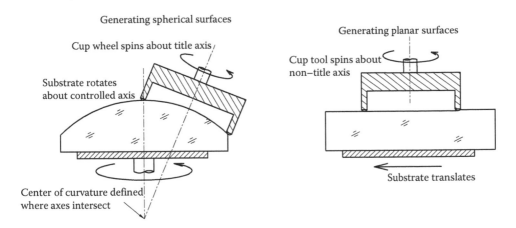

FIGURE 21.10 Generating with cup tools. Spherical surfaces are cut by tilting the axis of the cutting wheel so it intersects the axis of the rotation of the part. This will cut a spherical surface with the center of curvature at this intersection. Plano parts are milled by translating the optic in a direction perpendicular to the cup wheel.

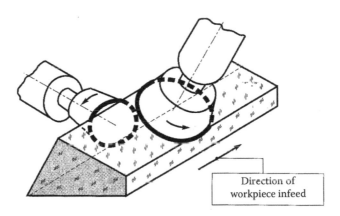

FIGURE 21.11 Generating a prism profile with two mill heads. (Courtesy of Karow 1993.)

Diamond tools are quite versatile and are used for many different operations. For example, a simple peripheral wheel can be used to cut the curvature into a part. The tool can be moved slowly across a spinning part. The shape of the cut is determined by the tool motion, which can be run on a numerically controlled (NC) machine, or it can be driven to follow a template.

A large emphasis has been placed on progress in both diamond tooling and in the machines that use them because accurate, fine cuts reduce the time spent in the grinding step that follows. Machining of the part on high-speed machines is very rapid, with removal rates up to several cubic inches per minute; this is two orders of magnitude faster than coarse loose-abrasive grinding.

Unfortunately, generating creates significant damage to the glass under the surface, which must be removed in subsequent grinding and polishing operations. One current area of interest is how diamond generating can produce finer surfaces and more accurate shapes. Abrasive action on glass occurs due to small fractures that form when an abrasive particle is pushed against it with enough force. When enough fractures intersect, small pieces of glass pop out leaving small pits. Underneath the pits are larger fractures that continue some distance, depending on the materials. The structure of glass as it is typically abraded is shown in Figure 21.12. Generally, smaller diamonds and softer matrix material impart less damage to the surface. However, finer diamonds lower the removal rate, and a softer matrix allows greater tool wear, which can reduce accuracy.

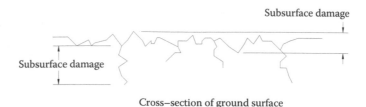

Cross–section of ground surface

FIGURE 21.12 An abraded optical surface consists of two components: the surface damage layer and a subsurface damage (cracked) layer. For loose abrasive grinding, the surface damage layer is about the size of the grit size and the maximum subsurface damage is about twice that. Diamond generated surfaces typically have less surface damage but a large subsurface component.

Most diamond surfacing methods use at least two different diamond wheels to rapidly produce a fine surface. Using a computer controlled machine, flat, spherical, and even aspherical parts can be rapidly surfaced. If the diamond tool and machine have sufficient accuracy, the tool can be brought to bear on the surface with a low force, so that the glass does not fracture. Material is removed by plastic flow with no subsurface damage, and the surfaces produced are specular. This process is called micromachining or ductile-regime machining (Bifano et al., 1988 [20], Golini et al., 1990, 1991, 1992, 1995 [21–24]).

Currently, surfaces ready for direct polishing can be routinely produced on production machines, bypassing all loose-abrasive grinding. However, most of these machines produce optics less than 100 mm in diameter. With larger parts, fine diamond machining is only performed on a few specialized machines, limited by the machine and mount stiffness. Microgrinding is just beginning to gain more widespread use in the industry. The combination of numerically controlled machines and diamond tooling will undoubtedly have a large impact on fabrication methods of the future.

Surfaces can also be fine generated using pel grinding, which uses a large tool covered with bound diamond in cylindrical pellet form [25]. The tool is made to match the shape to be generated, and is then driven at high speeds. The fine diamonds generate the surface directly, leaving little subsurface damage, allowing polishing without any subsequent fining operations.

The pellets are bonded to a curved tool to give the proper radius of curvature inverse to that of the part, as shown in Figure 21.13. The tool is rapidly rotated while the part is stroked over the tool in the same fashion as loose abrasive grinding. With higher speed and higher pressure, pel grinding quickly works the surface smooth enough for polishing. This method is very efficient for high-speed production of thousands of the same part, but it is costly to set up, since a new tool is required for each radius. Hence, more traditional loose abrasive grinding is used for low volume production.

21.3.4 Fining

As described above, the diamond machining leaves a smooth surface, below which is a layer of material riddled with fractures. These fractures, if left in the final polished part, are visible under bright illumination and cause the surface to scatter light. These fractures can be a hundred microns deep, so this amount of material must be removed before a good polish can be obtained (See figure 21.12) [26,27].

Grinding with loose abrasives is traditionally used to remove the damaged layer of glass. In this method, the part and the tool are rubbed together while an abrasive powder, usually in an aqueous solution, is maintained between them. The particles cause tiny fractures in the glass, which results in material removal as the fractures intersect. This abrasive action itself causes subsurface damage, but the sizes of the particles are chosen to reduce the amount of damage in a series of steps, generally reducing the damage by a factor of two with each grade. A rule of thumb for loose abrasive grinding is that the maximum depth of the pits after grinding is on the order of the size of the grain; subsurface fractures extend to about twice that. A typical sequence might be to diamond generate, remove 100 microns with

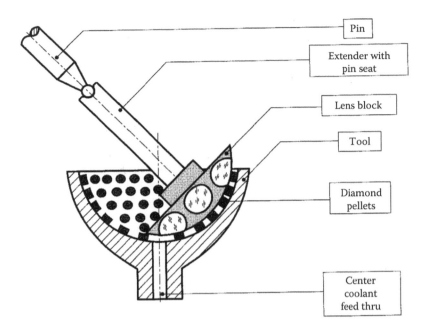

FIGURE 21.13 Typical configuration for using a pel grinder for working convex surfaces. (Courtesy of Karow 1993.)

a 40 μm abrasive, remove 50 microns with a 25 μm abrasive, and finally remove 25 microns with a 9 μm abrasive. This surface can be polished, but up 20 microns of material needs to be polished off to eliminate the remaining damage.

It is possible to remove all of the generating subsurface damage by polishing directly, skipping the step of loose abrasive grinding. It would, however, take an unacceptably long time to remove 100 microns of material by polishing. It only makes sense to polish a surface with a few microns of surface damage.

Other factors that contribute to the amount of damage produced include the hardness of the abrasive, the material being ground, the tool, and to some extent the shape of the abrasive grains. Harder abrasive grains or tools will remove material more rapidly, at the price of increased damage. The more plate-like grains found in modern aluminum oxides appear to produce less damage than their blockier counterparts, like garnet.

Tool materials for loose abrasive grinding range from cast iron, which is quite hard, to brass, glass, and aluminum on the softer side. The harder tools will grind faster and retain their form longer than the softer tools, at the cost of more subsurface damage. Most production tools are made from cast iron because they keep their shape well.

Loose abrasive grinding is used for fining large optical surfaces, and built-up or layered tools are the rule. Tools for larger optics are made from some soft, workable material, such as aluminum, wood, or plaster. The curvature of the tool is either machined or cast into the tool, which is then faced by bonding ceramic or glass tiles to the tool's surface. When the ceramic layer grinds down, it is replaced with another layer, or a fresh layer is bonded to the first.

The point at which grinding stops and polishing begins depends on a number of factors. While glass surfaces can be ground to a very fine finish—perhaps to a 1 μm grit size, which minimizes subsurface damage—other factors generally limit the final grit size to around 5 microns. For very small particle size, the intimacy between the part surface and the tool can cause the tool to seize on the part, which makes the two virtually inseparable without major forces being applied. The larger or more costly the part, the more this risk becomes unacceptable. Also, the risk of scratching the surface increases with small grit, especially when using very hard tools, such as cast iron or steel. On many surfaces it is a good compromise to perform the final grinding with a softer tool material, such as brass, aluminum, or glass, and use a slightly larger grit size. The softer tool material will result in less damage and reduce the risks.

Compared to generating with bound abrasive wheels, loose abrasive grinding is performed at much lower speeds. Very small parts are ground at a few hundred rpm whereas large parts are ground at only a few rpm. This lower speed accounts for the removal rate difference between the two methods. At higher speeds, the loose abrasive slurry mixture would be flung off the tool or part.

21.3.5 Polishing

Polishing an optical surface brings its surface figure, or form, into compliance with specification. At the same time, the surface finish or microroughness is reduced to an acceptable value. Polishing is a seemingly magical process, which uses a combination of mechanical motion and chemistry to produce surfaces smooth to molecular levels [28].

Most high quality optical surfaces are polished using a tool similar to the grinding tool, except that it is faced with visco-elastic pitch or polyurethane. This tool, called a lap, is stroked over the surface with an aqueous slurry of polishing compound. The surface is polished by a complicated chemical and mechanical interaction between the glass surface, the lap surface, and the slurry. Polishing is partially a chemical process, so different substances must be used to polish different materials; no one substance is ideal for all materials. Some polishing compounds for common optical materials are cerium oxide, zirconium oxide, alumina, and colloidal silica. These are available in proprietary mixtures from several suppliers.

Pitch laps are frequently used for high quality surfaces. Pitch is a generic term describing a group of substances made from the distillation of tar derived from wood or petroleum. It is very soft compared to glass so it will not scratch, and has a low melting temperature of about 50–100°C. Its viscosity, usually in the range of 10^8–10^{11} poise, allows the pitch to slowly flow at room temperature so that it takes the shape of the part being polished and remains in close contact.

Pitch laps give the best performance, but they require considerable maintenance. Production parts are polished using laps faced with polishing pads made of polyurethane. These synthetic pads work well with particular polishing compounds that have been optimized for use with the pad. Laps faced with these polishing pads are extremely stable, and they polish more quickly than pitch because they can be run at higher pressures and speeds. However, unlike pitch, these laps do not naturally flow to conform to the shape of the optic, so the pads must be applied to a precision-machined surface. This special tooling is efficient but expensive, and it will only work for a particular radius of curvature.

Metals, plastics, and crystals can be polished the same way as glass, but using different polishing pads and compounds. Metals are polished best with cloth polishers and polishing compound with very fine chrome oxide or diamond [29]. The quality of the finish depends on the hardness, porosity, and inclusions of the metal substrate. Plastic optics, such as acrylics, are polished with aluminum or tin oxide with soft synthetic polishing pads. Most crystals are polished using synthetic pads with a compound of colloidal silica, fine diamond, or alumina (aluminum oxide).

The macrotopography (surface figure) and microtopography (surface finish) are the two most difficult specifications to meet in the fabrication process and are the biggest cost drivers. The surface figure is commonly specified as an average (rms) or absolute value (peak-valley) height difference between the actual surface height and the ideal theoretical surface. This difference is usually specified in units of waves, or fractions of a wave, at the wavelength at which it is used or tested. Typical figure tolerances are 0.2–0.05 waves rms at the measurement wavelength of the He–Ne laser at 632.8 nm.

Control of the figure comes from the geometry of the polisher: how the lap is stroked and the table with the optic is rotated. High quality parts are time-consuming to make. They require the optician to measure the part, usually with interferometry, and to adjust the fabrication process to correct the errors in the surface. The cost of the optic will depend on the efficiency of the optician in converging to the final specification. The accuracy of the finished part will depend on both the residual errors that the optician measures and errors in the optical test.

Surface figure specifications are made as peak-to-valley (P-V) or root-mean-square (rms) departure of the surface from ideal. Peak-to-valley specifications are becoming less popular (particularly to the fabricators) because they only relate to very local regions of a surface. This specification makes sense for optical surfaces measured by inspection with a test plate. The optician uses a test plate to evaluate the large-scale distortions of the interferogram. He gives a limit to the irregularity that he sees, and he uses this visual assessment to qualify the P-V distortion in the surface.

For the case of computerized phase measuring methods, however, the P-V error is strictly the difference between the maximum point and the minimum point in the data. The high resolution of these instruments will provide surface maps with 30,000 points, so any two points are not statistically significant. In fact, the minimum and maximum of the data will usually be driven only by measurement noise. It is not uncommon to relate a P-V specification to an equivalent rms value by applying a simple rule of thumb—the allowable rms figure can be estimated by dividing the P-V specification by a factor of 5. Nonetheless, P-V specifications in the ½ to 1/20 wave are still common.

Clearly, higher quality surfaces require more time to make and are more expensive to produce. Flats and spheres can be produced by conventional methods down to 0.01 waves rms or better. They are made to 0.002 waves rms using special methods depending on the size and surface shape. Aspheres are considerably more difficult to figure and will be discussed in Section 3.

The finish is the local roughness of a surface compared to a perfectly smooth surface. It is usually specified as an average (rms) surface height irregularity over spatial scales of a few tens of microns. Unlike the figure, the surface finish comes from the process itself—the type of lap, polishing compound, pressures, and speeds. These processes are derived before starting the production parts, so the optician does not typically adjust the polishing based on measured results, as he does for the figure.

Interferometric techniques are now used to measure both the figure and the finish of optical surfaces to high precision. Using computer-controlled phase-shifting interferometry, surface figures can be measured to a few nanometers and surface finish to a few tenths of a nanometer. The ability to precisely measure these quantities has resulted in improved polishing processes that lead to better surfaces. However, the understanding of the polishing process, particularly of glass, remains mired in its great complexity.

High quality optical surfaces will generally be finished to less than 20 angstroms rms, down to a few angstroms rms. For most optics, the standard pitch polish, giving about 10 angstrom rms roughness, is more than adequate. Some applications require super-polished surfaces, with roughness below 2 angstroms rms. Special effort is required to produce such surfaces, and few fabricators have developed this capability.

Producing a high quality optical figure is perhaps the most cost sensitive aspect of fabrication. Various methods have been developed to create high quality surfaces on different types of surface shapes. Here, we will describe some of the methods used to produce flat surfaces, spherical surfaces, and aspheric surfaces. We look at established techniques, as well as some more modern methods under development.

21.3.5.1 Polishing of Spherical Surfaces

Spherical surfaces are the simplest of all to fabricate because of their symmetry. The grinding and polishing process tends to produce spherical surfaces. The fact that the tool and the workpiece are not full spheres, but are segments, allows variations in wear across the surfaces that can be used to change their radius of curvature. Most optical system designs utilize optics with only spherical surfaces, due to the relative ease of manufacture over aspherical ones, although aspherical surfaces can simplify the optical design.

Conventional methods for grinding and polishing spheres use one surface, usually the optic, to rotate on a turntable and the other, the lap, to move over it. Overarm machines, shown in Figure 21.14, stroke the tool over the part using an arm that attaches to the tool through a ball joint. Also, the roles can be reversed and the block with the optics can be attached to the arm and driven over a rotating tool. By adjusting the length of the stroke and the relative speeds of the rotation, as well as the length of the stroke, the radius of the two surfaces can be made to move longer, shorter, or stay the same. A skilled operator can make an accurate sphere with low surface roughness.

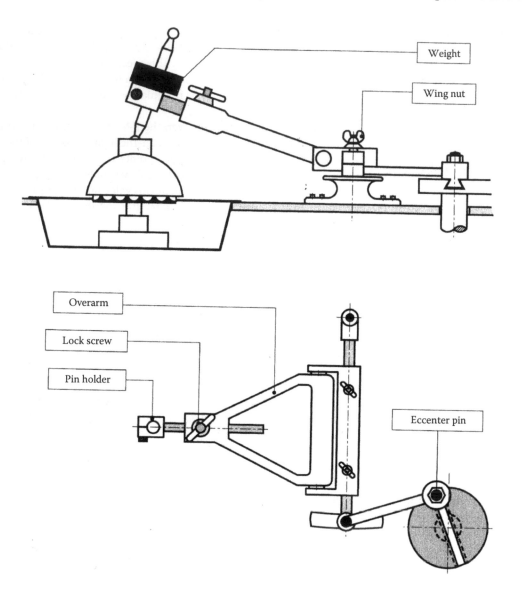

FIGURE 21.14 Overarm polishing machine. Production shops use machines with numerous spindles running simultaneously. (Courtesy of Karow 1993.)

As mentioned, various blocking methods can be utilized to increase production volume, such as the use of spot blocks or other multiple element blocking, as shown in Figure 21.15. Running the machine faster and automatically feeding the slurry can dramatically increase production rates. Also, high production volume can be achieved using diamond pellet tools for grinding and polyurethane pads for polishing. Fabrication of spherical surfaces where the process parameters have been finely tuned to produce predictable results allows economic fabrication of optics in large quantities. Most catalog items fall into this category of production optics, and optical designers should use these available parts whenever possible.

When custom optics are required, the story changes dramatically. Parts must be blocked individually, and tools and test plates may need to be fabricated for each surface. However, many optics houses keep a large range of both tools and test plates used in prior work. If catalog optics cannot be used in a design,

FIGURE 21.15 Numerous small parts with the same radius of curvature can be blocked together and processed simultaneously. (Courtesy of Newport Corp.)

it is always cost effective to choose radii for the spherical surfaces that are in the test plate and tooling inventory of a manufacturer.

Large spherical surfaces (>100 mm) are produced in the same way as small ones. The tooling and machines become proportionally large, but the basic method of rubbing two spheres together is the same. However, controlling the shape becomes increasingly difficult as the part diameter becomes larger. It is also increasingly difficult to handle the large tools. Generally, it is necessary to use a large tool (large meaning 60%–100% of the part diameter) after the part has been generated to smooth out errors in the surface. Following large tool work, smaller diameter tools are used to figure the surface to high accuracy. The use of smaller tools can have some effect on the surface slope errors, since a small tool is working locally and can leave behind local wear patterns. This becomes increasingly important as the tools get smaller. With skill and experience, an optician can keep these errors small by not dwelling too long at any one location on the surface.

21.3.5.2 Polishing of Flats

The production of a flat surface used to be difficult, due to the fact that the tolerance on the radius of the surface is the same as the tolerance on the irregularity; that is, power in the surface is an error to be polished out. This changed with the development of the continuous polishing (CP) machine (Figure 21.16).

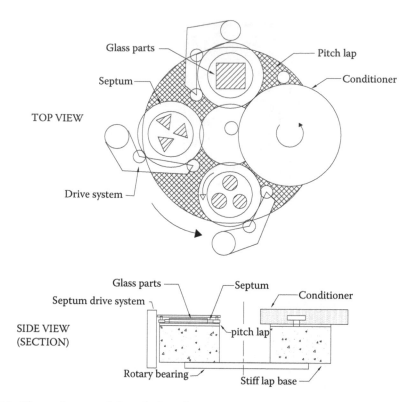

FIGURE 21.16 The continuous polishing (CP) machine can polish both flats and long radius spheres to very high surface figure quality and surface finish. As long as the parts do not pass over the edge of the lap and are rotated synchronously with the lap, they will experience uniform wear. The conditioner is a large disk that keeps the lap flat and also rotates synchronously with the lap.

A continuous polisher uses a large annular lap (at least three times the size of the part) that turns continuously. The parts to be polished are placed on the lap in holders, or septums, that are fixed in place on the annulus and are driven so they turn in synchronous motion with the lap. It can be shown that if the part is in synchronous rotation with the lap and always remains in full contact with the lap, then the wear will be uniform. By maintaining the flatness of the lap and providing uniform wear, any parts that are not initially flat very rapidly become so.

The lap of the continuous polisher is kept flat by the use of a large flat called a conditioner, or bruiser, having a diameter as large as the radius of the lap. The conditioner rides continuously on the lap, and is caused to rotate at a synchronous rate. By adjusting the conditioner's radial position, the lap can be brought to a flat condition that can be maintained for long periods. Slight adjustments in the position of the conditioner are made as parts are found to be slightly convex or concave. Careful attention must be paid to environmental control and slurry control to maintain consistent results. Since these machines run continuously, 24 hours a day, their throughput can be very large. Because the contact between lap and part is exceptionally good on these machines, they routinely produce excellent surfaces with no roll at the edge.

The uniform wear is not dependent on the shape of the part. This means that plano parts with highly unusual shapes can be fabricated to high quality right to their edges or corners. The only other variable that needs to be controlled to produce uniform wear is the pressure. Some parts with large thickness variations and low stiffness need to have additional weights added so that the pressure is nearly uniform across the part. If the figure is seen to be astigmatic, weights can be distributed on the back of the part to counteract any regions of decreased wear.

Instead of using pitch, the lap can be faced with grinding or polishing pads. Brass or other metal or ceramic surfaces are used for grinding. Polyurethane, or other types of synthetic pads, can be used for polishing. Pad polishers do not require as much maintenance as pitch laps and can produce excellent surfaces with the proper materials and conditions.

This technique has been extended to parts having two polished parallel faces, such as semiconductor wafers and various types of optical windows. Both faces are polished at the same time using what are called twin polishers. In this case, there is a lap on top and bottom, with the parts riding in septums in between. These machines rapidly grind and polish windows to high flatness, low wedge, and critical thickness.

Spherical parts can also be fabricated on a continuous polisher by cutting a radius into the lap and maintaining the radius with a spherical conditioner. In this way, numerous parts with exactly the same radius can be manufactured economically. This works well with parts whose radii are long compared to their diameters, that is, parts with large focal ratios. If the focal ratio becomes too small, the uniform wear condition is not valid due to an uncompensated angular velocity term in the wear equation. This term causes a small amount of spherical aberration in the part, which must be removed through pressure variation or some other means.

Continuous polishing machines have been built to 4 meters in diameter, capable of producing 1 meter diameter flats. To produce larger flats, a more conventional polishing machine is used, such as a Draper type, overarm type, or swing-arm type. In this case, the situation is reversed from a CP. The mirror is placed on a suitable support on the turntable of the polishing machine, and ground and polished with laps that are smaller than the part. This is a more conventional process, but it is difficult to achieve the smoothness and surface figure quality that the CP provides.

21.3.6 Centering and Edging

After polishing both sides of lenses, the edges are cut to provide an outer cylinder and protective bevels. The lenses are aligned on a rotary axis so both optical surfaces spin true, meaning that the centers of curvature of the spherical surfaces lie on the axis of rotation. This line, through the centers of curvature, defines the true optical axis of the lens. When the lens is rotated about the optical axis, the edge is cut with a peripheral diamond wheel. This ensures that the newly cut edge cylinder, which defines the mechanical axis of the part, is nominally aligned to the optical axis.

There are two common centering methods shown in Figure 21.17—one optical and the other mechanical. The lens can be mounted on a spindle that allows light to pass through the center. As the lens is rotated, any misalignment in the lens will show up as wobble for an image projected through the lens. The lens is centered by sliding the optic in its mount and watching the wobble. When the wobble is no longer discernable, the part is centered and can be waxed into place for edging.

Also, the centering can be automated using two co-axial cups that squeeze the lens. Here, the lens will naturally slide to the position where both cups make full ring contact, and will thus be aligned (at least as well as the alignment of the two cups). This method of bell chucking is self-centering, so it is naturally adapted to automated machines. It is important that the edges of the chucks are rounded true, polished, and clean so they will not scratch the glass surfaces.

When the optical element is centered and rotated about its optical axis, the outer edge is cut to the final diameter with a diamond wheel. This operation can be guided by hand, with micrometer measurements of the part, and it can also be performed automatically using numerically controlled machines.

When cutting the edge, a protective bevel should always be added to protect the corners from breakage. A sharp, non-beveled edge is easily damaged and the chips may extend well into the clear aperture of the part. A good rule of thumb for small optics is that bevels should be nominally 45°, with face width of 1% of the part's diameter.

Large optics, which are made one at a time, are frequently manufactured differently. The blanks are edged first, and then the optical surfaces are ground and polished, taking care to maintain the alignment

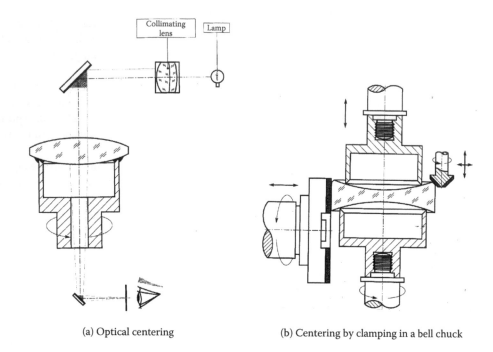

(a) Optical centering   (b) Centering by clamping in a bell chuck

FIGURE 21.17   Centering and edging of lenses. The lens can be centered on the chuck (a) optically by moving the element to null wobble of the image, or (b) automatically using a bell chuck. Once centered on the spindle, the edge and bevels are cut with diamond wheels. (Courtesy of Karow, 1993.)

of the optical axis with the mechanical axis. Also, optics with loose tolerance for wedge can be edged first, and then processed as described above.

21.3.7 Cleaning

The finished parts must be thoroughly cleaned to remove any residue of pitch, wax, and polishing compounds. The optics are typically cleaned in solvent baths with methyl alcohol or acetone. Optics can be cleaned one at a time by carefully wiping them with solvent-soaked tissues, or they can be cleaned in batches in large vapor degreasing units followed by an ultrasonic bath in solvent. Parts that were not edged after polishing tend to have stained bevels and edges from the polishing process. This can be difficult to clean and this residual compound can contaminate the coating chambers.

21.3.8 Cementing

Lenses and prisms are commonly bonded to make doublets or complex prisms. The bonded interface works extremely well optically as long as the cement layer is thin and nearly matches the refractive index of the glasses. The bonded surface allows two different glasses to be used to compensate for chromatic effects, and this interface introduces negligible reflection or scattering.

Most cementing of optics is performed using a synthetic resin, typically cured with UV light. The procedure for cementing lenses is first to clean all dust from the surfaces. Then, the cement is mixed and outgassed, and a small amount is dispensed into the concave surface. The mating convex surface is then gently brought in to press the cement out. Any air bubbles are forced to the edge and a jig is used to align the edges so that the lenses are centered with respect to each other. Excess cement is cleaned from the edge using a suitable solvent. When the lens is aligned, the cement is cured by illuminating with UV light, such as from a mercury lamp.

21.3.9 Current Trends in Optical Fabrication

Through the use of various types of motors, sensors, switching devices and computers, automation has begun to have a major impact on the productivity of fabrication equipment. Numerically controlled (NC) machining has made tooling and shaping of parts much more rapid and less costly. Generating has become more automated with the application of position encoders and radius measuring hardware and software. Grinding/polishing machines are slowly having most of their subsystems automated, although the basic process has remained as described above. For most precision optics made today, the optician's skill in the operation of the polishing machine still has a large impact on the results. However, automation is making the fabrication process less skill dependent and more "deterministic," a buzzword of modern optical fabricators.

New machines use advanced NC machining technology for efficiently and accurately manufacturing small optics. A single, high-precision machine can generate, grind, polish and shape a single lens. Metrology for each stage is integrated into the machine and corrections are applied automatically. Stiff, high-precision spindles with diamond wheels use shallow cuts to produce accurate surfaces with minimal subsurface damage. Ring tool polishers are used to bring the surfaces to final figure and finish. Although the machines are expensive compared with conventional labor-intensive methods, the future of production optics clearly includes this capability. The development of these machines has driven a wide range of deeper investigations into the grinding and polishing of glass. These will inevitably lead to further developments in the automation of optics production.

21.4 Fabrication of Aspheres and Non-Traditional Manufacturing Methods

In the previous section, we give the basic steps for making spherical and plano optics by following the conventional processes, although frequently these steps are made with advanced machinery. In this section, we describe the fabrication of aspheric surfaces and introduce a variety of methods that are in practice for making non-classical optics. Some aspheres are polished using direct extrapolations of spherical methods. Others rely on advanced, computer-controlled polishers. Aspheric surfaces can also be produced by methods other than polishing. Small optics are directly molded in glass and plastic. Aspheric and irregular surfaces are also replicated in epoxy, plastic, and electroformed metal.

Aspheric optical surfaces—literally any surfaces that are not spherical—are much more difficult to produce than the spheres and flats above. Since these non-spherical surfaces lack the symmetry of spheres, the method of rubbing one surface against another simply does not converge to the desired shape. Aspheric surfaces can be polished, but with difficulty, one at a time. The difficulty in making aspherics greatly limits their use, which is unfortunate since a single aspheric surface can often replace a number of spherical surfaces in a design.

21.4.1 Conventional Methods for Fabricating Aspherics

There is tremendous experience behind the traditional fabrication methods that were presented in the previous section. These methods can be applied for making aspheric surfaces, with a few adjustments. Since the methods work best for spheres, we define the difficulty of an asphere by its aspheric departure, or the difference between the aspheric shape and the closest fitting sphere.

Spherical surfaces are used for most optics because these surfaces are easy to describe, easy to manufacture, and easy to test. The spherical surface can be specified by a single parameter—its radius of curvature R. The spherical surface is the easiest to make because of its symmetry. The lap and the part tend to wear on the high spots, and since both are in constant motion about several axes, they will both tend to be spherical. Any other shape would present a misfit between the two, which would tend to be worn down. Testing of spherical surfaces also takes advantage of the symmetry.

When figuring optical surfaces by lapping, the optician uses two different effects to control the surface; natural smoothing and directed figuring. Small scale features, much smaller than the lap, tend to be removed by natural smoothing. This is the same process as using a sanding block to get a smooth texture in wood. As long as the block is rigid, any bumps in the wood will see large forces and will be removed quickly. This effect, for polishing and sanding, is diminished for features larger than the tool, or for the case where the tool is not rigid and easily conforms to the surface. Using good shop practices and large, rigid tools, optics can be finished spherical to about 0.2 µm of the ideal, using only natural smoothing. The symmetry of the spherical surface insures that the tool will fit the surface well everywhere.

Features on optical surfaces larger than the polishing tools can be shaped using directed figuring. This is simply controlling the process, based on surface measurements, to target the high areas on the optic and hit them directly. In its simplest form, an optician will use directed figuring by making a small tool and running it on the high regions of the optic, as determined by an optical test. In polishing, any combination of speed, dwell time, and pressure variation may be used, but the premise is the same.

21.4.1.1 Tools for Working Aspherics

The difficulty in polishing aspheric surfaces is due to the fact that a large rigid tool cannot fit everywhere on the surface. If the tool fits one place, it will not fit at a different position or orientation, and will lose the ability for natural smoothing. Opticians deal with this in two ways, both at the expense of large scale natural smoothing. They can make the tool smaller until the misfit is no longer important, or they can make the tool compliant so it will always fit. In fact, most opticians will use a combination of these for any single asphere.

For analysis of the tool misfit, we treat the case shown in Figure 21.18, with a circular lap, diameter $2a$, a distance b from the optical axis of the parent asphere. The misfit of the lap can be represented in several modes, which take the same form as optical aberrations. Power corresponds to a radius of curvature mismatch. Astigmatism gives the curvature difference in the two principal directions. Coma has a cubic form and spherical aberration (SA) has a quartic dependence on lap position.

We give the lap misfit for a few common conditions:

1. Lap fits a spherical surface with radius of curvature R.
2. Lap is revolving.
3. Lap is rotating a small amount $\Delta\theta$.
4. Lap is translating a small amount Δb.

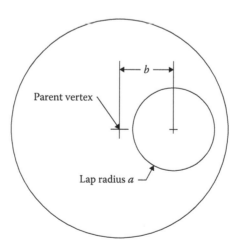

FIGURE 21.18 The lap misfit is calculated for a polishing surface with diameter $2a$, offset from the vertex of the parent asphere by an amount b.

Note that the spherical aberration term has no effect for the real cases (2,3,4). This is because the spherical aberration of the asphere is constant on the lap for any position. It is only the change of surface aberrations that affect polishing. Also, most of the terms for the aberrations in Table 21.4 can be neglected for two common cases. For a large tool with a small stroke near the axis, the coma term dominates. For a small tool, off-axis by an amount much larger than the tool size, the astigmatism and power terms dominate. The astigmatism and power are coupled so the P-V misfit for the case of the stroking tool will be equal to the sum of the power and astigmatism terms.

The relationships in Table 21.4 are used to design the equipment for grinding and polishing. In grinding, the shape errors should be less than the size of the grit in the grinding compound and, in polishing, the lap should fit to a few microns. (The better the lap fit, the better the finish.) The laps designed for aspheric surfaces use a combination of small size, small stroke, and compliance to maintain the intimate contact required.

21.4.1.2 Aspherizing

The traditional steps for making an aspheric surface are to first generate and grind to a spherical surface using the methods described in the previous section. Then, the surface is "aspherized" by grinding or polishing with a specially designed tool, stroke, or machine. For small departures of a few tens of microns, this can be polished. For steeper aspheres, it is generally ground into the surface and the entire surface is then polished with small or flexible tools.

There are a variety of methods for aspherizing. Full size compliant tools can be used with the contact area defined as petals that give the desired removal as the part is rotated underneath. Full size metal tools with the inverse aspheric curve are used for "plunge grinding" of small parts. Most commonly, smaller laps are used, and the dwell is adjusted based on the aspheric curve to be ground in. The aspherizing process is usually monitored with mechanical measurements, such as spherometry or profilometry.

21.4.1.3 Polishing

Once the part has been aspherized, it is polished and figured using a combination of large, semi-compliant tools and small tools. The optical test is critical for this process, as the optician will work the part based on the measurement. Unlike making spheres, there is no tendency for the process to give the correct shape. The optician iteratively measures the surface and works the surface until it meets the specification.

Mild aspheres have surface slopes that are only a few microns over the diameter of the part. In that instance, large tools can still be used to produce the asphere, and smooth aspheric surfaces can be made. When slopes become larger, say tens of microns over the part diameter, a single large tool cannot be used and small tools become the rule. For fast asphercis, where local slopes can become greater than several microns per millimeter, very small tools or other methods must be employed. The usual result from using a tool that is too large to fit the local surface is that the tool wears in a restricted region and produces ripples or zones in the surface. These zones can become quite sharp and are often difficult

TABLE 21.4 P-V Lap Misfit for the Cases Described Above

	Power	Astig	Coma	SA
1. Spherical lap	$\dfrac{Ka^4}{8R^3}+\dfrac{Ka^2b^2}{2R^3}$	$\dfrac{Ka^2b^2}{2R^3}$	$\dfrac{Ka^3b}{3R^3}$	$\dfrac{Ka^4}{32R^3}$
2. Revolving lap	0	$\dfrac{Ka^2b^2}{R^3}$	$\dfrac{2Ka^3b}{3R^3}$	0
3. Small rotation $\Delta\theta$	0	$\dfrac{Ka^2b^2}{R^3}\Delta\theta$	$\dfrac{Ka^3b}{3R^3}\Delta\theta$	0
4. Small translation Δb	$\dfrac{Ka^2b}{R^3}\Delta b$	$\dfrac{Ka^2b}{R^3}\Delta b$	$\dfrac{Ka^3}{3R^3}\Delta b$	0

to get rid of. Zones can be prevented or removed by using a properly sized tool, or by making the tool flexible enough to bend into the global shape of the surface, yet still retain some local stiffness. Much experience and knowledge has traditionally been required to produce high quality aspheres. However, more deterministic methods are being developed.

21.4.2 Modern Methods of Asphere Fabrication

21.4.2.1 Computer Controlled Polishing

Most aspheric surfaces are produced by highly skilled opticians using small tools and conventional machinery. There are, however, a number of methods being developed that integrate computer technology with radically different polishing methods, which can rapidly produce aspheric surfaces. The first of these is the computer controlled polishing (CCP) method [30–32]. This is essentially a traditional small tool method where the tool is driven in an orbital motion producing, on average, a known wear profile. This wear profile is applied to the measured errors in a surface to produce a tool path that essentially rubs longer on the high areas and less on the low areas, but in a precise relative way that can rapidly improve the figure. Sophisticated, proprietary computer algorithms are used to determine the optimal machine motions from the surface measurement and removal function.

Another method that radically departs from traditional polishing methods is the ion figuring method [33,34]. Here, the polished surface is bombarded by ions from an ion gun to remove material in a very deterministic way. The removal function of the ion gun is well established prior to use. Just like the CCP process, a tool path is developed from the measured surface errors to produce a dwell time function for the surface. The surface figure can be rapidly improved due to the high removal rate of the ion gun verses polishing. The process is highly deterministic, so many parts can be finished with a single run in the ion mill. Ion figuring is only used to remove about a micron from the surface, because it can degrade surface finish.

A figuring process that utilizes the etching of glass is the PACE or Plasma Assisted Chemical Etching method [35]. A small confined plasma, which is reactive with the glass substrate, is moved over the surface, and material is removed proportional to dwell time. By choosing a suitable tool path, the surface can be figured without introducing high spatial frequency errors into the surface. The tool size can be adjusted to produce the most appropriate removal profiles for the particular surface error. As with ion figuring, this method also demonstrates high removal rates and excellent figure convergence.

Another deterministic polishing method for small aspheres uses a lap made with a magnetorheological substance, which has viscosity that can be controlled by applying magnetic fields [24]. This tool gives a well-defined removal profile, which can be modulated with electromagnets. The parts are rotated under the lap and the magnetic field is adjusted under computer control according to the measured surface.

The finishing of optics with such computer controlled methods using small tools has been limited to large companies or research groups. These techniques provide excellent results when everything is worked out correctly. However, it takes many hours to polish a large optic with a small tool and, if something goes wrong in this process, the polisher can drive a small low region into the part. If this happens, the entire surface must be driven down to meet this low spot. One must have confidence in the process to use small tool figuring on production parts. Also, these methods rely on good, computer-acquired data, which is mapped carefully to the surface. If the polishing run is shifted slightly, relative to ideal, the polisher can drive low spots right next to the high spot it was intending to hit. Even with these difficulties, the large optics companies have developed excellent processes and equipment for computer controlled polishing.

While these methods are very efficient, they are expensive to implement and operate. The application of computer controlled polishing has been largely limited to special projects for defense or space-related work when more conventional methods would be nearly impossible to use. However, in recent years,

CCP processing is becoming widely used and will lead the way toward the integration of the computer and other high technologies with aspheric production methods.

Large-tool polishing is also possible for aspheric surfaces if the tool itself is controlled by the computer. Several groups have developed large, active tools that polish aspheric surfaces under computer control by changing the lap shape or force. The stressed lap polisher [36,37] uses a large, rigid polishing tool that is actively bent under computer control to take the shape of the aspheric surface. This retains the advantage of large tools to provide passive smoothing, even on steep aspheres [38]. A different concept has been demonstrated that uses a membrane lap with the polishing force dynamically controlled by computer [39]. This allows the use of a large tool, although there is little gain by passive smoothing.

One last semi-conventional method for making aspheres is the bend and polish technique developed by Schmidt [40] and applied elsewhere [41,42]. The substrate itself is carefully distorted by applying external forces or moments. The distortion is controlled and the part is polished spherical in its distorted state. The optic should then relax into the desired aspheric shape.

21.4.2.2 Molding

Many small aspheric lenses, such as camera lenses, are made by the direct molding of glass or plastic into an aspheric mold. The molds have the opposite shape of the finished asphere and are made from materials that can withstand the required high temperatures. These optics are readily mass produced by the millions with astonishingly good quality [43].

Small lenses are molded in glass using a method called Precision Glass Molding or PGM [44]. The lenses are formed into the final shape by being pressed into a die at high temperature. This method economically produces small (< 10 mm) spherical and aspherical optics in a variety of glasses, giving diffraction limited performance and excellent surface finish. These lenses are used in high-volume goods, such as pocket cameras. Larger condenser lenses for projectors, which have reduced requirements, are also made this way.

High quality plastic optics are mass produced by the process of injection molding [45]. Liquid plastic is forced into a heated mold cavity at high pressures. The plastic solidifies to the inverse shape of the mold. By carefully controlling the pressure and temperature profiles, high quality lenses up to 50 mm in diameter can be produced. The tooling to produce these lenses is quite expensive, but it enables a low cost process that produces lenses by the thousand. Plastic optics find use in the same type of applications as the molded glass lenses. Advantages to plastic optics are reduced weight and the ability to have complex mounting features integrated with the optic.

21.4.2.3 Replication of Optical Surfaces

In addition to molding, optical surfaces are created by replicating against a master. Compression molding of plastics is used to make large, flat optics, such as Fresnel lenses [46]. A thermoplastic blank is pressed between two platens and heated. Parts as large as 1.5 meters have been made using this method.

Optical surfaces, especially gratings, are often replicated into epoxy. Typically, the epoxy is cast between two glass surfaces, the master and the final substrate. A special chemical called a release agent is applied to the master surface so the epoxy will not stick to it. The result is a replicated inverse of the master, held fast to the final substrate. Diffraction limited accuracy can be obtained for parts made using a carefully controlled process.

Metal optics are electroformed against precision mandrels to make good, smooth optics. Electroforming is simply electroplating onto a surface with a suitable release. After completion, the thin metal "electroplate" can be removed and used as a reflective optic. Reflectors for high power light sources are made by electroforming a thin reflective layer of nickel or rhodium onto a convex mandrel. A layer of copper, several millimeters thick, is then electroformed on top to give the part structural rigidity. These optics are quite smooth, but can have large figure errors.

21.4.2.4 Single-Point Diamond Turning

In recent years, high performance machines have been produced that use sharp diamond tools to turn optical surfaces directly to finished tolerances. These machines use accurate motions and rigid mounts to cut the optical surface with a single diamond point, just as one would machine the part on a lathe. This has the obvious advantage that aspheric surfaces can be cut directly into the surface, without the need for special laps or metrology. In fact, some optical surfaces, such as axicons, would be nearly impossible with conventional processes. Single-point diamond turning (SPDT) is not new, but only in recent years has it become economical for production parts. Some references on the subject are in Arnold et al., 1977 [47], Gerchman 1986 [48], Rhorer and Evans 1995 [49], and Sanger 1987 [50].

A variety of materials have been fabricated using SPDT. The best results are for ductile metals such as aluminum, copper, nickel, and gold. Crystalline materials used for infrared applications, such as ZnSe, ZnS, and germanium, are also diamond turned with excellent results. Diamond turning does not work well for glass materials because they are brittle.

The surface structure obtained from diamond turning is different from conventional processes. Polished optics have no systematic structure in them, and they can be made perfect to a few angstroms. Diamond turned surfaces always have residual grooves from the diamond tool. These can be made quite small (10 nm) by making a final light cut with fine pitch. The surface scattering from these grooves limits the application of most diamond turned optics to infrared applications, which are not sensitive to such surface effects. In some cases, it is possible to post-polish the diamond turned part to smooth out these grooves [51].

There are two common configurations for diamond turning machines: as a precise lathe with the part spinning and the diamond bit carefully controlled, and as a fly cutter with the part fixed and the diamond bit moving on a rotating arm. The lathe-type machines produce both axisymmetric surfaces and off-axis optics (by mounting the optic off the axis of rotation.) The fly cutter geometry is used to produce flats, especially for crystals that are difficult to polish, and for multi-faceted prisms where the relative angle from one facet to the next can be controlled.

21.5 Conclusion

This chapter has given a summary of the most common fabrication methods in use today. Most optics are made by modern variants on classical methods, but the highest performance optics rely on more advanced techniques. Clearly, there are numerous fabrication methods for specialty optics that lie outside the scope of what has been presented here.

We present this information to the optical engineer to give some understanding of limitations and alternatives in the shop. An engineer who knows the basic issues can work directly with the fabricator to design cost effective systems. Clearly, the system cannot be optimized for either performance or cost if the fabricator is not involved in the decisions. Remember, without the fabricator, the optical engineer would have nothing but a pile of computer printouts and some sand!

References

1. Sumner R. Working Optical Materials, in *The Infrared Handbook*, Wolfe WK, Zissis GJ, Eds., Arlington, VA: Office of Naval Research, 1978.
2. Musikant S. *Optical Materials*, New York, NY: Marcel Dekker, Inc., 1985.
3. Willey RR, Parks RE. Optical Fundamentals, in *Handbook of Optomechanical Engineering*, Ahmad A, Ed. Boca Raton, FL: CRC Press, 1997.
4. Willey RR. The impact of tight tolerances and other factors on the cost of optical components, in *Optical Systems Engineering IV*, P. R. Yoder, Jr., Ed., Proc. SPIE **518**; 1984.

5. Willey RR, George R, Odell J, Nelson W. Minimized cost through optimized tolerance distribution in optical assemblies, in *Optical Systems Engineering III*, Taylor WH, Ed., Proc. SPIE **389**; 1983.

6. Parks RE. Optical specifications and tolerances for large optics, in *Optical Specifications: Components and Systems*, Smith WJ, Fischer RE, Ed., Proc. SPIE **406**; 1983.

7. Parks RE. Optical component specifications, in *International Lens Design Conference*, R. E. Fischer, Ed., Proc. SPIE **237**; 1980.

8. Smith WJ. Fundamentals of establishing an optical tolerance budget, in *Geometrical Optics*, R. E. Fischer, Price WH, Smith WJ, Eds., Proc. SPIE **531**; 1985.

9. Plummer JL. Tolerancing for economics in mass production optics, in *Contemporary Optical Systems and Components Specifications*, R. E. Fischer, Ed., Proc. SPIE **181**; 1979.

10. DeGroote-Nelson, Youngworth JRN, Aikens DM. The cost of tolerancing, Proc. SPIE **7433**; 2009.

11. Bennett JM, Mattsson L. *Introduction to Surface Roughness and Scattering*, Washington, DC: Optical Society of America, 1989.

12. Wilson RN. *Reflecting Telescope Optics II*, Heidleberg, Germany: Springer-Verlag, 1999.

13. Karow HH. *Fabrication Methods for Precision Optics*, New York, NY: Wiley, 1993.

14. Piscotty MA, Taylor JS, Blaedel KL. Performance Evaluation of Bound Diamond Ring Tools, in *Optical Manufacturing and Testing*, Doherty VJ, Stahl H, Eds., Proc. SPIE **2536**; 1995.

15. Brown NJ. Optical polishing pitch, Preprint UCRL-80301 (Lawrence Livermore National Laboratory, 1977).

16. Yoder PR. *Opto-Mechanical Systems Design*, 2nd Edition New York, NY: Marcel Dekker, 1993.

17. Ohmori H. Ultraprecision grinding of optical materials and components applying ELID "Electrolytic In-Process Dressing," in *International Conference on Optical Fabrication and Testing*, Kasai T, Ed., Proc. SPIE **2576**; 1995.

18. Stowers IF, et al. Review of precision surface generation processes and their potential application to the fabrication of large optical components, in *Advances in Fabrication and Metrology for Optics and Large Optics*, Arnold JB, Parks RE, Eds., Proc. SPIE **966**; 1988.

19. Horne DF. Loose abrasives, impregnated diamonds and electro-plated diamonds for glass surfacing, in *Advances in Optical Production Technology*, T. L. Williams, Ed., Proc. SPIE **109**; 1977.

20. Bifano TG, Dow TA, Scattergood RO. Ductile-regime grinding of brittle materials: Experimental results and the development of a model, in *Advances in Fabrication and Metrology for Optics and Large Optics*, Arnold JB, Parks RE, Eds., Proc. SPIE **966**; 1988.

21. Golini D, Jacobs SD. Transition between brittle and ductile mode in loose abrasive grinding, in *Advanced Optical Manufacturing and Testing*, Baker LR, Reid PB, Sanger GM, Eds., Proc. SPIE **1333**; 1990.

22. Golini D, Jacobs SD. The Physics of Loose Abrasive Microgrinding, *Applied Optics* **30**: 2761–2777; 1991.

23. Golini D, Czajkowski W. Center for Optics Manufacturing Deterministic Microgrinding, in *Current Developments in Optical Design and Optical Engineering II*, Fischer RE, Smith WJ, Eds., Proc. SPIE **1752**; 1992.

24. Golini D, Jacobs SD, Kordonsky W. Fabrication of glass aspheres using deterministic microgrinding and magnetorheological finishing, in *Optical Manufacturing and Testing*, Doherty VJ, Stahl H, Eds., Proc. SPIE **2536**; 1995.

25. Spira MW. Precision grinding with pellets and high-speed polishing by means of synthetic material, in *Advances in Optical Production Technology*, Williams TL, Ed., Proc. SPIE **109**; 1977.

26. Rupp W. Loose abrasive grinding of optical surfaces, *Applied Optics*, **11**, 2797–2810; 1972.

27. Rupp W. Surface structure of fine ground surface, *Optical Engineering*, **15** 392–396; 1976.

28. Holland L. *The Properties of Glass Surfaces*, New York, NY; John Wiley & Sons, Inc., 1964.

29. Brown NJ, Baker PC, Maney RT. Optical polishing of metals, in *Contemporary Methods of Optical Fabrication*, Stonecypher CL, Ed., Proc. SPIE **306**; 1981.

30. Bajuk DJ. Computer controlled generation of rotationally symmetric aspheric surfaces, *Optical Engineering* **15**, 401–406; 1976.

31. Jones RA. Grinding and polishing with small tools under computer control, Optical Engineering **18**, 390–393; 1979.

32. Jones RA, Rupp WJ. Rapid optical fabrication with CCOS, in *Advanced Optical Manufacturing and Testing*, Baker LR, Reid PB, Sanger GM, Eds. Proc. SPIE **1333**; 1990.

33. Meinel AB, Bushkin S, Loomis DA. Controlled figuring of optical surfaces by energetic ionic beams, *Applied Optics* **4**, 1674; 1965.

34. Allen LN, Keim RE. An ion figuring system for large optic fabrication, in *Current Developments in Optical Engineering and Commercial Optics*, Fischer RE, Pallicove HM, Smith WJ, Eds., Proc. SPIE **1168**; 1989.

35. Bollinger D, et al. Rapid, non-contact optical figuring of aspheric surfaces with plasma assisted chemical etching, in *Advanced Optical Manufacturing and Testing*, Baker LR, Reid PB, Sanger GM, Eds., Proc. SPIE **1333**; 1990.

36. Martin HM, Anderson DS, Angel JRP, Nagel RH, West SC, Young RS. Progress in the stressed-lap polishing of a 1.8-m f/1 mirror, in *Advanced Technology Optical Telescopes IV*, Barr LD, Editor, Proc. SPIE **1236**; 1990.

37. West SC, et al. Practical Design and Performance of the Stressed Lap Polishing Tool, Applied Optics **33**, 8094; 1994.

38. Burge JH. Simulation and optimization for a computer-controlled large-tool polisher, OSA Trends in Optics and Photonics *Vol. 24, Fabrication and Testing of Aspheres*, Taylor JS, Piscotty M, Lindquist A, eds, Washington, DC: Optical Society of America, 1999.

39. Korhonen T, Lappalainen T. Computer-controlled figuring and testing, in *Advanced Technology Optical Telescopes IV*, L. D. Barr, Ed., Proc. SPIE **1236**; 1990.

40. Schmidt B. *Mitt. Hamburg. Sternw.*, **7**, 15; 1932.

41. Everhart E. Making Corrector Plates by Schmidt's Vacuum Method, *Applied Optics* **5**: 713–715; 1966; and *Errata in Applied Optics*. **5**: 1360; 1966.

42. Lubliner J, Nelson J. Stressed Mirror Polishing, *Applied Optics* **19**, 2332–2340; 1980.

43. Aquilina T. Characterization of Molded Glass and Plastic Aspheric Lenses, in *Replication and Molding of Optical Components*, M. J. Riedl, Ed., Proc. SPIE **896**; 1988.

44. Pollicove HM. Survey of present lens molding techniques, in *Replication and Molding of Optical Components*, M. J. Riedl, Ed., Proc. SPIE **896**; 1988.

45. Hoff AM. Basic considerations for injection molding of plastic optics, in *Design, Fabrication, and Applications of Precision Plastic Optics*, Ning X, Hebert RT, Eds., Proc. SPIE **2600**; 1995.

46. Parks RE. Overview of optical manufacturing methods, in *Contemporary Methods of Optical Fabrication*, Proc. SPIE **306**; 1981.

47. Arnold JB, Sladky RE, Steger PJ, Woodall ND, Saito T. Machining nonconventional-shaped optics, *Optical Engineering* **16**: 347–354; 1977.

48. Gerchman M. Specifications and manufacturing considerations of diamond machined optical components, in *Optical Component Specifications for Laser-Based Systems and Other Modern Optical Systems*, Fischer RE, Smith WJ, Eds., Proc. SPIE **607**; 1986.

49. Rhorer RL, Evans CJ. Fabrication of optics by diamond turning, *Handbook of Optics*, Washington, DC: Optical Society of America, 1995.

50. Sanger GM. The Precision Machining of Optics, in *Applied Optics and Optical Engineering, Vol. 10*. New York, NY: Academic Press, Inc., 1987.

51. Bender JW, Tuenge SR, Bartley JR. Computer-controlled belt polishing of diamond-turned annular mirrors, in *Advances in Fabrication and Metrology for Optics and Large Optics*, J B Arnold and R E Parks, Eds., Proc. SPIE **966**; 1988.

Additional Suggested Reading

1. Fontane SD. *Optics Cooke Book, Optical Society of America*, 1991.
2. George RW, Michaud LL. Optical fabrication by precision electroform, in *Current Developments in Optical Engineering II*, Fischer RE, Smith WJ, Eds., Proc. SPIE **818**; 1987.
3. Horne DF. *Optical Production Technology*, Bristol, England: Adam Hilger, 1983.
4. Ingalls AG. *Amateur Telescope Making, Volumes* 1–3, Richmond, Virginia: Willmann-Bell, 1996.
5. Kumanin KG. *Generation of Optical Surfaces*, London, England: The Focal Library, 1962.
6. Malacara D. *Optical Shop Testing*, 2nd Edition, New York, NY: Wiley, 1992.
7. Parks R. Traditional optical fabrication methods, in *Applied Optics and Optical Engineering, Vol. X*, Shannon RR, Wyant JC, Eds., San Diego, CA: Academic Press, 1987.
8. Pollicove HM, Moore DT. Automation for Optics Manufacturing, in *Advanced Optical Manufacturing and Testing*, Baker LR, Reid PB, Sanger GM, Eds., Proc. SPIE **1333**; 1990.
9. Sanger GM. Editor, *Contemporary Methods of Optical Manufacturing and Testing*, Proc. SPIE **433**; 1983.
10. Scott RM. Optical Manufacturing, in *Applied Optics and Optical Engineering, Vol. III*, Lingslake R, Ed., New York, NY: Academic Press, 1965.
11. Taylor JS, Piscotty M, Lindquist A. eds, *Trends in Optics and Photonics Vol. 24, Fabrication and Testing of Aspheres*, Washington, DC: Optical Society of America, 1999.
12. Texereau J. *How to make a Telescope*, Richmond, Virginia: Willmann-Bell, 1984.
13. Twyman F. *Prism and Lens Making*, 2nd Edition, Bristol, England: Adam Hilger, 1988.
14. Zschommler W. *Precision Optical Glassworking*, London, England: MacMillan, 1984—also published as SPIE volume 472; 1984.

Web References

1. http://www.edmundoptics.com/technical-resources-center/optics/understanding-optical-specifications/
2. http://www.optimaxsi.com/innovation/optical-manufacturing-tolerance-chart/
3. http://www.cidraprecisionservices.com/scratch-dig-specs.html
4. http://www.prhoffman.com/pdf/MIL-Specs.pdf

Index

A

Abbe number *vs.* refractive index, 684, 685
Abbe value, 678
Aberrations, eye
  chromatic, 547–549
  correction of, 555
  higher-order, 552–554
  peripheral, 556
Abrasive grinding, 746
Absorption
  and fiber loss, 639
  filters, 615
  index, 679
AC interferometer, 170, 185
Acousto-optical scanners, 13
Acousto-optic filters, 617–618
Acousto-optic (AO) modulators, 453–457
Acrylamide polyvinylalcohol formulation, 376
Active optics system, 246–247
Actuators, in optical measuring systems, 105
Adaptive compensating system, 452
Adaptive optics system
  in astronomy, 250–253
  elements in, 246–247
  in ophthalmology, 253–256
  wavefront sensing in
    pyramid sensor, 248–250
    Shack-Hartmann sensor, 247–248
Adaptive quadrature filter, 130
Addition of color, 531–532
Adjacency effects, in film processing, 360
Agfa, 361
Airy distribution, 612
ALL, *see* Amplitude-locked loop
All-dielectric narrow band filters, 626
Allyldiglycol carbonate, 690
Aluminum, 690
Amplitude filter, 313, 314
Amplitude gratings, 265
Amplitude-locked loop (ALL), 196
Angle blocks, 172

Angle measurements, 171
  autocollimators and theodolites, 172–173
  divided circles and goniometers, 171–172
  interferometric methods, 181–182
  lateral shear interferometer for, 183
  level measurement, 175–176
  moiré methods in, 177–181
  in prisms, 173–175
Angle replication, of polygons, 174, 175
Anisotropic material
  crystalline materials and properties, 698–699
  crystals
    index ellipsoid, 707–708
    lattice of, 699
    linear electro-optic tensors, 701–704
    natural birefringence, 708
    point groups, and dielectric tensors, 700
    quadratic electro-optic tensors, 705–706
    wave surface, 708–710
    wavevector surface, 710–711
  electric fields, application of, 711–717
  electromagnetism, 696–698
  interaction of light with, 695
  magneto-optic effects, 718–719
Anisotropy, 660
Annealing, 683
Anti-counterfeiting, 624
Antireflection coatings, 628
AO modulators, *see* Acousto-optic modulators
APD, *see* Avalanche photodiode
Arc sources, 566–567
Area modulation, 347–350
Arithmetic (pixel-to-pixel) operations, 104
Arizona eye model, 545
Artificial light source, 561
Art, laser applications, 597–601
Aspheric optical surfaces, 729, 755
Astigmatism, 549
  axial and oblique, 549
  irregular, 550–551
Astronomical telescopes, 250
Astronomy, wavefront sensing in, 250–253
Asymmetrical arrangement, 226–227

Atmospheric seeing, 245
Attenuation, in optical fibers, 638–639
Attenuator, 659
Autocollimators, 159, 160, 172–173
Autofluorescence emission, 6
Avalanche gain, 414
Avalanche photodiode (APD), 414
    gain *vs.* electric field, 417
    staircase, 422
AΩ product, 610
Axial astigmatism, 549
Axicons, 731
AZEye15 model, 545
    cardinal points for, 547
    longitudinal spherical aberration of, 553, 555
    parameters of, 546, 547
Azoic colorants, 293

B

Baffles, 609
Bandpass quadrature filter, 129–130
Bandwidth-distance product, 641
Bandwidth, optical fiber, 639
Bar spherometer, 158
BCCD, *see* Buried channel CCD
Beam of light, 603, 604
Beam-scanning laser printer, 456
Beamsplitter, 12–13
Bending loss, 638
Beryllium, 689–690
Biaxial crystals, index ellipsoid for, 709
Biconic connector, 669
Binary amplitude grating, 269
Binary phase grating, 269
Binder optimization, 380
Biophotopolymer, 389
Birefringence, 680, 711–717
Bi-Ronchigram, 51
Bit error rate (BER), 388
Black-and-white films
    development process, 360
    image effects, 359–360
    image structure properties, 357–359
    manufacturers, 360–361
    photographic sensitivity, 354–357
Black body light sources, colors of, 529, 530
Blackbody radiator, 491–492
Blackbody simulators, 512, 513
Black light lamp, 572
Blanchard type geometry, 743
Bleaching process, 364
Blooming effect, 429
Blue LED, relative spectral radiance for, 575
Blue Ray Disk (BD), 386
Boolean functions, 332

Border effect, 360
Bragg's gratings
    diffraction efficiency of
        amplitude, 266
        phase, 265
    optical fibers, *see* Fiber Bragg gratings
    transmission and reflection, 267
Bragg's law, 263
BRDF function, 498–499
Broadband signals
    area modulation, 347–350
    optical processing of, 343–347
Bubbles, 680
Burch and Murty interferometers, 86–87
Buried channel CCD (BCCD), 426
Burrus LED, 573

C

Calcite, 708
Capillary splice, 670
Capsule endoscopy, 28
Carbon arc, 566
Cathode-ray tube (CRT), 429
Cathodoluminescent, 429
Cationic ring opening polymerization (CROP), 388
Cauchy-type dispersion formula, 546
CCD-addressed LCLV, 451, 452
Cemented doublet, 31
Cementing, of optics, 754
Centering methods, 753–754
Ceramic ferrules, 668
Ceramics, 689
Chalcogenide glass fibers, 646–647
Characteristic admittance, 604
Characteristic curve
    for color film, 362
    for negative photographic material, 355
Charge-coupled devices (CCDs) imagers, 511
    buried channel, 426
    camera, 137
        sub-Nyquist techniques, 145
    charge-injection devices charge, 429
    detector, in digital holographic interferometry, 221, 222
    energy-band diagram, MOS capacitor and, 423
    interline transfer and frame transfer, 428
    MOS readout scanner, 427
    overall MOS capacitance, 424
    surface potential equation, 425
Charge-injection devices (CIDs), 429
Chirped fiber Bragg gratings, 665–666
Cholesteric liquid crystal, 722
Cholesteric mesophase, 719
Christiansen filters, 617
Chromatic aberration, 547–549

Chromatic defect, 519
Chromatic dispersion, 678
Chromaticity CIE diagram
    applications of, 529
    chromaticity coordinates, 527, 528
    color matching functions, 526
    lines of constant hue, 529
    McAdam ellipses, 528, 529
    for r *vs.* g, 525
    with three colors, 528
Chromaticity coordinates, 524–525
Chromatic visual effects, 520
CIDs, *see* Charge-injection devices
CIELAB system, 535
CIELUV system, 535
Classical regularization, 107–110
Closed-fringes demodulation, 133
Coatings/films, 626–632, 692–693
Coaxial arrangement, 225
Coefficient of thermal expansion (CTE), 733
Coherence
    illumination, 304
    of light sources, for interferometers, 72–73
    in Young's interferometer, 75–76
Coherent optical image processing
    amplitude and phase filter, 313, 314
    image restoration, 315
    image subtraction, 316
    inverse filter transfer function, 312, 313
    output complex light distribution, 317
    relative degree of restoration, 314
Coherent optical signal processing, 302–304
Coherent spot projection system, 204
CO_2 Laser Doppler Velocimetry, 193
Color addition, 531–532
Color blindness, 519
Color defect, 519
Color evaluation boots, 571
Color films, subtraction process, 361–362
Colorimeters, 537
Color mixtures, 531–534
Color rendering index (CRI), 562
Color subtraction, 532–534
Color temperature, 562
Color theories, 518–519
Common path interferometers
    Burch and Murty interferometers, 86–87
    point diffraction interferometer, 87–88
Complex amplitude transmittance, 306
Compression molding, 759
Computer controlled polishing (CCP) method,
        758–759
Computer generated holograms (CGH)
    errors in, 280–282
    Fourier hologram, 278
    Fresnel hologram, 279

iterative computation for, 280
Computing with optics
    logic-based computing, 331–333
    systolic processor, 336–338
Concave surface, testing configuration for, 58, 60
Cones, human eye, 544
Confocal adaptive optics scanning ophthalmoscope, 256
Confocal cavity curvature measurement, 159, 160
Confocal microscopy, 9
    beamsplitter, 12–13
    illumination optics, 12
    principle of, 9–12
    relay system, 14
    scanners, 13
    types of, 14–16
Confocal-planar resonator, 588
Conic aspherical surface, 729, 731
Connectors
    for interconnection loss, 666–667
    performance, 668
    types, 669
Continuous polishing (CP) machine, 751, 753
Continuous wave (CW) lasers, 582
Contrast index, 356
Conventional relay system, 32, 33
Convex surfaces
    null screens for, 58
    testing configuration for, 59
Convolution integral, 300, 301
Convolution methods, 107, 108
Cornea, human eye, 543
Corning ULE™ titanium silicate, 689
Correlated color temperature, 562
Cosmetic properties, 736
Cotton–Mouton effect, 719
Couplers
    directional, 653
    fabrication of, 655–656
    2 x 2 coupler, 654, 655
CRI, *see* Color rendering index
Crossbar switch, 334
Crosslinker optimization, 380
CRT, *see* Cathode-ray tube
Crystal anisotropy, 710
Crystal growing (CG) process, 131, 132
Crystalline materials, 695, 760
Crystals, 699–710
    index ellipsoid of, 711
    liquid, 719–722
Curvature measurements, 157–160, 181
Curvature sensing, wavefront analysis by, 143–145
Cut-off frequency, of photodiodes, 413
Cut-off polarizer, 661
Cutoff wavelength, of optical fiber, 637
CW lasers, *see* Continuous wave lasers
Cyclic interferometers, 97–98

Cyclic multiple reflection interferometers, 96–97
Cylindrical refractive error, 549

D

Daughter wavelets, 325
DC interferometer, for distance measurement, 169
Decorative coating, 623–624
Deformation of gratings, 162, 163
Density, optical materials, 682
Dental photopolymers, 392
Detectors, for confocal microscopes, 14
Deterministic polishing method, 758
Development process, photographic emulsions, 360, 364
Diamond generating tools, 743–746
Dichro-mated gelatine (DCG), 377
Dichromatic LED, 573
Dielectric hollow waveguides, 647–649
Dielectric layers, 622–623
Dielectric permittivity tensor, 697
Dielectric system, 622, 679
Dielectric tensor, 695
Differential geometry, 195
Differential vibrometer, 202
Diffraction
    efficiency
        for thick gratings, 264
        of thin gratings, 270
    gratings
        frequency-shifting by, 194
        thick gratings characteristics, 264–267
        thin gratings characteristics, 267–271
        waves, interference, and grating criteria,
            261–263
    limited accuracy, 759
    by simple apertures
        Fresnel zone plate, 277–278
        multiple slits, 276
        pinhole, 275
        slit, 274–275
Diffraction-limited OTF, 4
Digital displays, 294
Digital holography
    advantages, 272
    interferometry, 221–224
Digital micro-mirror device (DMD), 15, 17
Digital sensors, 294
Digital Versatile Disk (DVD), 386
Diluent system, 375–376
Directional coupler, 653, 658
Directional radiator, 489–491
Direct phase detection system, 199
Direct-read-after-write optical disk system, 457
Discharge lamps, 568–570
Discrete chirp-Z transform (DCZT) matrix, 338
Discrete cosine transform (DCT), 337
    two-complement representation, 338, 339

Discrete Fourier transform (DFT), 337
Discrete Hilbert transform (DHT) matrix, 338
Discrete linear transformation (DLT) system, 337, 338
Discrete sine transform (DST), 337
Discretized sawtooth phase grating, 271
Dispersion, 604
    in optical fibers
        intermodal, 635
        material, 640
        waveguide, 641
Dispersive media, 604
Distance measurement, 183
DMP-128 system, 374
Doped polymer materials, 378–379
Doppler-broadened transition, 585
Doppler differential mode, 194
Doppler effect, 191
Doppler frequency shift, 190
Doppler signal processing, 196–197
Double-clad fibers, 644–646
Double-exposure hologram, 219
Double modulation system, 200–201
Double-slit interferometers; *see also* Young's double slit
    coherence in, 75–76
    stellar Michelson interferometer, 76–78
Dual beam vibrometer, 202
Ductile-regime machining, 746
Dupont film, 390
Dye-doped recording systems, 379
Dynamic Range, definition of, 386

E

EDFAs, *see* Erbium-doped fiber amplifiers
Edge-emitting LED, 575
Edging methods, 753–754
Edwin Land's color experiment, 520
Elasticity, optical materials, 682
Electric fields, application of, 711–717
Electric polarization, 696
Electrodes, 722
Electroforming, 759
Electromagnetic spectrum, 460
Electromagnetism, 696–698
Electronic devices, 293–294
Electronic industry, laser applications, 597
Electronic speckle pattern interferometry (ESPI)
    optical setups
        asymmetrical arrangement, 226–227
        coaxial arrangement, 225
        symmetrical arrangement, 227–229
    real-time vibration measurement, 231–232
    speckle formation, 224, 225
    video signal subtraction/addition, fringe formation
        by, 229–231
Electronic switches, 658
Electro-optic effect, 433, 434

Electro-optic (EO) modulators, 722
    equivalent circuit, 440
    Kerr cell light modulator, 445
    liquid-crystal light valve, *see* Liquid-crystal light
        valve
    Mach-Zehnder modulator, 442, 443
    phase modulators, 438
    Pockels amplitude modulator, 436, 437
    Pockels read-out optical modulator, 443, 444
    principle of, 433–436
    Q-switched laser system, 446
    transverse modulators, 439
Electro-optic tensor, 695
Embossed holograms, 292
Emissivity, 505
Endoscopy, 26
    objective lens, 29–32
    optical layout, 26–28
    relay lenses, 32–35
Energy level diagram, 418, 419
Engagement systolic array processor, 337
EO modulators, *see* Electro-optic modulators
Erasmus Bartholinus, 699
Erbium-doped fiber amplifiers (EDFAs), 643
Erbium-doped fibers, 643–644
Ethylene glycol dimethacrylate (EGDMA), 391
Exact eye model, 545
Excitation filter, 8
    in multiphoton microscopy, 9
Expanded beam connectors, 668
Exposure determination, 357
Extended range fringe pattern analysis, 137–147
External transmittance, 505–507
Extrinsic fluorescence, 6
Extrinsic fluorophores, 6
Extrinsic losses, 666, 668

F

Fabry-Perot etalon, 618
Fabry-Perot interferometer, 96
Face contact (FC) connector, 669
Faraday effect, 718
Fast Fourier transform (FFT), 126
FC connector, *see* Face contact connector
FEL lamp, 565
Ferroelectric liquid crystals, 446, 722
Ferrules, 668
Fiber Bragg gratings
    chirped, 665–666
    multiplexers, 664–665
    phase-mask fabrication process, 663–664
    as reflection filter, 664
Fiber confocal imaging, 15–16
Fiber connection, 666; *see also* Connectors
Fiber lasers, 581
Fiber-optic endoscope, 27

Fiber-optic splitters, 655
Field emission, 566
Field of view, 480
Filament, tungsten lamp, 562
Filling gas, 563
Film coatings, 692–693
Filtering operations, 104
Filters
    color, 530–531, 533
    fluorescence, 7–8
    holographic, 390–391
Finite schematic eye models, 545
"First-order building block," 119–121
Fizeau interferometer, 80–82
Flash lamps, 569, 570
Flexible endoscopes, 27
Fluorescence imaging systems, 6, 7–8
Fluorescence microscope, 5
    imaging systems, 6
        filters, 7–8
        multiphoton imaging, 8–9
        system consideration, 8
    process, 5–6
Fluorescent lamps, 539
    designation characteristics, 572
    electrical characteristics, 570
    optical characteristics, 571
    physical construction, 570
Fluoride glass fibers, 646
Focal length measurements, 157–160, 164–166
Foot-candle (ft-c), 472
Foot-lambert (ft-lb), 473
Formalism, 282–283
Foucaultgram, 38
    simulations, 39–41
Foucault test, 38–41
    and Schlieren techniques, 94
Fourier computer generated hologram, 278
Fourier decomposition process, 605
Fourier domain filtering, 305
Fourier method, fringe formation by, 186
Fourier optics
    convolution integral, 300, 301
        Fourier transformation by lens, 301, 302
    paraxial approximation, 300
Fourier-transform method, for open-fringes
        interferograms, 125–127
Four phase-shifted interferograms, 112
Four-step phase stepping, 116
Frame transfer CCD (FTCCD), 428
Fraunhofer diffraction integral, 273–274
Free electron lasers, 581
Freeform optical surfaces, 729
Freeform surfaces, 731
Frequency response, of regularized linear low-pass
        filters, 110
Frequency-shifting, 192

Frequency transfer function (FTF), 110, 113
  "first-order building block," 119–121
  for quadrature linear filter, 120
Frequency up-conversion, 649
Fresnel computer generated hologram, 279
Fresnel diffraction integral, 273
Fresnel losses, 499–501
Fresnel reflections, 679
Fresnel zone plate, 277–278
Fried parameter, 251, 252
Fringe counting, 168, 185
Fringe pattern analysis
  applicability of methods, 147
  intensity methods, 104
  phase methods, 105
  phase unwrapping, 133–137
  process, 106
  sinusoidal signal fluctuation, 102
  smoothing techniques, 107–110
  spatial phase-measuring methods
    regularized phase-tracking technique, 131–132
    robust quadrature filters, 129–130
    spatial carrier phase-shifting method, 127–129
  2π phase ambiguity, 105
  wavefront (extended) analysis
    by curvature sensing, 143–145
    with Hartmann screens, 141–143
    with lateral shearing interferometry, 139–141
    with linear gratings, 137–139
    screen-testing methods, 137
Fringe projection, 214
  measurement process, 215
  three basic configurations, 216
  3D profilometry, 217, 218
  triangulation principle, 215
FTCCD, *see* Frame transfer CCD
Fuji Film, 361
Full width at half maximum (FWHM), 10
Fundus cameras, 253
Fused quartz, 687, 688
Fusion splicing, 670

G

Gabor transform, 325
Gabor zone plate modulation, 277
Galium phosphide LEDs, 573
Galvanometer scanners, 13
Gamma, photographic sensitivity, 355
Gas-filled phototubes, 420
Gas lasers, 580
Gaussian beams, 593–594, 606, 610
Gaussian distribution of energy, 612
Gaussian lineshape function, 585
Geneva gauge, 158
Gerchberg-Saxton algorithm, 280, 281
Germicidal lamps, 572

GIPOF, *see* Graded-index plastic optical fiber
Glass ceramics, 689
Glass envelope, tungsten lamp, 564
Glass optical fibers, 634, 643
  vs. plastic fibers, 649, 650
Global interconnect, 334
Goniometers, 171–172
Graded-index plastic optical fiber (GIPOF), 635, 650
Gradient tests
  irradiance transport equation, 62–64
  Platzeck-Gaviola test, 60
  Roddier method, 64–66
Granularity measurement, of photographic emulsions, 358, 362
Grating criteria, 261–263
Grating monochromator, 616
Greivenkamp's approach, 146
Grid contours, 166, 168
Grinding/polishing machines, 728, 739, 746, 749
Gullstrand number 1 model, 545

H

Halfwaves, 621
Halogen, tungsten lamp, 563
Hard-Clad Silica (HCS) fiber, 643
Hardness of glass, 733
Hardness, of optical materials, 681–682
Harmonic wave, 604
Hartmanngram evaluation, 54–55
Hartmann screens, wavefront analysis with, 141–143
Hartmann test
  advantages of, 53
  characteristics of, 57
  experimental setup, 54
  screen design and Hartmanngram evaluation, 54–55
  Shack-Hartmann test, 55–56
HCS fiber, *see* Hard-Clad Silica fiber
Helium-neon (He-Ne) lasers, 72–73, 456
  Fizeau interferometer, 81–82
Helmholtz equation, 63, 261
He-Ne lasers, *see* Helium-neon lasers
Hering's color theory, 518
Hetero-association IPA neural network, 324
Higher-order aberrations, 552–555
High-intensity discharge lamps, 568
High-pressure sodium lamps, 568
HiLo microscopy, 20
Hole burning, 379
Hollow waveguides
  formation, mechanism of, 372, 373
  geometry, 369
  optical fibers, 647–649
  thickness, 369
Hologram(s)
  formalism, 282–283
  inline reflection, 284

inline transmission, 283–284
off-axis transmission, 285
rainbow, 286
recording and diffraction terms, 283
Holographic data storage (HDS), 385–389
Holographic diffractive optical element (HDOE), 390
Holographic filters, 390–391, 619
Holographic interferometry, 289–290
digital, 221–224
optical, 218–221
Holographic lenses (HLs), 390
Holographic polymer-dispersed liquid crystals (H-PDLCs), 383, 384
Holographic recording materials
dichromated gelatin, 291–292
electronic devices, 293–294
photochromic materials, 293
photopolymer, 292
silver halide, 290–291
Holographic stereogram, 288–289
Holography, silver halide emulsions for, 363–364
Homogeneity, 733
optical materials, 680
Hopfield neural network, 320–321, 334
Hopkins rod lens relay system, 32, 33
Hot isostatic pressing, 690
Hubble space telescope, 253
Human eye, 543
aberrations in, 254, 256
adaptive optics imaging, 253–256
axial astigmatism, 549
higher-order aberrations, 552–554
irregular astigmatism, 550–551
laser in situ keratomileusis, 555
longitudinal chromatic aberration, 547
ocular aberrometry, 551–552
peripheral aberrations, 556
schematic models, 545–547
transverse chromatic aberration, 548
Huygen's principle, 272, 299
Hybrid field-effect liquid crystal, 722
Hybrid ternary modulation (HTM), 388
Hypothetical optical processing system, 302

I

Illuminance, 472, 536
conversion factors, 473
Image intensifier, 432, 433
Image structure characteristics, of photographic emulsion, 357–359, 362
Image subtraction, 316, 319
Imaging equation, in frequency space, 3
Imaging tubes, 431
Incandescent lamp, 537
Inclusions, bubbles and, 680
Incoherent illumination, 304

Incoherent optical image processing, 317–319
Incoherent optical signal processing, 302–304
Incoherent spot projection system, 203–204
Index ellipsoid, 707–708, 711, 712
Indiana chromatic eye, 545
Induced fiber loss, 638
Infinitesimal angle, 477
Infrared materials, 691–692
Infrared optical fibers, 646–647
Inherent fiber loss, 638, 639
Inhomogeneous films, 693
Injection molding process, 759
Inline reflection hologram, 284
Inline transmission hologram, 283–284
Intensity phase conversion, 50
Interconnection losses, 666–667
Interconnection weight matrix (IWM), 320
Interference, 261–263, 618–620
Interferograms
in lateral shear interferometer, 92
in Twyman-Green interferometer, 85
wavefronts producing, 88
Interferometers, 618
Burch and Murty interferometers, 86–87
cyclic interferometers, 97–98
cyclic multiple reflection interferometers, 96–97
Fabry-Perot interferometer, 96
Fizeau interferometer, 80–82
helium-neon lasers for, 72–73
interferograms and Seidel primary aberrations, 88
lateral shear interferometers, 88
Mach-Zehnder interferometer, 84
Michelson interferometer, 78–79
Newton interferometer, 83
point diffraction interferometer, 87–88
Sagnac interferometer, 98–100
Shack-Fizeau interferometer, 81–82
Talbot interferometer, 92–93
Twyman-Green interferometer, 82
working principle, 71–72
Young's double slit, *see* Double–slit interferometers
Interferometric methods
angle measurements, 181–182
in small distance measurement, 168–171
techniques, 749
Interferometry, 739
Interline transfer CCD (ITCCD), 428
Intermediate frequency (IF) detection system, 199
Intermodal dispersion, optical fibers, 635
Internal transmittance, 505–507
Interpattern association (IPA) neural network, 322–325
Intrinsic fluorescence, 6
Intrinsic fluorophores, 6
Intrinsic losses, 666, 668
Invasive fiber polarizers, 660
Ionization coefficient, 415, 417
Iridescence, 624

Irradiance, 95, 530, 616
    function, 72, 96, 97
Irradiance transport equation, 62–64, 143
Irregular astigmatism, 550–551
ITCCD, *see* Interline transfer CCD

J

Jablonski energy diagram, 5
Joint transform correlator (JTC), 304, 306
Joint transform power spectrum (JTPS), 307

K

KDP modulator, 436
Kerr cell light modulator, 445
Kerr magneto-optic effect, 433, 719
Klein and Cook criteria, 263
Knife test, 38–41
    experimental setup for, 40
    Foucaultgram simulations, 39–41
Knoop hardness, 682
Kodak, 360–361

L

Laguerre polynomials, 591
Lambert's Law
    blackbody radiator, 491–492
    directional radiator, 489–491
    exitance *vs.* radiance, 493–496
    nonplanar sources, 492–493
    radiance of, 491
Large-distance optical measurements, 154
Large-tool polishing, 759
Laser, 579
    applications, 597–601
    for confocal microscopes, 12
    diffraction losses in, 588
    double-clad fiber, 645
    Helium-neon, 72–73, 81–82
    medium
        line-shape function, 584–585
        saturated gain coefficient, 585–586
        unsaturated gain coefficient, 581–584
    temporal operation, 581
    types of, 579–581
    up-conversion, ZBLAN fiber, 649
Laser Doppler Anemometry (LDA), 190
Laser Doppler displacement interferometer, 181
Laser Doppler velocimetry (LDV), 597
    Doppler signal processing, 196–197
    frequency shifting, 192–194
    photodetectors, 194–196
    physical principle, 190–191
    torsional geometry, 193

Laser Doppler vibrometers
    referenced vibrometer, 202
    scanning vibrometer, 203
    spot projection system, 203–204
Laser in situ keratomileusis (LASIK), 555
Laser interferometer, 184–185
Laser radar (LIDAR), 597
Laser speckle interferometry, 182, 183
Laser speckle photography (LSP), 185–189
Laser thermometry, 597
Laser Twyman-Green interferometer, 84
LASIK, *see* Laser in situ keratomileusis
Latent imaging polymers, 375
Lateral chromatic aberration, 30
Lateral resolution enhancement, 22
Lateral shearing interferometry, 88
    for angle measurement, 183
    theory, 47
    wavefront analysis, 139–141
Laws of Grassmann, 521
LCA, *see* Longitudinal chromatic aberration
LCD, *see* Liquid-crystal display
LCLV, *see* Liquid-crystal light valve
LCs, *see* Liquid crystals
LDA, *see* Laser Doppler Anemometry
LDV, *see* Laser Doppler Velocimetry
Leaching, 597
Least-squares technique, phase unwrapping using,
    133–134
LEDs, *see* Light-emitting diodes
Lens
    anatomy of, 735
    focal length of, 165, 166
    Fourier transformation by, 301, 302
    hydrostatic support of, 743
    system, 613
Level measurement, 175–176
Light-emitting diodes (LEDs), 333, 539
    applications, 576
    cross-section of, 572, 573
    electrical characteristics, 576
    in endoscopes, 28
    optical characteristics, 573–576
    plastic optical fibers, 650
    properties of, 574
Light energy, 608
Light modulation, 722
Light sources
    fluorescent lamps, 539
    and illuminants, 536
    incandescent lamp, 537
    sodium/mercury vapor lamps, 538
Lily pollen grain, 18
Linear arrays, 195
Linear distance measurements
    curvature and focal length, 157–160
    interferometric methods, 168–171

large distances, 154
  range finders and optical radar, 154–157
Linear electro-optic tensor, 433, 699, 701, 714
Linear grating, 161
  projected, 167, 168
Linear gratings, wavefront slope analysis with, 137–139
Linearly polarized (LPlm) modes, 636, 637
Linear polarization, 606, 660, 716
Linear system, 301
Line scanning confocal microscope, 14
Line-shape function, 584–585
Liou-Brennan eye model, 545
Liquid-core fibers, 648–649
Liquid-crystal display (LCD), 448
  and LCLV, 450
  system and characteristics, 449
Liquid-crystal light valve (LCLV)
  adaptive compensating system, 452
  CCD-addressed, 451, 452
  liquid crystals, 446, 447, 449
  schematic of, 451
  SLM structure, 448
  transmission characteristics of, 450
Liquid crystal on silicon (LCoS), 294
Liquid crystals (LCs), 446, 719–722
  field effects in, 447
  in photopolymerizable materials, 383–384
Liquid-crystal television (LCTV), 320
  logic functions using, 333
Liquid lasers, 580
Liquid level measurement, 164
Logic-based computing, 331–333
Long-discharge lamps, 569
Longitudinal chromatic aberration (LCA), 547, 548
Longitudinal mode modulator, 722, 723
Longitudinal modes, in He-Ne laser, 73
Longitudinal spherical aberration (LSA), 550, 553–555
Longpass filter, 7
Loose abrasive grinding, 746–748
Low current lamps, 563
Low-expansion glasses, 688
Low-pass filters, regularized linear, frequency response of, 110
Low-pressure discharge lamps, 569
LSA, *see* Longitudinal spherical aberration
LSP, *see* Laser speckle photography
Luminance conversion factors, 473
Luminous efficacy, 561
  for photopic and scotopic vision, 469, 471
  for standard photometric observer, 470
Luminous energy, 472
Luminous exitance, 472
Luminous flux (volume) density, 472
Luminous intensity, 473
Luminous light beam, 607
Lyot filter, 617, 618

M

Mach–Zehnder interferometer, 84, 658
Mach–Zehnder waveguide EO modulator, 442, 443
Magnetic field decomposition, 605
Magneto-optic effect, 718–719
Magnification, 2
Matching color functions, 520–523
Material dispersion, 640
Mating (alignment) sleeve, 668
Matrix-vector processors, 333–336
Maxwell's equation, 261
Mean wavelength, 11
Mechanical splices, 670–676
Medium distance measurements, *see* Moiré techniques
Mercury vapor lamps, 538, 568
Mesophase, 719
Metal-dielectric filters, 622, 624–626
Metal film polarizer, 660
Metal halide lamps, 567, 568
Metal ion/polymer systems, 377–378
Metallic hollow waveguides, 647–649
Meter setting, 357
Methacrylate acid (MAA), 391
Methyl methacrylate, 690
Michelson interferometer, 78–79, 168, 184
Microchannel spatial light modulator (MSLM), 452
Micro-electro mechanical system (MEMS), 294
Microgrinding, 746
Microlenses, fabrication of, 367
Micromachining, 746
Microroughness, 737
Microstructure fabrication, 392–393
Miller indices, 700
Minkowitz nonreacting interferometer, 169
Mirror materials; *see also* Optical glasses, 688
Mirror MEMS, 294
Mirror substrates, 734
MMFs, *see* Multimode fibers
Modern fabrication methods, 727
Modulation Transfer Function (MTF), 359, 362
Moharam and Young criteria, 263
Moh Hardness Scale, 681
Moiré deflectometry, 92–93, 139
Moiré effect, 160
Moiré fringes, 160, 161
  collimated/noncollimated light, 165
  displacement of, 164
Moiré (grating) interferometry, 234, 239–240
  conventional and photographic patterns, 237–238
  diffracted beams in, 236
  multiple-channel, 240
  three-channel system for, 240
  with tilted object, 239
  virtual grating in, 237

Moiré techniques
 angle measurements, 177
 medium distance measurement
  deformation of gratings, 162, 163
  focal length measurement, 164–165
  liquid level measurement, 164
  moiré fringes, 160, 161
  photoelectric fringe counting, 163–164
  Talbot effect, 164
 vibration analysis, 205–211
Monochromatic wave, 604
Monomer optimization, 379–380
Monomer/polymer system, 377
Moon ranging experiment, 156
Mother wavelet, 325
MSLM, *see* Microchannel spatial light modulator
Multicomponent crosslinking polymerization, 371
Multimode fibers (MMFs), 634–635
Multiphoton imaging, 6, 8–9
Multiple-channel grating interferometric systems, 240
Multiple-fiber confocal microscope, 16
Multiple reflection interferometers, 94–97
Multiple slits, diffraction by, 276
Multiple-wavelength interferometry, 171
Multiplexers, fiber Bragg grating in, 664–665
Munsell color space, 535
Murty's shearing interferometer, 91, 182
Mutual intensity function, 303

N

NA, *see* Numerical aperture
Nanocomposites, photopolymer, 380–381
Nanoparticles, photopolymer, 380–381
Narrowband filter, 625
Natural birefringence, 708
Navarro eye model, 545
Negative photographic material, 355
Nematic electro-optic effect, 721
Nematic liquid crystals, 446
Nematic mesophase, 719
Neural networks (NNs)
 architecture, 320
 definition, 319
 interpattern association, 322–325
 layout, 320
Newton interferometer, 82, 83
Night myopia, 550
Nipkow disk confocal microscope, 14–15
NNs, *see* Neural networks
Non-collinear type, 618
Noninvasive fiber polarizers, 661
Nonlinear SIM, 24–25
Non-local photopolymerization driven diffusion model (NPDD), 382–383

Normalized frequency, optical fiber, 636
NPDD, *see* non-local photopolymerization driven diffusion model
N-step least-squares phase-shifting algorithm (LS-PSA), 117, 119
Null screens, for convex surfaces, 58
Null Shack-Hartmann test, 57–58
Numerical aperture (NA), 2, 636, 666
Nyquist sampling theorem, 281

O

Objective lenses
 of confocal microscope, 13
 of endoscopes, 29–32
Objective speckles, 224, 225
Oblique astigmatism, 549, 556
Ocular aberrometry, 551–552
Ocular spherical aberration, 550
Off-axis sphere, 729
Off-axis transmission hologram, 285
Oligomer system, 375–376
Omnidex system, 374
1 x 2 switch, 657
OPDs, *see* Optical path differences
Open-circuit photovoltage, 412
Open-fringes interferogram, 125
 Fourier-transform method, 125–127
 synchronous demodulation of, 129
Ophthalmic glasses, 685, 687
Ophthalmology, adaptive optics in, 253–256
Optical alignment, 181
Optical cavity technique, 159
Optical coatings, 692–693
Optical collimation, 180–181
Optical contact method, 741
Optical designer, 732
Optical engineering, 727
Optical fabrication
 aspheric surfaces, 731
 capability, 728
 cementing, 754
 centering and edging methods, 753–754
 cleaning, 754
 conventional methods for, 755–758
 current trends in, 755
 diamond machining and generating, 743–746
 fining, 746–748
 material specifications, 732–735
 mechanical specifications, 735–739
 modern methods of, 758–760
 optical surface shapes, 728–731
 pitch polishing, 739
 polishing, 748–753
 process, 727
 single-point diamond turning, 760
 support methods for, 740–742

system tolerance analysis, 731
Optical fiber(s)
  attenuators, 659
  bandwidth, 639
  Bragg gratings, *see* Fiber Bragg gratings
  connection, *see* Connectors
  couplers, *see* Couplers
  double-clad, 644–646
  erbium-doped, 643–644
  fiber bundles, 650–652
  history of, 633–634
  hollow waveguides, 647–649
  infrared, 646–647
  linearly polarized modes, 636, 637
  losses, 638–639
  material dispersion, 640
  parameters, 636–638
  polarization controller, 662–663
  polarization splitter, 661–662
  polarizers, 660–661
  single-mode and multimode, 634–635
  splicers, 670–676
  structure, 634, 635
  switches, 656–659
  total internal reflection, 636
  types, 642–643
  waveguide dispersion, 641
  wavelength-division multiplexers, 656,
    657 658
  ZBLAN up-conversion lasers, 649
Optical glasses, 734
  Abbe number *vs.* refractive index, 684, 685
  bubble classes for, 680
  categories of, 732
  ceramic materials, 689
  low-expansion, 688
  physical properties for, 686
  refractive indices *vs.* wavelength, 683
Optical holographic interferometry, 218–221
Optical homogeneity, 680
Optical level, optical collimator, 181
Optical materials
  chemical properties, 682
  chromatic dispersion, 678
  coatings/films, 692–693
  cost of, 683
  Fresnel reflections, 679
  hardness, 681–682
  homogeneity, 680
  infrared and ultraviolet materials, 691–692
  plastics, 690–691
  refractive index, 677–678
  spectral transmission, 679–680
  thermal expansion, 680–681
  vitreous silica, 687–688
Optical metrology of diffuse objects, 213
  digital holographic interferometry, 221–224

fringe projection, 214
  measurement process, 215
  three basic configurations, 216
  3D profilometry, 217, 218
  triangulation principle, 215
grating (moiré) interferometry, 234, 239–240
  configuration, 236
  conventional and photographic patterns,
    237–238
  diffracted beams in, 236
  moiré pattern, 237
  multiple-channel, 240
  three-channel system for, 240
  with tilted object, 239
  virtual grating in, 237
optical holographic interferometry, 218–221
shearography, 232–234
Optical noise, 509–510
Optical path differences (OPDs), 72, 73, 78, 90
  in holographic interferometry, 219, 220
Optical plastics, 690–691
Optical radar, 154–157
Optical sectioning strength, 17–19
Optical sectioning structured illumination microscopy
    (OS-SIM), 16
  optical sectioning strength, 17–19
  principle, 17
  root-mean-square method, 19
  speed and solution problem, 19–20
  and SR-SIM, 25–26
Optical signal, in radiometry
  absolute measurements, 512–513
  imaging and non-imaging systems, 508–509
  optical noise and stray light, 509–510
  radiation detection and signal conditioning,
    510–511
  relative measurements, 513
  signal collection, 509
  signal-to-noise ratio, 510
Optical signal processing
  coherent and incoherent processing, 302–304
  exploitation of coherence, 309–312
  Fourier and spatial domain processing, 304–309
Optical surface shapes, 728–731
Optical transfer function (OTF), 4
Optical vibrometers, subcarrier system, 198
  direct phase detection system, 199
  double modulation system, 200–201
  intermediate frequency detection system, 199–200
Order-sorting filters, 616
Oscillation Project with Emulsion tRacking Apparatus
    (OPERA) experiment, 367
OS-SIM, *see* Optical sectioning structured
    illumination microscopy
OTF, *see* Optical transfer function
Oxide glass fibers, 646
Oximetry, 513

P

Parabolic equation, 63
Parallelepiped, 700
Parallelism measurement, 176–177
Parallel stacked mirror model, 263
Paraxial approximation, 300
Paraxial wave equation, 63
Particle image velocimetry (PVI), 185–189
PC connector, *see* Physical contact connector
PCM, *see* Phase conjugate mirror
PCS fiber, *see* Plastic-Clad Silica fiber
Peak-to-valley (P-V) specifications, 749
Pentaprism, 159, 160
Peripheral aberrations, 556
Permittivity tensor, 697
Phase conjugate mirror (PCM), 329
Phase contrast microscopy, 612
Phase demodulation, using Takeda's Fourier technique, 127
Phase filter, 313
Phase gratings, 265
Phase holograms, 293
Phase-locked loop (PLL), 157, 196
    Ronchi rulings using, 50
Phase-mask fabrication process, 663–664
Phase-shifting algorithms (PSAs), 111, 114
    eight-step, 123–125
    five-step, 116–117
    four-step, 116
    seven-step, 122–123
    six-step, 121
    three-step, 115–116
Phase-shifting interferometry (PSI), 111
    quadrature filters for, 118–119
    theory of, 111–114
Phase unwrapping, 133
    least-squares technique, 133–134
    regularized phase-tracking technique, 134–135
    temporal, 135–137
Phonon annihilation, 454
Photochemical repolymerization, 375
Photochromic materials, 293
Photoconductors, 409–411
Photocrosslinking polymers, 371–372, 376–378
Photodetectors, 194–196
Photodiodes, 196
    avalanche, *see* Avalanche photodiode
    cut-off frequency, 413
    equivalent circuit, 413
    ionization coefficient, 415, 417
    photocurrent and open-circuit photovoltage, 412
    vacuum, 419, 420
Photoelectric fringe counting, 163–164
Photoemissive tube technology, 420
Photographic emulsions, applications of, 366–367
Photographic sensitivity, 354–357

Photographic speed, 357
Photographic turbidity, 358–359
Photography, fluorescent lamps in, 571
Photometric brightness, 473
Photomultiplier tube (PMT), 420
    problems of, 421
    types of, 422
Photon exitance, 468
Photon flux, 467
Photonic switching, 658
Photon incidance, 467
Photon intensity, 468
Photon-phonon collison, 454
Photon spectral exitance, 468
Photon spectral incidance, 468
Photon spectral radiance, 468
Photopic vision, luminous efficacy for, 469, 471
Photopolymer recording materials
    polymerization process, 370–372
    state-of-the-art, 373
        photocrosslinkable polymers, 376–378
        post-treatment, systems requiring, 373–375
        self-processing materials, 375–376
Photopolymers, 292
    applications of
        holographic data storage, 385–389
        holographic filters, 390–391
        holographic optical elements, 389–390
        microstructure fabrication, 392–393
        sensors with polymers, 391–392
        waveguide fabrication, 390
    holographic recording systems, 368
        holograms, classification of, 369–370
        polymerization quantum yield, 368
        theoretical modeling of, 382–383
    nanocomposites, 380–381
    surface relief in, 381–382
Photorefractive crystals and polymers, 293
Photoresistors, 195, 292
    in optics, 366
    processing, 365–366
    types of, 365
Photosensitivity, 663
Photo sensors, 194
Phototransistors, 417, 418
Physical contact (PC) connector, 669
Pinhole, diffraction by, 275
Pinhole, in confocal microscope, 13
p-i-n photodiode, 414
Pitch
    laps, 748
    polishing, 739
Planar LED, 576
Plane wave propagating, 604
Plasma Assisted Chemical Etching method, 758
Plasmonics, 693
Plastic-Clad Silica (PCS) fiber, 643

Plastic optical fibers (POFs)
    core material, 643
    glass fibers *vs.*, 649, 650
Plastic optics, 748
Plastics, 690–691
Platzeck–Gaviola test, 60
PLL, *see* Phase-locked loop
Plumbicon, 432
PMT, *see* Photomultiplier tube
p-n photodiode, 414
Pockels effect, 433
Pockels EO amplitude modulator, 436, 437
Pockels read-out optical modulator (PROM), 443, 444
POFs, *see* Plastic Optical Fibers
Point-by-point approach, 187
Point diffraction interferometer, 87–88
Point scanning confocal microscope, 14
Point spread function (PSF), 3, 612
    of confocal microscope, 10, 11
Polarization, 604, 617, 620
    controller, 662–663
    density, 696
    effects, 621
    mechanism, 659–660
    modulation, 695, 711–717
    splitter, 661–662
Polarizers, 660–661
Polished fiber coupler, 655–657, 662
Polishing, 757–758
    aspheric surfaces, 756
    compound, 748
    of flat surface, 751–753
    of spherical surfaces, 749–751
Polycarbonate, 690
Polycrystalline fibers, 647
Polygonal scanners, 13
Polyhedral oligomeric silsesquioxane-methacrylate
    (POSS), 389
Polymer(s)
    sensors with, 391–392
    shrinkage, 384–385
Polymer dispersed liquid crystals, 383–384
Polymer liquid-crystal polymer slices (POLICRYPS)
    grating, 384
Polymethylmethacrylate (PMMA)
    hologram formation in, 373
    post-treatments, 375
Polynomial aspherics, 731
Polynomial fitting, 49
Polystyrene, 690
Polyurethane, 748, 753
Polyvinylalcohol/Cr(III) systems (DCPVA), 378
POSS, *see* Polyhedral oligomeric
    silsesquioxane-methacrylate
Post-image, 520
Potassium tantalate niobate (KTN), 445
Power, of radiation, 460–464

Power transfer across boundary, 496
    BRDF function, 498–499
    emissivity, 505
    external and internal transmittance, 505–507
    radiation absorbed in medium, 503–505
    radiation propagation, 501
    radiation scattering, 501–503
    specular and diffuse surface, 497–498
Power transfer equation, in integral form, 486–489
Precision Glass Molding (PGM), 759
Prepolymerized multicomponent systems, 376
Prisms
    angle error observation, 176
    angle measurements in, 173
    monochromators, 610, 616
Profilometric measurements, 50
Programmable logic array (PLA), 334
Projected linear grating, 167, 168
Projected solid angle, 481–482
Projection lamps, 565
PROM, *see* Pockels read-out optical modulator
PSAs, *see* Phase-shifting algorithms
PSI, *see* Phase-shifting interferometry
PVA/AA mixture, 388, 389, 391
    in waveguide fabrication, 390
PVA–acrylamide photopolymer, 388
PVI, *see* Particle image velocimetry
Pyramidal error, 173, 174
Pyramid wavefront sensor, 248–250

Q

Q-switched laser system, 446
Quadratic electro-optic
    effect, 433
    tensors, 701, 705, 716
Quadrature filters
    adaptive, 130
    bandpass, 129–130
    frequency transfer function for, 120
    for phase-shifting interferometry, 118–119
Quadrupole clad, 637
Quarterwave rule, 621, 627, 628
Quartz-halogen lamp, 563, 564

R

Radar imaging with optics
    image reconstruction, 342
    optical processing system, 343
    point-target problem, 340
    returned radar signals, 340, 341
    side-looking radar, 338–340
    signal recording format, 341
Radiance, 530
Radiant flux, 530
Radiating exitance, 531

Radiative power, 482
Radiative power transfer, in 3D-space
  geometrical factors
    angle, 477–479
    area, 475
    field of view, 480
    projected area, 476
    projected solid angle, 481–482
    solid angle, 480–481
    surface and area, 474
  power transfer equation, in integral form, 486–489
  radiative power, 482
  sources and collectors, 482–484
    line of sight, 483
Radiometry, 459
  biomedical applications of, 513
  in color studies, 530–531
  exitance and incidence, 464–465
  intensity, 466
  Lambert's law, directional radiator, 489–491
  photometric quantities and units, 468–473
  photon, 461
  photon-based quantities, 466–468
  power, 461–464
  radiance, 465–466
  wavelength, 460
Radius of curvature measurements, 157–160, 181
Rainbow hologram, 286
Raman–Nath grating, 455, 456
Range compensators, 155
Range finders, 154–157
Rayleigh distance, 92
Rayleigh scattering, 639
Rayleigh's criterion, 3, 4
Real-time holographic interferometry, 219
Real-time vibration measurement, 231–232
Reciprocity, 359–360
Reciprocity-law failure, 359, 360, 364
Reference beam mixing, 191
  pre-shifted, 192
Reference beam mode, 194
Referenced vibrometer, 202
Reflection filter, fiber Bragg grating as, 664
Reflection geometry, thick gratings characteristics, 265
Refractive index
  Abbe number *vs.*, 684, 685
  for fused quartz, 688
  of isotropic materials, 677–678, 687
  for optical glasses, 683, 685
  for optical plastics, 690
  for propagation, 707
  spectral filter, 623
Regularization methods, 107
Regularized linear low-pass filters, 110
Regularized phase-tracking (RPT) technique, 131–132, 143
  phase unwrapping, 134–135

Relative (normalized) index difference, 635
Relative radiometric measurements, 513
Relay system
  confocal microscopes, 14
  endoscopes, 32–35
ReRRCA, *see* Ronchigram recovery with random aberrations coefficients algorithm
Residual aberrations, 14
Resolution, 2–4
  enhancement, by structured illumination, 19
Resolving power, 358–359, 362
Retina
  adaptive optics imaging of, 254, 255
  of eye, 544
Retinal/fundus cameras, 253
Retinex theory of color, 520
Retrieval equation, 320
Retrofocus objective lens, 29
Reverse-biased *p-i-n* photodiode, 414
Reverse-biased *p-n* junction, avalanche in, 416
Right optical material, 732
Rigid endoscopes, 27
Ring spherometer, 158, 159
RMS, *see* Rotary mechanical splice
Robust quadrature filters, 118–119, 129–130
Rockwell hardness, 682
Roddier method, 64–66
Rods, human eye, 544
Ronchigram recovery with random aberrations coefficients algorithm (ReRRCA), 49, 50
Ronchi ruling, 137
Ronchi test, 92, 93, 137–139
  bi-Ronchigram, 51
  characteristics of, 57
  null test, 56–58
  Ronchigram evaluations, 49
  Ronchigram recovery with random aberrations coefficients algorithm, 49, 50
  Ronchigram simulations
    geometry used in, 46
    lateral shear interferometry theory, 47
    for parabolic mirror, 46
    procedure, 45
    for spherical wavefront, 44
    square Ronchi ruling, 49
    two crossed Ronchigrams, 47–49
  substructured Ronchi gratings, 51–53
Root-mean-square (RMS) method, 19
Rotary mechanical splice (RMS), 671
Rotational velocity, 184
Rugate filters, 631

S

Sagnac interferometer, 98–100
Sample Sensitivity, 386
SAPD, *see* Staircase avalanche photodiode

Saunders' method, 90
Sawtooth phase grating, 270
SBP, *see* Space-bandwidth product
Scalar theory of diffraction
　diffraction by simple apertures
　　Fresnel zone plate, 277–278
　　multiple slits, 276
　　pinhole, 275
　　slit, 274–275
　Fraunhofer diffraction integral, 273–274
　Fresnel diffraction integral, 273
　Kirchhoff diffraction integral, 272–273
Scanner, 13
Scanning vibrometer, 203
Scattering, 498
　fiber loss and, 639
　interferometer, 86, 87
　spectral filters, 617
SCCD, *see* Surface channel charge-coupled device
Schadt–Helfrich effect, 720
Schlieren techniques, 94
Schwider-Hariharan's PSA, 116–117
Scotopic vision, luminous efficacy for, 469, 471
SCPI, *see* Spatial carrier phase-shifting method
Screen-testing methods, 137
Se-based chalcogenide glasses, 647
Secondary emission, 566
Seidel primary aberrations, 88
Selenide glass fibers, 647
Self-processing materials, 375–376
Semiconductors, 615
Sensitizer/initiator optimization, 380
Sensitometry, 362
Sensors
　digital, 294
　with polymers, 391–392
Seven-step PSA, 122–123
Shack–Fizeau interferometer, 81–82
Shack-Hartmann test, 55–56
　characteristics of, 57
　null test, 57–58
Shack-Hartmann wavefront sensor, 247–248, 551, 552, 556
Shadow-casting logic array processor, 331–333
Sharpness measurement of, photographic emulsions, 359
Shearography, 232–234
Short arc lamps, 566–567
Shortpass filter, 7
Short-time Fourier transform (STFT), 325, 326, 330
Side-looking radar, 338–340
Signal conditioning, radiometric measurements, 510–511
Signal-to-noise ratio, 510
Silica fiber, attenuation in, 639
Silver halide, 290–291
　emulsions, 354
　　color films, 361–362
　　for holography, 363–364

image structure characteristics, 362
　sensitometry, 362
　spectral sensitivities of, 357, 358
Sine plate, 172
Single-crystalline fibers, 647
Single-fiber confocal microscopes, 16
Single-mode fibers (SMFs), 634
　multimode fibers *vs.*, 643
　refractive index profiles of, 637, 638
Single-monomer crosslinking polymerization, 371
Single-monomer linear polymerization, 370
Single refractive index, 699
Sinusoidal amplitude grating, 268
Sinusoidal phase grating, 269
Si photodiodes, 414
Six-step PSA, 121
SLD, *see* Super luminescent diode
"Slew rate," 196
Slit, diffraction by, 274–275
SLMs, *see* Spatial light modulators
Slope measurements, 181
Small distance measurement, 168–171
Smectic liquid crystals, 722
Smectic mesophase, 719
SMFs, *see* Single-mode fibers
Smirnov's system, 551
Smoothing techniques, 107–110
Snell's law, 636
Sodium lamps, high-pressure, 568
Solar telescope, 609
Solc filter, 617
Sol-gel glasses, 390, 693
Solid angle, 480–481
Solid-state sensors, 195
Space-bandwidth product (SBP), 280, 281
Spatial carrier phase-shifting method (SCPI), 127–129
Spatial domain processing, 304–309
Spatial filtering, 607–615
Spatial impulse response, 300
Spatial light modulators (SLMs), 305, 332, 333, 335, 386, 446
　configuration, 448
　confocal microscopes using, 15
　microchannel, 452
Spatially incoherent source, 304
Spatial phase-measuring methods
　open-fringes interferogram, 125
　　Fourier-transform method, 125–127
　　synchronous demodulation of, 129
　regularized phase-tracking technique, 131–132
　robust quadrature filters, 129–130
　spatial carrier phase-shifting method, 127–129
Speckle angle measurement, 182, 183
Speckle interferometers; *see also* Electronic speckle
　　pattern interferometry
　class 1and 2, 229
Spectral confocal imaging, 16
Spectral decomposition, 605

Spectral filtering, 607–609
    absorption filters, 615
    acousto-optic filters, 617–618
    all-dielectric filters and coatings, 626–632
    decorative coating, 623–624
    metal-dielectric filters, 624–626
    metals/dielectrics, coatings of, 622
    polarization, 617
    refraction and/or diffraction, 616–617
    scattering, 617
    sensitivity to tilting, 621–622
    surface admittance, 621
Spectral hole burning, 379
Spectral irradiance standard lamps, 565
Spectral lamps, characteristics for, 568
Spectral power, 463
Spectral radiance, 465–466, 490
Spectral radiant exitance, 464–465, 531
Spectral radiant incidance, 464–465
Spectral radiant intensity, 466
Spectral radiative power, 460
Spectral reflectance, 531
Spectral sensitivity curves, 362, 363
Spectral transmittance, 531
Spectrophotometers, 537
Speech spectrogram, 349
Spherical aberration, 550–551, 756
Spherical refractive error, 549
Spherical surfaces, 728, 729
    polishing of, 749–751
Splicers, fiber
    fusion splices, 670
    mechanical splices, 670–676
Spot block, 741
Spot projection system, 203
    coherent, 204
    incoherent, 203–204
Square cyclic interferometer, 98
SR-SIM, *see* Super-resolution structured illumination
    microscopy
Stage scanning, 13
Staircase avalanche photodiode (SAPD), 422
Standard human observer, radiative power detection
    by, 470
Standard type A source, 565
ST connector, *see* Straight tip connector
Stellar Michelson interferometer, 76–78
Step-index fiber, 635
Stereograms, holographic, 288–289
STFT, *see* Short-time Fourier transform
Stokes' shift, 430
Straight tip (ST) connector, 669
Stray light, 509–510
Stresses, 680
Stroboscopic lamps, 183–184
Structured illumination, 21
    resolution enhancement by, 19

Subjective speckles, 224, 225
Sub-Nyquist analysis, 145–147
Substructured Ronchi gratings, 51–53
Subtraction of color, 532–534
Subtraction process, 361–362
Sulfide glass fibers, 647
Super luminescent diode (SLD), 573
Super-resolution structured illumination microscopy
    (SR-SIM), 20
    comparative studies, 25
    instrumentation, 21–22
    nonlinear SIM, 24–25
    OS-SIM and, 25–26
    reconstruction algorithm, 22–24
    working principle, 21
Supporting bases, tungsten lamp, 564
Surface admittance, 621
Surface channel charge-coupled device (SCCD), 425
Surface relief, in photopolymers, 381–382
Susceptibility tensor, 697
Switches, 656–659
Symmetrical arrangement, 227–229
Symmetry, 699
Synchronous demodulation, of open-fringes
    interferogram, 129
Synchronous phase detection, 50
System tolerance analysis, 731
Systolic processing, 335, 336–338

T

Takeda's Fourier technique, 127
Talbot effect, 93, 164
Talbot interferometer, 92–93
TCA, *see* Transverse chromatic aberration
Templates, curvature measurements, 157
Temporal coherence, 310
Temporal phase-measuring methods
    "first-order building block," 119–121
    N-step least-squares phase-shifting algorithm, 117,
        119
    phase-shifting algorithms
        eight-step, 123–125
        five-step, 116–117
        four-step, 116
        seven-step, 122–123
        six-step, 121
        three-step, 115–116
    PSI theory, 111–114
    quadrature filters, 118–119
Temporal phase unwrapping, 135–137
Test object contrast (TOC), 358
Texas Instruments DMD, 294
Theodolites, 172–173
Thermal electron emission, 566
Thermal expansion, of optical materials, 680–681
Thermal repolymerization, 375

Thick gratings characteristics, 264–267
Thickness measurement, 166
thickness measurement, 166
Thin films, 692
    all-dielectric filters and coatings, 626–632
    coatings of metals and/or dielectrics, 622–623
    decorative coatings, 623–624
    metal-dielectric filters, 624–626
    quarterwaves, halfwaves and surface admittance, 621
    reflecting filter, 619
    sensitivity to tilting, 621–622
Thin gratings characteristics, 267–271
Three-step PSA, 115–116
Tilting, 621–622
Tilt measurement, 179–180
Time of flight method, 156
Tissue, 3D scattering, 503
Titanocene dichloride system, 375
TOC, *see* Test object contrast
Tooling, 727
Toroids, 731
Torsional geometry, 193
Total internal reflection, 636
Transfer holograms, 286
Transform kernel, 337
Transmission geometry, thick gratings characteristics, 264
Transmitted wavefront errors, 733
Transport equation, 143
Transverse aberration, 41, 42
Transverse chromatic aberration (TCA), 548
Transverse electro-optic modulators, 439
Transverse mode modulator, 723
Triangle index profile, 637
Triangular cyclic interferometer, 98
Triangulation principle, in fringe projection, 215
Trichromatic LEDs, 575
Trichromatic theory
    chromaticity coordinates, 524–525
    matching color functions, 520–523
    tristimulus values, 521–524
Tristimuli, colorimeters of, 537
Tristimulus values, 521–524
Tscherning aberroscope, 551, 552
Tunable grating filters, 665, 666
Tungsten lamp
    components of, 562–566
    designation, 564
    optical applications of, 565–566
Turbidity, of photographic emulsion, 358–359
Twin polishers, 753
Twisted nematic effect, 720, 721
Two crossed Ronchigrams, 47–49
Two interfering wavefronts, 72
Two-monomer linear polymerization, 371
Two-photon polymerization method, 392
2π phase ambiguity, 105

2 x 2 coupler, 654, 655
Twyman–Green interferometer, 82
    unequal path and interferogram, 89

U

Ultimate antireflection coating, 630
Ultraviolet materials, 691–692
Unaccommodated AZEye15 model, 547
Uniaxial crystal, 708
Uniaxial materials, 707
Unstable resonators, 613

V

Vacuum photodiode, 419, 420
Vander Lugt correlator (VLC), 304, 305
VCO, *see* Voltage-controlled oscillator
Velocity measurement, 182
    laser interferometer, 184–185
    LDV method, *see* Laser Doppler velocimetry
    particle image velocimetry, 185–189
    using stroboscopic lamps, 183–184
Very low-expansion glasses, 689
V-groove splice, 670, 671
VHOE, *see* Volume Holographic Optical Element
Vibration analysis
    interferometry for, 184
    laser Doppler vibrometers
        dual beam vibrometer, 202
        referenced vibrometer, 202
        scanning vibrometer, 203
        spot projection system, 203–204
    moiré techniques, 205–211
    optical vibrometers, 197–201
Vibrometers
    laser Doppler, *see* Laser Doppler vibrometers
    optical, 197–201
V-I characteristics, of photodiodes, 413
Video endoscope, 27
Video signal subtraction/addition, fringe formation by, 229–231
Vidicon, 431, 432
Virtual grating, 237
Visco-elastic pitch, 748
Visual and infrared thermometry, 513
Visual colorimeters, 537
Visual pigment, 519
Vitreous silica, 687–688
Voigt effect, 719
Voltage-controlled oscillator (VCO), 196
Volume Holographic Optical Element (VHOE), 390

W

Wavefront analysis
    by curvature sensing, 143–145

with Hartmann screens, 141–143
with lateral shearing interferometry, 139–141
with linear gratings, 137–139
screen-testing methods, 137
Waveguide dispersion, 641
Waveguide fabrication, photopolymers in, 390
Wavelength-division multiplexer (WDM), 643, 656, 657–658
Wavelet transform processing
optical implementations, 329
principle of wavelet transform, 325–329
scalegram, 330
Wave surface, 708–709
WDM, *see* Wavelength-division multiplexer
Weakly Interacting Massive Particles (WIMP), 367
"Whiffle-tree" supports, 742
White LEDs, 575
Wide-angle scattering, 737
Wide-field microscopy
disadvantage of, 9
optical layout, 1–2
resolution, 2–4

Wireless endoscopy, 28
Wire stems, 563
Wire test, 41–43

X

Xenon lamps, 566, 567, 570

Y

Young-Helmholtz color theory, 518–519
Young's double slit; *see also* Double-slit interferometers
interference in, 74
light source and observing plane, 75
locus of fringes, 74, 75
Young's fringe formation, 187, 188

Z

ZBLAN fibers, 649
Zernike polynomials, 553, 731